Basic Radiation Oncol

Murat Beyzadeoglu · Gokhan Ozyigit
Cüneyt Ebruli

Basic Radiation Oncology

Second Edition

Murat Beyzadeoglu
Gulhane Faculty of Medicine
Department of Radiation Oncology
University of Health Sciences Turkey
Ankara, Turkey

Gokhan Ozyigit
Department of Radiation Oncology
Hacettepe University Faculty of Medicine
Ankara, Turkey

Cüneyt Ebruli
Department of Radiation Oncology
Tepecik Training and Research Hospital
University of Health Sciences Turkey
Izmir, Turkey

1st edition: © 2010, University of Health Science Turkey,Gülhane Basımevi, Turkey 2010
Original Turkish Edition published by University of Health Sciences Turkey, Gülhane Basimevi, 2018

ISBN 978-3-030-87310-3   ISBN 978-3-030-87308-0   (eBook)
https://doi.org/10.1007/978-3-030-87308-0

© The Editor(s) (if applicable) and The Author(s), under exclusive license to Springer Nature Switzerland AG 2010, 2022
This work is subject to copyright. All rights are solely and exclusively licensed by the Publisher, whether the whole or part of the material is concerned, specifically the rights of reprinting, reuse of illustrations, recitation, broadcasting, reproduction on microfilms or in any other physical way, and transmission or information storage and retrieval, electronic adaptation, computer software, or by similar or dissimilar methodology now known or hereafter developed.
The use of general descriptive names, registered names, trademarks, service marks, etc. in this publication does not imply, even in the absence of a specific statement, that such names are exempt from the relevant protective laws and regulations and therefore free for general use.
The publisher, the authors and the editors are safe to assume that the advice and information in this book are believed to be true and accurate at the date of publication. Neither the publisher nor the authors or the editors give a warranty, expressed or implied, with respect to the material contained herein or for any errors or omissions that may have been made. The publisher remains neutral with regard to jurisdictional claims in published maps and institutional affiliations.

This Springer imprint is published by the registered company Springer Nature Switzerland AG
The registered company address is: Gewerbestrasse 11, 6330 Cham, Switzerland

Foreword to First Edition

Revolutionary advances have been taking place in radiation oncology as our world has entered a new millennium. Developments in radiological and functional imaging techniques over the last two decades have enabled us to delineate tumors more accurately in the three spatial and the fourth (temporal) dimensions. More powerful computerized planning systems are facilitating accurate three-dimensional dose calculations as well as inverse planning processes. We have even started to use robotic technology to track our targets in real time for more precise delivery of radiotherapy.

All of those high-tech machines sometimes cause us to spend many hours in front of detailed displays of serial computerized tomography (CT) slices. However, we are still treating our patients with the same types of ionizing radiation discovered more than a century ago. Therefore, every member of a radiation oncology team should know the interactions of ionizing radiation with matter at the atomic level, and be familiar with its effects in biological systems. In addition, every radiation oncologist should have an essential knowledge of evidence-based clinical oncology in relation to the indications and technical aspects of radiotherapy at major cancer sites.

Basic Radiation Oncology is an up-to-date, bedside-oriented textbook that integrates radiation physics, radiobiology, and clinical radiation oncology. The book includes the essentials of all aspects of radiation oncology, with more than 300 practical illustrations and color figures. The layout and presentation is very practical and enriched with many eye-catching conceptual highlights. The first two chapters review crucial ideas in radiation physics and radiobiology as well as the terminology of clinical radiation oncology. Basic descriptions of all high-tech radiotherapy machines are also given. The remaining 11 clinical chapters describe anatomy, pathology, general presentation, treatment algorithms, and the technical aspects of radiotherapy for major cancer sites. The 2010 (seventh) edition of the AJCC Staging System is provided for each tumor type. Practical details about key studies, particularly randomized ones, and available RTOG consensus guidelines for the determination and delineation of targets are also included at the end of each clinical chapter.

Basic Radiation Oncology meets the need for a practical and bedside-oriented radiation oncology textbook for residents, fellows, and clinicians of radiation, medical and surgical oncology, as well as for medical students, physicians, and medical physicists interested in clinical radiation oncology.

K. S. Clifford Chao
Vice President, China Medical University, Taiwan

Preface

It is with great pleasure that we introduce the second edition of basic radiation oncology. The aim of writing *Basic Radiation Oncology* was to provide a structured overview of the theory and practice of radiation oncology, including the principles of radiation physics, radiation biology, and clinical radiation oncology. We have encompassed the fundamental aspects of radiation physics, radiobiology, and clinical radiation oncology. In the last two decades, there have been many technical and conceptual advances in both treatment planning systems and radiation delivery systems. However, there are minor changes in the basic interactions of radiation with atoms or cells. Therefore, basic concepts that are crucial to understanding radiation physics and radiobiology are reviewed in depth in the first two sections. The third section describes radiation treatment regimens appropriate for the main cancer sites and tumor types according to the eighth edition of the American Joint Committee on Cancer Staging System. Many "pearl boxes" are used to summarize important information, and there are more than 350 helpful illustrations. *Basic Radiation Oncology Second Edition* meets the need for a practical radiation oncology book. It will be extremely useful for residents, fellows, and clinicians in the fields of radiation, medical, and surgical oncology, as well as for medical students, physicians, and medical physicists with an interest in clinical oncology.

Evidence-based data are also available at the end of the section on each clinical subsite. However, every clinician should be aware of the fact that there is a very fine line between evidence-based and probability-based medicine. Cancer is a highly complex subject, and it is impossible to fit it into a simple mathematical formula or "p" value. Therefore, we must not throw away experience-based data during our clinical decision-making procedures. We extend our most sincere gratitude to Professor Cevdet Erdol, the Rector of University of Health Sciences Turkey, as well as to our families for their understanding as we worked to meet our publication deadlines.

We hope that second edition of *Basic Radiation Oncology* continues to serve as a practical, up-to-date, bedside oriented high-yield learning *Radiation Oncology* book.

Ankara, Turkey	Murat Beyzadeoglu
Ankara, Turkey	Gokhan Ozyigit
Izmir, Turkey	Cüneyt Ebruli

Acknowledgments

The authors are indebted to Mr Sushil Kumar Sharma, Daniela Heller and Rosemarie Unger from Springer for their assistance in preparing *Basic Radiation Oncology* 2nd Edition. Special thanks are extended to Professor K.S. Clifford Chao for honoring us with his foreword. We extend our most sincere gratitude to Gulhane Medical Faculty Dean Professor Mehmet Ali Gulcelik and to our colleagues and friends at University of Health Sciences Turkey, Gulhane Faculty of Medicine and the Faculty of Medicine at Hacettepe University.

Contents

1 Radiation Physics 1
   1.1 Introduction and Atom 1
   1.2 Radiation 1
       1.2.1 Photon 2
   1.3 Ionizing Radiation 2
       1.3.1 Ionizing Electromagnetic Radiation 3
       1.3.2 Ionizing Particulate Radiation 7
   1.4 The Interaction of Radiation with Matter 8
       1.4.1 Photoelectric Effect 9
       1.4.2 Compton Effect 9
       1.4.3 Pair Production 10
       1.4.4 Coherent Effect (= Rayleigh Scattering, = Thomson Scattering) 11
   1.5 Specific Features of X-Rays 11
   1.6 Specific Features of Electron Energies 14
   1.7 Ionizing Radiation Units 14
   1.8 Radiotherapy Generators 16
       1.8.1 Kilovoltage Machines (<500 kV) 16
       1.8.2 Megavoltage Therapy Machines (>1 MV) 17
       1.8.3 Cobalt-60 Teletherapy Unit 18
       1.8.4 Treatment Head and Collimator 19
       1.8.5 Gantry 20
       1.8.6 Linear Accelerator (Linac) 20
       1.8.7 Magnetron 20
   1.9 Measurement of Ionizing Radiation 22
       1.9.1 Portable Measuring Equipment 23
       1.9.2 Ionization Chamber 23
       1.9.3 Geiger–Müller Counter (GM Counter) 23
       1.9.4 Film Dosimeters 24
       1.9.5 Thermoluminescence Dosimeters (TLD) 24
       1.9.6 Other Measuring Equipment 25
   1.10 Radiation Dosimetry 25
       1.10.1 Phantom 25
       1.10.2 Definition of Beam Geometry 25
       1.10.3 Build-Up Region 26
       1.10.4 Half-Value Layer (HVL) 26
       1.10.5 Percentage Depth Dose (PDD) 27

		1.10.6	Isodose Curves	29
		1.10.7	Ionization Chamber in a Water Phantom	31
		1.10.8	Dose Profile	31
		1.10.9	Penumbra	32
		1.10.10	Inverse Square Law	33
		1.10.11	Backscatter Factor (BSF)	33
		1.10.12	Tissue to Air Ratio (TAR)	34
		1.10.13	Tissue Maximum Ratio (TMR)	34
		1.10.14	Scatter Air Ratio (SAR)	35
		1.10.15	Collimator Scattering Factor ($S\,c$)	35
		1.10.16	Phantom Scattering Factor ($S\,p$)	35
		1.10.17	Monitor Unit (MU) Calculation in a Linear Accelerator	35
		1.10.18	Treatment Time Calculation in a Co-60 Teletherapy Unit	37
	1.11	Beam Modifiers		37
		1.11.1	Bolus	37
		1.11.2	Compensating Filters	37
		1.11.3	Wedge Filters	37
		1.11.4	Shielding Blocks	38
		1.11.5	Multileaf Collimator (MLC)	39
	1.12	Pearl Boxes		41
	References			45
2	**Radiobiology**			47
	2.1	Cell Biology and Carcinogenesis		47
		2.1.1	Cell Structure	47
		2.1.2	Cell Types and Organelles	47
		2.1.3	Cell Cycle	49
		2.1.4	Carcinogenesis and the Cell Cycle	52
		2.1.5	Features of Cancer Cells	53
	2.2	Cellular Effects of Radiation		54
		2.2.1	The Direct Effect of Radiation at the Molecular Level	55
		2.2.2	The Indirect Effect of Radiation at the Molecular Level	57
	2.3	Factors Modifying the Biological Effects of Ionizing Radiation		58
		2.3.1	Characteristics of the Radiation	58
	2.4	Target Tissue Characteristics		59
		2.4.1	Whole Body Dose Equivalent	59
		2.4.2	Effective Dose	60
		2.4.3	Relative Biological Effect (RBE)	61
		2.4.4	Overkill Effect	61
	2.5	Target Theory		61
	2.6	Cell Survival Curves		62
		2.6.1	Surviving Fraction	63
		2.6.2	Exponential Survival Curves	65

		2.6.3	Linear–Quadratic Model (LQ Model)	67
		2.6.4	Types of Cellular Damage Due to Radiation	70
		2.6.5	Factors Affecting the Cell Survival Curve	70
	2.7	Tissue and Organ Response to Radiation		73
		2.7.1	Bergonie and Tribondeau Law	75
		2.7.2	Factors Determining Radiation Damage According to Ancel and Vintemberger	75
		2.7.3	Rubin and Casarett Tissue Sensitivity Classification	76
	2.8	Stochastic and Deterministic Effects		76
		2.8.1	Deterministic Effect	77
		2.8.2	Stochastic Effects	77
	2.9	Tumor Response to Radiation		78
		2.9.1	Therapeutic Index	78
		2.9.2	Tumor Control Probability (TCP)	79
		2.9.3	Normal Tissue Complication Probability (NTCP)	81
	2.10	The Five R's of Radiotherapy		82
		2.10.1	Repopulation	84
		2.10.2	Repair	85
		2.10.3	Redistribution (= Reassortment)	86
		2.10.4	Reoxygenation	87
		2.10.5	Radiosensitivity (Intrinsic Radiosensitivity)	88
	2.11	Fractionation		88
	2.12	Radiobiology of SBRT/SABR		90
	2.13	Radiation Protection		90
	2.14	Pearl Boxes		92
		2.14.1	Radiation Hormesis	92
		2.14.2	Abscopal Effect	93
		2.14.3	Bystander Effect	93
		2.14.4	Avalanche Phenomenon	93
		2.14.5	Radiation Recall Phenomenon	93
		2.14.6	Normalized Total Dose (NTD$_{2\,Gy}$)	94
		2.14.7	Normal Tissue Toxicity After Stereotactic Body Radiation (SBRT)	95
	References			95
3	**Clinical Radiation Oncology**			99
	3.1	Introduction and History		99
		3.1.1	Radiotherapy Types According to Aim	99
		3.1.2	Radiotherapy Types According to Timing	100
		3.1.3	Radiotherapy Types According to Mode	101
	3.2	The Radiotherapy Procedure		102
		3.2.1	Simulation	102
		3.2.2	Treatment Planning	108
		3.2.3	Target Volume Definitions	109
		3.2.4	Setup and Treatment	115
		3.2.5	Quality Assurance	115
		3.2.6	Treatment Fields in Radiotherapy	117
	References			122

4 Central Nervous System Tumors ... 123
- 4.1 Introduction ... 123
- 4.2 Anatomy ... 123
  - 4.2.1 Brain ... 123
  - 4.2.2 Brain Ventricles ... 124
  - 4.2.3 Spinal Cord ... 124
- 4.3 General Presentation and Pathology ... 125
  - 4.3.1 Spinal Cord Tumors ... 126
  - 4.3.2 Brain Tumors ... 126
- 4.4 Staging ... 130
- 4.5 Diagnostic Imaging ... 131
- 4.6 Treatment ... 131
  - 4.6.1 Low-Grade CNS Tumors ... 131
  - 4.6.2 High-Grade CNS Tumors ... 134
  - 4.6.3 Primary CNS Lymphoma (PCNSL) ... 136
  - 4.6.4 Meningiomas ... 136
  - 4.6.5 Pituitary Tumors ... 137
  - 4.6.6 Arteriovenous Malformations ... 139
  - 4.6.7 Vestibular Schwannoma ... 139
  - 4.6.8 Primary Medulla Spinalis Tumors ... 139
  - 4.6.9 Medulloblastoma ... 139
  - 4.6.10 Brain Metastasis ... 141
- 4.7 Radiotherapy ... 141
  - 4.7.1 External Radiotherapy ... 141
  - 4.7.2 Craniospinal RT ... 143
  - 4.7.3 Symptomatic Treatments ... 145
  - 4.7.4 Side Effects of CNS Radiotherapy ... 145
- 4.8 Selected Publications ... 149
  - 4.8.1 Low-Grade Glial Tumors ... 149
  - 4.8.2 Radiotherapy Timing ... 151
  - 4.8.3 Grade III Glial Tumors ... 151
  - 4.8.4 High Grade (III–IV) Glial Tumors ... 152
  - 4.8.5 RT Dose ... 152
  - 4.8.6 RT Fields (Whole Brain Vs. Localized Field) ... 153
  - 4.8.7 Fractionation ... 153
  - 4.8.8 Temozolomide ... 153
- 4.9 Pearl Boxes ... 154
  - 4.9.1 Good Prognostic Factors in Low-Grade Glial Tumors ... 154
  - 4.9.2 Prognostic Factors in High-Grade Tumors ... 154
  - 4.9.3 Tumors Capable of Subarachnoid Seeding via CSF ... 154
  - 4.9.4 CNS Tumors That Can Metastasize Outside the CNS ... 154
- References ... 154

Contents

5 Head and Neck Cancers 159
- 5.1 Tips for Delineating the Neck 163
- 5.2 Target Volume Determination and Delineation for the Neck 163
- 5.3 Pharyngeal Cancers 163
  - 5.3.1 Nasopharyngeal Cancer 163
  - 5.3.2 Oropharyngeal Cancer 174
  - 5.3.3 Hypopharyngeal Cancer 186
- 5.4 Laryngeal Cancer 192
  - 5.4.1 Pathology 192
  - 5.4.2 General Presentation 193
  - 5.4.3 Staging 194
  - 5.4.4 Treatment Algorithm 197
  - 5.4.5 Radiotherapy 198
  - 5.4.6 Selected Publications 200
- 5.5 Oral Cavity Cancers 202
  - 5.5.1 Pathology 203
  - 5.5.2 General Presentation 204
  - 5.5.3 Staging 204
  - 5.5.4 Treatment Algorithm 206
  - 5.5.5 Radiotherapy 206
  - 5.5.6 Selected Publications 210
- 5.6 Sinonasal Cancers 210
  - 5.6.1 Pathology 210
  - 5.6.2 General Presentation 211
  - 5.6.3 Staging 212
  - 5.6.4 Treatment Algorithm 215
  - 5.6.5 Radiotherapy 216
  - 5.6.6 Selected Publications 219
- 5.7 Major Salivary Gland Tumors 221
  - 5.7.1 Pathology 222
  - 5.7.2 General Presentation 223
  - 5.7.3 Staging 223
  - 5.7.4 Treatment Algorithm 224
  - 5.7.5 Radiotherapy 225
  - 5.7.6 Selected Publications 228
- 5.8 Thyroid Cancer 230
  - 5.8.1 Pathology 231
  - 5.8.2 General Presentation 232
  - 5.8.3 Staging 233
  - 5.8.4 Radiotherapy 235
  - 5.8.5 Selected Publications 236
- 5.9 Radiotherapy in Unknown Primary Head–Neck Cancers 238
- 5.10 Selected Publications for Head and Neck Cancers 240
- 5.11 Pearl Boxes 246
- References 248

6 Lung Cancer ... 251
- 6.1 Introduction ... 251
  - 6.1.1 Nonsmall Cell Lung Cancer (NSCLC) ... 254
- 6.2 Small Cell Lung Cancer (SCLC) ... 271
  - 6.2.1 Pathology ... 271
  - 6.2.2 General Presentation ... 271
  - 6.2.3 Staging ... 272
  - 6.2.4 Treatment Algorithm ... 272
  - 6.2.5 Radiotherapy ... 272
  - 6.2.6 Selected Publications ... 272
- References ... 274

7 Breast Cancer ... 277
- 7.1 Pathology ... 277
- 7.2 General Presentation ... 282
- 7.3 Staging ... 282
- 7.4 Treatment Algorithm ... 294
- 7.5 Radiotherapy ... 296
- 7.6 Selected Publications ... 298
- References ... 312

8 Genitourinary System Cancers ... 313
- 8.1 Prostate Cancer ... 313
  - 8.1.1 Pathology ... 315
  - 8.1.2 General Presentation ... 316
  - 8.1.3 Staging ... 318
  - 8.1.4 Treatment Algorithm ... 322
  - 8.1.5 Radiotherapy ... 325
  - 8.1.6 Selected Publications ... 327
- 8.2 Testicular Cancer ... 336
  - 8.2.1 Pathology ... 337
  - 8.2.2 Treatment Algorithm ... 340
  - 8.2.3 Radiotherapy ... 341
- 8.3 Bladder Cancer ... 345
  - 8.3.1 Pathology ... 345
  - 8.3.2 General Presentation ... 347
  - 8.3.3 Staging ... 347
  - 8.3.4 Treatment Algorithm ... 348
  - 8.3.5 Radiotherapy ... 350
  - 8.3.6 Selected Publications ... 350
- References ... 354

9 Gynecological Cancers ... 357
- 9.1 Cervical Cancer ... 357
  - 9.1.1 Pathology ... 357
  - 9.1.2 General Presentation ... 358
  - 9.1.3 Staging ... 358
  - 9.1.4 Treatment Algorithm ... 360
  - 9.1.5 Radiotherapy ... 362

		9.1.6 Selected Publications................................... 372
	9.2	Endometrial Cancer... 379
		9.2.1 Pathology.. 380
		9.2.2 General Presentation 381
		9.2.3 Staging.. 381
		9.2.4 Treatment Algorithm for Endometrial Cancer............ 383
		9.2.5 Radiotherapy 385
		9.2.6 Selected Publications....................................... 389
	9.3	Vaginal Cancer... 394
		9.3.1 Pathology.. 395
		9.3.2 General Presentation 395
		9.3.3 Staging.. 395
		9.3.4 Treatment Algorithm 397
		9.3.5 Radiotherapy 398
		9.3.6 Selected Publications....................................... 399
	9.4	Vulvar Cancer.. 402
		9.4.1 Pathology.. 402
		9.4.2 General Presentation 402
		9.4.3 Staging.. 403
		9.4.4 Treatment Algorithm 403
	9.5	50 Gy for cN0, 54 Gy for cN+.. 403
		9.5.1 Radiotherapy 404
		9.5.2 Selected Publications....................................... 404
	References... 408	
10	**Gastrointestinal System Cancers**................................. 411	
	10.1	Esophageal Cancer... 411
		10.1.1 Pathology.. 411
		10.1.2 General Presentation 413
		10.1.3 Staging.. 413
		10.1.4 Treatment Algorithm 415
		10.1.5 Radiotherapy 420
		10.1.6 Selected Publications..................................... 422
	10.2	Gastric Cancer.. 425
		10.2.1 Pathology.. 426
		10.2.2 General Presentation 426
		10.2.3 Staging.. 426
		10.2.4 Treatment Algorithm 429
		10.2.5 Radiotherapy 433
		10.2.6 Selected Publications..................................... 435
	10.3	Pancreatic Cancer.. 436
		10.3.1 Pathology.. 438
		10.3.2 General Presentation 438
		10.3.3 Staging.. 438
		10.3.4 Treatment Algorithm 439
		10.3.5 Radiotherapy 441
		10.3.6 Selected Publications..................................... 442
	10.4	Rectal Cancer... 444

		10.4.1 Pathology.. 445
		10.4.2 General Presentation 446
		10.4.3 Staging.. 447
		10.4.4 Treatment Algorithm 447
		10.4.5 Radiotherapy 449
		10.4.6 Selected Publications......................... 451
	10.5	Anal Cancer.. 452
		10.5.1 Pathology.. 453
		10.5.2 General Presentation 453
		10.5.3 Staging.. 453
		10.5.4 Treatment Algorithm 453
		10.5.5 Radiotherapy 453
		10.5.6 Selected Publications......................... 455
	References.. 458
11	**Soft Tissue Sarcoma** 463
	11.1	Introduction 463
	11.2	Pathology.. 463
	11.3	General Presentation 466
	11.4	Staging... 467
	11.5	Treatment.. 469
	11.6	Radiotherapy 471
		11.6.1 Three-Dimensional Conformal/IMRT............ 472
	References.. 475
12	**Non-melanoma Skin Cancers**... 477
	12.1	Introduction 477
	12.2	Pathology.. 478
	12.3	General Presentation 479
	12.4	Staging... 480
	12.5	Treatment.. 481
	12.6	Radiotherapy 484
		12.6.1 2D Conventional Radiotherapy 484
	References.. 486
13	**Lymphomas and Total Body Irradiation** 487
	13.1	Hodgkin's Lymphoma 487
		13.1.1 Pathology/General Presentation................. 487
		13.1.2 Clinical Signs................................. 489
		13.1.3 Staging.. 489
		13.1.4 Treatment Algorithm 490
		13.1.5 Radiotherapy 491
	13.2	Selected Publications..................................... 493
		13.2.1 Early-Stage HL 493
		13.2.2 Advanced-Stage HL........................... 497
		13.2.3 Ongoing Trials................................ 500
		13.2.4 Nodular Lymphocyte-Predominant HL........... 500
	13.3	Non-Hodgkin's Lymphoma 500
		13.3.1 Pathology/General Presentation................. 500

		13.3.2	Staging................................... 501
		13.3.3	Treatment Algorithm......................... 502
		13.3.4	Radiotherapy............................... 503
		13.3.5	Selected Publications........................ 503
	13.4	Cutaneous Lymphoma............................. 507
		13.4.1	Treatment Algorithm......................... 509
		13.4.2	Total Skin Irradiation (TSI)................... 509
		13.4.3	Selected Publications........................ 510
	13.5	Total Body Irradiation (TBI)......................... 512
		13.5.1	Selected Publications........................ 514
	References.. 519
Index... 521

Authors and Contributors

About the Authors

Murat Beyzadeoglu graduated from Cerrahpasa Medical School at the University of Istanbul in 1982 before going on to specialize in radiation oncology. He worked at the Department of Clinical Oncology at the Royal Marsden Hospital, London, and then became an Assistant Professor at Gulhane Military Medical School in Ankara. After working as the Turkish National Military Representative at SHAPE Medical Center, Mons, Belgium, and the Jules Bordet Cancer Institute in Brussels, he became first Assistant Professor, then Associate Professor, and currently Professor and Chairman at the University of Health Sciences Turkey, Gulhane Faculty of Medicine, Department of Radiation Oncology. From 2008 to 2010 he also served as President of the Balkan Military Medical Committee. In 2016, he became Chairman at the Department of Radiation Oncology at Gulhane Medical Faculty at Health Sciences University Turkey. Professor Beyzadeoglu is the editor of three previous Springer books and has authored numerous articles in peer-reviewed journals and book chapters.

Gokhan Ozyigit graduated from Hacettepe University Faculty of Medicine (English) in 1996 and then obtained ECFMG Certification in the USA. In 2001, he completed his Residency at the Department of Radiation Oncology at Hacettepe Univ1ersity. He later became a Research Fellow at the Departments of Radiation Oncology at Washington University and the University of Texas M.D. Anderson Cancer Center between 2001–2003. In 2004, he was appointed Assistant Professor at the Department of Radiation Oncology at Hacettepe University and in 2012 became Professor there. He is currently Chair at the Department of Radiation Oncology and Director of Advanced Cancer Technologies Research & Application Center at Hacettepe University. Dr. Ozyigit is an editorial board member of several international journals. He is also a reviewer or advisor for a number of other international journals, co-editor of ten previous books, and has authored numerous articles in peer-reviewed journals and book chapters. He received several national awards in the field of Oncology. He is currently President of Turkish Society for Radiation Oncology.

Cüneyt Ebruli graduated from Akdeniz University Faculty of Medicine (Antalya/Turkey) in 2000. In 2006, he completed his Residency at the Department of Radiation Oncology at Dokuz Eylul University (Izmir/Turkey). He later became Staff Radiation Oncologist at the Departments of Radiation Oncology at Kocaeli State Hospital. In 2013, he was appointed as Staff Radiation Oncologist to the Dr. Suat Seren Chest Diseases and Training and Research Hospital (Izmir/Turkey) and in 2018 transferred as Senior Registrar to Health Sciences University Turkey, Izmir Tepecik Training and Research Hospital. Dr. Cüneyt Ebruli is interested in aerodigestive cancers including head and neck cancers and upper cervical esophageal cancers; as an extra-medical, Lycia/Caria Archeology is in the field of interest and he continues his education in Dokuz Eylül University, Faculty of Fine Arts, Department of Photography. He received several national awards in the field of Radiation Oncology and has authored numerous articles in peer-reviewed journals and book chapters and is a co-editor of the first edition of Springer *Basic Radiation Oncology*.

Contributors

Selcuk Demiral, MD Department of Radiation Oncology, University of Health Sciences Turkey, Gulhane Faculty of Medicine, Ankara, Turkey

Ferrat Dincoglan, MD Department of Radiation Oncology, University of Health Sciences Turkey, Gulhane Faculty of Medicine, Ankara, Turkey

Selenge Bedük Esen, MD Department of Radiation Oncology, Hacettepe University, Faculty of Medicine, Ankara, Turkey

Ferhat Eyiler, MD Department of Radiation Oncology, University of Health Sciences Turkey, Tepecik Training and Research Hospital, Izmir, Turkey

Hakan Gamsız, MD Department of Radiation Oncology, University of Health Sciences Turkey, Gulhane Faculty of Medicine, Ankara, Turkey

Melis Gultekin, MD Department of Radiation Oncology, Hacettepe University, Faculty of Medicine, Ankara, Turkey

Pervin Hurmuz, MD Department of Radiation Oncology, Hacettepe University, Faculty of Medicine, Ankara, Turkey

Omer Sager, MD Department of Radiation Oncology, University of Health Sciences Turkey, Gulhane Faculty of Medicine, Ankara, Turkey

Sezin Yüce Sarı, MD Department of Radiation Oncology, Hacettepe University, Faculty of Medicine, Ankara, Turkey

Bora Uysal, MD Department of Radiation Oncology, University of Health Sciences Turkey, Gulhane Faculty of Medicine, Ankara, Turkey

Introduction and History

Roentgen was working on Crook's Vacuum tube on November 8th, 1895. He suddenly realized that shadows of his wife's finger bones and ring in her finger appeared on the palette. This was the discovery of X-rays and the beginning of radiation history (Fig. 1).

Henri Becquerel opened his drawer in his laboratory on March 1896. He was greatly surprised when he saw blackened photo glasses despite their being kept in a totally dark medium. This was the discovery of natural radioactivity (Fig. 2).

It has been more than a century since the discovery of X-rays by Roentgen and that of natural radioactivity by Becquerel, and in that time the field of radiation oncology has seen enormous changes due to the now-standard use of extraordinarily complex systems and high-technology products to treat cancers. The first use of X-rays in war front. A shrapnel piece in the hand of

Fig. 1 The first X-ray (an X-ray of the hand of Anna Berthe Roentgen)

Fig. 2 The blurred photo glasses of Becquerel

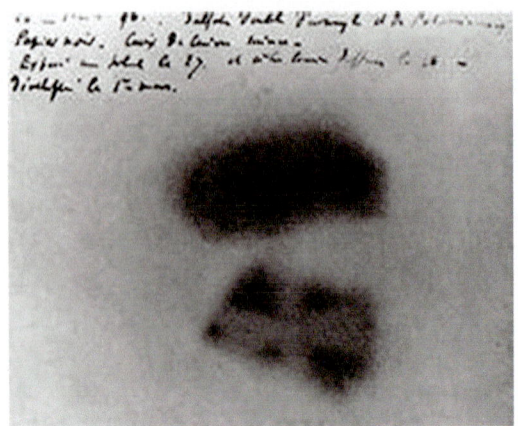

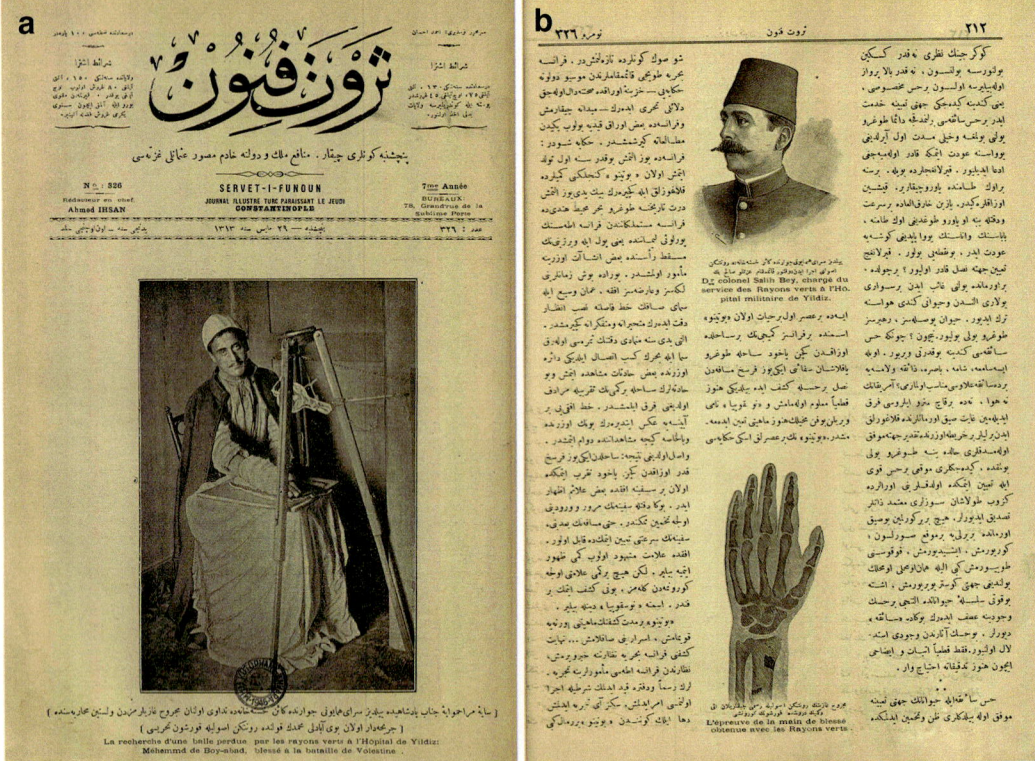

Fig. 3 (**a, b**) The first use of X-rays in war front. A shrapnel piece in the hand of an Ottoman soldier Mehmed at the Greco-Turkish war (10 June 1897)

Fig. 4 The first radiation oncologist, Emil Grubbe

an Ottoman soldier Mehmed at the Greco-Turkish war (10 June 1897) (Fig. 3). Indeed, just 2 months after the discovery of X-rays (i.e., on 1 Jan 1896), a medical student named Emil Grubbe (Fig. 4) used X-rays to treat a 65-year-old female patient named Rosa Lee with recurrent breast carcinoma at a lamp factory in Chicago.

We can summarize the most important developmental steps in the field of radiation oncology chronologically as follows:

1895: Discovery of X-rays by Wilhelm Conrad Roentgen (Germany)
1895: Use of X-rays in breast cancer by Emil Grubbe (Chicago, USA)
1896: Use of X-rays in nasopharyngeal cancer and in pain palliation by Voigt J. Ärztlicher Verein (Germany)
1896: Discovery of natural radioactivity by Henri Becquerel (Paris, France)
1896: Use of X-rays in the treatment of gastric cancer by Despeignes (France)
1896: Use of X-rays in the treatment of skin cancer by Léopold Freund (Austria)
1897: Discovery of electrons (Thompson)
1898: Discovery of radium by Pierre and Marie Curie (France)
1899: Definition of the alpha particle (E. Rutherford)
1901: The first use of radium in skin brachytherapy (Dr. Danlos, France)
1903: Publications showing the efficacy of radiotherapy in lymphoma (Senn & Pusey)
1905: Discovery of the sensitivity of seminoma to radiation (A. Béclère, France)
1905: Discovery of the photoelectric effect by A. Einstein (Germany)
1906: Discovery of characteristic X-rays (G. Barkla)
1922: Demonstration of the Compton effect (Arthur H. Compton)
1931: First cyclotron (Ernest O. Lawrence, USA)
1932: Discovery of neutrons (Sir James Chadwick, UK)

Fig. 5 The first rotational therapy in Radiation Oncology (Prof. Friedrich Dessauer, 1933, Istanbul-Turkey)

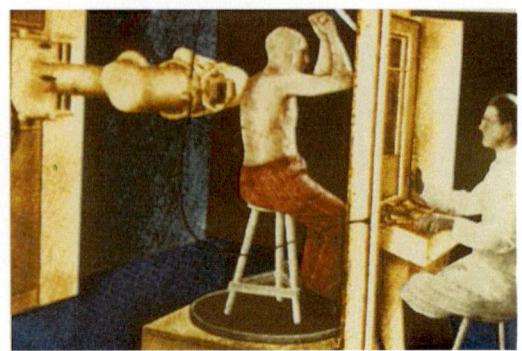

1934:   Discovery of artificial radioelements (Irène and Frédéric Joliot-Curie, France)
1934:   23% cure rate in head and neck cancer (Henri Coutard)
1934:   The first rotational therapy in Radiation Oncology in Istanbul, (Prof. Friedrich Dessauer Turkey) (Fig. 5)
1934:   Death of Mrs. Marie Curie due to pernicious anemia (myelodysplasia)
1940:   The first betatron (Donald W. Kerst)
1951:   The first cobalt-60 teletherapy machine (Harold E. Johns, Canada)
1952:   The first linear accelerator (linac) machine (Henry S. Kaplan, USA)
1968:   Discovery of the gamma knife (Lars Leksell)
1971:   The first computerized tomography (CT) (G.N. Hounsfield, UK)
1973:   The first MRI machine (Paul C. Lauterbur, Peter Mansfield)
1990:   The first use of computers and CT in radiotherapy (USA)
1994:   The first clinical IMRT treatment (USA)
1996:   FDA approval of the first IMRT software
2001:   FDA approval of robotic radiosurgery
2002:   FDA clearance of spiral (helical) tomotherapy
2003:   The first use of image-guided radiation therapy (IGRT) technology
2017:   MR Linear accelerator (UMC Utrecht-Holland)

Radiation Physics

1.1 Introduction and Atom

The word "atom" derives from the Greek word "atomos," which means indivisible; an atom was the smallest indivisible component of matter according to some philosophers in Ancient Greece [1]. However, we now know that atoms are actually composed of subatomic particles: protons and neutrons in the nucleus of the atom, and electrons orbiting that nucleus (Fig. 1.1).

The total number of protons and neutrons in a nucleus (p + n) (i.e., the total number of "nucleons") is termed the mass number of that atom, symbolized by A [1]. The total number of protons is called the atomic number and is symbolized by Z. The atomic number and the mass number of an element X are usually presented in the form $_Z^A X$ (Fig. 1.2).

Electrons are negatively charged particles.

Protons are positively charged particles. The mass of a proton is about 1839 times greater than that of an electron.

Neutrons are uncharged (neutral) particles. The mass of a neutron is very slightly larger than that of a proton.

Protons and neutrons form the nucleus of an atom, and so these particles are also called nucleons.

The diameter of an atom is about 10^{-8} cm, whereas the diameter of the atomic nucleus is 10^{-13} cm.

Nuclide → if an atom is expressed in the form $_Z^A X$, it is called a nuclide (e.g., $_2^4 He$).

Radionuclide → if the atom is expressed in the form $_Z^A X$ and is radioactive, it is called a radionuclide.

1.2 Radiation

The propagation of energy from a radiative source to another medium is termed radiation. This transmission of energy can take the form of particulate radiation or electromagnetic radiation (i.e., electromagnetic waves). The various forms of radiation originating from atoms, which include (among others) visible light, X-rays, and γ-rays, are grouped together under the terms "electromagnetic radiation" [1] or "the electromagnetic spectrum" [1, 2]. Radio waves, which have the longest wavelengths and thus the lowest frequencies and energies of the various types of electromagnetic radiation, are located at one end of the electromagnetic spectrum, whereas X-rays and γ-rays, which have the highest frequencies and energies, are situated at the other end of this spectrum.

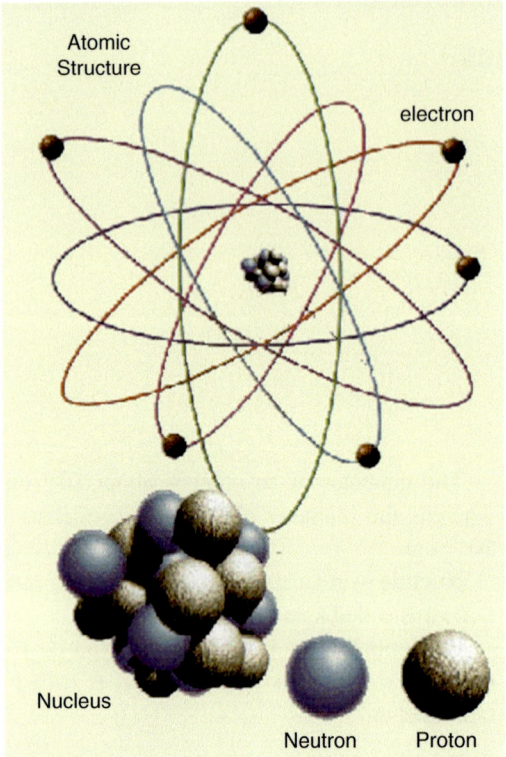

Fig. 1.1 The structure of an atom

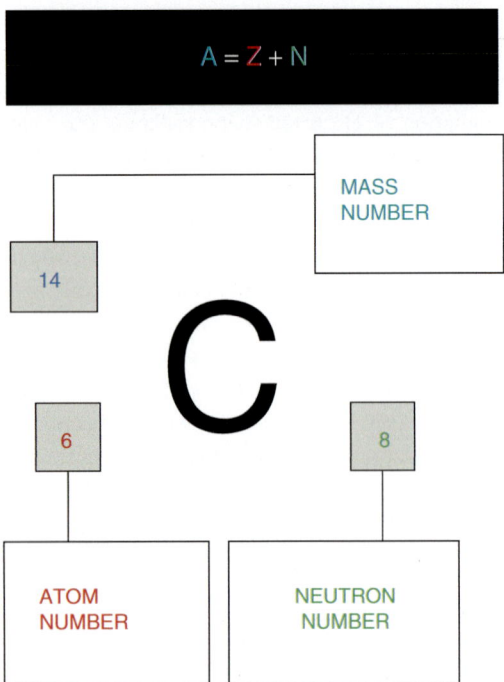

Fig. 1.2 Writing an element in nuclide format (carbon-14 is used as an example)

1.2.1 Photon

- If the smallest unit of an element is considered to be its atoms, the photon is the smallest unit of electromagnetic radiation [3].
- Photons have no mass.

1.2.1.1 Common Features of Electromagnetic Radiation [4, 5]

- It propagates in a straight line.
- It travels at the speed of light (nearly 300,000 km/s).
- It transfers energy to the medium through which it passes, and the amount of energy transferred correlates positively with the frequency and negatively with the wavelength of the radiation.
- The energy of the radiation decreases as it passes through a material, due to absorption and scattering, and this decrease in energy is negatively correlated with the square of the distance traveled through the material.

Electromagnetic radiation can also be subdivided into ionizing and nonionizing radiations. Nonionizing radiations have wavelengths of $\geq 10^{-7}$ m. Nonionizing radiations have energies of <12 electron volts (eV); 12 eV is considered to be the lowest energy that an ionizing radiation can possess [4].

Types of Nonionizing Electromagnetic Radiation [5]
- Radio waves
- Microwaves
- Infrared light
- Visible light
- Ultraviolet light

1.3 Ionizing Radiation

Ionizing (high-energy) radiation has the ability to remove electrons from atoms; i.e., to ionize the atoms. Ionizing radiation can be electromagnetic or particulate radiation (Fig. 1.3). Clinical radiation oncology uses photons (electromagnetic) and electrons or (rarely) protons or neutrons (all

1.3 Ionizing Radiation

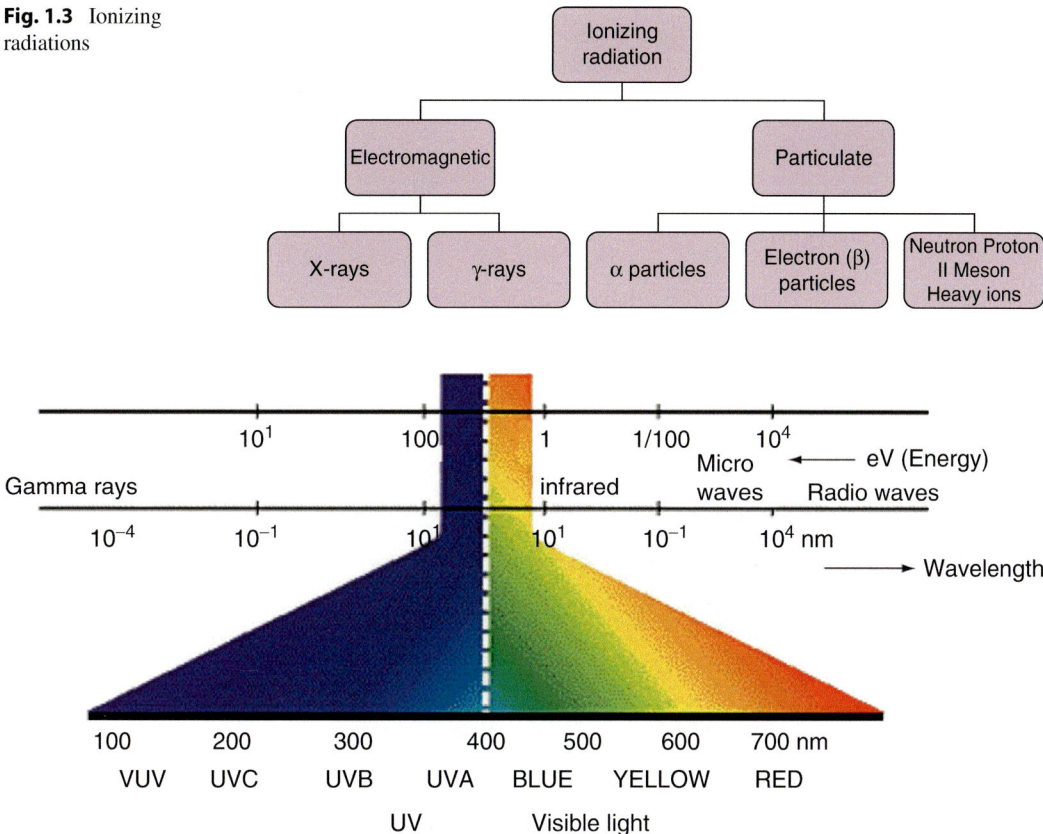

Fig. 1.3 Ionizing radiations

Fig. 1.4 Electromagnetic spectrum

three of which are particulate) as radiation in the treatment of malignancies and some benign conditions [6].

1.3.1 Ionizing Electromagnetic Radiation

The electromagnetic spectrum comprises all types of electromagnetic radiation, ranging from radio waves (low energy, long wavelength, low frequency) to ionizing radiations (high energy, short wavelength, high frequency) (Fig. 1.4) [7].

Electrons are knocked out of their atomic and molecular orbits (a process known as ionization) when high-energy radiation interacts with matter [8]. Those electrons produce secondary electrons during their passage through the material. A mean of energy of 33.85 eV is transferred during the ionization process, which in atomic and molecular terms is a highly significant amount of energy. When high-energy photons are used clinically, the resulting secondary electrons, which have an average energy of 60 eV per destructive event, are transferred to cellular molecules.

1.3.1.1 X-Rays

X-rays were discovered by the German physicist Wilhelm Conrad Roentgen in 1895 [9]. The hot cathode Roentgen tube, which was developed by William David Coolidge in 1913, is a pressured (to 10^{-3} mmHg) glass tube consisting of anode and cathode layers between which a high-energy (10^6–10^8 V) potential is applied (Fig. 1.5a, b). Electrons produced by thermionic emission in the cathode are accelerated toward the anode by the potential. They thus hit the anode, which is a metal with high melting temperature. X-rays are

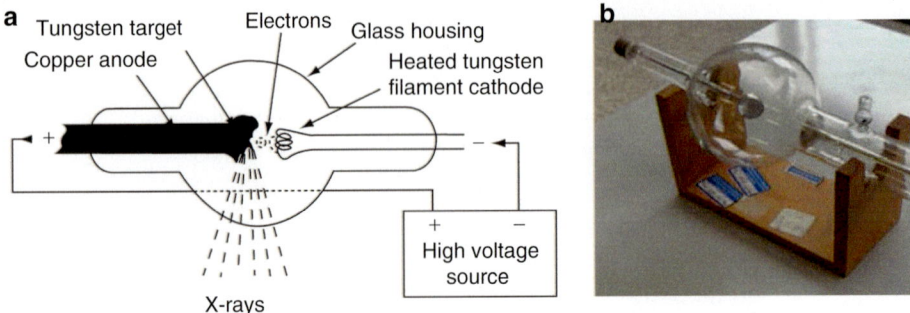

Fig. 1.5 (**a**) Schematic representation of an X-ray tube; (**b**) photograph of an X-ray tube

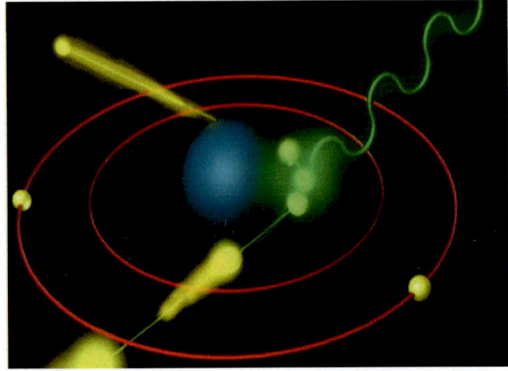

Fig. 1.6 Bremsstrahlung process

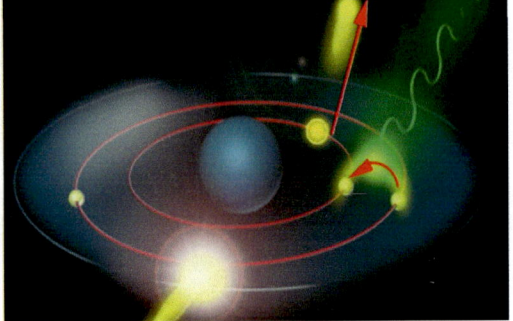

Fig. 1.7 Characteristic X-ray generation

produced by the sudden deceleration of these electrons due to Coulomb interactions with nuclei in the anode (this sudden deceleration of fast-moving electrons is known as bremsstrahlung; Fig. 1.6). The energy and the wavelength of the X-rays depend on the atomic number of the target (anode) metal, as well as the velocity and the kinetic energy of the electrons. This process is used to produce medical radiation in diagnostic X-ray units, linear accelerators (linacs), and betatrons.

X-rays are produced by extranuclear procedures. Two kinds of X-rays are created by X-ray tubes [10, 11]. The first type corresponds to the bremsstrahlung X-rays mentioned above. The second type occurs because an electron in an inner atomic orbital is knocked out by an incoming electron, and the resulting space in the orbital is filled by another electron that moves from an outer atomic orbital (Fig. 1.7). This electron must shed energy to move in this manner, and the energy released is radiated as characteristic X-rays [12]. They are characteristic due to the fact that their energy depends on the target metal onto which the electrons are accelerated.

X-rays produced by bremsstrahlung have a broad energy spectrum (→ heterogeneous), while characteristic X-rays are monoenergetic beams.

1.3.1.2 Gamma (γ) Rays

Gamma rays are physically identical to X-rays, but they are emitted from atomic nuclei (intranuclearly). An unstable atomic nucleus sheds its excess energy in the form of either an intranuclear electron (e⁻) (beta particle) or a helium nucleus (an "alpha particle") (Fig. 1.8). If it still possesses excess energy after that, gamma rays are emitted in order to reach its steady state (Fig. 1.9).

Gamma rays have well-defined energies. For instance, two monoenergetic gamma rays with a mean energy of 1.25 MV (1.17 and 1.33 MV) are emitted after beta rays of 0.31 MV energy have

1.3 Ionizing Radiation

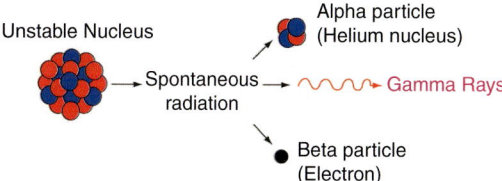

Fig. 1.8 Alpha particle generation

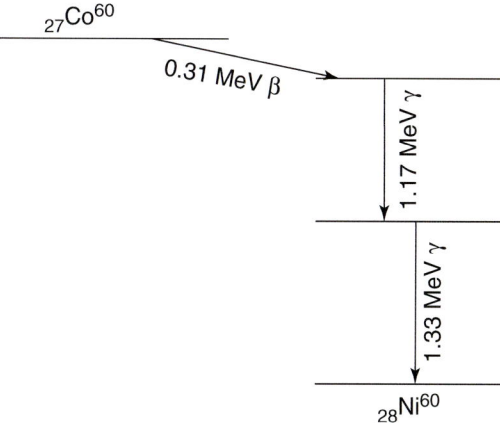

Fig. 1.9 Co-60 decay

been emitted during the decay of ^{60}Co (cobalt-60; Co-60). Through this process, ^{60}Co transforms into a final, stable decay product, ^{60}Ni (nickel-60; Ni-60). There is actually a stable naturally occurring form of cobalt: ^{59}Co. ^{60}Co is created through neutron bombardment in nuclear reactors, and has a half-life of 5.26 years. One gram of ^{60}Co has an activity of 50 Ci (1.85 terabecquerels) [13, 14].

The half-life of a radioisotope is the time required for its activity to half [15].

The activity of a radioisotope is the number of decays per second, and is defined in becquerels or curies.

– Becquerel (Bq): the standard unit of (radio) activity; it is defined as one disintegration (decay) per second.
– Curie (Ci): an older unit of (radio)activity, corresponding to 3.7×10^{10} disintegrations per second.

The decay of a radioactive nucleus is a spontaneous process. There are three forms of radioactive decay. Alpha or beta particles are emitted during the alpha and beta decays of an unstable nucleus in order to reach a stable nucleus. A gamma decay occurs without any change in the form of the nucleus.

There are also three types of beta decay. In all of them, the mass number of the nucleus remains constant during the decay, while the numbers of protons and neutrons change by one unit. Furthermore, the emission of some massless, uncharged particles called neutrinos and antineutrinos is observed during each beta decay process. The existence of these particles was first suggested by Pauli in 1930, although it was Fermi that provided the name "neutrino" [16].

Alpha Decay [16]. An alpha particle consisting of two protons and two neutrons is emitted if a nucleus is unstable because it has an excessive number of both protons and neutrons (Fig. 1.10).

After alpha decay, the alpha particle possesses most of the energy, due to the conservation of momentum and the fact that the alpha particle is much less massive than the residual nucleus. Although the $^{4}_{2}$He nucleus is very energetic, it does not travel very far compared to most forms of radiation, due to its relatively heavy mass. Alpha decay is usually observed in nuclei with mass numbers of more than 190. The energy spectrum of alpha decay is not continuous, and varies between 4 and 10 MeV. Alpha particles strongly interact with the electrons of the matter through which they pass, since they are charged particles.

Beta Decay [17]. There are three types of beta decay.

If a radionuclide is unstable because it has an excess number of neutrons in its nucleus, it transforms one of the neutrons into a proton and an electron in order to reduce the amount of energy in its nucleus (Fig. 1.11). The electron is rapidly propelled out of the nucleus, while the proton remains. This high-speed electron is called a β^- particle or negatron, and the process is termed β^- decay. The atomic number of the radionuclide increases by one, and thus it changes into the next element in the periodic table. Note that the mass

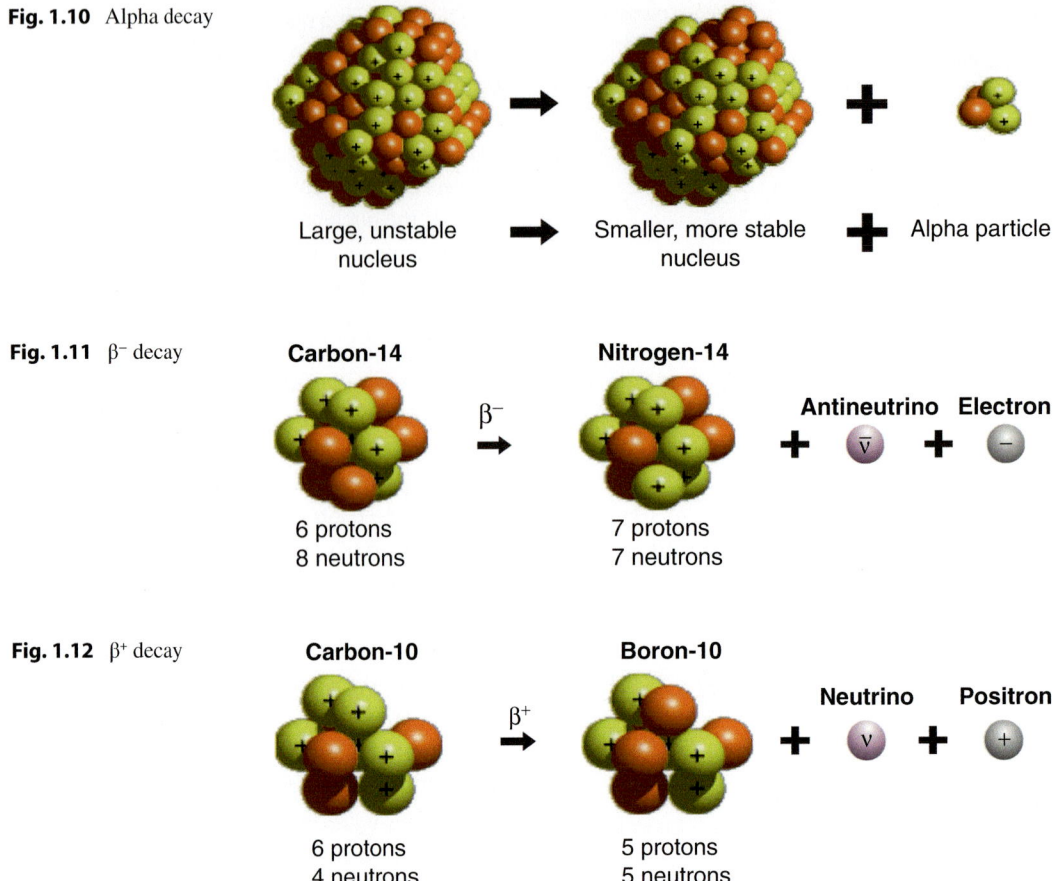

Fig. 1.10 Alpha decay

Fig. 1.11 β⁻ decay

Fig. 1.12 β⁺ decay

number does not change (it is an "isobaric" decay) [16, 17].

If a radionuclide is unstable due to an excess amount of protons or a lack of neutrons, one of the protons transforms into a neutron and a small positively charged particle called a positron in a process termed *β⁺ decay* [17]. The neutron stays in the nucleus while the positron is propelled out of it (Fig. 1.12). The atomic number of the radionuclide that emits the positron decreases by one, and thus it changes into the preceding element in the periodic table. Again, note that the mass number does not change.

If the nucleus is unstable due to an excess amount of protons, one of the electrons close to the atomic nucleus, such as an electron in a K and L orbital, is captured by the nucleus (Fig. 1.13a–c). This electron then combines with a proton, yielding a neutron and a neutrino. This process is called *electron capture* [16]. Note that no particle is emitted from the nucleus, but the atomic number decreases by one, as in positron decay. Yet again, the mass number does not change. The space in the inner orbital is filled by an electron from an outer orbital, resulting in the emission of characteristic X-rays.

Instead of being released in the form of an X-ray, the energy may occasionally be transferred to another electron which is ejected from the atom later. This is referred to as the "Auger effect" and the ejected electron is called the "Auger electron" (Fig. 1.13a–c).

β⁻ decay, β⁺ decay, electron capture phenomenon are summarized in Fig. 1.14.

Gamma Emission [13, 14, 16]. A nucleus is not always fully stable (i.e., at its basal energy level) just after it decays; sometimes, the nucleus will be in a semi-stable state instead

1.3 Ionizing Radiation

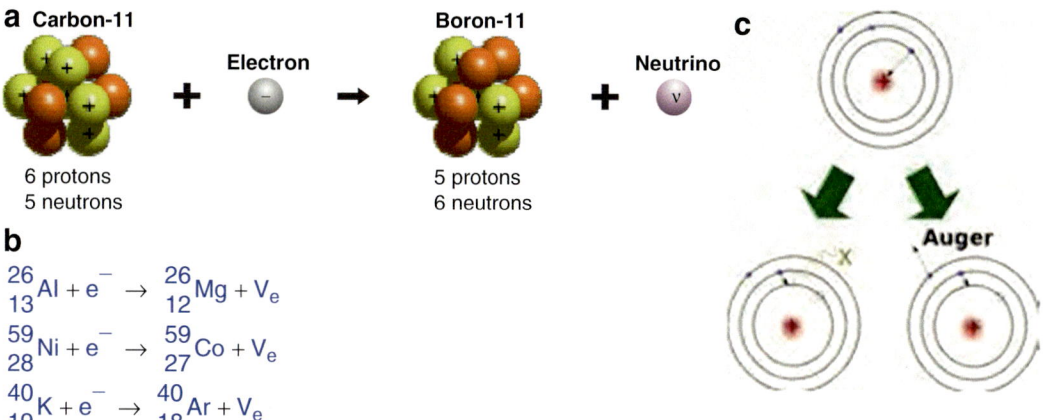

Fig. 1.13 (a–c) Electron capture phenomenon

$$^A_ZX \rightarrow\, ^A_{Z+1}X' + e^- + \bar{v}_e \quad \bullet\ \beta^- \text{ decay}$$

$$^A_ZX \rightarrow\, ^A_{Z-1}X' + e^+ + v_e \quad \bullet\ \beta^+ \text{ decay}$$

$$^A_ZX + e^- \rightarrow\, ^A_{Z-1}X' + e^+ + v_e \quad \bullet\ \text{Electron capture}$$

Fig. 1.14 β⁻ decay, β⁺ decay, electron capture phenomenon

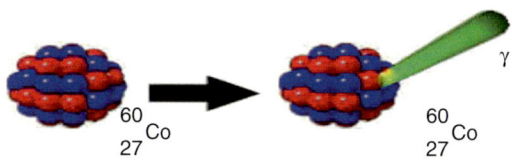

Fig. 1.15 Gamma emission

(Fig. 1.15). The excess energy carried by the nucleus is then emitted as gamma radiation. There is no change in the atomic or mass number of the nucleus after this decay, so it is termed an "isomeric" decay.

The half-lives of gamma radiation sources are much shorter than sources of other types of decay, and are generally less than 10^{-9} s. However, there are some gamma radiation sources with half-lives of hours or even years. Gamma energy spectra are not continuous.

Isotope [18]. Atoms with the same atomic number but different mass numbers are called isotopes (e.g., $^{11}_{6}C$, $^{12}_{6}C$, $^{13}_{6}C$).

Isotone. Atoms with the same number of neutrons, but different numbers of protons are called isotones (e.g., $^{9}_{3}Li$, $^{10}_{4}Be$, $^{11}_{5}B$, $^{12}_{6}C$).

Isobar. Atoms with the same number of nucleons but different numbers of protons are called isobars (e.g., $^{12}_{5}B$, $^{12}_{6}C$, $^{12}_{7}N$).

Isomer. Atoms with the same atomic and mass numbers but which are in different energy states are called nuclear isomers (Tc^{99m}).

1.3.2 Ionizing Particulate Radiation

Electrons, protons, alpha particles, neutrons, pi mesons, and heavy ions are all forms of ionizing particulate radiation (Figs. 1.16, 1.17, and 1.18) [19]. Electrons are particles that are generally used in routine clinics. Other particles are only used in specific clinics worldwide.

Electrons, due to their negative charge and low mass, can be accelerated to high energies in linacs or betatrons.

Electrons are normally bound to a (positively charged) nucleus. The number of electrons is equal to the number of protons in a neutral atom. However, an atom can contain more or less electrons than protons, in which case it is known as a negatively or positively charged ion, respectively. Electrons that are not bound to an atom are called

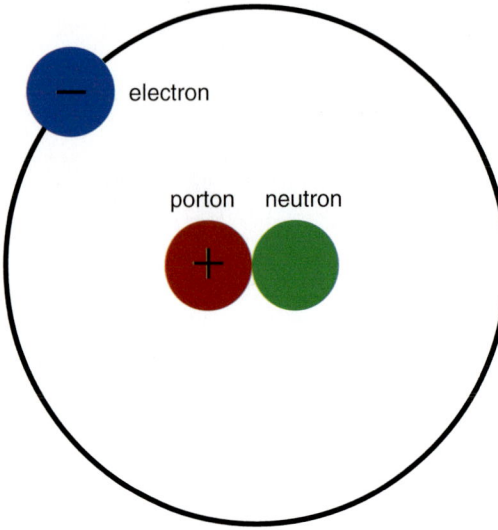

Fig. 1.16 Atom and particles

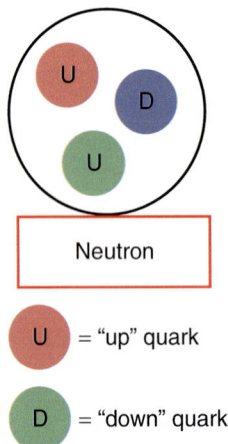

Fig. 1.17 Neutron

free electrons; free electrons can be produced during nuclear decay processes, in which case they are called beta particles.

Neutrons are the neutrally charged particles that enable the formation of stable large atomic nuclei (Fig. 1.17) by decreasing the repulsion between the protons in the nucleus. However, neutrons, like protons, actually consist of particles called quarks; a neutron is one up quark and two down quarks, while a proton (Fig. 1.18) is two up quarks and one down quark.

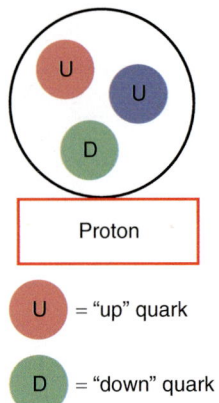

Fig. 1.18 Proton

The mass of an electron is 9.109 3826(16) × 10^{-31} kg.

The electrical charge of an electron is −1.602 176 53(14) × 10^{-19} C.

Electrons have much smaller ranges (i.e., they travel smaller distances) in matter than gamma and X-rays, and can be absorbed by plastics, glass, or metal layers (Fig. 1.19).

1.4 The Interaction of Radiation with Matter

Radiation is scattered and absorbed when it passes through tissue [19, 20]. The intensities of monoenergetic X-rays or gamma rays attenuate exponentially within tissues. In other words, the intensity of radiation constantly decreases as it propagates within tissues. This decrease depends on the type of tissue and its thickness. If the wavelength stays constant, the intensity of the radiation passing through a tissue can be calculated by the following formula:

As seen in the above formula, the intensity of the radiation decreases exponentially with the absorbent thickness, and the intensity of the outgoing radiation depends on the tissue absorption coefficient and its thickness.

$$\left[I = I_0 \cdot e^{-\mu t} \right] \quad (1.1)$$

I = intensity of outgoing radiation beam,
I_0 = intensity of incoming radiation beam,

1.4 The Interaction of Radiation with Matter

Fig. 1.19 Penetration ranges of various ionizing radiations

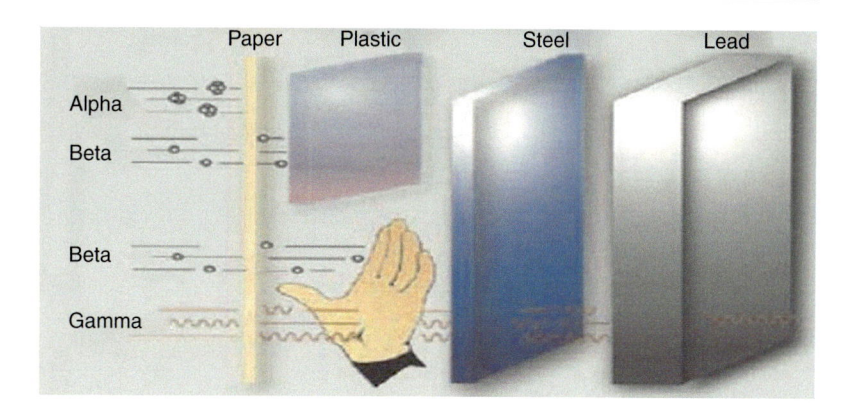

μ = absorption coefficient (which is positively correlated with the fourth power of the atomic number of the penetrated tissue and the third power of the wavelength of the radiation), t = tissue thickness.

Three of the five types of interaction of radiation with matter determine the absorption coefficient → photoelectric effect, Compton effect, and pair production.

1.4.1 Photoelectric Effect

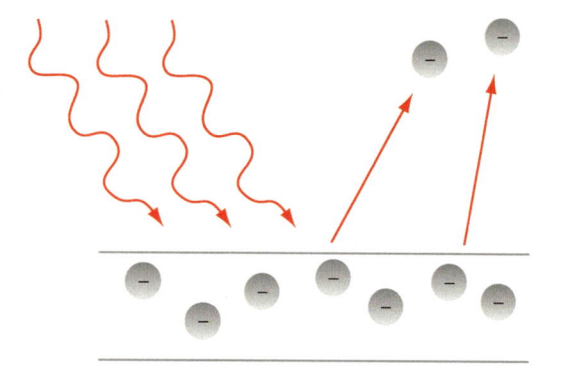

Fig. 1.20 Photoelectric effect

This phenomenon, which was theorized by Albert Einstein in 1905, was actually first observed by Heinrich Rudolf Hertz in 1887, and was therefore also known as the Hertz effect [21]. To define it simply, when any electromagnetic radiation reaches a surface (generally a metallic surface), it transfers its energy to the electrons of that surface, which are then scattered. At the atomic level, the incoming radiation knocks an electron from an inner atomic orbital, propelling it from the atom (Fig. 1.20).

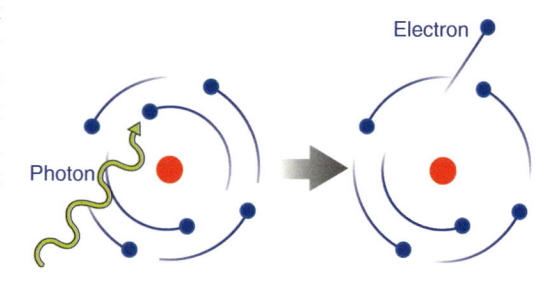

Fig. 1.21 The illustration of photoelectric effect

1.4.1.1 Photoelectric Effect (Fig. 1.20)
This is the basic interaction in diagnostic radiology (Fig. 1.21).

It is dominant at energies of less than 35 kV, and in atoms with high atomic numbers (Z).

Since the atomic number of bone is higher than that of soft tissue, bone absorbs more radiation than soft tissue. This absorption difference is the basis of diagnostic radiology.

This effect also explains why metals with high atomic numbers (e.g., lead) are used to absorb low-energy X-rays and gamma rays.

1.4.2 Compton Effect

In the Compton effect, a photon collides with an electron in an outer orbital, and the photon and electron are scattered in different directions

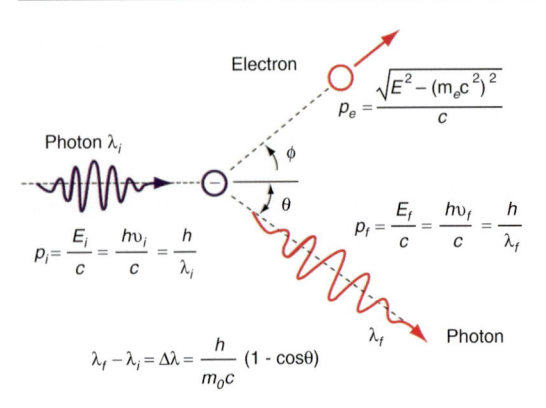

Fig. 1.22 Math associated with the Compton effect

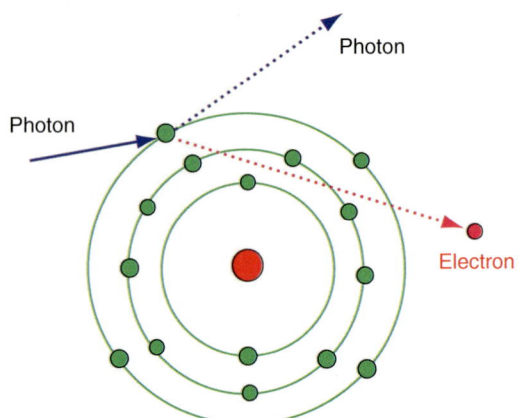

Fig. 1.23 Illustration of the Compton effect

(where q is the angle between the directions) [21]. The energy of the incoming photon is transferred to the electron in the form of kinetic energy. The scattered electron also interacts with the outer orbital electrons of other atoms. After the interaction, the photon has a lower energy than it did beforehand (Fig. 1.22).

1.4.2.1 Compton Effect (Fig. 1.23)

This is the main mechanism for the absorption of ionizing radiation in radiotherapy.

It is the dominant effect across a wide spectrum of energies, such as 35 kV–50 MV.

It has no dependency on the atomic number (Z) of the absorbent material, but it does depend on the electron density of the material.

The absorption of incoming radiation is the same for bone and soft tissues.

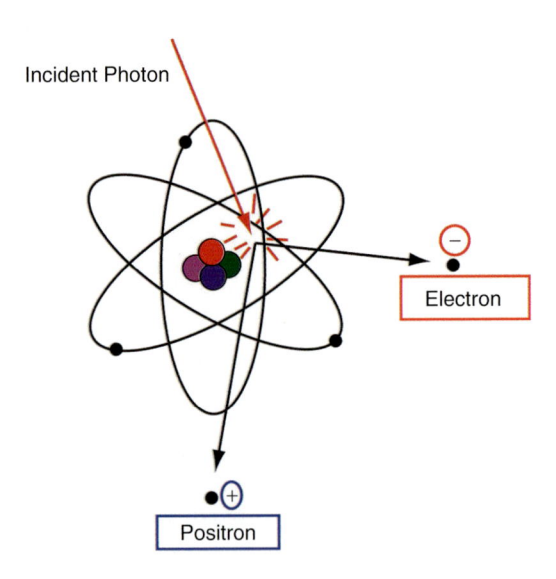

Fig. 1.24 Pair production

1.4.3 Pair Production

This is a relatively rare effect. In it, a photon transforms into an electron and a positron near a nucleus (Fig. 1.24) [21]. The electron sheds all of its energy by the absorption processes explained above. On the other hand, the positron propagates through the medium ionizing atoms until its energy has dropped to such a low level that it pulls a free electron close enough to combine with it, in a process called annihilation. This annihilation causes the appearance of a pair of photon moving in opposite directions, and each with 0.511 MeV of energy. These annihilation photons are absorbed through either photoelectric or Compton events.

1.4.3.1 Pair Production

The threshold photon energy level for pair production is 1.02 MeV; below this, pair production will not occur.

The probability of pair production occurring increases as Z increases.

Pair production is more frequently observed than the Compton effect at energies of more than 10 MeV (Fig. 1.25).

1.4.4 Coherent Effect (= Rayleigh Scattering, = Thomson Scattering)

Here, an electron is scattered when an electromagnetic wave or photon passes close to it [21]. This type of scattering is explained by the waveform of the electromagnetic radiation. There are two types of coherent scattering: Thomson scattering and Rayleigh scattering (Fig. 1.24). The wave/photon only interacts with one electron in Thomson scattering, while it interacts with all of the electrons of the atom in Rayleigh scattering. In Rayleigh scattering, low-energy radiation interacts with an electron, causing it to vibrate at its own frequency. Since the vibrating electron accelerates, the atom emits radiation and returns to its steady state. Thus, there is no overall transfer of energy to the atom in this event, so ionization does not occur. The probability of coherent scattering is high in heavy (i.e., high-Z) matter and for low-energy photons (Fig. 1.26).

The probabilities of the various photon–matter interactions are summarized in Table 1.1.

The differences between bremsstrahlung X-rays and characteristic X-rays are described in Table 1.2.

- The coherent effect is the reason that the sky is blue during the day and reddish at sunset (Fig. 1.27)
- X-rays and γ-rays show the same intrinsic effects, but are produced in different ways (X-rays are generated extranuclearly, and gamma rays intranuclearly)
- Low-energy (nonionizing) radiation cannot ionize atoms because it does not have sufficient energy to do so, and thus it only causes excitation
- The target in mammography tubes is molybdenum, which produces characteristic X-rays of 17 keV.

Electron volt (eV): this is the amount of kinetic energy gained by an electron when it is accelerated by a potential difference of 1 V (1 eV = $1.60217646 \times 10^{-17}$ erg = $1.60217646 \times 10^{-19}$ J).

1.5 Specific Features of X-Rays

X-rays are a type of electromagnetic radiation with wavelengths of 10–0.01 nm, frequencies of 30–30,000 pHz (10^{15} Hz), and typical photon energies of 100 eV–100 keV (Table 1.3).

X-rays are generally produced in either X-ray tubes or linacs. X-ray tubes are the main source of X-rays in laboratory instruments. In such a tube, a focused electron beam is accelerated under high voltage within a glass vacuum tube, they hit to a fixed or rotating target. When the electrons approach target atoms, Coulomb interactions with the nuclei cause the electrons to be suddenly deflected from their previous paths and slowed. During this braking process, energy in the form of X-rays is produced in a continuous spectrum (→ bremsstrahlung X-rays). High-energy electrons hit inner orbital electrons and knock them out of the atom during the ionization process. Free electrons from outer orbits then fill the empty spaces in the inner orbitals, and X-rays

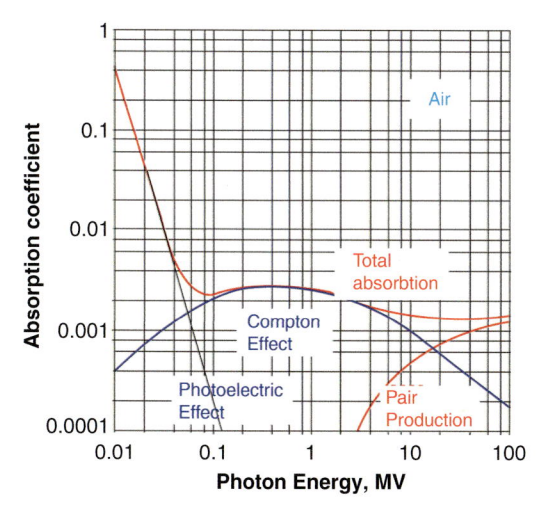

Fig. 1.25 Relationship between photon energy and absorption coefficient for various types of interaction with matter (in air)

Fig. 1.26 Rayleigh scattering

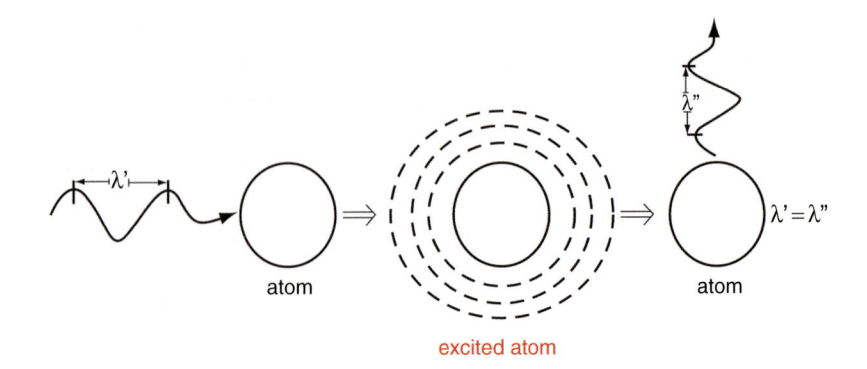

Table 1.1 Interaction probabilities for various photon energies in the photoelectric effect, the Compton effect, and pair production in water

Photon energy (MeV)	Interaction probability (%)		
	Photoelectric effect	Compton effect	Pair production
0.01	95	5	0
0.026	50	50	0
0.060	7	93	0
0.150	0	100	0
4.00	0	94	6
10.00	0	77	23
24.00	0	50	50
100.00	0	16	84

with energies that are characteristic of the target are produced (→ characteristic X-rays) (Fig. 1.28).

Early in the history of radiotherapy, the X-ray beams used had energies of only between 250 and 400 kV. The penetrative abilities of those X-rays were rather poor. Most of the energy of the X-rays produced can be explained by the X-ray energy spectrum. Unfiltered low-energy X-rays can easily be absorbed by superficial tissues in the human body, and so excessive dose

Table 1.2 The differences between bremsstrahlung X-rays and characteristic X-rays

Bremsstrahlung	Characteristic X-rays
Also known as white radiation or braking radiation	Since it is produced by the movement of an outer orbital electron to an inner orbital, the energy of the X-ray is equal to the difference in the binding energies of the two orbits
Photon energy spectrum is equal to the initial electron energy	Produced X-rays are monoenergetic (constant energy)
Occurrence probability increases with the square of the target's atomic number	Characteristic X-rays comprise 30% of the X-rays used in diagnostic radiology, but only 3% of those used in radiotherapy
Used to create most X-rays >100 keV	
For <100 keV, the angle between the photon produced by bremsstrahlung and the outgoing electron is 90°; this angle shortens when the beam energy is >100 keV	Occurrence probability increases with the square of the target's atomic number
Produced X-rays are heterogeneous in energy	

1.5 Specific Features of X-Rays

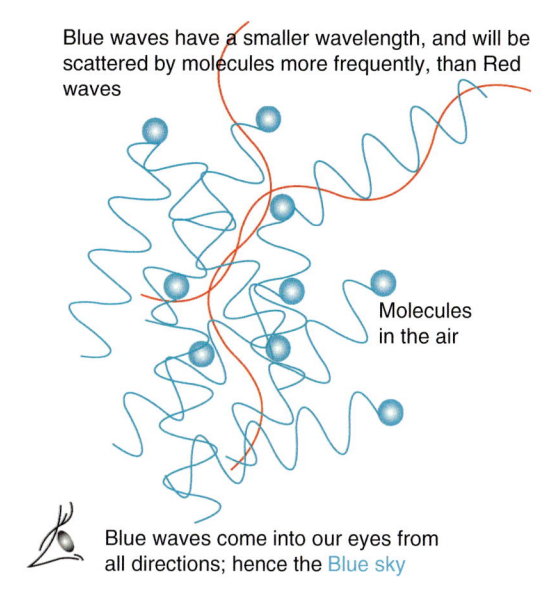

Blue waves have a smaller wavelength, and will be scattered by molecules more frequently, than Red waves

Molecules in the air

Blue waves come into our eyes from all directions; hence the Blue sky

Fig. 1.27 Why the sky is blue

accumulation can occur in these areas, causing serious dermal reactions during the treatment of deeply seated tumors.

X-rays are used in various applications in industry and medicine (Table 1.4).

The energy of an X-ray photon depends on its wavelength:

$$\left[E = h\frac{c}{l} \right] \quad (1.2)$$

Here, h is the Planck constant (6.626 × 10^{34} J/s), c is the speed of light, and λ is the wavelength.

Table 1.3 Features of X-rays of various energies

Photon energy		Frequency (Hz)	Wavelength (pm, =10^{-12} m)	Half-value layer (HVL)			
(keV)	(J)			Cement	Lead	Human body	Aluminum
1	1.602 × 10^{-16}	2.418 × 10^{17}	1240	0.87 μm	0.117 μm	1.76 μm	2.17 μm
10	1.602 × 10^{-15}	2.418 × 10^{18}	124	147 μm	4.68 μm	1220 μm	97.9 μm
100	1.602 × 10^{-14}	2.418 × 10^{19}	12.4	17.3 mm	0.110 mm	38.6 mm	15.1 mm
1000	1.602 × 10^{-13}	2.418 × 10^{20}	1.24	46.4 mm	8.60 mm	93.3 mm	41.8 mm
10,000	1.602 × 10^{-12}	2.418 × 10^{21}	0.124	132 mm	12.3 mm	298 mm	111 mm

Fig. 1.28 X-ray spectrum

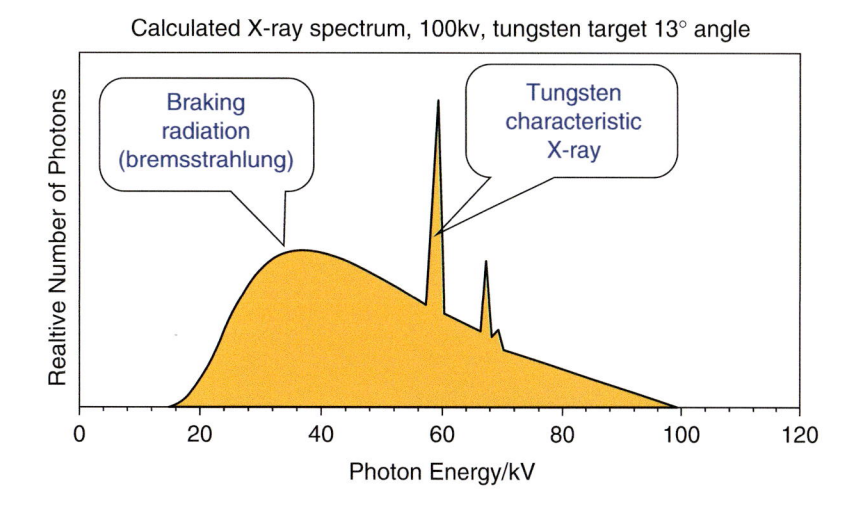

Calculated X-ray spectrum, 100kv, tungsten target 13° angle

Braking radiation (bremsstrahlung)

Tungsten characteristic X-ray

Realtive Number of Photons

Photon Energy/kV

Table 1.4 The various uses of X-rays and their features

Usage		Acceleration potential	Target	Source type	Mean photon energy
X-ray crystallography		40 kV 60 kV	Cooper Molybdenum	Tube	8–17 keV
Diagnostic radiology	Mammography	26–30 kV	Rhodium Molybdenum	Tube	20 keV
	Dentistry	60 kV	Tungsten	Tube	30 keV
	Roentgen	50–140 kV	Tungsten	Tube	40 keV
	CT	80–140 kV	Tungsten	Tube	60 keV
Airport, customs	Airport	80–160 kV	Tungsten	Tube	80 keV
	Customs	450 kV–20 MV	Tungsten	Tube/linear accelerator	150 keV–9 MeV
Structural analysis		150–450 kV	Tungsten	Tube	100 keV
Radiotherapy		10–25 MV	Tungsten/high atomic number matter	Linear accelerator	3–10 MeV

1.6 Specific Features of Electron Energies

Electrons with energies of 2–10 MeV are low-energy electrons, whereas electrons with energies of 10–42 MeV are high-energy electrons. Low-energy electrons can also be produced by some radioactive isotopes such as yttrium-90. Medium- to high-energy electrons are produced in linacs or betatrons. They are used in superficial treatments due to their limited penetration ability.

Energy is selected according to target depth and field size in radiotherapy (Fig. 1.29). Co-60 can be selected for thin regions such as the head and neck and extremities, whereas high-energy X-rays are preferred for thicker regions such as the abdomen, pelvis, and thorax. Electrons are generally used to treat skin cancers and superficially located tumors.

1.7 Ionizing Radiation Units

The amount of radiation delivered needs to be known in order to determine possible harmful biological effects and to reach definite conclusions in studies that use ionizing radiation. Specific units are required for radiation measurement.

Units of radiation measurement have changed dramatically over the years, and some units have been completely abandoned (e.g., the pastille), while other units have been introduced (Fig. 1.30).

Activity unit. This is the number of spontaneous nuclear disintegrations (N) per unit time (t) ($A = N/t$), as measured in becquerels (Bq). Note that an older system of units, the curie (Ci), is also often encountered.

Kerma (kinetic energy released in the medium). This is the sum of the initial kinetic energies of all of the charged particles liberated by uncharged ionizing radiation (neutrons, protons) in a sample of matter divided by the mass of the sample. The kerma is measured in the same units as absorbed dose (Gy).

Absorbed dose. The basic quantity associated with radiation measurement in radiotherapy is the absorbed dose. This defines the amount of energy absorbed from a radiation beam per unit mass of absorbent material. It is measured in grays (Gy), although an older unit, the rad, is also still used.

Exposure. This is the amount of ionization produced by photons in air. Since it is impossible to directly measure the absorbed dose in tissue, the measurement of radiation is performed in air.

1.7 Ionizing Radiation Units

Fig. 1.29 Relative depth dose distributions of photons and electrons

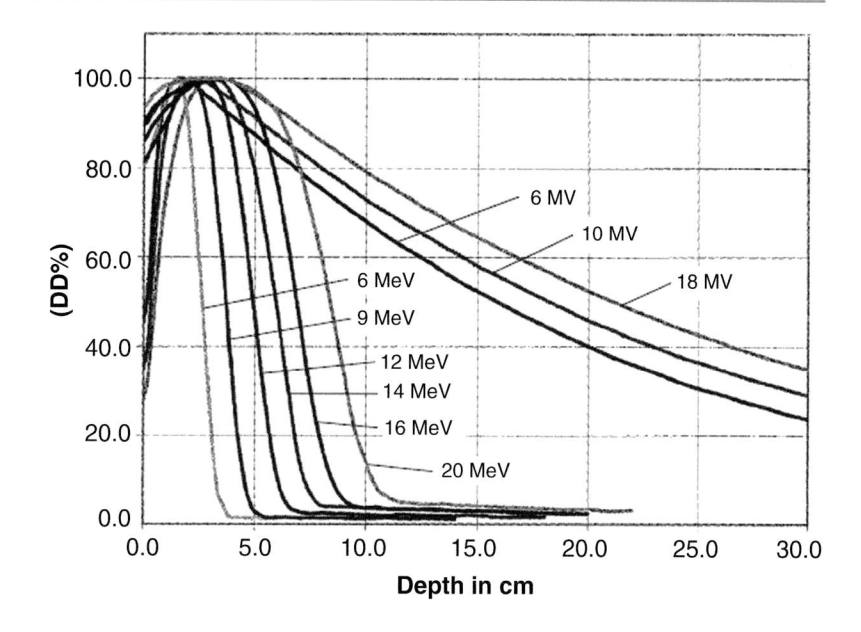

Fig. 1.30 Scientists whose names are used as the units in which radiation-associated quantities are measured

W.K. Röntgen   H. Gray   H. Becquerel   R. Sievert   M. Curie

The exposure is the amount of radiation required to liberate a positive or negative charge of one electrostatic unit of charge (esu) in 1 cm^3 of dry air at standard temperature and pressure (this corresponds to the generation of approximately 2.08×10^9 ion pairs). It is measured in coulombs per kilogram (C/kg), although the old unit of the roentgen (R) is also commonly encountered.

Equivalent dose. Since different radiations have different harmful effects on human tissues, the basic dosimetric unit of absorbed dose (→ Gy) is not sufficient for studies of radiation protection. Thus, the absorbed dose in tissue must be multiplied by a radiation-weighting factor that depends on the type of radiation employed. The resulting dose is called the equivalent dose, and it is measured in sieverts (Sv), although an older unit, the rem (roentgen equivalent man), is often used too.

$$H = D \times \mathrm{WR} \qquad (1.3)$$

H = equivalent dose (Sv), WR = radiation-weighting factor (no unit), D = dose (Gy)

1 Sv = 1 J/kg = 100 rem.

The measured quantities associated with ionizing radiation can be briefly summarized as follows (Fig. 1.31):

Source → activity units

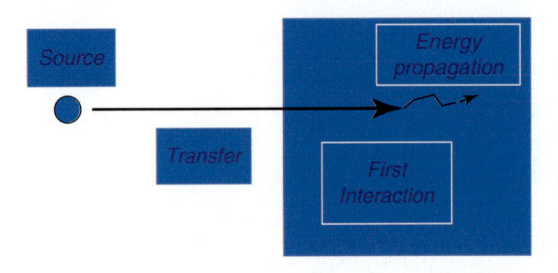

Fig. 1.31 Three points associated with ionizing radiation measurement

The first interaction point → kinetic energy released in matter (kerma)
Matter → absorbed dose

Radioactivity. This is the transition of an unstable nucleus to a steady state through the emission of particulate or electromagnetic radiation from the nucleus.

Curie (Ci). This is an activity of 3.7×10^{10} disintegrations per second.

Becquerel (Bq). This is an activity of one disintegration per second.

1 Ci = 33.7×10^{10} Bq.
1 Bq = 2.7×10^{-11} Ci

The *reference air kerma* is used to define the visible activity. It is the dose delivered in 1 h to air 1 m away from a source with an activity of 1 MBq. Its units are 1 µGy–1. m^2 = 1 cGy. h^{-1}. cm^2.

Rad. This is the amount of radiation that causes one erg (of energy) to be absorbed per gram of irradiated material (rad = radiation absorbed dose).

1 rad = 100 erg/g.

Gray (Gy). This is the amount of radiation that causes one joule to be absorbed per kilogram of irradiated material.

1 Gy = 1 J/kg.
1 Gy = 100 cGy = 100 Rad.

Roentgen (R). In normal air conditions (0°C and 760 mmHg pressure), this is the amount of X-radiation or gamma radiation that produces 2.58×10^{-4} coulombs of electrical charge (in the form of ions) in 1 kg of air.

C/kg. In normal air conditions, this is the amount of radiation that produces one coulomb of electrical charge (in the form of ions) in 1 kg of air.

Integral dose. This is the total energy absorbed in the treated volume (in J = kg × Gy).

The roentgen and C/kg are only used for photonic radiation (X-rays and gamma rays), not for particulate radiation.

The energies of therapeutic or diagnostic gamma rays and X-rays are in the kilovolt (kV) or megavolt (MV) range, while the energies of therapeutic electrons are in the megaelectronvolt (MeV) range.

1.8 Radiotherapy Generators

Kilovoltage X-rays are generally used in the treatment of skin cancers and superficial tumors, whereas megavoltage X-rays are used in the management of deeply seated tumors [22]. Megavoltage electrons, on the other hand, are used in the treatment of superficial tumors.

1.8.1 Kilovoltage Machines (<500 kV)

Contact therapy machines
40–50 kV
Filtered with 0.5–1.0 mm aluminum
SSD = 2 cm
50% depth dose is 5 mm
Superficial therapy machines (Fig. 1.32)

Fig. 1.32 Superficial therapy machine

50–150 kV
Filtered with 1–4 mm aluminum
SSD = 20 cm
50% depth dose is 1–2 cm
Orthovoltage therapy machines (Fig. 1.33)
150–500 kV
Filtered with 1–4 mm copper
SSD = 50 cm (for a field size of 20 × 20 cm)
50% depth dose is 5–7 cm.
Basic therapy machine before 1950
Supervoltage therapy machines
500–1000 kV
Filtered with 4–6 mm copper
50% depth dose is 8–10

1.8.2 Megavoltage Therapy Machines (>1 MV)

Van de Graaf generator (Fig. 1.34)
  Also known as an electrostatic generator
  Produces energies of up to 25 MV
Cobalt-60 teletherapy unit (Co-60)
  Manufactured in 1951
  Two gamma rays with energies of 1.17 and 1.33 MV
  Dose rate >150 cGy/min
  SSD = 80–100 cm (for field sizes of 35 × 35 or 40 × 40 cm)
  50% depth dose is 10 cm
  The half-life of Co-60 is 5.27 years

Source is 2 cm in diameter, and has a large penumbra in comparison to a 4 MV linac.
Betatron (Fig. 1.34)
Developed in 1940
Developed for the circular induction acceleration of electrons and light particles. The magnetic field guide is increased over time in order to keep the particles in a constant-diameter circle
Mean energy is 45 MeV (maximum energy 300 MV)
Not used after the advent of linacs, due to their large dimensions, high costs, and low dose rates
Linear accelerator (linac)

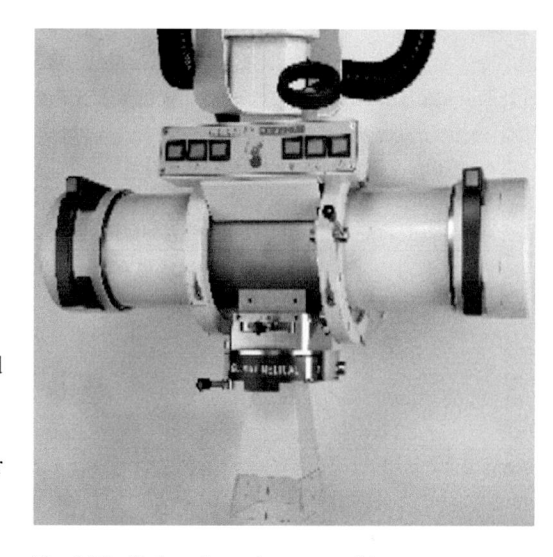

Fig. 1.33 Orthovoltage therapy machine

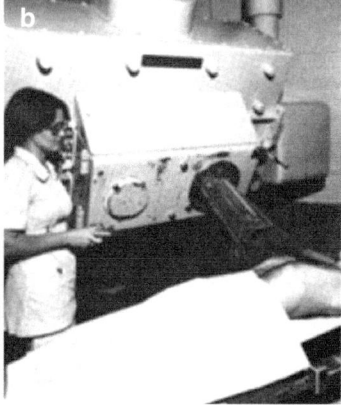

Fig. 1.34 (**a**) Van de Graaf generator; (**b**) betatron; (**c**) microtron

Entered into routine clinical service in 1953

SSD = 100 cm (for a field size of 40 × 40 cm)

Produces several photon and electron energies

Microtron (Fig. 1.34)

Entered clinics in 1972

Combination of a linac and a cyclotron

A circular particle accelerator for accelerating electrons to energies of several MeVs

Has a simple structure and it is easy to select the appropriate energy. Small compared to other linacs. Just one microtron generator can provide electrons for more than one treatment room

Produces energies of up to 50 MeV

Cyclotron

A circular particle accelerator in which charged subatomic particles generated at a central source are accelerated spirally outward in a plane perpendicular to a fixed magnetic field by an alternating electrical field. A cyclotron is capable of generating particle energies of between a few million and several tens of millions of electron volts

Orthovoltage treatment machines [22]

These machines have similarities with machines used in diagnostic radiology. X-rays are produced by accelerating electrons into tungsten. These X-rays are shaped by collimators before they reach the tumor tissue. Essentially all of the dose is directed onto the surface of the skin; the percentage depth dose (PDD) curve drops sharply. Thus, deeply seated tumors cannot be treated using these machines. Orthovoltage treatment machines are currently only found in a few centers, and are only used to treat skin cancers or some very superficial lesions in the head and neck region.

Megavoltage treatment machines [22]

Megavoltage treatment machines are the foundation of modern radiotherapy techniques. Photon beams are produced in megavoltage treatment machines. Their one difference from orthovoltage machines is the skin-sparing effect (i.e., the maximum energy is delivered in the subepidermal region). The skin-sparing effect is directly proportional to the energy of the photons. The maximum dose point distance from the skin is 0.5 cm for Co-60, 1.5 cm for 6 MV photon beams, and 2.5 cm for 10 MV photon beams. In addition, these machines have more suitable depth doses for deeply seated tumors and a sharper dose distribution at the field edge due to reduced side scattering. When electrons are used, deeper healthy tissues are well protected in the treatment of superficially located tumors by controlling the penetration depth.

1.8.3 Cobalt-60 Teletherapy Unit

Natural cobalt is a hard, stable, bluish-gray, easily breakable metal with properties similar to iron and nickel. Its atoms contain 27 protons, 32 neutrons, and 27 electrons [23]. Nonradioactive cobalt can be found mixed with various minerals in nature, and has been used to impart a blue color to glass and ceramics for thousands of years.

In 1735, a Swedish scientist, George Brandt, showed that the bluish color in colored glasses was due to a previously unknown element that he named cobalt. The melting point of cobalt is 1495 °C, its boiling point is 2870 °C, and its density is 8.9 g/cm^3.

The well-known isotope of cobalt is unstable radioactive Co-60. This isotope was discovered by Glenn Seaborg and John Livingood at California Berkeley University in 1930. Co-60 is now produced commercially in nuclear reactors.

The decay of Co-60 starts with a β^- decay, and then two gamma emissions with energies of 1.17321 and 1.33247 MV are observed (Fig. 1.35).

Co-60 teletherapy units have a cylindrical source 2 cm in diameter. The activity of the source is generally between 5000 and 15,000 Ci (Fig. 1.36). A source with an activity of less than 3000 Ci is replaced with a new one; this is necessary after 5–7 years of use.

Co-60 teletherapy units provide good performance for tumors with depths of <10 cm. Thus, the use of a linac is recommended for more deeply seated tumors.

The half-life ($t_{1/2}$; i.e., the time required for the activity of the source to half) of Co-60 is 5.27 years. For practical purposes it is considered

1.8 Radiotherapy Generators

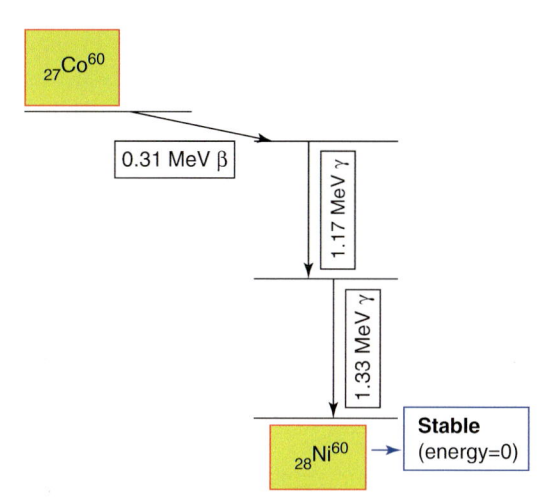

Fig. 1.35 Cobalt-60 decay scheme

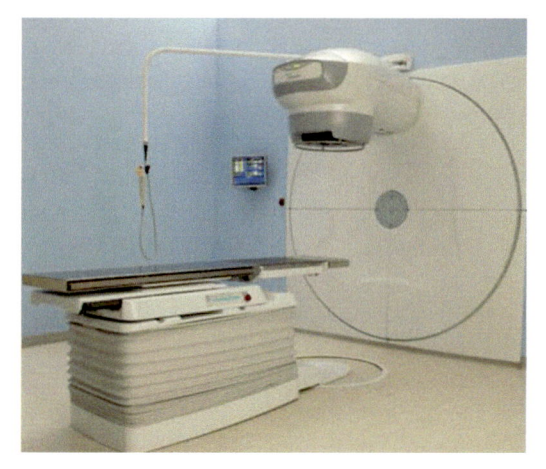

Fig. 1.36 Cobalt-60 teletherapy unit

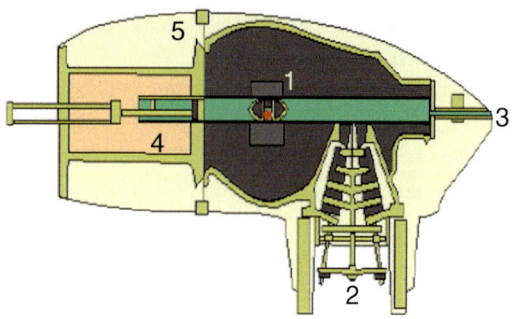

Fig. 1.37 Treatment head of a cobalt-60 teletherapy unit. The cobalt source (*orange*) is situated in a drawer, and surrounded by lead (*1*). When the device is in the resting position, the source is protected by layers of enriched uranium. The source is then pushed by a pneumatic system (*4*) to the treatment position. (*2*) The collimator system; (*3*) manual system that can pull the source to the resting position in case of emergency; (*5*) link between the head and the rotating part of the machine that is used to change the source when its activity is no longer sufficient for treatment

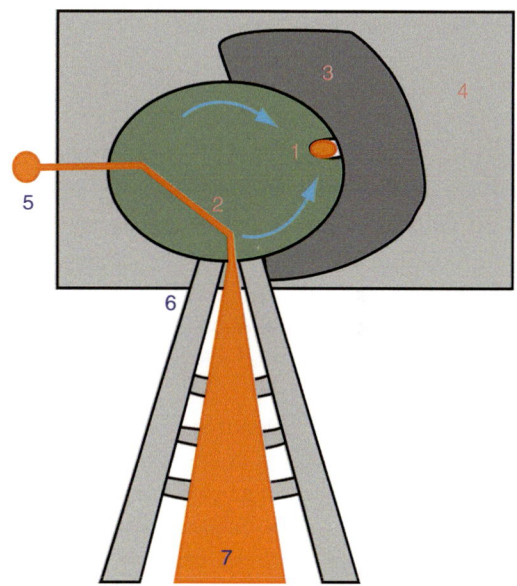

Fig. 1.38 Cobalt-60 head. (*1*) Cobalt-60 source; (*2*) tungsten cylinder; (*3*) enriched uranium; (*4*) lead; (*5*) laser source; (*6*) collimator; (*7*) γ-rays

harmless and inactive after ten half-lives. Thus, Co-60 should be stored safely for approximately 53 years.

1.8.4 Treatment Head and Collimator (Figs. 1.37 and 1.38)

This has the capacity to take a source with an activity of 10.000 Roentgens per hour at a meter (RHm) (165 Roentgens per minute at a meter (Rmm)).

The leakage from the treatment head is not more than 2 mR/h at 1 m.

The drive mechanism for the source within the treatment head has a very simple linear structure, and returns to its parked position spontaneously in emergencies (even during electric interruptions).

It has the property of interlocking with the source head at an angle of 0°.

The source–skin distance (SSD) is 80–100 cm.

The rotational movement of the collimator is continuous, and it can rotate 360° about its own axis.

An optical distance indicator showing the SSD is present on the treatment head of the system.

The collimator system can move to any position when the gantry is rotated.

1.8.5 Gantry

The source–isocenter distance (SAD) is 80–100 cm.

The rotational movement of the gantry is motorized and controlled in two directions continuously; its rotation speed can be adjusted.

The gantry can rotate by 360°.

1.8.6 Linear Accelerator (Linac)

There are two types of accelerator that are used for radiation treatment [23]. Betatron and linac electron accelerators comprise 99% of all current accelerator machines used in radiation treatment. Cyclotrons, on the other hand, are heavy particle accelerators that are used for proton or neutron treatments.

Free electrons emitted from a metal wire via thermionic emission (as in the case of the X-ray tube) are accelerated in an electromagnetic field to increase their kinetic energies. These accelerated high-energy electrons can either be used directly for radiotherapy (generally for superficial therapy), or they are directed into a target and high-energy X-rays are produced (for deeply seated tumors). In this way, X-rays with energies of 4–25 MV are produced by electrons with energies of 4–25 MeV. It is impossible to accelerate the electrons to more than 400 kV within conventional X-ray tubes. Thus, high-frequency magnetic wave chambers are used in linac machines, and the negatively charged electrons are accelerated by the magnetic fields in such machines, thus gaining kinetic energy.

The operational principle of electron accelerators. An electric impulse is deposited in the modulator. A specific control mechanism sends this impulse simultaneously to the electron gun and to the section responsible for microwave production (called the klystron or magnetron) at certain intervals (frequency: 50–200 Hz). The electrons liberated by the pulses are sent to the accelerator tube. An automatic frequency control module generates electromagnetic waves in the accelerator tube with the same frequency.

Microwave chamber. This consists of cylindrical, conductive metal chambers (8 cm in diameter), and it produces 3000 MHz electromagnetic waves.

Electron acceleration. High-frequency electromagnetic waves occurring within the chamber move into the canal in the middle of cylinder, and electrons are accelerated linearly by passing from one chamber to other one within this canal. The velocity of an electron exiting this tube is equal to the sum of the velocities gained by the electron within each chamber.

Electrons produced in the electron gun are sent to the accelerator tube with energies of 50 keV. The electrons ride on top of the electromagnetic waves, so they are accelerated and their energies are boosted to the MeV level. They reach their maximum energies at the end of the accelerator tube.

Electrons exiting through the accelerator tube are then diverted at an angle of 90 or 270° and guided to the head, where the beam exits.

1.8.7 Magnetron (Fig. 1.39) [24]

This is an oscillator that produces hundreds of microwaves per second. The frequency of the microwaves is 3000 MHz.

1.8 Radiotherapy Generators

The exit power of a low-energy linac magnetron (lower than 6 MV) is 2 MW.

Klystron (Fig. 1.40) [24]

This does not produce microwaves; it is a microwave amplifier.

Microwaves produced in low-power oscillators are sent to the klystron in order to gain power (energy).

The klystrons used in a high-energy linac can boost the energy to 25 MV, leading to an exit power of 5 MW.

The dose stabilities of klystrons are much better than those of magnetrons.

- The electron gun used in a linac is a hot-wire filament (Fig. 1.41)
- The main purpose of the flattening filter in a linac is to collect low-energy X-rays and enable the passage of high-energy X-rays (i.e., it "flattens" the beam)
- In some linacs an electromagnetic wave transmitter is used instead of a scattering foil

The Electa SL 25 linac treatment head consists of (in order): a tungsten target, a primary collimator, a flattening filter (main filter) containing a mixture of tungsten and aluminum that is used to adjust the spatial modulation of the beam intensity, a scattering foil to spatially spread the electron beams, two ion chambers, a motorized 60° wedge filter and multileaf collimator (Fig. 1.42a).

Fig. 1.39 Magnetron

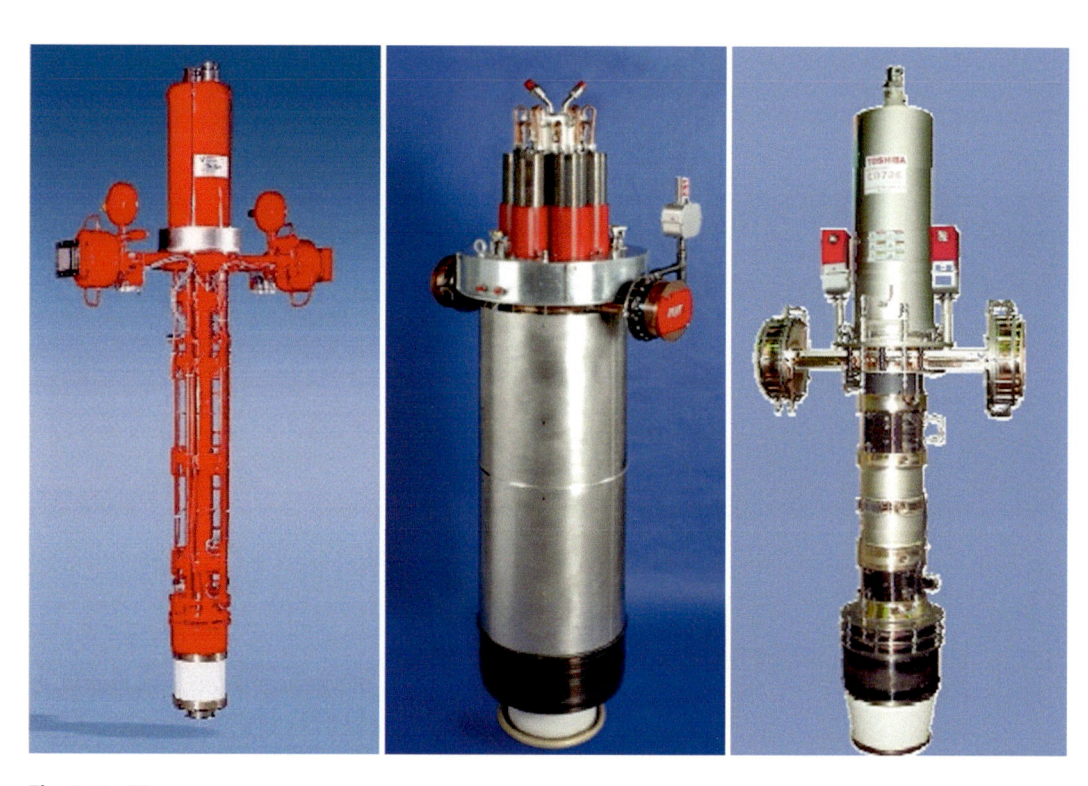

Fig. 1.40 Klystron

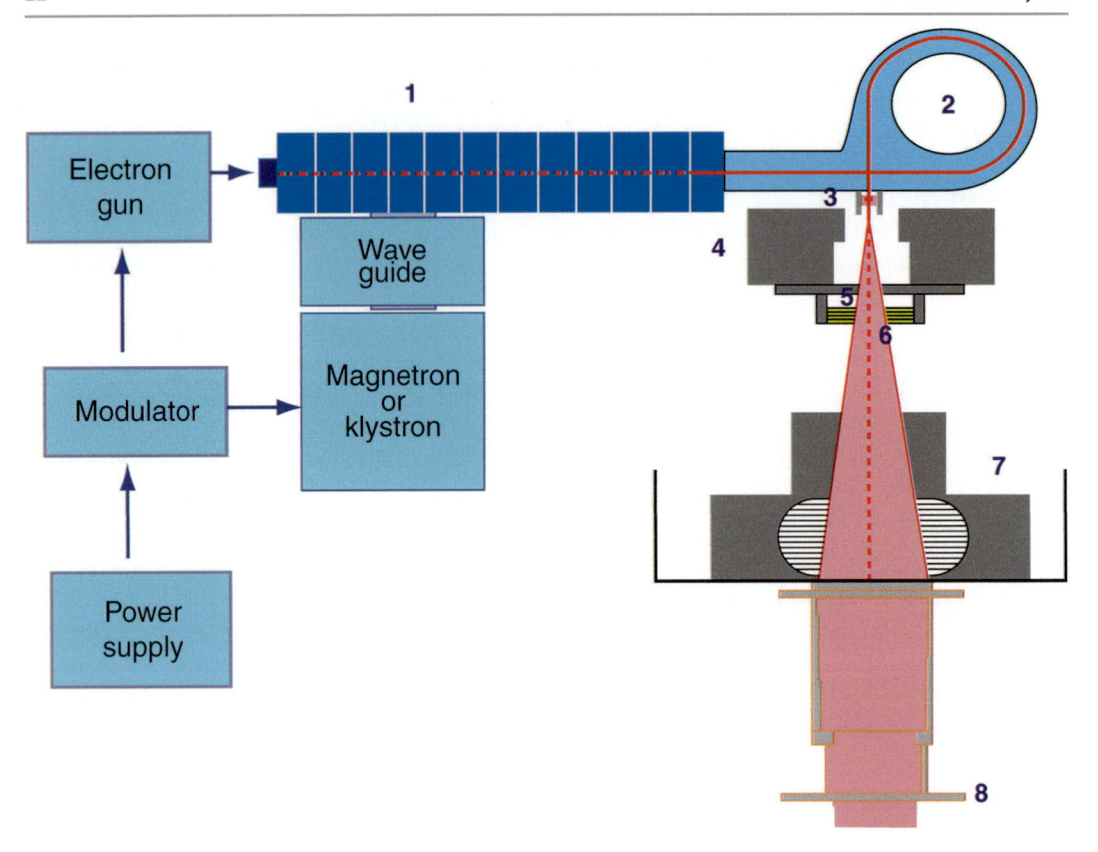

Fig. 1.41 General illustration of a linear accelerator. (1) The production and the acceleration of electrons, (2) The 270° bending of electrons, (3) Target and primary filter, (4) Primary collimators, (5) Main filter, (6) Ionizing chamber, (7) Multileaf collimator, (8) Electron applicator

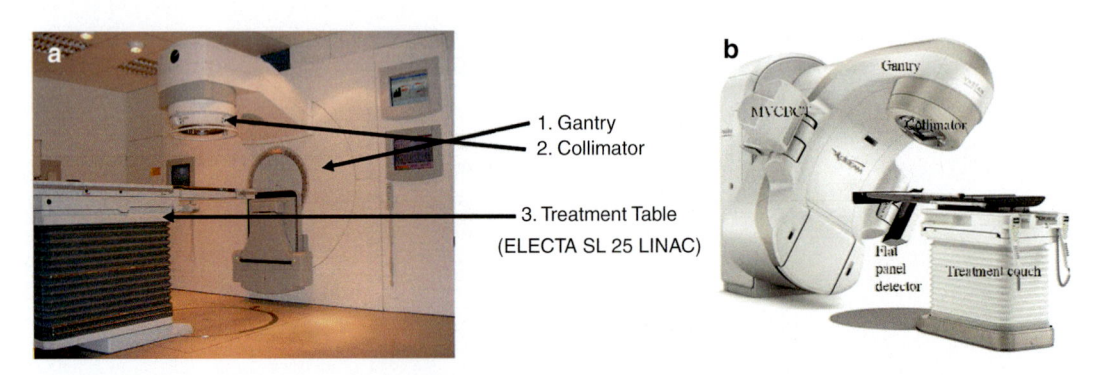

Fig. 1.42 Electa SL 25 (**a**) and Varian VitalBeam (**b**) linear accelerator (linac)

1.9 Measurement of Ionizing Radiation

The maximum allowable dose is limited to 20 mSv per year for people working with radiation. This limit is 1 mSv for the normal population. The effects of irradiation on an organism may change according to the dose, the type of contamination, and the features of the radiation source. It is crucial to perform measurements on the radiation generators used for diagnostic or

1.9 Measurement of Ionizing Radiation

therapeutic purposes. The measurement of radiation is called dosimetry, and the equipment used for dosimetric procedures is called a dosimeter or a detector [25].

Detector Types
1. Detectors according to the principle of operation:
   (a) Pulse type
   (b) Current type
2. Detectors according to their structures:
(a) Gas-filled detectors:
   - Ionization chambers
   - Proportional detectors (with or without a window)
   - Geiger–Müller detectors (with or without a window)
   - Gas scintillation detectors
(b) Solid-state detectors:
- Crystal detectors
  – Scintillation detectors
  – Semiconductive detectors
- Plastic detectors (solid or liquid)
- Glass detectors
3. Film detectors
4. Dosimeters:
(a) Electron spin resonance (ESR)/alanine
(b) Thermoluminescence (TLD)
5. Chemical detectors
6. Neutron detectors

1.9.1 Portable Measuring Equipment

The basic principle of a portable measuring device is ionization inside a gas-filled detector. Radiation creates ion pairs, and these ion pairs are collected and converted into an electrical signal (impulse or current) when they pass through an electrical field. This signal is used to determine whether and how much radiation is present.

There are several types of detector, and they all work on the same principle. Ion chambers and Geiger–Müller counters are the two main types of measuring instrument.

1.9.2 Ionization Chamber [25]

This is designed to measure the dose rate of ionizing radiation in mR/h or R/h. The detector usually takes the form of a cylinder that is filled with air (Fig. 1.43). When the radiation interacts with the air in the detector, ion pairs are generated, and collecting them results in a small current. The charge occurring within the air defines the dose rate.

- Ion chambers are used for the measurement of X-rays, gamma rays, and beta particles.

1.9.3 Geiger–Müller Counter (GM Counter)

A GM counter consists of a tube filled with "Q-gas" (98% helium and 1.3% butane; see Fig. 1.44). As in an ion chamber, the detector records every interaction instead of measuring the average current that occurs after several reactions. In other words, one ionizing event will produce a pulse or a count in the GM tube. This does not take into account the number of pairs that started the pulse; all pulses are the

Fig. 1.43 Basic design of a Farmer-type ionization chamber

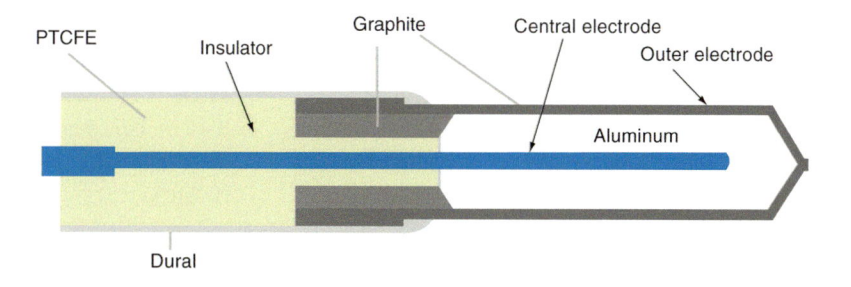

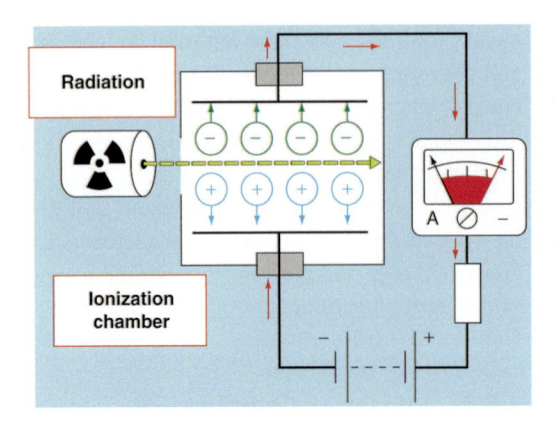

Fig. 1.44 Simple illustration of a Geiger–Müller counter

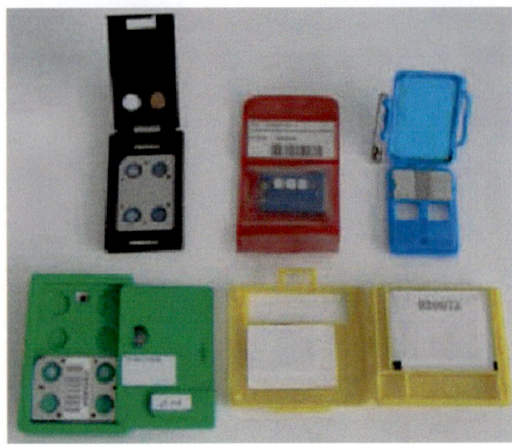

Fig. 1.45 Personal dosimeters

same. Thus, a GM counter does not differentiate between types of radiation or their energies. For this reason, most GM counters are calibrated to give counts per minute (CPM). GM counters are usually used to simply detect the presence of radioactive material.

- GM counters are used to detect low-energy X and gamma rays.

1.9.4 Film Dosimeters [25]

A film dosimeter is the oldest and most popular system that is used to determine the dose taken a person working with radiation (Fig. 1.43). It is based on the effect of radiation on the optical density of developed photographic film. A dosimeter consists of two main parts: the film and its holder. There are filters of various thickness and types in order to have the optical density resulting from the radiation on the film to be independent from radiation type and energy. The dose calculation is done by measuring optic densities behind the filters on the film (Fig. 1.45).

- The radiation doses taken from beta, gamma, and X-ray sources can be measured by film dosimeters

1.9.5 Thermoluminescence Dosimeters (TLD) [25]

Thermoluminescence is the phenomenon where a material luminesces (emits significant amounts of light glows) when it is heated. In a solid crystalline material, there is forbidden energy zone between the valance band and conduction band where no electron can exist. When the crystal is excited by radiation, traps appear, and electrons coming from the valance band or returning from the conductive band are caught in these traps (Fig. 1.46). Thus, some of the energy transferred from the radiation to the crystal is deposited. If this crystal is heated to a certain temperature, the electrons gain enough energy to release themselves from these traps and return to the valance band, and the energy they release when they do so is emitted as light (i.e., luminescence). This luminescence can be measured by an electrometer. Crystals that exhibit this phenomenon include LiB_4O_7, LiF, and $CaSO_4$, and these are used in TLDs.

- TLDs can separately measure gammas, X-rays, beta particles, and thermal neutrons with energies of between 10 keV and 10 MeV.

Fig. 1.46 Energy level diagram for a TLD crystal

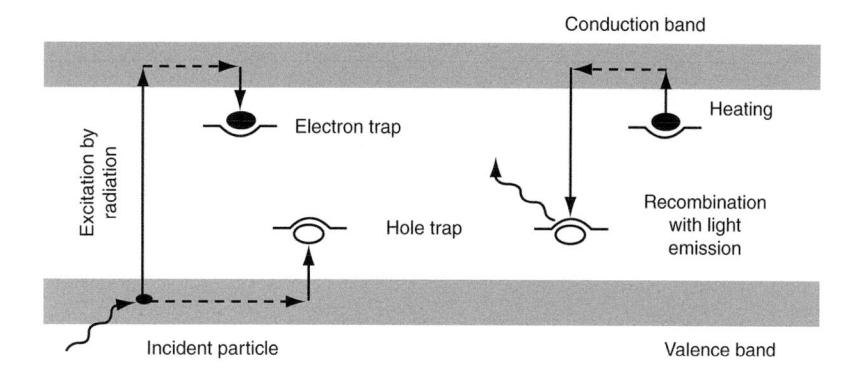

1.9.6 Other Measuring Equipment

1.9.6.1 Electron Spin Resonance (ESR)/ Alanine Dosimeter [25]

One device that can be used to measure high doses is the ESR/alanine dosimeter. Here, powdered alanine crystals are mixed with a defined ratio of combiners. When the alanine is exposed to radiation, free radicals are generated. The number of radicals is determined by the ESR technique, and this can then be used to measure the absorbed dose. Alanine is a tissue-equivalent material due to its characteristics (e.g., composition, density, and effective atomic number). ESR/alanine dosimeters give sensitive, reliable, and repeatable results for a wide range of doses.

Neutron Dosimeter [25]

These dosimeters are sensitive to low-energy thermal neutrons with energies of between 0.02 and 50 eV. They utilize lithium and copper TLD crystals, and can be used to measure neutron doses of between 0.005 mSv and 0.5 Sv.

1.10 Radiation Dosimetry

The quality of the radiation produced can differ between radiotherapy machines. The quality of radiation depends on the type of radiation, its energy, and its penetrating ability. These characteristics should be experimentally measurable and confirmable. Several physical measurements, measuring techniques, and units help to increase radiotherapy efficacy, in accordance with the principle of *primum non nocere*. Dosimetric measurements are performed in patients, and water or tissue-equivalent phantoms and physical treatment parameters are determined.

1.10.1 Phantom

Phantoms are models constructed from tissue-equivalent material and used to determine the radiation absorption and reflection characteristics of a human body or a specific organ (Fig. 1.47) [26]. Phantom materials are equivalent to human tissues in terms of their characteristics under X-ray and electron irradiation. Soft tissue, bone, and lungs are equivalent of real density. Soft tissues are mimicked by heat-hardened plastic material. The effective atomic number of soft tissue is 7.30 ± 1.25%, and its density is 0.985 ± 1.25 g/cm^3. Although lung has the same atomic number as soft tissue (7.30), its density is 0.32 ± 0.01 g/cm^3. The bones used in a phantom are real human bones, and the same cavities exist as in the human body. The phantom is divided into slices of 2.5 cm thickness. There are drilled holes of 2.5–6 mm diameter for TLD insertion in each slice.

1.10.2 Definition of Beam Geometry

The accurate delivery of a radiation dose to a patient depends on the precise positioning of the

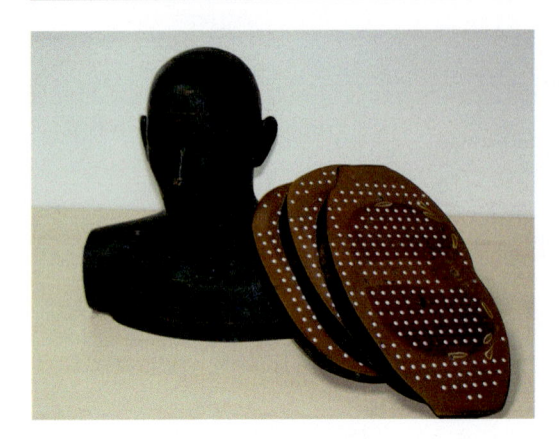

Fig. 1.47 An Alderson RANDO phantom

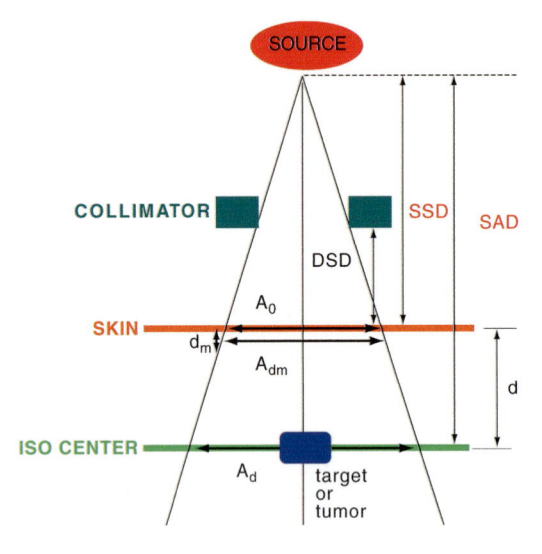

Fig. 1.48 Geometric parameters between the source and target

patient and radiation source. The geometric parameters linking the source and target (tumor at treatment) are described below (Fig. 1.48).

A_0 and A_{d_m} can be found mathematically using the geometric parameters between source and the target:

$$A_d = \frac{A_0 \times (\text{SSD} + d)}{\text{SSD}} = A_0 \times \frac{\text{SAD}}{\text{SSD}} \quad (1.4)$$

$$A_{d_m} = A_0 \frac{(\text{SSD} + d_m)}{\text{SSD}} \quad (1.5)$$

SSD: skin to source distance, SAD: skin to axis distance, DSD: diaphragm to skin distance, A_0: field on skin, A_d: field at D_0 depth (field at the location of the tumor): field size at the point of dose maximum (field at build-up point), d: depth (tumor depth), d_m: depth at dose maximum ($D_{\max}$).

SSD and SAD are geometric parameters that can be changed, and the ability to change them leads to two different planning and treatment set-up modalities (Fig. 1.49):

- *Constant SSD technique* [27]. Here, the isocenter (the point at which all of the radiation beams cross) of the treatment machine (or simulator) is on the patient's skin (= nonisocentric technique).
  – Field and dose are defined according to A_0.
- *Constant SAD technique* [27]. Here the isocenter of the treatment machine (or simulator) is in the patient (in the tumor) (= isocentric technique; Fig. 1.50).
  – Field and dose are defined according to A_d.

1.10.3 Build-Up Region

The region between the skin and the depth at dose maximum ($D_{\max}$) is called the build-up region (Fig. 1.51) [27, 28]. This region between the surface and depth d is referred to as the dose build-up region in megavoltage beams, and it results from the kinetic energy deposited in the patient by secondary charged particles (which have relatively long ranges) released inside the patient by photon interactions (photoelectric effect, Compton effect, pair production) (Tables 1.5 and 1.6).

1.10.4 Half-Value Layer (HVL)

A radiotherapy machine is characterized by the penetrating ability of its radiation. The thickness of material that decreases the intensity of the incoming radiation to half of its initial value (50%) is called the half-value layer (HVL), and quoted in mm or cm of absorbent material (Fig. 1.52) [29]. The absorbent materials that are

1.10 Radiation Dosimetry

Fig. 1.49 Isodose curves in SSD and SAD techniques

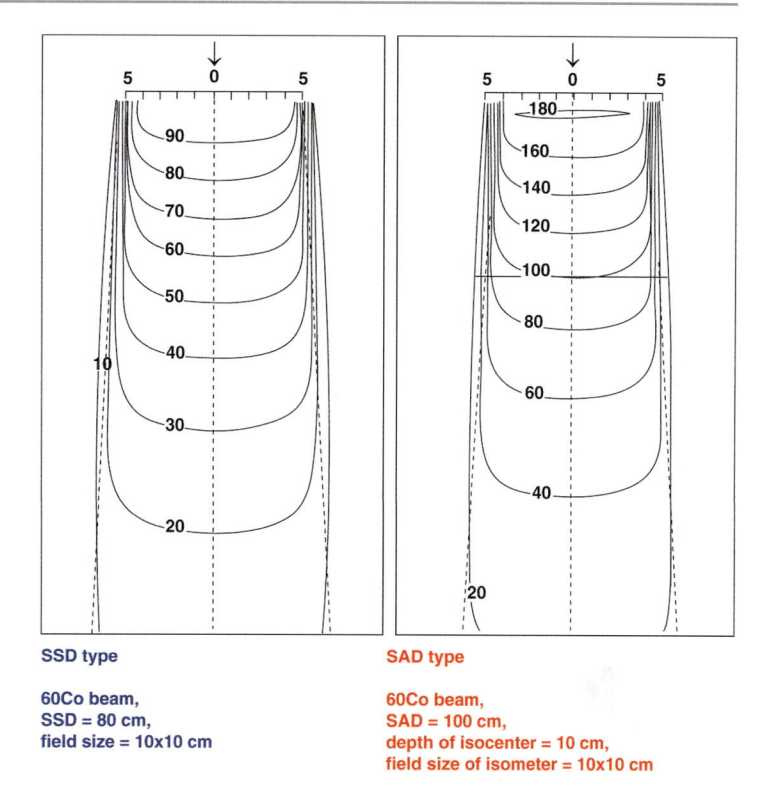

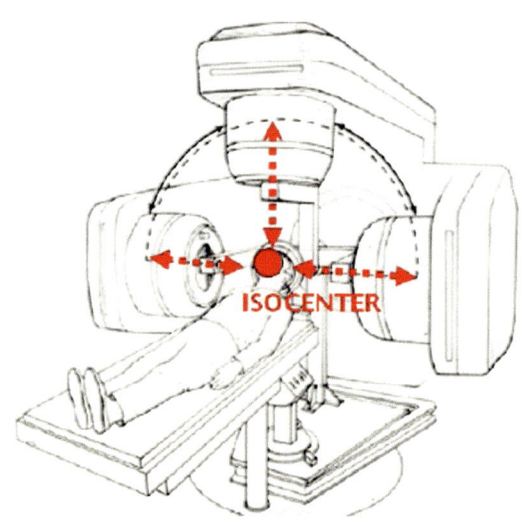

SSD type

60Co beam,
SSD = 80 cm,
field size = 10x10 cm

SAD type

60Co beam,
SAD = 100 cm,
depth of isocenter = 10 cm,
field size of isometer = 10x10 cm

Fig. 1.50 Isocenter

generally used in this context are aluminum, copper, and lead (Table 1.7).

The HVL is generally used to characterize low-energy X-ray machines.

A high-energy radiation beam is characterized by its maximum energy and the depth at the 50% isodose curve.

Blocks with thicknesses of 4–5 HVLs are used in radiotherapy. A block five HVLs thick transmits 3.125% of the incoming radiation.

1.10.5 Percentage Depth Dose (PDD)

The ratio (in percent) of the dose absorbed at a predefined depth (D_x) to D_{max} (the dose maximum) for a predefined SSD and field size is termed the percentage depth dose (PDD or DD%) [30]:

$$DD_x\% = 100 \times \frac{D_x}{D_{max}} \quad (1.6)$$

Synonyms for D_{max}:

- Given dose
- Entrance dose

Fig. 1.51 Build-up region

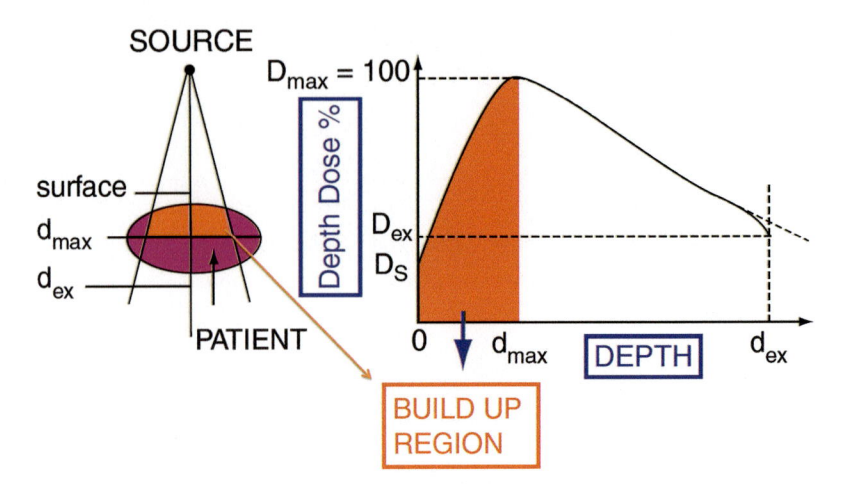

D_S = surface dose at the beam entrance side
D_{ex} = surface dose at the beam exit side
D_{max} = dose maximum often normalized to 100

Table 1.5 Build-up points (depth of D_{max}) for various photon energies

	Superficial	Orthovoltage	Co-60	4 MV	6 MV	10 MV	18 MV	25 MV
Depth of D_{max} (cm)	0	0	0.5	1	1.5	2.5	3.5	5

Field size = 5 × 5 cm

Table 1.6 Build-up points (depth of D_{max}) for various electron energies

	6 MeV	9 MeV	12 MeV	15 MeV	18 MeV	22 MeV
Depth of D_{max} (cm)	1.2	1.56	2.1	1.7	1.5	1.49

Fig. 1.52 Transmission of an X-ray beam through an aluminum absorbent

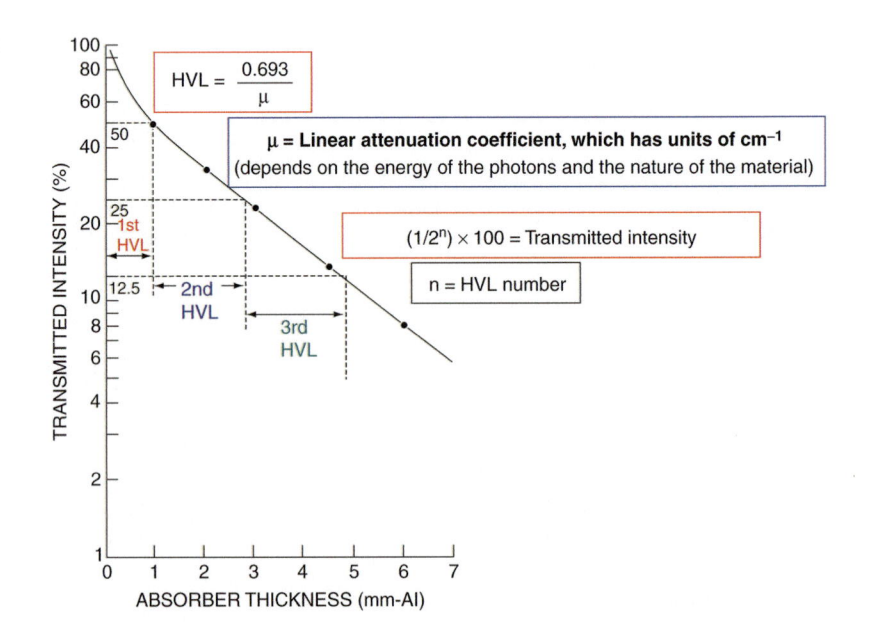

$$HVL = \frac{0.693}{\mu}$$

μ = **Linear attenuation coefficient, which has units of cm^{-1}**
(depends on the energy of the photons and the nature of the material)

$(1/2^n) \times 100$ = Transmitted intensity

n = HVL number

DD% is also defined as the dose at a specific depth as a function of distance, field, and energy in a water phantom.

The percentage depth dose curve provides information on the quality of the radiation and its energy.

The depth at dose maximum can be calculated.

The most probable energy at the surface of the phantom can be found by calculating the range of the electrons. This can give information on X-ray contamination.

Dose: Energy transferred per unit mass of target tissue, and its unit is Gray (Gy).

- The dose in radiotherapy is normalized to the D_{max} calculated in the phantom.

PDD curves are created by plotting DD% values at different depths from the surface of the phantom (Figs. 1.53, 1.54, and 1.55).

1.10.6 Isodose Curves

Isodose curves are prepared by combining the points in the phantom or target tissue that receive the same dose (Fig. 1.56) [31]. They are

Table 1.7 Half-value layers of various radioisotopes

Radioisotopes	Half-value layer (cm)		
	Lead	Iron	Cement
Tc-99m	0.02	–	–
I-131	0.72	–	4.7
Cs-137	0.65	1.6	4.9
Ir-192	0.55	1.3	4.3
Co-60	1.1	2.0	6.3

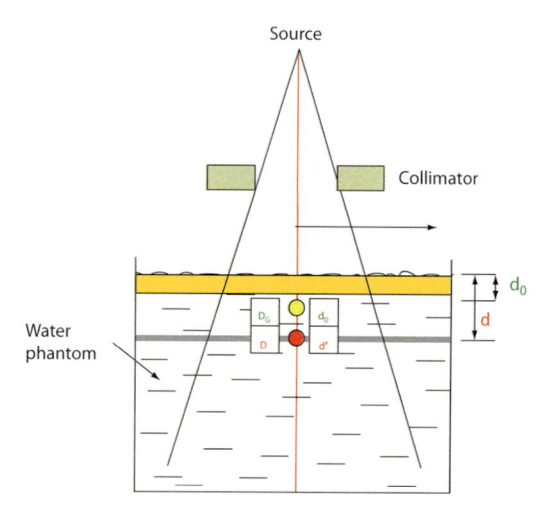

Fig. 1.53 Percentage depth dose (PDD) in a water phantom

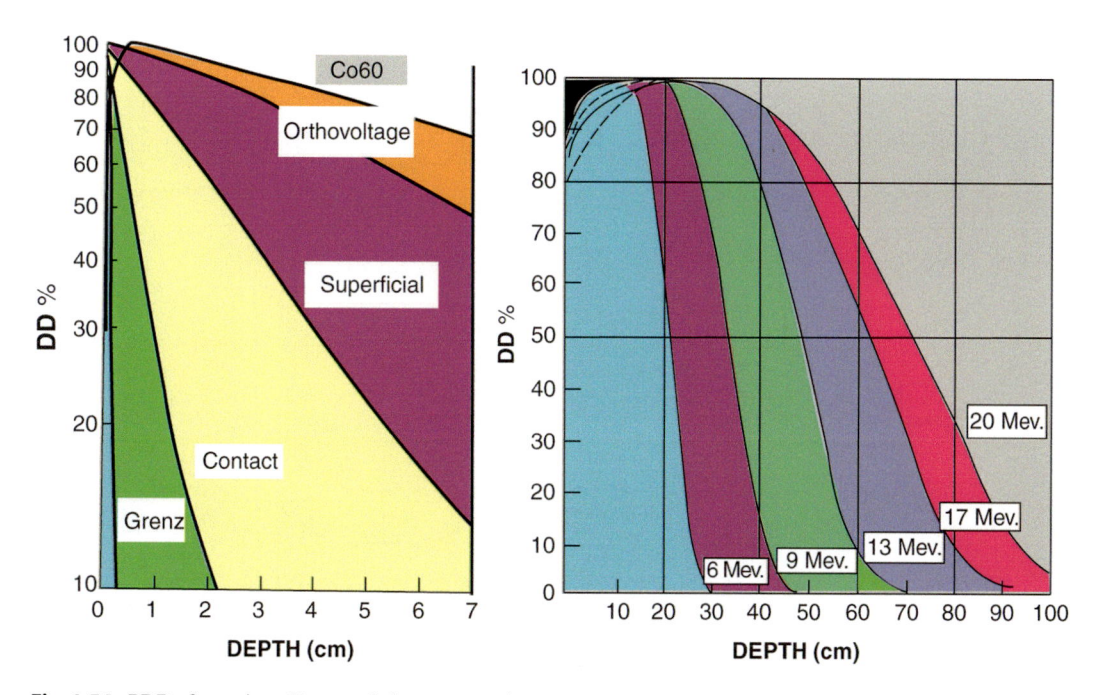

Fig. 1.54 PDDs for various X-ray and electron energies

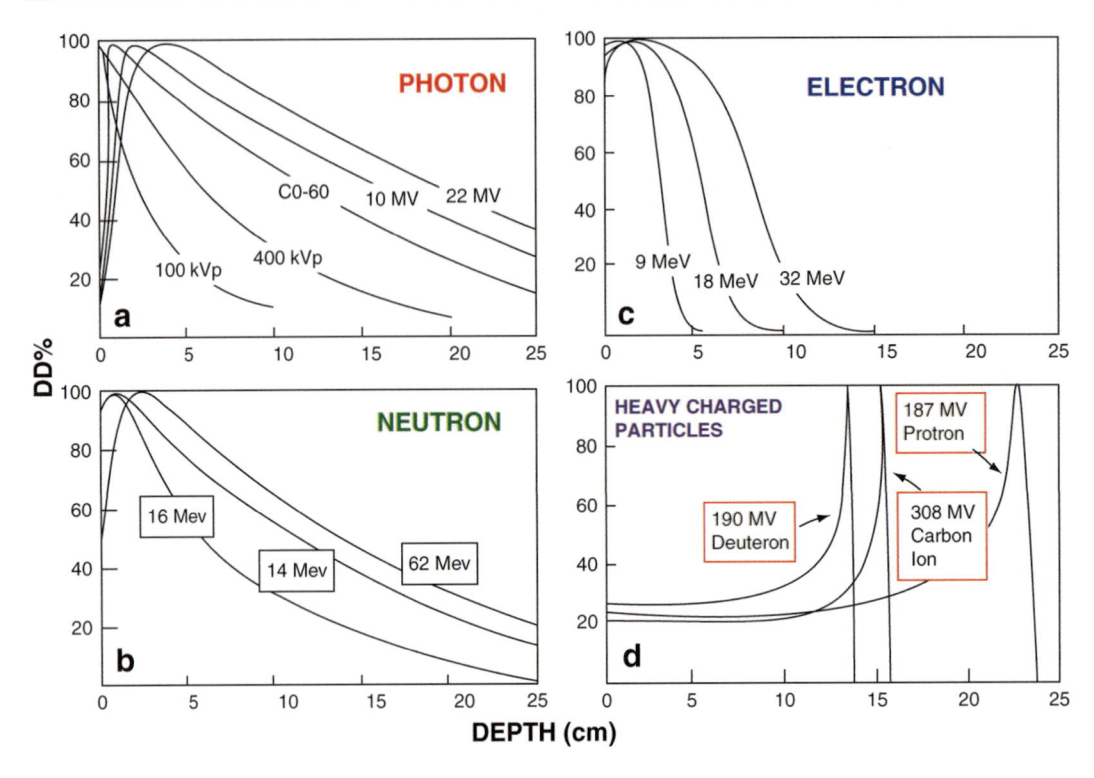

Fig. 1.55 PDD curves for photon, electron, neutron, and heavy charged particles

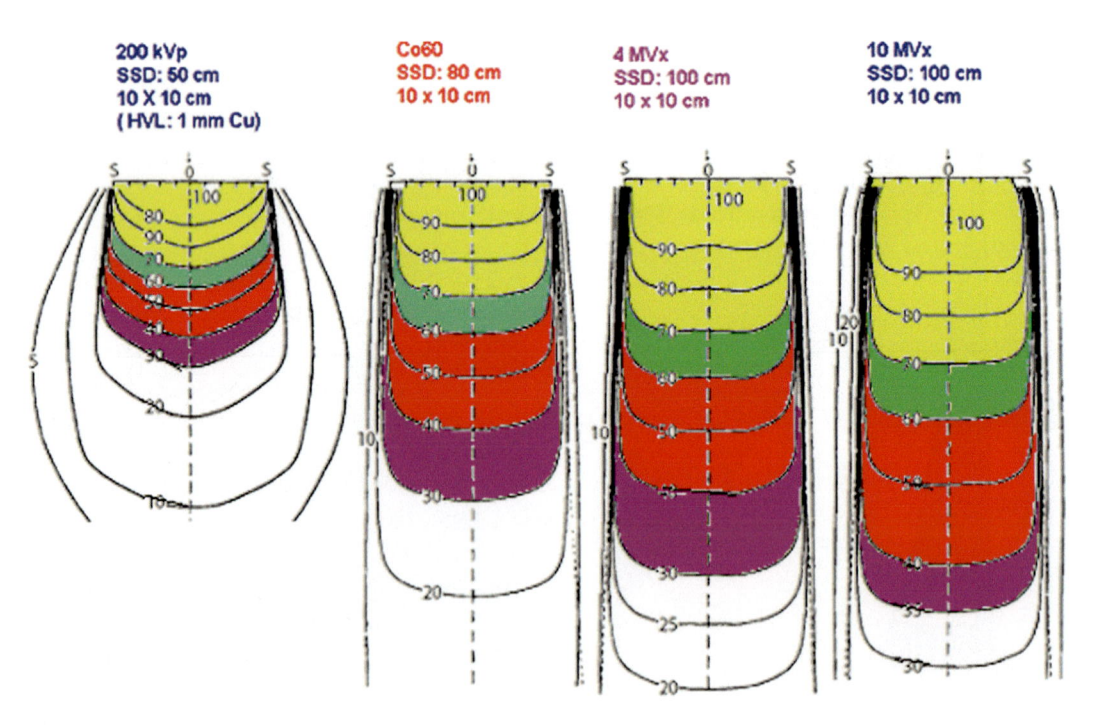

Fig. 1.56 Isodose curves for various X-ray energies

1.10 Radiation Dosimetry

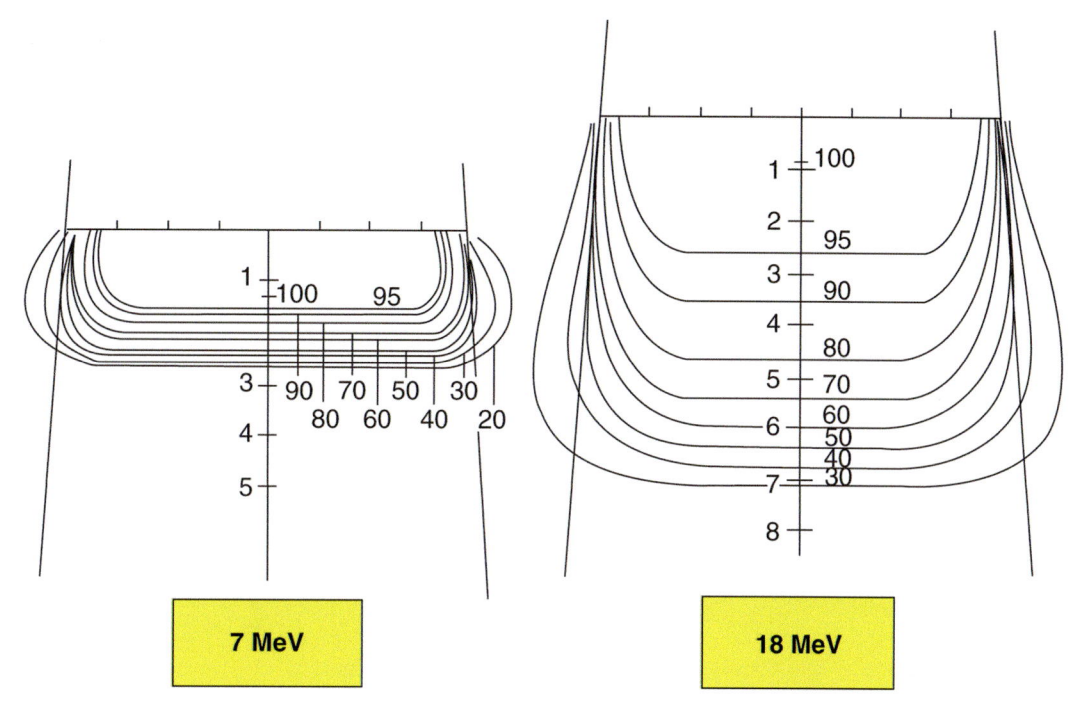

Fig. 1.57 Isodose curves for various electron energies

calculated by various dosimetric measurements, and the highest dose is considered 100%. The curves are placed in percentage order, and then used to create the dose distribution graphics for the target tissue and the energy of interest (Fig. 1.57). By using the isodose curves during treatment planning, the dose distribution of the radiation delivered to the target tissue and neighboring structures can be seen from different angles.

- In a plot of isodose curves, the *y*-axis shows the depth below the surface of the skin, while the *x*-axis shows the range of the field

1.10.7 Ionization Chamber in a Water Phantom

- The *isodose distribution* along the central axis is determined for that energy
- The *dose profile* perpendicular to the central axis (i.e., parallel to earth) is determined for that energy

1.10.8 Dose Profile

The characteristics of the delivered radiation can be determined by performing measurements in ionization chamber within a water phantom (Fig. 1.58). These characteristics are the flatness, symmetry, and penumbra for that energy (Fig. 1.59) [32].

1.10.8.1 General Features of a Water Phantom System

The system can perform three-dimensional computerized controlled analyses of the dose, depth dose, dose ratios [TMR, TPR (measurement and calculation)], and isodose calculations.

It has an air scanner capable of making measurements in air, along with an ion chamber, and build-up blocks. There is also a mechanism in these equipments which can be used to mount to the head of actual treatment machines.

The system has a specific test apparatus that can be used to test the isocenter point for the treatment machine (an "isocheck").

It has an apparatus that can check the monitor unit controls of a linac teletherapy unit (a "monicheck").

It also has a "linear array" apparatus for dynamic wedge and MLC measurements.

Solid water phantom (RW 3). This is used for absorbed dose measurements of photon and electrons. It has dimensions of 40 × 40 × 40 cm, and utilizes layers with thicknesses of 1 mm, 0.5, 1 cm.

1.10.9 Penumbra

The penumbra is defined as the region of steep dose rate decrease at the edge of radiation beam, noting that the dose rate decreases as a function of the distance from the central axis (Fig. 1.60) [33]:

$$P = \frac{s(\text{SSD} + d - \text{SCD})}{\text{SCD}} \quad (1.7)$$

P = penumbra, S = source diameter, SCD = SDD: source–collimator distance (=source–diaphragm distance), d = depth.

Field size does not affect penumbra.

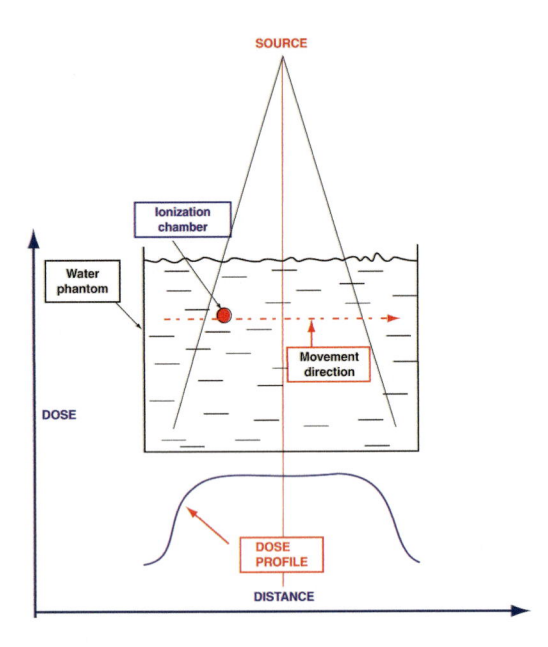

Fig. 1.58 Measurement of dose profile in a water phantom

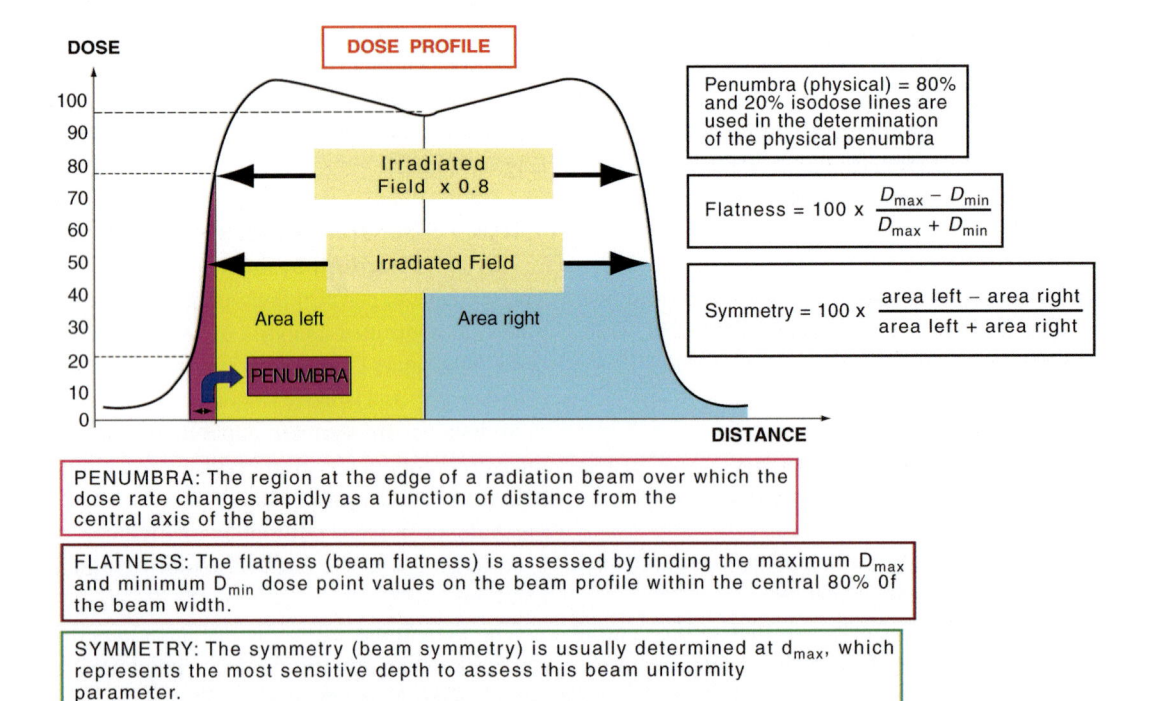

Fig. 1.59 Dose profile and its components

1.10 Radiation Dosimetry

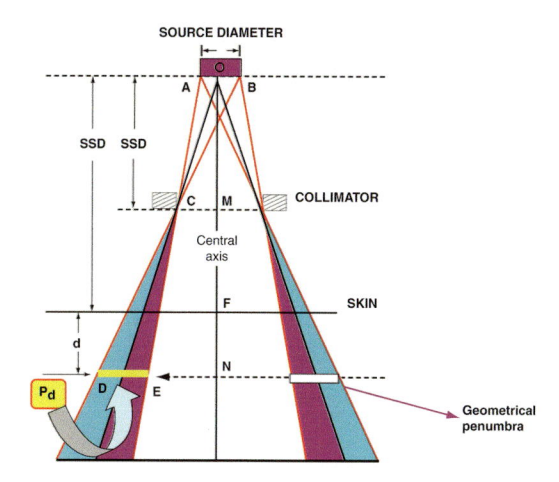

Fig. 1.60 Penumbral parameters and the calculation of the penumbra

1.10.9.1 Types of Penumbra

The *physical penumbra* is the penumbra measured in the dose profile. It is the distance between the points at which the 20 and 80% isodose curves cross the *x*-axis at D_{max}

There are several components to the physical penumbra:

- **Geometrical penumbra.** This occurs due to the size of the source; large sources have larger geometrical penumbras.
- **Transmission penumbra.** This occurs due to the beam emerging from the edges of blocks or collimators. It can be decreased by making sure that the shapes of the focalized blocks take into account the beam divergence.

Factors that increase the penumbra:
Increase in SSD
Increase in source diameter
Increase in SDD (SCD)
Factors that decrease the penumbra:
Decrease in SSD
Decrease in source diameter
Increase in SDD (SCD)

1.10.10 Inverse Square Law

This is the decrease in radiation intensity with the square of distance from the source (Fig. 1.61) [34]. For instance, when the distance from the source triples, the surface dose decreases ninefold (Fig. 1.62). In tissues, the depth into the tissue thickness is another factor that must be considered in addition to the distance from the source, and the dose decreases exponentially with depth into the tissue.

The inverse square law is very important in both radiotherapy and protection from radiation. Since radiotherapy treatments are applied at a short distance from the source, the dose drops off rapidly with distance due to the inverse square law. This situation is observed for both brachytherapy and external radiotherapy.

Most external radiotherapy is delivered as teletherapy (at a distance of 80–120 cm). Thus, the dose fall-off due to distance is relatively small.

Brachytherapy isodose curves show a rapid dose decrease → rapid fall-off

- Here, the isodose curves are narrow, so isodose distances are short

Teletherapy isodose curves show a slow dose decrease → slow fall-off

- Here, the isodose curves are wide, so isodose distances are long

Isocentric treatment (constant SAD) is affected more by the inverse square law than the SSD technique.

1.10.11 Backscatter Factor (BSF)

In a phantom, the ratio of the dose maximum to the dose in air at the same depth is called the backscatter factor (BSF) [35].

$$\text{BSF} = \frac{D_{max}}{D_{air}} \quad (1.8)$$

BSF
- Increases as the energy increases (gets closer to 1)

Fig. 1.61 The relationship between intensity and distance from the source

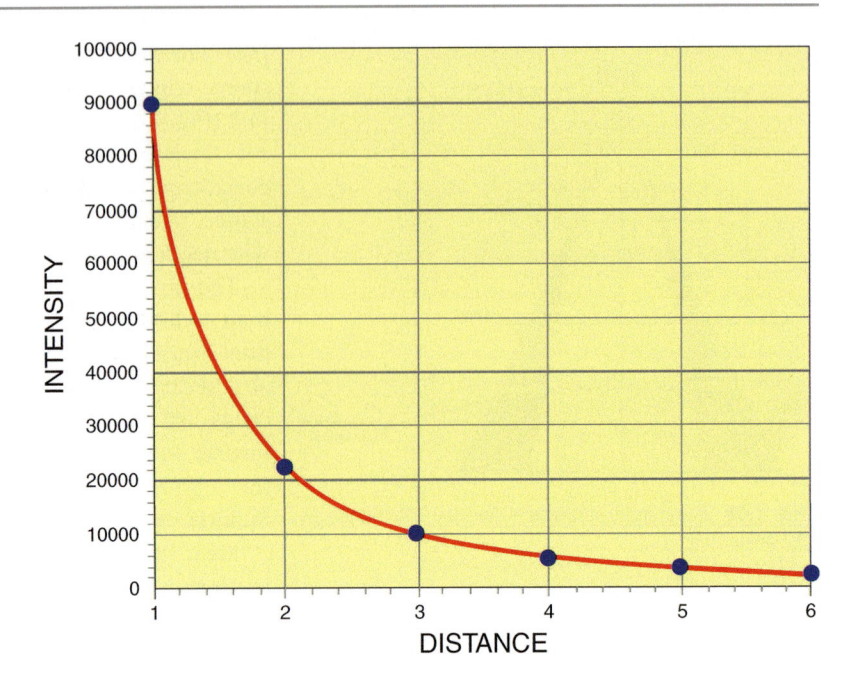

- Increases as the field size increases (gets closer to 1)
- Is independent of SSD

Since the energy of the scattering photon increases as the energy increases, BSF increases.

- At >2 MV, the BSF approaches 1
- The depth at which the BSF is measured depends on the energy
- The BSF measurement depth at energies below that of Co-60 is the surface, since D_{max} is close to the surface

1.10.12 Tissue to Air Ratio (TAR)

The ratio of the dose at depth d (D_d) in a phantom to the dose at the same depth in air (D_{air}) for the distance used in SAD is defined as the tissue to air ratio (TAR) [35]:

$$\text{TAR} = \frac{D_d}{D_{d\,air}} \quad (1.9)$$

The BSF is only defined at D_{max}, whereas TAR can be defined at any depth.

- When $d = D_{max}$, TAR = BSF.

TAR

- Increases as the energy increases
- Increases as the field size increases
- Is independent of SSD at low megavoltage energies
- Is dependent on SSD at high megavoltage energies (due to electron contamination)
- The BSF includes primary radiation and scattered radiation; the TAR only includes scattered and absorbed radiation

If $D_d = D_{max}$ in the TAR formula → *peak scatter factor (PSF)*.

1.10.13 Tissue Maximum Ratio (TMR)

The ratio of the dose measured at a depth d (D_d) to D_{max} in a phantom is defined as the tissue maximum ratio (TMR) [36]:

$$\text{TMR} = \frac{D_d}{D_{max}} \quad (1.10)$$

1.10 Radiation Dosimetry

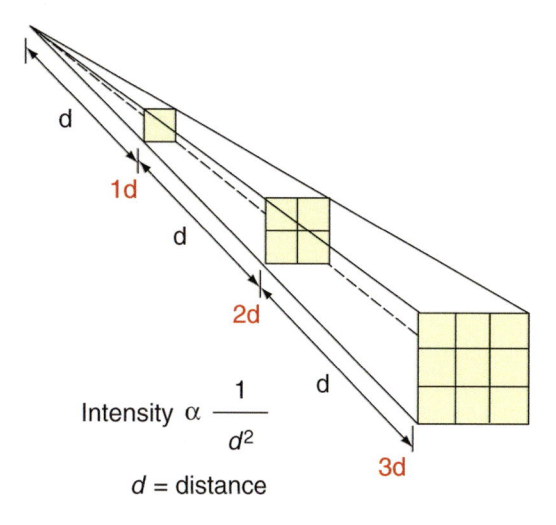

Fig. 1.62 Inverse square law

It is defined by performing two measurements in the phantom (i.e., D_d and D_{max} are measured).

The TMR is normalized to D_{max}, in contrast to the TAR.

$$\text{TMR} = \frac{\text{TAR}}{\text{BSF}} \quad (1.11)$$

TMR
- Increases as energy increases
- Increases as the field size increases
- Is independent of SSD at low megavoltage energies
- Is dependent on SSD at high megavoltage energies (due to electron contamination)

The Differences Between TAR and TMR
- TAR uses the dose in air
- TMR uses the dose at D_{max} in phantom
- TAR is used in isocentric treatment technique
- TMR calculation is done instead of TAR at energies more than 3 MV

1.10.14 Scatter Air Ratio (SAR)

The ratio of dose measured at d depth ($D_{d\text{-phantom}}$) in phantom to the dose measured at the same depth in air ($D_{d\text{-air}}$) is defined as SAR [37].

$$\text{SAR} = \frac{D_{d-\text{phantom}}}{D_{d-\text{air}}} \quad (1.12)$$

It is used for the calculation of mean scattered dose.

It is independent on SSD as TAR, but dependent parameter on energy, depth, and field size.

1.10.15 Collimator Scattering Factor (S_c)

The ratio of the dose measured in any field at a depth d in air to D_{max} measured in a reference field (10 × 10 cm²) in air is called the collimator scattering factor (S_c or CSF; Fig. 1.63) [38].

The collimator scattering factor is also termed the output factor.

S_c is measured in an ion chamber with a build-up cap.

It is:

- Correlated with field size
- Correlated with energy (scattering increases as the field size and energy increase)

1.10.16 Phantom Scattering Factor (S_p)

The ratio of the dose measured in a definite field size at a depth d to D_{max} measured in a reference field (10 × 10 cm²) is defined as the phantom scattering factor (S_p) (Fig. 1.64) [37, 38].

$$\text{Total scattering factor} = S_c + S_p \quad (1.13)$$

The ratio of the BSF calculated in any field at a depth d to the BSF in a reference field in a phantom is another way of defining S_p.

S_p is important for determining scattered radiation from a phantom.

1.10.17 Monitor Unit (MU) Calculation in a Linear Accelerator

Monitor units are the units in which the output of a linac is measured. Linacs are calibrated to give

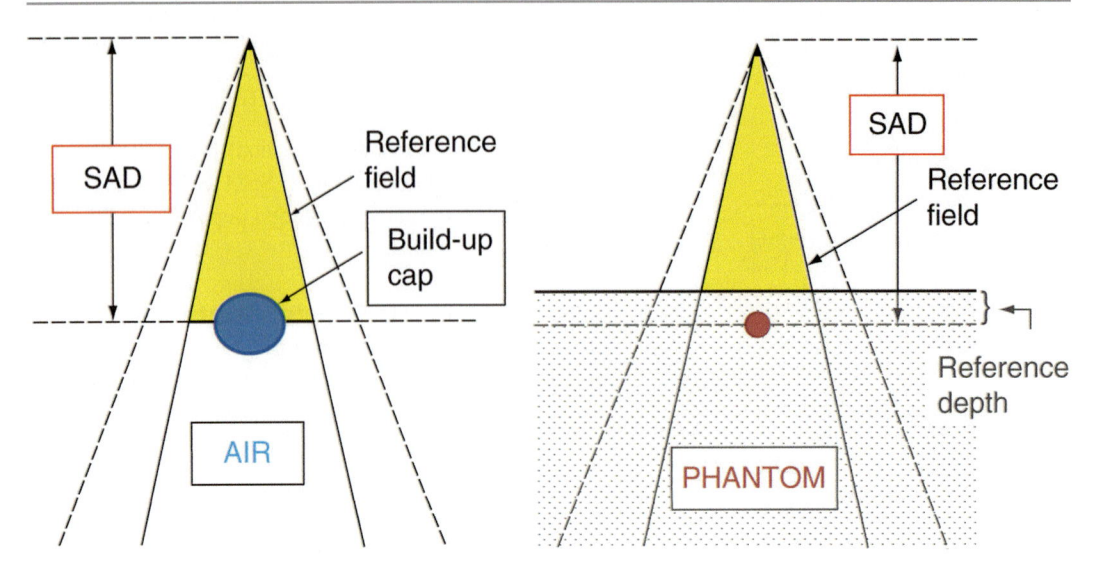

Fig. 1.63 Calculation of the collimator and phantom scattering factor (Fig. 5.3, p 71 of [42])

1 cGy at a SAD distance of 100 cm, for a field size of 10 × 10 cm, and at the depth corresponding to D_{max}, and this calibration dose is defined as one monitor unit (MU).

MU Calculation in the SSD Technique [39] (Nonisocentric Technique)

$$MU = \frac{TD \times 100}{K \times (DD\%)_d \times S_c(r_c) \times S_p(r) \times SSD_{factor}}$$

(1.14)

If a tray or wedge is used, a tray factor (TF) or a wedge factor (WF) is added to the denominator as a multiplier.

$$r_c = r\frac{SAD}{SSD}$$

(1.15)

$$SSD_{factor} = \left(\frac{SCD}{SSD+t_0}\right)^2$$

(1.16)

TD: fraction dose, K: 1 cGy/MU, t_0: reference depth, S_c: collimator scattering factor, S_p: phantom scattering factor, DD%: percentage depth dose, r: collimator field size, SCD: source–collimator distance, SSD: source–skin distance.

MU Calculation in the SAD Technique [39] (Isocentric Technique)

$$MU = \frac{ID}{K \times TMR(d,r_d) \times S_c(r_c) \times S_p(r_d) \times SAD_{factor}}$$

(1.17)

If a tray or wedge is used, a tray factor (TF) or a wedge factor (WF) is added to the denominator as a multiplier.

$$r_c = r\frac{SAD}{SSD}$$

(1.15)

$$SSD_{factor} = \left(\frac{SCD}{SAD}\right)^2$$

(1.18)

ID: fraction dose, K: 1 cGy/MU, t_0: reference depth, S_c: collimator scattering factor, S_p: phantom scattering factor, TMR: tissue maximum ratio, r: collimator field size, SCD: source–collimator distance, SSD: source–skin distance.

Fig. 1.64 The measurement of CSF and S_p, and how they relate to the field size

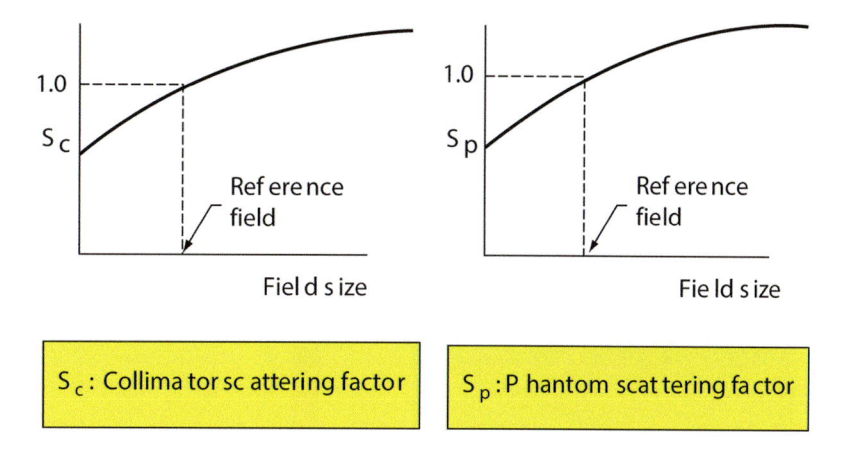

S_c: Collimator scattering factor

S_p: Phantom scattering factor

1.10.18 Treatment Time Calculation in a Co-60 Teletherapy Unit

$$\text{Time}(\min) = \frac{TD \times 100}{D_0 \times (DD\%)_d \times S_c(r_c) \times S_p(r) \times SSD_{factor}} \quad (1.19)$$

$$r_c = r\frac{SAD}{SSD} \quad (1.15)$$

$$SSD_{factor} = \left(\frac{SCD}{SSD + t_0}\right)^2 \quad (1.16)$$

TD: fraction dose, D_0: dose rate (specific SAD in phantom at D_{max}), t_0: reference depth, S_c: collimator scattering factor, S_p: phantom scattering factor, DD%: percentage depth dose, r: collimator field size, SCD: source–collimator distance, SSD: source–skin distance.

1.11 Beam Modifiers

1.11.1 Bolus

Bolus is used for tissue compensation, and is put on the skin at right angles to the beam axis. It is made from a tissue-equivalent density material (Fig. 1.65) [40]. Bolus use leads to increased effects of radiation scattered into the skin. Thus, the entrance dose to the skin increases. Secondary electrons produced by the bolus also increase the skin dose, since the bolus is in contact with skin (→ the depth corresponding to D_{max} gets close to the surface).

1.11.2 Compensating Filters

The dose distribution is not homogeneous if the surface of the patient is not flat. Therefore, a compensating filter is positioned between the beam source and the skin to reduce the dose delivered to the area with thinner tissue in order to achieve a homogeneous dose distribution in the irradiated volume (Fig. 1.66) [40]. Compensating filters are made of aluminum–tin or copper–tin mixtures, and are individually designed to compensate for tissue irregularities.

1.11.3 Wedge Filters

Metal wedge filters can be used to even out the isodose surfaces for photon beams delivered onto flat patient surfaces at oblique beam angles (Fig. 1.67) [40]. These wedges can be static,

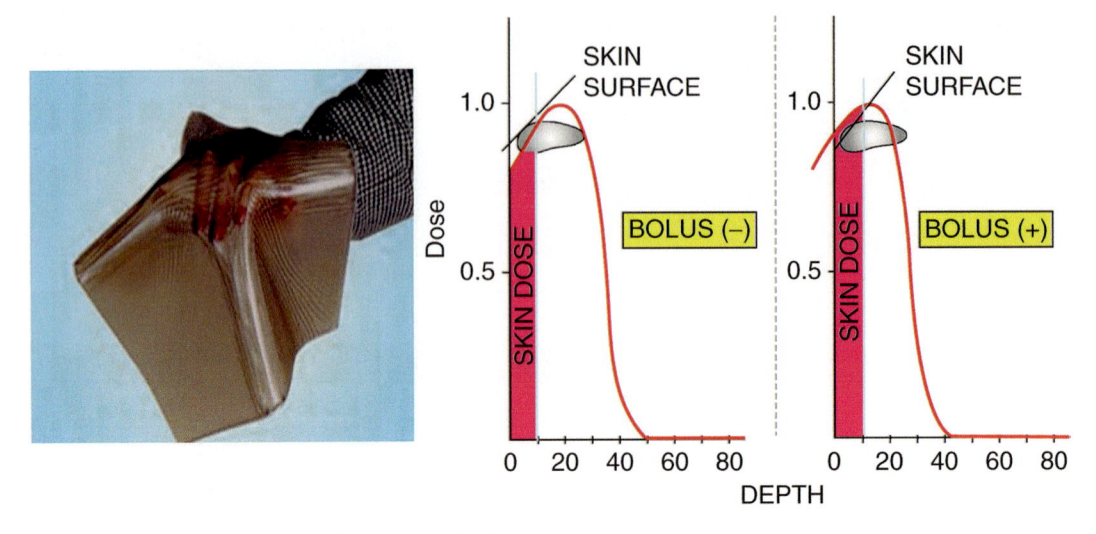

Fig. 1.65 Bolus material and its effect on the skin dose

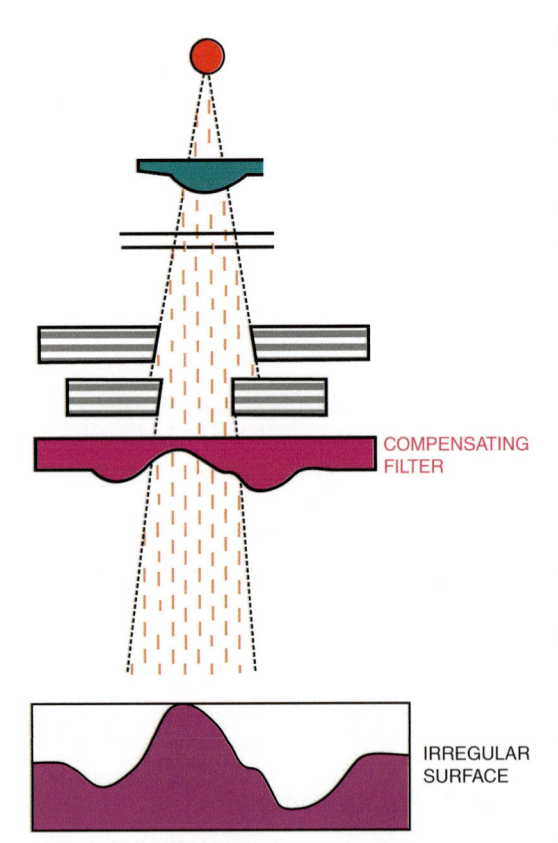

Fig. 1.66 Compensating filter (Fig. 4.11, p 73 of [42])

dynamic, or motorized. They are most commonly used in tangential irradiation (e.g., the breast, head, and neck regions), and they prevent hot spots in vital organs and cold spots in the radiation field. They provide a more homogeneous dose distribution.

Wedge angle (q): the angle between the plane of the ground and the point at which the central axis crosses the 50% isodose line (if $E > 6$ MV, at 10 cm] (Fig. 1.68).

$$\text{Hinge angle } (\Phi) = 180 - 2\theta \quad (1.20)$$

1.11.4 Shielding Blocks

These are manufactured in order to shield the normal critical structures in radiotherapy portals. There are two types of shielding block. Standard blocks come with the teletherapy unit, and have various sizes and shapes. They are designed according to the area that needs to be protected. On the other hand, focalized blocks are individually made in mold rooms to shield the areas of the field that need protecting (according to the simulation procedure). Standard blocks are only used in emergencies.

Focalized blocks have the advantages of providing divergence, a very close fit to the region that needs protecting, and easy setup. However, they are

1.11 Beam Modifiers

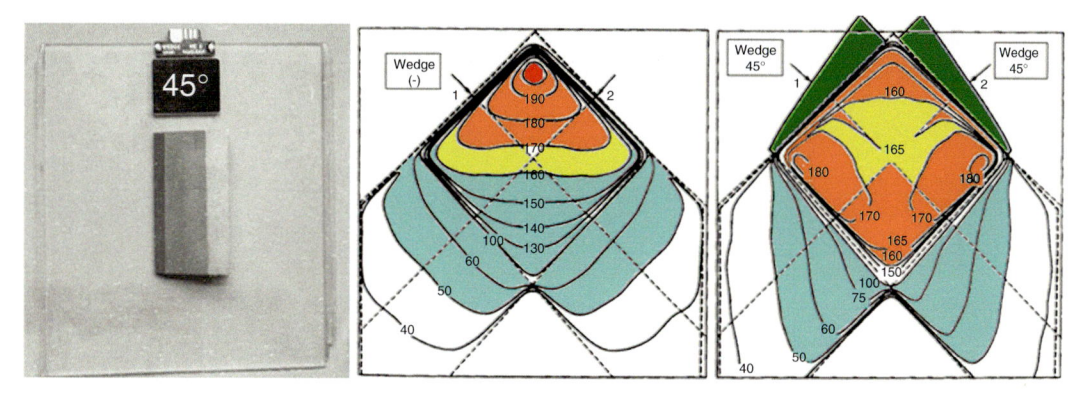

Fig. 1.67 Change in isodose profile resulting from the use of a static wedge filter

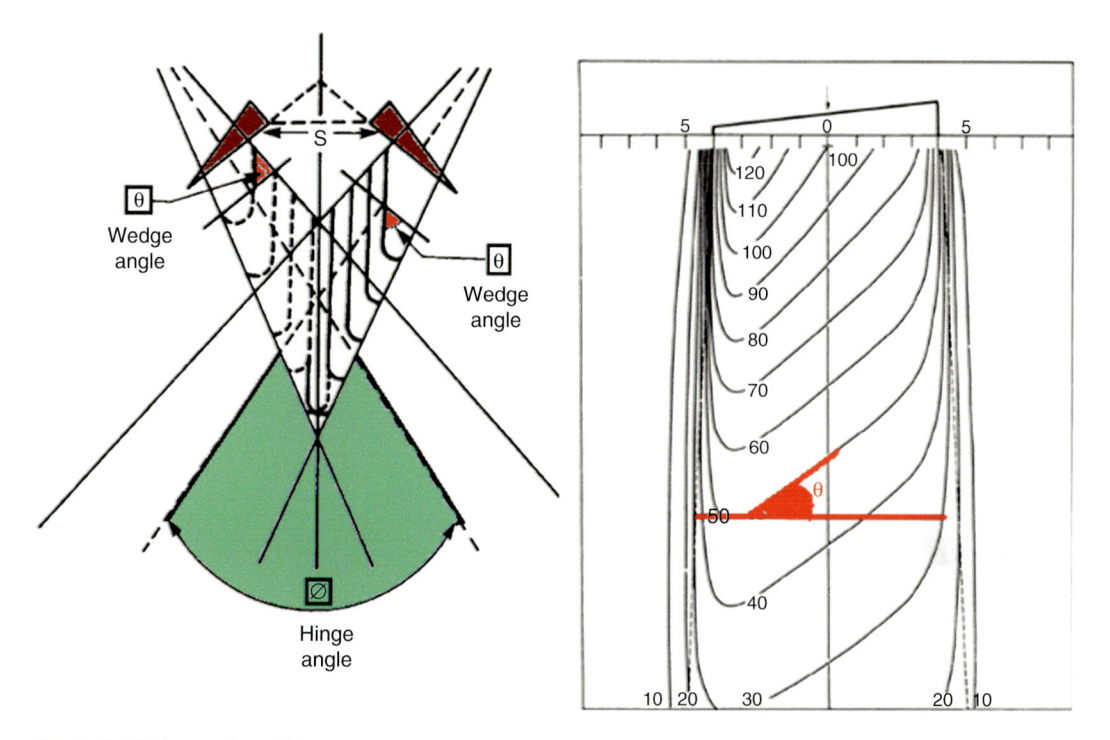

Fig. 1.68 Wedge angle and hinge angle

time-consuming to make and use, result in an increased workload, and are expensive (Fig. 1.69).

Focalized blocks are made up of lead or Cerrobend® (Fig. 1.70). Cerrobend is a mixture of lead (26.7%), bismuth (50%), zinc (13.3%), and cadmium (10%) that melts at 70 °C and has an HVL of 1.3 cm [40].

1.11.5 Multileaf Collimator (MLC)

Irregular fields cannot be shaped without focalized blocks in conventional radiotherapy machines. The collimator systems in Co-60 and old linac machines provide only a rectangular field. Multileaf collimators, on the other hand,

Fig. 1.69 Process of preparing focalized blocks. (**a**) Area to be protected is delineated. (**b**) Block cutter is adjusted according to the parameters of the treatment machine (SSD, SAD). (**c**) Block thickness is adjusted according to the energy and the treatment machine. (**d**) Block mold is cut by hot-wire. (**e**) Cerrobend is poured into block mold. (**f**) Focalized blocks after cooling

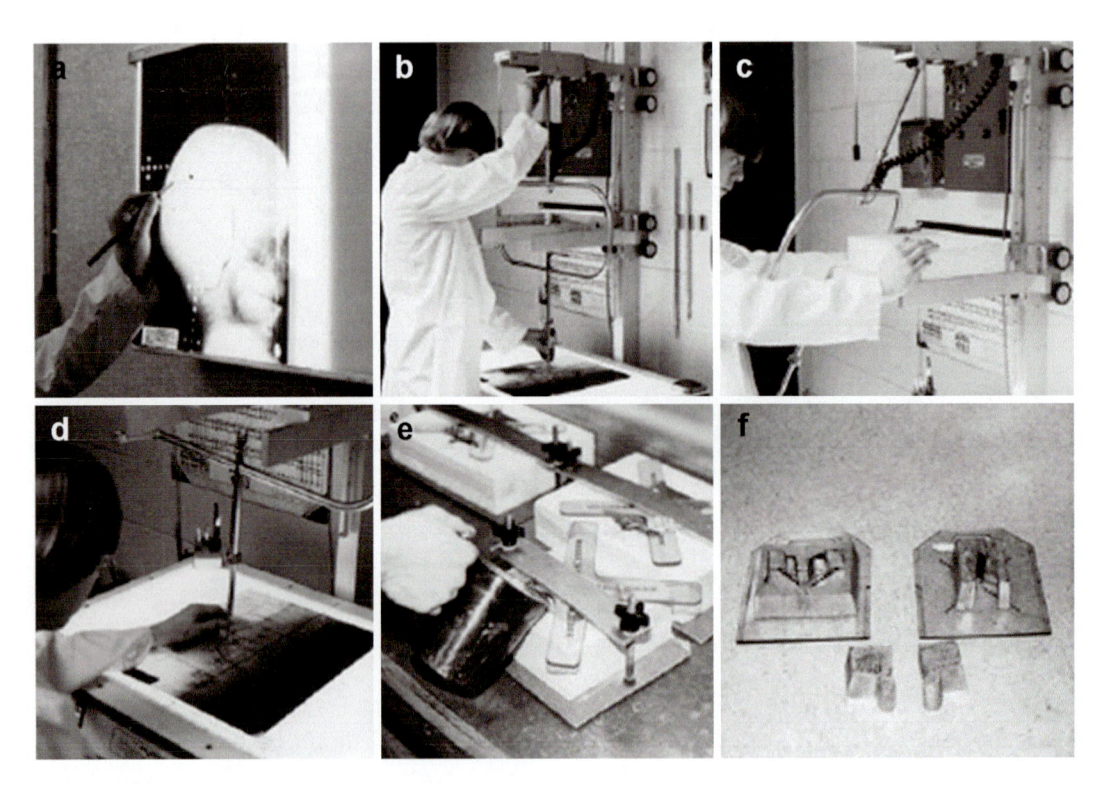

Fig. 1.70 Block cutter, Cerrobend, and focalized blocks

are composed of many leaves, and each leaf can move independently (Fig. 1.71) [41]. By modulating those leaves appropriately, irregular fields can easily be radiated without using blocks (Tables 1.8, 1.9, and 1.10).

Fig. 1.71 Various types of multileaf collimator systems

Table 1.8 Some radiation-associated units and their features

Parameter	Radiation type	Medium in which measurement is performed	Unit	Specific unit
Exposure	X-rays and gamma rays	Air	C/kg	R (roentgen) = 2.58 × 10^{-4} C/kg
Absorbed dose	All radiation types	Everywhere	erg/g, J/kg	1 rad = 100 erg/g 1 Gy = 1 J/kg
Equivalent dose	All radiation types	Human	erg/g, J/kg	1 REM = 100 erg/g × WR 1 Sv = 1 J/kg × WR

WR weighting factor, formerly known as the quality factor (*Q*)

Table 1.9 Summary of photon–matter interactions

Photon–electron interactions	Photon–nucleus interactions
Interaction with bound electrons Photoelectric effect Coherent effect	Direct interaction with nucleus Photodisintegration
Interaction with free electrons Compton effect	Coulomb interaction with nucleus Pair production

Note that bremsstrahlung is not a photon–matter interaction; it is the interaction of a charged particle (an electron) with matter

1.12 Pearl Boxes

The skin-sparing effect and penetrating ability increase at high energies.

The depth (in cm) of the 10% isodose line for electrons is nearly half of their initial energy in MeV → $E/2$. This is known as the *penetration range* of electrons. It is important for determining an energy that cannot reach spinal cord while selecting a boost dose in head and neck radiotherapy.

The penetration range of electrons triples in lung tissue → because of the air present!

The depth (in cm) of the 80% isodose line for electrons is nearly one-third of their energy in MeV → $E/3$. This is known as the *therapeutic range* of electrons.

The depth (in cm) of the 90% isodose line for electrons is nearly one-quarter of their energy in MeV → $E/4$.

The build-up point (i.e., that corresponding to the dose maximum, d_{max}, in cm) for electrons corresponds practically to $E/6$.

- For example, for 6 MeV, $D_{max} = 6/6 = 1$ cm.

This formula is valid up to 16 MeV; at 16 MeV and over d_{max} is £ 1.5 cm).

Common features of X-rays and gamma rays in their interaction with tissue are the production of ionization and high-speed electrons, depositing energy, and undergoing scattering.

The skin dose increases as the electron energy increases.

The skin dose decreases as the photon energy increases.

Table 1.10 Advantages and disadvantages of various radiations used in external radiotherapy

Radiation	Advantages	Disadvantages
Photon	Wide range of use Skin-sparing effect	Entrance dose > tumor dose High-dose region during the exit of the radiation from the patient
Electron	Sparing of normal tissues beyond the tumor due to their limited ranges	Large penumbra due to scattering Only used in superficial tumors
Proton	No dose beyond tumor Very low dose proximal to the tumor	Large penumbra at 20 cm depth Limited use Expensive

The TAR is difficult to measure at Co-60 and 6 MV photon energies, so the TPR is used instead of TAR in dose calculations.

An advantage of TLD over ion chamber is that it can be used in vivo.

Silver bromide crystals exposed to radiation in film dosimeters change color, and this color change is measured through optic densitometry.

The kerma is greatest at the surface, and decreases with depth. The absorbed dose increases until the depth at D_{max}, and decreases as the depth increases beyond that.

The maximum block thickness for electrons is 1 cm.

Block thicknesses (lead):
For Co-60, → 5 cm
For 6 MV X-rays → 6 cm
For 25 MV X-rays → 7 cm

5 HVL: 3.125% of radiation passes under block in Co-60. For a linac, block thickness → 4 HVL.

The old unit of exposure, the roentgen, cannot be used for particulate radiation. It is only used for exposure to X-rays and gamma rays.

An electron loses 2 MeV of energy as it propagates through each centimeter of soft tissue.

Frequencies:
Co-60 (mean 1.25 MV) → 3×10^{20} cycles/s (Hz).
6 MV X-rays → 12×10^{20} cycles/s (Hz).
25 MV X-rays → 60×10^{20} cycles/s (Hz).

Photons travel at the speed of light, since they have no mass; electrons are slower, since they have mass.

Tyratron → this is present in all linacs, where it converts DC current to pulse current. It activates the klystron or the magnetron and the filament of the electron gun.

Energy	Dmax (cm)	50% isodose (cm)
250 keV	On the surface	7
1.25 MeV	0.5	11
4 MeV	1	14
6 MeV	1.5	16
18 MeV	3	21
25 MeV	3.5	23

Since electrons show a rapid fall-off and a finite range, tissues beyond the target are spared and a uniform superficial dose is provided when electrons are used.

Delta electrons → electrons produced by the interactions of secondary electrons with atoms.

Trimmer bars → satellite collimators used in teletherapy machines to decrease the penumbra.

Hardening the X-ray beam → eliminating low-energy photons within the beam using selectively absorbing filters.

TVL (tenth-value layer) → thickness of absorbent that decreases the incoming radiation dose by 90% (TVL = 3.32 HVL).

Divergence → the tendency of a radiation beam to widen as it propagates farther from the source.

Electron energies are calibrated in an ion chamber at the reference depth of the 50% isodose line.

Binding energy of a nucleus is described as the energy needed to separate nucleons from the nucleus. Binding energy of electron in a shell is expressed as the energy needed to separate an electron from that shell.

Transient equilibrium (TE) and secular equilibrium (SE) are radioactive equilibrium types. TE takes place if the parent's half-life is greater than the daughter's half-life. Following the occurrence of TE, daughter's activity exceeds the parent's activity. An example of TE is the production of technetium (^{99m}Tc) out of the parent nuclide molybdenum (^{99}Mo)'s decay.

SE takes place if parent's half-life is much more greater than daughter's half-life and when the isotopes reside in a closed environment. Following the occurrence of SE, daughter's activity equates the parent's activity. A representative example of SE is the production of radon (^{222}Rn) from radium (^{226}Ra).

Grenz rays are referred to as very low energy X-rays produced in the range of 10–20 kVp at X-ray tubes.

Attenuation occurs by scattering or absorption of photons by the matter.

X-rays produced by electrons on X-ray tube are forward peaked and radiation beam profile is not flat which renders them unsuitable for use in radiation treatment.

Decay of source strength over years results in reduction of dose output and increase in treatment time for Co 60 which necessitates regular replacement of sources at every 5–6 year periods.

When compared with photons, electrons are more prone to the effects of tissue heterogeneities.

The "f factor" is used for converting exposure or exposure rate in the air to dose or dose rate in a medium.

Calibration of high energy photons are performed in a water phantom at 10 cm depth with an ion chamber as per the AAPM TG-51 protocol.

Calibration of electrons is performed in water at a reference depth specified by the 50% depth dose using an ion chamber as per the AAPM TG-51 protocol.

Factors affecting PDD include the field size, energy, depth, and SSD.

BSF is dependent on field size and energy, but it is independent of SSD.

BSF corresponds to TAR at the dmax depth.

Among the factors of energy, SSD, field size, total dose, and tissue thickness; dmax is most dependent on energy.

Main sources of scattered radiation affecting the patient dose include patient scatter and the collimator scatter.

Collimator scatter is measured by use of the dose in the air for varying collimator jaw settings.

Transmission penumbra:

- Penumbra caused by collimation blocks.
- This type of penumbra can be reduced if the blocks are individualized.

Possible scenarios in which an atomic particle hits a target:

- Nuclear reaction
- Elastic scattering

In elastic scattering, the particles after collision are identical to those before the collision.

The total kinetic energy of the particles does not change during the collision.

A small fraction of the total kinetic energy is transferred to the target.

Radiation is not produced in elastic scattering.

- Inelastic scattering

In inelastic scattering, the particles after the collision are identical to those before the collision.

Most of the total kinetic energy is transferred to the target.

Radiation is produced in inelastic collisions.

Possible Scenarios When a Photon Enters the Human Body (Fig. 1.72a–c)

- No interactions occur (a)
- Photon–matter interactions occur, which divert the photon from its original path (b)
- Scattered photons may produce secondary photon–matter interactions (c)

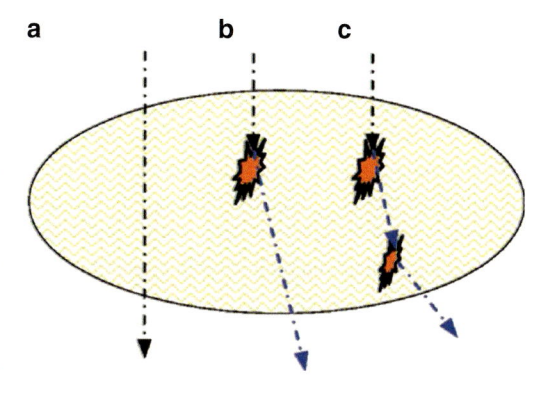

Fig. 1.72 (**a–c**) Possible scenarios when a photon enters the human body

Fig. 1.73 Bragg peak

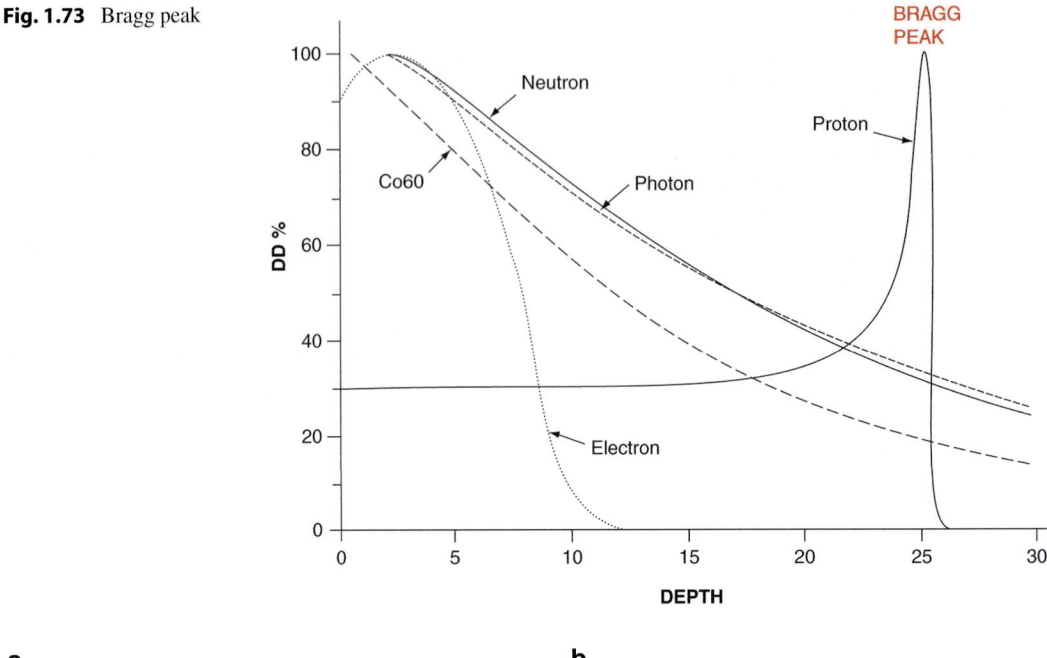

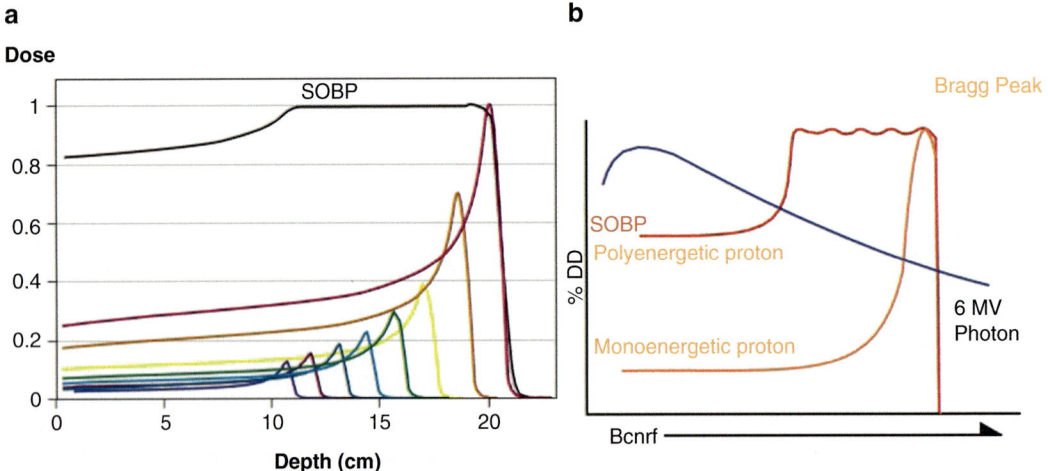

Fig. 1.74 (**a**, **b**) Bragg peak for different proton energies (**a**) and SOBP for monoenergetic and polyenergetic protons (**b**)

Bragg Peak (Fig. 1.73)

Protons with energies of 200 MV enter the body with a speed of 180,000 km/s. Their range in the body is only 25–30 cm. The energy transferred by the protons moving through the tissue is inversely proportional to their velocities. Thus, the protons lose most of their energy in the region 1–4 mm before they are stopped completely, and create a kind of "energy explosion" (known as the Bragg peak) at the target point. This peak in the dose delivered to tissues in the Bragg peak region can be seen in Fig. 1.73. On the other hand, the most effective dose is delivered just beneath the surface in classical radiotherapy using X-rays, and the dose decreases further into the tissue. Thus, healthy tissues before and beyond the target are exposed to unwanted radiation. The Bragg peak width contains the entire target volume and is defined as Spread Out Bragg Peak (SOBP) (Fig. 1.74).

- This phenomenon was first described by Sir William Henry Bragg.

References

1. Khan Faiz M (2003) Physics of radiation therapy, 3rd edn. Lippincott Williams & Wilkins, Philadelphia, pp 3–4
2. Podgorsak EB (2005) Radiation oncology physics: a handbook for teachers and students. International Atomic Energy Agency, Vienna, pp 3–7
3. Peres A (1958) Photons, gravitons and the cosmological constant. Il Nuovo Cimento (1955–1965) 8(4):533–538
4. Kano Y (1966) The fluctuation formula for the photon number in stationary electromagnetic fields. Il Nuovo Cimento B (1965–1970) 43(1):1–5
5. Potzel W, van Bürck U, Schindelmann P, Hagn H, Smirnov GV, Popov SL, Gerdau E, Yu Shvyd'ko V, Jäschke J, Rüter HD, Chumakov AI, Rüffer R (2003) Interference effects of radiation emitted from nuclear excitons. Hyperfine Interact 151–152(1–4):263–281
6. Bentzen S, Harari P, Tome W, Mehta M (2008) Radiation oncology advances. Springer, New York, p 1
7. Kaul A, Becker D (2005) Radiological protection. Springer, Berlin, p 24
8. Khare SP (1992) K-shell ionisation of atoms by positron and electron impacts. Hyperfine Interact 73(1–2):33–50
9. Kostylev VA (2000) Medical physics: yesterday, today, and tomorrow. Biomed Eng 34(2):106–112
10. Ulrich A, Born M, Koops HWP, Bluhm H, Justel T (2008) Vacuum electronics components and devices. Springer, Berlin, p 5
11. Barouni M, Bakos L, Papp Zemplén É, Keömley G (1989) Reactor neutron activation analysis followed by characteristic X-ray spectrometry. J Radioanalytical Nuclear Chem 131(2):457–466
12. Khan Faiz M (2003) Physics of radiation therapy, 3rd edn. Lippincott Williams & Wilkins, Philadelphia, p 33
13. Podgorsak EB (2005) Radiation oncology physics: a handbook for teachers and students. International Atomic Energy Agency, Vienna, p 21
14. Podgoršak E (2007) Radiation physics for medical physicists, 1st edn. Springer, Berlin, pp 262–265
15. Tatjana J (2005) Nuclear principles in engineering. Springer, Berlin, pp 127–171
16. Hooshyar MA, Reichstein I, Malik Bary F (2005) Nuclear fission and cluster radioactivity. Springer, Berlin, pp 153–173
17. Hobbie Russell K, Roth Bradley J (2007) Intermediate physics for medicine and biology. Springer, Berlin, pp 481–513
18. Magill J, Galy J (2005) Radioactivity radionuclides radiation. Springer, Berlin, pp 117–123
19. Dietze G (2005) Radiological protection. In: Kaul A, Becker D (eds) Radiological protection. Springer, Berlin, pp 355–368
20. Stabin Michael G (2008) Radiation protection and dosimetry. Springer, Berlin, pp 244–308
21. Fasso A, Göbel K, Höfert M, Ranft J, Stevenson G (2006) Shielding against high energy radiation. Springer, Berlin, pp 265–266
22. Podgoršak E (2007) Radiation physics for medical physicists, 1st edn. Springer, Berlin, pp 107–114
23. Podgorsak EB (2005) Radiation oncology physics: a handbook for teachers and students. International Atomic Energy Agency, Vienna, p 153
24. Khan Faiz M (2003) Physics of radiation therapy, 3rd edn. Lippincott Williams & Wilkins, Philadelphia, pp 44–45
25. Anatoly Rosenfeld B (2006) Semiconductor detectors in radiation medicine. In: Tavernier S, Gektin A, Grinyov B, Moses WW (eds) Radiation detectors for medical applications. Springer, Berlin, pp 111–147
26. Khan Faiz M (2003) Physics of radiation therapy, 3rd edn. Lippincott Williams & Wilkins, Philadelphia, p 160
27. Levitt SH, Purdy JA, Perez CA, Vijayakumar S (2006) Physics of treatment planning in radiation oncology. In: Levitt SH, Purdy JA, Perez CA, Vijayakumar S (eds) Technical basis of radiation therapy, 4th edn. Springer, Berlin, pp 69–106
28. Podgorsak EB (2005) Radiation oncology physics: a handbook for teachers and students. International Atomic Energy Agency, Vienna, p 171
29. Podgorsak EB (2005) Radiation oncology physics: a handbook for teachers and students. International Atomic Energy Agency, Vienna, p 599
30. Khan Faiz M (2003) Physics of radiation therapy, 3rd edn. Lippincott Williams & Wilkins, Philadelphia, p 179
31. International Commission on Radiation Units and Measurements (1973) Measurement of absorbed dose in a phantom irradiated by a single beam of x or gamma rays. Report No. 23. National Bureau of Standards, Washington, DC
32. Webster EW, Tsien KC (eds) (1965) Atlas of radiation dose distributions. In: Single-field isodose charts, vol 1. International Atomic Energy Agency, Vienna
33. Khan Faiz M (2003) Physics of radiation therapy, 3rd edn. Lippincott Williams & Wilkins, Philadelphia, p 53
34. VanderLinde J (1993) Classical electromagnetic theory. Springer, Berlin, pp 269–311
35. Cunningham JR, Johns HE, Gupta SK (1965) An examination of the definition and the magnitude of back-scatter factor for cobalt 60 gamma rays. Br J Radiol 38:637
36. Holt JG, Laughlin JS, Moroney JP (1970) Extension of concept of tissue-air ratios (TAR) to high energy x-ray beams. Radiology 96:437
37. Khan Faiz M (2003) Physics of radiation therapy, 3rd edn. Lippincott Williams & Wilkins, Philadelphia, p 175
38. Khan FM, Gerbi BJ, Deibel FC (1986) Dosimetry of asymmetric X-ray collimators. Med Phys 13:936
39. Khan Faiz M (2003) Physics of radiation therapy, 3rd edn. Lippincott Williams & Wilkins, Philadelphia, pp 183–185

40. Eric K, Sasa M, James P (2006) Treatment aids for external beam radiotherapy. In: Levitt SH, Purdy JA, Perez CA, Vijayakumar S (eds) Technical basis of radiation therapy, 4th edn. Springer, Berlin, pp 167–177
41. Ehrgott M, Hamacher HW, Nußbaum M (2007) Decomposition of matrices and static multileaf collimators: a survey. In: Carlos JS, Alves Panos M (eds) Pardalos and Luis Nunes Vicente. Optimization in medicine. Springer, Berlin, pp 25–46
42. Goitein M (2008) Radiation oncology: a physicist's-eye view. Springer, New York

Radiobiology

2.1 Cell Biology and Carcinogenesis

Radiobiology, in general terms, is the science that evaluates the effects of radiation in living organisms. In the field of radiation oncology, it is defined as the science that investigates the interactions between ionizing radiation and living systems and the consequences of these interactions.

2.1.1 Cell Structure

Atoms form molecules, molecules make macromolecules, macromolecules build complex organic structures, and then cells—which are the main structural component of tissues and reflect all features of life—are formed. All cells have generally similar structures. However, they specialize according to the location of the tissue (i.e., according to the functions of that tissue). The basic structures of all organisms are formed from these cells. In humans, there are approximately 10^{14} cells [1].

All cells are surrounded by a cell membrane, and they have a liquid-like cytoplasm and organelles within the membrane (Fig. 2.1).

2.1.2 Cell Types and Organelles

Cells can be simply divided into two categories: prokaryotic cells and eukaryotic cells [2].

- **Prokaryotic cells.** Bacteria and blue–green algae belong in this group. These cells have no nucleus and are surrounded by a nuclear membrane. In addition, they do not have any membranous organelles, such as mitochondria. DNA—genetic material—is scattered within the cytoplasm. These cells do have ribosomes. The vital functions of this type of cell are performed in the cytoplasm and cellular membrane.
- **Eukaryotic cells.** These cells have membranous organelles, so their nuclear material is not scattered within the cytoplasm, and they have a real nucleus. Eukaryotic cells are more highly developed than prokaryotic cells—the cells of animals, plants, fungi, and protists are eukaryotic. These cells have organelles in their cytoplasm. Chromosomes, consisting of DNA and proteins, are located in the cell nucleus. Eukaryotic cells divide by mitosis.

Cytoplasm. The cytoplasm is a semifluid matrix that fills the space between the cell membrane and the nucleus. All vital events occur in the cytoplasm of a living organism. It generally forms a homogeneous transparent mass.

Mitochondria. These are ellipsoidal or cudgel-shaped organelles of length 2–3 µm and diameter 0.5 µm. They are the energy-generating unit of cells. The citric acid cellular respiration cycle (the Krebs cycle) occurs in this organelle.

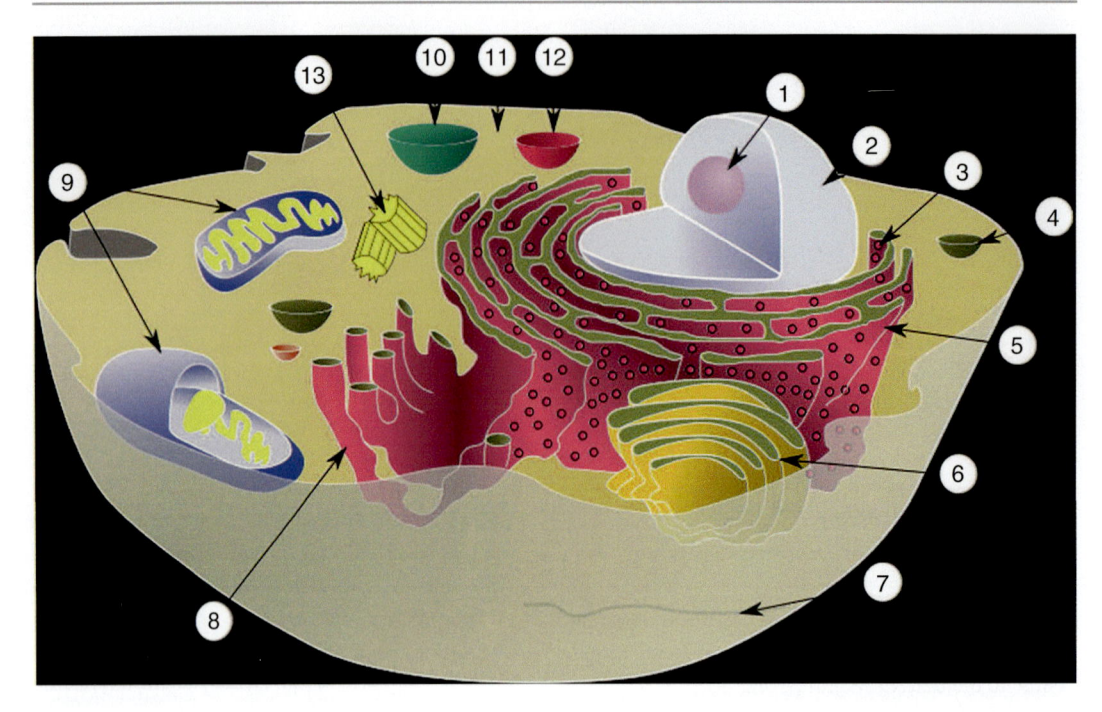

Fig. 2.1 Schematic illustration of a eukaryotic cell. (*1*) Nucleolus; (*2*) nucleus; (*3*) ribosome; (*4*) vesicle; (*5*) rough endoplasmic reticulum (ER); (*6*) Golgi apparatus; (*7*) cytoskeleton; (*8*) smooth endoplasmic reticulum; (*9*) mitochondria; (*10*) vacuole; (*11*) cytoplasm; (*12*) lysosome; (*13*) centrioles within a centrosome

The energy released by breaking the chemical bonds of organic molecules is transformed into ATP within mitochondria.

Lysosome. This is a round organelle surrounded by a membrane, and it contains hydrolytic enzymes. They perform the function of digestion in the cell, clearing away excessive or harmful intracellular structures.

Golgi apparatus. The Golgi apparatus or complex consists of a combination of membranous tubes or saccules. It is generally close to the nucleus and is particularly conspicuous in actively secreting secretory cells. Its main function is believed to be the storage of proteins secreted by the cell. It performs the functions of secretion and packing.

Endoplasmic reticulum. The endoplasmic reticulum ensures that nutrition circulates in the cytoplasm and synthesizes lipids and hormones. It is a complex serial channel system located between the cell membrane and the nuclear membrane. If it does not contain ribosomes it is called "smooth endoplasmic reticulum." This secretes steroid hormones in steroid-secreting cells and performs detoxification in others.

Ribosome. Ribosomes are located along the channels of the endoplasmic reticulum and are found scattered within the cytoplasm. They perform protein synthesis. They are approximately 150 Å in diameter. Their structures are composed of 65% RNA (ribonucleic acid) and 35% protein. Proteins synthesized by ribosomes are sent to either intracellular or extracellular regions with the aid of the endoplasmic reticulum.

Cell nucleus. This has a granular and fibrous structure (Fig. 2.2). Most of the genetic information of the cell is located in the cell nucleus within chromosomes, which are long, linear, folded DNA molecules formed through the collection of many proteins like histones. The genes located in these chromosomes comprise the nuclear genome of the cell. The role of the cell nucleus is to maintain the integrity of these genes and to control cellular functions by arranging gene expression.

2.1 Cell Biology and Carcinogenesis

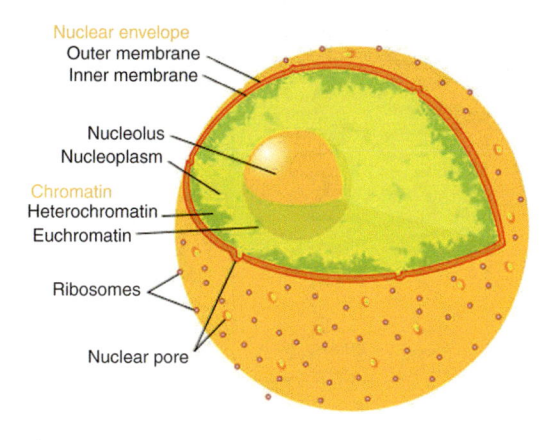

Fig. 2.2 Eukaryotic cell nucleus

Nuclear membrane. This is the outer covering of the nucleus. Ribosomes are stuck in the nuclear membrane, and it contains pores.

Chromatin. This is the structure that transforms into chromosomes during division and when moving towards poles.

Nucleolus. This is the center of the nucleus, and it synthesizes proteins and ribosomes.

Nuclear matrix. This fills the space between the chromatin and the nucleus and contains proteins and ions.

Chromosome [3]. The nucleus contains most of the genetic material in the cell (Fig. 2.3). This genetic material actually consists of multiple linear DNA molecules called chromosomes. Chromosomes are found in the form of a complex combination of DNA and proteins termed chromatin during most of the cell cycle, and chromatin forms the chromosomes of one karyotype during division (Fig. 2.4).

A small number of cellular genes are found in mitochondria:

1. Chromatid
2. Centromere
3. Short arm
4. Long arm

2.1.3 Cell Cycle

During its life, a cell generally exhibits a long period or phase (interphase) during which no division occurs and a division phase (mitosis).

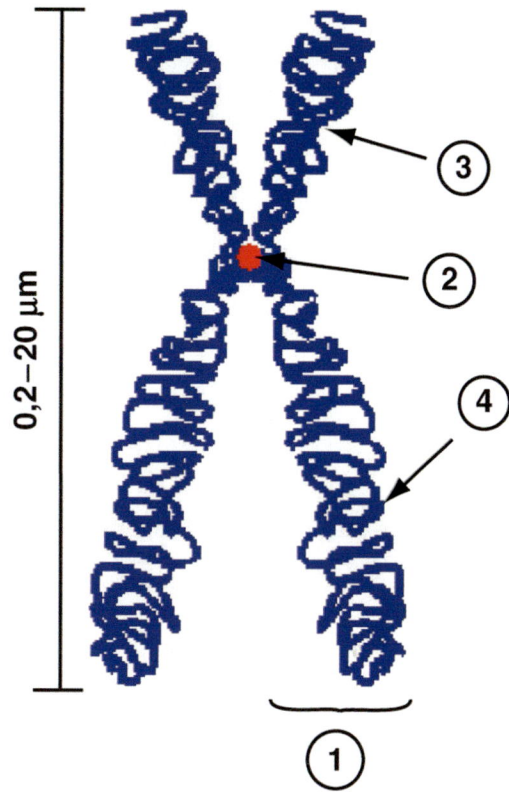

Fig. 2.3 The structure of a chromosome

This is called the cell cycle [4]. The cell cycle is repeated at each cellular stage, and the length of time corresponding to a cell cycle varies with cell type (Fig. 2.5). The interphase is very long in some cells, and these types of cells never divide during the life period of the organism (an example is a neuron). Generally, cells grow until they reach a certain size, then they divide.

The radiosensitivity of a cell is fourfold greater during the mitotic phase than during the interphase. Radioresistance is high in the S, late G_1, and G_0 phases. The resistance of the S phase is due to the large amounts of synthesis enzymes present, which have the ability to rapidly repair DNA.

M: Mitosis
Prophase
Metaphase
Anaphase
Telophase
I: Interphase
G_1

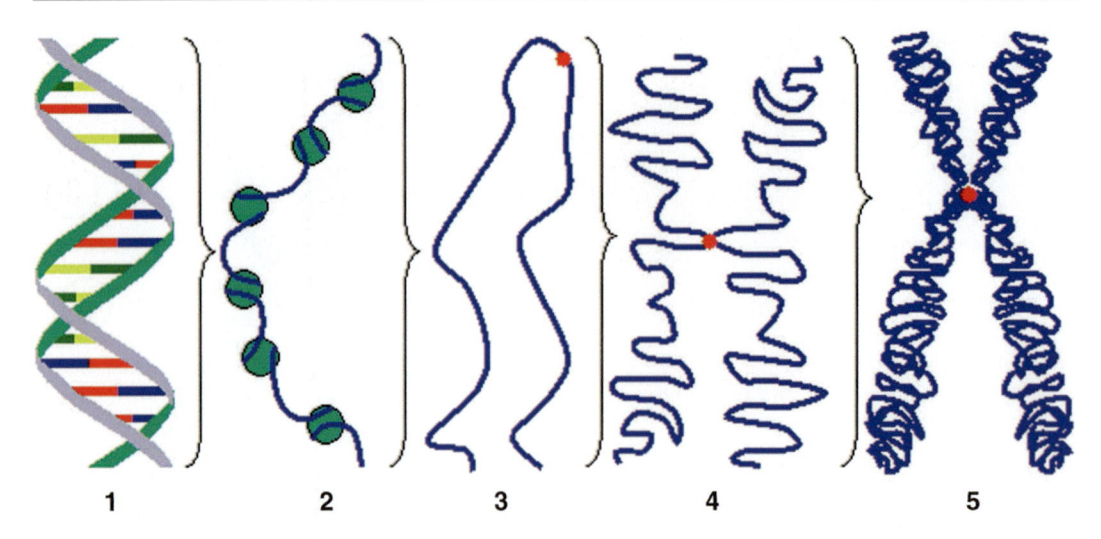

Fig. 2.4 Relationship between DNA and chromosomes. (*1*) DNA; (*2*) DNA + protein; (*3*) chromatin; (*4*) chromatids; (*5*) chromosome. DNA molecules combine and make proteins, proteins form chromatins, chromatins mate with each other during division and become chromatids, and chromatids combine and forms chromosomes

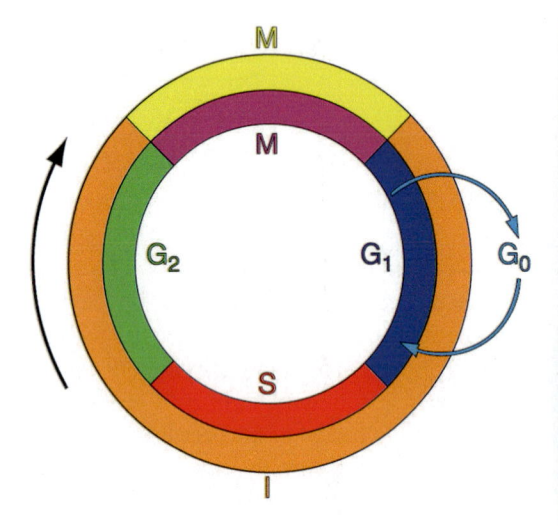

Fig. 2.5 Cell cycle and its stages

S
G₂
G₀

2.1.3.1 The Stages of Mitosis [4]

Mitosis is the division of a cell into two cells through the mating of its genome. Mitosis is only observed in eukaryotic cells. Somatic cells are formed via mitosis, whereas germ cells are formed by meiotic division.

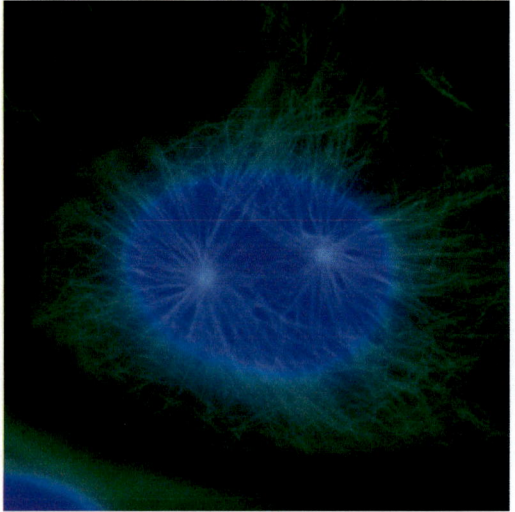

Fig. 2.6 Prophase

Prophase (Fig. 2.6)
The nuclear membrane and endoplasmic reticulum disappear.
　The chromosomes shorten and thicken.
　Centrosomes move towards opposite poles.
　The nucleolus disappears.
　Spindle cells form from the poles to the center.

2.1 Cell Biology and Carcinogenesis

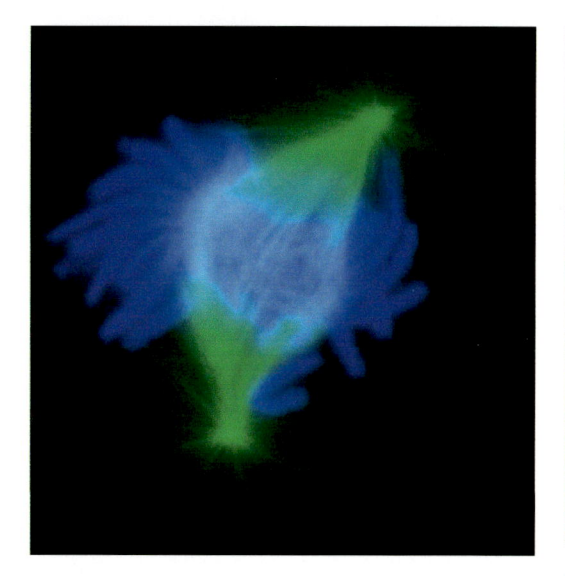

Fig. 2.7 Metaphase

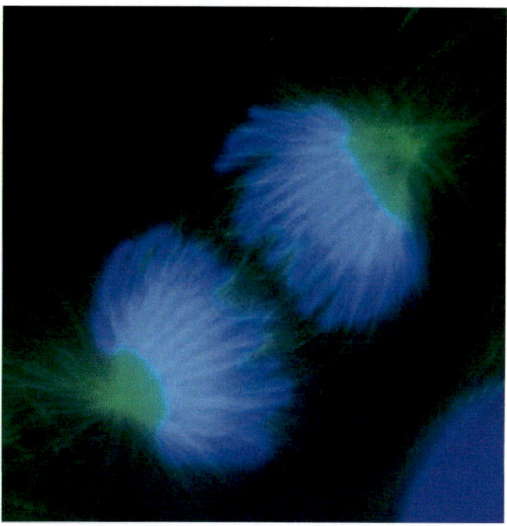

Fig. 2.8 Anaphase

Metaphase (Fig. 2.7)
The chromosomes shorten and thicken further.
　Sister chromatids are kept together using centromeres.
　The chromosomes are arranged side-by-side in a row in the equatorial plane.
　The chromosomes hold on to spindle cells with their centromeres.

Anaphase (Fig. 2.8)
The contraction and relaxation movements of spindle cells break the centromeres that lock the chromatids together.
　The sister chromatids are separated from each other and are moved to opposite poles.

Telophase (Fig. 2.9)
The chromosomes stop moving.
　The chromosomes unwind their helices and become chromatins.
　The nucleolus reappears.
　RNA and protein syntheses start.
　Spindle cells disappear.
　The nuclear membrane forms, and the endoplasmic reticulum takes on a shape again.
　Vital events restart in the cell.
　Cytogenesis occurs, and division finishes.

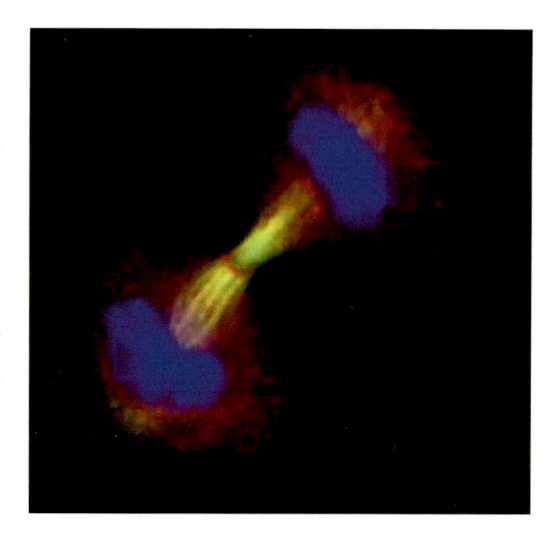

Fig. 2.9 Telophase

2.1.3.2 Interphase and Its Stages [4, 5]

The interphase is the preparation phase for the redivision of a divided cell. It is the longest phase of the eukaryotic cell cycle. For instance, the interphase of a human skin cell is about 22 h, while the cell cycle of such a cell lasts approximately 24 h. The interphase is divided into three stages.

$G_1 \rightarrow (G: Gap_1)$

This occurs just after cytogenesis.

Metabolic events continue intensely.

This is the stage where matter transportation, synthesis, lysis reactions, organelle production, RNA synthesis, and tissue functions continue at their highest levels.

It is the longest stage. Dividable cell growth occurs during this stage.

Cells that lose their ability to divide continue with their functions and life activities (e.g., muscle and nerve cells still function at this stage).

$S \rightarrow (S: Synthesis)$

DNA is duplicated, and the number of chromatins doubles ($\rightarrow$ replication).

The most intense protein synthesis is performed at this stage.

The order of centromere duplication is observed.

$G_2 \rightarrow (G: Gap_2)$

Enzymes related to division are synthesized.

The number of organelles increases.

DNA synthesis finishes, but RNA synthesis continues.

Centrosome synthesis finishes, and these centrosomes start moving towards opposite poles.

G_0 phase

Cells have a natural mechanism that protects them during difficult developmental conditions. Under these conditions, the cells transiently stop their cellular activities. This phase is called the G_0 phase. In the G_0 phase, some genes in the DNA are covered with various proteins; i.e., the DNA is programmed.

The most radiosensitive stages during the cell cycle are the early G_2 and M stages (Fig. 2.10).

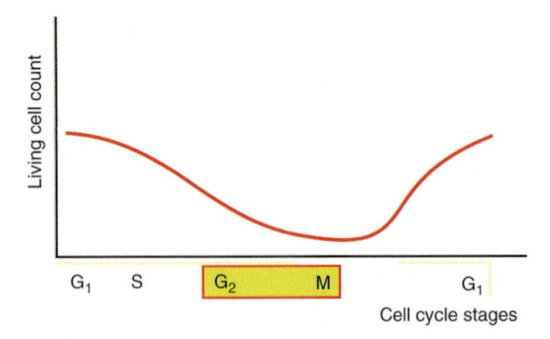

Fig. 2.10 Relationship between cell cycle stages and living cell count

2.1.4 Carcinogenesis and the Cell Cycle

Cell proliferation in tissues is a normal function in organisms. Decreasing cell proliferation or increasing the death rate prevents any excessive increase. The replication of a cell into two similar cells is prompted by extrinsic biochemical signals, and a series of phases regulated by inner or outer growth factors occur. Some oncogenes and proteins specific to the cell cycle are activated synchronously throughout the cell cycle and are then inactivated.

The development of cancer at the cellular level is termed *carcinogenesis*. The combination of mutations that affect biological events such as cell survival, growth control, and differentiation is the basis for carcinogenesis. Tumor cells gain several phenotypic features during the development of cancer. Those changes cause the rapid and uncontrolled proliferation of tumor cells, as well as their spread to surrounding tissues. In addition, those cells can survive independently in specific microenvironments and have the ability to metastasize.

Cells with proliferative capacity normally stop at certain checkpoints. The most important of these is the first, which occurs just before DNA synthesis, and the second, which occurs just prior to mitosis. These histologic resting periods probably occur due to the decreased activity of cyclin-dependent kinases and tumor suppressor proteins. Actually, since cells in these phases of the cell cycle synthesize the proteins used in the next phases, they are biochemically active. At these checkpoints, any genetic defects are repaired. In summary, while the cell progresses through the cycle, it stops at two checkpoints and is controlled. Normal cells have mechanisms for detecting errors in the DNA sequence. A group of repair mechanisms replace damaged nucleotides with normal molecules when the DNA is damaged. These mechanisms ensure that the genetic material in each of the two daughter cells is the same as that of the mother cell.

A small proportion of the normal cell population consists of immortal cells (i.e., they have the capacity for unlimited division). These cells can renew themselves based on signals from other parts of the organism; they also mature and differentiate into new cells that perform the required functions of the organism. However, only a few tissue types can differentiate; most lose their survival abilities, passing into the resting period after aging, and consequently die. Eukaryotes have four populations of cells.

Oncogenes. Genes that are mutated or synthesized in abnormally excessive amounts and easily transform normal cells into cancer cells are termed oncogenes.

Cyclins [6]. These are specific proteins that activate various phases of the cell cycle. Most cells with proliferative abilities divide as a response to external signals like growth factors, some hormones, and antigen histocompatibility complexes that affect cell surface receptors. These cell surface receptors transmit the received signal to the nucleus and the cell divides. Tyrosine kinases are an important component of cascade reactions, which are initiated by proliferative signals from extracellular growth factors, and propagate to the nucleus. Cyclins combine with specific tyrosine kinases called cyclin-dependent kinases, activating them as well as regulating their effects. Various cyclins are synthesized throughout the different phases of the cell cycle, and their levels either increase or decrease synchronously during each phase of the cell cycle.

The first checkpoint of the cell cycle [7]. This is located in the late G_1 phase just prior to the S phase. DNA should be error-free before it exits from G_1, and even extracellular signals specific for DNA synthesis and all of the mechanisms should work properly. If any damage is detected, the cells try to either repair the damage or die by apoptosis. The protein p53 plays a prominent role in checking for DNA damage at this checkpoint.

The second checkpoint of the cell cycle [7]. This is located just prior to the M phase. Cell cycle inhibitors stop the cell cycle until they are sure that the new daughter cells will have perfect genetic copies of the DNA in the original cell. If DNA replication does not finish entirely and correctly, or all of the proteins, spindle cells, and other materials needed for mitosis are not formed completely, the cell cycle stops at this checkpoint until all errors have been corrected. It then enters the M phase.

2.1.4.1 Eukaryotic Cell Populations [8]

Germ cells. These have the capacity for unlimited proliferation, probably due to meiotic division. Unlike cancer cells, these cells form immortal cell lines through meiotic division.

Stem cells. These cells have two functions. The first is proliferation; the second is differentiation and then to carry out the required specific functions of the organism. Unlike cancer cells, these cells can only pass through a limited number of cell cycles.

Partially differentiated cells. These have a limited capacity for proliferation, and their daughter cells are fully differentiated with no proliferative ability.

Fully differentiated cells. These cells never proliferate.

- Differentiated normal cells, in contrast to immortal cancer cell lines, have a biological timer that counts the number of cell divisions. When a certain number of divisions have occurred, the cell cannot divide any further. For instance, human fibroblast cells divide approximately 50 times into cell lines. After that, they cannot divide, regardless of the nutritive conditions present.

2.1.5 Features of Cancer Cells

Cancer is a disorder characterized by the continuous proliferation of cells [9]. This event happens when the increase in the number of excessively proliferating cells is not balanced by normal cell loss. These cells continuously invade and damage the organs of organisms. Although cancer cells die more quickly than the normal cells they derive from, new cell formation occurs so quickly that the cancer cells accumulate. This imbalance arises from both the genetic abnormalities of cancer cells and the inability of the organism to recognize and destroy these cells.

2.1.5.1 Unique Features of Cancer Cells
[9, 10]

Clonal origin. Most cancer cells originate from just one abnormal cell. However, some cancers arise from more than one malign clone. These clones are formed because of either field damage (tissue cells exposed to more than one carcinogen) or heritable defects in some genes.

Immortality. Most normal cells can undergo a limited number of divisions. On the other hand, cancer cells can undergo an unlimited number of divisions and form endless numbers of cells. One of the mechanisms for immortality is associated with telomeres, which are the tips of chromosomes. During normal cell differentiation, these telomeres shorten. However, the telomeres are renewed by the effect of the enzyme telomerase in cancer cells and stem cells. Telomerase activity normally decreases during cell differentiation. Since the cell loses its capacity for proliferation, fully differentiated cells enter a resting state and consequently die. However, telomerase retains its efficacy in several cancer types, or it is reactivated. Therefore, the telomere length remains constant in these cells, and they proliferate indefinitely (they become immortal).

Genetic instability. This situation is caused by defects in the DNA repair process and in DNA mismatch recognition, which results in the heterogeneity of cancer cells. Cancer cells form clones that gradually respond less and less to the proliferation control mechanism. The ability of these clones to survive in foreign environments also gradually increases, and they gain the ability to metastasize.

Loss of contact inhibition. Normal cells growing in a culture medium cannot divide if they do not stick to the bottom layer. Normal cells also lose their ability to divide when they form a layer across the whole surface. They do not divide, even in the presence of all of the required growth factors and other nutritional elements in the Petri dish. Cancer cells, however, divide independently without needing to stick to the bottom layer of the Petri dish. Furthermore, they continue to grow even when they have formed more than one layer in the cell culture.

Continuous increase in proliferation. This situation is a characteristic of cancer cells in a culture medium. Although cancer cells consume the required nutrition factors, they continue to grow, and they actually end up killing themselves.

Metastasis. This feature is not found in benign tumors and normal cells. Metastasis occurs because of the loss of cellular proteins responsible for adherence to the extracellular matrix, intercellular interaction defects, abnormalities in cell adherence to the basal membrane, abnormalities in basal membrane production, or the destruction of the basal membrane by enzymes like metalloproteases.

2.2 Cellular Effects of Radiation

Ionizing radiation injects energy into a material as it passes through it, like a microscopic bullet, until the radiation is stopped by the material due to absorption [11]. In addition, radiation breaks the molecular bonds of the material in its path and changes the structure of the material. If the material consists of long molecular chains, the chains that are broken by the radiation form new bonds at random. In other words, radiation cuts long molecules at various positions, like a welding flame, and reconnects them in different ways.

Living cells commonly consist of long protein chains, and some of these molecules can be broken by exposing the cell to radiation. The molecular fragments can then rebond in various ways, resulting in new molecules. These new molecules cannot function like the original molecules, and so they need to be repaired. Otherwise, these defective molecular structures will accumulate in the cell, changing the cell's metabolism; if the defective molecule is DNA, it can result in the formation of a cancer cell. Cells have certain repair mechanisms that they can employ for this type of damage. Cells in developed organisms can even check their molecules one by one, and they prefer to rebuild these molecules at certain intervals rather than repairing any damage. However, the capacity for cell repair is limited, and if this limit is exceeded, the damaged mole-

2.2 Cellular Effects of Radiation

cules will start to accumulate and affect the vital survival functions of the cell.

There is no such thing as a fully radioresistant cell. Structures that form the cell, such as the nucleus and particularly chromosomes undergoing division, are more radioresistant than the cell cytoplasm. One of the most prominent effects of radiation at the cellular level is the suppression of cell division. The growth of cells that are exposed to radiation, particularly during cell division (mitosis), is interrupted.

Ionizing radiation can cause breaking, sticking, clamping, and curling in chromosomes. Broken chromosomes can reorganize, remain the same, or combine with other chromosomes. All of these events result in mutations or in eventual cell death (Fig. 2.11).

Although all molecules can be damaged by radiation, DNA molecules that carry genetic information related to cell division and growth are the most probable targets. Radiation may damage or change a small part of the DNA molecule (e.g., only one gene); it can break one or several locations on the DNA helix. The damage is repaired in most cases, but cell death or transformation is observed in some circumstances, and this may result in malign transformations and cause cancer. Dead cells are normally eliminated by the organism. However, if the number of cell deaths exceeds a certain limit, they will affect the proper functioning of the organism and can kill it.

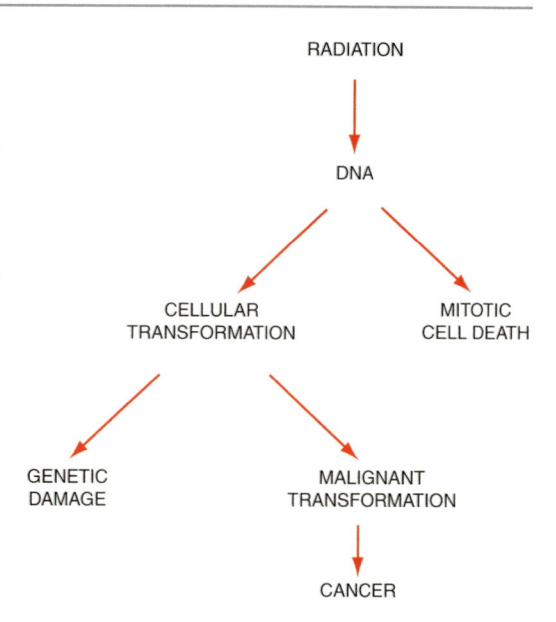

Fig. 2.12 Relationships between dose and mutated cell, living cell, and mutated living cell counts

Radiation can have direct and indirect effects on DNA molecules.

The presence of ionizing radiation increases mutation frequency. Mutation frequency is linearly related to dose (Fig. 2.12). Since low mutation rates occur with low doses, this relation is not linear at low radiation doses. Dose rate (i.e., the amount of radiation received over a specific amount of time) does not affect mutation frequency. In other words, the total number of mutations is the same regardless of the period of exposure to the radiation [12].

2.2.1 The Direct Effect of Radiation at the Molecular Level

The direct effect of radiation is to ionize molecules in its path. While the direct effect of low linear energy transfer (LET) radiation is largely insignificant (e.g., in terms of DNA damage), the direct effect dominates for high-LET radiation [13].

LET → loss of energy per unit tract length.

A quarter to a third of the damage produced in cellular macromolecules by radiation is due to its direct effect. This means that most of the damage

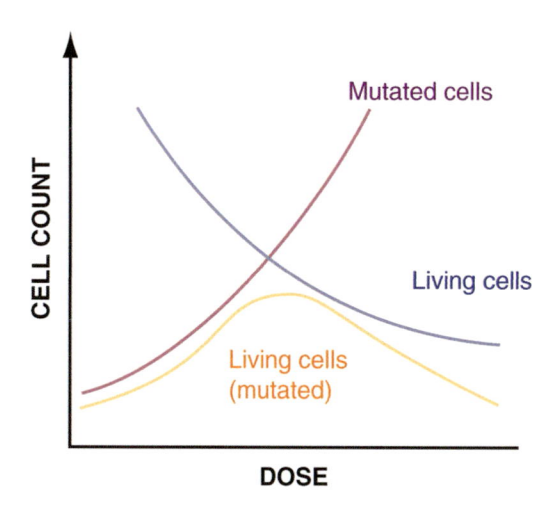

Fig. 2.11 Radiation and carcinogenesis

is caused by the indirect effect of the radiation. Damage to cellular proteins following irradiation at biologically relevant doses appears to be of relatively minor importance.

Radiation directly affects DNA molecules in the target tissue (Fig. 2.13a) [13]. The direct ionization of atoms in DNA molecules is the result of energy absorption via the photoelectric effect and Compton interactions. If this absorbed energy is sufficient to remove electrons from the molecule, bonds are broken, which can break one DNA strand or both (Fig. 2.13b, c). A single broken strand can usually be repaired by the cell, while two broken strands commonly result in cell death.

A dose of 2 Gy of X-rays is equal to an energy of 2 J/kg. Since 1 J/kg is equal to 6.25×10^{18} eV/kg [13], 2 Gy is equal to 12.5×10^{18} eV/kg. Since the minimum energy required for ionization is 33 eV, the number of ions per kilogram is calculated by dividing 12.5×10^{18} eV/kg by 33 eV, which yields 4×10^{17} ions/kg. If we apply two doses to the whole body (we know that there are 9.5×10^{25} atoms/kg in the human body), the number of atoms in the whole body ionized by a dose of 2 Gy can be found by

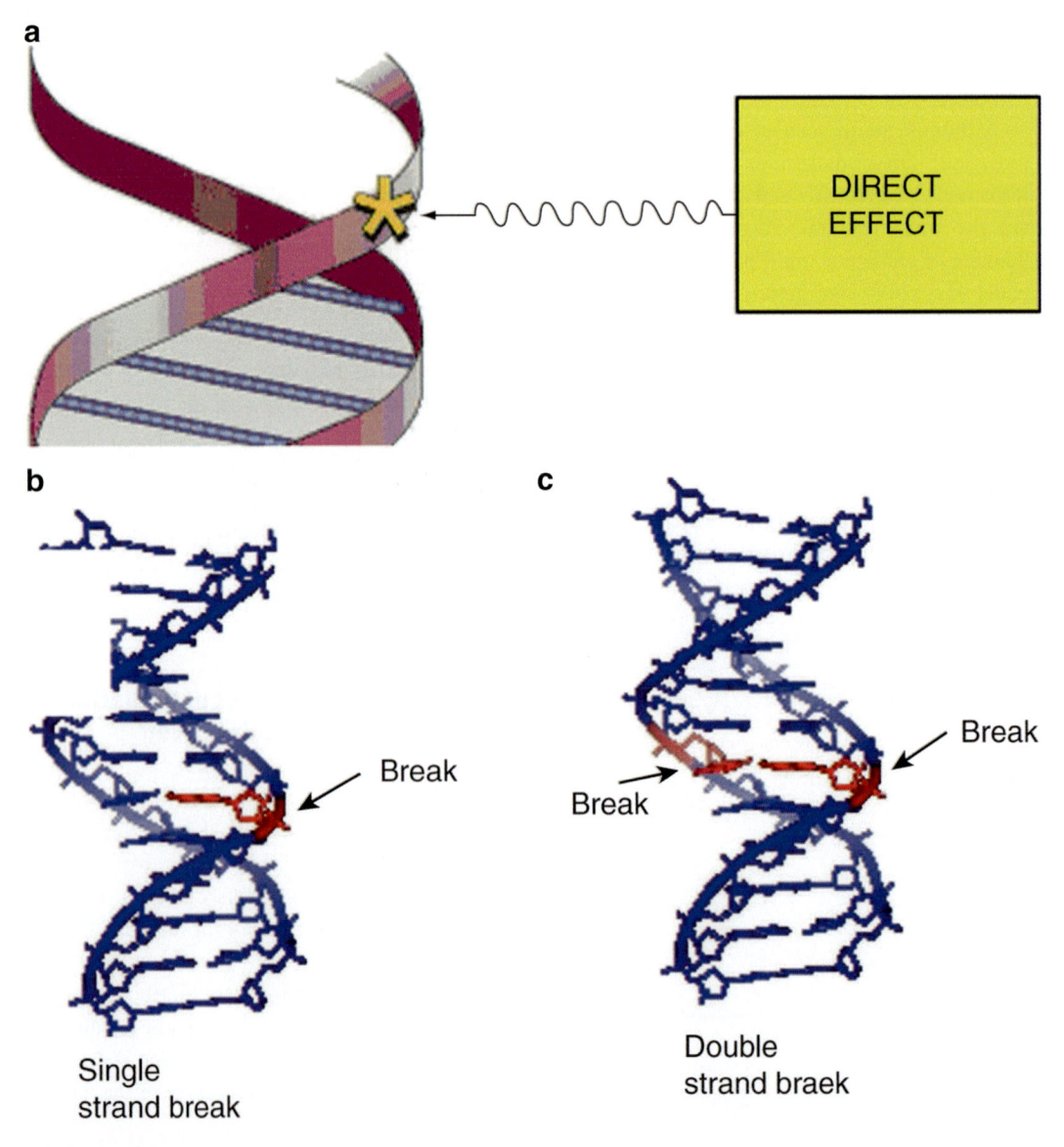

Fig. 2.13 (a) Direct effect of radiation. (b) Single-strand DNA break. (c) Double-strand DNA break

2.2 Cellular Effects of Radiation

dividing the ions/kg by the atoms/kg. The result is nearly 1×10^{-8} (one in a hundred million), which means that the direct effect of X-rays in terms of DNA damage in tissue is relatively minor.

When normal cell DNA is damaged by radiation provided in the kinds of doses normally used in radiotherapy, the cell cycle is stopped by the protein p53. The DNA is repaired; the cell then reenters the cell cycle and continues to proliferate. If the DNA cannot be repaired, the cell enters apoptosis—the programmed cell death pathway. At high radiation doses, the molecules utilized by the DNA repair mechanisms are damaged, so repair is not possible, the cell loses its ability to divide, and it subsequently dies.

2.2.2 The Indirect Effect of Radiation at the Molecular Level

The indirect effect of radiation on molecules includes the formation of free radicals by energy transfer from radiation and the resulting molecular damage caused by the interactions of these free radicals with DNA (Fig. 2.14) [13, 14]. This phenomenon is most probably due to the interaction of radiation with water molecules, since the human body is approximately 70% water. Free radicals are electrically neutral atoms that contain "free" (i.e., unbound) electrons. They are highly electrophilic and reactive.

Fig. 2.14 Indirect effect of radiation (Fig. 1.1 of [64])

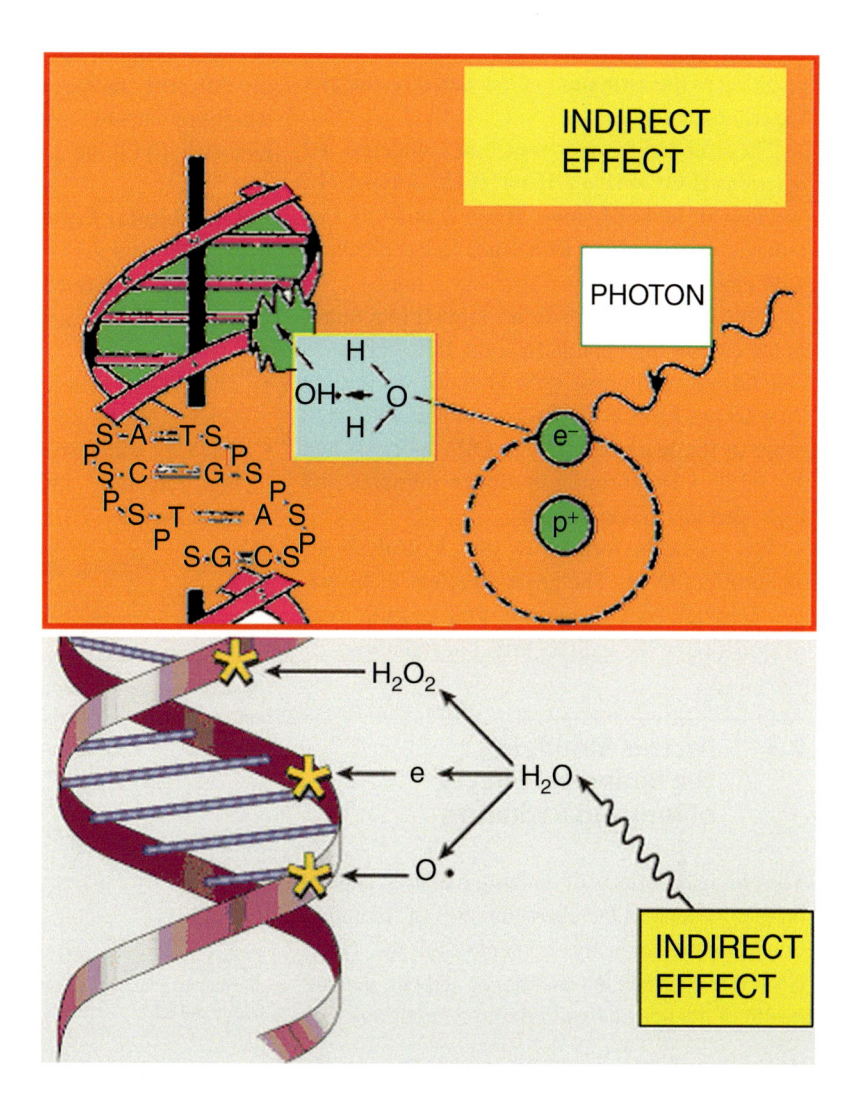

Simple free radicals (H or OH) have very short lifetimes (10^{-10} s), and this time span is too short for them to travel from the cytoplasm to the nucleus, where the DNA is located. Therefore, the H combines with O_2 and transforms into a more potent and lethal free radical with a longer lifetime, called hydrogen dioxide (HO_2) [14]. Although hydrogen peroxide, H_2O_2, has an even longer lifetime (10^{-5} s), it cannot move from one place to another. It oxidizes the surroundings of the cells close to where it is formed and prevents the nutrition of neighboring tissues or cells. This results in cell death through nutritive deficiency or the isolation of these cells from other tissues.

Water (H_2O) is ionized when exposed to radiation, and as $H_2O \rightarrow H_2O^+ + e^-$, a positively charged water molecule and a free electron are formed [14].

This free electron (e^-) interacts with another water molecule in the reaction $e^- + H_2O^+ \rightarrow H_2O^-$, resulting in the formation of a negatively charged water molecule.

These charged water molecules undergo the reactions $H_2O^+ \rightarrow H^+ + OH$ and $H_2O^- \rightarrow H + OH^-$, yielding H^+ and OH^- ions. These H and OH free radicals may combine with other free radicals or with other molecules.

If the LET of the radiation is high (particularly in the case of alpha particles), the free OH^- radicals do not recombine with H^+ radicals, and so they do not form H_2O. They combine with each other in the reactions $OH^- + OH^- \rightarrow H_2O_2$ and $H^+ + H^+ \rightarrow H_2$, forming hydrogen peroxide and hydrogen gas molecules [14].

Free radicals formed by the hydrolysis of water affect DNA. The negative effect of hydrogen peroxide on cell nutrition may be employed as evidence of the indirect effect of radiation.

2.3 Factors Modifying the Biological Effects of Ionizing Radiation

The biological effects of ionizing radiation depend on factors such as the characteristics of the radiation (energy, intensity, content) and the target (structure of irradiated tissue; age, gender, general health of the person exposed to the radiation).

2.3.1 Characteristics of the Radiation

The potential harm to biological materials caused by their irradiation is directly proportional to the efficacy with which the radiation deposits energy in the material. Proton, neutron, and alpha particles lose their energies over much shorter distances than X-rays and gamma rays with the same energy.

Since high-LET radiation (particulate radiation) transfers more energy per unit length of material, the probability of causing DNA damage in a short period of time is high. Thus, a dose of high-LET radiation is more destructive than the same dose of low-LET radiation (electromagnetic radiation).

Radiation weighting factors (W_R) are determined in order to compare the biological effects of different radiation types [18, 19]. These weighting factors are also called radiation quality factors (QF) (Table 2.1).

2.3.1.1 Linear Energy Transfer (Table 2.2) [15]

The energy transferred to the tissue by ionizing radiation per unit tract length is called the LET.

- The LET is a function of the charge and the velocity of the ionizing radiation.
- The LET increases as the charge on the ionizing radiation increases and its velocity decreases.

Table 2.1 Weighting factors for various radiation types (ICRP 1991)

Radiation type	Energy interval	Weighting factor = quality
Photon (gamma and X-rays)	All energy levels	1
Electron	All energy levels	1
Neutron	<10 kV	5
Neutron	10–100 kV	10
Neutron	100 kV–2 MV	20
Neutron	2–20 MV	10
Neutron	>20 MV	5
Proton	>20 MV	5
Alpha particles, heavy nuclei	All energy levels	20

2.4 Target Tissue Characteristics

Table 2.2 LET values for various radiation types [15]

Radiation	Energy	Relative LET value (keV/μm)
250 kV X-ray	250 kV	3
3 MV X-ray	3 MV	0.3
Cobalt 60	1.17–133 MV	0.3
Beta 10 kV	10 kV	2.3
Beta 1 MV	1 MV	0.25
Neutron 2.5 MV	2.5 MV	20
Neutron 19 MV	19 MV	7
Proton 2 MV	2 MV	16
Alpha 5 MV	5 MV	100

- Alpha particles are slow and positively charged. Beta particles, on the other hand, are fast and negatively charged. Therefore, the LET of an alpha particle is higher than that of a beta particle.
- Lethal effects increase as the LET increases.
- The units of the LET are keV/μm.

Absorbed dose [16]. The basic quantity of radiation measurement in radiotherapy is the "absorbed dose." This term defines the amount of energy absorbed from a radiation beam per unit mass of absorbent material. The unit of absorbed dose is the Gray (Gy). It changes continuously along the path of the radiation because the radiation slows down. In addition, secondary radiation energies occur due to secondary scattering from the particle's path in tissue. The type and effects of each form of radiation type should be known exactly in order to define the total effect of the radiation.

Equivalent dose (dose equivalent) [16, 17]. Different radiations cause different damages in human tissues. The *absorbed dose* ($\rightarrow$ Gy) is not adequate for studies of radiation protection. Thus, the absorbed dose in tissue should be multiplied by the radiation weighting factor for this radiation type. The calculated result is defined as the equivalent dose, measured originally in roentgen equivalent in man units (REMs), but now measured in Sieverts (Sv).

If the mean absorbed radiation dose (Gy) in a tissue or organ is multiplied by the appropriate radiation weighting factor (W_R), the equivalent dose (H_T) is found.

Sv is a large unit, so doses are frequently expressed in millisieverts (mSv) or microsieverts (μSv) for practical purposes.

2.3.1.2 Dose Rate [20]

This is the dose delivered per unit of time. If a radiation dose that causes irreparable damage when delivered over a short time period is delivered over longer periods, the cell or organism may survive.

2.4 Target Tissue Characteristics

Different tissues have different radiosensitivities. Cells that divide frequently (e.g., blood-forming cells in the bone marrow) are frequently affected by radiation more than rarely dividing cells (e.g., connective and fat tissue). Metabolic factors such as the oxygen concentration in the irradiated volume are also important.

The International Commission on Radiological Protection (ICRP) defined a mean reference human in order to estimate the absorbed doses at certain places on the body (Fig. 2.15). In general, the results on radiation absorption obtained in this way can be related to real irradiation. A further simplification was recommended by the ICRP in 1977. The recommended limits on dose equivalence are based on regular irradiation of the whole body. Irradiation of the whole or only part of the body can be expressed as the equivalent whole-body irradiation by taking into account weighting factors for certain organs (Table 2.3) [18, 19].

The durations of physical, chemical, and biological events after radiation has penetrated the cell are shown in Fig. 2.16.

2.4.1 Whole Body Dose Equivalent [17]

$$H_{wb} = \Sigma_T W_T H_T \quad (2.1)$$

Fig. 2.15 Reference human according to the ICRP (Fig. 4.21 of [65])

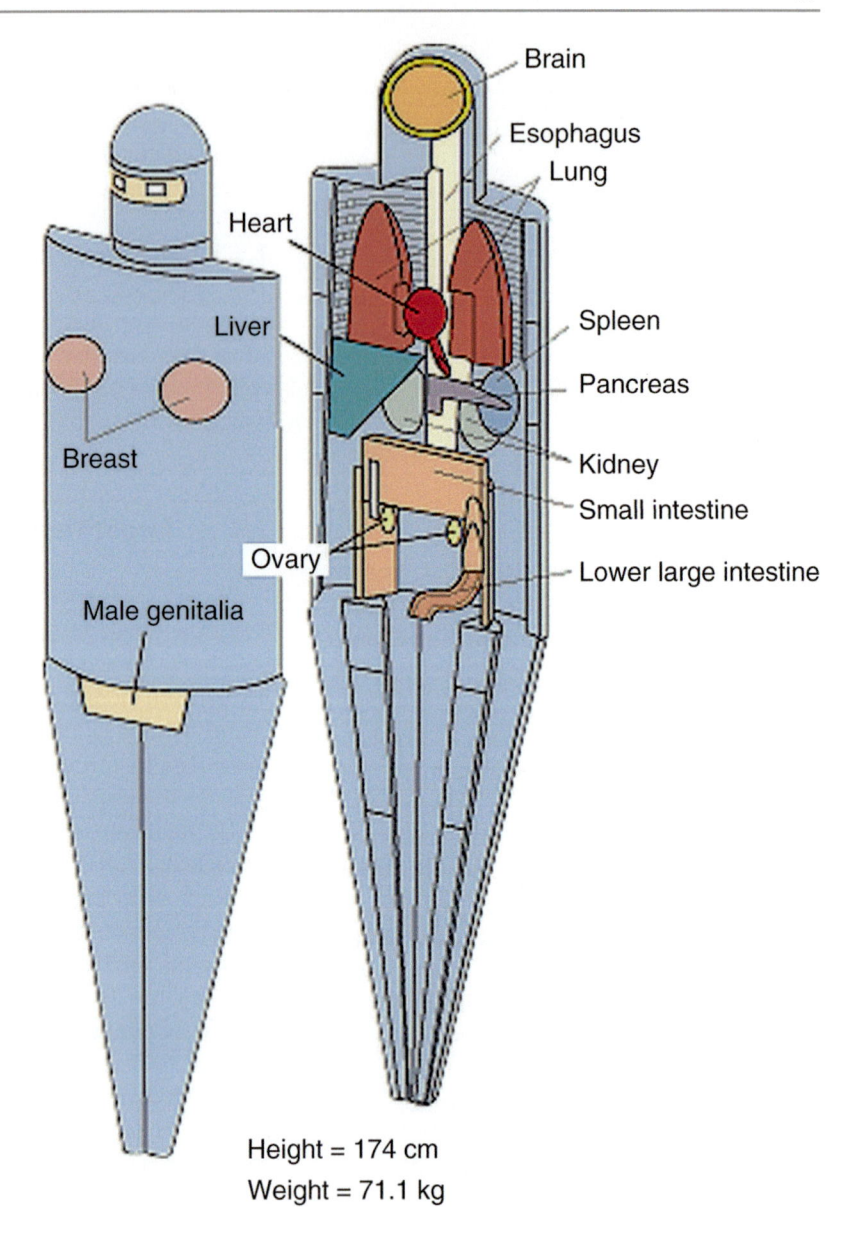

Height = 174 cm
Weight = 71.1 kg

Table 2.3 Tissue weighting factors

Tissue	Tissue weighting factor (W_T)
Gonads	0.20
Lung, bone marrow, stomach, colon	0.12
Thyroid, liver, esophagus, breast, bladder	0.05
Bone surface	0.01
Skin	0.01
Other organs	0.05
Total	1

H_T: dose equivalent for tissue, W_T: weighting factor for tissue.

2.4.2 Effective Dose [17]

This is the dose calculated by multiplying the equivalent dose by the tissue weighting factor (W_T) (Fig. 2.17).

- The units of effective dose are sieverts, just like equivalent dose.

2.5 Target Theory

$$WT = \frac{\text{Risk at organ or tissue (depending on stochastic effects)}}{\text{Total risk at body (depending on stochastic effects)}} \quad (2.2)$$

2.4.3 Relative Biological Effect (RBE) [21, 22]

The RBE is the ratio of the 250 kV X-ray dose that produces a specific biological effect to the test dose of any radiation that produces the same effect. The RBE is related to the LET.

$$RBE = \frac{\text{The 250 kV X–ray dose required for a specific effect}}{\text{Tested dose of any radiation required for a specific effect}} \quad (2.3)$$

2.4.4 Overkill Effect

The decrease in the curve of RBE vs. LET at LET values of above 100 keV/μm has been interpreted as an "overkill effect," where the ionization density within a single cell is greater than the two ionization events needed to inactivate the cell (Fig. 2.18). In other words, any dose beyond that needed to produce the two events per cell is, in effect, wasted. Densely ionizing radiation is inefficient at producing the maximum amount of cell death.

2.5 Target Theory

The number of DNA or critical target cells "hit" by the radiation depends on random events in target theory, and has no direct relation to the ionizing radiation dose [23]. Therefore, there is no threshold at which the effects of the radiation are observed. Whatever the delivered radiation dose, there is always a chance of it hitting DNA or cells and producing harmful effects. The phenomenon where the effects of the radiation do not depend on dose is known as the "stochastic effect."

Target theory explains the cell damage caused by radiation based on the principles of probability. It assumes that there are certain critical molecules or critical targets within cells that need to

Fig. 2.16 Durations of physical, chemical, and biological stages

Fig. 2.17 Relationship between exposure and equivalent dose

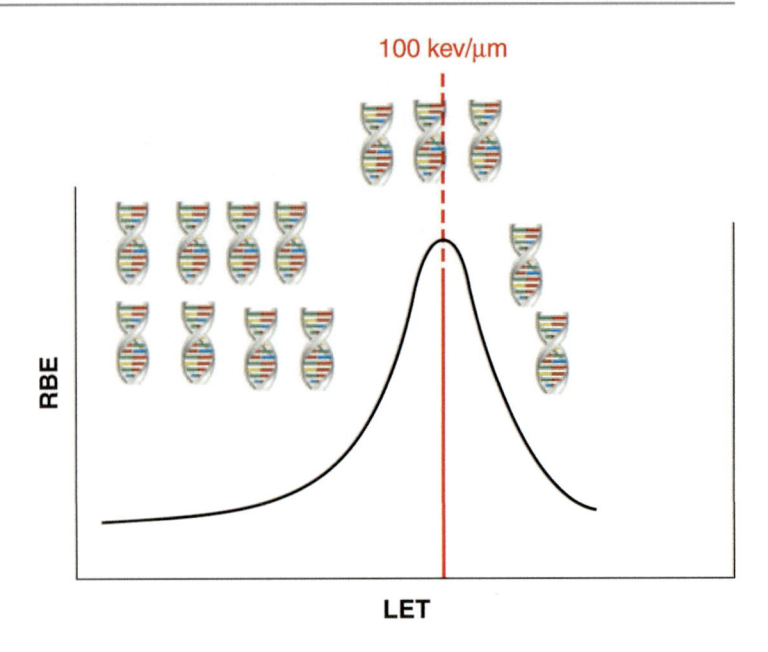

be hit or inactivated by the radiation to kill the cell.

Single Target–Single Hit [23, 25]:

Here, there is only one target in the cell that is associated with cell death, and a single hit on this target is adequate to inactivate the target.

- This is a valid assumption for viruses and some bacteria.

Multiple Target–Single Hit [23, 24]:

Here, there is more than one target per cell, and a single hit of any of these targets is required for cell death.

Not all targets are hit; some of them are killed, while others are damaged by low doses. This type of damage is called sublethal damage (SLD). Cells with SLD may repair themselves during interfractional periods.

This is a valid assumption for mammalian cells.

2.6 Cell Survival Curves

When cell culture lines are exposed to radiation, some of them lose their capacity to divide and cannot form colonies (→ reproductive cell death), some only divide to a small degree and form small colonies, some divide slowly and form colonies over longer periods, some lose their capacity to divide but continue to grow and become giant cells, and still others degenerate and die. The remaining cells are not affected by the radiation, and they represent the surviving fraction (SF) after irradiation of the cell culture (→ SF) [26].

If the SF is calculated for various doses, then it can be presented as a cell–dose plot. Combining the points on the plot leads to a cell survival curve.

Curves showing the relation between the radiation dose and SF are termed cell survival curves. If the dose is plotted on the y-axis and the SF (as a percentage of the original number of cells in the culture) is plotted on the x-axis, a sigmoid curve is obtained (Fig. 2.19a). If the logarithm of the SF is plotted on the x-axis, a semilogarithmic curve is obtained (Fig. 2.19b) [27].

LD50 value can be obtained from a sigmoid survival curve (LD50 is the dose that kills 50% of cells → lethal dose).

The number of cells in cell lines within cell cultures can increase in one of two ways: either arithmetically or exponentially (geometrically).

The number of cells increases linearly (by a constant number) with each generation in an arithmetic increase. In an exponential increase, the number of cells doubles with each generation,

2.6 Cell Survival Curves

Fig. 2.18 Overkill effect

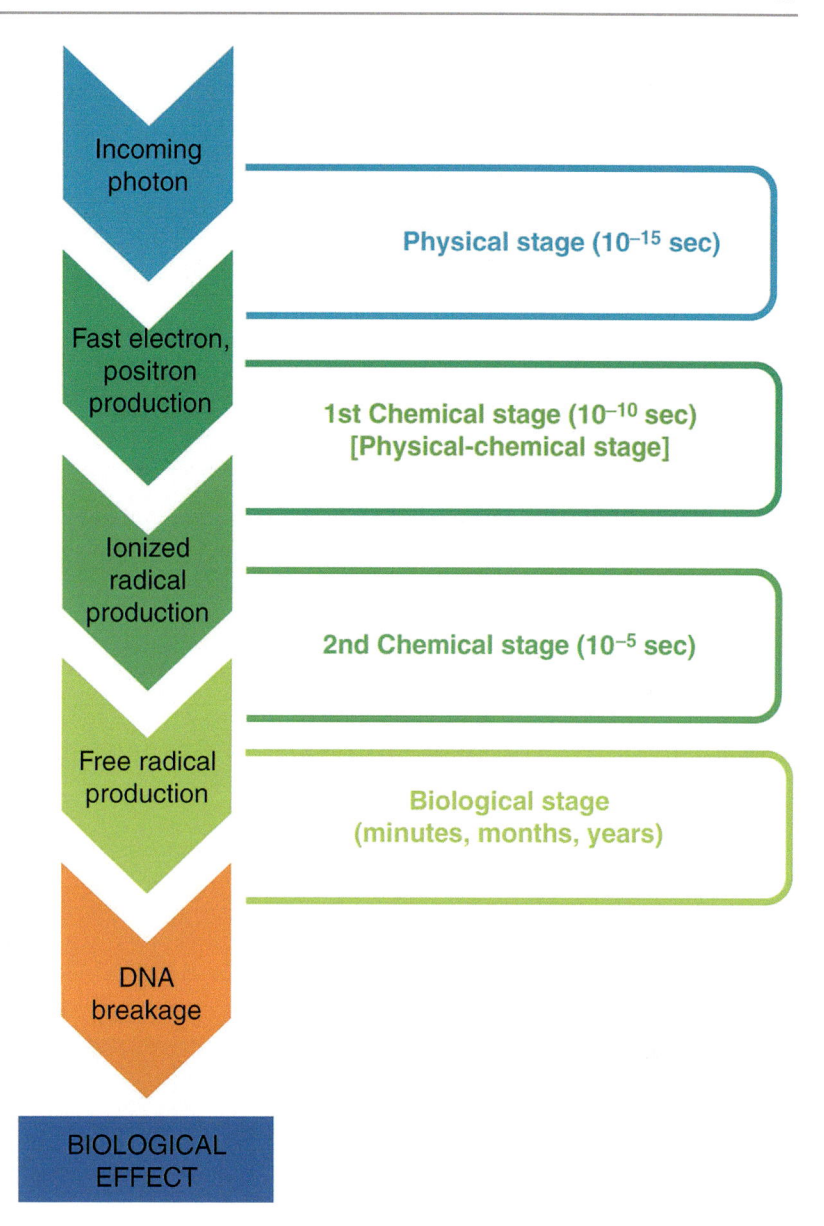

and so exponential growth is faster than arithmetic growth (Fig. 2.20).

2.6.1 Surviving Fraction [26, 27]

The ratio of the number of cells that form colonies to the number of seed cells under normal conditions (i.e., no irradiation) in a cell culture is termed the plating efficiency (PE). The same ratio obtained under irradiated conditions and divided by the PE is called the surviving fraction (SF):

$$\text{Surviving fraction}\,(\text{SF}) = \frac{\text{Colony number}_{\text{rad}}}{\text{Seed cell number}_{\text{rad}} \cdot \text{PE}} \quad (2.4)$$

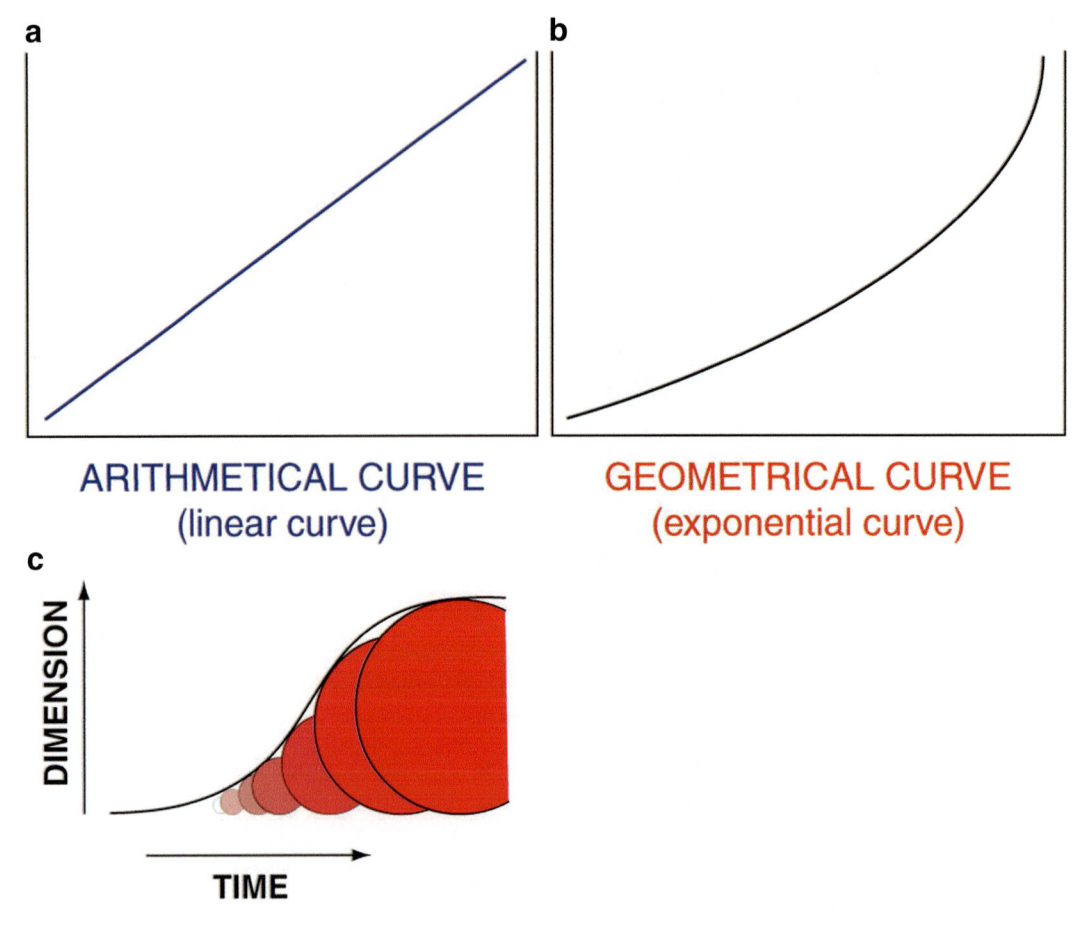

Fig. 2.19 (**a**) Sigmoid curve; (**b**) semilogarithmic curve; (**c**) time-dimension curve

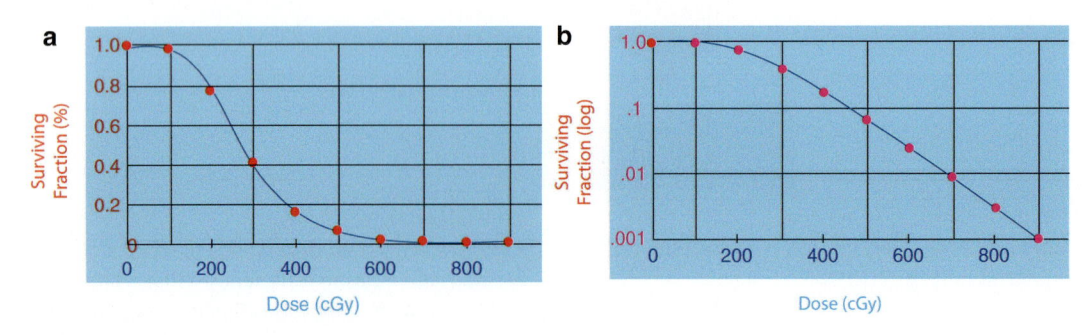

Fig. 2.20 Survival curves: (**a**) arithmetical; (**b**) geometrical (exponential) number

- For example, if 100 cells are seeded into an unirradiated culture and ten colonies are formed, then the PE is 10/100. If there are five colonies after a 450 cGy dose of radiation, the SF is 5/[100 × 10/100] = 1/2. Thus, the SF of 450 cGy is 50%.

Survival curves are radiobiologically defined using semilogarithmic curves, and these curves provide information on some parameters such as the number of cells killed by the radiation and cell radiosensitivity.

2.6.2 Exponential Survival Curves

As the value of D_0 decreases → $1/D_0$ increases → slope increases → *radiosensitive cell*.

As the value of D_0 increases → $1/D_0$ decreases → slope decreases → *radioresistant cell*.

These are the survival curves resulting from the single target–single hit hypothesis of target theory (Fig. 2.21) [16, 26–28]. They show that cell death due to irradiation occurs randomly. At certain doses with one unit increase, both same number of cell deaths and same proportion of cell death occur.

D_0 = dose that decreases the surviving fraction to 37%.

This is the dose required to induce an average damage per cell.

A D_0 dose always kills 63% of the cells in the region in which it is applied, while 37% of the cells will survive.

$1/D_0$ = the slope of the survival curve.

After 100 radiation "hits," the probability that one of the hits will be a target → e^{-1} (e ≈ 2.718 …).

- e^{-1} is approximately 37%. In other words, 63% of the targets will be hit after 100 hits, while 37% of the targets will survive.
- This corresponds to the survival curve observed for viruses and some bacteria.
- Some cells are observed to be very sensitive to radiation (for example, germ cells also show this behavior).
- It may also be observed at very low dose rates and for high-LET radiation.

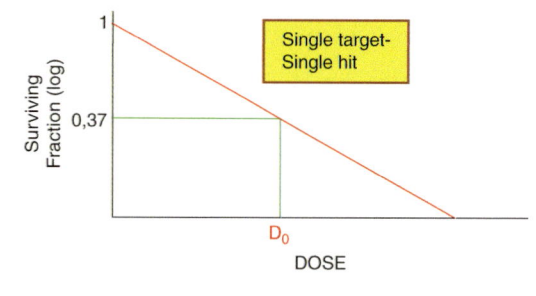

Fig. 2.21 Single target–single hit hypothesis

For exponential survival curves, SF is given by

$$SF = e^{-D/D_0} \quad (2.5)$$

This can be used to determine the proportion of the original cells that will survive if a dose D is delivered.

2.6.2.1 Shouldered Survival Curves with Zero Initial Slope

These survival curves are based on the multiple target–single hit hypothesis of target theory [16, 26–28]. They are produced by the hypothesis of requiring multiple targets per cell, and only one of these targets needs to be hit to kill the cell (Fig. 2.22).

SF for shouldered survival curves with a zero initial slope:

$$SF = 1\left[1 - e^{-D/D_0}\right]^n \quad (2.6)$$

This gives the proportion of the original cells that survive if a dose D is delivered.

D_0: the dose that yields a surviving fraction of 37%.

D_q: half-threshold dose → the region of the survival curve where the shoulder starts (indicates where the cells start to die exponentially) (= quasi-threshold dose).

n: extrapolation number (the number of D_0 doses that must be given before all of the cells have been killed).

D_q → the width of the shoulder region.

$$D_q = D_o \log_n 2.7 \quad (2.7)$$

If n increases → D_q increases → a wide shouldered curve is observed.

If n decreases → D_q decreases → a narrow shouldered curve is observed.

If D_q is wide and D_0 is narrow, the cell is radioresistant.

The D_0 and D_q values for the tumor should be smaller than those of normal tissue to achieve clinical success.

2.6.2.2 Shouldered Survival Curves with Nonzero Initial Slope

If we carefully examine the shouldered survival curve with an initial slope of zero, the curve is

Fig. 2.22 Multiple target–single hit hypothesis

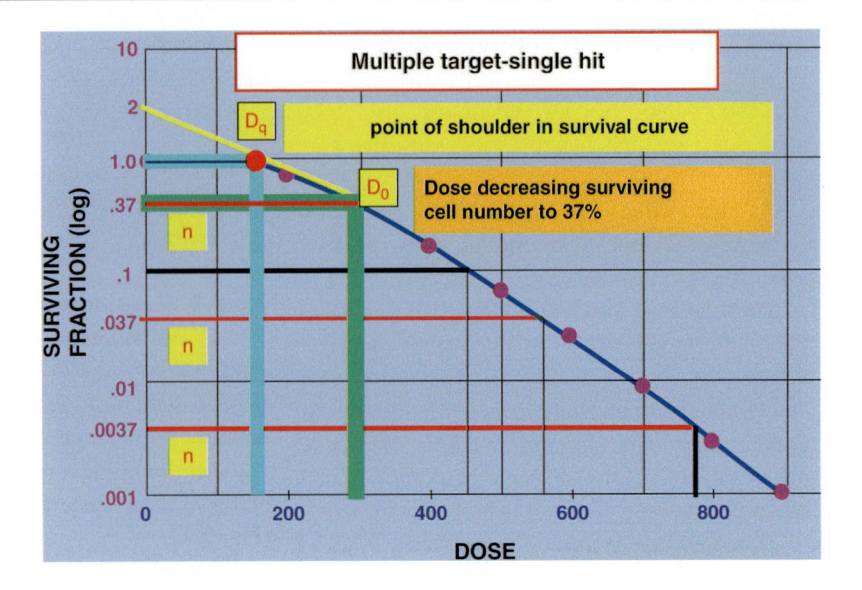

straight when the dose is small [26–28]. This indicates that there is a threshold dose where the radiation starts to exert an effect. However, studies have demonstrated that radiation has an effect regardless of the radiation dose. The model that takes this observed behavior into account has two components with a nonzero initial slope.

SF for shouldered survival curves with a nonzero initial slope:

$$\mathrm{SF} = e^{-D/D_1}\left[1-\left(1-e^{-D/D_0}\right)^n\right] \quad (2.8)$$

Components of Shouldered Survival Curves with Nonzero Initial Slope
(Fig. 2.23) [16, 27, 28]
- Component corresponding to the single target–single hit model (blue in the figure).

  This shows lethal damage.
  This shows the cells killed by the direct effect of the radiation.
  This shows the effect of high-LET radiation.

- Component corresponding to the multiple target–single hit model (red in the figure).

  This shows the accumulation of SLD.
  This shows the cells killed by the indirect effect of the radiation.

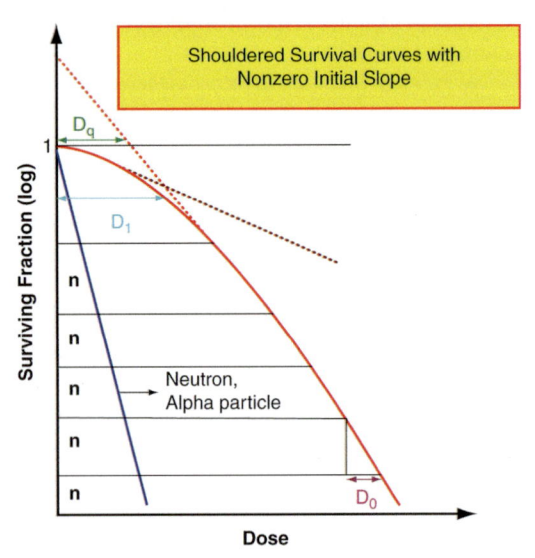

Fig. 2.23 Shouldered survival curves with a nonzero initial slope

This shows the effect of low-LET radiation.

$1/D_1$: the slope of the component corresponding to multiple target–single hit (the slope of the initial region).

D_q: the dose at which the shoulder starts for the multiple target–single hit component (the quasi-threshold dose).

$1/D_0$: the slope of the terminal region of the multiple target–single hit component.

n: extrapolation number.

2.6.3 Linear–Quadratic Model (LQ Model)

In this model, developed by Douglas and Fowler in 1972, it was assumed that cell death due to ionizing radiation has two components (Fig. 2.24) [29].

If we transform the cell death probability curve into an SF curve, the linear–quadratic model assumes a linear–quadratic relation between fraction dose and fraction number.

The first component
Directly proportional to dose → D

- Linear component

The second component
Directly proportional to the square of the dose → D^2

- Quadratic component

If the effect with one radiation hit is $p1$, then

- $p1 = \alpha D$.

α → initial slope of the survival curve (low-dose region)
α → linear coefficient (Fig. 2.25).

- Corresponds to the cells that cannot repair themselves after one radiation hit.
- Important for high-LET radiation.

Apoptotic and mitotic death are dominant.
If the effect of two radiation hits is $p2$, then

- $p2 = \beta D^2$

β → quadratic coefficient.

- Corresponds to cells that stop dividing after more than one radiation hit but can repair the damage caused by the radiation.
- Important for low-LET radiation.
- Mitotic death is dominant.

$$\text{Total effect } p1 + p2 = \alpha d + \beta d^2 \quad (2.9)$$

$$SF = e^{-(\alpha d + \beta d^2)} \quad (2.10)$$

α → shows the intrinsic cell radiosensitivity, and it is the natural logarithm ($\log_e$) of the proportion of cells that die or will die due to their inability to repair radiation-induced damage per Gy of ionizing radiation.

β → reflects cell repair mechanisms, and it is the natural logarithm of the proportion of repairable cells due to their ability to repair the

Fig. 2.24 Components of the linear–quadratic model

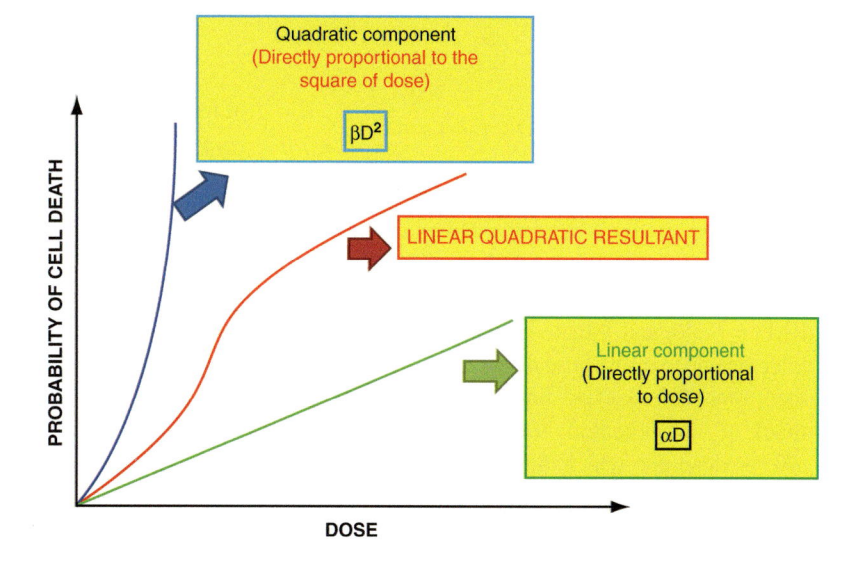

Fig. 2.25 Relation I between dose and surviving fraction in the LQ model

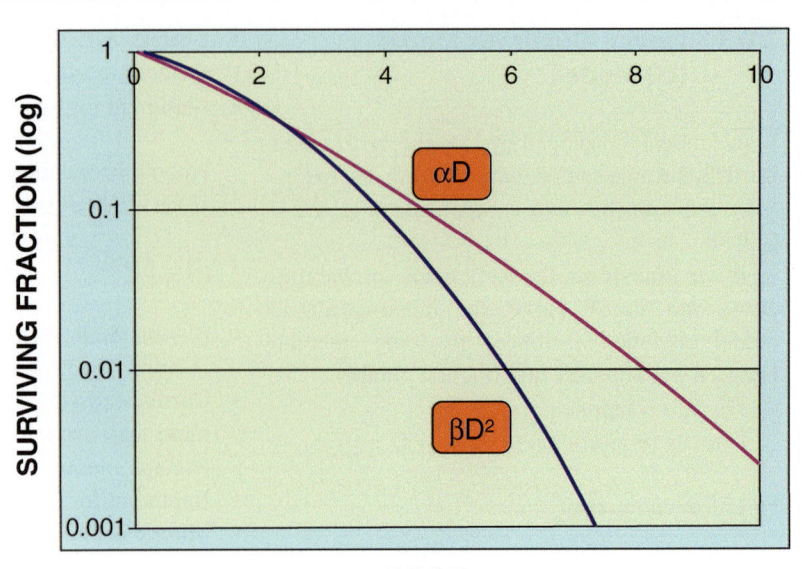

radiation-induced damage per Gy of ionizing radiation.

What is the LQ model used for?

To formulate equivalent fractionation schemes.

To calculate additional doses after breaks from radiotherapy.

To get information on acute and late responses.

$$E = n(\alpha d + \beta d^2) \quad (2.11)$$

$$E = nd(\alpha + \beta d) \quad (2.12)$$

$$E/\alpha = nd\left(1 + \frac{d\beta}{\alpha}\right) \quad (2.13)$$

$$E/\alpha = nd\left(1 + \frac{d}{\alpha/\beta}\right) \quad (2.14)$$

$$\text{BED} = \frac{E}{\alpha} \quad (2.15)$$

$$\text{BED} = nd\left(1 + \frac{d}{\alpha/\beta}\right) \quad (2.16)$$

$$E = nd(\alpha + \beta d) + \log_e 2(T - T_k)/T \quad (2.17)$$
$$(\log_e 2 = 0.693)$$

$$\text{BED} = E/\alpha = nd\left(1 + \frac{d}{\alpha/\beta}\right) - \frac{0.693}{\alpha T_p}(T - T_k) \quad (2.18)$$

$E = \log_e$ of the total cell number, including irreparable cells (α) or partially repairable cells (β), n = fraction number, d = fraction dose, BED = biological effective dose = extrapolated tolerance dose = response dose, T = overall treatment time, T_k = kickoff time (repopulation start time), T_p = potential tumor doubling time, $\alpha/\beta \rightarrow$ dose for which the number of acutely responding cell deaths is equal to the number of late-responding cell deaths (the dose for which the linear and quadratic components of cell death are equal) (Fig. 2.26).

Tumor response and acute effects in normal tissues → α/β = 10 Gy.

Late effects in normal tissues → α/β = 3 Gy.

The α/β ratio may differ among tumor types; e.g., it is 1.5 for melanoma and 1.5–3.5 for prostate adenocarcinoma.

2.6 Cell Survival Curves

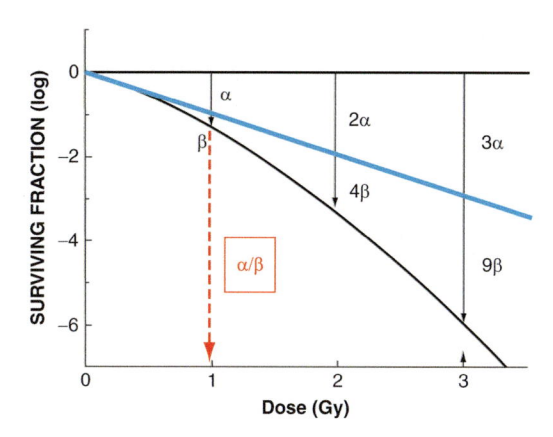

Fig. 2.26 Relation II between dose and surviving fraction in the LQ model (Fig. 1.1, p 9 of [66])

- This model does not take into account the effect of treatment time.

2.6.3.1 Models Used Before the LQ Model

1. **Strandqvist model** [30, 31]. This was developed by Magnus Strandqvist in 1944. Here, the relationship of skin tolerance to radiation dose for a particular skin cancer treatment time is plotted using a logarithmic curve. The slope of this curve is constant and equal to 0.22. Cohen showed that this slope value was valid for skin cancer, but he observed that the slope was 0.33 for skin erythema. In summary, this model assumed that the tolerable fraction dose was related to the treatment time T as $T^{0.33}$.

2. **Ellis model** [31, 32]. This was developed by Ellis in 1966. While only the total dose is important in the Strandqvist model, the dependence of the tolerable dose on the number of fractions and the overall treatment time is accounted for in this model. The dose obtained using this model is termed the nominal standard dose (NSD), and so the model is known as the NSD model.
   - The NSD is the dose required to cause maximum tumor damage without exceeding the tolerance levels of healthy tissues.

$$D = NSD \times N^{0.24} \times T^{0.11} \quad (2.19)$$

$$NSD = D \times N^{-0.24} \times T^{-011} \quad (2.20)$$

D: total dose at skin level, NSD: nominal standard dose, N: fraction dose, T: overall treatment time

3. **Orton–Ellis model** [31]. This is a modified form of the NSD model. It is also known as the TDF (time–dose factor) model. It can be summarized as:

$$TDF = d^{1.538} \times X^{00.169} \times 10^{-3} \quad (2.21)$$

X = treatment time/fraction number, d = fraction number.

If a total dose of 66 Gy is given in 30 fractions comprising 2 Gy daily fraction doses for 6 weeks, what are the BED values for acute effects, tumor response, and late effects ($\alpha/\beta = 3$ for late effects, $\alpha/\beta = 10$ for acute effects and tumor response)?

BED = $nd\,(1 + d/[\alpha/\beta])$.
$BED_{10} = 2 \times 30 \times (1 + 2/10) = 72$ Gy.
$BED_{10} = 72$ Gy for acute effects and tumor response.
$BED_3 = 2 \times 30 \times (1 + 2/3) = 100$ Gy.
$BED_3 = 100$ Gy for late effects.

The BED formulation is less reliable for fraction doses of more than 3 Gy.

2.6.3.2 In Head and Neck and Lung Cancers

BED calculation for tumor response:
$T_k = 21$ days and $T_p = 3$ days.
BED calculation for normal tissue:
$T_k = 7$ days and $T_p = 2.5$ days.
In the prostate for late effects, $\alpha/\beta = 1.5$ Gy.
In the CNS and kidney for late effects, $\alpha/\beta = 2$ Gy.

- Tissues that respond early to irradiation (radiosensitive).

These die linearly.
The α/β ratio is large.

- Tissues that respond late to radiation (radioresistant).

These die quadratically.
The α/β ratio is small.

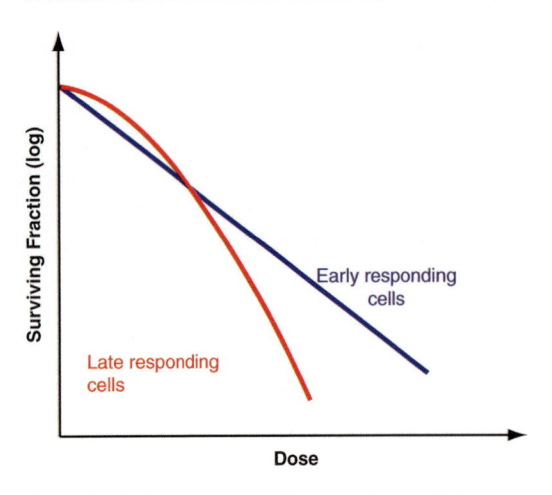

Fig. 2.27 Relation between SF and radiosensitivity

This hypothesis is valid for both tumor and normal tissues (Fig. 2.27).

The BED formula utilizing the LQ model can be employed to compare two different radiotherapy schedules:

$$n_2 d_2 = n_1 d_1 \frac{\alpha/\beta + d_1}{\alpha/\beta + d_2} \qquad (2.22)$$

$n_1, d_1 \rightarrow$ fraction dose and number of fractions in the first scheme. $n_2, d_2 \rightarrow$ fraction dose and number of fractions in the second scheme.

2.6.4 Types of Cellular Damage Due to Radiation

1. **Lethal damage** [33, 34]. This is irreversible, irreparable damage, resulting in cell death.
   - This usually results from the direct effect of radiation.
   - Double strand breakage in DNA (+).
   - Particularly observed in high-LET radiation.
2. **SLD** [33, 34]. SLD can be repaired within hours under normal conditions, unless an additional radiation dose is given (inducing further SLD).
   - This generally occurs due to the indirect effect of radiation.
   - Single strand breakage in DNA (+).
   - Observed in low-LET radiation.
3. **Potentially lethal damage** [34]. This is repairable, depending on the changes in the cell environment after exposure to radiation.
   - Under normal conditions, this type of damage is lethal to cells undergoing mitosis that are exposed to radiation.
   - However, such damage can be repaired in suboptimal environmental conditions after exposure to radiation because the cell gets the signal that suboptimal conditions that are not suitable for mitosis are present. The cell then prefers to repair this potential damage rather than initiate mitosis.

2.6.5 Factors Affecting the Cell Survival Curve

1. *Cell cycle*. Duration of each phase in the human cell cycle: G1 = 1.5–14 h, S = 6–9 h, G2 = 1–5 h, M = 0.5–1 h.
   - The responses of cells in different phases to radiation vary (Fig. 2.28).
   - The most radiosensitive cell phases are late G2 and M.
   - The most radioresistant cell phases are late S and G_1.
2. **LET**. Radiosensitivity increases with high-LET radiation (Fig. 2.29).

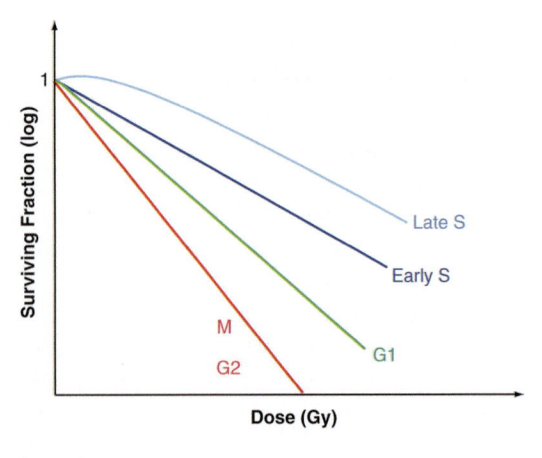

Fig. 2.28 Cell cycle and SF

2.6 Cell Survival Curves

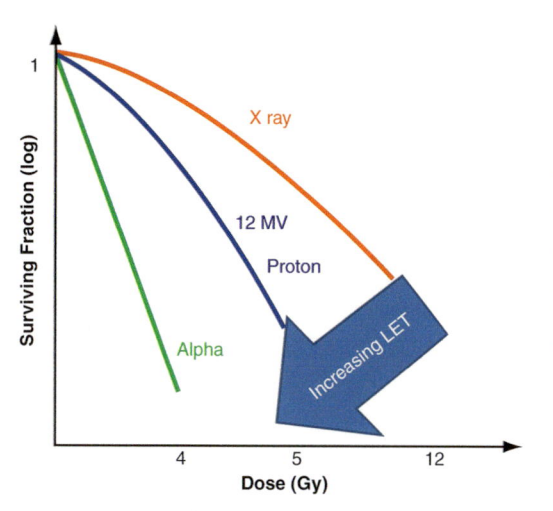

Fig. 2.29 Linear energy transfer and SF

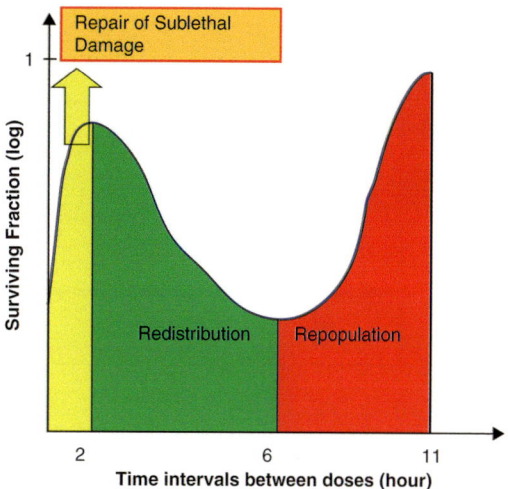

Fig. 2.30 Repair of sublethal damage and SF

- The slope of the survival fraction (SF) curve ($1/D_0$) is large for high-LET radiation.
- The slope of the SF curve ($1/D_0$) is small for low-LET radiation.

3. **Repair of sublethal damage (SLDR)** [35]. SLD is usually repaired 2–6 h after the delivery of radiation (Fig. 2.30).
   - SLD is not fatal, but the second dose increases radiosensitivity.
   - It can be lethal if there is an insufficient repair period between two fractions.
   - Repair abilities differ among normal tissues and tumors.
   - Inhibition of SLDR is the rationale for the additive effect of chemoradiotherapy.
   - SLDR depends on dose rate, and it is evident between dose rates of 0.01–1 Gy/min.

4. Repair of potentially lethal damage (PLDR) [36].
   Some damage that is lethal during normal growth can be repaired under suboptimal conditions (Fig. 2.31).

- The first human DNA repair gene to be discovered is located in the 18th chromosome.
- Mitomycin C, which selectively affects hypoxic tumor cells, acts through this gene and inhibits PLDR.

5. **Dose rate**. Cell survival is @@greater for a delivered radiation dose if the dose rate is decreased (Fig. 2.32).
   - This is due to the proliferation of undamaged living cells and SLD repair during radiotherapy.
   - This effect is very important in brachytherapy applications. The dose rate in external therapy is 100 cGy/min. Low dose rates are used in brachytherapy, and high doses can be given due to normal tissue repair and repopulation.

6. **Oxygenation** [37]. Soluble oxygen in tissues increases the stability and toxicity of free radicals. The increase in the effect of radiation after oxygenation is defined as the oxygen enhancement ratio (OER) (Fig. 2.33).

$$\text{OER} = \frac{\text{Required dose under hypoxic conditions}}{\text{Required dose under oxygenated conditions}} \quad (2.23)$$

The maximum value of the OER is 3. Oxygenation can modify the indirect effect of free radicals. However, the OER plays no role in the direct effect of high-LET radiation; OER is 1 in this case.

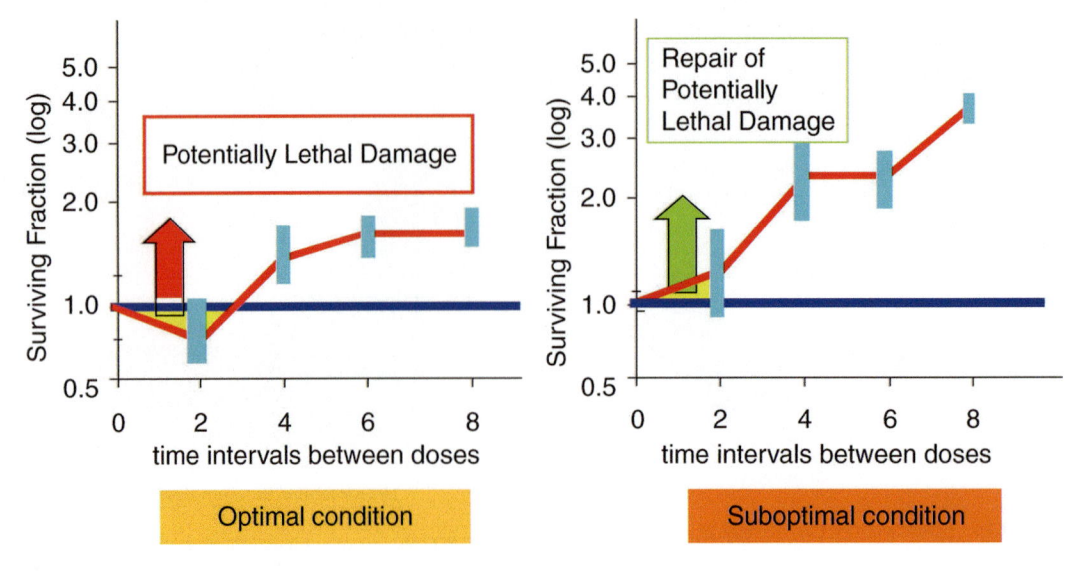

Fig. 2.31 Repair of potentially lethal damage and SF

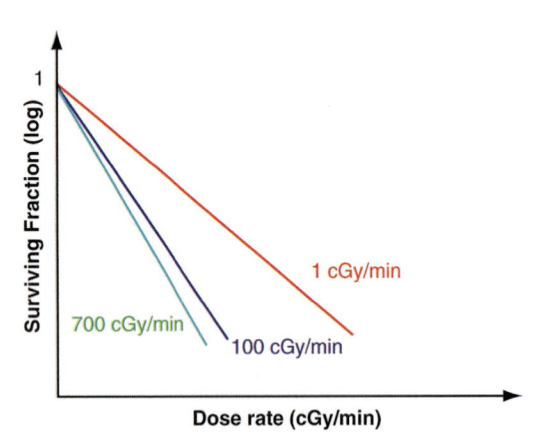

Fig. 2.32 Dose rate and SF

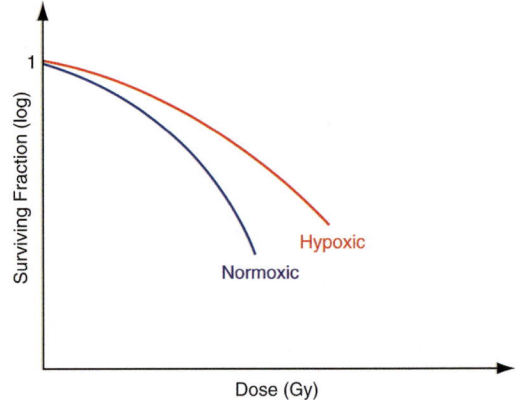

Fig. 2.33 Oxygenation and SF

Tumors become less hypoxic during fractionated radiation schedules.

7. **Temperature**. Most cells are more sensitive to radiation at high temperatures. However, there are more chromosome aberrations at low temperatures (probably due to the suppression of the DNA repair process at low temperatures).
8. **Chemical agents**
   (a) *Radioprotective agents* [38]. Free radical scavengers are radioprotective agents.
   - Thiol compounds, sulfhydryl amines like cysteine, cystamine, and isothiouronium, dimeric compounds containing excess sulfhydryl (SH) radicals, and antioxidants like vitamins A, C, and E can decrease radiation damage.
   - These compounds protect cells by neutralizing free radicals, producing hypoxic conditions, and forming disulfide bonds in proteins, strengthening protein structure.
   - Thiols, on the other hand, transiently inhibit DNA synthesis, giving the cell time to repair SLD using repair enzymes. However, they are not used prophylactically as radioprotective agents due to their side effects.

- Alcohol, morphine, and tranquilizing agents decrease respiration, thus increasing radioresistance.
- *Amifostine (WR-2721)*. The results of phase III trials have confirmed the safety and efficacy of amifostine as a radioprotective agent that reduces xerostomia in patients with head and neck cancers who are receiving radiotherapy. It is also a cytoprotectant that prevents cisplatin-induced renal toxicity and neutropenia in patients with ovarian cancer.

(b) *Radiosensitizers* [39]. Oxygen is the leading radiosensitizer. Oxygen mimetic agents with electron affinity (metronidazole, misonidazole, nitroimidazoles, etanidazole (SR-2508)), DNA analogs (actinomycin D, adriamycin, methotrexate, 5-fluorouracil), and caffeine can increase the damaging effects of radiation.

Table 2.4 Radiosensitivities of various tissues

The most sensitive	Lymphocyte
	Immature hematopoietic cells
	Intestinal epithelium
	Spermatogonia
	Ovarian follicle cells
Sensitive	Bladder epithelium
	Esophagus epithelium
	Gastric mucosa
	Epidermal epithelium
	Optic lens epithelium
Moderately sensitive	Endothelium
	Growing bone and cartilage
	Fibroblast
	Glial cells
	Mammary gland epithelium
	Lung epithelium
	Renal epithelium
	Hepatic epithelium
	Pancreas epithelium
	Thyroid epithelium
	Surrenal gland epithelium
Less sensitive	Mature erythrocyte
	Muscle cell
	Mature connective tissue
	Mature bone and cartilage
	Ganglion cell

2.7 Tissue and Organ Response to Radiation

Tissue is defined as a collection of similarly functioning cells that have the same origin and are similar in shape and structure. Tissues form organs. The response of a tissue to radiation is determined by its precursor cells (Table 2.4).

Flexure dose (D_f). This is the dose calculated by multiplying the α/β ratio of acutely responding tissue by 0.1 in the LQ model.

It refers to the maximum dose at which F-type tissues are protected.

It is important in hyperfractionation.

D_f is the maximum dose at which late-responding tissues are protected but early-responding tissues die. It is the dose attained just before the death of the first cell.

If D_f is high, late side effects decrease and acute side effects increase. The total dose may increase. If D_f is high, the suitability for hyperfractionation increases.

Well-differentiated cells show a reduced capacity to divide compared to undifferentiated ones. This indicates that undifferentiated cells accrue more damage from radiation. For instance, bone marrow cells, intestinal crypt cells, and basal skin cells are undifferentiated cells; these are damaged early and at low doses.

The α/β ratio for acute-responding tissues is high (10).

The α/β ratio for late-responding tissues is low (3).

The α/β ratio in human tumors varies between 1 and 25.

Tissue and organ radiation tolerance is an important clinical parameter. The tolerance doses of normal tissue and organs surrounding the tumor are very important in radiotherapy planning. The tolerance dose depends on the delivered fraction dose and the irradiated tissue volume.

Dose-limiting organs are classified into three classes according to their radiation tolerances: I, II, and III (Tables 2.5, 2.6, and 2.7).

Tolerance doses are determined for 2 Gy daily fraction doses and for 5 days/week.

Table 2.5 Class I organs: radiation damage is morbid and/or highly fatal [44]

	Damage	TD 5/5 (Gy)	TD50/5 (Gy)	Irradiated field size or volume
Bone marrow	Aplasia, pancytopenia	25	4.5	Total
		35	40	Segmental
Liver	Acute and chronic hepatitis	25	40	Total
		15	20	Total thin band
Stomach	Perforation, bleeding	45	55	100 cm^2
Small intestine	Perforation, bleeding	45	55	400 cm^2
		50	65	100 cm^2
Brain	Infarction, necrosis	60	70	Total
		70	80	25%
Spinal cord	Infarction, necrosis	45	55	10 cm
Heart	Pericarditis, pancarditis	45	55	100 cm^2
Lung	Acute and chronic pneumonia	30	35	60%
		15	25	Total
Kidney	Acute and chronic nephrosclerosis	15	20	Total
		20	25	Whole thin band
Fetus	Death	2	4	Total

Table 2.6 Class II organs: radiation damage is of low–moderate morbidity, or occasionally fatal [44]

	Damage	TD5/5 (Gy)	TD50/5 (Gy)	Irradiated field size or volume
Oral cavity and pharynx	Mucositis, ulceration	60	75	50 cm^2
Skin	Acute and chronic dermatitis	55	70	100 cm^2
Esophagus	Esophagitis, ulceration	60	75	75 cm^2
Rectum	Ulcer, stricture	60	80	100 cm^2
Salivary glands	Xerostomia	50	70	50 cm^2
Bladder	Contracture	60	80	Total
Ureters	Stricture	75	100	5–10 cm
Testis	Sterilization	1	2	Total
Ovaries	Sterilization	2–3	6–12	Total
Growing cartilage	Growth retardation	10	30	Total
Child bone	Dwarfism	10	30	10 cm^2
Adult cartilage	Necrosis	60	100	Total
Adult bone	Fracture, sclerosis	60	100	10 cm^2
Eye Retina Cornea Lens	Retinopathy Keratopathy Cataract	55 50 5	70 60 12	Total Total Total/partial
Endocrine glands Thyroid Surrenal Hypophysis	Hypothyroidism Hypoadrenalism Hypopituitarism	45 60 45	150 20–30	Total
Peripheral nerves	Neuritis	60	100	Total
Ear Middle ear Vestibule	Serous otitis Meniere's syndrome	50 60	70 70	Total

2.7 Tissue and Organ Response to Radiation

Table 2.7 Class III organs: radiation damage is not morbid or is reversibly morbid [44]

	Damage	TD5/5 (Gy)	TD50/5 (Gy)	Irradiated field size or volume
Muscle	Fibrosis	60	80	Total
Lymphatics	Atrophy, sclerosis	50	70	Total
Large artery–vein	Sclerosis	80	100	10 cm^2
Joint cartilage	–	500	500	
Uterus	Perforation, necrosis	100	200	Total
Vagina	Ulcer, fistule	90	100	Total
Breast (adult)	Atrophy, necrosis	50	100	Total

2.7.1 Bergonie and Tribondeau Law [40]

The radiosensitivity of a tissue depends on:

- The excess amount of less-differentiated cells in the tissue
- The excess amount of active mitotic cells
- The duration of active proliferation of the cells

According to the Bergonie and Tribondeau law, the effect of radiation on undifferentiated divided cells with high mitotic activity is much greater than the effect of radiation on undivided differentiated cells.

2.7.1.1 Michalowski Tissue Sensitivity Classification [41]

Hierarchical tissues. These tissues are divided into two compartments containing differentiated and undifferentiated cell groups, respectively. The cells in these tissues are H-type cells.

- These are cells that can continuously divide, such as stem cells and intestinal epithelial cells.
- They respond acutely to radiation.

Flexible tissues. These are tissues that are not divided into compartments containing different cells. The cells die together during tissue damage, and they are F-type cells.

- These are tissues that consist of cells that divide if necessary, such as liver and thyroid cells.
- They are late-responding tissues.

Many tissues respond to radiation in a manner that can be considered a hybrid of these two tissue types. The response of a tissue to radiation derives from both parenchymal cells and vascular stromal cells. Cells that cannot renew themselves (such as central nervous system (CNS) cells and striated muscle cells) are less sensitive to radiation, and radiation damage is most likely due to the effect on vascular stroma. A radiation dose that kills most of the stem cells in the parenchymal compartment of a tissue activates the repopulation of functional mature cells, and mature cells originating from stem cells play an important role in re-establishing tissue function after irradiation.

2.7.2 Factors Determining Radiation Damage According to Ancel and Vintemberger [42]

Biological stress in the cell.
Biological stress is important in cell division. While radiation damage to rapidly dividing cells is observed early, damage to slowly dividing cells is seen in late.

Cell status before and after radiation dose.
This indicates the environmental conditions of the cell → the radiation response of a cell changes in optimal and suboptimal conditions:

- Radiation response increases in optimal conditions.
- Radiation response decreases in suboptimal conditions.

2.7.3 Rubin and Casarett Tissue Sensitivity Classification [43]

This classifies tissues according to proliferation kinetics:

Tissues consisting of vegetative intermitotic cells (VIM).
These consist of undifferentiated cells.
These cells have a very short cell cycle.

- Examples include stem cells and intestinal stem cells.
  - These have short lifetimes but can continuously repopulate.
  - These are the most radiosensitive tissues.

Tissues consisting of differentiated intermitotic cells (DIM).
These consist of cells with a partial proliferative capacity.
Their mitotic activity stops when they become mature.

- An example is spermatogonia.

Multipotential connective tissues (MPC).
These consist of cells with relatively long lifetimes.
These cells divide at irregular intervals.

- The most prominent example is the fibroblast.

Tissues consisting of reverting postmitotic cells (RPM).
These cells do not divide under normal conditions; they only divide if necessary.
These tissues consist of cells with long lifetimes.

- Examples include liver parenchymal cells, pulmonary cells, and renal cells.

Tissues consisting of fixed postmitotic cells (FPM).
These cells never divide.
These tissues consist of cells with very long lifetimes.

- Examples include CNS cells, muscle cells, and erythrocytes.
- These are the most radioresistant tissues.

The radiosensitivities of tissues are variable, except in the cases of VIM and FPM.
TD5/5: this defines the minimum tolerance dose, which is the dose that yields a complication rate of less than 5% over 5 years.
TD50/5: this defines the maximum tolerance dose, which is the dose that yields a complication rate of 50% over 5 years.
Serial organs [45].
The functional subunits (FSUs) of serial organs are structured serially (Fig. 2.34).

- If critical damage due to radiation occurs in any functional subunit, complications are observed in the whole organ.
- Examples include the spinal cord, esophagus, rectum, and coronary arteries.

2.7.3.1 Parallel Organs [45]
Their FSUs are parallel in structure (Fig. 2.35).

- If critical damage due to radiation occurs in any functional subunit, complications are only observed in that subunit, and the organ continues its function.
- Examples include lungs, liver, and myocardium.

2.8 Stochastic and Deterministic Effects

The effects of radiation on tissues and organs can be classified into three groups: acute, subacute, and chronic [46].
Acute effects: changes that occur in the first 6 months.
If the radiation dose is high enough, the organ's parenchymal tolerance is exceeded and organ death occurs. If the dose is low, the organ continues to function fully or partially, even in the presence of parenchymal damage.
Subacute effects: changes that occur between 6 and 12 months.

2.8 Stochastic and Deterministic Effects

Fig. 2.34 Response to radiation in serial organs (Fig. 5.7, p 101 of [20])

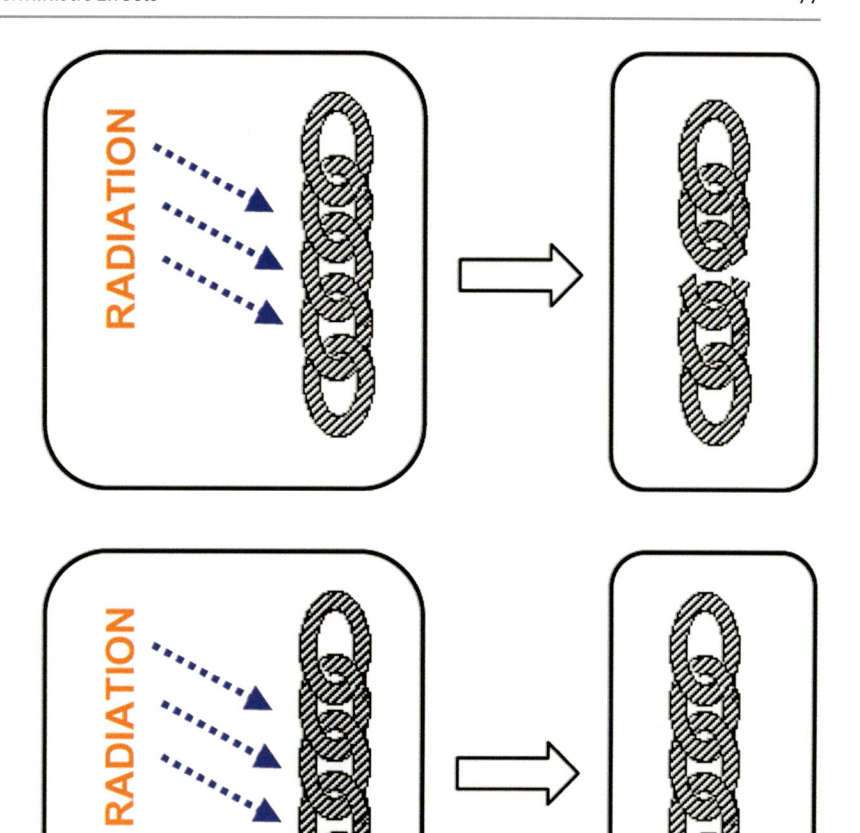

Fig. 2.35 Response to radiation in parallel organs (Fig. 5.8, p 102 of [20])

Secondary parenchymal degeneration resulting in decreased resistance to radiation is observed.

Chronic effects: changes that occur after 12 months.

Carcinogenesis, genetic mutations, and chromosomal aberrations occur.

2.8.1 Deterministic Effect [46]

- The acute and subacute effects of radiation are known as deterministic effects (nonstochastic effects) (Fig. 2.36). The intensities of these effects are directly proportional to the dose.

They have a specific threshold dose.
Effects appear at higher doses than the threshold dose.

There is a relationship between dose and individual effects.

Cataract, skin erythema, sterility, radiation myelitis, and fibrosis are all examples of deterministic effects.

For example, if the total body irradiation dose is >5 Gy, bone marrow suppression is observed, but this suppression is not observed for a dose of <5 Gy.

2.8.2 Stochastic Effects [46]

- The chronic effects of radiation are known as stochastic effects (Fig. 2.37).

These are statistically measurable effects.
There is no threshold dose for these effects.
There is no relationship between the dose and individual effects.

Fig. 2.36 Deterministic and stochastic effects of radiation

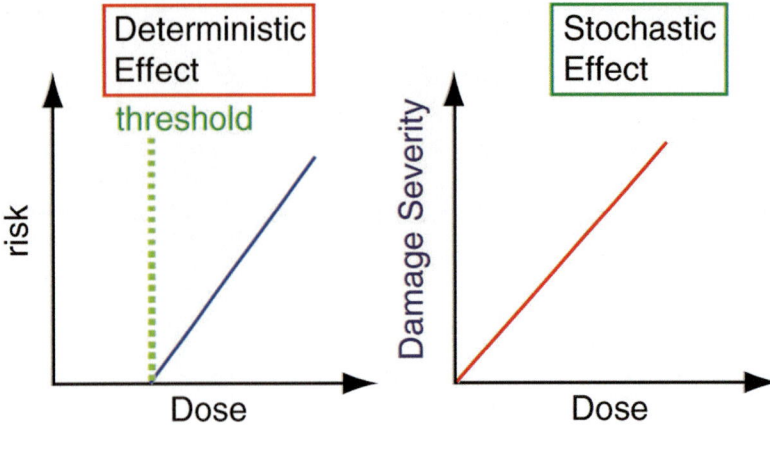

Fig. 2.37 TCP and NTCP curves

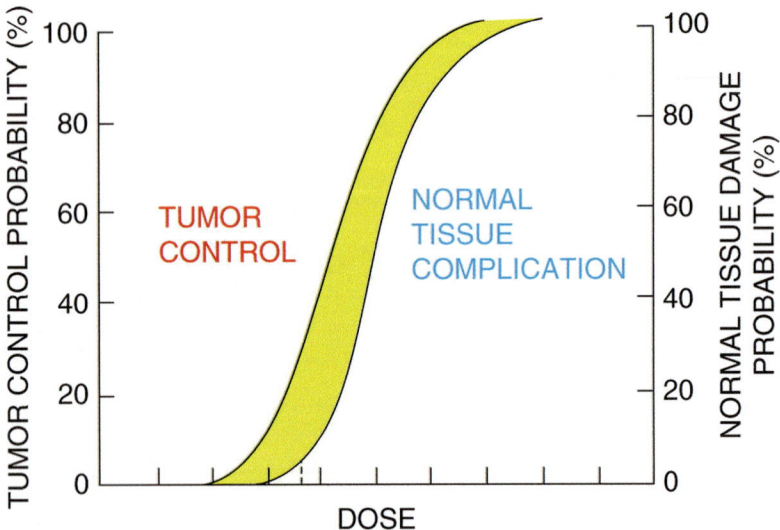

Carcinogenesis, genetic mutations, and chromosome aberrations are all stochastic effects.

Stochastic incidence → the incidence of cancer is 250 cases per one million for 1 REM radiation.

2.9 Tumor Response to Radiation

The aim of radiotherapy is to annihilate the tumor tissue while minimizing damage to the normal surrounding tissues. Thus, the radiosensitivities of the tumor and its surrounding tissues are important considerations when determining the best treatment. It is well known that cells in tumor tissues have chaotic growth patterns and various radiosensitivities. Furthermore, tumor cells exhibit a variety of sizes, chromosome structures, and cytoplasms. However, this is not true of neighboring normal tissue cells. The principle of *primum non nocere* is always valid in radiotherapy. In light of this principle, several concepts for destroying the tumor while protecting healthy tissues have been developed in the field of radiation oncology.

2.9.1 Therapeutic Index

The therapeutic index defines how the tumor control probability (TCP) relates to the normal tissue complication probability (NTCP) for different doses [47]. Normal tissues may get damaged by the dose required to control the tumor; on the other

2.9 Tumor Response to Radiation

hand, the tumor may not receive an adequate dose if the normal tissues require protection. Achieving the optimal balance between TCP and NTCP is a basic aim of radiotherapy. All new technologies are directed towards this aim.

TCP and NTCP curves are sigmoid in shape. The purpose of treatment is to move the TCP curve to the left and the NTCP curve to the right (Fig. 2.37).

- The therapeutic index (= therapeutic window) increases if the region between two curves becomes large, and the expected benefit from treatment increases.

When the fraction dose is increased from 2 to 2.5 Gy, the total dose to control the tumor decreases. Since the maximum tolerable dose is constant, the total dose received by normal tissue increases, and the therapeutic window narrows. Therefore, the treatment scheme in the second graphic in Fig. 2.38 is unacceptable compared to that in the first graphic.

2.9.2 Tumor Control Probability (TCP)

The efficacy of radiotherapy treatment is evaluated by the locoregional TCP and the treatment-related NTCP [48].

TCP is directly proportional to the dose and inversely proportional to the number of cells in the tissue (or the volume of the tumor). The total dose required to control the subclinical disease in epithelial cancers is 40–50 Gy, whereas it is 60–70 Gy for clinically observable gross disease. The most important dose-limiting factor is the tolerance of the surrounding tissues to radiation.

Local tumor control. This is the destruction of tumor cells, where they are determined. It is also defined as the death of the last clonogenic cancer cell.

Radiation affects tumor cells in a very similar way to normal tissue and organs; its effect is nonspecific.

Tumoral factors affecting the TCP:

- Intrinsic radiosensitivity
- Location and size of tumor
- Cellular type of tumor
- Effect of oxygen

Treatment-related factors affecting the TCP:

- Dose–time fractionation
- Radiation quality (RBE, LET)
- Dose rate
- Use of radiosensitizers
- Combination of radiotherapy with surgery and/or chemotherapy
- Technique (e.g., small field sizes)
- Treatment modality (e.g., brachytherapy, conformal RT, IMRT, IGRT, targeted RT

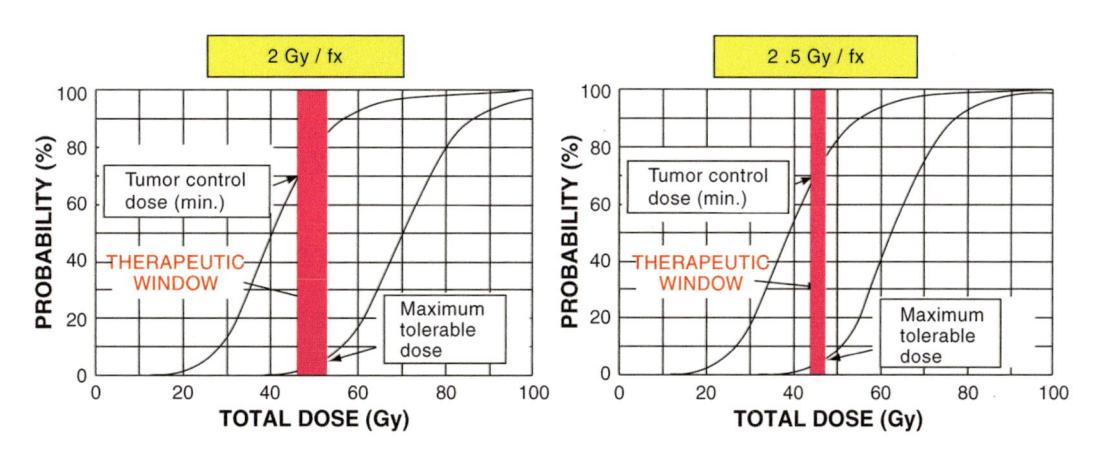

Fig. 2.38 Relationships between fraction dose, total dose, and therapeutic window

The tumor volume decreases if a dose d_1 is delivered to a volume v_1. If we assume that a second dose d_2 is delivered to a new volume v_2, the total TCP will be as follows:

$$\mathrm{TCP} = \mathrm{TCP}(d_1,v_1) \cdot \mathrm{TCP}(d_2,v_2) \mathrm{TCP}(d_3,v_3)\ldots = \prod_{i=1}^{n} \mathrm{TCP}(d_i,v_i) \quad (2.24)$$

$$\mathrm{TCP}(D) \approx e^{N \cdot e^{-(\alpha d + \beta d^2)}} \quad (2.25)$$

$$\mathrm{TCP} = e^{-(SF \times N)} \quad (2.26)$$

where SF = surviving fraction and N = clonogenic cell number.

$$P(D) = \frac{1}{1+\left(\dfrac{D_{50}}{D}\right)^k} \quad (2.27)$$

k = slope of dose–response curve, D = total dose, D_{50} = tolerance dose, $P(D)$ = expected probability of cure for the given total dose (%), $\mathrm{TCD}_{50} = D_{50} = \mathrm{ED}_{50} = \mathrm{TD}_{50}$

- The tissue tolerance dose is equal to the dose that kills 50% of clonogenic cells (Fig. 2.39).

$$(\mathrm{TCD}_{50} = \mathrm{TD}_{50})$$
$$\mathrm{TCD}_{50} = \mathrm{TCP}(1-\mathrm{NTCP}) \quad (2.28)$$
$$\mathrm{TCD}_{50} = D_{50} = \mathrm{ED}_{50} = \mathrm{TD}_{50}$$

- The dose needs to be increased threefold to increase the TCP from 10 to 90%.

If the TCP and NTCP curves are close to each other → tumor is radioresistant.

If the TCP and NTCP curves are far from each other → tumor is radiosensitive (Fig. 2.40).

Fig. 2.39 Relationship between TCP, NTCP, and TCD50

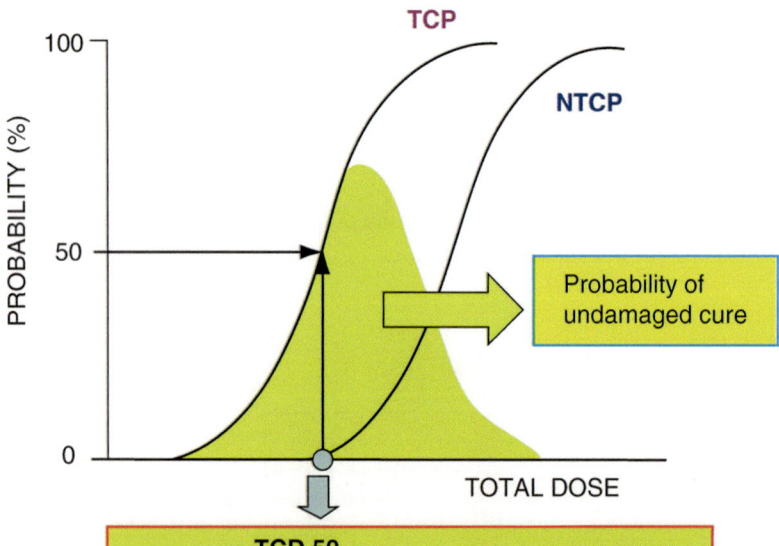

TCD 50
the radiation dose required to control 50% of tumours
or
the radiation dose required to kill 50 of clonogenic cells

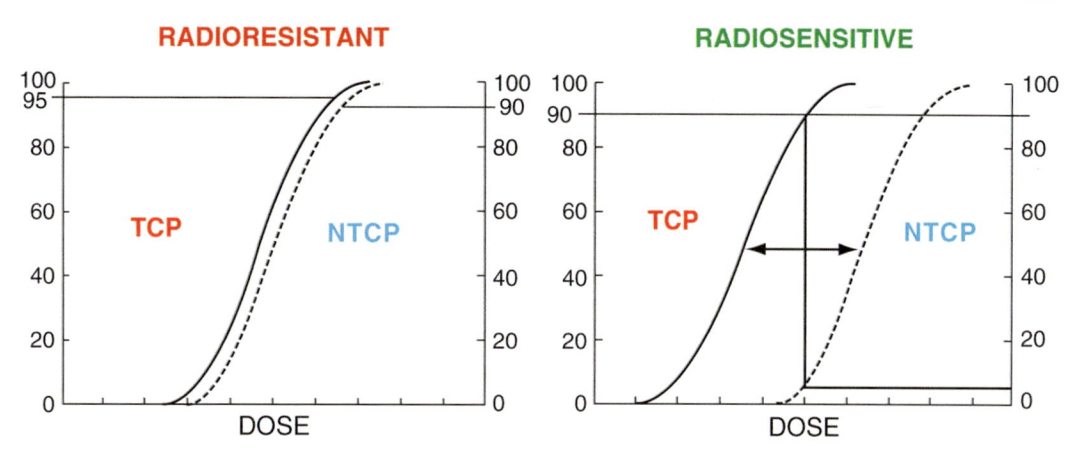

Fig. 2.40 Relationship between TCP, NTCP, and radiosensitivity

2.9.3 Normal Tissue Complication Probability (NTCP)

The TCP is a function of the total dose, fraction dose, irradiated volume including the whole tumor, and treatment reproducibility [48, 49]. The NTCP is a function of the total dose, fraction dose, fraction number, and the volume of tissue exposed to the radiation [48, 49].

The Lyman model is used in NTCP calculations [50]. This model is highly complex. It is a mathematical model of biological effects based on the use of algebraic definitions.

- The volume of irradiated normal tissue and its radiosensitivity make this model more complex.
- In addition, tissue types (parallel or serial organs) as well as the FSUs of tissues are added to this formulation, making the model even more complex [50].

$$P(D) = \sum_{k=M+1}^{N} \binom{k}{N} \cdot P_{FSU}^{k} \left(1 - P_{FSU}\right)^{N-k} \quad (2.29)$$

- However, the most important issue is not the formulation, but the knowledge of TD50, and not to exceed this limit during planning.

$$NTCP = \frac{1}{1 + \left(\dfrac{D_{50}}{D}\right)^{k}} \quad (2.30)$$

Factors affecting NTCP [50]:

Factors related to organ tissue

- Tissue radiosensitivity
- The volume of organ tissue within the radiotherapy portal
- Organ type: serial or parallel

Factors related to treatment

- Dose–time fractionation
- Quality of radiation (RBE, LET)
- Dose rate
- Use of radioprotectors
- Combination of RT with surgery and/or chemotherapy
- Technique (e.g., addition of boost field)
- Treatment modality (e.g., brachytherapy, conformal RT, IMRT, IGRT, targeted RT)
- SF2 = surviving fraction after 2 Gy irradiation → as SF2 increases, TCP decreases.

TCP and NTCP calculations are mechanistic models. The *critical volume model* was developed by Niemierko in 1997. This model is used empirically for 3D treatment plans that involve calculating the *equivalent uniform dose* (EUD). This model is based on the hypothesis that clonogenic cells that have the same survival curves can be irradiated with the same uniform dose. The α/β ratios, clonogenic cell number, dose, number of fractions, type of tissue and type of tumor, as well as the SF2 are fed into this formulation. The EQD2 (equivalent dose at 2 Gy) is derived in

Fig. 2.41 Relationship between TCP and EUD

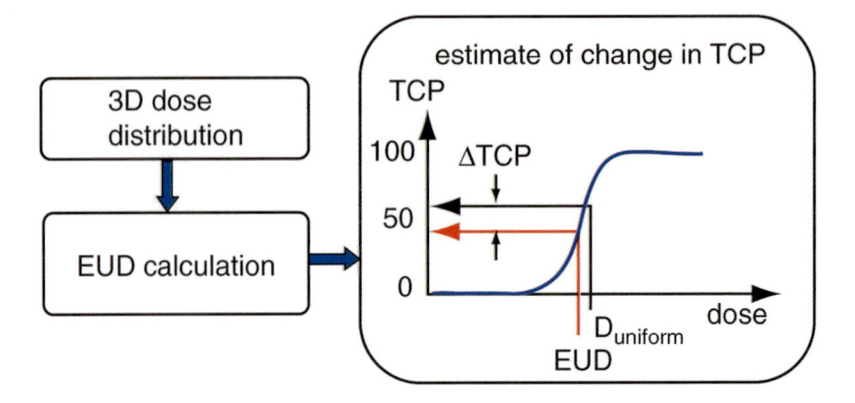

addition to the SF2, and this is used to compare different fractionation schemes [51].

- SF2 = surviving fraction after 2 Gy irradiation
  → as SF2 increases, TCP decreases

$$EQD_2 = D \times \left[d + \frac{\alpha/\beta}{2} + \alpha/\beta \right] \quad (2.31)$$

where D = total dose and d = fraction dose.

The value of the EUD lies between the minimum dose and the mean dose for tumor control (Fig. 2.41). A decrease in the TCP dose was observed when the EUD was calculated and used.

$$D_{min} \text{ " EUD " } D_{mean}$$

As the irradiated volume increases, the normal tissue tolerance dose decreases due to the increase in functional subunit number.

The essential parameters of the NTCP are the irradiated tissue volume and the dose delivered (Figs. 2.42 and 2.43).

Major references for NTCP estimation models:

- Lyman model: Burman C, Kutcher GJ, Emami B, Goitein M (1991) Fitting of normal tissue tolerance data to an analytic function. Int J Radiat Oncol Biol Phys 21:123–135.
- Critical volume model: Stavrev P, Stavreva N, Niemierko A, Goitein M (2001) The application of biological models to clinical data. Phys Medica 17(2):71–82.

2.9.3.1 The Clinical Importance of TCP and NTCP [49–51]

- The TCP and NTCP can be used to estimate the treatment success and side effects in particular.
- Dose–volume histograms created by treatment planning systems (particularly 3D-conformal radiotherapy and IMRT) as well as TCP and NTCP mathematical modeling are very useful for graphically demonstrating normal tissue damage ratios within the treated tumor volume, and can be used to guide clinicians during treatment planning (Fig. 2.43).

2.9.3.2 Therapeutic Ratio

TR = normal tissue tolerance dose/tumor control dose.

2.10 The Five R's of Radiotherapy

The delivery of radiation in small daily fractions is known as fractionated radiotherapy. Fractionated radiotherapy studies were started after it was realized that single, high-dose radiotherapy is ineffective for tumor control and has serious side effects. Claudius Regaud observed that ram spermatogenesis decreased after fractionated doses, but there were no side effects on the scrotal skin.

Although Regaud started the first fractionation studies, the first radiation oncologist, Grubbe, treated a breast cancer patient (Mrs.

2.10 The Five R's of Radiotherapy

Fig. 2.42 The essential parameters of NTCP and its relationships (Fig. 5.4, p 97 of [20])

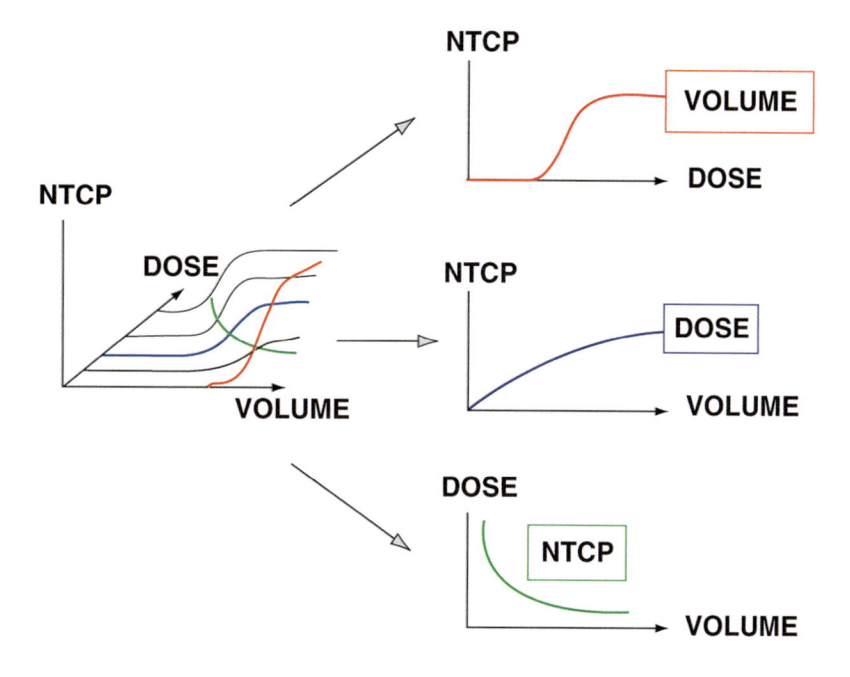

Fig. 2.43 DVH curves for the spinal cord in two different treatment plans (Fig. 5.6, p 100 of [20])

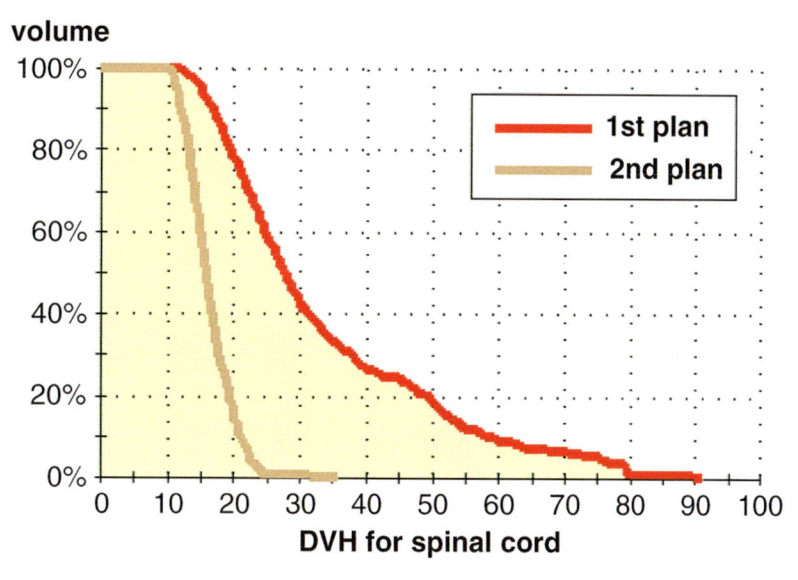

Rosa Lee) with radiotherapy for 1 h each day for 18 days in 1896. He found beneficial responses, but did not publish them since he was just a medical student. However, Grubbe published two scientific articles 50 years after his first use of radiotherapy.

Grubbe EH (1946) X-ray treatment; its introduction to medicine. J Am Inst Homeopath 39(12):419–422.

Grubbe EH (1947) The origin and birth of X-ray therapy. Urol Cutaneous Rev. 51(5):375–359.

Coutard, a colleague of Regaud, applied fractionated radiotherapy to head and neck cancers in 1934, and obtained successful results [52]. He emphasized the importance of the time–dose concept in radiotherapy (Fig. 2.44).

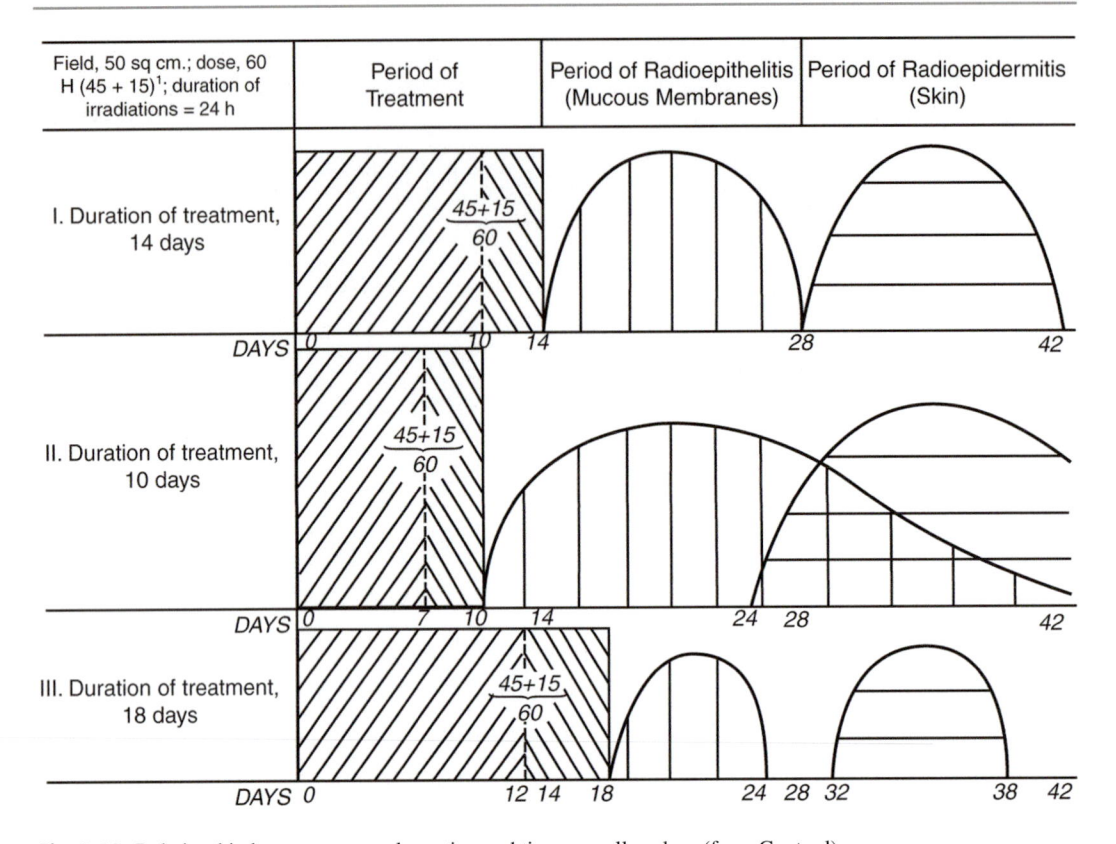

Fig. 2.44 Relationship between mucosal reaction and time as well as dose (from Coutard)

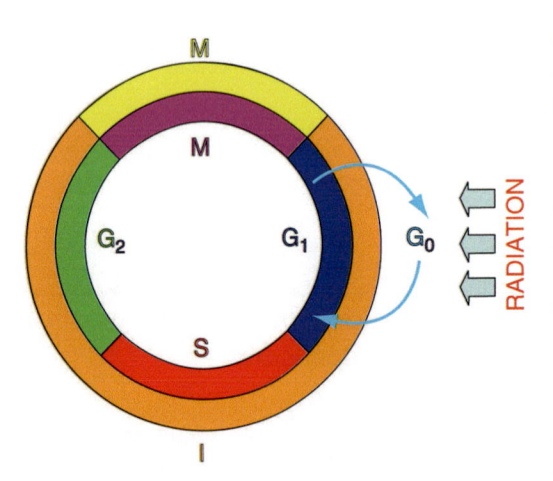

Fig. 2.45 Cell cycle and repopulation

therapy were defined as repair, reassortment (redistribution), repopulation, and reoxygenation by Withers in 1975. A fifth "R," radiosensitivity, was added to this list by Bernard Fertil in 1981. All of these are important for estimating the responses of normal or tumor tissues to radiotherapy.

Fractioned radiotherapy is founded on five main features:

- Repopulation
- Repair
- Redistribution (= reassortment)
- Reoxygenation
- Radiosensitivity (intrinsic radiosensitivity)

Coutard observed that skin and mucosal reactions seen during radiotherapy of pharyngeal and laryngeal cancers depend on the dose and the duration of radiotherapy.

Biological factors that affect the responses of normal and tumor tissues in fractionated radio-

2.10.1 Repopulation

Both tumor and healthy normal cells continue to proliferate even when they are exposed to radiation [53, 54]. This proliferation is a physiological

response of tumor and normal tissues to decreases in cell number.

This repopulation enables tumor cells to partially resist the lethal effects of radiotherapy. The time required for the tumor cell number to double is known as the "tumor doubling time," T_p. This doubling time is less than 2 days for most tumors. This period can also be considered the repopulation time, and it varies during radiotherapy. Repopulation is slow at the beginning of radiotherapy, but it speeds up after the first doses of radiation therapy. This increase in repopulation rate is termed "accelerated repopulation," and the time taken for it to begin is termed the "kickoff time" (T_k). This accelerated repopulation becomes even faster if the treatment is interrupted after the tumor doubling time for any reason. Normal tissues also repopulate during radiotherapy; this issue is important for the repair of acute side effects. Therefore, radiotherapy schemes should be arranged so as to allow normal tissues to repopulate.

2.10.1.1 The Consequences of Proliferation

- Increases the number of tumor cells to be destroyed → against treatment
- Increases the number of normal tissue cells following irradiation → in favor of treatment

Resting cells in the G_0 phase enter the cell cycle in order to compensate for the cells killed by radiotherapy, and they undergo mitosis → repopulation (Fig. 2.45).

Early-responding tissues repopulate faster than the tumor during interfraction periods.

If the overall treatment time becomes longer than the period required, the tumor enters the accelerated repopulation mode and its response to radiation decreases due to tumoral proliferation.

- Accelerated repopulation begins after 28 days of treatment for head and neck tumors (Fig. 2.46) [54, 55].

Radiotherapy should be completed as soon as possible, within the tolerance limits of acutely responding normal tissues, due to the risk of accelerated repopulation.

2.10.2 Repair

Radiotherapy causes lethal damage to tumor cells and SLD in normal tissues. The application of radiotherapy in fractionated doses allows normal tissues time to repair [55, 56].

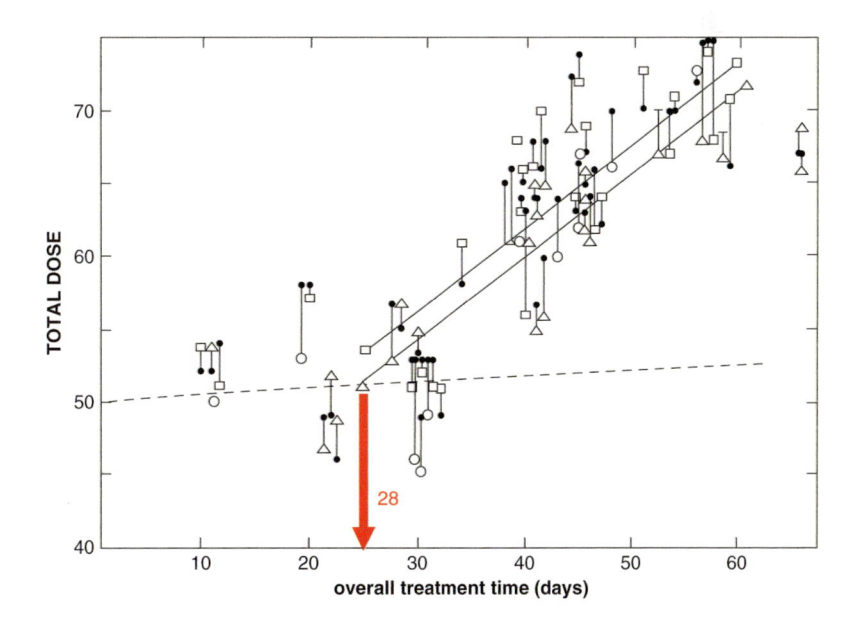

Fig. 2.46 Accelerated repopulation

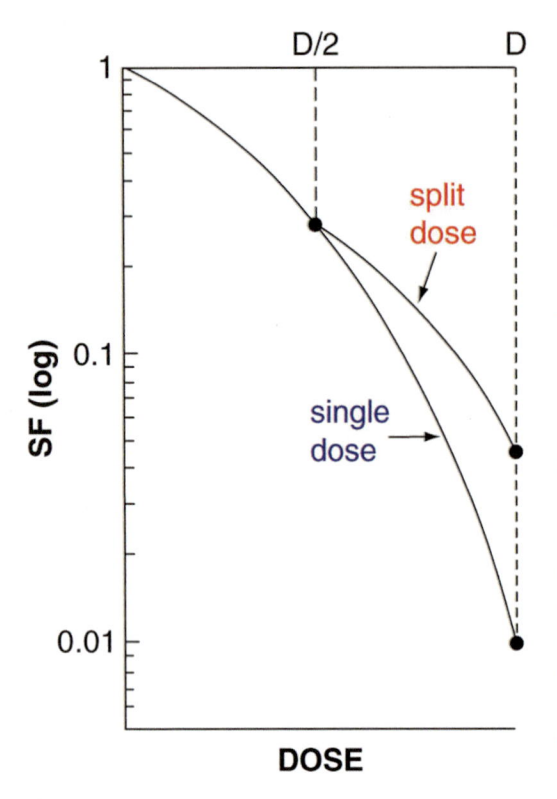

Fig. 2.47 Fractionated radiotherapy and the cell survival curve

One parameter used in this context is half the time required for cell repair after radiation damage ($t_{1/2}$), and the value of this parameter can be minutes to hours. Therefore, interfraction intervals should be at least 6 h in order to allow normal tissue cells to repair radiation damage.

If an optimal interval is left between fractions (6–12 h), normal tissue cells responding late to radiation have the capacity for faster repair than tumor cells.

According to multiple target–single hit theory, SLD occurs in mammalian cells at low doses, and this damage is repaired during interfraction intervals.

The repair of SLD in the spinal cord is much slower than that in other normal tissues. Thus, the interfraction interval should be at least 8 h in spinal cord irradiation.

Tumor cell SLD repair starts at the initial point of the shoulder (D_q) in the survival curve of the LQ model. Fractionating the next dose prevents this sublethal repair, shifting the dose away from the shoulder. Normal tissue cells, on the other hand, start to repair SLD before D_q, and so are not affected by the use of fractionation (Fig. 2.47).

Repopulation and repair → more important for normal tissues than tumor tissues.

As the protection of normal tissues increases, radioresistance increases.

Redistribution and reoxygenation → more important for tumor tissues than normal tissues; as more tumor tissue dies, its radiosensitivity increases.

When the total radiation dose is applied by dividing it into small fractions, and if the interval between two fractions is long enough (>6 h), normal tissues can protect themselves from radiation through SLD repair and repopulation.

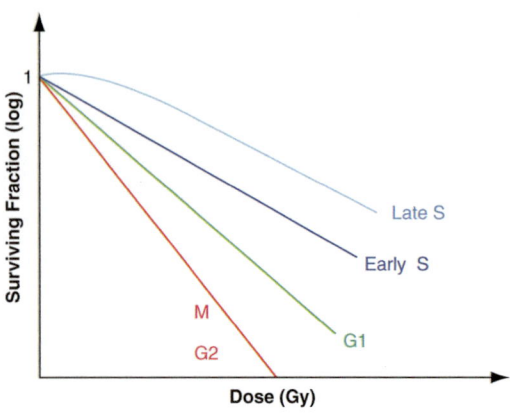

Fig. 2.48 Phases of the cell cycle and the survival curve

2.10.3 Redistribution (= Reassortment)

The radiosensitivities of cells vary with the phase of the cell cycle (Fig. 2.48) [54, 57]. The most sensitive phases are M and G_2, while the most resistant is the S phase. Cells in resistant phases of the cell cycle may progress into a sensitive phase during the next dose fraction. Therefore, the probability that tumor cells will be exposed to radiation during a sensitive phase increases, and this probability will continue to increase over the

Fig. 2.49 Oxygenation and the cell survival curve

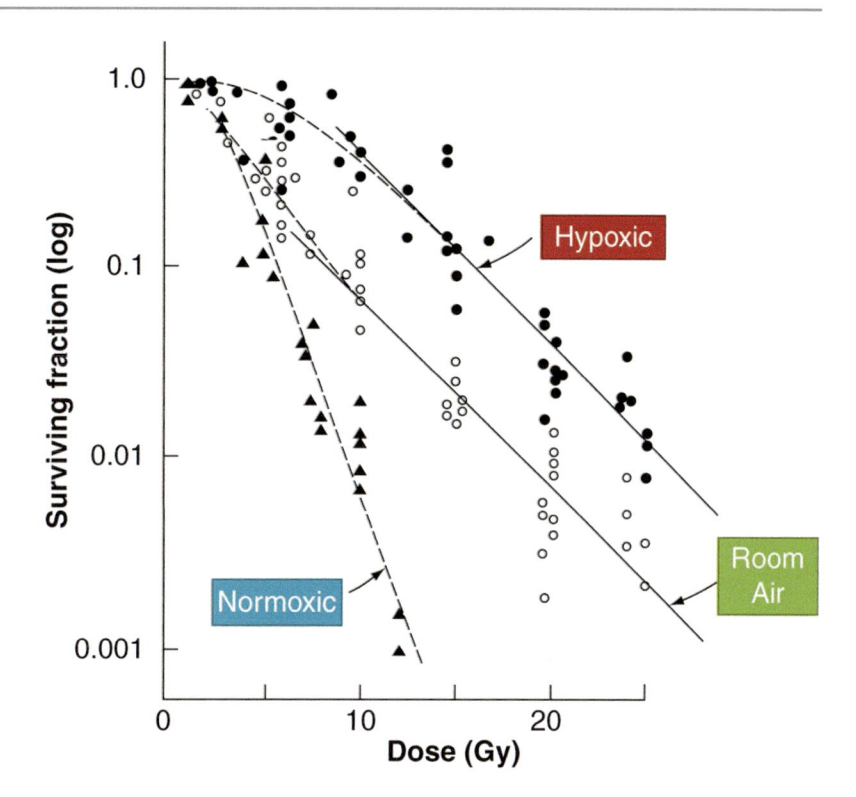

course of the treatment, and so the benefit of the radiation will also increase.

The durations of cell cycle phases: G_1 = 1.5–14 h, S = 6–9 h, G_2 = 1–5 h, M = 0.5–1 h

- The most sensitive: M and G_2
- The most resistant: S

2.10.4 Reoxygenation

As the tumor volume increases through the proliferation of tumor cells, the vascularity of the tumor tissue becomes insufficient to meet its requirements, and hypoxic–necrotic regions begin to occur within the tumor tissue [58, 59]. Hypoxic cells are 2–3 times more resistant to radiation (→ oxygen is required for the indirect effect to occur) (Fig. 2.49). Well-oxygenated cells that are radiosensitive die over the full course of fractionated radiotherapy. Therefore, since the oxygen supply is constant, the hypoxic cells gradually obtain much better vascularity and oxygenation, and their radiosensitivities increase (Fig. 2.50).

Oxygenating a tumor from the hypoxic state

- If hemoglobin is low, a blood transfusion may be given.
- High-pressure oxygen or carbogen may be applied during radiotherapy.
- The patient may be prevented from using hypoxic materials like cigarettes during radiotherapy.
- Hypoxic radiosensitizers may be used (e.g., metronidazole).

If the time interval between fractions is t and T is the overall treatment time, then for:

Reoxygenation	T should be minimum
Redistribution	t should be minimum
Repair	T should be minimum for normal tissues
Repopulation	T should be minimum for the tumor

Fig. 2.50 Fractionated radiotherapy and reoxygenation

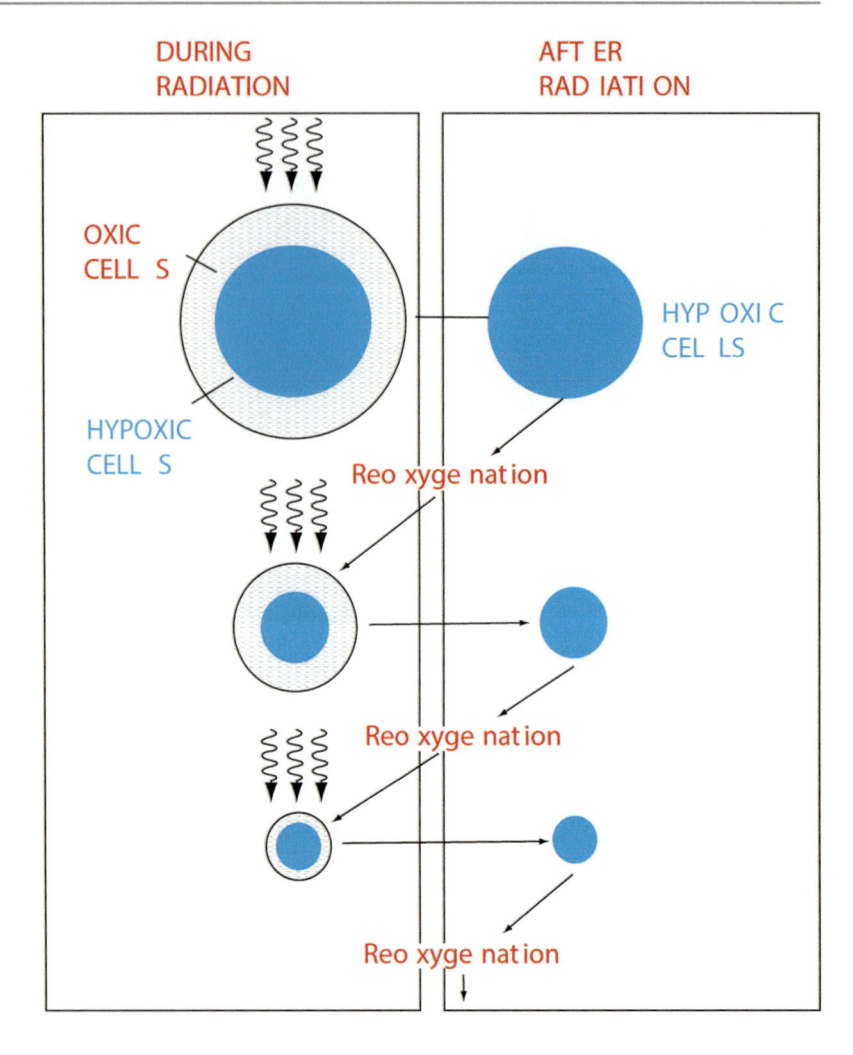

2.10.5 Radiosensitivity (Intrinsic Radiosensitivity)

Radiosensitivity (the fifth R of radiotherapy) is a concept that involves multiple components [51, 59]. Radiosensitivity may be affected by environmental conditions. The term "radiosensitivity" was first defined by Bergonie and Tribendau in 1907; they suggested that radiosensitivity was directly proportional to mitosis and inversely proportional to differentiation. Since radiosensitivity may be affected by external conditions, the term SF_2 was introduced by Fertil in 1981.

SF_2 = surviving cell fraction after a radiation dose of 2 Gy.

As SF_2 increases, radiosensitivity decreases.

SF_2 is represented graphically for some tumor cell lines in Fig. 2.51.

Radiosensitizers are used to decrease SF_2.

2.11 Fractionation

The five R's of radiotherapy form the basis for fractionation [47, 60, 61]. The total dose cannot be given in just one fraction, since this would produce serious adverse reactions in normal tissues. Therefore, it is necessary to divide the total dose into fractions. Normal cells can protect themselves from the radiation through repair and repopulation during the interfraction periods, whereas tumor cells are sensitized to

2.11 Fractionation

Fig. 2.51 SF$_2$ in some tumor lines

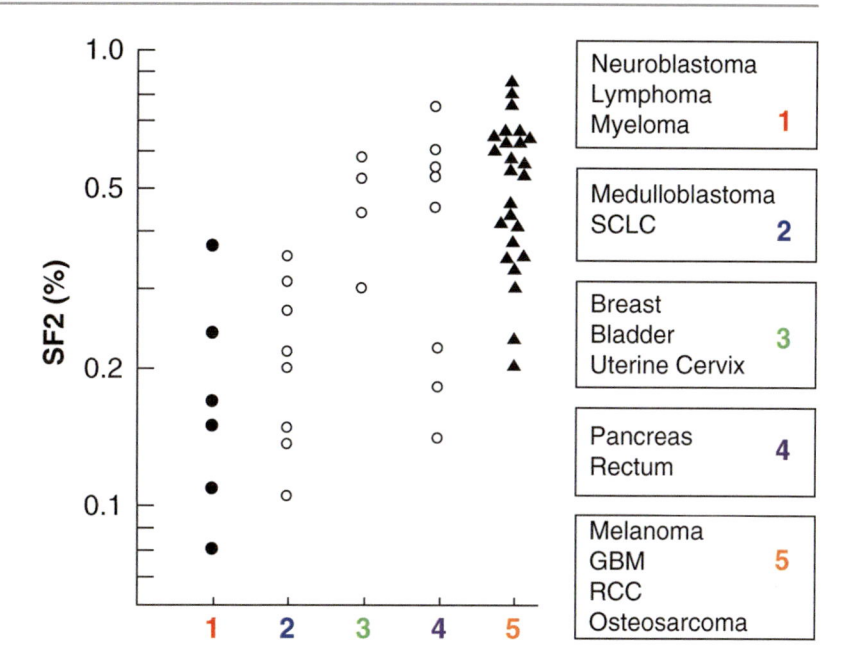

Group	Tumors
1	Neuroblastoma, Lymphoma, Myeloma
2	Medulloblastoma, SCLC
3	Breast, Bladder, Uterine Cervix
4	Pancreas, Rectum
5	Melanoma, GBM, RCC, Osteosarcoma

the radiation through reoxygenation and redistribution.

Conventional Fractionation

Fraction dose	1.8–2 Gy
Number of fractions per day	1
Number of fractions per week	5
Number of fractions per treatment	25–35
Total dose	45–70 Gy

Hyperfractionation

Fraction dose	1.1–1.2 Gy
Number of fractions per day	>2
Number of fractions per week	10
Number of fractions per treatment	60–70
Total dose	45–70 Gy or >10%

Aims of hyperfractionation

To decrease the fraction dose and increase the total dose

To increase local control

To decrease late effects in normal tissues

In hyperfractionation

Acute side effects are similar or slightly increased compared to conventional fractionation.

Late side effects are decreased compared to conventional fractionation.

Accelerated Fractionation

Fraction dose	1.1–2 Gy
Fraction number/day	>1
Fraction number/week	>5
Fraction number/treatment	25–35
Total dose	45–70 Gy or less

The aim of accelerated fractionation

To decrease overall treatment time, to decrease accelerated repopulation

In accelerated fractionation

Early side effects are more than conventional fractionation.

Late side effects are the same as conventional fractionation.

Treatment may be stopped early or the total dose may be decreased due to the excess amount of early side effects → this may cause a decrease in local control.

Hypofractionation

Fraction dose	>2 Gy
Number of fractions per day	≤1
Number of fractions per week	≤5
Number of fractions per treatment	≤25–35
Total dose	<45–70 Gy

The aim of hypofractionation

Hypofractionation is generally used for palliative purposes in the management of metastatic tumors. Late side effects are undervalued and palliation is provided over a very short time period.

In hypofractionation

Early side effects are similar to those associated with conventional fractionation.

Late side effects are increased compared to conventional fractionation.

Split Course. This refers to a fractionated treatment regimen that includes a planned interruption in order to decrease acute side effects. It is no better than conventional fractionation with regard to treatment efficacy.

Concomitant Boost. Here, two fractions are given daily in the fourth week of radiotherapy in order to prevent accelerated repopulation in head and neck cancers. The boost dose is given to the primary tumor site.

Hyperfractionation

Standard regimen:

70 Gy/35 fractions/7 weeks

$BED = D(1 + d/\alpha/\beta)$

$BED_3 = 70 (1 + 2/3) = 116$ Gy

$BED_{10} = 70 (1 + 2/10) = 84$ Gy

If we want to give the same total dose using two fractions per day:

$BED_{10} = 84 = X(1 + 1.2/10)$

$X = 75$ Gy

$BED_3 = 75(1 + 1.2/3) = 105$ Gy

Note that late side effects will increase.

However, 75 Gy/62 fractions = 6 weeks.

Accelerated Fractionation

Standard regimen:

70 Gy/35 fractions/7 weeks

$BED = D(1 + d/\alpha/\beta)$

$BED_3 = 70(1 + 2/3) = 116$ Gy

$BED_{10} = 70(1 + 2/10) = 84$ Gy

If we give 70 Gy/45 fractions/5 weeks with 1.6 Gy fractions and 3 fractions per day:

$BED_3 = 70(1 + 1.6/3) = 107$

$BED_{10} = 70(1 + 1.6/10) = 81$ (total treatment time will decrease).

- Early side effects depend on the total dose.
- Late side effects depend on the fraction dose.

2.12 Radiobiology of SBRT/SABR

SBRT (Stereotactic Body Radiotherapy) or SABR (Stereotactic Ablative Body Radiotherapy) is a "hypofractionated" radiation therapy regimen used in the treatment of primary or metastatic disease either alone or in combination with a systemic agent. The radiobiological principles of SBRT/SABR are not yet clearly defined. It is well known that newly formed tumor blood vessels are fragile and highly sensitive to ionizing radiation. Various evidence suggests that irradiation of tumors at high doses per fraction (>10 Gy per fraction) not only kills tumor cells but also causes significant damage to tumor vasculature. Vascular damage and disruption of the intratumor environment cause ischemic or indirect/secondary tumor cell death within a few days after radiation exposure, indicating that vascular endothelial damage plays an important role in the response of tumors to SBRT [64].

It is hypothesized that extensive tumor cell death due to the direct effect of radiation on tumor cells and secondary effect through vascular damage may trigger an anti-tumor immune response by causing the mass release of tumor-associated antigens and various proinflammatory cytokines. However, the precise role of the immune attack on tumor cells in SBRT and SRS has not yet been fully defined.

The "5 R" valid for conventional fractionated radiotherapy cannot explain effective tumor control with SBRT and SRS since it does not involve indirect cell death. The linear-quadratic model is for cell death caused by DNA breaks; therefore, the usefulness of this model for ablative high-dose SBRT and SRS is limited.

2.13 Radiation Protection

The ICRP, the International Atomic Energy Agency (IAEA), and similar organizations have published many recommendations relating to protection from ionizing radiation over the last

50 years. Although those recommendations are not enforceable by law, many countries have adopted these recommendations and put them into practice.

In principle, protection from all radiation sources is required. However, no practical measure can be taken to protect ourselves against the normal levels of radiation resulting from natural radiation sources and radioactive fallout due to previous nuclear tests. Nevertheless, we can control whether or how nuclear tests will occur in the future. The use of radiation in medicine is a decision that should be made clinically. In this field, it is not suitable to limit personal doses that are delivered for diagnostic or therapeutic reasons. Thus, collective doses are very high in these circumstances. Medical personnel must carefully follow related laws and ICRP recommendations [62].

On the other hand, the principles of radiation protection should be applied fully to all medical and industrial uses of radiation, as well as by all other users of nonnatural radiation, radiation workers, and the normal population.

The working conditions for radiation personnel who are pregnant should be arranged to ensure that the dose received by the fetus is as low as possible. The fatal dose for the remaining period of pregnancy should not exceed 1 mSv. Lactating personnel should not be allowed to work in places that incur a risk of radiation contamination.

- Patient visitors and volunteers (during diagnosis/treatment): <5 mSv.
- Children visiting patients: <1 mSv

The essential principles of radiation protection, as published in ICRP report 60 and the IAEA document BSS-115 (*Basic Protection Standards*) are:

- *Justification*. Radiation should never be delivered with no benefit, considering its harmful effects.
- *Optimization*. The dose should be as low as possible, considering personal doses, the number of irradiated persons, and economical and social factors, except when medical irradiation is performed for therapeutic purposes. This is known as the ALARA (as low as reasonably achievable) principle [62, 63].
- *Dose limits*. Radiation dose to normal individuals should not exceed the permissible organ and tissue equivalent doses.

Dose limits for radiation personnel [62]:

- For the whole body, the effective dose limit for five consecutive years: 100 mSv (i.e., a mean of 20 mSv per year).
- For the whole body, the effective dose limit in any 1 year: 50 mSv.
- For lens, the equivalent dose limit per year: 150 mSv.
- For hands, food, and skin, the equivalent dose limit per year: 150 mSv.

Dose limits for the general public [62]:

- For the whole body, the effective dose limit for five consecutive years: 25 mSv (i.e., a mean of 5 mSv per year).
- For the whole body, the effective dose limit in any 1 year: 1 mSv.
- For lens, the equivalent dose limit per year: 15 mSv.
- For hands, food, and skin, the equivalent dose limit per year: 50 mSv.

Dose limits for 16–18-year-old students and trainees (i.e., in those using radiation for educational purposes) [62]:

- For the whole body, the effective dose limit in any year: 6 mSv.
- For lens, the equivalent dose limit per year: 50 mSv.
- For hands, foot, and skin, the equivalent dose limit per year: 150 mSv.

Note that radiation workers must be at least 18 years old.

Radiation protection. Methods of protection from radiation can be grouped into two classes:

1. **Protection from internal radiation.** Internal contamination occurs via the entry of radioactive materials during respiration, digestion, or through damaged mucosa or skin. The radioactive material will radiate throughout the period when it is in the body. Therefore, precautions should be taken to prevent internal radiation from entering the body through media, foods, clothes, respiration, and skin that are already contaminated with radioactive materials. These precautions include the use of special respiration equipment, a full face mask, and filters, wearing protective clothes, blocking respiratory entry using towels when such equipment is absent, and banning the consumption of food and water in contaminated regions.
2. **Protection from external radiation.** There are three methods that are used to protect from external radiation:
   (a) *Distance.* As radiation intensity is inversely proportional to the square of the distance from the radiation source (inverse square law), increasing the distance from the source is a good protective measure. For instance; if the dose rate is 100 mSv/h at 1 m, it will be 1 mSv/h at 10 m.
   (b) *Time.* As the amount of radiation delivered is directly proportional to the time spent close to the radiation source, the time spent close to the source should be as short as possible. For instance, if the dose rate is 100 mSv/h, and someone stays in this field for 1 h, they will receive a dose of 100 mSv, but the dose will be 1000 mSv if they stay for 10 h.
   (c) *Shielding.* The most effective method of protecting from external radiation is shielding, and protective barriers with suitable features should be placed between the radiation source and the person in order to decrease the dose received. Shielding can be made from highly protective materials like soil, concrete, steel, and lead.

2.14 Pearl Boxes

2.14.1 Radiation Hormesis

Although the harmful effects of high-dose ionizing radiation are accepted, some preclinical and clinical data show that low-dose radiation actually stimulates some biological functions. The word "hormesis" derives from the Greek word "hormone," which means "stimulate." Radiation hormesis refers to the stimulating effect of ionizing radiation. This concept is generally defined as the physiological benefits gained from exposure to low-LET radiation in the total absorbed dose range of 1–50 cGy.

- In 1996, Yonezawa et al. observed that 30% of 21-ICR rats lived for 30 days after 8 Gy X-ray irradiation, whereas their life spans increased by 70% after a 5 cGy dose.

Adequate time and the optimal medium (adequate nutrition, oxygen) should be provided between each fraction to allow the repair of SLD (→ for normal tissues!).

- SLD is repaired in 0–2 h.
- There is a cell cycle phase change every 2–10 h (→ reassortment = redistribution).
- Repopulation starts after 12 h in tumor tissue.

Basic Advantages of Fractionation

- Protects normal tissues from early side effects.
- Provides reoxygenation in tumor cells.

Basic Disadvantages of Fractionation

- Overall treatment time increases.
- Tumor cell proliferation increases during treatment breaks due to increased duration.

Molecular oxygen should be present in the medium during radiotherapy to ensure maximum cell death through the fixation of free radicals. Therefore, hypoxia decreases cellular radiosensitivity. This condition provides the basis for some applications, such as the delivery of oxygen at 3 atm pressure (hyperbaric) to patients during radiotherapy, the use of hypoxic cell sensitizers like nitroimidazole, and the use of high-LET radiation that decreases the inverse effect of hypoxia on tumor cells.

2.14.2 Abscopal Effect

Abscopal effect is the occurrence of systemic side effects in the regions that are far away from the irradiated area. Increasing the fraction size over 2 Gy per fraction increases the abscopal effect in parallel tissues. For many years, the abscopal effect due to conventional radiation has been poorly studied [65, 66]. However, with the development and utilization of immunotherapy strategies that include radiotherapy and targeted immune modulators and immune checkpoint blockade, the abscopal effect is gaining more importance in less immunogenic tumors such as breast cancer. In recent years, the abscopal effect has been investigated by many researchers and it has been shown that the immune system probably leads to immunogenic tumor cell death with critical mediators including the dendritic cells, T regulatory cells, and suppressor cells [67–69].

2.14.3 Bystander Effect

The bystander effect is where cells that are not exposed to radiation, but which are in close proximity to cells that are, show similar biological effects to the cells exposed directly to radiation.

- This effect is thought to occur due to intercellular communication and cytokines.

2.14.4 Avalanche Phenomenon

The avalanche phenomenon refers to the fact that, when the number of flexible tissue cells has dropped to a certain level (point F) after the cells have been exposed to radiation, the remaining cells die more rapidly. This effect is more prominent at high doses and with additional treatments (chemotherapy, surgery). This dose dependency is due to not only cell turnover but also the entry of other lethally irradiated cells into the cell proliferation mode, and cell death occurs as an avalanche or cascade.

2.14.5 Radiation Recall Phenomenon

This is a severe reaction that is observed in particular in the skin of the irradiated area during chemotherapy after the end of external radiotherapy. Skin edema, erythema, rashes, and discoloration are seen. The mechanism that causes this is the interaction of basal layers of the irradiated skin with cytotoxic agents secreted from dead cells due to chemotherapy. In general, this effect is observed in patients receiving chemotherapy after a long period of radiation (weeks, months). The chemotherapeutic agents that most often cause this phenomenon are actinomycin, doxorubicin, methotrexate, fluorouracil, hydroxyurea, and paclitaxel.

The half-life of a free radical is longer (microseconds) than that of an ion radical. Free radicals can diffuse and so damage regions outside of the primary radiation pathway. They play an important role in the oxygen effect. They are the basis for the indirect effect of radiation.

Two hundred meV proton radiation and 1.1 MV gamma radiation are examples of low-LET radiation, like 250 kVp X-rays.

The components of cosmic rays are high-LET radiations, like Fe ions and carbon ions, while solar flares are composed largely of energetic protons (which are a low-LET radiation).

Telomeres shorten during each period of mitosis due to the low telomere activity in primary human fibroblasts and most other somatic cells.

- This shortening generally causes a loss of replicative potential and G_1 arrest, which is also known as replicative cell aging.
- Primary human fibroblast aging can occur due to various mechanisms, including exposure to ionizing radiation. Some somatic cells may

undergo additional divisions before a proliferative block, and this condition is termed a crisis. The crisis includes extensive genetic instability and ends in cell death in most situations.

The quadriradial-type aberrant chromosome number increases in Fanconi aplastic anemia. An increase in this type of chromosome is also observed upon exposure to radiation. A prominent increase is also seen upon the use of chemotherapeutic agents that create cross-linkages, like mitomycin.

SLD repair and the value of the β parameter in the cell survival curve decreases as the dose rate decreases.

If there is no shoulder in the cell survival curve, $n = 1$ and $D_{37} = D_0$.

In high-LET radiations, the extrapolation number (n) decreases until it reaches 1 as LET increases.

D_q and n are more predictive regarding the effect of fractionation on survival than D_0.

As n increases with dose rate, D_{10}, D_0, and α/β all decrease.

- D_{10} is the dose that decreases the SF to 10%.
- A dose rate of 1–0.1 Gy/min allows cellular repair.
- A dose rate of 0.1–0.01 Gy/min allows redistribution.
- A dose rate of 0.01–0.001 Gy/min allows repopulation.

In cases of ataxia telangiectasia, the α component of the cell survival curve is larger than it is in the normal population.

Inhibiting SLD selectively increases double-hit kill and the β component of the cell survival curve increases.

Skin infections occur due to microvessel damage during and after radiotherapy; infections in oral mucosa arise due to decreased saliva, loss of the antibacterial effect of saliva, and increases in oral acidity.

In cell culture, 90% of cells growing exponentially die after receiving a D_{10} dose. The remaining cells are cells in the S phase, which is the radioresistant phase of the cell cycle.

The differences in intrinsic radiosensitivity among several types of cancer cells are mainly due to variations in the α component. Variations in the β component do not affect radiosensitivity so much.

According to animal experiments, the most sensitive fetal phase to radiation is just after conception and before the implantation of the embryo into the uterus.

Irradiation in the early fetal period, corresponding to weeks 8–15 of human gestation, mostly increases the risk of mental retardation.

The main risks of irradiation during preimplantation, organogenesis, and the late fetal period are prenatal death, congenital malformation, growth retardation, and carcinogenesis, respectively.

The effects of irradiation during gestation (other than inducing cancer) are deterministic, not stochastic. Since there is a threshold dose for deterministic effects, the severity of such an effect depends on the dose. Since the developing embryo/fetus is particularly radiosensitive, the dose received by the fetus should be minimized. Unfortunately, 36% of the 40 million radiological tests performed per year were performed in fertile women until the 1980s.

The most radiosensitive phase of human gestation regarding neonatal deaths is 2–6 weeks, the phase of organogenesis. Some congenital malformations that can occur at this stage are not compatible with life.

The time required for SLD repair in most tissues ($T_{1/2}$) → 1.5 h.

2.14.6 Normalized Total Dose (NTD$_{2\,Gy}$)

This is used to convert a BED to an LQ equivalent dose in 2 Gy fractions. The NTD is very useful for deriving hypofractionated regimens in LQ equivalent doses (stereotactic radiotherapy, hypofractionation in prostate cancer and breast cancer, etc.). NTD$_{2\,Gy}$ is the total dose in 2 Gy fractions that would give the same log cell kill as the schedule being analyzed.

$$RE = 1 + \frac{d}{\alpha/\beta}$$

Here, RE = relative effectiveness, d = fraction dose, and α/β = alpha/beta ratio.
- Prostate tumor α/β = 1.5 Gy.
- RE for 2 Gy fractions: 1 + 2/1.5 = 2.33.
- Divide $BED_{1.5\,Gy}$ by this RE (2.33) to get $NTD_{2\,Gy}$.
- For example, 7.30 × 5 fractions = 36.5 Gy.
- $BED_{1.5\,Gy}$ = 36.5(1 + 7.3/1.5) = 214.13.
- $NTD_{2\,Gy}$ = 214.13/2.33 = 91.9 Gy.

Reference: Fowler JF (2005) The radiobiology of prostate cancer including new aspects of fractionated radiotherapy. Acta Oncolo 44:265–276.

2.14.7 Normal Tissue Toxicity After Stereotactic Body Radiation (SBRT)

SBRT doses per fraction for serial organs:
Spinal cord: 8–10 Gy/1 fraction, 5–6 Gy/3 fractions, 4–5 Gy/5 fractions
 Trachea and bronchi: 7–9 Gy/5 fractions
 Brachial plexus: 8–10 Gy/5 fractions
 Esophagus: 6–8 Gy/5 fractions
 Chest wall and ribs: 10–15 Gy/3 fractions, 6–8 Gy/5 fractions
 Small bowel: 10–12 Gy/1–2 fractions, 6–8 Gy/5 fractions

Note that therapeutic or close to therapeutic doses with 1–3 fractions are not recommended for trachea, bronchi, brachial plexus, and esophagus.

SBRT doses per fraction for parallel organs:
Lung: 20 Gy/1 fraction, 20 Gy/3 fractions, 8–10 Gy/5 fractions
Liver: 25 Gy/1 fraction, 20 Gy/3 fractions, 8–10 Gy/5 fractions

SBRT dose–volume limits for parallel organs:
Lung: 700–1000 mL of the lung not involved with gross disease, V_{20} of 25–30%
Liver: 700–1000 mL of the liver not involved with gross disease, two-thirds of normal liver <30 Gy
Kidney: minimize the region receiving >20 Gy, two-thirds of one kidney <15 Gy (assuming that there is another functional kidney).

Reference: Milano MT et al (2008) Normal tissue toxicity after small field hypofractionated stereotactic body irradiation.

Radiation-induced cataract given as a single acute exposure threshold for humans is around 200 rem (2 Gy). The cataract formation latent period range is between 6 months to many years. Radiation-induced cataract differs from senile, diabetic, or trauma-induced cataracts. The early phase of radiation-induced cataract progression is seen in the posterior subcapsular region as an opacified small ring surrounding a clear center. As the clear center opacification progresses, the radiation-induced cataract is indistinguishable from other cataract forms.

References

1. de Pouplana LR (ed.) (2005) The genetic code and the origin of life. Springer, pp 75–91
2. Pollard TD, Earnshaw WC (2007) Cell biology. Saunders, pp 20–47
3. Sobti RC, Obe G (eds.) (2002) Some aspects of chromosome structure and function. Springer, pp 112–115
4. Moeller SJ, Sheaff RJ (2006) G1phase: components, conundrums, context. In: Kaldis P (ed.). Cell cycle regulation. Springer, pp 1–29
5. Hartwell LH, Culotti J, Pringle JR et al (1974) Genetic control of the cell division cycle in yeast. Science 183:46
6. Harper JW, Adams PD (2001) Cyclin-dependent kinases. Chem Rev 101:2511
7. Zinkel SS, Korsmeyer SJ (2005) Apoptosis. In: DeVita VT, Hellman S, Rosenberg SA (eds) Cancer: principles & practice of oncology, 7th edn. Lippincott Williams & Wilkins, Philadelphia, pp 95–98
8. Jékely G (Ed.) (2007) Eukaryotic membranes and cytoskeleton. Springer, pp 35–40
9. Lenhard Rudolph K (2007) Telomere shortening induces cell intrinsic checkpoints and environmental alterations limiting adult stem cell function. In: Gutierrez LG, Ju Z. Telomeres and telomerase in ageing, disease, and cancer, part II. Springer, pp 161–80

10. Bignold L, Coghlan B, Jersmann H (2006) Cancer morphology, carcinogenesis and genetic instability: a background. In: Bignold LP (ed.) Cancer: cell structures, carcinogens and genomic instability. Springer, pp 1–25
11. Bodansky D (2007) Effects of radiation exposures. In: Bodansky D (ed.) Nuclear energy. Springer, pp 85–121
12. Kaul A, Becker D (eds.) (2005) Radiological protection. Springer, pp 5–40
13. Reith W (2016) Radiation biology and radiation protection. In: Vogl T, Reith W, Rummeny E (eds) Diagnostic and interventional radiology. Springer, Berlin, p 13
14. Lewanski CR, Gullick WJ (2001) Radiotherapy and cellular signaling. Lancet Oncol 2:366
15. Podgorsak EB (2005) Radiation oncology physics: a handbook for teachers and students. International Atomic Energy Agency, Vienna, p 486
16. Saw CB, Celi JC, Saiful HM (2006) Therapeutic radiation physics primer. Hematol Oncol Clin North Am 20(1):25–43. Review
17. Michael G (2008) Stabin: quantities and units in radiation protection. In: Stabin MG (ed.) Radiation protection and dosimetry. Springer, pp 67–74
18. Podgorsak EB (2005) Radiation oncology physics: a handbook for teachers and students. International Atomic Energy Agency, Vienna, p 556
19. Joseph M (2005) Galy Jean radioactivity, radionuclides, radiation. Springer, pp 117–123
20. Goitein M (2008) Radiation oncology: a physicist's— eye view. Springer, pp 5–6
21. Beck-Bornholdt HP (1993) Quantification of relative biological effectiveness, dose modificationfactor and therapeutic gain factor. Strahlenther Onkol 169(1):42–47
22. Joseph M (2005) Galy Jean radioactivity radionuclides radiation. Springer, pp 102–3
23. Katz R, Cucinotta FA (1999) Tracks to therapy. Radiat Meas 31(1–6):379–388. Review
24. Blackstock A, McMullen K (2005) Radiotherapy and chemotherapy. In: Jeremic B (ed.) Advances in radiation oncology in lung cancer. Springer, p 158
25. Hobbie RK, Roth BJ (2007) Intermediate physics for medicine and biology. Springer, p 463
26. Bond VP (1995) Dose, effect severity, and imparted energy in assessing biological effects. Stem Cells 13(Suppl 1):21–29. Review
27. Podgorsak EB (2005) Radiation oncology physics: a handbook for teachers and students. International Atomic Energy Agency, Vienna, p 492
28. Michael G (2008) Stabin: quantities and units in radiation protection. In: Stabin MG (ed.) Radiation protection and dosimetry. Springer, pp 100–2
29. Fowler JF (2006) Practical time-dose evaluations, or how to stop worrying and learn to love linear quadratics. In: Levitt SH, Purdy JA, Perez CA, Vijayakumar S (eds.) Technical basis of radiation therapy, 4th rev. edn. Springer, pp 444–46
30. Strandqvist M (1944) Studien uber die cumulative Wirkung der Rontgenstrahlen bei Fraktionierung. Erfahrungen aus dem Radiumhemmet an 280 Haut und Lippenkarzinomen. Acta Radiol 55(Suppl):1–300
31. Thames HD Jr (1988) Early fractionation methods and the origins of the NSD concept. Acta Oncol 27(2):89–103. Review
32. Ellis F (1969) Dose, time and fractionation: a clinical hypothesis. Clin Radiol 20:1–7
33. Goitein M (2008) Radiation oncology: a physicist's— eye view. Springer, pp 3–4
34. Podgorsak EB (2005) Radiation oncology physics: a handbook for teachers and students. International Atomic Energy Agency, Vienna, pp 485–491
35. Dan Garwood L, Cho C, Choy H (2006) Clinical principles and applications of chemoirradiation. In: Levitt SH, Purdy JA, Perez CA, Vijayakumar S (eds.) Technical basis of radiation therapy, 4th rev. edn. Springer, pp 40–41
36. Little JB, Hahn GM, Frindel E, Tubiana M (1973) Repair of potentially lethal radiation damage in vitro and in vivo. Radiology 106:689
37. The effect of oxygen on impairment of the proliferative capacity of human cells in culture by ionizing radiations of different LET. Int J Radiat Biol Relat Stud Phys Chem Med 1966;10(4):317–2
38. Grdina DJ, Murley JS, Kataoka Y (2002) Radioprotectants: current status and new directions. Oncology 63(Suppl 2):2–10
39. Thomas CT, Ammar A, Farrell JJ, Elsaleh H (2006) Radiation modifiers: treatment overview and future investigations. Hematol Oncol Clin North Am 20(1):119–139
40. Bergonie J, Tribondeau L (1906) Interprétation de quelques résultats de la radiothérapie et essaide fixation d'une technique rationelle. C R Acad Sci 143:983–985
41. Michalowski AS (1992) Post-irradiation modification of normal-tissue injury: lessons from the clinic. BJR Suppl 24:183–186. Review
42. Ancel P, Vintemberger P (1928) C R Soc Biol 99, 832
43. Rubin P, Casarett GW (1968) Clinical radiation pathology as applied to curative radiotherapy. Cancer 22(4):767–778
44. Emami B, Lyman J, Brown A et al (1991) Tolerance of normal tissue to therapeutic irradiation. Int J Radiat Oncol Biol Phys 21(1):109–122
45. Withers HR, Taylor JM, Maciejewski B (1988) Treatment volume and tissue tolerance. Int J Radiat Oncol Biol Phys 14(4):751–759
46. Awwad HK (2005) Normal tissue radiosensitivity: prediction on deterministic or stochastic basis? J Egypt Natl Canc Inst 17(4):221–230. Review
47. Willers H, Held KD (2006) Introduction to clinical radiation biology. Hematol Oncol Clin North Am 20(1):1–24. Review
48. Baumann M, Petersen C, Krause M (2005) TCP and NTCP in preclinical and clinical research in Europe. Rays 30(2):121–126. Review

References

49. Baumann M, Petersen C (2005) TCP and NTCP: a basic introduction. Rays 30(2):99–104. Review
50. Lyman JT (1992) Normal tissue complication probabilities: variable dose per fraction. Int J Radiat Oncol Biol Phys 22(2):247–250
51. Niemierko A (1997) Reporting and analyzing dose distributions: a concept of equivalent uniform dose. Med Phys 24(1):103–110
52. Coutard H (1937) The result and methods of treatment of cancer by radiation. Ann Surg 106(4):584–598
53. Tubiana M, Dutreix J, Wambersie A (1990) Introduction to radiobiology. Taylor & Francis, London, pp 119-123-131-135
54. Baumann M, Dörr W, Petersen C et al (2003) Repopulation during fractionated radiotherapy: much has been learned, even more is open. Int J Radiat Biol 79(7):465–467
55. Baumann M, Liertz C, Baisch H et al (1994) Impact of overall treatment time of fractionated irradiation on local control of human FaDu squamous cell carcinoma in nude mice. Radiother Oncol 32(2):137–143
56. Willers H, Dahm-Daphi J, Powell SN (2004) Repair of radiation damage to DNA. Br J Cancer 90(7):1297–1301
57. Trott KR (1982) Experimental results and clinical implications of the four R's in fractionated radiotherapy. Radiat Environ Biophys 20(3):159–170. Review
58. Popple RA, Ove R, Shen S (2002) Tumor control probability for selective boosting of hypoxic subvolumes, including the effect of reoxygenation. Int J Radiat Oncol Biol Phys 54:921–927
59. Podgorsak EB (2005) Radiation oncology physics: a handbook for teachers and students. International Atomic Energy Agency, Vienna, pp 499–505
60. Lee CK (2006) Evolving role of radiation therapy for hematologic malignancies. Hematol Oncol Clin North Am 20(2):471–503. Review
61. Thames HD, Ang KK (1998) Altered fractionation: radiobiological principles, clinical results, and potential for dose escalation. Cancer Treat Res 93:101–128
62. ICRP (2006) Assessing dose of the representative person for the purpose of radiation protection of the public. ICRP publication 101. Approved by the Commission in September 2005. Ann ICRP. 36(3):vii–viii, 5–62
63. Prasad KN, Cole WC, Haase GM (2004) Radiation protection in humans: extending the concept of as low as reasonably achievable (ALARA) from dose to biological damage. Br J Radiol 77(914):97–99. Review
64. Song CW, Kim MS, Cho LC, Dusenbery K, Sperduto PW (2014) Radiobiological basis of SBRT and SRS. Int J Clin Oncol 19(4):570–578. https://doi.org/10.1007/s10147-014-0717-z. Epub 2014 Jul 5
65. Mole RH (1953) Whole body irradiation; radiobiology or medicine? Br J Radiol 26(305):234–241
66. Nobler MP (1969) The abscopal effect in malignant lymphoma and its relationship to lymphocyte circulation. Radiology 93:410–412
67. Demaria S, Ng B, Devitt ML et al (2004) Ionizing radiation inhibition of distant untreated tumors (abscopal effect) is immune mediated. Int J Radiat Oncol Biol Phys 58:862–870
68. Kepp O, Senovilla L, Vitale I et al (2014) Consensus guidelines for the detection of immunogenic cell death. Onco Targets Ther 3:e955691
69. Marin A, Martin M, Linan O et al (2015) Bystander effects and radiotherapy. Rep Pract Oncol Radiother 20(1):12–21

Clinical Radiation Oncology

3.1 Introduction and History

Radiation was first applied in a clinical setting by an American medical student, Emil Grubbe, on 29th January 1896. Grubbe, the son of German immigrants, introduced the term "radiotherapy" in 1903: "He had been interested in comparing the effects of phototherapy and radiotherapy" [1]. Since then, the field of radiation oncology has developed rapidly in parallel with technology, and has become indispensable in the treatment of cancers.

Initially, naturally radioactive radium was applied in every situation where medicine was deficient; radium extracts were assigned a trademark and sold in markets (Fig. 3.1). However, the serious side effects of their use were understood, and so the application of radium was regulated by strict rules and laws. Finally, the fields of the diagnostic use and the therapeutic use of radiation were separated, and that of radiation oncology, a branch of internal medicine that uses ionizing radiation of various types and energies for the treatment of cancer and some benign diseases was established.

Radiation oncology made use of several treatment modalities and machines in its infancy. All of these new treatment modalities blazed a trail in the medical community (Fig. 3.2); this process of development continues even today in the field of radiation oncology.

Approximately 50–60% of all cases of cancer require radiotherapy at some stage during their treatment. The radiation oncologist decides whether radiation therapy is indicated. However, it is best to take a multidisciplinary approach (surgery, medical oncology, nuclear medicine, radiology) when deciding on the final treatment in clinical practice.

3.1.1 Radiotherapy Types According to Aim

Curative radiotherapy. This is the application of radiotherapy alone to cure. Used in cases of early-stage Hodgkin's lymphoma, nasopharyngeal cancer, some skin cancers, and early glottic cancers (curative radiotherapy = definitive radiotherapy) for example.

Palliative radiotherapy. This is the alleviation of cancer symptoms by applying palliative doses of radiation. Used in cases of brain and bone metastases and superior vena cava syndrome for example.

Palliative radiotherapy with curative doses: It is defined as the administration of high doses of radiotherapy in cases where other treatment modalities cannot be applied for different reasons. Used in cases of inoperable lung/laryngeal cancer and brain tumors for example.

Prophylactic (preventive) radiotherapy. This is the prevention of possible metastases or recurrences through the application of radiotherapy. An example is whole-brain radiotherapy for

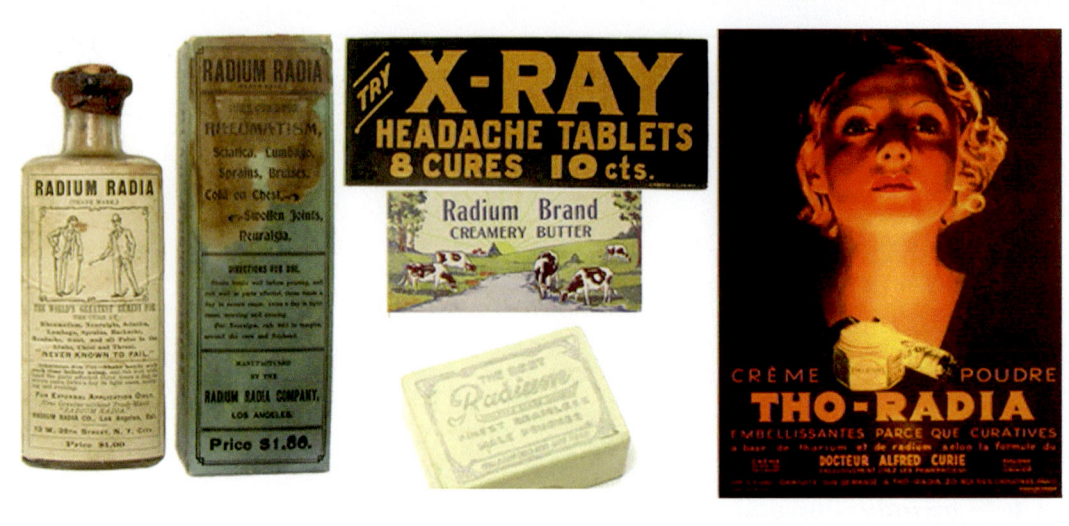

Fig. 3.1 Various industrial compounds that included radium from the beginning of the twentieth century

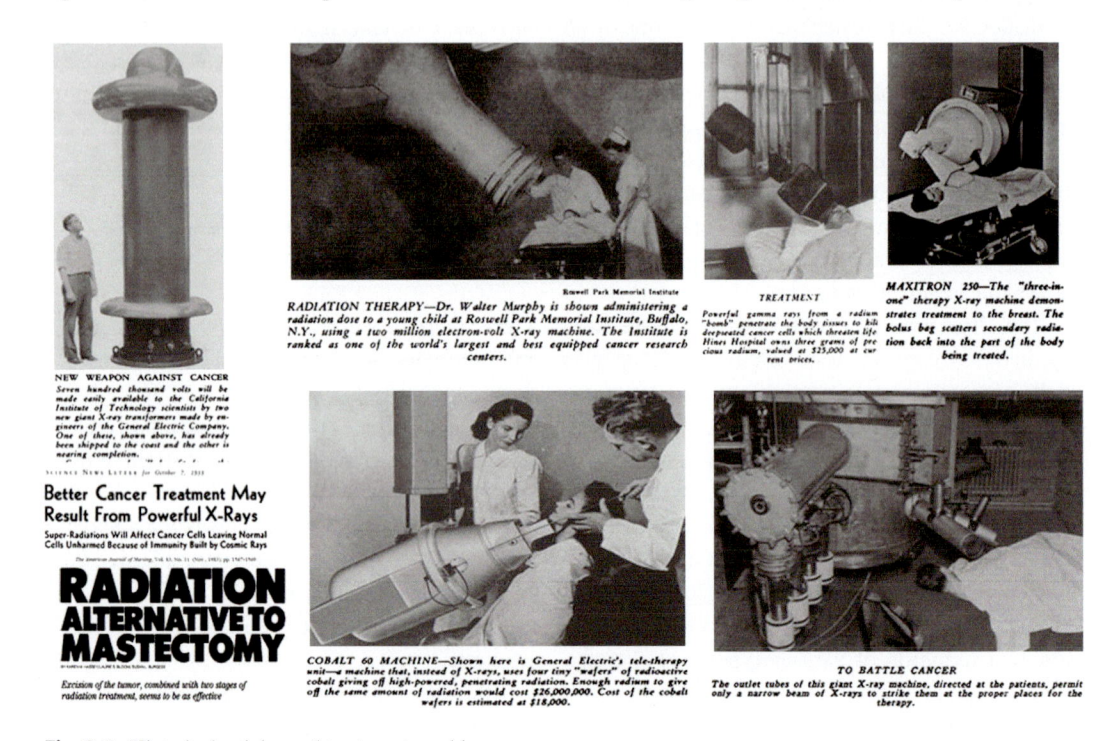

Fig. 3.2 Historical articles and treatment machines

acute lymphoblastic leukemia and small cell lung cancer.

Total body irradiation. This is the ablation of bone marrow by radiation in order to suppress the immune system, eradicate leukemic cells, and clear space for transplant cells during bone marrow transplantation conditioning.

3.1.2 Radiotherapy Types According to Timing

Adjuvant radiotherapy. Radiotherapy given after any kind of treatment modality.

- If given after surgery → *postoperative radiotherapy*

Neoadjuvant radiotherapy. Radiotherapy given before any kind of treatment modality.

- If given before surgery → *preoperative radiotherapy*

Radiochemotherapy (chemoradiotherapy). Radiotherapy given concurrently with chemotherapy.

3.1.3 Radiotherapy Types According to Mode

External radiotherapy (teletherapy/external beam radiotherapy). Radiotherapy applied to the body externally using a treatment machine.

Brachytherapy (endocurietherapy/sealed-source radiotherapy). Radiotherapy performed by placing temporary or permanent radiation sources into body cavities.

Intraoperative radiotherapy (IORT). Radiotherapy given under intraoperative conditions, usually by electron beams or low-energy X-rays. It is delivered to the tumor bed just after the resection of the primary tumor, and external radiotherapy is generally required afterward.

Stereotactic radiotherapy (SRT). Radiotherapy delivered by several beams that are precisely focused on a three-dimensionally localized target. A special frame or a thermoplastic mask is used for CNS tumors, while a body frame may or may not be used for extracranial sites.

- If given in one-five fractions in ablative doses → *stereotactic radiosurgery (SRS)*

Stereotactic body radiotherapy (SBRT): It is fractionated radiotherapy (in a single or limited number of fractions) for well-defined extacranial malignant/benign lesions.

Three-dimensional conformal RT (3D-CRT). A radiotherapy technique where the dose volume is made to conform closely to the target through the use of 3D anatomical data acquired from CT or MRI imaging modalities. The aim is to apply the maximum dose to the target while sparing neighboring structures as much as possible with the aid of advanced computer software and hardware.

Intensity-modulated radiotherapy (IMRT). A highly developed form of 3D-CRT. IMRT provides a highly conformal dose distribution around the target through the use of nonuniform beam intensities. This is achieved through using either static or dynamic segments. The isodose distribution can then be matched closely to the target by modulating the intensity of each subsegment.

Image-guided radiotherapy (IGRT). The integration of various radiological and functional imaging techniques in order to perform high-precision radiotherapy. The main aims are to reduce setup and internal margins, and to account for target volume changes during radiation therapy, such as a tumor volume decrease or weight loss.

Photodynamic therapy. It is a relatively tissue-specific treatment and hematoporphyrin derivatives are used. Mechanism of action is that hematoporphyrin derivatives persist for several days in tumoral tissues, although they are excreted a few hours after ingestion into normal tissues. Hematoporphyrin derivatives create cytotoxic effects when exposed to light of certain wavelengths.

Tomotherapy. It is the IMRT type in which the irradiation head rotates 360° around the treatment volume, similar to the CT. Includes 64 MLCs and treatment verification is easier due to daily CT scans.

Cyberknife® (robotic radiosurgery). A type of SRT/radiosurgery technique. It provides frameless treatment of tumors at both cranial and extracranial sites, and utilizes a 6 MV linac mounted on a robotic arm.

Boron neutron capture therapy. Here, a boron compound that is selectively absorbed by brain tumor cells is given to the patient. The tumor tissues that absorb the boron are then irradiated with slow neutrons. The boron atoms react with these neutrons to generate alpha radiation, which damages DNA via ionization events.

Hyperthermia This prevents tumoral repair by utilizing a supraadditive (synergistic) effect with radiation: tumor tissues get colder more slowly than normal tissues. Hyperthermia is more effective under hypoxic and acidic conditions. The critical temperature for hyperthermia is 43 °C.

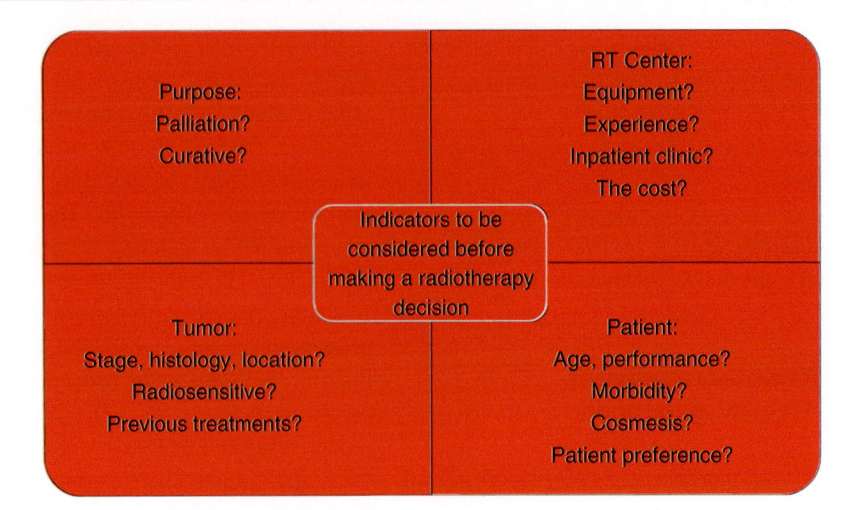

[Figure: Indicators to be considered before making a radiotherapy decision — Purpose: Palliation? Curative?; RT Center: Equipment? Experience? Inpatient clinic? The cost?; Tumor: Stage, histology, location? Radiosensitive? Previous treatments?; Patient: Age, performance? Morbidity? Cosmesis? Patient preference?]

Patients should be informed about the treatment (procedures, duration, acute and late side effects, cost, and alternative modalities) after radiotherapy has been decided upon, and informed consent should be obtained for legal purposes.

3.2 The Radiotherapy Procedure

Radiotherapy is performed by a group of team members, including the radiation oncologist, radiation physicist, radiotherapist, dosimetrist, nurse, psychologist, and/or social workers. The treatment success in radiotherapy is highly dependent on adequate technical equipment.

After the decision to use radiotherapy has been taken in a multidisciplinary meeting, the radiation oncologist needs to evaluate reports related to the patient's history, physical exams, and particularly pathology reports, radiological imaging studies, as well as nuclear imaging reports, and add such details to the patient's chart. Adding a short summary to the end of chart may prove useful after registry. If there are any prominent and apparent physical signs, they should be photographed and added to the chart, since they may be useful for evaluating the success or failure of treatment. With the patient's consent, such pictures can also be used for academic studies.

Procedures and data that are not recorded on the chart are accepted as non-existent.

The patient should be informed about his disease and treatment according to his education level just after clinical examination. The steps involved in the radiotherapy procedure and their duration, as well as acute and late effects, their starting times, and preventive measures for them should also be told to the patient. The consent form should then be signed by the patient and physician, and one copy should be given to the patient while the other should be added to the patient's chart. After that, the first step in radiotherapy—simulation—is scheduled (immobilization, imaging, and tumor localization) (see Fig. 3.3).

Basic steps of radiotherapy procedure may be shown in a chain. Any failure in the rings of this chain leads to treatment failure.

3.2.1 Simulation

Simulation is radiotherapy field determination using a diagnostic X-ray machine with similar physical and geometrical features to the actual teletherapy machine. The patient is immobilized before simulation and then the tumor is localized either in a direct scopy X-ray machine or in serial CT slices. The simulation can be performed by CT, MRI, or rarely by PET–CT (Fig. 3.4).

The simulation performed by a conventional simulator is a real-time simulation procedure, since it is done directly in the patient (Figs. 3.5

3.2 The Radiotherapy Procedure

Fig. 3.3 The steps involved in radiotherapy procedure

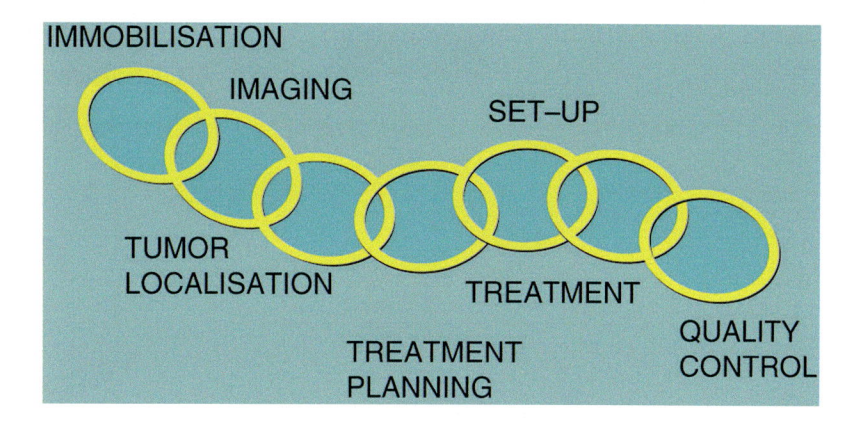

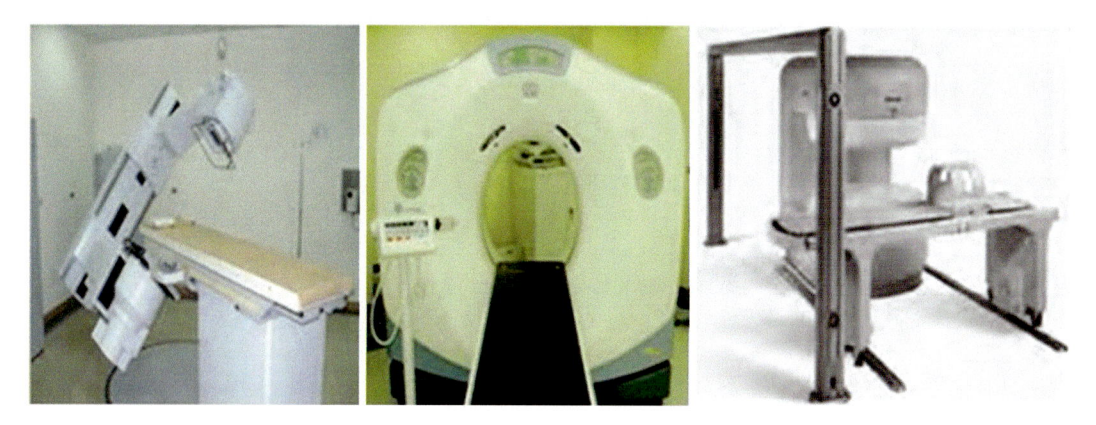

Fig. 3.4 (a–c) Various simulator types: conventional simulator, CT simulator, and MRI simulator (a, b: courtesy of Gulhane Medical Faculty, Ankara)

and 3.6). However, the simulation performed by a CT or MRI is a virtual simulation since the tumor is localized digitally.

3.2.1.1 Conventional Simulation Steps

Immobilization. The patient should be immobile during therapy. Movements cause changes in the treatment area and increase side effects, thus affecting treatment success. The patient should be positioned in the most comfortable, easily reproducible way that is suitable for the irradiated region of interest.

- Various types of apparatus are used for immobilization. The most frequently used apparatus is the thermoplastic mask (Fig. 3.7). Such a mask should not only be tight; there should be no space between the patient's skin and the mask. The mask should be checked during every setup procedure for tightness or looseness (due to edema or weight loss), and should be remade if necessary.

Patient positioning. The patient's treatment position should be recorded on both the patient's chart and the simulation film (e.g., supine, prone, hands up, hands on side) (Fig. 3.8). The gantry is adjusted according to the actual treatment machine's source–axis distance (SAD) before simulation. After the patient has been positioned on the couch, the desired SSD value (usually 80–100 cm) is achieved by adjusting the couch height. The SSD is reduced by half the thickness of the patient. Finally, the irradiation field is determined according to the chosen technique (fixed SSD or fixed SAD).

Imaging and tumor localization. The patient is placed on the simulator couch in the required position. The mask, base plate, T-arm, breast board, knee support, or any other similar immo-

Fig. 3.5 A conventional simulator and its movement abilities [Levitt et al. 8, p 108, Fig. 6.1]

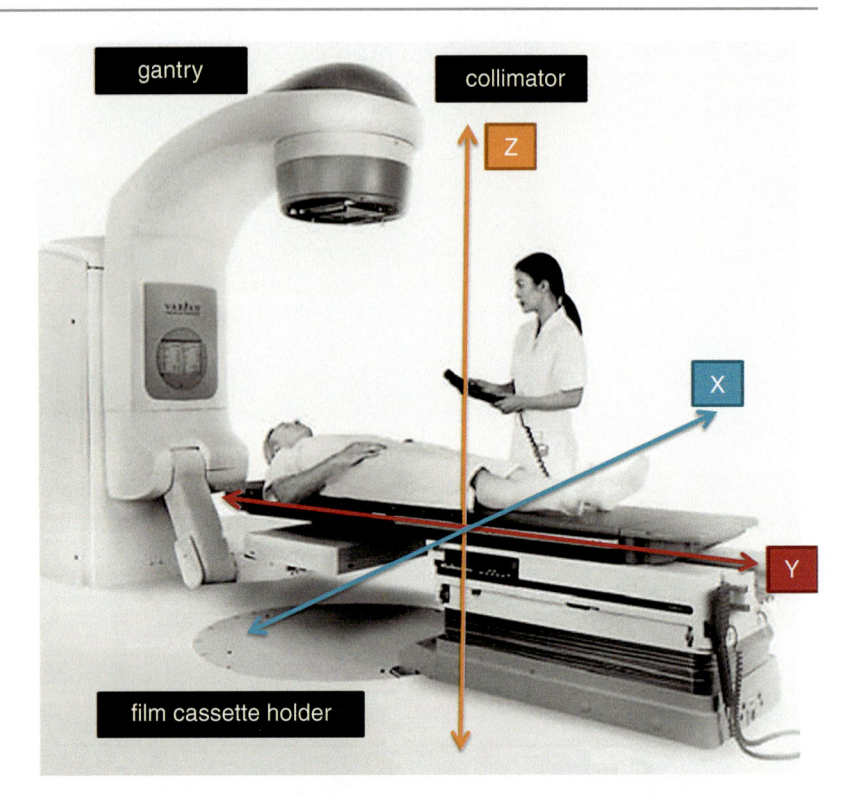

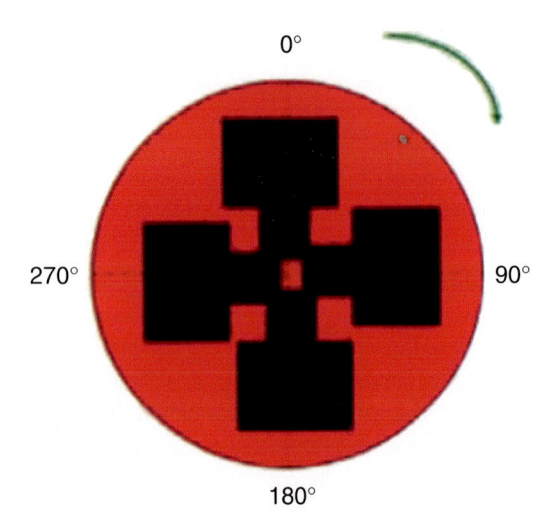

Fig. 3.6 Gantry angles and positions

bilization device is positioned accurately. The probable radiotherapy fields are determined and SSDs are calculated according to patient thickness. Gantry angles, field sizes, and collimator angles are arranged by the simulator software under intermittent X-ray scopy (Fig. 3.9).

Any scar, palpable mass, and drain site is marked with flexible wires; lead markers should be placed in lateral epicanthal regions for head and neck irradiation. These markings are easily seen during the simulation procedure.

Oral or IV contrast material can be used if required. Therefore, the contrast material should be prescribed before simulation; first aid equipment and drugs (e.g., adrenaline, atropine) should be available in the simulator room.

- An oral solution containing barium is generally used in gastrointestinal irradiation
- IV nonionic radiopaque materials are used for the imaging of the kidneys and bladder
- Radiopaque rectal and vaginal applicators or tamponades with contrast material can be used in the pelvic region

Treatment fields and block regions are marked on the patient's mask or skin after determining the radiotherapy fields. Simulation radiographs are then taken, and protected areas are marked

3.2 The Radiotherapy Procedure

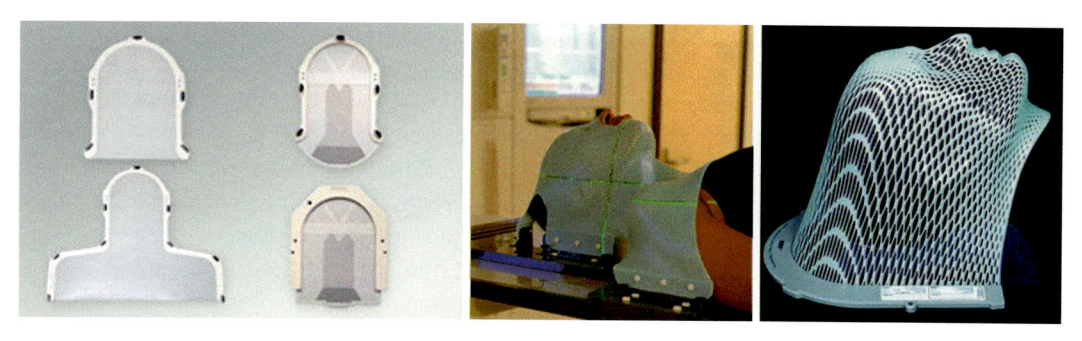

Fig. 3.7 Thermoplastic mask

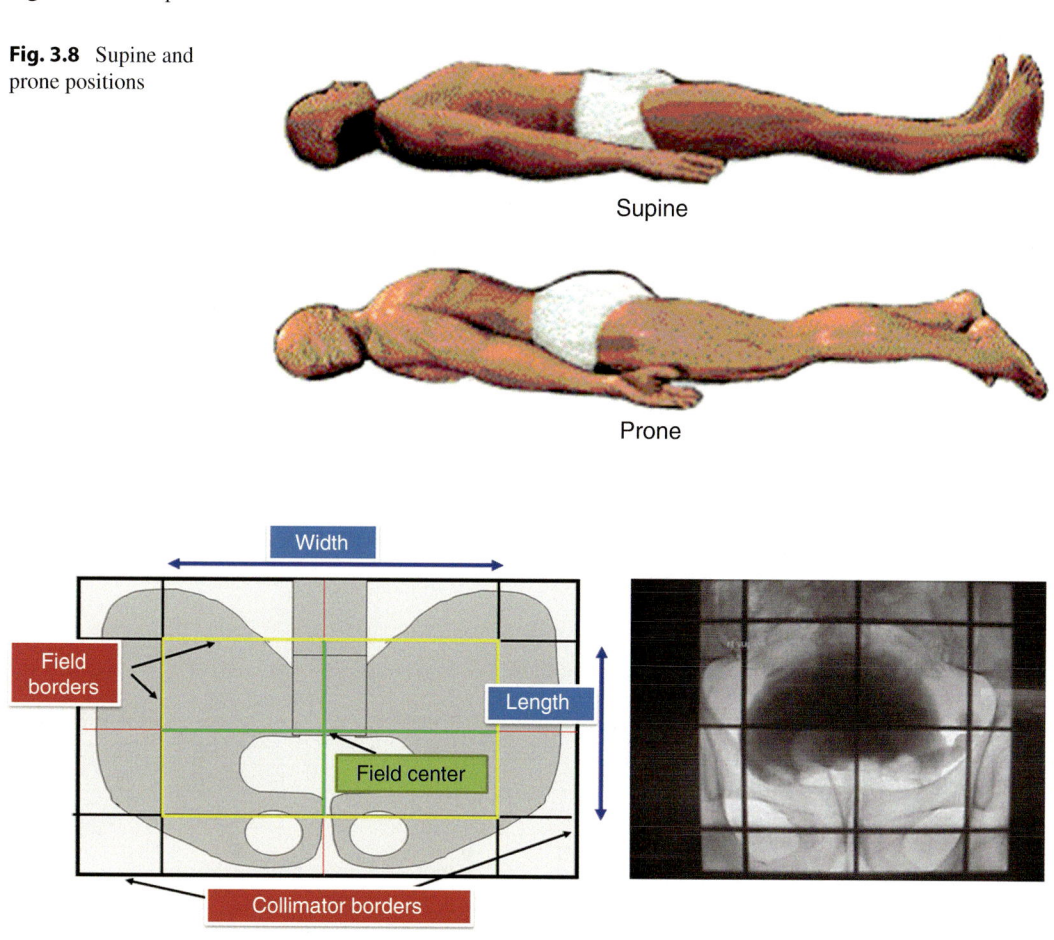

Fig. 3.8 Supine and prone positions

Fig. 3.9 Treatment field in a conventional simulator, and a simulation film

on this film, which is sent to the block-cutting room. Focalized blocks are then fabricated. The patient is sent home at this point since block fabrication is a lengthy procedure. If no block is used, or if standard blocks are used, the patient's chart is sent to the physics room for dose calculations, and the patient can be treated on the same day.

If there is no block-cutting unit, protected areas are determined by wires, and these wires are verified in the simulation film. These areas are spared by standard blocks.

- If focalized blocks are used, the blocks can be checked on X-ray scopy by special trays mounted on the simulator's gantry. Minor errors can be corrected, but the blocks should be refabricated when there are major errors.
- Breast contours are observed by dosimetrists or radiation physicists in conventional breast simulations and then used for treatment planning, wedge and dose calculations.
- The daily dose fraction and the total dose are determined, and the patient's chart is sent to the medical physics room for dose calculations after all of these procedures. Schedules are arranged for the treatment machine.

3.2.1.2 Parameters that Should Be Written on the Simulation Film

Patient name
   Simulation date
   Field size
   Gantry angle
   Collimator angle
   SSD
   SAD
   Depth
   Magnification factor
   Physician name
   Technician name

Patient position
Right and left signs
Patient data and all other data associated with the disease, as well as all parameters related to the simulation, should be recorded on the treatment chart. The treatment fields should also be presented graphically along with the date, since these data may be needed for possible future irradiation.

3.2.1.3 CT Simulation (Fig. 3.10)

The mask and other required equipment are made on the day of CT simulation by the radiotherapist, under the supervision of the radiation oncologist, for the patient who is to receive conformal radiotherapy (Fig. 3.11).

The patient is sent to the nurse for an IV route before CT simulation if an IV contrast material is to be used. Then the patient is positioned on the CT couch, and the mask, knee support, alpha cradle, or any other similar device is fitted on the CT couch if required.

The lasers are turned on and they are positioned at the midline according to the region of interest.

Reference points are determined by radiopaque markers located at the cross-sections of the lasers (Fig. 3.12).

Fig. 3.10 CT simulation and 3D-conformal RT steps

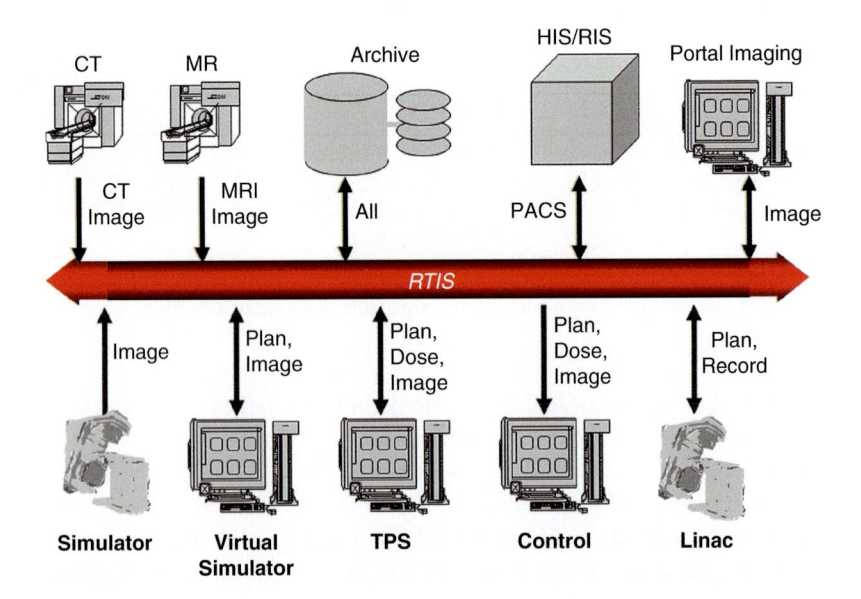

3.2 The Radiotherapy Procedure

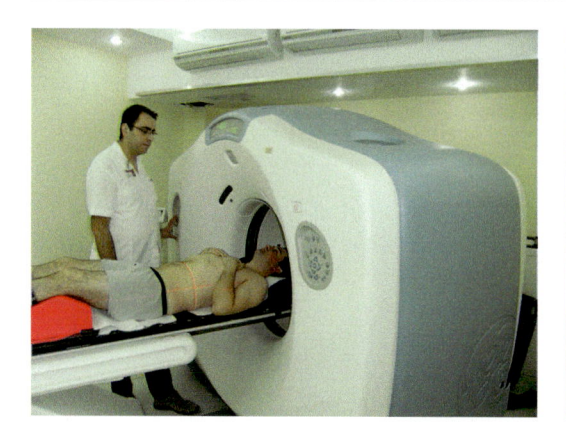

Fig. 3.11 Patient on a CT simulator couch

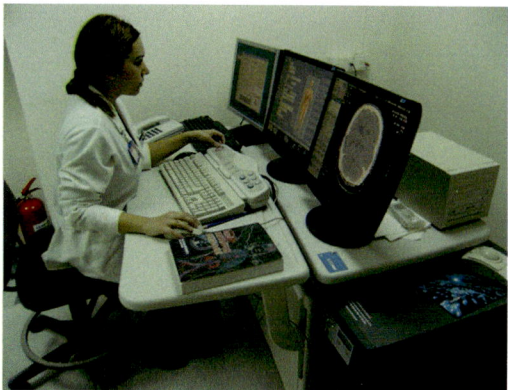

Fig. 3.13 CT command room

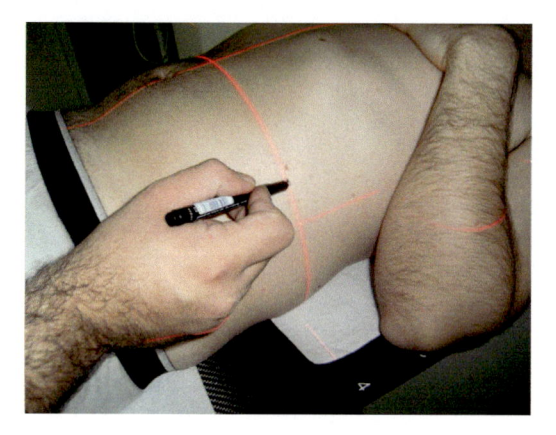

Fig. 3.12 Determination of reference points

Reference points are predetermined locations for each region of the body. There are three reference points: one is craniocaudal and the others are on the right and left lateral sides.

Contrast material is given intravenously by the nurse, if required. Adequate measures should be taken for any possible anaphylactic reactions.

Any required adjustments are performed by the CT technician in the CT command room (Fig. 3.13). The region of interest (that for which serial CT slices are to be taken) is determined by the radiation oncologist. The slice thickness is also determined. All of these data are transferred to the CT computer. After the region of interest has been verified on screen, serial slices are taken.

These slices are sent online to the treatment planning room via the network (a radiotherapy information system utilizing PACS) (Fig. 3.14).

The patient should then rest for 20 min after CT to check for any possible adverse reactions. After this, they can be sent home.

- If no mask is used, particularly in the case of body simulations, reference points should be permanently tattooed on, otherwise they will easily disappear.

A virtual digital simulation is performed after the CT slices have been transferred to the network. The acquired CT slices should be transferred to the treatment planning computer from the network for treatment planning. The radiation oncologist then performs contouring on the acquired CT slices (Fig. 3.15). External body contours are usually delineated automatically. The radiation oncologist then contours the GTV, CTV, PTV, and the organs at risk.

The slice with the reference points determined during CT simulation is recorded as the reference slice.

Contours are carefully checked and recorded, and the planning phase then begins.

Radiotherapy portals are determined during the first stage of planning according to the PTV (Fig. 3.16). Beams are placed according to the PTV using the beam's eye view (Fig. 3.17).

Delineation of the multileaf collimator (MLC) or the blocks is performed, and critical structures are spared. The penumbral region should be taken into account during MLC or block placing.

Data on the energy, fraction number and dose, and the treatment machine are entered into the

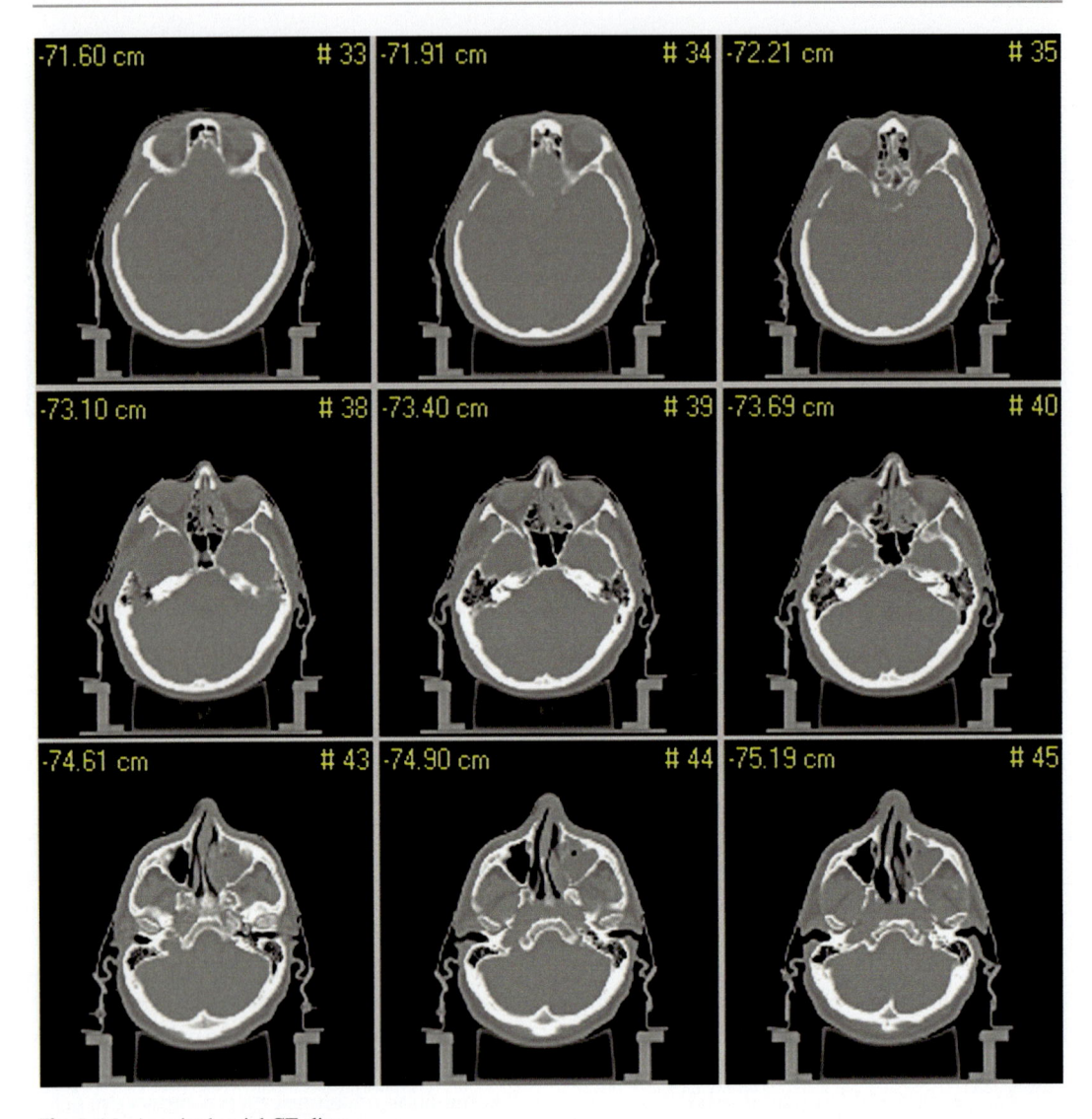

Fig. 3.14 Acquired serial CT slices

planning computer. The treatment planning system (TPS) starts its calculations, and the final dose distribution and dose–volume histograms are formed.

The doses for the target and the organs at risk are evaluated with the DVH, and the isodoses are carefully checked in each slice. The reference isodose or isocenter is determined for the dose prescription, and the final treatment plan should be verified by the radiation oncologist.

The final verified treatment plan is sent online to the treatment machine via the network, treatment parameters are recorded on the patient's chart, and isodose curves and DVH printouts are also attached to the chart (Fig. 3.18).

3.2.2 Treatment Planning

3.2.2.1 Conventional Planning

The energy is selected according to tumor depth and the surrounding normal tissues after determining the radiotherapy fields through simulation. High energies are selected for deeply seated tumors, while lower energies or electron beams are selected for superficially located tumors.

3.2 The Radiotherapy Procedure

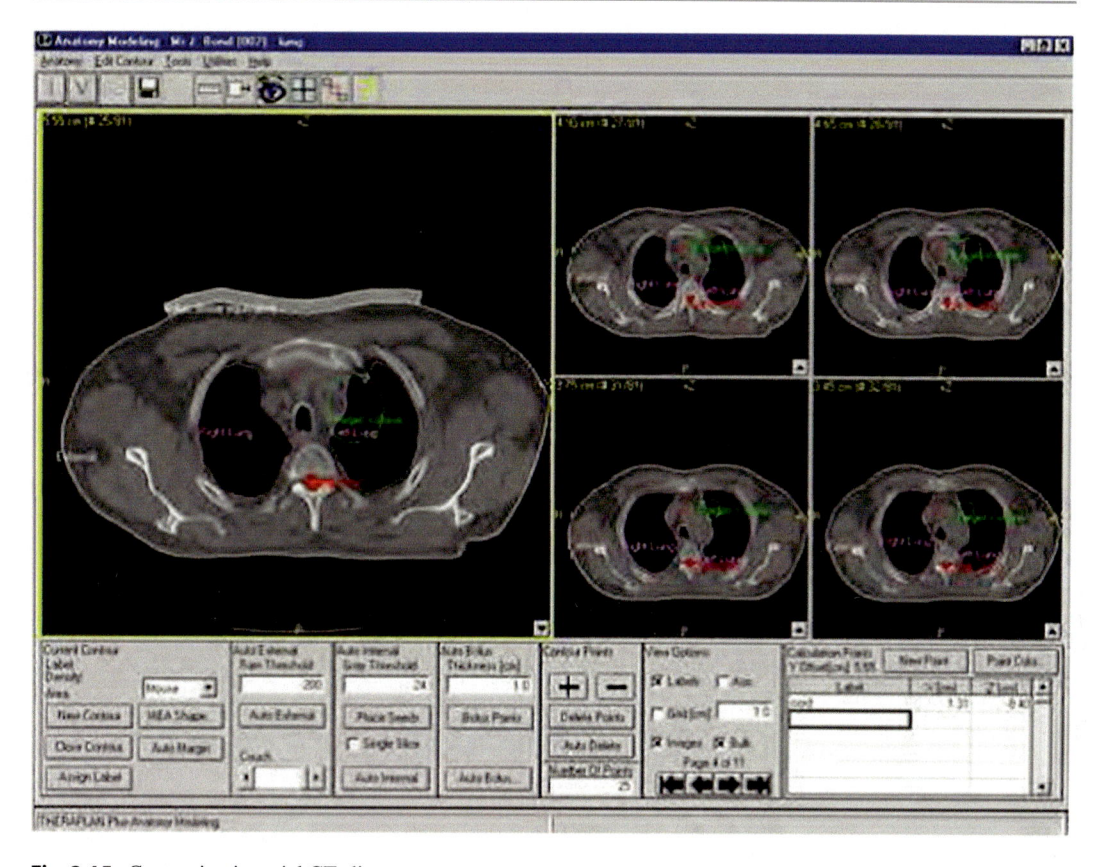

Fig. 3.15 Contouring in serial CT slices

Therapies Utilizing Megavoltage Energy

Energy is selected according to target depth.

- High-energy X-rays are preferred for thick regions like the thorax and abdomen.
- Co-60 can be used for thin regions like the head and neck, or extremities.
- Target volume, organs at risk, and depth are the main factors in energy selection.

Therapies Utilizing Electron Energy

These are preferred for superficial tumors. The target volume should be the 80–95% isodose region of the selected energy, depending on the distance from the skin and underlying tissues.

- Deeply seated normal tissues are spared in electron therapies.
- The depth of the 80% isodose corresponds to one-third of the electron energy (E/3).
- Bolus can be used to increase the superficial dose if required.

3.2.2.2 Conformal Planning

The basic idea here is quite different from that of conventional planning. The treatment volume is determined virtually after contouring the target volume and the organs at risk in serial CT slices. Blocks or the MLC and isodoses can be seen and selected digitally in the TPS.

3.2.3 Target Volume Definitions

Successful radiotherapy treatment depends on determining the optimal technique for the target volume and surrounding normal tissues as well as the accurate application of this technique. Therefore, cells that form tumors are destroyed

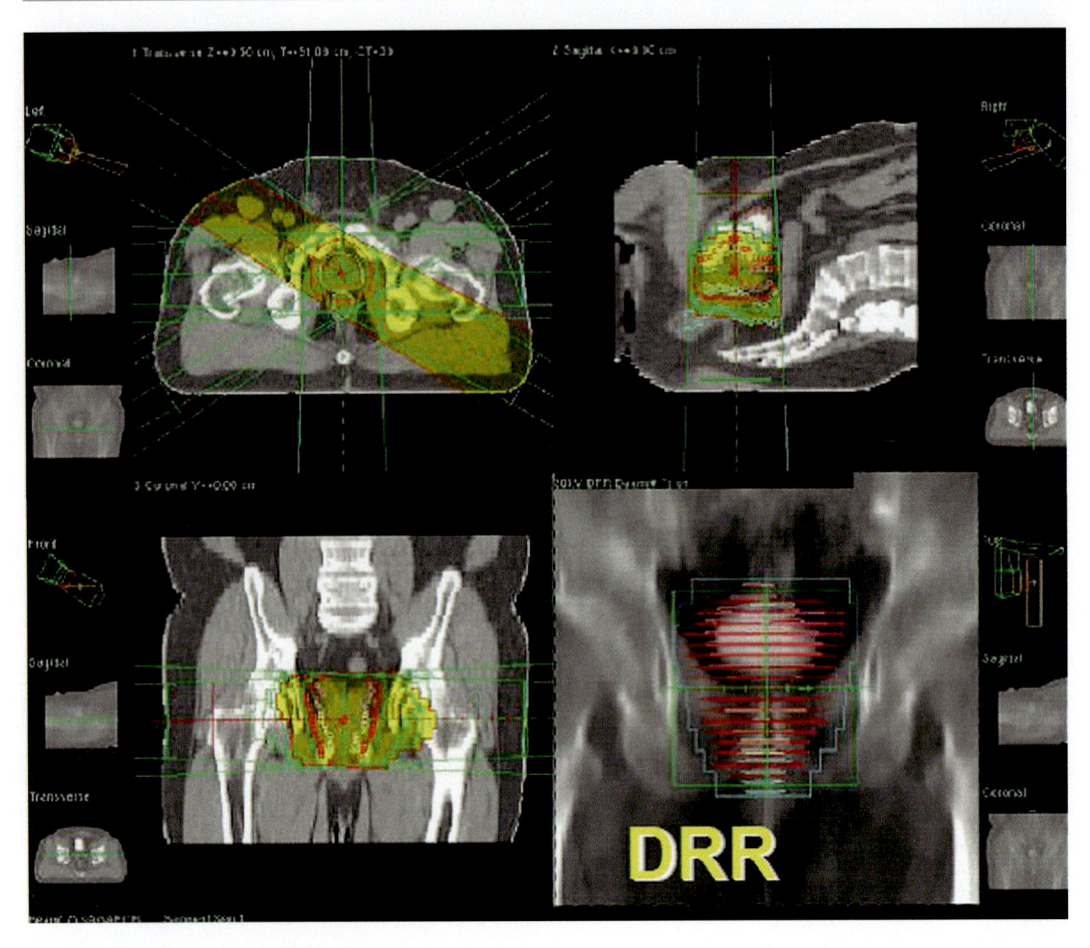

Fig. 3.16 Anterior radiotherapy field of a prostate cancer case

while causing minimal damage to the surrounding normal critical structures [2, 3].

The use of a common terminology is very important, since it allows procedures to be understood as well as comparisons to be made between therapeutic results after planning and between records at different institutions. The International Commission on Radiation Units and Measurements (ICRU) has published many reports that are used to determine treatment parameters and define target volumes so that radiotherapy can be accurately planned. These reports include ICRU 50 and 62 on photon energies of external treatments, ICRU 71 on electron energies, and ICRU 38 (1985), ICRU 58 (1958), and ICRU 72 (2004) on brachytherapy treatment. The ICRU also published ICRU 78 on proton therapy in 2007.

The ICRU 50 and 62 reports define the target volumes and organs at risk. Accurate volume definition plays a very important role in radiotherapy [4, 5].

3.2.3.1 Volume Definitions According to ICRU 50 (Fig. 3.19) [4]

The *gross tumor volume (GTV)* is the macroscopic volume of the tumor. The GTV defines the tumor volume determined by clinical exam and imaging modalities (visible, palpable).

- Visible or palpable tumor volume, clinical volume

The *clinical target volume (CTV)* encompasses the possible regions into which the microscopic disease may extend, or regions with a high

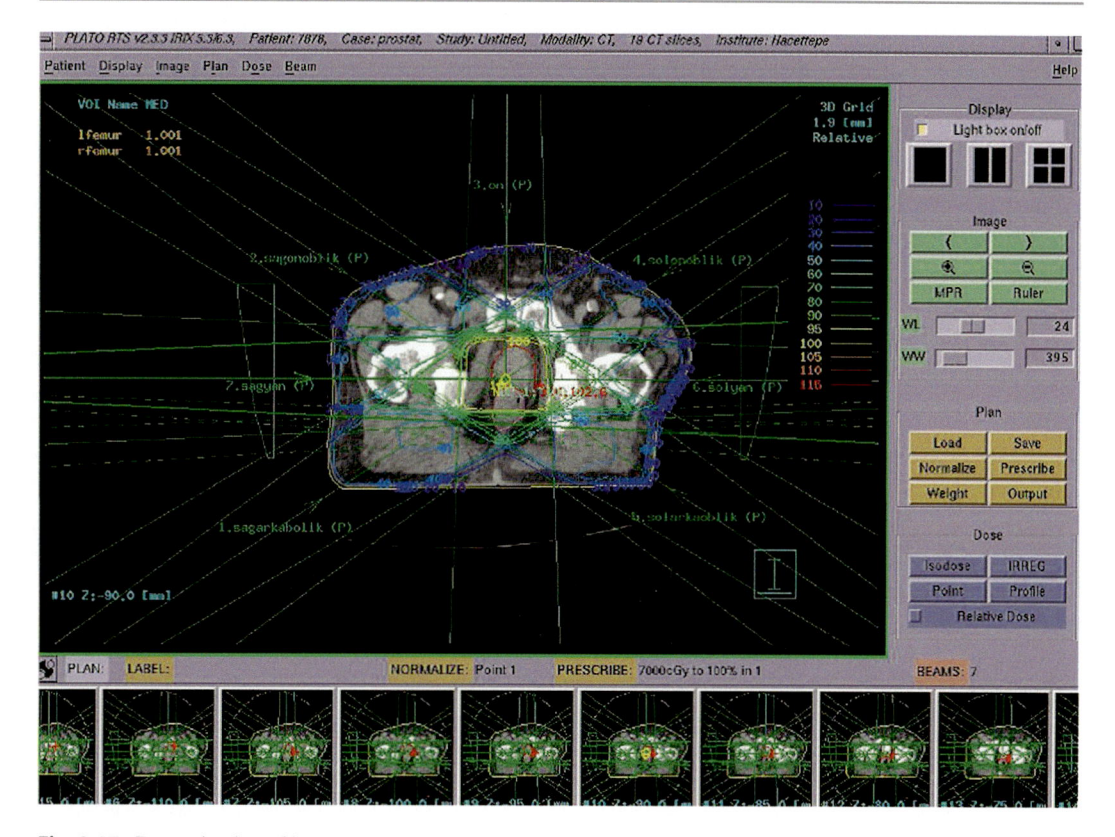

Fig. 3.17 Determination of beams

risk of involvement based on clinical experience (invisible tumor).

- Subclinical volume and clinical volume

The planning target volume (PTV) defines the volume formed when the CTV is extended due to physiological organ movements or technical reasons.

- Physiological movements like respiration, bladder, and rectum fullness
- Technical reasons like patient movement, mask movement, or couch movement

The treatment volume (TV) is the volume, including the reference isodose, that has the minimum probability of incurring complications.

The irradiated volume (IV) is the volume that receives a significant dose, based on normal tissue tolerance doses.

3.2.3.2 Volume Definitions According to ICRU 62 (Fig. 3.20) [5]

In addition to the volumes defined by the ICRU 50 report, two new volumes termed the internal target volume (ITV) and the planning organs at risk volume (PRV) were added. Internal margin (IM), setup margin (SM), organ at risk (OAR), and conformity index (CI) were also defined.

The internal margin (IM) defines physiological organ movements.

- Respiration, swallowing, bladder fullness, lung movements, cardiac movement, and intestinal movement

The *setup margin (SM)* defines movements relating to the treatment and technique, and daily changes in setup position.

- Patient movement, treatment machine movement, setup errors

Fig. 3.18 Dose–volume histogram

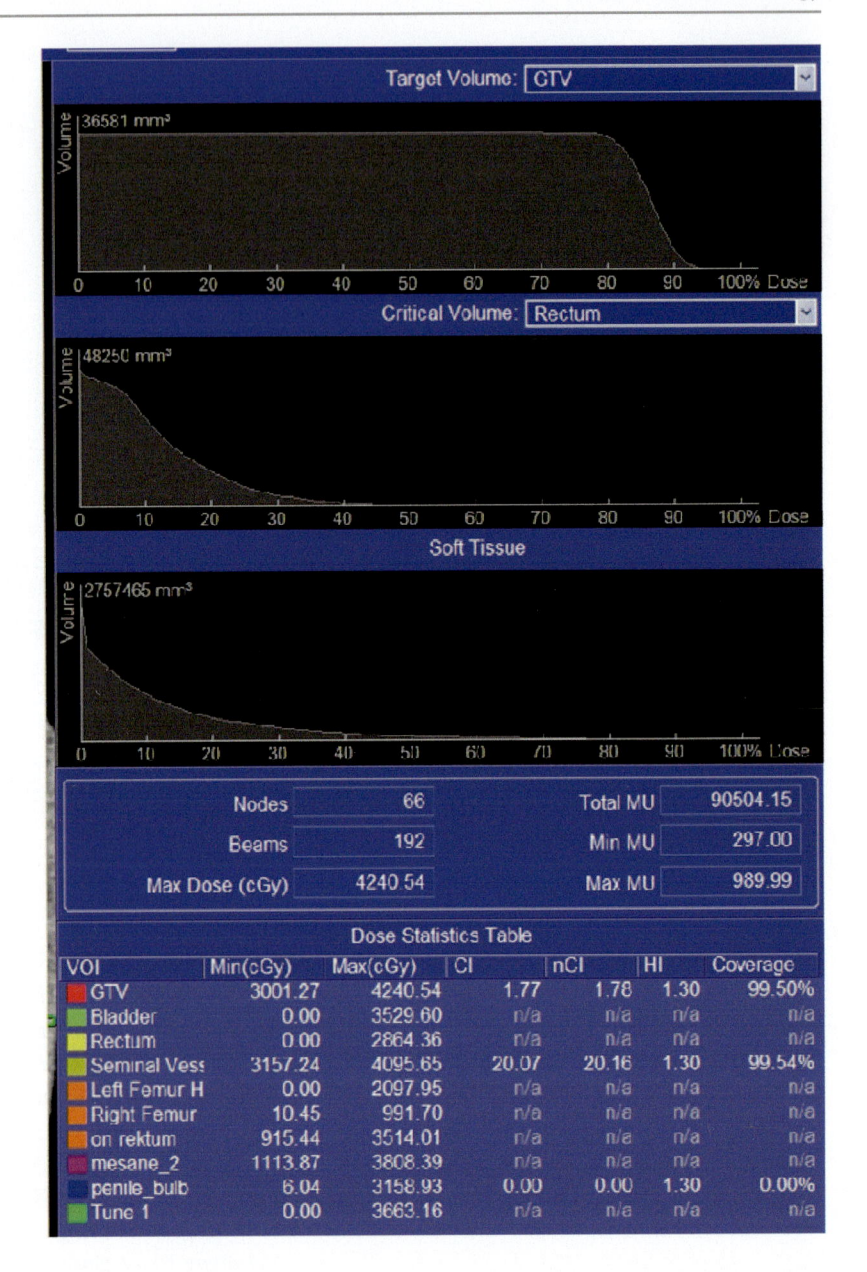

An OAR is an organ that may remain in the treatment field, and can cause changes to treatment plans and doses (spinal cord, heart, lungs, kidney, eye, etc.).

The internal target volume (ITV) is the combined volume of the CTV and IM.

The planning organ at risk volume (PRV) defines the volume of the OAR that may reside in the PTV during treatment.

CI = TV/PTV.

- The TV should include PTV (CI ratio)
- This is necessary for dose homogeneity within the PTV

The *ICRU reference point and dose* [5] is a point outside the rapid dose change region that determines the PTV; it is easy to define and is dose-definable physically.

- It is a point on the central axis or close to it.
- It is defined separately for each treated region.

3.2 The Radiotherapy Procedure

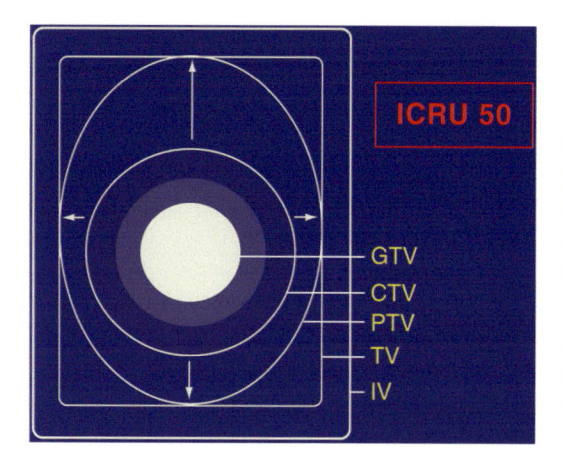

Fig. 3.19 Treatment volumes according to the ICRU-50 report [4]

D_{max} is the maximum dose point within the PTV and the organ at risk.

D_{min} is the minimum dose point within the PTV.

Hot spots are high-dose regions in the treated volume.

- These are points that receive more than 100% of the total dose.
- They are acceptable in the GTV or CTV, but are unacceptable in organs at risk, and require technical correction.

The dose distribution should be homogeneous within the PTV, and an isodose heterogeneity of −5 to 7.5% is acceptable.

3.2.3.3 Uncertainties in the Volume Definitions in ICRU 62 [6]

The most important uncertainty is in the delineation of the CTV. The CTV is not just the region that has a high risk of the tumor spreading to it, because the whole body is at risk. Therefore, the CTV is the region that has a high risk of bearing tumor cells and that can be included within a reasonable radiotherapy field. The data used to determine CTV borders are generally based on weak and anecdotal information. Therefore, CTV is practically just an automatic enlargement of the GTV to a certain margin, and this approach is too weak to define the CTV.

The PTV defines a physical volume, while the GTV and CTV define biological volumes [6].

The PTV does not contain information about the radiation beam penumbra and thus does not provide any information on the actual treatment field size either. Therefore, it is very important to enlarge the treatment field to include a beam penumbra that is larger than the PTV.

The PTV is not useful for defining the field size since the relationship of the penumbra to the field size is weak for charged particles like protons.

3.2.3.4 ICRU 71 [7]

This report is similar to reports 50 and 62 in terms of general volume definitions.

- It recommends that the point of D_{max} dose of electron energy selected within PTV at ICRU reference point is determined at D_{max} of the selected electron energy for treatment.
- The changes in OAR and PRV with energy and patient movement in electron therapies are explained with examples.
- It provides information about some special electron therapy techniques (total skin irradiation, intraoperative RT).

3.2.3.5 ICRU 78 (Fig. 3.21) [6, 7]

This report on proton therapies was published in 2007.

A new volume called the remaining volume at risk (RVR) was added to the volume definitions.

- RVR is examined in two parts: the region of the PRV and OAR included within the imaged field, and the region of the PRV and OAR remaining outside of the PTV.
- The dose received by the RVR is important when estimating late side effects (particularly secondary malignancies).
- It specifies that RVR should always be delineated [6].
- It recommends that OARs should be delineated during planning [6].

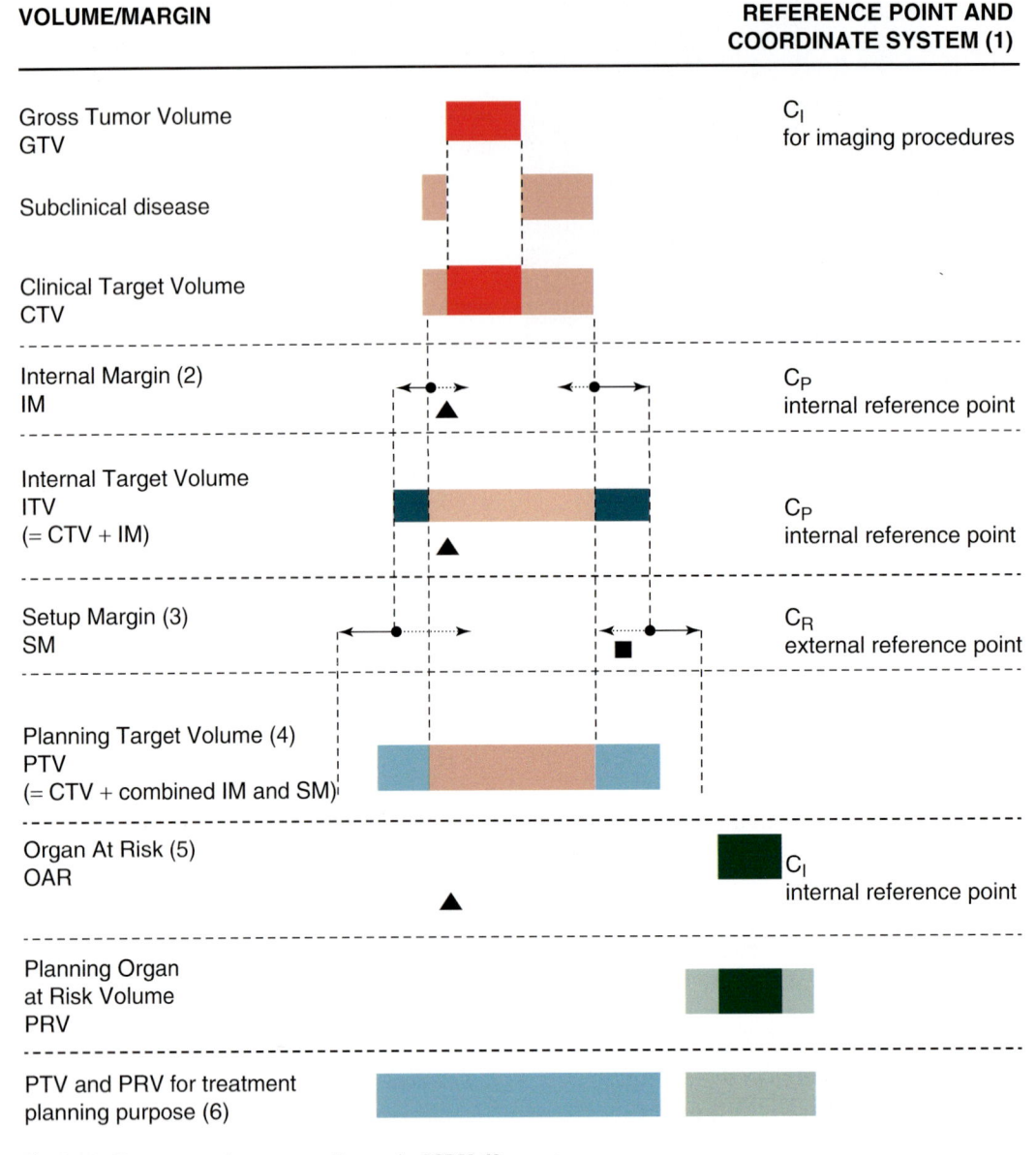

VOLUME/MARGIN		REFERENCE POINT AND COORDINATE SYSTEM (1)
Gross Tumor Volume GTV		C_I for imaging procedures
Subclinical disease		
Clinical Target Volume CTV		
Internal Margin (2) IM		C_P internal reference point
Internal Target Volume ITV (= CTV + IM)		C_P internal reference point
Setup Margin (3) SM		C_R external reference point
Planning Target Volume (4) PTV (= CTV + combined IM and SM)		
Organ At Risk (5) OAR		C_I internal reference point
Planning Organ at Risk Volume PRV		
PTV and PRV for treatment planning purpose (6)		

Fig. 3.20 Treatment volumes according to the ICRU-62 report

Other concepts that are pointed out are:

Volume of interest. This is any volume that needs to be defined. It can be generically used to define special volumes such as the PTV and OAR.

Surface of interest. This is a flat or curved plane or a particular surface.

Point of interest. This is any point in space.

The IAEA report in 2000 considers proton dosimetry, and the importance of the value of w, as measured in an ion chamber, is mentioned [6].

The definition of the RBE-weighted dose should be mentioned in proton therapy, and it is shown that this dose can be found by multiplying the total delivered dose (according to the photon energy of Co-60) by 1.1. This dose is called the bioeffective dose. It was mentioned that the bioeffective dose can change with depth, dose, and fraction dose.

The old term for the RBE-weighted dose was the "cobalt gray equivalent (CGE)"; however, the

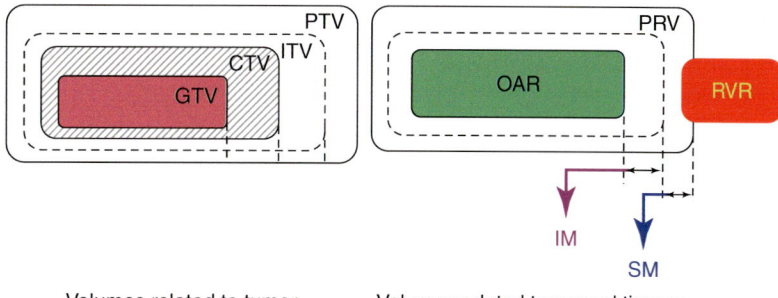

3.2 The Radiotherapy Procedure

Fig. 3.21 Treatment volumes according to the ICRU-78 report

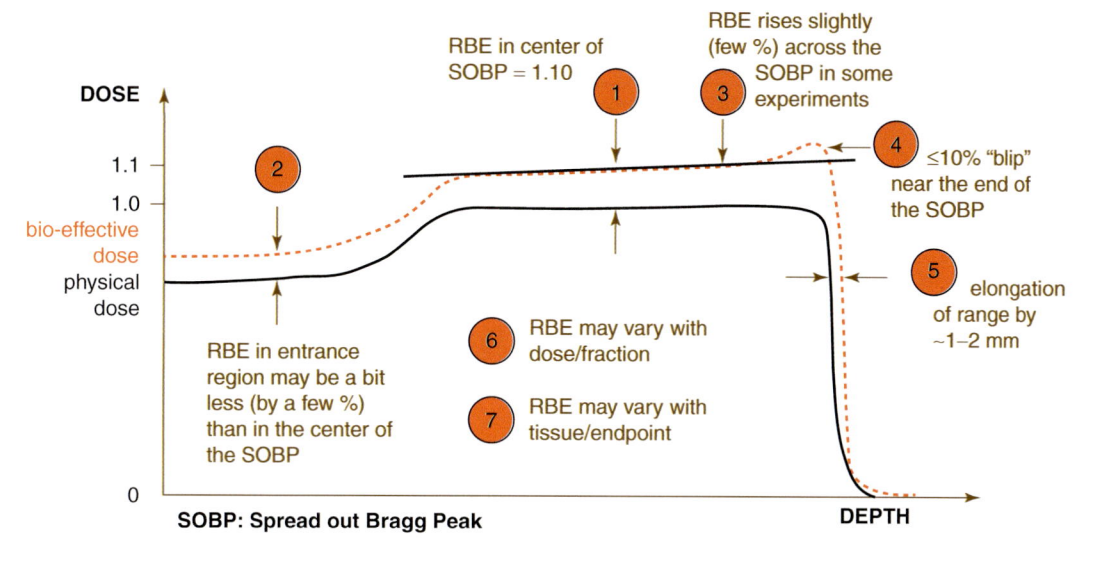

Fig. 3.22 Spread out Bragg peak [6]

CGE is not recommended, since it is not given in SI units (Fig. 3.22) [6].

- The CTV can be used in proton therapy beam design (Fig. 3.23) [6].
- The CTV can be used when generating the PTV, but it can vary [6].
- The PTV can be used for both proton therapy and photon therapy, thus enabling uniformity of reporting within institutions (Table 3.1) [6].

3.2.4 Setup and Treatment

The radiation oncologist should attend the first patient treatment (i.e., during setup) and see the patient at regular intervals during radiotherapy. This is true of both conventional and advanced radiotherapies. The applicability of a well-designed therapy plan requires meticulous review of the physician.

Port films are used to control actual treated fields if they are exactly the same as the planned ones. These films are regularly taken during therapy when the patient is on the treatment couch and compared with the simulation films (Fig. 3.24). The patient's position is then adjusted accordingly.

Digitally reconstructed radiographies created by the TPS are compared with images taken by an electronic portal imaging device in more developed techniques (e.g., conformal RT, IMRT) (Fig. 3.25).

3.2.5 Quality Assurance

Quality assurance (QA) is absolutely essential in radiotherapy (Fig. 3.26). QA includes all aspects of the therapy, from recording the patient's data

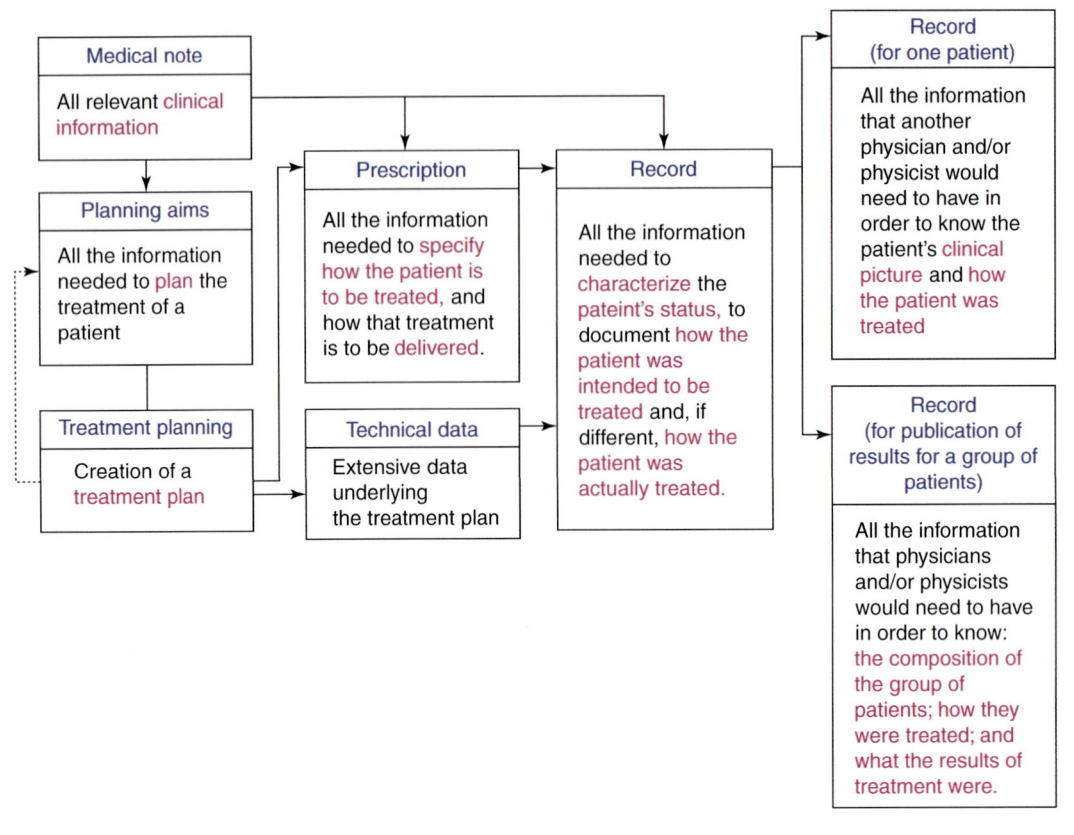

Fig. 3.23 The common algorithm for photon and proton therapies according to the ICRU-78 report [6]

Table 3.1 Similarities and differences between proton therapies and photon therapies according to ICRU 78 [6]

	Same or different?
Imaging modalities	Same or different?
GTV, CTV delineation	Same
PTV delineation	Different
Electron densities of tissues	Different
Treatment planning tools	Same
Treatment planning design	Same
Weights of planned beams	Different
Calculations of field irregularities	Quite different
Selection of plan	Same
Final prescription	Same
Simulation of plan	Same
Delivery of treatment and corrections	Different
Plan check during treatment	Same
Documentation and reporting	Same
Follow-up after treatment	Same

to follow-up procedures after treatment. A problem at any stage of this overall procedure affects all functions. Therefore, control mechanisms should be standardized and continuously monitored if they are to be successful.

Single Treatment Fields (Fig. 3.27). These are generally used for superficially located tumors (e.g., skin cancers), superficial lymphatic regions (e.g., supraclavicular lymphatics), metastatic nodules, and vertebral metastases.

Advantages include the fact that treatment planning is not usually required, dose calculations are rapid and easy, and setup is easy.

Factors determining the dose include subcutaneous tissue, heterogeneity in tumor volume, and normal tissues under the tumor.

Multiple Treatment Fields (Fig. 3.28). These are used for deeply seated tumors or large tumors to provide a homogeneous dose distribution and to spare normal tissues.

Factors that determine the dose include the size of the irradiated volume, the treatment machine, the cross-section of the treatment field, and the selected energy.

3.2 The Radiotherapy Procedure

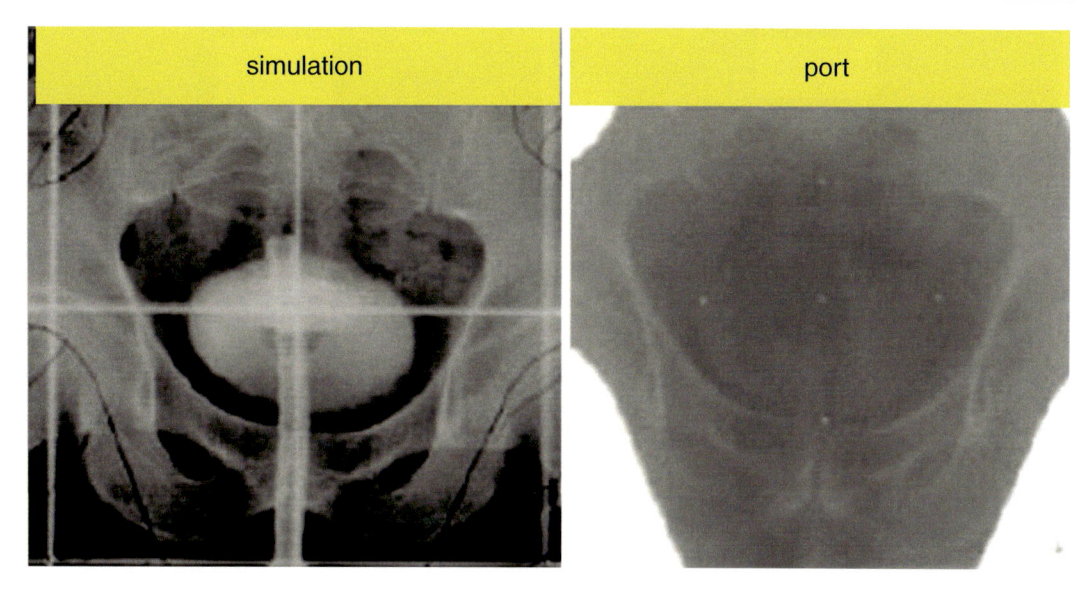

Fig. 3.24 Simulation and port films

Special multiple treatment field examples include *tangential beams*, where parallel opposed tangential fields are used for convex-shaped regions such as the chest wall (Fig. 3.29), and *oblique fields (double-wedge technique)*, where nonparallel and nonopposed fields are used for convex-shaped regions such as parotid gland tumors (Fig. 3.30).

3.2.6 Treatment Fields in Radiotherapy

Treatment fields in radiotherapy and beams are generally classified into two groups: single and multiple fields.

1. **Intensity Modulated Radiation Therapy (IMRT)**
   - In 3D conformal radiotherapy, diaphragm aperture is designed so that the radiation beam geometrically conforms to the target volume.
     - The radiation beam generates a convex dose distribution.
     - Beam's eye view (BEV) is generally used for sparing of distant critical structures geometrically.
   - In IMRT, radiation beam modulation is performed before designation of diaphragm aperture.
     - Radiation beamlets have different intensities and frequently show a grill-like intensity mapping pattern (Fig. 3.31).
       Dose intensity of the radiation beamlets is designed by inverse planning and dose optimization is provided.
       This mathematical optimization provides the desired dose and dose uniformity at target volume and dose reduction at non-target tissues by use of different inputs.
       Ideal target dose and maximal sparing of critical organs are achieved by accumulation of dose from multiple beamlets with different dose intensities.
       Another feature of IMRT is its ability to generate concave-shaped dose distributions.
   - This feature of IMRT allows for sparing of the spinal cord and parotid glands in head and neck irradiation
   - Different techniques may be used for IMRT delivery depending on the capacity of equipment such as the linear accelerator

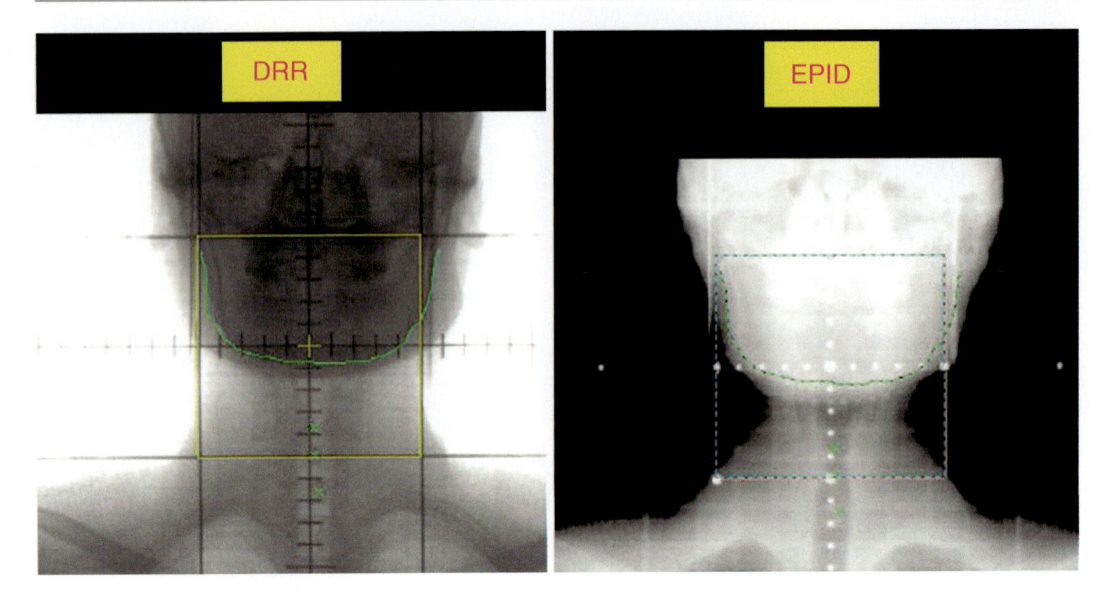

Fig. 3.25 Imaging of DRR and EPID

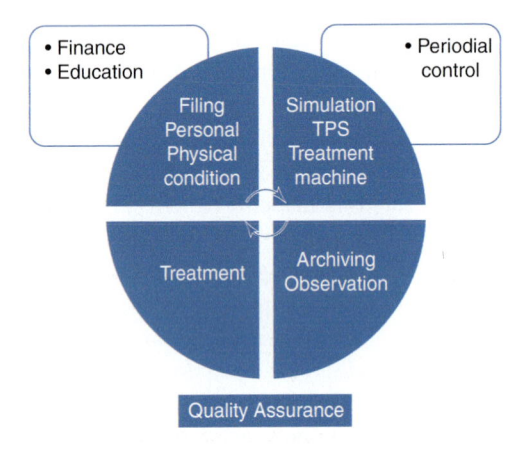

Fig. 3.26 Quality assurance and its components

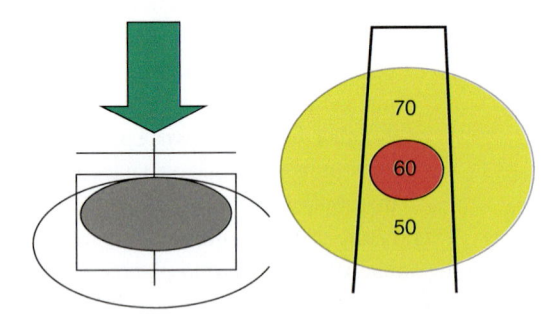

Fig. 3.27 Single RT field and change in isodose

and treatment planning systems. Most common techniques for IMRT.

- Step and Shoot IMRT:
  - Typically 5 to 9 beam segments are used in this technique [8].
  - Gantry angle is constant.
  - At each segment angle, dose with modulated intensity is sent to the target volume by use of computer-controlled MLCs.
  - MLC leaves move when the beam is on.
  - "Beam off" position is activated when the desired monitor unit dose is reached, and the MLC leaves take the position of the next segment. Modulated dose for that segment is sent when the "beam on" position is activated again.
  - This cascade continues till the completion of all segments.
  - In this technique, the desired fraction dose is composed of multiple static components.
  - The treatment time is prolonged since the radiation dose is delivered by small segments.
- Sliding Window IMRT:
  - In this technique, gantry angle is constant likewise, however, the MLC leaves move at both "beam off" and "beam on"

3.2 The Radiotherapy Procedure

Fig. 3.28 Multiple RT fields and change in isodose

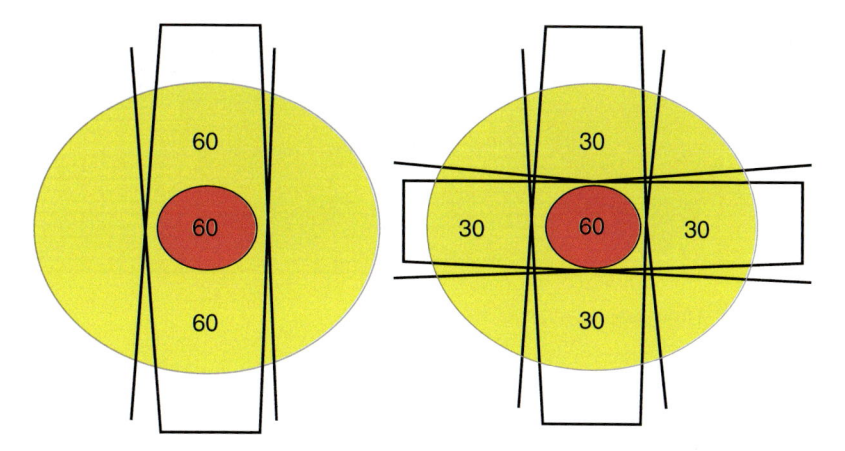

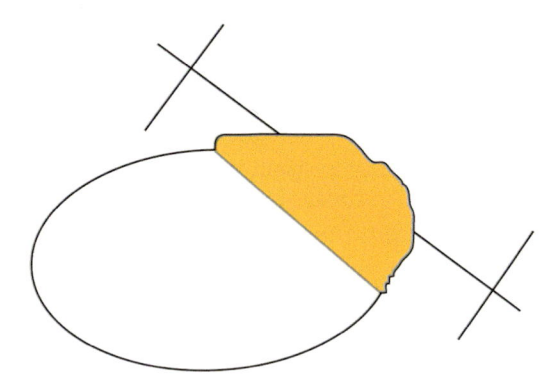

Fig. 3.29 Tangential fields

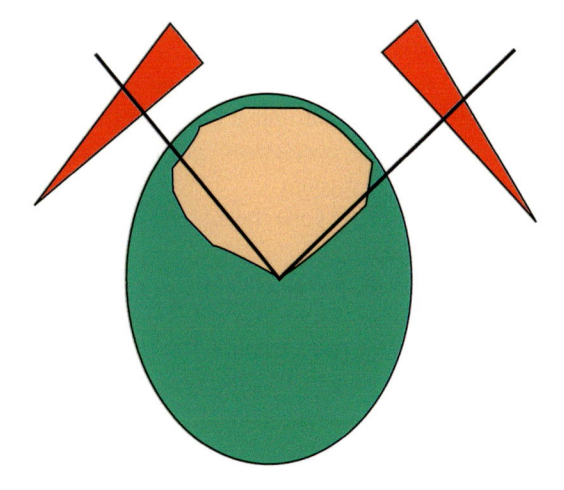

Fig. 3.30 Oblique fields (double-wedge technique)

positions with radiation delivered during MLC movements [9].
- Volumetric intensity arc therapy (VMAT/Rapid Arc)
  - VMAT is a type of computer-controlled arc treatment.
  - MLC leaves move under computer control with rotation to generate the intensity-modulated arc therapy [10].
  - Dose rate, gantry speed, and other parameters are computer-controlled.
  - Superposition of multiple arcs results in highly conformal dose distributions [11].
  - This technique typically allows for shorter treatment times with lower monitor units.
  - Treatment time may be decreased to 1/3 of step and shot IMRT treatment time.
- IMRT is in improvement with regard to improvements in the accuracy of dose calculation algorithms.
  - With improved computer systems, deformable image registration and tracking of breathing motion and anatomical changes, it is possible to perform 4D therapies with desired target doses along with optimal normal tissue sparing.
2. **Image-Guided Radiation Therapy – IGRT**
   - Changes in tumor size and position during the course of radiotherapy may be accounted for by use of IGRT to improve the accuracy of RT [12].
   - IGRT improves target localization and treatment accuracy which allows for safer administration of radiation dose to the tar-

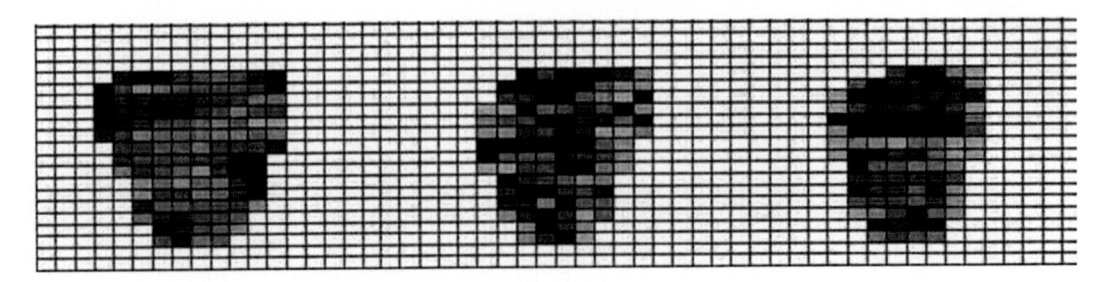

Fig. 3.31 Dose mapping of IMRT beamlets

get with optimal normal tissue sparing to achieve improved clinical outcomes.
- Image-guided treatment refers to the use of 2D/3D imaging techniques for imaging at various stages including pretreatment imaging, on-treatment imaging, and post-treatment imaging for optimal treatment.
- IGRT may utilize CT simulation images, cone beam CT (CBCT) images, planar kilovoltage (kV) imaging or planar megavoltage (MV) imaging, and digitally reconstructed radiographs (DRRs).
  – Registration of CBCT images acquired at the treatment room and planning CT images acquired at the CT simulation may be an example of volumetric IGRT.
- Importance of IGRT is increasing with wider utility of SRS and SBRT [13].
- Image-guided contouring of target volume and critical organs:
  – Accurate contouring of target volume and critical organs is crucial in RT.
  – Computed tomography (CT);
    Leading imaging modality for treatment simulation and planning is CT which is used for delineation as well as for calculation of tissue heterogeneity [14].
    CT offers high spatial integrity, high spatial resolution, superior visualization of bony structures, and allows for calculation of radiation dose based on the electron density.
  – Magnetic resonance imaging (MRI) [15] and positron emission tomography (PET) [16]
    Fusion of the planning CT images with MRI or PET images may improve contouring of target volume and critical organs
- Implementation of IGRT
  – Available IGRT technologies may be categorized as follows:
    Conventional IGRT technologies
      Portal imaging
      Fluoroscopic imaging
      Ultrasonic imaging
    In-room CT-based IGRT technologies
      CT-on-Rail
        includes a diagnostic CT system integrated with the linear accelerator at the treatment room
- Kilovoltage cone beam CT (kV-CBCT) (Fig. 3.32)
- Megavoltage cone beam CT (MV-CBCT) (Fig. 3.33)
- Megavoltage fan beam CT (MVCT)
  – Real-time tracking systems
    Exac-Trac system
    CyberKnife (Fig. 3.34)
    ViewRay system
    Calypso 4D localizing system.

3.2 The Radiotherapy Procedure

Fig 3.32 kV-CBCT-based IGRT system [Elekta] (courtesy of Gulhane Medical Faculty, Ankara)

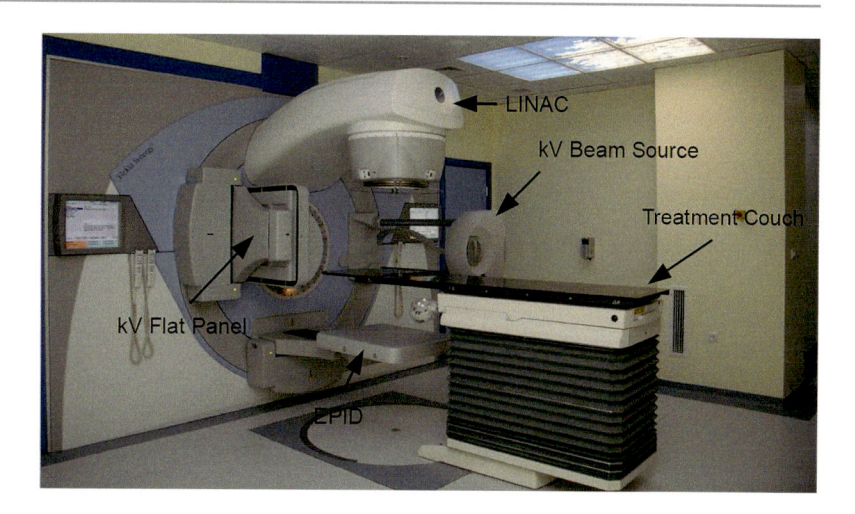

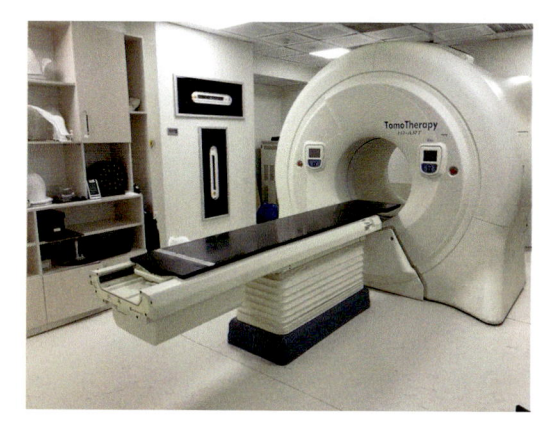

Fig. 3.33 MVCT-based IGRT Tomotherapy (courtesy of Gulhane Medical Faculty, Ankara)

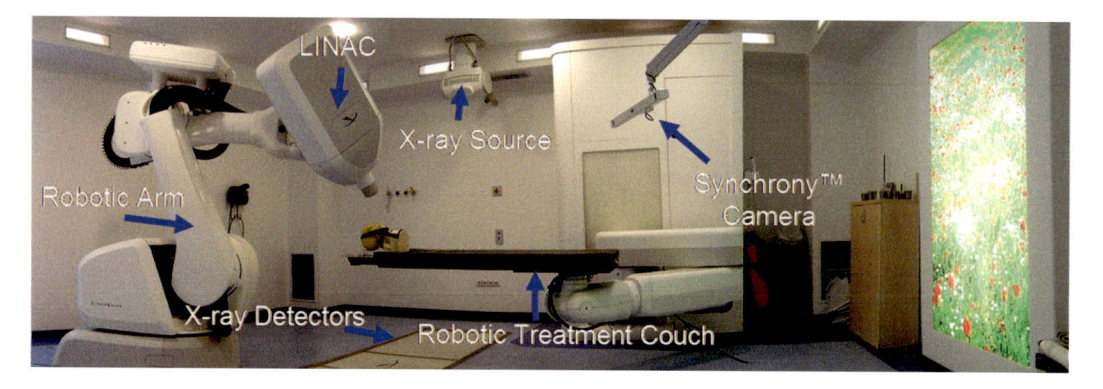

Fig 3.34 Cyberknife (courtesy of Hacettepe University, Ankara)

References

1. Del Regato JA (1972) The American Society of Therapeutic Radiologists. Introduction. Cancer 29(6):1443–1445
2. Purdy JA (2004) Current ICRU definitions of volumes: limitations and future directions. Semin Radiat Oncol 14(1):27–40. Review
3. Kantor G, Maingon P, Mornex F, Mazeron JJ (2002) Target-volume contours in radiotherapy. General principles. Cancer Radiother 6(Suppl 1):56s–60s. Review
4. International Commission on Radiation Units and Measurements (ICRU) (1993) Report 50. Prescribing, recording, and reporting photon beam therapy. ICRU, Bethesda
5. International Commission on Radiation Units and Measurements (ICRU) (1999) Report 62. Prescribing, recording, and reporting photon beam therapy (Supplement to (ICRU) Report 50). ICRU, Bethesda
6. Goitein M (2008) Designing a treatment beam. In: Radiation oncology: a physicist's-eye view. Springer, pp 111–37
7. International Commission on Radiation Units and Measurements (ICRU) (2004) Report 71. Prescribing, recording, and reporting electron beam therapy. ICRU, Bethesda
8. Bortfeld T, Boyer AL, Schlegel W, Kahler DL, Waldron TJ (1994) Realization and verification of three-dimensional conformal radiotherapy with modulated fields. Int J Radiat Oncol Biol Phys 30(4):899–908
9. Williams PC (2003) IMRT: delivery techniques and quality assurance. Br J Radiol 76(911):766–776
10. Yu CX (1995) Intensity-modulated arc therapy with dynamic multileaf collimation: an alternative to tomotherapy. Phys Med Biol 40(9):1435–1449
11. Otto K (2008) Volumetric modulated arc therapy: IMRT in a single gantry arc. Med Phys 35(1):310–317
12. Bissonnette JP, Craig T (1999) X-ray imaging for verification and localization in radiation therapy in modern technology of radiation oncology (suppl. 1). Modern technology of radiation oncology. Medical Physics Pub, Madison. isbn:0-944838-38-3
13. Lo SS, Fakiris AJ, El C et al (2010) Stereotactic body radiation therapy: a novel treatment modality. Nat Rev Clin Oncol 7(1):44–54
14. Kijewski PK, Bjarngard BE (1978) The use of computed tomography data for radiotherapy dose calculations. Int J Radiat Oncol Biol Phys 4:429
15. Mundt AJ, Roeske JC (2006) In: Bortfeld T, Schmidt-Ullrich R (eds) Image- guided radiation therapy. Springer, Berlin
16. Bailey DL, Townsend DW, Valk PE (2005) Positron emission tomography: basic sciences. Secaucus, Springer. isbn:185233-798

Central Nervous System Tumors

4.1 Introduction

According to SEER (Surveillance Epidemiology and End Results) data, central nervous system (CNS) tumors are seen in 7.7/100000 in men and 5.4/100000 in women, and account for 2% of all cancers in the USA [1]. Approximately 13% of cases are seen in the pediatric population (<20 years of age) and the prevalence is bimodal in the fourth and fifth decades, with a median age of diagnosis at 57 years [2]. CNS tumors tend to behave differently depending on age, histology, and location.

Tentorium cerebelli. This is an extension of the dura mater that covers the top of the cerebellum (Fig. 4.1). The top portion is called the supratentorium, and that below is known as the infratentorium.

- Supratentorial location is frequent in adults.
- Infratentorial location is frequent in children.

4.2 Anatomy

4.2.1 Brain (Fig. 4.2) [2]

- Embryologically, the brain consists of the prosencephalon (forebrain), the mesencephalon (midbrain), and the rhombencephalon (hind brain) [3].
- The prosencephalon forms important parts of the brain including the two hemispheres (telencephalon) and the diencephalon (interbrain) during the later stages of embryonic life.
- White matter is located inside the structure, while gray matter surrounds the white matter like a shell.
- The shell of gray matter is called the cerebral cortex. There are important functional centers at particular positions in the cerebral cortex.
- The white matter of the brain does not contain any neuronal cells. It consists solely of neuronal pathways.
- The part that links the two hemispheres of the brain is the interbrain (diencephalon).
- The midbrain (mesencephalon) is the smallest part of the brain. It acts as a connection between the spinal cord and the neuronal pathways of the brain. It also contains important and vital functional cranial nerves (e.g., the oculomotor nerve or N. oculomotorius, and the trochlear nerve or N. trochlearis) and nuclei (e.g., the nucleus ruber and the substantia nigra) [3].
- The hindbrain (rhombencephalon) consists of three important parts: the pons, bulbus, and cerebellum. An important brain cavity, the fourth ventricle, is located between these three parts. The pons and the bulbus (also known as the medulla oblongata or simply "medulla") contain important start and end nuclei for some cranial nerves.

Brain and spinal cord is covered by three membranes, called meninges: the dura mater, pia mater, and arachnoid mater (Fig. 4.3) [4].

- There is a cavity filled with fluid between the pia mater and arachnoid mater. This fluid protects the brain from excessive pressure. It also protects against external mechanical trauma.
- The dura mater and pia mater are very sensitive to pain.
- The neuronal network of the arachnoid mater is very poorly developed.

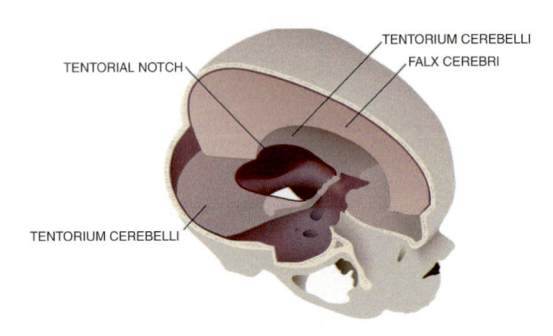

Fig. 4.1 Tentorium cerebelli

4.2.2 Brain Ventricles [2]

- The *first and second ventricles* lie within the hemispheres of the brain (telencephalon).
- The *third ventricle* is located in the interbrain (diencephalon).
- The *fourth ventricle* is the space between the pons, bulbus, and cerebellum. Its base, which is formed by the pons and bulbus, is rhomboidal in shape, so it is called the rhomboid fossa [3].

Cerebrospinal Fluid (CSF)

There is a vascular network in the ventricles of the brain that is called the choroid plexus, which secretes the CSF found in the ventricles. Brain ventricles normally produce 300–400 mL of CSF daily. In addition, CSF is found in the subarachnoid space located between the arachnoid mater and pia mater [4].

4.2.3 Spinal Cord (Fig. 4.4) [3]

The spinal cord runs through the vertebral canal between the top portion of the atlas and the intervertebral disc of L1–2 in adults.

Fig. 4.2 Brain lobes

Fig. 4.3 Brain meninges and subarachnoid space

- The spinal cord is not a straight pipe; it exhibits enlargements in two places.
- The first enlargement is between the C4 and T1 segments.
- The second enlargement is between the L1 and S3 segments.

The spinal cord gets thinner and ends at the intervertebral disc of the L1 and L2 vertebra. This end region of the spinal cord is called the conus medullaris. A bond originating from the conus medullaris connects the spinal cord to the tip of the coccyx. This bond is known as the filum terminale, and is approximately 20 cm in length [4].

4.3 General Presentation and Pathology

Sixty percent of all primary brain tumors are glial tumors, and two-thirds of these are clinically aggressive, high-grade tumors (Fig. 4.5) [1].

In the WHO 2016 classification, tumors are classified based on histopathological and molecular parameters [5] (Table 4.1).

Newly Added Entities in the WHO 2016 Classification
- Epithelioid glioblastoma

Diffuse leptomeningeal glioneuronal tumor
Diffuse midline glioma, H3K27M-mutant
Embryonal tumor with multilayered rosettes
Ependymoma RELA fusion-positive
Medulloblastoma—WNT

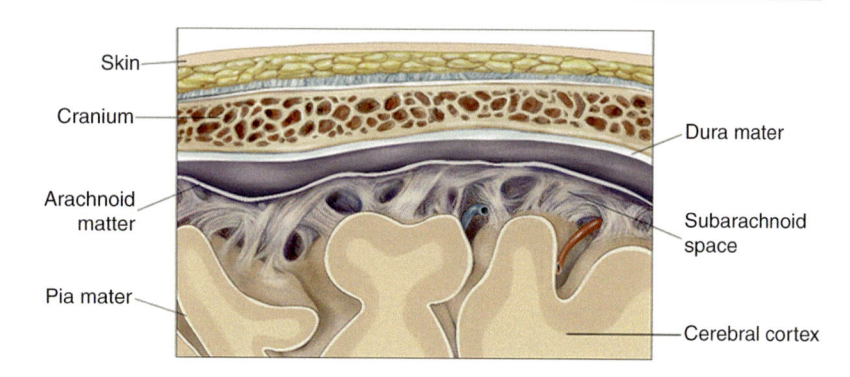

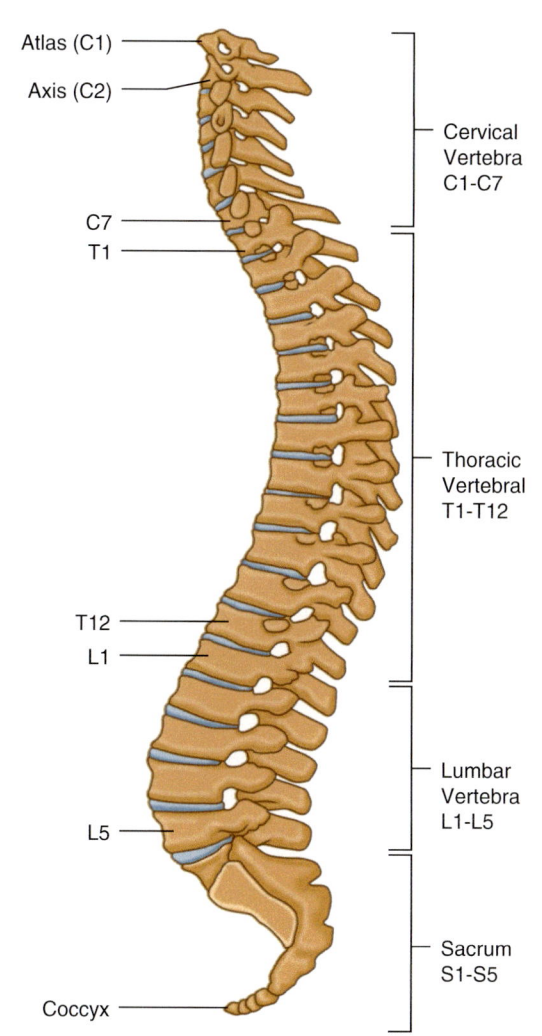

Fig. 4.4 Sections of spinal cord

- It may extend to the L3 vertebral body in small children.
- It is 45 cm long, 30 g in weight, and 1 cm in diameter.

Fig. 4.5 Survival distributions for brain tumors according to SEER data

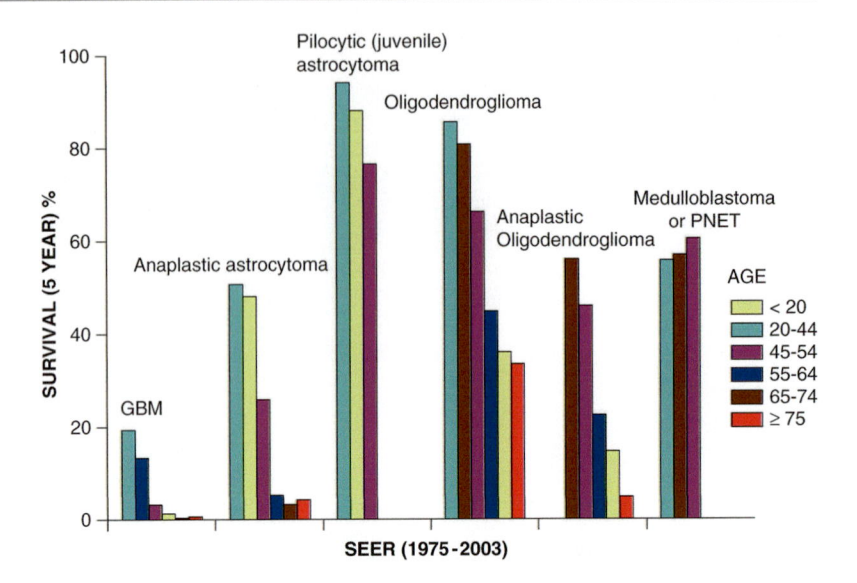

WHO 2016 CNS Tumor Grading
- **Grade I:** They are lesions with low proliferative potential and may typically be treated by surgical resection alone.
- **Grade II:** Despite low mitotic activity, they generally contain infiltrative atypical cells. They recur more frequently than grade I tumors after local treatment. Some grade II tumors tend to upgrade.
- **Grade III:** They contain "nuclear atypia/anaplasia" and "increased mitotic activity", which are considered as histological evidence of malignancy. These lesions have anaplastic histology and infiltrative capacity. Generally, intensive adjuvant radiotherapy and/or chemotherapy is used for management.
- **Grade IV:** Lesions are mitotically active, prone to necrosis, and are usually associated with neovascularity and a tendency to infiltrate into surrounding tissue along with craniospinal spread, leading to rapid postoperative progression and fatal outcomes. Lesions are usually treated with aggressive adjuvant therapy (typically Stupp protocol combined chemoradiotherapy) [5].

4.3.1 Spinal Cord Tumors [5]

Spinal tumors are classified as either extradural, intradural extramedullary, or intradural intramedullary. Among all spinal tumors, 55% are extradural, 40% are intradural extramedullary, and 5% are intradural intramedullary [6].

4.3.2 Brain Tumors [1]

Symptoms and findings depend on tumor histology, location, and age. Brain tumors cause neurological disorders by either infiltrating normal CNS structures or obstructing CSF pathways and subsequently increasing intracranial pressure. This increase in intracranial pressure results in the early symptoms of brain tumors: headaches, vomiting, and lethargy.

- Headaches are frequently seen, and they occur more frequently in the morning.
- Papilledema is present in 25% of cases.
- Endocrine tumors have hyperfunctional or hypofunctional findings.
- Tumors located in the motor area generally cause hemiparesis or hemiplegia, while tumors of the cerebellum cause balance and coordination disorders.
- Frontal tumors cause personality and memory disorders.
- Infratentorial tumors cause obstructive hydrocephalus, cerebellar dysfunctions, and cranial nerve disorders.
- Hypophyseal tumors compress the optic chiasm and cause visual disorders.

Table 4.1 2016 WHO classification of CNS tumors

Diffuse astrocytic and oligodendroglial tumors WHO grade II	Diffuse astrocytoma • IDH-mutant • Gemistocytic astrocytoma • IDH-wildtype • NOS
	Oligoastrocytoma—NOS
	Oligodendroglioma • IDH-mutant, 1p19q co-deleted • Oligodendroglioma NOS
Diffuse astrocytic and oligodendroglial tumors WHO grade III	Anaplastic astrocytoma • IDH-mutant • IDH-wildtype • NOS
	Anaplastic oligoastrocytoma
	Anaplastic oligodendroglioma • IDH-mutant, 1p19q co-deleted • Anaplastic oligodendroglioma NOS

Diffuse astrocytic and oligodendroglial tumors WHO grade IV	Glioblastoma
	• IDH wildtype
	    Giant cell glioblastoma
	    Gliosarcoma
	    Epithelioid glioblastoma
	• IDH mutant
	• NOS

	Diffuse midline glioma, H3K27M-mutant
Other astrocytic tumors	WHO grade I • Pilocytic astrocytoma • Subependymal giant cell astrocytoma
	WHO grade II • Pilomyxoid astrocytoma • Pleomorphic xanthoastrocytoma
	WHO grade III • Anaplastic pleomorphic xanthoastrocytoma
Other gliomas	WHO grade I • Angiocentric glioma
	WHO grade II • Chordoid glioma of the third ventricle
	WHO (grade not yet assigned) • Astroblastoma
Ependymal tumors	WHO grade I • Subependymoma • Myxopapillary ependymoma
	WHO grade II • Ependymoma • Papillary ependymoma • Clear cell ependymoma • Tanycytic ependymoma • RELA fusion (+) ependymoma
	WHO grade III • Anaplastic ependymoma
Choroid plexus tumors	WHO grade I • Choroid plexus papilloma
	WHO grade II • Atypical choroid plexus papilloma
	WHO grade III • Choroid plexus carcinoma

(continued)

Table 4.1 (continued)

Neuronal and mixed neuronal-glial tumors	WHO grade I • Desmoplastic infantile astrocytoma and ganglioglioma • Dysembryoplastic neuroepithelial tumor (DNET) • Dysplastic gangliocytoma of the cerebellum—(Lhermitte-Duclos) • Gangliocytoma • Ganglioglioma • Papillary glioneuronal tumor • Paraganglioma of the filum terminale • Rosette-forming glioneuronal tumor of the fourth ventricle
	WHO grade II • Central neurocytoma • Extraventricular neurocytoma • Cerebellar liponeurocytoma
	WHO grade III • Anaplastic ganglioglioma
	WHO grade unknown • Diffuse leptomeningeal glioneuronal tumor

Embryonal tumors	WHO grade IV
	• Medulloblastoma
	  Genetically defined
	    WNT-activated
	    SHH-activated
	    Medulloblastoma SHH-activated & TP53—wildtype
	    Group 3
	    Group 4
	  Histologically defined
	    Classic
	    Desmoplastic/nodular
	    Extensive nodular
	    Large cell/anaplastic
	    NOS
	• CNS neuroblastoma
	• CNS ganglioneuroblastoma
	• Embryonal tumors with multilayered rosettes
	  C19MC-altered
	  NOS
	• Medulloepithelioma
	• Atypical teratoid/rhabdoid tumor
	• CNS embryonal tumor with rhabdoid tumor
	• CNS embryonal tumor

Tumors of pineal region	WHO grade I • Pineocytoma
	WHO grade II or III • Papillary tumor of pineal region • Pineal parenchymal tumor of intermediate differentiation
	WHO grade IV • Pineoblastoma

Tumors of sellar region	WHO grade I
	• Craniopharyngioma
	  Adamantinomatous
	  Papillary
	• Granular cell tumor
	• Pituicytoma
	• Spindle cell oncocytoma

4.3 General Presentation and Pathology

Table 4.1 (continued)

Tumors of cranial and paraspinal nerves	WHO grade I
	• Schwannoma (neurilemoma, neurinoma)
	    Cellular schwannoma
	    Plexiform schwannoma
	    Melanotic schwannoma
	• Neurofibroma
	    Atypical neurofibroma
	    Plexiform neurofibroma
	• Perineurioma

	WHO grade II, III, or IV
	• Malignant peripheral nerve sheath tumor (MPNST)
	    Epithelioid
	    With perineural differentiation

Tumors of meningothelial cells	WHO grade I • Meningioma • Meningothelial meningioma • Fibrous meningioma • Microcystic meningioma • Psammomatous meningioma • Angiomatous meningioma • Secretory meningioma • Metaplastic meningioma • Lymphoplasmacyte-rich meningioma
	WHO grade II • Atypical meningioma • Clear cell meningioma • Chordoid meningioma
	WHO grade III • Anaplastic meningioma • Papillary meningioma • Rhabdoid meningioma
Mesenchymal, nonmeningothelial tumors	WHO grade I, II, or III • Solitary fibrous tumor of the dura/hemangiopericytoma
	WHO grade I • Angiolipoma • Chondroma • Desmoid-type fibromatosis • Hemangioblastoma • Hemangioma • Hibernoma • Leiomyoma • Lipoma • Myofibroblastoma • Osteochondroma • Osteoma • Rhabdomyoma
	WHO grade III • Epithelioid hemangioendothelioma • Angiosarcoma • Chondrosarcoma • Ewing sarcoma/PNET • Fibrosarcoma • Kaposi sarcoma • Leiomyosarcoma • Liposarcoma (intracranial) • Osteosarcoma • Rhabdomyosarcoma • Undifferentiated pleomorphic sarcoma/malignant fibrous histiocytoma

(continued)

Table 4.1 (continued)

Melanocytic lesions	Primary melanocytic tumors of the CNS • Meningeal melanocytosis • Meningeal melanocytoma • Meningeal melanomatosis • Meningeal melanoma
Lymphomas	• Diffuse large B-cell lymphoma of the CNS • Immunodeficiency-associated CNS lymphomas 　AIDS-related diffuse large B-cell lymphoma 　EBV-positive diffuse large B-cell lymphoma, NOS 　Lymphomatoid granulomatosis • Intravascular large B-cell lymphoma • Low-grade B-cell lymphomas of the CNS • T-cell and NK/T-cell lymphomas of the CNS • Anaplastic large cell lymphoma 　ALK-positive 　ALK-negative • Dural MALT lymphoma
Histiocytic tumors	• Erdheim-Chester disease • Histiocytic sarcoma • Juvenile xanthogranuloma • Langerhans cell histiocytosis • Rosai-Dorfman disease
Germ cell tumors	• Choriocarcinoma • Embryonal carcinoma • Germinoma • Mixed germ cell tumors Teratoma 　Mature 　Immature 　　with malignant transformation • Yolk sac tumor

Spinal Cord Tumors [1]

Pain is the earliest and most frequently observed symptom. Pain occurs in 80% of cases, while motor weakness, sphincter problems, and sensory disorders are seen in 10% of cases. Pain has a tendency to increase with movement and at night. Neurological deficits are generally seen during the terminal periods of the disease and with late diagnosis.

Extradural spinal tumors

- Metastatic extradural tumors
- Primary extradural tumors

Intradural extramedullary spinal tumors

- Meningioma
- Schwannoma/neurofibroma

Intradural intramedullary spinal tumors

- Astrocytoma
- Ependymoma

Other

- Hemangioblastoma

4.4 Staging

CNS tumors have no TNM designation, and no stage grouping applies. Although there was an old AJCC staging system in brain tumors, it is not used in practice [6]. Therefore, AJCC did not recommend any staging system for CNS tumors. Although staging was available in previous AJCC guidelines (1992), attempts to develop a TNM-based classification and staging system for CNS tumors were neither practical nor feasible. The main reasons for this are as follows.

- For T category, there is no relationship between tumor size and histology.
- For the N category, there is lack of a lymphatic system in the brain and medulla spinalis.
- For category M, most CNS tumors do not metastasize.

Prognostic factors and biomarkers are more important than staging for CNS tumors.

Glial tumors constitute 60% of primary brain tumors and 2/3 of these glial tumors are clinically aggressive high-grade tumors [5].

4.5 Diagnostic Imaging

Magnetic resonance imaging (MRI) is the most common and well-established imaging modality for evaluation of intracerebral neoplasms. Computed tomography (CT) has a much lower resolution and is primarily used for patients who may not undergo MRI, or for thorough assessment of bony structures, or in case of emergencies.

Non-neoplastic conditions that can be confused with high-grade glioma or contrast-enhancing tumors on MRI include subacute infarction, subacute hemorrhage, vascular malformations, and abscess. Demyelination, inflammation, and infections that can be seen with low-grade tumors can also be confused with contrast-enhancing tumors.

- High-grade gliomas are seen as a contrast-enhancing mass lesion with surrounding edema.
- Malignant gliomas are 5% multifocal.
- High-grade gliomas are typically hypointense masses on T1-weighted MR images.
  - Vasogenic edema is common and is seen as a hyperintense signal abnormality on the white matter on T2/FLAIR weighted MRI images
  - Glioblastomas usually show central necrosis or central deletion with cystic changes and a thick ringing (rim) along the edges of the tumor.
- Low-grade gliomas are usually seen as hyperintense lesions with both cortex and underlying white matter on T2/FLAIR-weighted images
  - Vasogenic edema is generally absent.
  - Low-grade gliomas often do not enhance with contrast, but the presence or absence of this condition is not a reliable indicator.
  - Although calcification may be seen occasionally in oligodendrogliomas, it is not specific for this histology.

Open biopsy, endoscopic biopsy, and stereotactic biopsy can be performed for histological diagnosis of intracranial tumors. In addition, both open and closed surgical procedures may be used for diagnosis and treatment.

4.6 Treatment

4.6.1 Low-Grade CNS Tumors

Low-grade gliomas are rare primary tumors of the central nervous system. They often follow an indolent course but may result in death for the majority of patients. Surgery, radiotherapy (RT), and chemotherapy are useful in selected situations, but there are many areas of controversy regarding the optimal management of these patients.

- Patients having a large mass presenting with severe neurological symptoms may be treated with upfront surgery. This allows for establishing the histological diagnosis along with tumor debulking.
- Surgical resection is recommended for patients with transient symptoms and small tumors that do not have a mass effect on adjacent critical structures. However, close follow-up may also be an alternative approach.
- Surgical resection is indicated if there are signs and symptoms of accelerated tumor growth, transformation to high-grade glioma, persistent seizures, or progressive neurological impairment. In surgery, maximal safe resection is recommended rather than debulking [7].

Surgery alone may not be sufficient in patients with low-grade glioma, and additional radiotherapy and chemotherapy treatments may be used. However, optimal timing of additional therapy is controversial. Presence of tumor-related symptoms and the spectrum of molecular prognostic factors should be considered in decision-making for patient selection for immediate postoperative treatment. Individual prognostic factors (including age limit 40 years) are relative indications.

Younger patients (40–45 years) with good prognostic molecular features (for instance, those with isocitrate dehydrogenase [IDH] mutation and 1p/19q-codeletion) who have undergone complete surgery may be followed after surgery [8]. Additional therapy for these patients is indicated in the setting of recurrence and progression.

- Early postoperative treatment is recommended for patients with IDH-wild type tumor or for those with other poor prognostic factors (for instance, residual disease, age > 40–45, neurological deficit, large lesion with massive effect) [8]

In patients with low-grade glioma, radiotherapy and chemotherapy are recommended as the postoperative treatment rather than chemotherapy or RT alone. As adjuvant chemotherapy; there is no consensus on the choice of temozolomide or PCV (procarbazine, lomustine, and vincristine), and treatment selection should be individualized. Particularly for oligodendroglial tumors, there is data supporting the use of PCV (proven to improve survival in a randomized study) [9].

However, there is also data in favor of temozolomide (it is easier to administer, better tolerated, and has shown efficacy in a randomized study of patients with grade III anaplastic glioma without 1p/19q codeletion) [10].

Surgery, radiotherapy, and chemotherapy treatments may be used for the management of recurrent disease after upfront treatment. Selected treatment should be individualized based on the extent of recurrence along with previously received treatments.

Pilocytic astrocytomas have a more favorable prognosis than diffuse astrocytomas. Surgical resection is used for diagnosis and initial treatment, and gross total resection can be curative.

- Radiotherapy is usually indicated in unresectable or progressed disease after surgery [11].

Low-grade or well-differentiated gliomas, especially in children and young adults, constitute a heterogeneous group of intracranial and spinal tumors. Differentiation, pleomorphism, hyperchromasia, cellularity, mitotic index, necrosis, and vascular proliferation are used for the diagnosis and determination of tumor grade.

- One of the most important factors for prognosis is tumor proliferation rate (measured using Ki-67 staining).
- Ki67 index below 1% is a very good prognostic factor and probably a better predictor of quality of life than tumor grade [12]. However, Ki-67 labeling index does not predict which patients should benefit from postoperative radiotherapy [13].
- 1p/19q deletion and isocitrate dehydrogenase mutations are favorable prognostic markers for low-grade glioma.

Seizure is the most common symptom in patients with low-grade glioma.

- Patients who do not have other neurological symptoms other than seizures have a better prognosis than patients presenting with mental state changes or other focal neurological deficits [14].

Age at diagnosis is an important prognostic factor in patients with low-grade glioma.

- Many studies have shown age at diagnosis >40 years to be a poor prognostic factor [15].

Other important poor prognostic factors include astrocytic histology, tumor size ≥6 cm, tumors crossing the midline, and persistence of neurological deficits after resection [15].

Prognostic factors.

- Age ≥ 40
- Astrocytic histology
- Tumor size ≥6 cm
- Tumors crossing the midline
- Persistence of neurological deficits after resection

Low Risk Group: 0–2 prognostic factors (+)
High Risk Group: 3–5 prognostic factors (+)
The median life expectancy in low-risk patients is 7.8 years compared with 3.7 years for the high-risk group [15].

4.6.1.1 Radiotherapy Timing for Low-Grade CNS Tumors

The role of early postoperative radiotherapy is controversial. Postoperative management by radiotherapy or "wait-and-see" for low-grade glioma is poorly defined. The European Organization for Cancer Research and Treatment (EORTC) 22845 study is the largest study investigating the role of postoperative radiotherapy in low-grade glioma [16]. This study has shown that salvage radiotherapy is as effective as early radiotherapy and suggests that selected patients can be followed up safely after tumor resection [16].

4.6.1.2 Radiotherapy Dose for Low-Grade CNS Tumors

There are two randomized phase III studies regarding the dose of radiotherapy for low-grade gliomas.

1. EORTC 22845 study [16].
2. Intergroup study conducted by the North Central Cancer Treatment Group (NCCTG), Radiation Therapy Oncology Group (RTOG), and Eastern Cooperative Oncology Group (ECOG) [17].

According to the results of these studies, the decision when to initiate radiotherapy should be made after a thorough assessment of the patient's risk factors, clinical status, and the risks and potential benefits of treatment. Conformal radiotherapy techniques (including IMRT) are recommended to avoid damage to normal brain tissue.

- A dose of 1.8 Gy/fraction and 50.4–54 Gy is recommended.

4.6.1.3 Chemotherapy for Low-Grade CNS Tumors

RTOG 9802 trial investigated the role of PCV (procarbazine, lomustine, vincristine) chemotherapy in patients receiving radiotherapy for low-grade gliomas [18]. Patients with high-risk disease (age $\geq$ 40 years or subtotal resection) were randomized to 54 Gy radiotherapy in 30 fractions or RT + six cycles of PCV [18]. No difference was demonstrated in five-year OS between the treatment arms (RT; 63%, RT + PCV 72%, $p = 0.13$). However, PFS benefit was found in the PCV arm (RT; 46%, RT + PCV 63%, $p = 0.06$). In the latest update performed after 11.9 years of follow-up, both median survival (13.3 vs 7.8 years, $p = 0.03$) and PFS were longer in the arm treated with PCV (10.4 years vs 4.0 years, $p = 0.002$).

Temozolomide (TMZ) has been shown to be useful for patients with recurrent low-grade gliomas and is being investigated for its potential role as an integrated upfront therapy for low-grade gliomas. Radiation Therapy Oncology Group (RTOG) 0424 was a phase II study of a high-risk low-grade glioma (LGG) population (age 40 years, astrocytoma histology, bilateral hemispheric involvement) who were treated with temozolomide (TMZ) and radiation therapy (RT), and outcomes were compared to those of historical controls [19]. Patients with LGGs with 3 or more risk factors for recurrence (age $\geq$ 40 years, astrocytoma histology, bihemispherical tumor, preoperative tumor diameter of $\geq$6 cm, or a preoperative neurological function status of >1) were treated with RT (54 Gy in 30 fractions) and concurrent and adjuvant TMZ. The 3-year OS rate was 73.1% (95% confidence interval: 65.3–80.8%), which was significantly improved compared to that of prespecified historical control values ($p < 0.001$). Three-year progression-free survival was 59.2%. These results suggest the utility of temozolamide for selected patients with high risk for early recurrence.

4.6.1.4 Chemotherapy Alone for Low-Grade CNS Tumors

Chemotherapy alone has also been investigated as an upfront approach in low-grade oligodendrogliomas [20].

- In the EORTC 22033–26033 study, a total of 477 high-risk WHO grade II patients (aged>40 years, progressive disease, tumor size>5 cm, tumor crossing the midline, or neurological symptoms) were evaluated, and eligible patients were randomly assigned (1:1) to receive either conformal radiotherapy (up to 50.4 Gy; 28 doses of 1.8 Gy once daily, 5 days per week for up to 6.5 weeks) or dose-dense

oral temozolomide (75 mg/m² once daily for 21 days, repeated every 28 days [one cycle], for a maximum of 12 cycles).
- The median progression-free survival was 46 months (95% CI 40–56) with radiotherapy and 39 months (35–44) with temozolomide (unadjusted HR 1.16 [95% CI 0.9–1.5], log-rank p = 0.22). Analyses showed a median progression-free survival of 62 months (95% CI 41–not reached) for IDHmt/codel patients, 48 months (41–55) for IDHmt/non-codel patients, and 20 months (12–26) for IDHwt patients. Patients with IDHmt/non-codel tumors had longer progression-free survival if treated with radiotherapy than with temozolomide.

Classification of low-grade gliomas by evaluating the genomic mutation or genetic changes of gliomas has a prognostic and predictive role in patient assessment. The isocitrate dehydrogenase gene (IDH), 1p/19q, and TP53 combinations are important predictors of tumor behavior. The indolent progression characteristic of low-grade gliomas is more common in tumors with IDH mutations [21].

4.6.2 High-Grade CNS Tumors

Diffuse high-grade gliomas include anaplastic astrocytoma (AA), anaplastic oligodendroglioma (AO), anaplastic oligoastrocytoma (AOA), and glioblastoma (GBM).

- These tumors represent the majority of primary CNS tumors in adults and are 10 times more common in adults than children.
- GBM is estimated to account for 82% of high-grade gliomas [22].

Surgery is the treatment of choice for initial management of high-grade glioma.

- Maximum surgical resection allows for rapid reduction of the mass effect.
- In many retrospective analyzes, maximal surgical resection has been shown to improve outcomes [23].
- If resection cannot be performed safely due to tumor location or comorbid diseases, a biopsy should be performed to establish the diagnosis.
- However, if the tumor is in an area such as the brain stem that will almost certainly result in neurological deterioration, a biopsy is not indispensable. It is also important to rule out a primary brain tumor on MR images from a solitary metastasis, infection, or other disease processes. In addition, patients' age, performance status, and basal neurological function are also important in the selection of patients for surgery [24].

In recent years, a number of prognostic biomarkers have been identified:

- The most common chromosomal abnormality in tumors with oligodendroglial component (AO and AOA), loss of 1p/19q heterozygosity is associated with better treatment response and survival [25].
- O6-methylguanine methyltransferase (MGMT) has been identified as a particularly important prognostic marker in both grade III and grade IV tumors.
- Silencing of this gene by promoter methylation improves survival [26].
  - The importance of this prognostic marker has been further recognized with the development of the alkylating agent temozolamide (TMZ), which is used as the standard chemotherapy in many malignant gliomas.
  - Patients without gene methylation may benefit from TMZ, but the effect is greater in patients with promoter methylation [26].

Mutations in IDH1 and IDH2 have been identified as strong predictors of survival in patients with malignant gliomas as well as in low-grade tumors [27].

- Radiotherapy is recommended after maximal surgical resection for almost all patients with GBM. Despite the aggressive nature of GBM, adjuvant radiotherapy achieves more than twice the median survival compared with supportive care alone [28].
- Radiotherapy should usually commence immediately after adequate wound healing has been achieved (within 2 weeks after surgery).
- MRI-CT fusion is now considered standard for radiotherapy treatment planning [29].

4.6 Treatment

- The typical approach for high-grade glial tumor radiotherapy is to deliver 45–50 Gy of radiotherapy with a 1–2 cm margin [including potential microscopic disease (CTV)] to T2 hyperintense areas. Subsequently, in T1 post-contrast weighted MRI series, the total dose is increased to 60 Gy with boost to the contrast-enhancing area.
  - 3D conformal or IMRT planning can be used depending on the size and location of the tumor.
  - Long-term radiotherapy is not well-tolerated in elderly patients with low-performance status.
  - Radiotherapy alone has been shown to provide a significant survival advantage with minimal side effects compared to supportive care alone [30].
  - In a randomized study comparing 60 Gy in 30 fractions to 40 Gy in 15 fractions, no significant difference was found for patients aged 60 years or older [31].
  - In the prospective randomized NOA-08 study, the option of using TMZ alone instead of the standard 60 Gy RT was also investigated [32]. Temozolomide alone has been found to be noninferior to radiotherapy alone in the treatment of elderly patients with malignant CNS tumors in this study. In addition, the MGMT promoter methylation seems to be a useful biomarker for outcomes by treatment and may assist in decision-making.
- Given the radioresistant nature of malignant gliomas, the role of dose escalation with stereotactic radiosurgery or particle radiotherapy is under investigation.
- Until now, there is little convincing evidence that increasing the dose with these techniques provides a significant advantage [33].
  - Further studies are in progress investigating sophisticated techniques such as carbon ion radiotherapy and boron neutron capture radiotherapy [34].
- A very striking change in management of GBM is the introduction of the alkylating agent TMZ.
  - In a study published in 2005, Stupp et al. showed that for patients with GBM, the use of TMZ concurrently with RT and adjuvant increased the median survival time from 12.1 to 14.6 months and two-year OS from 10 to 26% [35].

    While the median survival time was 23.4 months in patients with (+) MGMT methylation, it was 12.6 months in those with MGMT methylation (−).

    TMZ has been shown to provide a survival advantage in all subgroups, except for patients over 70 years of age [35].
- The therapeutic approach to anaplastic (grade III) gliomas is generally the same as for GBM. Before TMZ, a combination of procarbazine, lomustine, and vincristine (PCV) was frequently used [36].
  - Most treatment centers abandoned PCV chemotherapy due to higher toxicity compared with TMZ.
  - In a study that randomized all forms of anaplastic gliomas, no significant difference was found between upfront radiotherapy, PCV, and TMZ arms with regard to OS or PFS [37].

    However, it is important to note that in this study, 78% of the patients in the chemotherapy arm received salvage radiotherapy and 48% of the patients in the radiotherapy arm received salvage chemotherapy [37].
- Important factors affecting treatment response and survival in high-grade gliomas:
  - Age.
  - Performance status
  - Histological tumor type (anaplastic glioma/glioblastoma)
  - Molecular factors
      O6-methylguanine-DNA methyl transferase (MGMT) methylation
      1p/19q codeletion (mainly in oligodendroglial tumors)
      Isocitrate dehydrogenase type 1 (IDH1) or type 2 (IDH2) mutations
- Median survival times in high-grade gliomas [38]
  - Glioblastoma,
      10–12 months
  - Anaplastic astrocytoma,
      IDH-wild: 2–3 years

- Anaplastic astrocytoma, IDH-mutant: 8–10 years
- Anaplastic oligodendroglioma, IDH-mutant, 1p/19q-codeleted: 15–20 years.

4.6.3 Primary CNS Lymphoma (PCNSL)

Primary CNS lymphoma (PCNSL) is defined as a non-Hodgkin lymphoma in the craniospinal axis (brain, leptomeninx, eyes, and medulla spinalis) without evidence of systemic involvement. It is completely different from systemic lymphoma with CNS involvement which is treated differently. Multiple prognostic factors for PCNSL have been defined including advanced age, poor performance status, increased lactate dehydrogenase (LDH), increased CNS protein levels, and deep brain involvement. Depending on the number of risk factors present, the median survival time can vary from more than 5 years to less than 1 year [39]. PCNSL treatment historically included whole-brain radiotherapy, but despite the high initial response rates, the results were poor. In the 1990s, RTOG investigated the feasibility of using neoadjuvant multi-agent chemotherapy including high-dose methotrexate (MTX) before RT. Although the toxicity rate is high, especially in elderly patients, the addition of chemotherapy provides a significant advantage [40]. The Memorial Sloan Kettering Cancer Center (MSKCC) published its experience using a regimen of reduced-dose whole-brain radiotherapy (23.4 Gy) for complete responders while administering standard-dose whole-brain radiotherapy (45 Gy) to those who responded partially to induction chemotherapy containing MTX and rituximab [41]. In this study with a three-year median follow-up, two-year OS and PFS were reported as 67% and 57%, respectively [41].

Median survival with RT is 9 months in the setting of primary chemotherapy failure [42]. Intrathecal chemotherapy may be considered in patients with multifocal disease [43]. High-dose systemic MTX-based induction therapy is recommended rather than whole-brain RT alone for patients who are eligible candidates for chemotherapy. For patients with good performance status (e.g., ECOG ≤3), a MTX-based combination regimen is recommended instead of MTX chemotherapy alone. In addition to chemotherapy, rituximab administration is also recommended. In patients with complete response, optimal consolidation therapy is not clear, at least half of these patients eventually relapse. Consolidative therapy options include nonmyeloablative chemotherapy (etoposide, cytarabine), autologous hematopoietic cell transplantation (HCT) and high-dose chemotherapy, or whole-brain RT. Selection among therapeutic options is based on multiple factors such as the patient's age, performance status, comorbidity, and patient preferences. In case of recurrence, therapeutic options include re-treatment with high-dose MTX (prior complete remission with this agent), alternative chemotherapy regimens (cytarabine and etoposide), hematopoietic cell transplantation, and whole-brain RT.

4.6.4 Meningiomas

Meningiomas constitute 15–20% of primary brain tumors in the USA and are the most common benign CNS tumors. Spinal meningiomas are rare, representing only 10–12% of all meningiomas. The incidence of meningioma increases with age and peaks around 70 years of age. Meningiomas are more common in women compared to men. This can be attributed to the presence of hormone receptors, such as the progesterone receptor, on the tumor cell surface in women [44]. Indeed, the use of hormone replacement therapy is associated with the risk of developing meningioma [45]. Other risk factors include a history of NF-2 and RT. Meningiomas are typically seen as extra-axial masses on CT and MRI images. They appear isointense in T1-weighted MRI series and show high contrast enhancement with intravenous contrast.

Grad I or benign meningiomas constitute 70–90% of all meningiomas and disease-free survival is achieved in >80% of patients with appropriate treatment modalities. Grade II or atypical meningioma constitutes 15–25% of all cases and the probability of recurrence in 5 years is 7–8 times higher than benign meningiomas. Only 1–2% of meningiomas are anaplastic (grade III) and have a very poor prognosis despite aggressive treatment. All meningiomas, most of which are detected incidentally and rarely become symptomatic, do not require treatment intervention. Surgical resections are curative in the majority (except 20–30%) of patients with benign meningiomas that cause neurological symptoms or are at high risk of developing neurological symptoms [46].

The extent of resection is assessed using the Simpson scale.

Grade I, complete removal including resection of underlying bone and associated dura.

Grade II, complete removal and coagulation of dural attachment.

Grade III, complete removal w/o resection of dura or coagulation.

Grade IV, subtotal resection.

Grade V, simple decompression with or without biopsy [47].

Ten-year recurrence rates are 9% for a Simpson grade I resection and 40% for grade IV. Radiotherapy is frequently used for unresectable, partially resected, or recurrent lesions. The benefit of adding adjuvant RT to partially resected meningiomas has been demonstrated [48]. Radiotherapy doses in the range of 54–60 Gy have been shown to be safe and effective [49]. RT provides long-term local control rates exceeding 90% for 5 years with minimal morbidity [50]. Stereotactic radiosurgery (SRS) is a technique that may be used safely for the management of relatively smaller meningiomas. Encouraging treatment results have been achieved by the use of stereotactic irradiation with typical single-fraction doses ranging from 12 to 20 Gy [51]. With proton RT, high RT doses can be achieved for the tumor that can better cover the tumor tissue in areas close to critical structures, and normal tissue sparing may be improved. Asymptomatic, small-sized meningiomas without signs of edema can be closely followed up diligently without upfront surgery or radiotherapy (RT).

For symptomatic, large, infiltrated, or severely edematous meningiomas in surgically accessible areas, maximal safe resection is recommended instead of RT or follow-up. In addition to providing local tumor control, surgery also provides alleviation of the mass effect and confirms histological diagnosis and tumor grade. It is important to distinguish atypical (WHO grade II) and malignant (WHO grade III) meningiomas from benign (WHO grade I) meningiomas according to specific histopathological criteria, since higher grade tumors have significantly higher rates of local recurrence, morbidity, and mortality.

Diagnosis of atypical or malignant meningioma requires confirmation by histopathology obtained at the time of surgical resection or biopsy.

Neuro-imaging features are not specific enough to differentiate low-grade and high-grade meningiomas without histopathological verification. Imaging features that are suspicious for a higher grade tumor include intratumoral cystic change, peritumoral edema, a decrease in apparent diffusion coefficient, and an increase in cerebral blood volume.

The role of adjuvant RT is dependent on tumor grade and extent of resection. Adjuvant RT is recommended for all malignant meningiomas regardless of the Simpson grade. Radiotherapy rather than postsurgical observation is recommended for atypical meningiomas undergoing complete resection (or biopsy). Potential benefit of RT for atypical meningiomas with gross total resection is related to the risks of adverse effects and late toxicities. Factors that increase the risk of radiation-induced complications include advanced age, low-performance status, wide RT portal, and proximity to critical structures. While adjuvant RT is favored for patients with a low risk of RT-induced complications, close follow-up may also be considered when there is high risk for RT toxicity.

4.6.5 Pituitary Tumors

Pituitary gland tumors are typically benign tumors associated with multiple endocrine symptoms and clinical syndromes. However, tumors require intervention in the presence of mass lesions that cause increased intracranial pressure, effects on the optic apparatus (causing bitemporal hemianopsia and other visual impairments with compression of the optic chiasm) or clinical symptoms due to a disorder in one or more endocrine signaling pathways.

Neuroendocrine abnormalities are typically associated with tumors originating from the anterior pituitary of Rathke's sac origin. These types of tumors make up about 80% of adult pituitary adenomas and may cause abnormal hormone release such as prolactin, corticotropin, follicle-stimulating hormone (FSH), luteinizing hormone LH), growth hormone (GH), and thyroid-stimulating hormone (TSH).

Approximately one-third of pituitary tumors are classified as nonfunctional adenomas and do not secrete detectable levels of pituitary hormones. The standard surgical strategy for resection of pituitary adenomas is the transsphenoidal approach and has supplanted frontotemporal craniotomy [52]. Transsphenoidal approach provides decompression and allows for tissue sampling for detailed pathological analysis. Surgical cure rates approach 90% with this approach using the modern techniques [53]. However, tumors involving the cavernous sinus, clivus, or suprasellar region can rarely be surgically treated. Therefore, adjuvant treatment methods such as external RT or radiosurgery should be considered. RT should also be considered when endocrine abnormalities cannot be controlled with medical treatments such as bromocriptine, cabergoline, or octreotide.

Various techniques can be used for radiotherapy including 3D conformal RT, IMRT, and stereotactic radiotherapy [54]. Nonfunctional adenomas can be treated with 45–50.4 Gy daily in fractions of 1.8 Gy, and functional adenomas with 50.4–54 Gy [54]. Recommended doses for SRS are 15–18 Gy for nonfunctional tumors and 20 Gy for functional tumors (although higher doses of up to 30 Gy can also be used) [55].

Compared to fractionated RT with SRS, the risk of damage to the optic pathway is significantly lower. This toxicity depends on the dose received by the optical apparatus. Particle RT has also emerged as a useful method for the treatment of pituitary adenomas since it can reduce the overall volume of irradiated tissue and decrease the exposure of normal structures in the vicinity of the target volume [56].

For functional tumors, it may take several years for hormones to normalize, and patients often need to continue medical therapy until normalization occurs. In patients receiving pituitary irradiation, secondary pituitary dysfunction requiring hormonal replacement develops in 20% of cases within 5 years [54].

If the pituitary adenoma is smaller than 3 cm and not close to the optic pathway (at least 3 to 5 mm away), SRS is recommended. For adenomas larger than 3 cm and/or closer to the optic pathway, fractionated RT or hypofractionated radiotherapy is recommended.

For SRS of clinically nonfunctioning pituitary adenomas:

- A dose of 18 Gy (14–20 Gy range) is recommended.
- The optical pathway should not receive more than 8–10 Gy.

For fractionated RT for clinically nonfunctional pituitary adenomas:

- A dose of 45–50.4 Gy is recommended (1.8 Gy/fx).
- The optical pathway should not receive more than 54 Gy.

For SRS of clinically functioning pituitary adenomas:

- A dose of 20 Gy (20–25 Gy range) is recommended.
- The optical pathway should not receive more than 8–10 Gy.

For fractionated RT of clinically functioning pituitary adenomas:

- A dose of 50–54 Gy is recommended (1.8 Gy/fx).
- The optical pathway should not receive more than 54 Gy.

While the purpose of RT for nonfunctional pituitary adenomas is to control tumor growth, purpose of RT for functional tumors is biochemical control as well as tumor growth control.

4.6.6 Arteriovenous Malformations

Intracranial arteriovenous malformations (AVMs) are rare vascular anomalies that occur when abnormally large arteries and vessels communicate directly without the capillary bed. Because of this irregular structure, venous drainage is under the effect of a very large hydrodynamic pressure. This pressure creates a 2–4% risk of bleeding per year and leads to catastrophic situations of up to 30% at the first bleeding [57]. Spontaneous hemorrhage is the most common finding and can cause seizures, headaches, and focal neurological signs. The therapeutic goal in AVMs is to prevent bleeding and minimize the side effects associated with therapy. If feasible, surgical resection or embolization is preferred. In deeply located lesions, depending on the size of the lesion, radiotherapy (especially SRS) is an effective alternative for nonsurgical candidates. In SRS, a single dose of 15–25 Gy RT is typically administered, depending on the size and location of the nidus [58]. It is thought that with SRS, a cytokine cascade is released resulting in endothelial proliferation and the vascular abnormality collapses with the eventual obliteration of the vessels.

4.6.7 Vestibular Schwannoma

Vestibular Schwannomas, also known as acoustic neuromas, are tumors that originate from myelin-producing Schwann cells that develop along the vestibular nerve. Except for NF-2, they are always unilateral (for NF-2, bilaterality is pathognomic). Although they account for less than 8% of intracranial tumors, they constitute 80–90% of cerebellopontine angle tumors. Hearing loss is present in 95% of the cases (2/3 of them have tinnitus). Mild imbalance is a common symptom. Facial pain and paralysis may be seen with larger lesions and are among less frequent symptoms. Close follow-up (especially in elderly patients) is a viable option for small and minimally symptomatic tumors. Rapid tumor growth (>2.5 mm/year) rather than tumor size may be a useful indicator for treatment selection. Although surgical resection is possible for majority of lesions, hearing may be preserved in less than 50% of patients and the risk of facial nerve palsy is higher compared to radiotherapy [59]. SRS and fractionated radiotherapy are currently considered as safe and effective primary therapy modalities with local control rates approaching 100% in several studies [60]. Single-fraction doses of 12–14 Gy may be delivered with SRS, and doses up to 54 Gy in 30 fractions may be used for fractionated RT.

4.6.8 Primary Medulla Spinalis Tumors

Primary spinal tumors account for 10–15% of all primary tumors in the CNS. Intradural, extramedullary tumors include meningiomas, neurilemmomas, and chordomas. Those other than chordoma are usually controlled only by surgery and rarely require postoperative radiotherapy. Intramedullary tumors include ependymomas, astrocytomas, oligodendrogliomas, and hemangioblastomas. These tumors often cannot be cured by surgery alone and postoperative radiotherapy is usually required. Doses of 45–50 Gy are typically used considering the radiosensitivity of the spinal cord [61]. Medulla spinalis ependymomas are relatively rare, and differently from their brain counterpart, they are more common in adults than in children. Mean age of onset is >40 years of age and these tumors are typically low grade [62].

Myxopapillary ependymomas typically arise in the lumbosacral region near the filum terminale. Gross total resection is preferred and long-term local control can be achieved in 50% of cases. After subtotal resection, 90–100% local control can be achieved with adjuvant RT [63]. However, these tumors tend to recur elsewhere in the CNS.

4.6.9 Medulloblastoma

Medulloblastoma, an embryonal tumor located in the cerebellum, is a relatively common pediatric brain tumor. The median age at diagnosis is 5–6 years, but medulloblastoma can be seen from infancy to adulthood [64]. Medulloblastoma is more common in males than in females. Medulloblastomas often occur within the midline cerebellar vermis and as they grow they compress and invade the fourth ventricle. As a result, the tumor causes symptoms of increased intracranial pressure by blocking cerebrospinal fluid (CSF) flow in the Sylvian aqueduct and fourth ventricle. Dysmetria and ataxia are more likely to occur in adolescents and young adults as the tumor usually develops more lateral to the cerebellar midline. In the postoperative period, children may develop "posterior fossa syndrome", which includes dysphagia, truncal ataxia, mutism, and rarely respiratory failure. Medulloblastoma is associated with a relatively high risk of CNS spread at diagnosis (11–43%), so extensive preoperative staging is required prior to initiation of therapy [64]. Imaging should be performed within 24–48 h postoperatively to evaluate the resection cavity. Gadolinium-enhanced spinal MRI and a lumbar puncture (LP) for CSF cytology should be performed 10–14 days after surgery to detect metastatic involvement for complete staging.

The extent of the disease is classified as standard risk or high risk [65]:

- Standard risk group is defined as
  – ≥ 3 years of age,
  – postoperative residual tumor ≤1.5 cm²,
  – M0 disease (no spinal involvement detected by MRI and LP).
- High-risk group is defined as
  – postoperative residual tumor>1.5 cm2 and/or, M1
  – age > 3 years

Although not included in this risk classification, anaplastic histology has been shown to have a poorer prognosis compared to patients with classical medulloblastoma, thus standard-risk patients with anaplastic histology should be evaluated and treated like high-risk patients [63]. Medulloblastomas consist of small-round blue cells with many histological variants (e.g., desmoplastic/nodular, large cell, extensive nodularity, anaplastic). Medulloblastomas are highly cellular tumors with molecular genetic classification [5].

With regard to prognosis, in 2012, the medulloblastoma was divided into four main subgroups [66]:

1. Medulloblastoma, WNT-activated
2. Medulloblastoma, SHH-activated, TP53-mutated
3. Group 3 (Group 3 tumors express high MYC levels and have the poorest prognosis of the MB subgroups)
4. Group 4 (Group 4 tumors account for ~35% of all MBs, and are the most frequent subgroup. These tumors exhibit an intermediate prognosis, similar to that of the SHH subgroup)

While cure rates exceed 80% for standard-risk patients receiving craniospinal RT (CSRT) and chemotherapy, 60–70% disease-free survival can be achieved in high-risk patients with more intensive chemotherapy and high-dose RT.

- In standard-risk medulloblastoma therapy, the entire craniospinal axis is treated to 23.4 Gy craniospinal RT (CSRT) [usually in combination with vincristine]. Then, with an additional (boost) of 30.6 Gy to the tumor bed the total dose of RT in this region is increased up to 54 Gy [67].
- In high-risk medulloblastoma or other cerebral embryonal tumors (such as pineoblastoma, ependymoblastoma, and medulloepithelioma), the total RT dose is increased to 55.8 Gy with a boost dose of 19.8 Gy to the posterior fossa after 36 Gy craniospinal RT with concurrent chemotherapy [67].
- The gross tumor volume (GTV) is contoured according to the gadolinium-enhanced T1-weighted signal changes. GTV includes the postoperative resection cavity, all preoperative tumor contact areas, and residual tumor mass.
- The clinical target volume (CTV) is typically defined as a 1.5 cm margin around the GTV and does not include bone or tentorium.
- Planning target volume (PTV) is generated by adding a margin of 0.3–0.5 cm around the CTV.

In cases of medulloblastoma under 3 years of age, postoperative craniospinal radiotherapy (CSRT) may pose significant risk for severe neurological disorders and growth retardation.

Although CSRT can improve disease control in patients under 3 years of age, this therapy should not be used to prolong survival in young children. In this context, initial therapy should include combined chemotherapy for patients under 3 years of age, and CSRT should be deferred until 3 years of age [68].

Medulloblastoma is rare in adults, and there is no randomized study to dictate treatment recommendations especially with regard to the role of chemotherapy. Like children, adults are usually stratified into risk groups based on the extent of surgical resection and the presence or absence of disseminated or metastatic disease, and the treatment strategy is determined accordingly [69].

- Adults with standard-risk disease are recommended combined therapy, including both CSRT and adjuvant multiple chemotherapy. Alternative therapy for patients with low-performance status or multiple comorbidities is CSRT.
- For adults with metastatic, unresectable, or recurrent medulloblastoma, treatment strategy is not clear. These patients should be included in clinical trials whenever possible. Outside of a clinical trial, combined chemotherapy is recommended following CSRT rather than CSRT alone in the majority of patients.

4.6.10 Brain Metastasis

Metastases are the most common intracranial tumors in adults, accounting for more than one-half of all intracranial tumors. Symptoms of brain metastases can be effectively alleviated in selected patients with aggressive local therapy. As life expectancy is often limited by extracranial disease, aggressive treatment of brain metastases is appropriate for those who are expected to survive longer. The strongest predictors of improved survival include good performance status, younger age (under 65 years), and controlled extracranial disease [70]. As more effective systemic therapies emerge, the underlying cancer histology and genotype are gaining increased importance. Management of brain metastases has become increasingly individualized as advanced systemic therapies offer greater potential for both systemic and intracranial disease control for some types of cancer and genotypes.

- Whole-brain RT (WBRT) continues to be the mainstay for the management of patients with good performance who are not suitable for SRS or surgery due to large metastases or multiple metastases.
- Patients with a single brain metastasis are recommended cavity radiosurgery after surgical resection or stereotactic radiosurgery only (SRS) rather than whole-brain RT alone (WBRT) [71].
- Surgical resection is usually performed for large, symptomatic metastases or if there is uncertainty about diagnosis of the primary tumor [72]. SRS is a viable alternative to surgery for small or inaccessible single metastasis.
- In most patients undergoing surgical resection, SRS or fractionated SRT directed at the surgical cavity is recommended rather than adjuvant WBRT or observation [73].
- For limited brain metastases with a size of less than 3 cm, SRS may be preferred instead of SRS + WBRT. Although addition of WBRT increases intracranial control, it does not improve overall survival and may impair quality of life due to adverse effects and neurocognitive deterioration.
- Treatment options for recurrent brain metastases include re-irradiation, surgery (in selected patients), and systemic treatments.

4.7 Radiotherapy

4.7.1 External Radiotherapy

Conventional Simulation

The tumor is located in simulation films by CT and MRI of the patient (preoperative imaging modalities if the patient is operated on).

Accessory equipment: thermoplastic mask (Fig. 4.6), special RT headrest (Fig. 4.7).

- *Patient position*: supine for brain tumor, prone for spinal tumor.

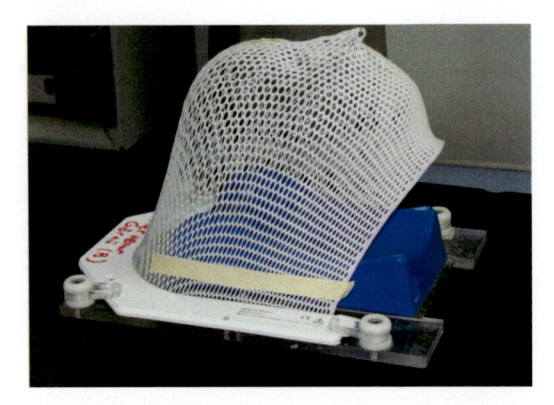

Fig. 4.6 Thermoplastic mask

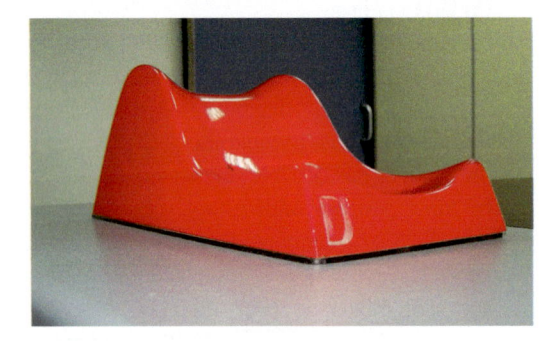

Fig. 4.7 Special RT cushion for head and neck area

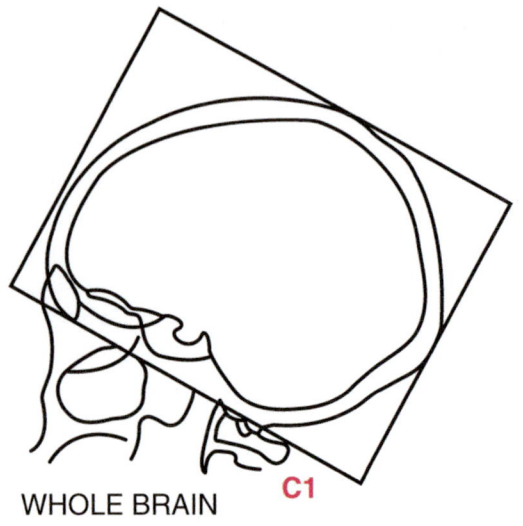

Fig. 4.8 Illustration of whole-cranium RT field

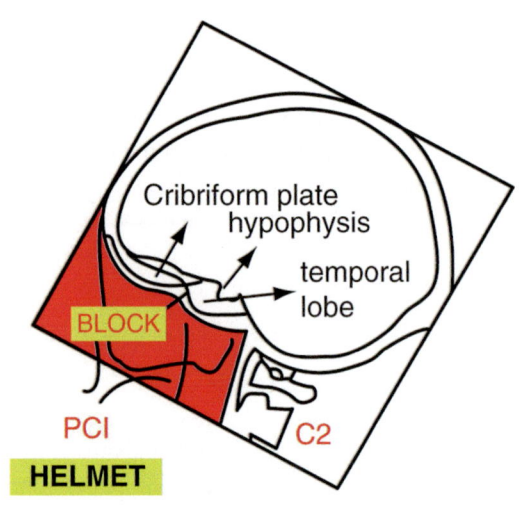

Fig. 4.9 Illustration of helmet-type whole-cranium RT field

4.7.1.1 Brain Tumors
Whole-brain RT (Fig. 4.8)

- Superior: skin fall off
- Anterior: skin fall off
- Posterior: skin fall off
- Inferior: cranial base

In the case of prophylactic cranial irradiation (PCI), the inferior border is the bottom of the C2 vertebra → helmet field.

Helmet field (Fig. 4.9). Anterior border is 2 cm posterior to lens; inferior border is the bottom of the C2 vertebra and 0.5 cm inferior to the cribriform plate in PCI (helmet type → similar to a German helmet in World War II).

CSF circulates around the optic nerve. Therefore, the orbital apex should be included in the helmet field. Corpus vertebrae C1 and C2 should not be spared due to a high risk of recurrence.

Local brain RT (Fig. 4.10)

- After tumor is located, 1–3 cm margin is given.
- For boost: tumor +1 cm.

4.7.1.2 Conformal Planning
GTV → contrast-enhanced field in T1 MRI/CT
  Low-grade tumors

- CTV = GTV + 1.5 cm (if tumor is benign CTV = GTV)
- PTV = CTV + 0.5 cm

High-grade tumors

- Local cranial field
- CTV1 = GTV + 2.5 cm (or high-intensity field in T2 MRI, edema field)

- PTV1 = CTV1 + 0.5 cm
- Boost field
- CTV2 = GTV + 1 cm
- PTV2 = CTV2 + 0.5 cm

Contrast material: 50–70 cc IV bolus should be given in CT simulator.
Technique

Conventional planning → two lateral fields are used.

Conformal planning → varies according to tumor localization, but generally more than two fields are used.

4.7.1.3 Dose
- Low-grade tumor
    Phase I → 40 Gy
    Phase II → 54 Gy
- High-grade tumor
    Phase I → 40 Gy
    Phase II → 60 Gy

Energy → Co-60, 4–18 MV photon energies.

4.7.2 Craniospinal RT

The aim is to irradiate all of the subarachnoid space. The patient is simulated in the prone position and with a thermoplastic head and neck mask (anesthesia may be required in children) (Fig. 4.11).
Cranial field

- Two lateral fields are simulated.
- Inferior border of the field includes cervical vertebrae as much as possible.
- Oral cavity, neck outside spinal cord, nasopharynx is protected.

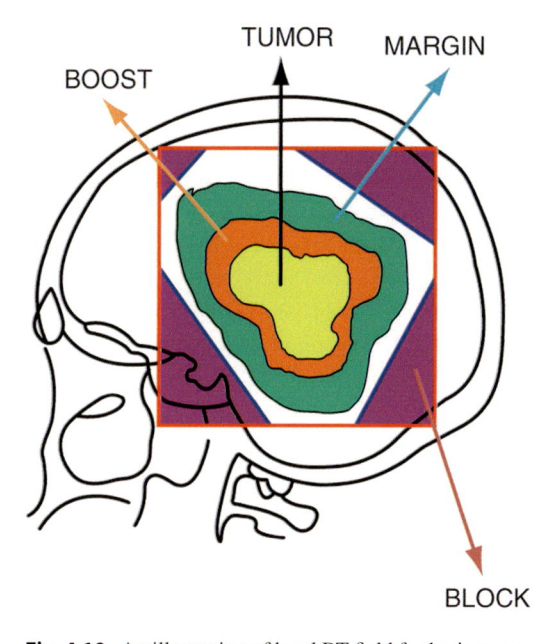

Fig. 4.10 An illustration of local RT field for brain tumor

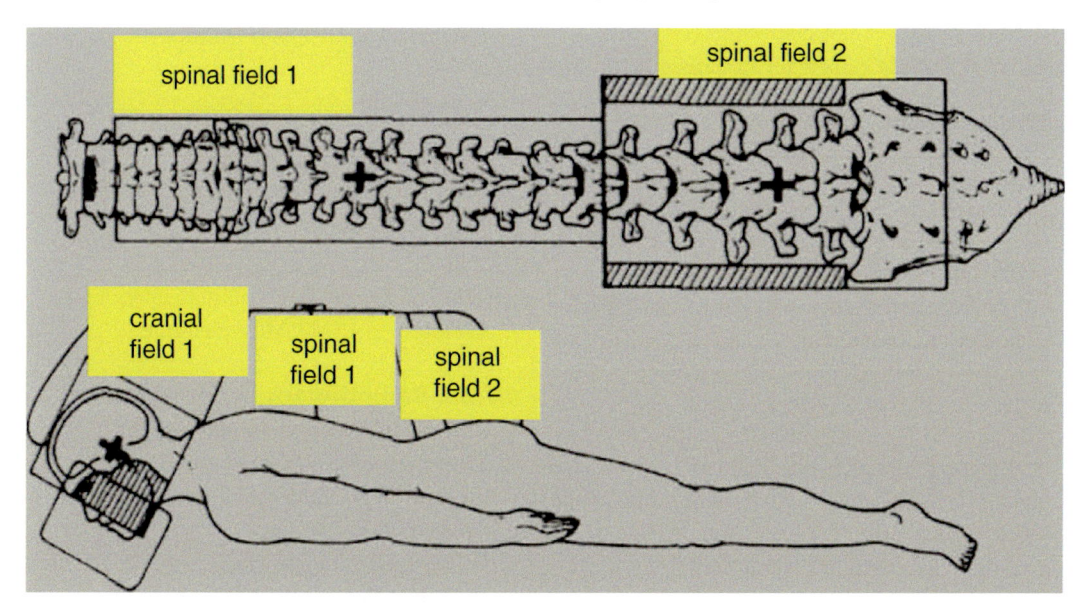

Fig. 4.11 Craniospinal irradiation technique

4.7.2.1 Gap Calculation

$$S1 = 1/2L1\left(\frac{D1}{SSD1}\right) \quad (4.1)$$

$$S2 = 1/2L2\left(\frac{D2}{SSD2}\right) \quad (4.2)$$

$$S = S1 + S2 \quad (4.3)$$

$L1$, $L2$ = length of each field, $D1$, $D2$ = depth of each field, and S = gap length.

Craniospinal simulation for conformal treatments (Fig. 4.14)

 CTV_1: whole brain
 CTV_{1B}: boost field (posterior fossa for medulloblastoma)
 $CTV_{cervical}$: the first spinal field
 CTV_1: the second spinal field
 0.5 cm is added to the CTV for PTV

Dose in craniospinal irradiation

- Whole-brain and spinal field

  1.8 Gy daily fraction dose, total 30–36 Gy

- Posterior fossa boost dose in medulloblastoma

  1.8 Gy daily fraction dose; total 18 Gy posterior fossa boost dose (making the total posterior fossa dose 54 Gy)

Energy in craniospinal irradiation

- Cranial field

  Co-60, 6 MV photon energies

- Spinal field

  Adult → Co-60, 6 MV photon energies
  Children → 15–21 MeV electrons

4.7.2.2 Radiotherapy in Spinal Tumors
- Primary treatment is surgery.
- Primary tumors are generally low-grade glial tumors, but metastatic tumors are more frequent.
- Postoperative radiotherapy may be beneficial for subtotally resected low-grade tumors, but radiotherapy is indicated in all high-grade tumors.

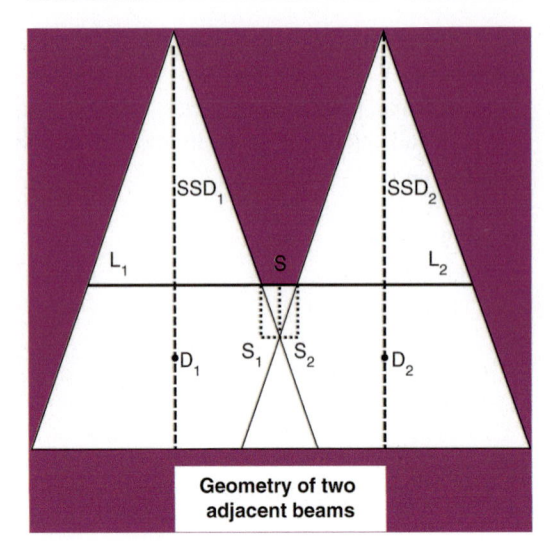

Fig. 4.12 Gap calculation for abutting fields

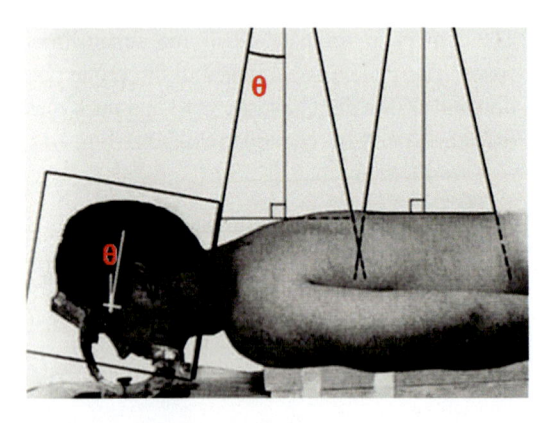

Fig. 4.13 Collimator angle (θ) for cranial field (from [12], reproduced with the permission from Springer Science and Business Media)

Spinal field

- Two separate single fields (single field in children).
- The superior border starts from the inferior border of cranial field.
- Second field ends at the S2 vertebra.
- Field widths vary between 4 and 6 cm.

A gap calculation is needed in order to find the space required between the cranial and spinal fields (Fig. 4.12).

Collimator angle for cranial field (Fig. 4.13):

$$\tan^{-1}\theta = \frac{\text{spinal field length}}{2 \times \text{SSD}} \quad (4.4)$$

4.7 Radiotherapy

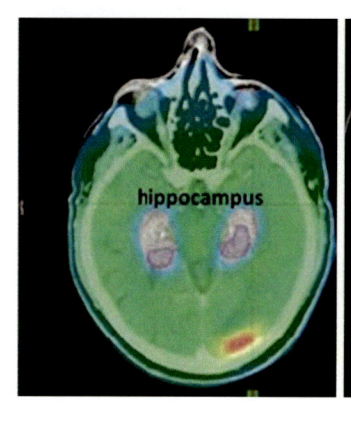

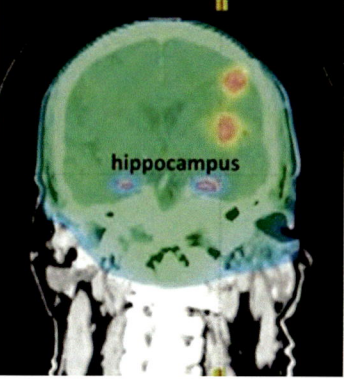

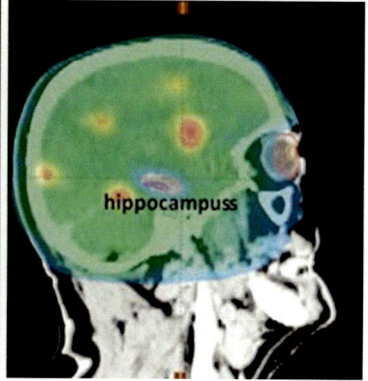

Fig. 4.14 The contouring of hippocampus

- Single field in prone position.
- Superior and inferior borders are one corpus vertebrae corpus from the location of the lesion.
- A suitable energy is selected according to field depth (generally photon energies).
- 1.8–2 Gy daily fractions to a total of 45–50 Gy.
- Metastatic spinal tumors: 30 Gy in ten fractions of palliative RT.

4.7.2.3 3-Dimensional Conformal RT/IMRT (Tables 4.2, 4.3, 4.4, 4.5, 4.6, 4.7, 4.8, 4.9, 4.10, and 4.11)

4.7.3 Symptomatic Treatments

4.7.3.1 Symptoms of Increased Intracranial Pressure and Spinal Cord Compression

16 mg/day dexamethasone is started and continued throughout radiotherapy; the dose is tapered and stopped after RT (an antiulcer medication should always be given with steroids).

- The dose can be increased to 32 mg in the case of spinal cord compression.

Epileptic seizures or prophylaxis for them
- Phenytoin sodium 100 mg 2 × 1 (the dose is adjusted according to blood level).
- Carbamazepine 200 mg 1 × 1 (the dose is adjusted according to blood level).
- Valproate sodium 200 mg 1 × 1 (the dose is adjusted according to blood level).

4.7.3.2 Temozolomide Concurrent with RT

Indicated in high-grade glial tumors
  GBM, anaplastic astrocytoma
  Seventy-five milligrams per square meter per day P.O. during radiotherapy
  Complete blood count should be examined weekly

4.7.3.3 Temozolomide After RT (Adjuvant)

Indicated in high-grade glial tumors
  GBM, anaplastic astrocytoma
  One hundred and fifty milligrams to 200 mg per square meter once every 28 days/for 5 days, P.O. after radiotherapy
  Complete blood count should be examined before each cycle

4.7.4 Side Effects of CNS Radiotherapy

Skin. In the early period: hyperemia, dry squamation, wet squamation, and ulceration. In the late period: telangiectasia, fibrosis, tissue necrosis, and achromia.

Hair. Alopecia starts on the seventh day (temporary at 7 Gy, permanent at >7 Gy).

Table 4.2 Adult supratantorial low grade infiltrative astrocytoma/oligodendroglioma treatment algorithm (excluding pilocytic astrocytoma)

Adult supratentorial low-grade infiltrative astrocytoma/oligodendroglioma (excluding pilocytic astrocytoma)			
Primary brain tumor finding on MRI (+)			
1-maximal safe resection possibility (+)	Low risk (Gross total resection, Age ≤ 40)		• Follow up • RT • Chemo (PCV–TMZ)
	High risk (Subtotal resection, Age > 40)		• RT + adjuvant PCV • RT + adjuvant TMZ • RT + concomitant TMZ + adjuvant TMZ
2-maximal safe resection possibility(−)	Subtotal resection/open biopsy/stereotactic biopsy		• RT + adjuvant PCV • RT + adjuvant TMZ • RT + concomitant TMZ + adjuvant TMZ
3- follow-up			
RT: 3D conformal RT, IMRT (SRS has no significant role in low-grade gliomas) Dose: 45–54 Gy total, 1.8–2 Gy/fraction			
Tumor volume should be contoured according to pre- and postoperative MR imaging. GTV contouring is defined according to the FLAIR and/or T2-weighted image. CTV = GTV + 1–2 cm			
Follow-up: Brain MRI imaging every 3–6 months for the first 5 years, then annually			

Table 4.3 Progressive or recurrent adult supratontorial low grade infiltrative astrocytoma/oligodendroglioma treatment algorithm (excluding pilocytic astrocytoma)

Adult supratentorial low-grade infiltrative astrocytoma/oligodendroglioma (excluding pilocytic astrocytoma)				
Recurrence or progression				
Previous RT history (+)	Resectability (+)	Surgery	Cranial MRI	Chemo
	Resectability (−)	Biopsy (+)	Biopsy (−)	Chemo
Previous RT history (−)	Resectability (+)	Surgery	Cranial MRI	• RT + adjuvant PCV • RT + adjuvant TMZ • RT + concomitant TMZ + adjuvant TMZ
	Resectability (−)			• RT • Chemo
3- İzlem				
RT: 3D conformal RT, IMRT (SRS has no significant role in low-grade gliomas)				
Dose: 45–54 Gy total, 1.8–2 Gy/fraction				
Tumor volume should be contoured according to pre- and postoperative MR imaging. GTV contouring is defined according to the FLAIR and/or T2-weighted image CTV = GTV + 1–2 cm				
Follow-up: Brain MRI imaging every 3–6 months for the first 5 years, then annually				

Mucosa. Erythema, edema, patchy mucositis, and confluent mucositis (mucositis starts at >3 Gy).

Eye. In the early period: conjunctivitis, decreased teardrops, infection. In the late period: necrosis in retina, dry eye, cataract. Optic nerve and optic chiasma damage at >50 Gy, total blindness at >70 Gy.

CNS. Acute side effects depend on edema and dose. Late effects: transient demyelination (somnolence syndrome) and brain necrosis 2–3 years after high dose of radiation. *Spinal cord (L'Hermitte's sign).* Transient demyelination in the spinal cord. Patient feels a sensation like an electrical shock starting from the arms and spreading throughout the body during head flex-

Table 4.4 Anaplastic glioma treatment algorithm

Anaplastic glioma			
Anaplastic Oligodendroglioma (1p19q co-deleted)	RT + PCV (neoadjuvant or adjuvant)		• RT: 3D conformal RT, IMRT
	RT + concomitant TMZ + adjuvant TMZ		• FLAIR and/or T2 weighted sequences are used in GTV contouring
• Anaplastic oligoastrositoma • Anaplastic Astrositoma (good performance status)	RT + PCV (neoadjuvant or adjuvant)		• CTV = GTV + 1–2 cm • PTV = 3–5 mm (less margins with IGRT) • Boost volume includes gross residual tumor and resection cavity
	RT + concomitant TMZ + adjuvant TMZ		
• Anaplastic Astrositoma (poor performance status, KPS < 60)	Hypofractionated RT (preferably)		• RT dose: 60 Gy total, 2 Gy/fx or 59.4 Gy total, 1.8 Gy/fx or 55.8–59.4 Gy total, 1.8 Gy/fx or 57 Gy total, 1.9 Gy/fx
	Standard RT		
Hypofractionated RT (in poor performance or elderly patients)			• 34 Gy/10 fx • 40.5 Gy/15 fx • 50 Gy/20 fx • 25 Gy/5 fx (in elderly/frail patients with small tumors)

Table 4.5 Treatment algorithm for progressed or recurrent anaplastic glioma

Anaplastic glioma			
Recurrence/progression			
Diffuse or multiple	• Inclusion of eligible cases in clinical trials • Palliative care/supportive therapy • Systemic chemotherapy • Surgery (symptomatic lesions) • Alternating electric field therapy		
Localized	Resectable	Cranial MRI	• Inclusion of eligible cases in clinical trials • Palliative care/supportive therapy • Systemic chemo • Re-irradiation • Alternating electric field therapy
	Unresectable		

ion. This appears after 8 weeks of radiotherapy, and heals spontaneously. Myelitis may occur when tolerance doses are exceeded in the spinal cord (Table 4.12).

4.7.4.1 Hippocampal-Sparing Whole-Brain Radiotherapy

The hippocampus, which is the most critical structure of episodic memory function and associated with deep amnestic syndromes when damaged, may be spared in 3D conformal RT/IMRT in contrast with conventional 2-dimensional radiotherapy [79–81] (Fig. 4.15).

- The hippocampus should be contoured by fusion of T1-weighted MRI images with CT-simulation images.
- The contouring of the hippocampus begins at the most caudal part of the temporal horn, which is shaped like a crescent.
- Continue posterocranially along with the medial edge of the temporal horn.
- Uncal recess separates the hippocampus from cerebral gray matter, and the hippocampus is located superior and anterior to the uncal recess.
- The posterocranial extension of the hippocampus is antero-medial to the lateral ventricular atrium.
- Contouring is finished at the lateral edges of the quadrigeminal cystern (just before the fornix crus emerges).

Table 4.6 Treatment algorithm and RT principles in glioblastoma

Glioblastoma				
Age < 70	KPS > 60	MGMT methylation (+)		• RT + concomitant TMZ + adjuvant TMZ
		MGMT methylation (−) or unknown		• RT
	KPS < 60	MGMT methylation (+)		• RT (hypofractionated/standard)
		MGMT methylation (−) or unknown		• Temozolamide • Palliative care/supportive therapy
Age > 70	KPS > 60	MGMT methylation (+)		• Hypofractionated RT + concomitant TMZ + adjuvant TM
		MGMT methylation (−) or unknown		• Temozolamide • Hypofractionated RT
	KPS < 60	MGMT methylation (+) or unknown		• Standard RT + concomitant TMZ + adjuvant TMZ
		MGMT methylation (−) or unknown		• Standard RT + concomitant TMZ + adjuvant TMZ • Hypofractionated RT • TMZ • Palliative care/supportive therapy

Radiotherapy
- RT: 3D conformal RT, IMRT
- Tumor volume should be contoured according to pre- and postoperative MR imaging
- FLAIR and/or T2-weighted series are used in GTV contouring
- CTV = GTV + 2–2.5 cm
- PTV = 3–5 mm
- Boost volume includes gross residual tumor and resection cavity

Standard fractionated RT	• 60 Gy total, 2 Gy/fx or 59.4 Gy total, 1.8 Gy/fx or 55.8–59.4 Gy total, 1.8 Gy/fx or 57 Gy total, 1.9 Gy/fx
Hypofractionated RT (in poor performance or elderly patients)	• 34 Gy/10 fx • 40.5 Gy/15 fx • 50 Gy/20 fx • 25 Gy/5 fx (in elderly/frail patients with small tumors)

Follow up (MRI: 2–4 weeks after RT, every 2–4 months for the next 3 years, then every 6 months)

Table 4.7 Treatment algorithm in progressed or recurrent glioblastoma

Glioblastoma			
Recurrence/progression			
Diffuse or multiple	• Inclusion of eligible cases in clinical trials • Palliative care/supportive therapy • Systemic chemo • Surgery (symptomatic lesions) • Alternating electric field therapy		
Localized	Resectable	Cranial MRI	• Inclusion of eligible cases in clinical trials • Palliative care/supportive therapy • Systemic chemo • Re-irradiation • Alternating electric field therapy
	Unresectable		

Table 4.8 Treatment algorithm for complete/incomplete resected meningioma

Meningioma			
Complete or incomplete resected			
• Complete or incomplete resected meningioma		Grade I	• SRS (12–16 Gy/fx) (in selected cases)
		Grade II	• 54–60 Gy RT (1.8–2 Gy/fx)
• Any grade III meningioma or incompletely resected grade II meningioma		Grade III	• 59.4–60 Gy RT (1.8–2 Gy/fx)

Grade II CTV: GTV (if present) + tumor bed + 1–2 cm margin
Grade III CTV: GTV (if present) + tumor bed + 2–3 cm margin

4.8 Selected Publications

4.8.1 Low-Grade Glial Tumors

Adjuvant RT dose

EORTC 22844 → randomized phase III. 343 cases. Ages 16–65. Histology: astrocytoma (except totally resected pilocytic astrocytoma), oligodendroglioma, mixed oligoastrocytoma. Randomized after surgery to 45 vs. 59.4 Gy. RT field was contrast-enhanced area + 2 cm until 45 Gy, contrast-enhanced area + 1 cm between 45 and 54 Gy, and only the contrast-enhanced area between 54 and 59.4 Gy. Median follow-up was 6.2 years.

- Five-year OS: low dose 58% vs. high dose 9%; 5-year PFS: low dose 47% vs. high dose 50%. No significant difference.
- Conclusion: 45 Gy is adequate.

Karim AB et al. (1996) A randomized trial on dose-response in radiation therapy of low-grade cerebral glioma: European Organization for Research and Treatment of Cancer (EORTC) Study 22844. Int J Radiat Oncol Biol Phys 36(3):549–556.

Table 4.9 Treatment algorithm in unresectable meningioma

Meningioma Unresectable		
• Asymptomatic/symptomatic • Unresectable meningioma • Risk of neurological complications if left untreated	Grade I	• SRS (12–16 Gy/fx) (close to critical structures) • 45–54 Gy RT (1.8–2 Gy/fx)
	Grade II	• 54–60 Gy RT (1.8–2 Gy/fx)
	Grade III	• 59.4–60 Gy RT (1.8–2 Gy/fx)

Grade II CTV: GTV + tumor + 1–2 cm margin
Grade III CTV: GTV + tumor + 2–3 cm margin

Table 4.10 Treatment algorithm for resected symptomatic grade i, complete resected grade ii meningioma

Meningioma Resected symptomatic grade I, complete resected grade II		
• Resected symptomatic grade I • Complete resected grade II	Grade I	• SRS (12–16 Gy/fx) (close to critical structures) • 45–54 Gy RT (1.8–2 Gy/fx
	Grade II	• 54–60 Gy RT (1.8–2 Gy/fx)

Grade II CTV: GTV + tumor bed + 1–2 cm margin

Table 4.11 Treatment algorithm for adult medulloblastoma

Adult medulloblastoma			
• Resected or unresectable • Craniospinal metastasis in MRI/LP (−) • High risk of recurrence (unresectable tumor or residual tumor >1.5 cm², dissemination beyond neuroaxis, large cell/anaplastic histology or supratentorial region)	• Gross total or partial resection • Craniospinal metastasis in MRI/LP (−) • Standard risk (residual tumor <1.5 cm², classical or dermoplastic histology)	• Local brain recurrence • Maximal tumor resection	• Metastatic
High risk	Standard risk	Recurrence/Progression	• Metastatic
• Craniospinal RT + adjuvant chemo	• Standard-dose craniospinal RT • Reduced-dose craniospinal RT + adjuvant Chemo	• Chemotherapy and/or re-irradiation (including SRS)	• Focal RT
	Craniospinal: 36 Gy boost: 54–55.8 Gy		
• Craniospinal: 36 Gy boost: 54–55.8 Gy	• Craniospinal: 23.4 Gy boost: 54–55.8 Gy		
IMRT or proton RT to reduce craniospinal RT toxicity		• SRS should be considered after resection	
Chemotherapy dose may be changed during concomitant chemo + RT treatment. Vincristine may need to be removed from the chemotherapy protocol			

Table 4.12 Tolerance doses of intracranial structures

Organ	Volume	Max. volume	Max. point dose	Clinical endpoint/toxicity ≥ grade 3	Reference
Brain	1/3	TD5/5: 60 Gy TD50/5: 75 Gy		Necrosis/infarct	Emami [74]
	2/3	TD5/5: 50Gy TD50/5: 65 Gy		Necrosis/infarct	
	3/3	TD5/5: 45 Gy TD50/5: 60 Gy		Necrosis/infarct	
	1/3	TD5/5: 72Gy		Necrosis/infarct	Lawrence [75]
Optic nerve and chiasma	1/3	TD5/5: 50 Gy TD50/5: 65 Gy		Visual loss	Emami
	2/3	TD5/5: 50 Gy TD50/5: 65 Gy		Visual loss	
	3/3	TD5/5: 50 Gy TD50/5: 65 Gy		Visual loss	
			55 Gy	Visual loss	Mayo [76]
Brainstem			≤54 Gy	Neuropathy	Daly [77]
	<1 mL	55 Gy		Neuropathy	Schoenfeld [78]
	1/3	TD5/5: 60 Gy TD50/5: 75 Gy		Neuropathy	Emami
	2/3	TD5/5: 53 Gy TD50/5: 70 Gy		Neuropathy	
	3/3	TD5/5: 50 Gy TD50/5: 65 Gy		Neuropathy	
Retina	1/3	TD5/5:45 Gy TD50/5: 65 Gy		Blindness	Emami
	2/3	TD5/5: 45 Gy TD50/5: 65 Gy		Blindness	
	3/3	TD5/5: 45 Gy TD50/5: 65 Gy		Blindness	
Lens	1/3	TD5/5: 10 Gy TD50/5: 18 Gy		Cataract	Emami
	2/3	TD5/5: 10 Gy TD50/5: 18 Gy		Cataract	
	3/3	TD5/5: 10 Gy TD50/5: 18 Gy		Cataract	

TD5/5: Radiation dose that would result in 5% risk of severe complications(≥grade 3) within 5 years after irradiation. TD50/5: Radiation dose that would result in 50% risk of severe complications (≥grade 3) within 5 years after irradiation

4.8 Selected Publications

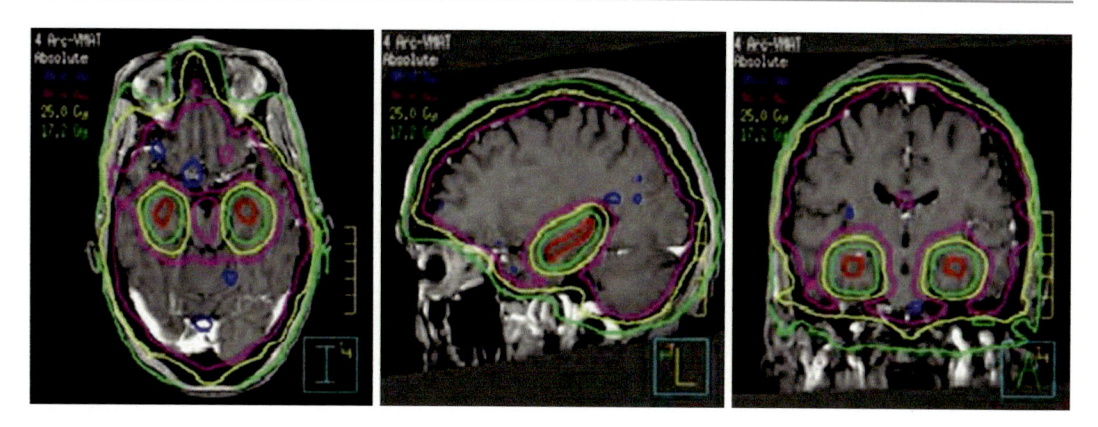

Fig. 4.15 Hippocampus localization in hippocampus-sparing whole-brain radiotherapy

Intergroup NCCTG/RTOG/ECOG → randomized. 203 cases. Ages >18. Histology: astrocytoma (except totally resected pilocytic astrocytoma), oligodendroglioma, mixed oligoastrocytoma. Randomized after surgery to 50.4 vs. 64.8 Gy. RT field was contrast-enhanced area + 2 cm until 50.4 Gy and contrast-enhanced area + 1 cm between 50.4 and 64.8 Gy.

- Five-year OS: low dose 72% vs. high dose 65%; 5-year PFS: low dose 58% vs. high dose 52%. No significant difference.
- Toxicity was greater in the high-dose arm (5 vs. 2.5%).
- Conclusion: high dose is unnecessary and is more toxic.

Shaw E et al. (2002) Prospective randomized trial of low- vs. high-dose radiation therapy in adults with supratentorial low-grade glioma: initial report of a North Central Cancer Treatment Group/Radiation Therapy Oncology Group/Eastern Cooperative Oncology Group study. J Clin Oncol 20(9):2267–2276.

4.8.2 Radiotherapy Timing

EORTC 22845 → randomized. 314 cases. Ages 16–65. Supratentorial low-grade astrocytoma (except for totally resected pilocytic astrocytoma, optic nerve glioma, brain stem glioma, third ventricle glioma, and infratentorial glioma), oligodendroglioma, and mixed oligoastrocytoma. Randomized after surgery to early radiotherapy 50.4 Gy vs. late radiotherapy 50.4 Gy. RT field was contrast-enhanced area + 2 cm until 45 Gy and contrast-enhanced area + 1 cm between 45 and 50.4 Gy. Contrast CT was performed every 4 months until progression.

- Median PFS: 5.3 years in early RT arm; 3.4 years in late RT arm.
- Five-year PFS: 55% in early RT arm; 35% in late RT arm (significant).
- Median OS: 7.4 years in early RT arm; 7.2 years in late RT arm (not significant).
- One-year epilepsy control: 41% in early RT arm; 25% in late RT arm (significant).
- No difference in terms of toxicities.
- Conclusion: early RT increases PFS and controls epileptic seizures, but has no effect on OS.

van den Bent MJ et al. (2005) Long-term efficacy of early vs. delayed radiotherapy for low-grade astrocytoma and oligodendroglioma in adults: the EORTC 22845 randomized trial. Lancet 366(9490)985–990.

4.8.3 Grade III Glial Tumors

EORTC 26591 → randomized phase III. 368 cases. Histologically confirmed anaplastic oligodendroglioma and anaplastic oligoastrocytoma (>25% included oligodendroglial elements).

Median age 49. Randomized after surgery and postoperative RT (45 + 14.4 Gy = 59.4 Gy) to surveillance and six cycles of a PCV (procarbazine 60, lomustine 110, vincristine 1.4 mg/m²) chemotherapy regimen.

- Conclusion: adjuvant PCV increases PFS but has no effect on OS.

van den Bent MJ et al. (2006) Adjuvant procarbazine, lomustine, and vincristine improve progression-free survival but not overall survival in newly diagnosed anaplastic oligodendrogliomas and oligoastrocytomas: a randomized European Organisation for Research and Treatment of Cancer phase III trial. J Clin Oncol 24(18):2715–2722.

4.8.4 High Grade (III–IV) Glial Tumors

RT vs. best supportive care.
 2002 Review → it was demonstrated in six randomized trials that postoperative RT was superior to best supportive care in terms of OS.

- Relative risk was 0.81 in postoperative RT.

Laperriere N (2002) Radiotherapy for newly diagnosed malignant glioma in adults: a systematic review. Radiother Oncol 64(3):259–273.
 SGSG, 1981 → randomized. 118 cases of Grade III/IV histology. Randomized after surgery to (1) 45 Gy whole-brain RT + bleomycin, (2) 45 Gy whole-brain RT, and (3) best supportive care.

- Median OS: 10.8 months in 45 Gy whole-brain RT + bleomycin arm vs. 10.8 months in whole-brain RT arm and 5.2 months in best supportive care arm.
- RT is dramatically superior to best supportive care.
- No effect of bleomycin.

Kristiansen K (1981) Combined modality therapy of operated astrocytomas grade III and IV. Confirmation of the value of postoperative irradiation and lack of potentiation of bleomycin on survival time: a prospective multicenter trial of the Scandinavian Glioblastoma Study Group. Cancer 47(4):649–652.

4.8.5 RT Dose

BTSG, 1979 → 420 cases.

- Median OS: 18 weeks in no RT arm, 28 weeks in 50 Gy RT arm, 36 weeks in 55 Gy RT arm, 42 weeks in 60 Gy RT arm.
- Conclusion: increasing the dose improves survival in malignant gliomas.

Walker MD (1979) An analysis of the dose–effect relationship in the radiotherapy of malignant gliomas. Int Radiat Oncol Biol Phys 5(10):1725–1731.
 RTOG 74–01/ECOG 1374, 1988 → 420 cases. Four arms: 60 Gy whole-brain RT; 60 Gy whole-brain RT + 10 Gy boost; 60 Gy + BCNU; 60 Gy + CCNU + dacarbazine.

- Median OS: 9.3 months in 60 Gy whole-brain RT, 8.2 months in 60 Gy whole-brain RT + 10 Gy boost
- Conclusion: no effect of increasing the dose beyond 60 Gy.

Nelson DF et al. (1988) Combined modality approach to treatment of malignant gliomas—reevaluation of RTOG 7401/ECOG 1374 with long-term follow-up: a joint study of the Radiation Therapy Oncology Group and the Eastern Cooperative Oncology Group. NCI Monogr (6):279–284.
 RTOG 98-03 → phase I/II. 209 cases. Dose escalation with 3D-CRT: 66, 72, 78, 84 Gy (stratification was done according to tumor volume).

- PTV1: GTV + 15 + 3 mm (2 Gy/day until 46 Gy).
- PTV2: GTV + 3 mm (boost) 66, 72, 78, 84 Gy (2 Gy/day).
- Conclusion: dose escalation was possible; there was no dose-limiting toxicity.

Werner-Wasik M et al. (2004) Phase I/II conformal three-dimensional radiation therapy dose escalation study in patients with supratentorial glioblastoma multiforme: report of the Radiation Therapy Oncology Group 98-03 Protocol. ASTRO 2004 Abstr 2769.

Canada 2004 → randomized. 100 cases; >60 years. Two arms: 60 Gy/30 fractions, overall treatment time > 6 weeks; 40 Gy/15 fractions, overall treatment time > 3 weeks. Standard RTOG fields were used.

- Median OS: 5.1 months in the first arm, 5.6 months in the second arm (no significance).
- Corticosteroid requirements were less in the second arm.
- Conclusion: low-dose RT with a shorter overall treatment time can be used in older patients with malignant gliomas.

Roa W (2004) Abbreviated course of radiation therapy in older patients with glioblastoma multiforme: a prospective randomized clinical trial. J Clin Oncol 22(9):1583–1588.

4.8.6 RT Fields (Whole Brain Vs. Localized Field)

Anderson (1991) → retrospective. 60 cases: GBM (39), AA (21), treated between 1982 and 1986. 53 cases with small RT field, 7 cases with whole brain.

- Conclusion: no significant difference between the two RT fields.

Garden AS (1991) Outcome and patterns of failure following limited-volume irradiation for malignant astrocytomas. Radiother Oncol 20(2):99–110.

BTCG 80-01, (1989) → randomized. 571 cases treated between 1982 and 1983. Randomized to two RT arms and three CT arms. RT arms: whole-brain 60.2 Gy and whole brain 43 + 17.2 Gy boost field.

- No difference in OS, but whole brain + boost field is more effective than whole brain.

Shapiro WR (1989) Randomized trial of three chemotherapy regimens and two radiotherapy regimens and two radiotherapy regimens in postoperative treatment of malignant glioma. Brain Tumor Cooperative Group Trial 8001. J Neurosurg 71(1):1–9.

4.8.7 Fractionation

RTOG 83-02 → 786 cases. Phase I/II. AA + GBM. Dose escalation trial of combined hyperfractionation (HF; 1.2 Gy b.i.d. 64.8, 72, 76.8 and 81.6 Gy) RT with BCNU and combined accelerated HF (AHF; 1.6 Gy b.i.d. 48, 54.4 Gy) RT with BCNU.

- Median OS: 9.6 months in AHF, 11 months in HF (no difference in OS in terms of doses).
- OS was better in GBM cases receiving higher RT doses (76.8 and 81.6 Gy).

Werner-Wasik M et al. (1996) Final report of a phase I/II trial of hyperfractionated and accelerated hyperfractionated radiation therapy with carmustine for adults with supratentorial malignant gliomas. Radiation Therapy Oncology Group Study 83-02. Cancer 77(8):1535–1543.

4.8.8 Temozolomide

EORTC/NCI Canada 2005 → randomized. 573 GBM patients. Randomized to RT alone (60 Gy) vs. RT concurrent with daily temozolomide (75 mg/m^2, 7 days/week), then six cycles of adjuvant temozolomide (150–200 mg/m^2 for 5 days during each 28-day cycle). RT field: CTV = GTV + 2–3 cm. Median PFS 4 months, median OS 6.1 months.

- Median follow-up was 28 months. Median survival: 14.6 months in chemoradiotherapy (CRT) arm, 12.1 months in RT-alone arm (significant).
- Two-year OS: 26.5% in CRT arm vs. 10.4% in RT-alone arm.
- Mortality decreased 37% in CRT arm.

- Progression was 85% in CRT arm vs. 94% in RT-alone arm.
- Conclusion: current standard therapy for GBM is RT + temozolomide.

Stupp R et al. (2005) Radiotherapy plus concomitant and adjuvant temozolomide for glioblastoma. N Engl J Med 352(10):987–996.

4.9 Pearl Boxes

4.9.1 Good Prognostic Factors in Low-Grade Glial Tumors

- Young age
- Good performance status
- Oligodendroglial component
- Gross total resection

Oligodendrogliomas are chemosensitive → BCNU (carmustine), procarbazine, vincristine.

4.9.2 Prognostic Factors in High-Grade Tumors

Tumor- and patient-related factors

- Age
- KPS
- Histology
- Symptom duration
- Tumor localization
- Tumor size

Treatment-related factors

- Status of surgery extent
- Postoperative residue size
- Radiotherapy dose
- Chemotherapy usage

4.9.3 Tumors Capable of Subarachnoid Seeding via CSF

- Medulloblastoma
- Ependymoma (high grade)
- Pinealoblastoma
- Germ cell tumors

Bone marrow: 20% in cranium, 25% in spinal column, and 10% in spongy bone. Therefore, approximately 50% of the bone marrow is irradiated in craniospinal irradiation.

Posterior fossa syndrome This syndrome may be observed after medulloblastoma surgery, and is characterized by mutism, dysphagia, and truncal ataxia lasting more than a month. This is not a contraindication for craniospinal RT of medulloblastoma.

The most frequently seen genetic mutation in brain tumors is p53 mutation (17p).

CNS radiotherapy actually acts on oligodendrocytes and glial cells. Although radiation affects neurons, this effect does not cause any actual CNS damage.

RT can be given to some brain tumors without any histopathological diagnosis, including:

- Optic glioma
- Diffuse pontine glioma
- Meningioma
- Choroidal melanoma

4.9.4 CNS Tumors That Can Metastasize Outside the CNS

- Glioblastoma
- Medulloblastoma
- Ependymoma
- Meningioma
- Hemangiopericytoma

The mean tumor doubling time is 39.5 days in malignant gliomas.

References

1. Siegel R, Miller K, Jemal A (2017) Cancer treatment and survivor statistics, 2017. CA Cancer J Clin 67(1):7–30
2. Fletcher CD, Hogendoorn P, Mertens F, Bridge J (2013) WHO classification of tumours of soft tissue and bone, 4th edn. IARC Press, Lyon

References

3. Tillmann B (2005) Atlas der Anatomie des Menschen. Springer, Heidelberg, pp 120–122
4. Tillmann B (2005) Atlas der Anatomie des Menschen. Springer, Heidelberg, pp 240–242
5. Louis DN, Perry A, Reifenberger G et al (2016) The 2016 World Health Organization classification of tumors of the central nervous system: a summary. Acta Neuropathol 131(6):803–820
6. Tonn JC, Westphal M, Rutka JT, Grossman SA (eds.) (2006) Neurooncology of CNS tumors. Springer, pp 620–36
7. Pouratian N, Schiff D (2010) Management of low-grade glioma. Curr Neurol Neurosci Rep 10(3):224
8. Leu S, von Felten S, Frank S (2013) IDH/MGMT-driven molecular classification of low- grade glioma is a strong predictor for long-term survival. Neuro Oncol. 15(4):469–479. Epub 2013 Feb 13
9. Buckner JC, Pugh SL, Shaw EG, et al. (2014) Phase III study of radiation therapy (RT) with or without procarbazine, CCNU, and vincristine (PCV) in low-grade glioma: RTOG 9802 with Alliance, ECOG, and SWOG (abstract). J Clin Oncol 32:5s (suppl; abstr 2000). http://meetinglibrary.asco.org/content/127483-144
10. Weller M, van den Bent M, Tonn JC (2017) European Association for Neuro-Oncology (EANO) guideline on the diagnosis and treatment of adult astrocytic and oligodendroglial gliomas. Lancet Oncol 18(6):e315. Epub 2017 May 5
11. Armstrong GT, Conklin HM, Huang S (2011) Survival and long-term health and cognitive outcomes after low-grade glioma. Neuro Oncol 13(2):223. Epub 2010 Dec 22
12. Montine TJ, Vandersteenhoven JJ, Aguzzi A et al (1994) Prognostic significance of Ki-67 proliferation index in supratentorial fibrillary astrocytic neoplasms. Neurosurgery 34(4):674–678; discussion 678–679
13. Fisher BJ, Naumova E, Leighton CC et al (2002) Ki-67: a prognostic factor for low-grade glioma? Int J Radiat Oncol Biol Phys 52(4):996–1001
14. McCormack BM, Miller DC, Budzilovich GN, Voorhees GJ, Ransohoff J (1992) Treatment and survival of low-grade astrocytoma in adults—1977–1988. Neurosurgery 31(4):636–642; discussion 642
15. Pignatti F, van den Bent M, Curran D et al (2002) Prognostic factors for survival in adult patients with cerebral low-grade glioma. J Clin Oncol 20(8):2076–2084
16. van den Bent MJ, Afra D, de Witte O et al (2005) Long-term efficacy of early versus delayed radiotherapy for low-grade astrocytoma and oligodendroglioma in adults: the EORTC 22845 randomised trial. Lancet 366(9490):985–990
17. Shaw E, Arusell R, Scheithauer B et al (2002) Prospective randomized trial of low versus high-dose radiation therapy in adults with supratentorial low-grade glioma: initial report of a North Central Cancer Treatment Group/Radiation Therapy Oncology Group/Eastern Cooperative Oncology Group study. J Clin Oncol 20(9):2267–2276
18. Shaw EG, Wang M, Coons SW et al (2012) Randomized trial of radiation therapy plus procarbazine, lomustine, and vincritine chemotherapy for supratentorial adult low-grade glioma: initial results of RTOG 9802. J Clin Oncol 30(25):3065–3070
19. Fisher BJ, Hu C, Macdonald DR et al (2015) Phase 2 study of temozolomide-based chemoradiation therapy for high-risk low-grade gliomas: preliminary results of Radiation Therapy Oncology Group 0424. Int J Clin Oncol Biol Phys 91(3):497–504
20. Pace A, Vidiri A, Galie E et al (2003) Temozolomide chemotherapy for progressive low-grade glioma: clinical benefits and radiological response. Ann Oncol 14(12):1722–1726
21. Cancer Genome Atlas Research Network, Brat DJ, Verhaak RG et al (2015) Comprehensive, integrative genomic analysis of diffuse lower-grade gliomas. N Engl J Med 372(26):2481–2498
22. Gunderson LL, Tepper JE (2007) Clinical radiation oncology, 2nd edn. Elsevier Churchill Livingstone, Philadelphia
23. Mirimanoff RO, Gorlia T, Mason W et al (2006) Radiotherapy and temozolomide for newly diagnosed glioblastoma: recursive partitioning analysis of the EORTC 26981/22981-NCIC CE3 phase III randomized trial. J Clin Oncol 24(16):2563–2569
24. Gorlia T, van den Bent MJ, Hegi ME et al (2008) Nomograms for predicting survival of patients with newly diagnosed glioblastoma: prognostic factor analysis of EORTC and NCIC trial 26981-22981/CE.3. Lancet Oncol 9(1):29–38
25. Bello MJ, Leone PE, Vaquero J et al (1995) Allelic oss at 1p and 19q frequently occurs in association and may represent early oncogenic events in oligodendroglial tumors. Int J Cancer 64(3):207–210
26. Hegi ME, Diserens AC, Gorlia T et al (2005) MGMT gene silencing and benefit from temozolomide in glioblastoma. N Engl J Med 352(10):997–1003
27. Parsons DW, Jones S, Zhang X et al (2008) An integrated genomic analysis of human glioblastoma multiforme. Science 321(5897):1807–1812
28. Kristiansen K, Hagen S, Kollevold T et al (1981) Combined modality therapy of operated astrocytomas grade III and IV. Confirmation of the value of postoperative irradiation and lack of potentiation of bleomycin on survival time: a prospective multicenter trial of the Scandinavian Glioblastoma Study Group. Cancer 47(4):649–652
29. Thornton AF Jr, Sandler HM, Ten Haken RK et al (1992) The clinical utility of magnetic resonance imaging in 3-dimensional treatment planning of brain neoplasms. Int J Clin Oncol Biol Phys 24(4):767–775
30. Keime-Guibert F, Chinot O, Taillandier L et al (2007) Radiotherapy for glioblastoma in the elderly. N Engl J Med 356(15):1527–1535
31. Roa W, Brasher PM, Bauman G et al (2004) Abbreviated course of radiation therapy in older patients with glioblastoma multiforme: a prospective randomized clinical trial. J Clin Oncol 22(9):1583–1588
32. Wick W, Platten M, Meisner C et al (2012) Temozolomide chemotherapy alone versus radiotherapy alone for malignant astrocytoma in the elderly:

the NOA-08 randomised, phase 3 trial. Lancet Oncol 13(7):707–715
33. Tsao MN, Mehta MP, Whelan TJ et al (2005) The American Society for Therapeutic Radiology and Oncology (ASTRO) evidence-based review of the role of radiosurgery for malignant glioma. Int J Clin Oncol Biol Phys 63(1):47–55
34. Mizumoto M, Tsuboi K, Igaki H et al (2010) Phase I/II trial of hyperfractionated concomitant boost proton radiotherapy for supratentorial glioblastoma multiforme. Int J Clin Oncol Biol Phys 77(1):98–105
35. Stupp R, Hegi ME, Mason WP et al (2009) Effects of radiotherapy with concomitant and adjuvant temozolomide versus radiotherapy alone on survival in glioblastoma in a randomised phase III study: 5-year analysis of the EORTC- NCIC trial. Lancet Oncol 10(5):459–466
36. Stupp R, Mason WP, van den Bent MJ et al (2005) Radiotherapy plus concomitant and adjuvant temozolomide for glioblastoma. N Engl J Med 352(10):987–996
37. Wick W, Hartmann C, Engel C et al (2009) NOA-04 randomized phase III trial of sequential radiochemotherapy of anaplastic glioma with procarbazine, lomustine, and vincristine or temozolomide. J Clin Oncol 27(35):5874–5880
38. Ostrom QT, Gittleman H, Xu J et al (2016) CBTRUS statistical report: primary brain and other central nervous system tumors diagnosed in the United States in 2009–2013. Neuro Oncol 18(suppl_5):v1
39. Abrey LE, Ben-Porat L, Panageas KS et al (2006) Primary central nervous system lymphoma: the Memorial Sloan-Kettering Cancer Center prognostic model. J Clin Oncol 24(36):5711–5715
40. Nelson DF, Martz KL, Bonner H et al (1992) Non-Hodgkin's lymphoma of the brain: can high dose, large volume radiation therapy improve survival? Report on a prospective trial by the Radiation Therapy Oncology Group (RTOG): RTOG 8315. Int J Clin Oncol Biol Phys 23(1):9–17
41. Shah GD, Yahalom J, Correa DD et al (2007) Combined immunochemotherapy with reduced whole-brain radiotherapy for newly diagnosed primary CNS lymphoma. J Clin Oncol 25(30):4730–4735
42. Nguyen PL, Chakravarti A, Finkelstein DM, Hochberg FH, Batchelor TT, Loeffler JS (2005) Results of whole-brain radiation as salvage of methotrexate failure for immunocompetent patients with primary CNS lymphoma. J Clin Oncol 23(7):1507–1513
43. Korfel A, Schlegel U (2013) Diagnosis and treatment of primary CNS lymphoma. Nat Rev Neurol 9(6):317–327
44. Donnell MS, Meyer GA, Donegan WL (1979) Estrogen-receptor protein in intracranial meningiomas. J Neurosurg 50(4):499–502
45. Blitshteyn S, Crook JE, Jaeckle KA (2008) Is there an association between meningioma and hormone replacement therapy? J Clin Oncol 26(2):279–282
46. Mirimanoff RO, Dosoretz DE, Linggood RM, Ojemann RG, Martuza RL (1985) Meningioma: analysis of recurrence and progression following neurosurgical resection. J Neurosurg 62(1):18–24
47. Simpson D (1957) The recurrence of intracranial meningiomas after surgical treatment. J Neurol Neurosurg Psychiatry 20(1):22–39
48. Miralbell R, Linggood RM, de la Monte S, Convery K, Munzenrider JE, Mirimanoff RO (1992) The role of radiotherapy in the treatment of subtotally resected benign meningiomas. J Neurooncol 13(2):157–164
49. Goldsmith BJ, Wara WM, Wilson CB, Larson DA (1994) Postoperative irradiation for subtotally resected meningiomas. A retrospective analysis of 140 patients treated from 1967 to 1990. J Neurosurg 80(2):195–201
50. Solda F, Wharram B, De Ieso PB, Bonner J, Ashley S, Brada M (2013) Long-term efficacy of fractionated radiotherapy for benign meningiomas. Radiother Oncol 109(2):330–334
51. Pollock BE, Stafford SL, Utter A, Giannini C, Schreiner SA (2003) Stereotactic radiosurgery provides equivalent tumor control to Simpson grade 1 resection for patients with small- to medium-size meningiomas. Int J Clin Oncol Biol Phys 55(4):1000–1005
52. Mortini P, Losa M, Barzaghi R, Boari N, Giovanelli M (2005) Results of transsphenoidal surgery in a large series of patients with pituitary adenoma. Neurosurgery 56(6):1222–1233; discussion 1233
53. Jane JA Jr, Thapar K, Kaptain GJ, Maartens N, Laws ER Jr (2002) Pituitary surgery: transsphenoidal approach. Neurosurgery 51(2):435–442; discussion 442-444
54. Loeffler JS, Shih HA (2011) Radiation therapy in the management of pituitary adenomas. J Clin Endocrinol Metab 96(7):1992–2003
55. Sheehan JP, Pouratian N, Steiner L, Laws ER, Vance ML (2011) Gamma knife surgery for pituitary adenomas: factors related to radiological and endocrine outcomes. J Neurosurg 114(2):303–309
56. Wattson DA, Tanguturi SK, Spiegel DY et al (2014) Outcomes of proton therapy for patients with functional pituitary adenomas. Int J Radiat Oncol Biol Phys 90:532
57. Stefani MA, Porter PJ, terBrugge KG, Montanera W, Willinsky RA, Wallace MC (2002) Large and deep brain arteriovenous malformations are associated with risk of future hemorrhage. Stroke 33(5):1220–1224
58. Engenhart R, Wowra B, Debus J et al (1994) The role of high-dose, singlefraction irradiation in small and large intracranial arteriovenous malformations. Int J Clin Oncol Biol Phys 30(3):521–529
59. Karpinos M, The BS, Zeck O et al (2002) Treatment of acoustic neuroma: stereotactic radiosurgery versus microsurgery. Int J Clin Oncol Biol Phys 54(5):1410–1421
60. Andrews DW, Werner-Wasik M, Den RB et al (2009) Toward döşe optimization for fractionated stereotactic radiotherapy for acoustic neuromas: comparison of two dose cohorts. Int J Clin Oncol Biol Phys 74(2):419–426

References

61. Marks LB, Yorke ED, Jackson A et al (2010) Use of normal tissue complication probability models in the clinic. Int J Clin Oncol Biol Phys 76(3 Suppl):S10–S19
62. Hanbali F, Fourney DR, Marmor E et al (2002) Spinal cord ependymoma: radical surgical resection and outcome. Neurosurgery 51(5):1162–1172; discussion 1172–1174
63. McLaughlin MP, Marcus RB Jr, Buatti JM et al (1998) Ependymoma: results, prognostic factors and treatment recommendations. Int J Clin Oncol Biol Phys 40(4):845–850
64. Gajjar A, Chintagumpala M, Ashley D et al (2006) Risk-adapted craniospinal radiotherapy followed by high-dose chemotherapy and stem-cell rescue in children with newly diagnosed medulloblastoma (St Jude Medulloblastoma-96): longterm results from a prospective, multicentre trial. Lancet Oncol 7(10):813–820
65. Polkinghorn WR, Tarbell NJ (2007) Medulloblastoma: tumorigenesis, current clinical paradigm, and efforts to improve risk stratification
66. Taylor MD, Northcott PA, Korshunov A et al (2012) Molecular subgroups of medulloblastoma: the current consensus. Acta Neuropathol 123(4):465–472
67. Packer RJ, Sutton LN, Elterman R et al (1994) Outcome for children with medulloblastoma treated with radiation and cisplatin, CCNU, and vincristine chemotherapy. J Neurosurg 81(5):690–698
68. Duffner PK, Horowitz ME, Krischer JP, Friedman HS, Burger PC, Cohen ME, Sanford RA, Mulhern RK, James HE, Freeman CR (1993) Postoperative chemotherapy and delayed radiation in children less than three years of age with malignant brain tumors. N Engl J Med 328(24):1725
69. Friedrich C, von Bueren AO, von Hoff K, Kwiecien R, Pietsch T, Warmuth-Metz M, Hau P, Deinlein F, Kuehl J, Kortmann RD, Rutkowski S (2013) Treatment of adult nonmetastatic medulloblastoma patients according to the paediatric HIT 2000 protocol: a prospective observational multicentre study. Eur J Cancer 49(4):893–903. Epub 2012 Nov 19
70. Borgelt B, Gelber R, Kramer S, Brady LW, Chang CH, Davis LW, Perez CA, Hendrickson FR (1980) The palliation of brain metastases: final results of the first two studies by the Radiation Therapy Oncology Group. Int J Radiat Oncol Biol Phys. 6(1):1
71. Patchell RA, Tibbs PA, Walsh JW, Dempsey RJ, Maruyama Y, Kryscio RJ, Markesbery WR, Macdonald JS, Young B (1990) A randomized trial of surgery in the treatment of single metastases to the brain. N Engl J Med 322(8):494
72. Sawaya R, Hammoud M, Schoppa D, Hess KR, Wu SZ, Shi WM, Wildrick DM (1998) Neurosurgical outcomes in a modern series of 400 craniotomies for treatment of parenchymal tumors. Neurosurgery 42(5):1044
73. Brown PD, Ballman KV, Cerhan JH et al (2017) Postoperative stereotactic radiosurgery compared with whole brain radiotherapy for resected metastatic brain disease (NCCTG N107C/CEC·3): a multicentre, randomised, controlled, phase 3 trial. Lancet Oncol 18(8):1049. Epub 2017 Jul 4
74. Emami B, Lyman J, Brown A et al (1991) Tolerance of normal tissue to therapeutic irradiation. Int J Radiat Oncol Biol Phys 21:109–122
75. Lawrence YR, Li XA, el Naqa I et al (2010) Radiation dose–volume effects in the brain. Int J Radiat Oncol Biol Phys 76:S20–S27
76. Mayo C, Martel MK, Marks LB et al (2010) Radiation dose-volume effects of optic nerves and chiasm. Int J Radiat Oncol Biol Phys 76:S28–S35
77. Daly ME, Chen AM, Bucci MK et al (2007) Intensity-modulated radiation therapy for malignancies of the nasal cavity and paranasal sinuses. Int J Radiat Oncol Biol Phys 67:151–157
78. Schoenfeld GO, Amdur RJ, Morris CG et al (2008) Patterns of failure and toxicity after intensity-modulated radiotherapy for head and neck cancer. Int J Radiat Oncol Biol Phys 71:377–385
79. Dickerson BC, Eichenbaum H (2010) The episodic memory system: neurocircuitry and disorders. Neuropsychopharmacology 35:86–104
80. Chera BS, Amdur RJ, Patel P et al (2009) A radiation oncologist's guide to contouring the hippocampus. Am J Clin Oncol 32:20–22
81. Gondi V, Pugh SL, Tome WA et al (2014) Preservation of memory with conformal avoidance of the hippocampal neural stem-cell compartment during whole-brain radio-therapy for brain metastases (RTOG 0933): a phase II multi- institutional trial. J Clin Oncol 32:3

Head and Neck Cancers

5

Head and neck cancers constitute approximately 10% of all cancers [1]. 600,000 patients are diagnosed with head and neck cancer every year worldwide. More than half of these patients present with a locally advanced disease. Head and neck cancers are more common in males than females. They are mostly seen in elder patients (50–70 years) but can also be seen in children.

Many risk factors for the development of head and neck cancers have been defined. Certain cancer types are more frequently observed in certain geographic regions (e.g., nasopharyngeal cancer in the Far East Asia). Smoking and alcohol use have a very close association with head and neck cancers. Chewing tobacco and tobacco-like substances also increase the risk of oral cavity cancers [2]. Genetic tendency is another important risk factor. A previous history of head and neck cancer as well as a history of another cancer in the first-degree family members increases the risk. Besides, multiple head and neck cancers can be seen simultaneously or metachronously in the same patient. Exposure to radiation as well as to the Sun (ultraviolet radiation) is closely related to an increased risk of head and neck cancer [3]. Nutritional disorders and vitamin deficiencies are other risk factors. Bad nutritional habits and iron deficiency anemia particularly in women can cause head and neck cancers [4]. Poor oral hygiene, use of inappropriate prostheses, chronic infections, gastroesophageal reflux, and particular viral infections (Epstein-Barr virus [EBV], human papilloma virus [HPV]) are additional risk factors [5].

Table 5.1 Head and neck lymphatics

Level	Lymphatics
Ia	Submental lymphatics
Ib	Submandibular lymphatics
II	Upper jugular lymph nodes
III	Mid-jugular lymph nodes
IV	Lower jugular lymph nodes (transverse cervical)
V	Spinal accessory chain lymph nodes (posterior triangle)
VI	Prelaryngeal, pretracheal, paratracheal lymph nodes
VII	Upper mediastinal lymph nodes

The head and neck region has a rich lymphatic network (Table 5.1). This network is divided into sublevels defined by the American Joint Committee on Cancer (AJCC) for the purposes of neck dissection and radiotherapy (Fig. 5.1) [6]. A more comprehensive guideline was proposed by a joint of study groups experienced in head and neck cancer radiotherapy in 2014 which suggested new neck levels [7] (Table 5.2). However, the AJCC levels are used universally for practical concerns. The borders for neck node levels are shown in Table 5.3. Certain lymphatics have special names; Virchow's node is used for supraclavicular; Delphian's node is used for the precricoid node; and Rouviere's node is the most superior node in the retropharyngeal region.

Each site-specific cancer has its specific route for lymph node metastasis (Fig. 5.2).

Fig. 5.1 Lymphatic levels of neck for head and neck cancers

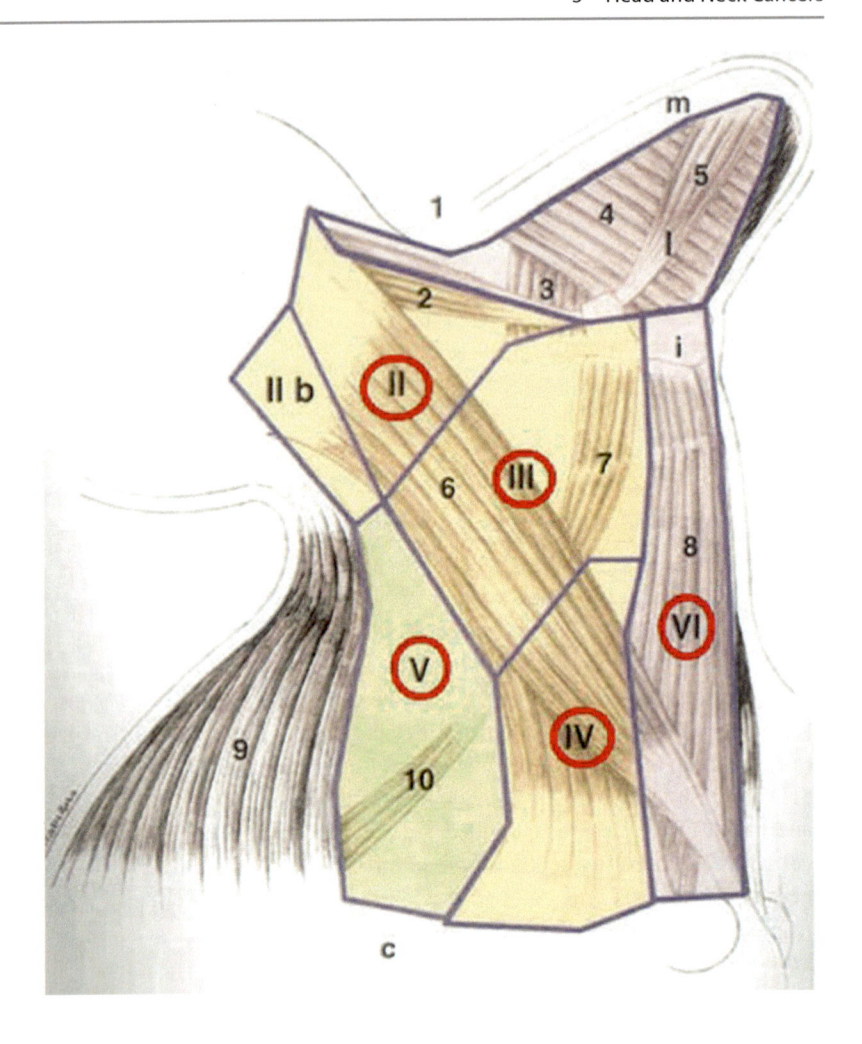

Table 5.2 Updated neck node levels [7]

Level	Terminology
Ia	Submental group
Ib	Submandibular group
II	Upper jugular group
III	Middle jugular group
IVa	Lower jugular group
IVb	Medial supraclavicular group
Va	Upper posterior triangle nodes
Vb	Lower posterior triangle nodes
Vc	Lateral supraclavicular group
VIa	Anterior jugular nodes
VIb	Prelaryngeal, pretracheal, and paratracheal nodes
VIIa	Retropharyngeal nodes
VIIb	Retrostyloid nodes
VIII	Parotid group
IX	Bucco-facial group
Xa	Retroauricular and subauricular nodes
Xb	Occipital nodes

Over 30% of all head–neck cancer cases show clinical lymph node positivity (Table 5.3) [9]:
- Pharyngeal wall cancer: 50%
- Pyriform sinus cancer: 49%
- Supraglottic laryngeal cancer: 39%

Head–neck cancers with clinical neck lymph node (−) but pathological lymph node (+) (Table 5.4) [9]:
- Pyriform sinus cancer: 59%
- Pharyngeal wall cancer: 37%
- Tongue cancer: 33%
- Supraglottic laryngeal cancer: 26%
- Floor of mouth cancer: 21%
- Glottic laryngeal cancer: 15%

Table 5.3 Lymphatic involvement ratios in various head–neck cancers (%)

Region	Level I	Level II	Level III	Level IV	Level V	RP
Nasopharynx	17	94	85	19	61	86
Glottic larynx	6	61	54	30	6	
Supraglottic larynx	6	61	54	30	6	4
Piriform sinus	2	77	57	23	22	9
Pharyngeal wall	11	84	72	40	20	21
Oral tongue	39	73	27	11	0	
Floor of mouth	72	51	29	11	5	
Alveolar ridge and retromolar trigone	38	84	25	10	4	
Base of tongue	19	89	22	10	18	6
Tonsil	8	74	31	14	12	12
Thyroid	0	87	100	100	10	

RP retropharyngeal

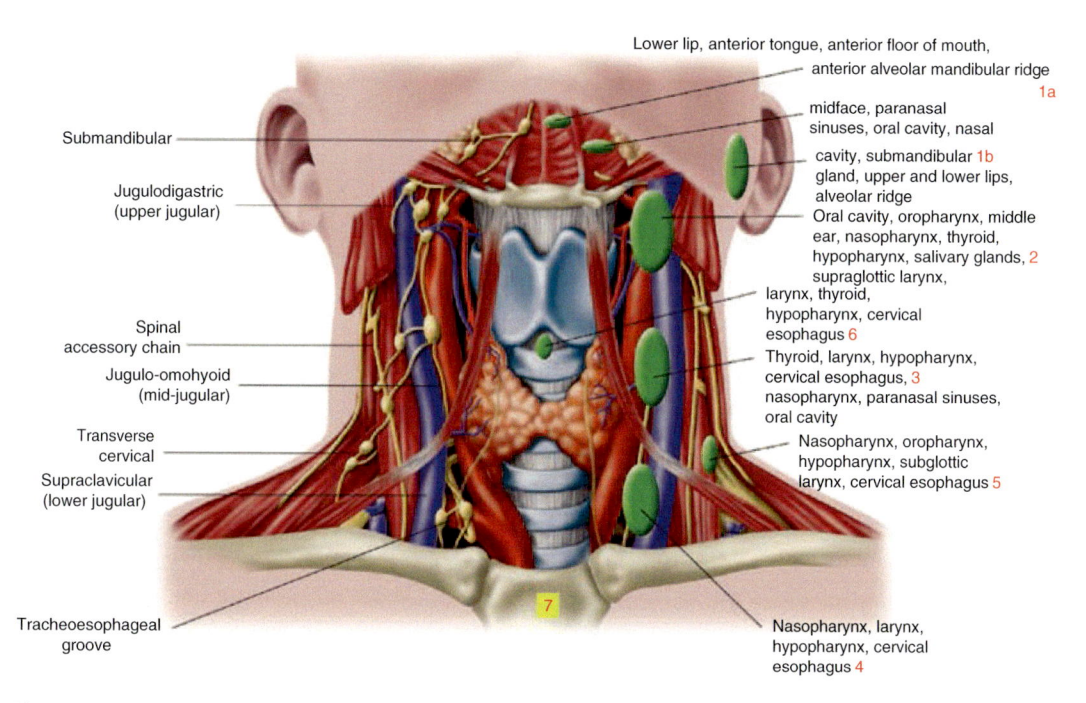

Fig. 5.2 Route of lymphatic involvement for head and neck cancers (from [8], p 18, Fig. 2.1, reproduced with the permission from Springer Science and Business Media)

Table 5.4 Recommendations for target volume dose prescriptions

Study	CTV1	CTV2	CTV3
	Total dose (Gy)/fraction dose (Gy)		
Chao et al. [10]	70 / 2	59.4 / 1.8	56 / 1.6
Butler et al. [11]	60 / 2.4	–	50 / 2
RTOG-0022 [12]	66 / 2.2	60 / 2	54 / 1.8
Lee et al. [13]	70 / 2.12	59.4 / 1.8	–

CTV clinical target volume, *RTOG* Radiation Therapy Oncology Group

Intensity-modulated radiotherapy (IMRT) is the treatment of choice in patients to undergo RT. The IMRT technique has no advantage in regard to survival or local control rates but it significantly reduces the rate of toxicity, particularly xerostomia [14, 15].

The Radiation Therapy Oncology Group (RTOG)/European Organization for Research and Treatment for Cancer (EORTC) group has

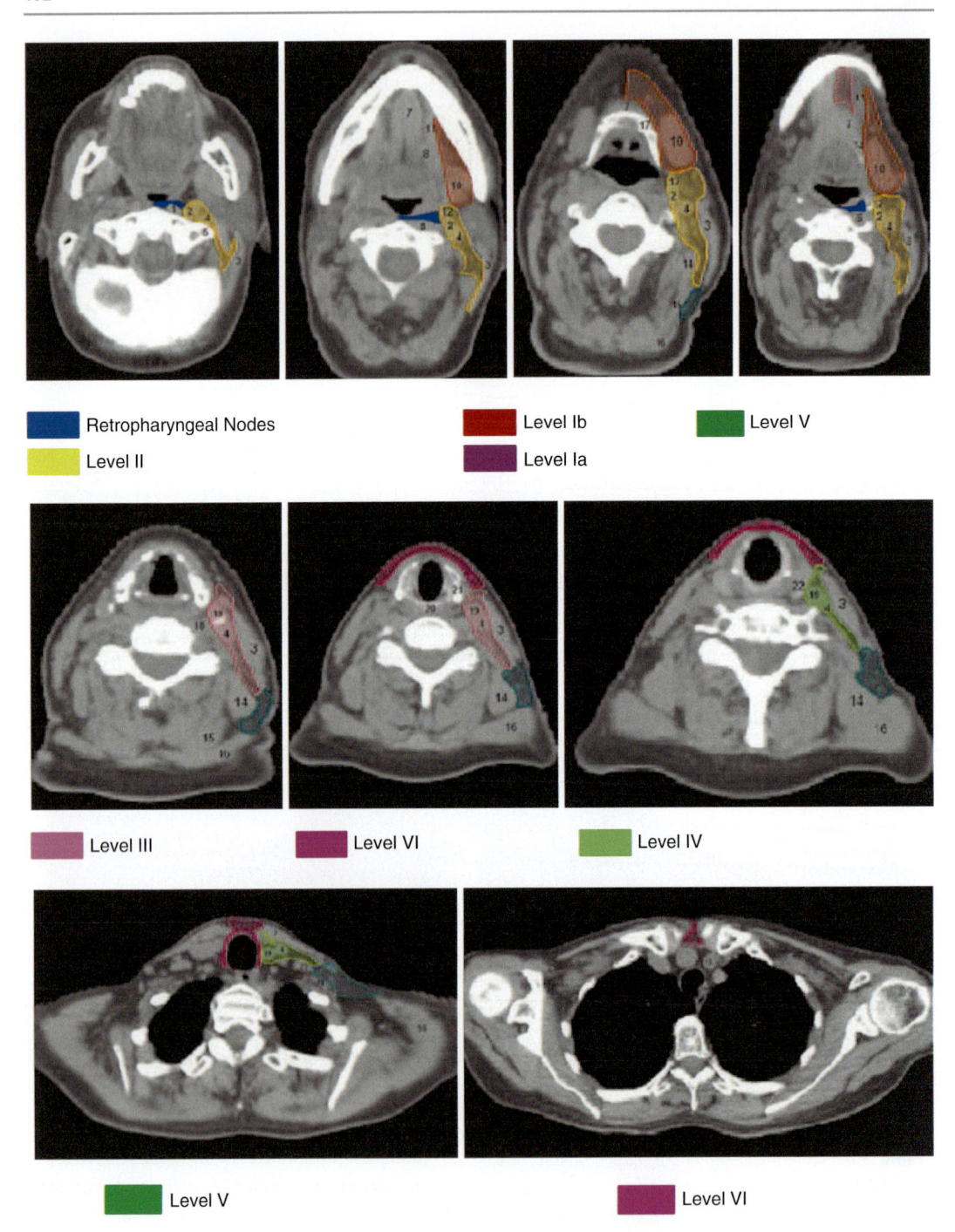

Fig. 5.3 Consensus guidelines for the delineation of N0 (elective) neck nodes [16]

published a consensus document on the determination and delineation of N0 neck nodes for conformal RT of head and neck cancers (Fig. 5.3) [16].

5.1 Tips for Delineating the Neck

- If level II is positive, the cranial border is the skull base.
- If level IV is positive, the caudal border is the sternoclavicular junction.
- In patients with neck dissection and no extracapsular extension (ECE), clinical target volume (CTVn) should include wider margins than negative necks, and a 2–3 mm of skin sparing is necessary to decrease skin toxicity.
- In patients with neck dissection and ECE, CTVn should have wider margins (including the sternocleidomastoid and/or paraspinal muscles), and in the regions where there is ECE, the skin is more generously included in CTV. If the muscular fascia is invaded, the entire muscle should be delineated as CTV.

5.2 Target Volume Determination and Delineation for the Neck

- -Gross tumor volume (GTVn) is the grossly involved lymph nodes on computed tomography (CT), magnetic resonance imaging (MRI), and positron emission tomography (PET)/CT. In postoperative cases, GTVn is not stated as it is assumed to be grossly resected.
- -CTVn:
  - Definitive IMRT:
    CTV1: GTVn + 0.5–1 cm
    CTV2: Positive lymph node levels
    CTV3: Uninvolved lymph node levels (elective)
- Adjuvant IMRT:
  CTV1: Positive lymph node levels with ECE
  CTV2: Positive lymph node levels without ECE
  CTV3: Uninvolved lymph node levels (elective)

Different dose prescriptions have been used in studies for CTV1, CTV2, and CTV 3 (Table 5.4).

5.3 Pharyngeal Cancers

The superior border of the pharynx is the cranial base (the base of the sphenoid sinus), and it extends inferiorly to the cricopharyngeal sphincter, becoming thinner as it goes down. It is located between the C1 and C6 vertebrae, and is 12–13 cm long in adults. It has a muscular structure, and is covered with mucosa [17].

> **Subsites of the Pharyngeal Region (Fig. 5.4)**
> Superior part continuous with the nasal cavity: *nasopharynx* or *epipharynx*. Middle part continuous with the oral cavity: *oropharynx* or *mesopharynx*. Inferior part continuous with the larynx: *laryngopharynx* or *hypopharynx*.

5.3.1 Nasopharyngeal Cancer

The nasopharynx extends from the cranial base to the soft palate and is continuous with the nasal cavity via the choanae anteriorly (Fig. 5.5).

> **Normal Anatomical Structures in the Nasopharynx [17]**
> Base of sphenoid sinus and cranial base superiorly Nasal cavity and choanae superior–anteriorly Pharyngeal side of soft palate inferior–anteriorly
> Laterally: ostium of eustachian tubes, tubal tonsils (Gerlach tonsils)

5.3.1.1 Pathology

The most common malignant histopathology in the nasopharynx is squamous cell carcinoma (SCC). Recent studies have shown that the histological subtype of nasopharyngeal SS is an important prognostic factor. The World Health Organization (WHO) classification is the most

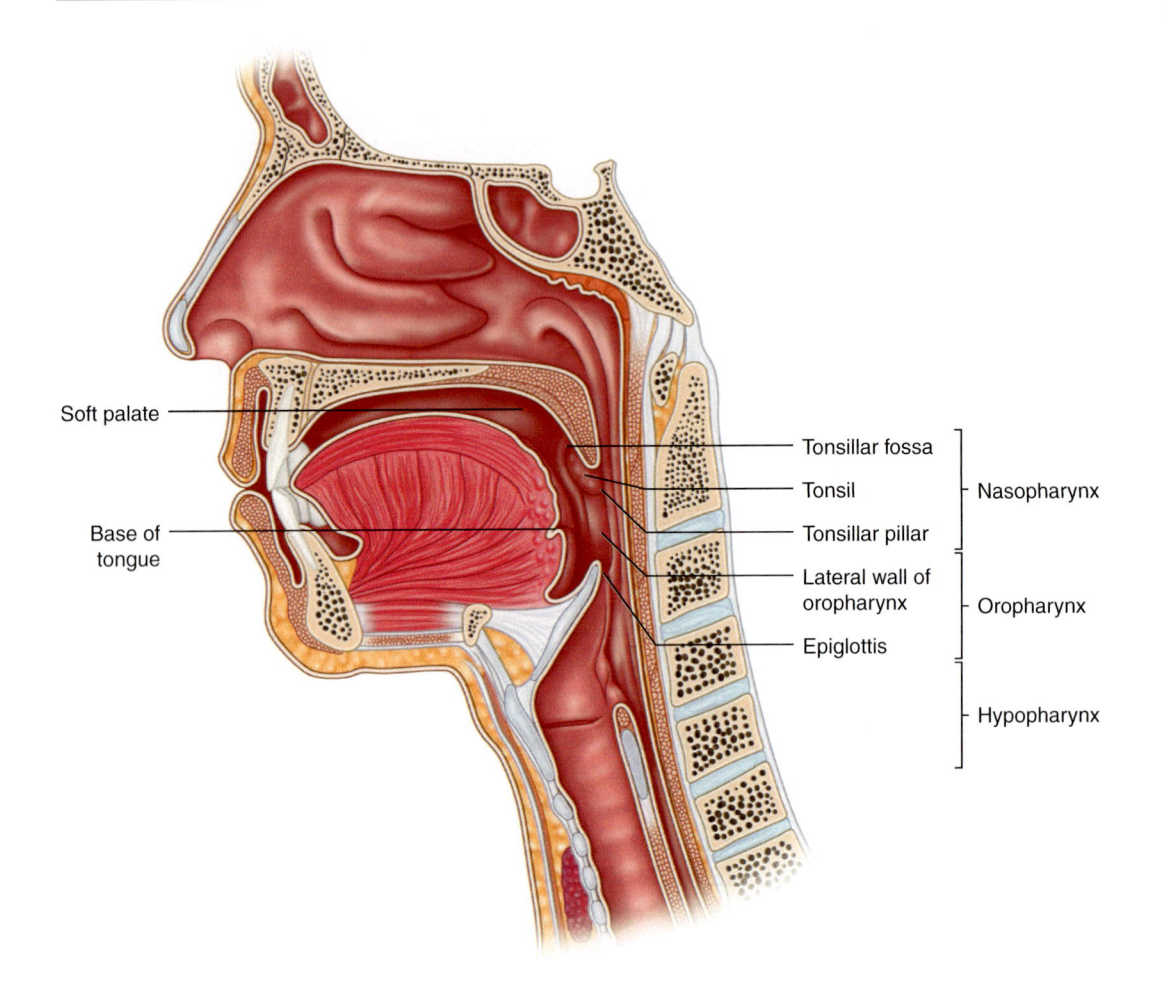

Fig. 5.4 Subsites of the pharyngeal region (from [18], p 28, Fig. 4.1, reproduced with the permission from the American Joint Committee on Cancer)

widely used histopathological classification system. This system essentially differentiates between tumors that do or do not produce keratin. In the 1978 version, three types were present as keratinizing, nonkeratinizing, and undifferentiated SCC. In 2017, the version was updated.

> **WHO Classification, 2017 [19]**
> Keratinizing SCC
>   Nonkeratinizing carcinoma
>
>   – differentiated subtype
>   – undifferentiated subtype
>
> Basaloid SCC

Keratinizing SCC: seen in older patients, less frequently associated with Epstein-Barr virus (EBV), and has the best prognosis (represents 20% of all nasopharyngeal cancer cases).

Nonkeratinizing SCC: includes most nasopharyngeal cancers.

- Differentiated (30–40%)
- Undifferentiated (40–50%)
  – lymphoepithelioma or lymphoepithelial carcinoma is a morphologic variant

Basaloid SCC: has the worst prognosis, very rare (<1%).

5.3 Pharyngeal Cancers

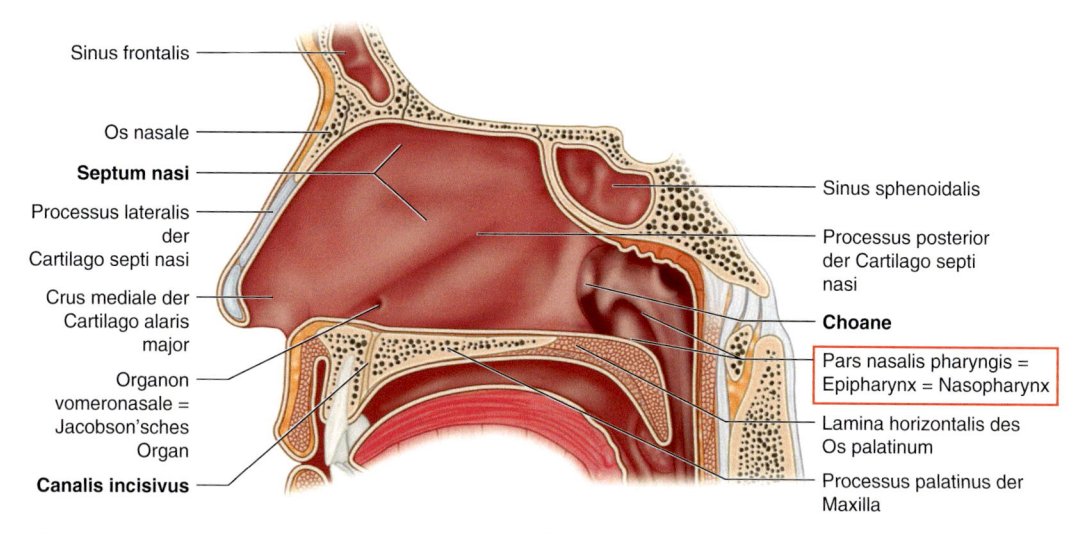

Fig. 5.5 Nasopharyngeal anatomy

Labels: Sinus frontalis; Os nasale; **Septum nasi**; Processus lateralis der Cartilago septi nasi; Crus mediale der Cartilago alaris major; Organon vomeronasale = Jacobson'sches Organ; **Canalis incisivus**; Sinus sphenoidalis; Processus posterior der Cartilago septi nasi; **Choane**; Pars nasalis pharyngis = Epipharynx = Nasopharynx; Lamina horizontalis des Os palatinum; Processus palatinus der Maxilla

5.3.1.2 General Presentation

Symptoms appear to be related to the localization and extent of the tumor. These symptoms can be grouped into four classes:

1. Nasal and nasopharyngeal symptoms are seen as a result of nasopharyngeal localization and tumor extension into the nasal cavity anteriorly. These symptoms are nasal obstruction and nasal bleeding.
2. Otological symptoms are due to the entrance to a Eustachian tube becoming filled and obstructed. These are the symptoms and findings of serous otitis media: hearing loss, tinnitus, and otophonia.
3. Cervical symptoms are neck lymph node (LN) metastases that are the extension of the nasopharyngeal cancer into first the parapharyngeal and retropharyngeal (RP) LN and then the jugular and spinal accessory chain. Fifty percent of patients have bilateral while 80–90% have bilateral neck metastases at presentation. Neck mass is the first sign in 40% of cases.
4. Ophthalmoneurological symptoms are observed due to the extension of the tumor into the cranial base and then into the foramen lacerum, various other foramens and fissures, and the cranial nerves. Dry eye due to the invasion of the greater superficial petrosal nerve; hypo- or hyperesthesia in the face due to the invasion of the trigeminal nerve; and ophthalmoplegia and diplopia due to extension into the cavernous sinus and superior orbital fissure, resulting in III, IV, and VI cranial nerve invasion, are all important findings. Parapharyngeal lymphatic involvement or extensive invasion of the lower cranial nerves (IX, X, XI, and XII) causes nasolali, aspiration, stridor, shoulder pain, motor weakness, and speaking disorders. Horner's syndrome may develop as a result of invasion of the cervical sympathetic chain located around large vessels.

5.3.1.3 Staging

The 2017 AJCC staging system has some variations from the 2010 system. For the T staging, a

T0 stage was added for an EBV-positive pathologic cervical LN with an unknown primary. T2 tumor includes the adjacent muscle involvement. As cranial nerve involvement has a worse prognosis than the skull base involvement, the prior was included in T4 and the latter in T3. Extension to the masticator space and infratemporal fossa was replaced by a more specific soft tissue involvement. For the N staging, N3b of supraclavicular fossa extension was switched to lower neck involvement (i.e., below the caudal border of the cricoid cartilage). Also, N3a and N3b were merged into a single N3. For total staging, previous stages IVA and IVB were merged to IVA and the previous IVC is now IVB.

T Staging (Fig. 5.6) [20]
- TX: Primary tumor cannot be assessed
- T0: No evidence of primary tumor
- T1: Tumor confined to the nasopharynx, or extends to the oropharynx and/or nasal cavity without parapharyngeal extension
- T2: Tumor with parapharyngeal and/or adjacent soft tissue (medial pterygoid, lateral pterygoid, prevertebral muscles) extension
- T3: Tumor invades bony structures at skull base, cervical vertebra, pterygoid structures, and/or paranasal sinuses
- T4: Tumor with intracranial extension and/or involvement of cranial nerves, hypopharynx, orbit, and parotid gland and/or extensive soft tissue infiltration beyond the lateral surface of the lateral pterygoid muscle

N Staging (Fig.5.7)
- Nx: Regional lymph nodes cannot be assessed
- N0: No regional lymph node metastasis
- N1: Unilateral metastasis in cervical lymph node(s) and/or unilateral or bilateral retropharyngeal lymph node(s), not more than 6 cm in greatest dimension, above the caudal border or cricoid cartilage
- N2: Bilateral metastasis in cervical lymph node(s), not more than 6 cm in greatest dimension, above the caudal border of cricoid cartilage
- N3: Unilateral or bilateral metastasis in cervical lymph node(s), larger than 6 cm and/or extension below the caudal border of cricoid cartilage. Metastasis in lymph node(s) larger than 6 cm

AJCC Stage Groups
- Stage 0: TisN0M0
- Stage I: T1N0M0
- Stage II: T0N1M0, T1N1M0, T2N0M0, T2N1M0
- Stage III: T0N2M0, T1N2M0, T2N2M0, T3N0M0, T3N1M0, T3N2M0
- Stage IVA: T4N0M0, T4N1M0, T4N2M0, AnyTN3M0
- Stage IVB: Any T, AnyNM1

MRI is the preferred imaging technique in the staging evaluation of nasopharyngeal carcinoma. The sensitivity of MRI is significantly superior to CT for detecting skull base involvement, intracranial involvement, RP LN involvement, and prevertebral muscle infiltration. Thin slices (3 mm) of MRI should be used for accurate staging. In patients with advanced locoregional disease, detection of distant metastasis is recommended. PET/CT is the preferred diagnostic tool with a 100% sensitivity and 90.1% specificity in these patients.

EBV infection has recently been shown to be associated with nasopharyngeal carcinoma independent of geography and ethnicity. Titers of IgA antiviral capsid antigen (VCA) and IgG anti–early antigen (EA) antibodies can be used for screening, early detection, and diagnosing

5.3 Pharyngeal Cancers

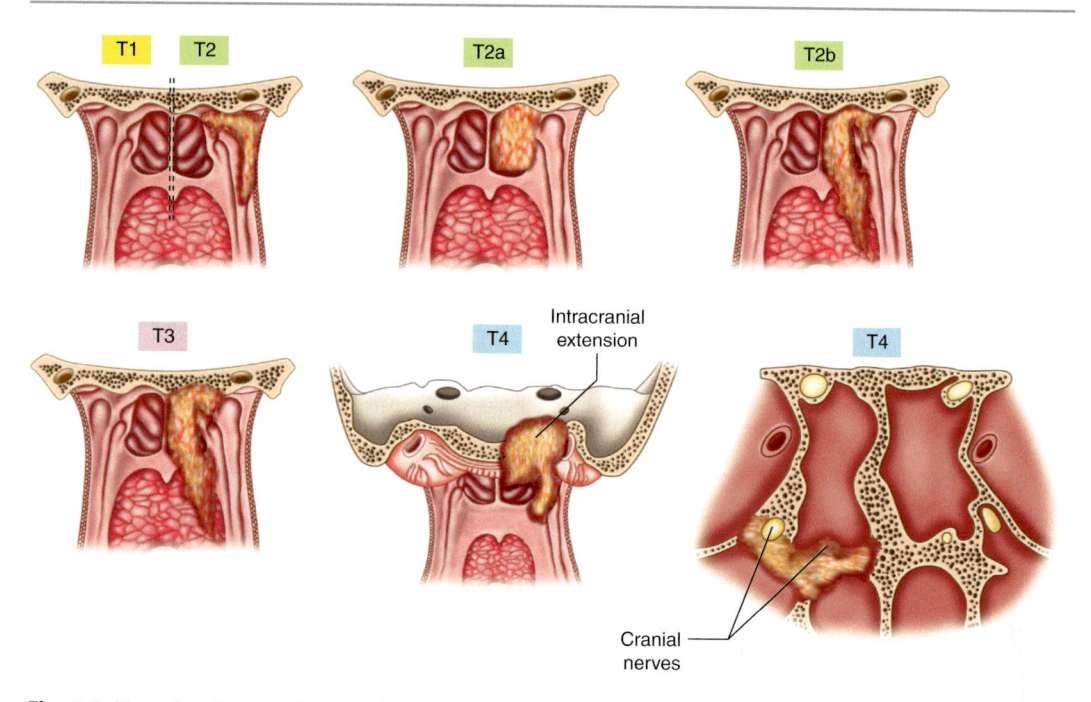

Fig. 5.6 T staging for nasopharyngeal cancers (from [20], reproduced with permission from the American Joint Committee on Cancer)

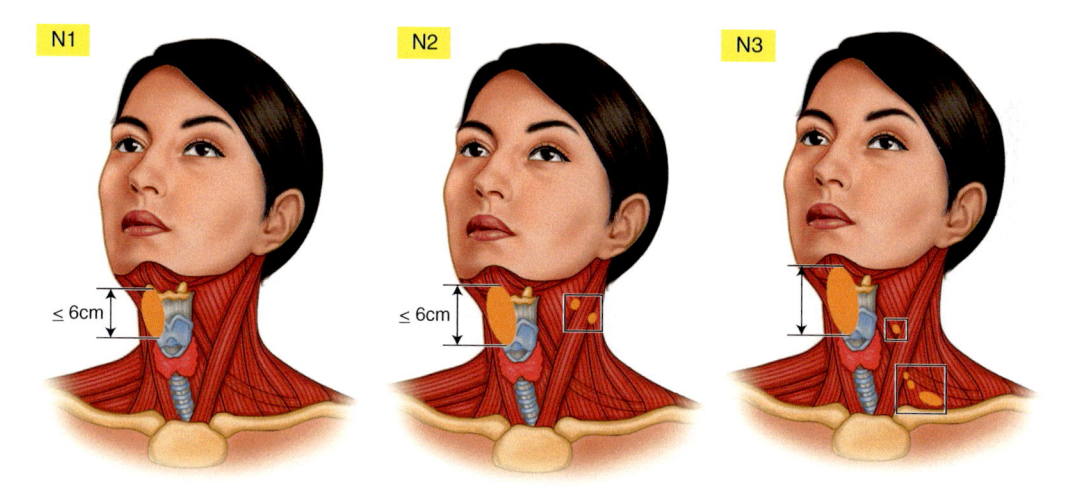

Fig. 5.7 N staging for nasopharyngeal cancers (from [20], reproduced with the permission from the American Joint Committee on Cancer)

nasopharyngeal carcinoma. Also, high circulating EBV DNA levels are associated with advanced stages, poor prognosis, and tumor recurrence. Besides, the clearance rate of plasma EBV DNA during early treatment can predict tumor response and overall survival in patients with metastatic/recurrent nasopharyngeal carcinoma.

Fig. 5.8 Neck node metastasis ratio for nasopharyngeal cancers

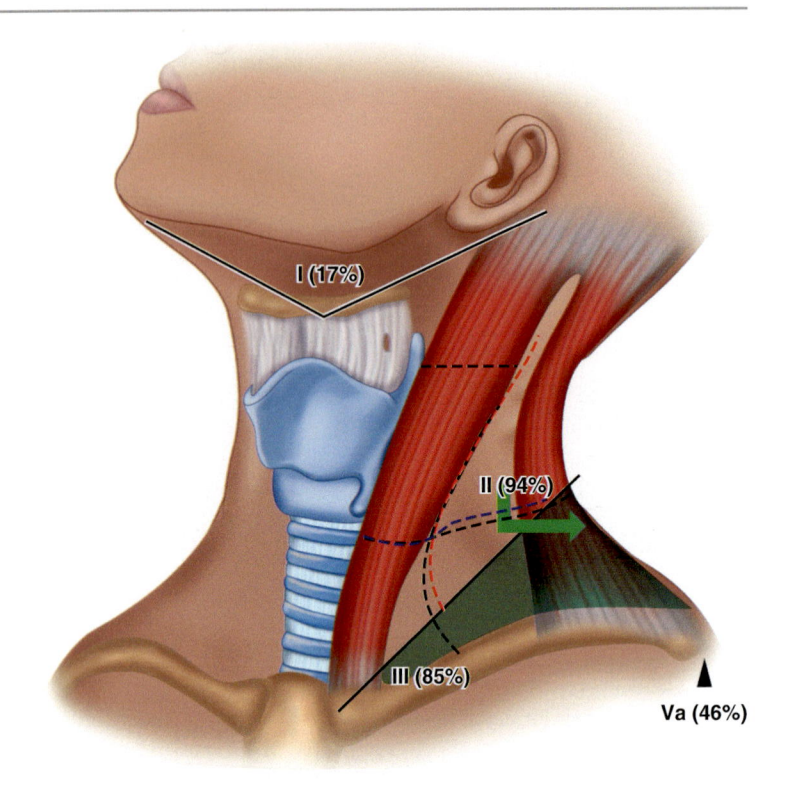

I (17%)
II (94%)
III (85%)
Va (46%)

The most important prognostic factor in nasopharyngeal cancer is the nodal status (N) followed by T stage (Fig. 5.8). While an advanced T category is associated with a worse local control (LC) and overall survival (OS), an advanced N category predicts increased risk of distant metastasis (DM) and worse survival:

- N0 → 17% distant metastasis (+). N3 → 78% distant metastasis (+).
- DM is very rare, and the most frequent sites are the lungs, bones and liver.

5.3.1.4 Treatment Algorithm
The treatment algorithm for nasopharyngeal cancers is summarized in Fig. 5.9.

Surgery does not play a major role in the management of nasopharyngeal cancer, except when diagnostic biopsies and neck dissections are performed in certain circumstances.

5.3.1.5 Radiotherapy
IMRT is the treatment of choice [13] (Figs. 5.10 and 5.11, Table 5.5).

5.3 Pharyngeal Cancers

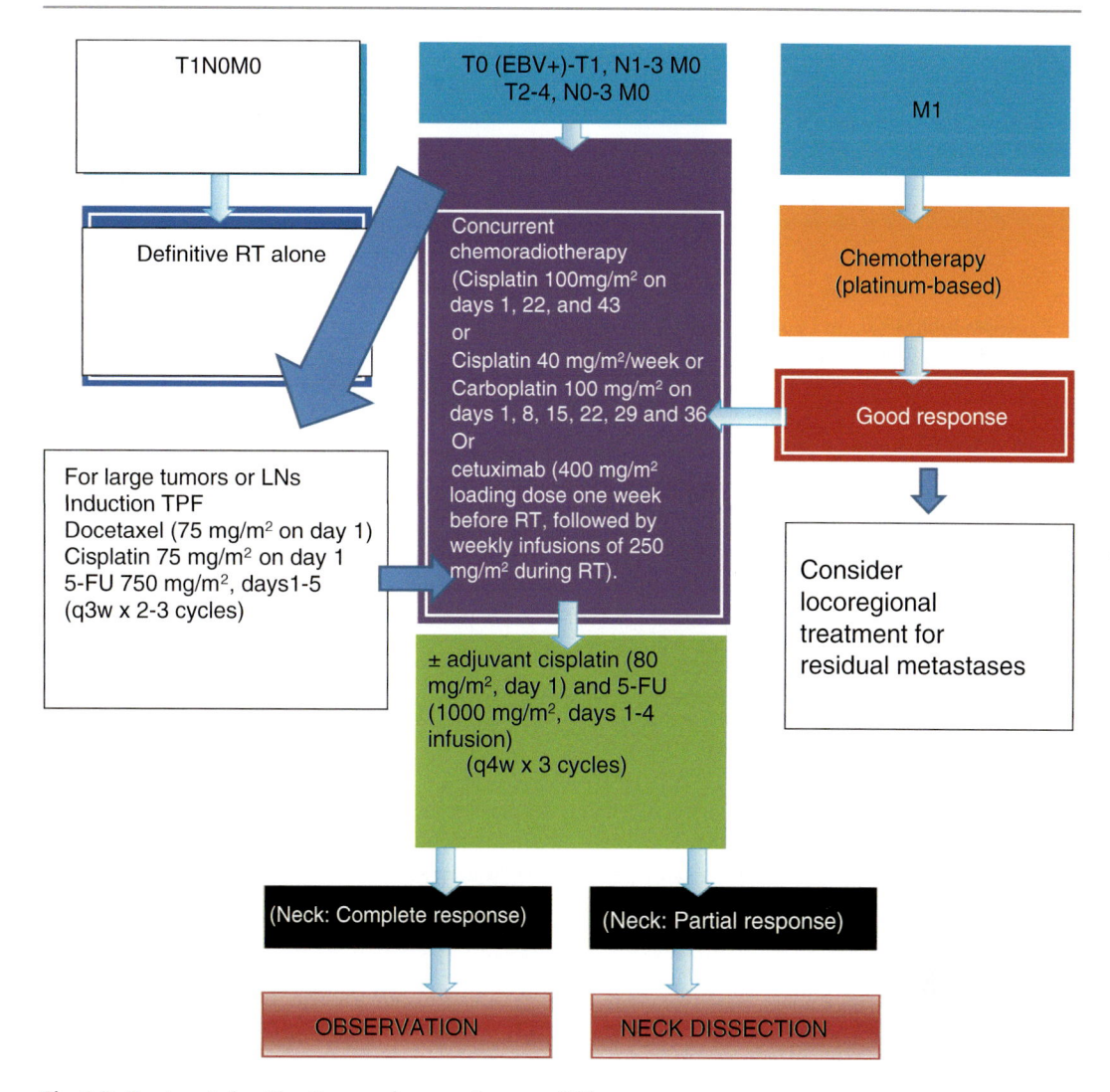

Fig. 5.9 Treatment algorithm for nasopharyngeal cancers [21]

- GTVp = Primary nasopharyngeal tumor
- GTVln = Gross metastatic LNs (>1 cm or with central necrosis)
- *In case of induction chemotherapy, targets should be determined by the pre-chemotherapy extent.
- CTV1 = GTV + 5–10 mm (for both the primary tumor and LNs)

- *Margins can be decreased to 1 mm in critical regions near the brainstem or spinal cord.
- CTV2p = Entire nasopharynx, anterior half of the clivus (entire clivus if involved), skull base (should include bilateral foramen ovale and rotundum), pterygoid fossae, bilateral upper deep

jugular and parapharyngeal spaces, inferior sphenoid sinus (entire sphenoid sinus if T3–T4), posterior third of the nasal cavity and maxillary sinuses. Also, the cavernous sinus should be included in high-risk patients (T3, T4, bulky disease involving the roof of the nasopharynx).
- CTV2ln = Retropharyngeal nodal regions and positive lymph node levels
- CTV3 = Bilateral level Ib (in cN+ neck), II, III, IV, V, and supraclcaviular fossa (SCF)
- Planning target volume (PTV) = CTV + 3–5 mm
- PTV1 = 70 Gy in 33 fractions (2.12 Gy per fraction)
- PTV2 = 59.4–63 Gy in 33 fractions (1.8–1.9 Gy per fraction)
- PTV3 = 54–57 Gy in 33 fractions (1.64–1.73 Gy per fraction)

Side Effects Related to Radiotherapy

Skin → early period: hyperemia, dry desquamation, wet desquamation, and ulceration; late period: telangiectasia, fibrosis, and tissue necrosis.

Hair → alopecia starts after seven days (temporary until 7 Gy, permanent >7 Gy).

Mucosa → erythema, edema, patchy mucositis, and confluent mucositis (mucositis starts after >3 Gy).

Esophagus → early period: retrosternal pain due to esophagitis (10–12 days). This heals in 1–2 weeks.

Late effects: muscular and epithelial changes, fibrosis, ulceration, and swallowing difficulty.

Salivary gland → decrease in saliva starts in 24 h, and its viscosity decreases. pH becomes acidic, bicarbonate and IgA levels decrease. If irradiated volume of salivary gland is more than 50%, marked xerostomia is seen.

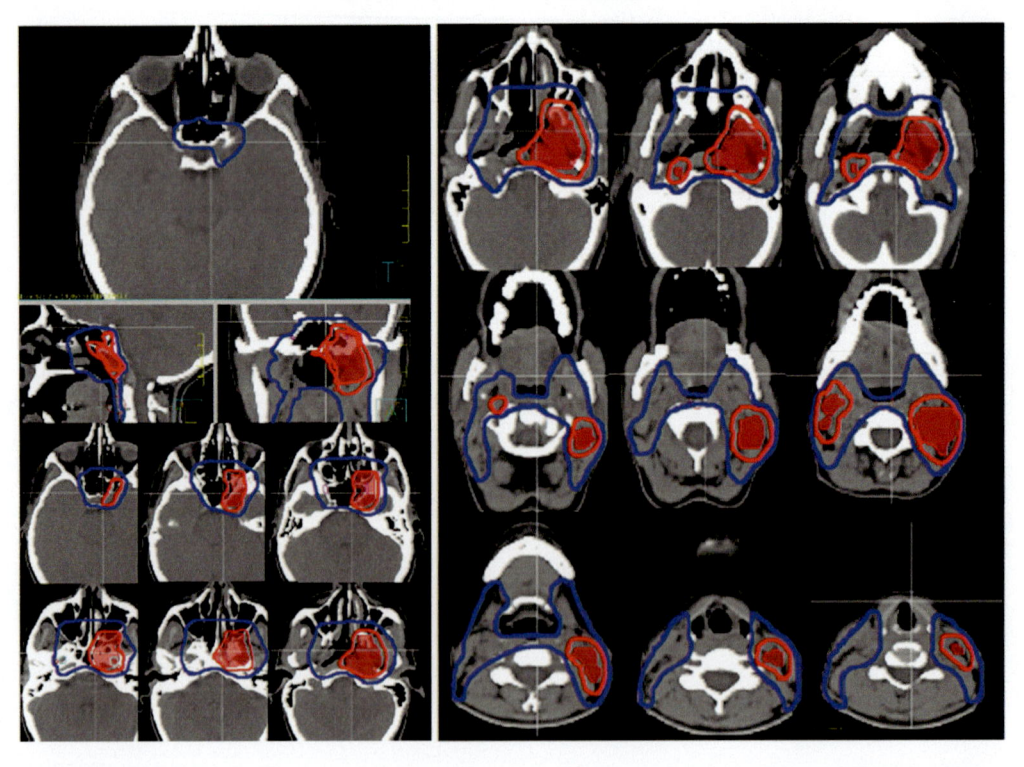

Fig. 5.10 Contouring of a patient with T4N2M0 nasopharyngeal cancer. CTV70 (red), CTV63 (blue), CTV57 (not shown-lower neck)

5.3 Pharyngeal Cancers

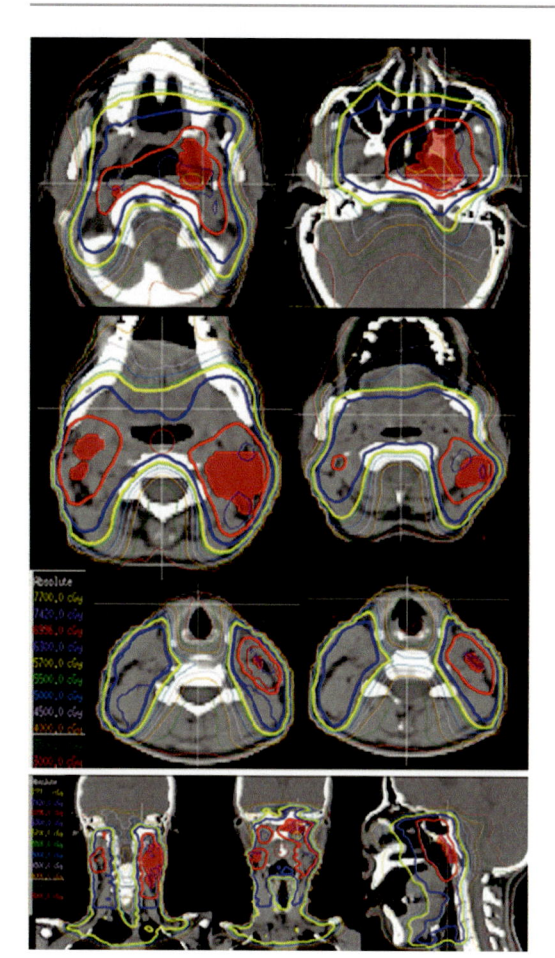

Fig. 5.11 IMRT plan of the same patient

Table 5.5 Critical organ doses in head and neck radiotherapy

Organ	Constraints
Brain	Mean <50 Gy
Brainstem	Max < 54 Gy or 1% of PTV cannot exceed 60 Gy
Optic nerves and chiasm	Max <54 Gy or 1% of PTV cannot exceed 60 Gy
Spinal cord	Max <45 Gy or 1 cc of the PTV cannot exceed 50 Gy
Brachial plexus	Max <66 Gy
Temporal lobes	Max <60 Gy or 1% of PTV cannot exceed 65 Gy
Lens	Max <10 Gy, try to achieve <5 Gy
Eyes	Mean < 35 Gy, max <50 Gy
Mandible and temporomandibular joints	Max <70 Gy or 1 cc of the PTV cannot exceed 75 Gy
Oral cavity	Mean < 40 Gy
Parotid gland	Mean < 26 Gy (in at least one gland) or At least 20 cc of both parotid glands <20 Gy or At least 50% of one gland <30 Gy (should be achieved in at least one gland)
Cochlea	V55 < 5%
Glottic larynx	Mean < 36–45 Gy
Esophagus	Mean < 45 Gy

5.3.1.6 Selected Publications

Radiotherapy alone

RTOG, 1980 → 109 patients with nasopharynx cancer were randomized to a split-course RT (two cycles of 30 Gy in 3-Gy fractions in 2 weeks, with a rest period of 3 weeks) or continuous RT (60–66 Gy in 2–2.2 Gy fraction dose). Complete response 96% for T1, 88% for T2, 81% for T3, and 74% for T4. For N (+), 5-year survival is 71–93%. Acute and late toxicity, DM rate, and tumor response rate were comparable in the two arms.

Marcial VA et al. (1980) Split-course radiation therapy of carcinoma of the nasopharynx: results of a national collaborative clinical trial of the Radiation Therapy Oncology Group. Int J Radiat Oncol Biol Phys. 6(4):409–14

Sanguinetti et al., 1997 → RT alone is efficacious for T1 tumors. Five-year LC: 93% vs. 79% for T2, 68% for T3, and 53% for T4 tumors. LC is decreased with higher T stage, squamous histology, and cranial nerve deficits

Sanguineti G, Geara FB, Garden AS, Tucker SL, Ang KK, Morrison WH, Peters LJ (1997) Carcinoma of the nasopharynx treated by radiotherapy alone: determinants of local and regional control. Int J Radiat Oncol Biol Phys 37(5):985–996

Altered fractionation

Teo et al, 2000 → 159 patients with nasopharyngeal cancer received 2.5 Gy/fraction once-daily for 8 fractions and then randomized to 1.6 Gy BID for an additional 32 fractions vs. 2.5 Gy once-daily for another 16 fractions. The trial was terminated early because of excessive neurologic toxicities and no improvement in tumor control in the accelerated fractionation arm.

Teo PM et al. (2000) Final report of a randomized trial on altered-fractionated radiotherapy in nasopharyngeal carcinoma prematurely terminated by significant increase in neurologic complications. Int J Radiat Oncol Biol Phys 48(5):1311–1322.

Lee et al., 2011 → 189 patients with T3-4N0-1M0 nasopharyngeal carcinoma randomized to conventional fractionation (CF) RT (5 fractions/week) vs. CF RT + chemotherapy (concurrent cisplatin and adjuvant cisplatin and 5-fluorouracil (FU)) vs. accelerated fractionation (AF) RT (6 fractions/week) vs. AF RT + chemotherapy. Five-year failure-free rate (FFR) was significantly higher in the AF + chemotherapy arm (88%) compared to all other arms (63% for CF RT, 56% for AF RT, 65% for CF RT+ chemotherapy). Late toxicity was similar with a borderline increase in OS.

Lee AWM et al. (2011) A randomized trial on addition of concurrent-adjuvant chemotherapy and/or accelerated fractionation for locally-advanced nasopharyngeal carcinoma. Radiother Oncol 98:15–22.

Concurrent chemoradiotherapy

RTOG 88-17/Intergroup 0099, 1998 → The most important randomized trial showing the survival benefit of chemoradiotherapy (CRT) in stage III and IV disease. 147 cases. Median follow-up: 2.7 years. The randomization arm included concurrent cisplatin with adjuvant cisplatin/5-FU. Three-year progression-free survival (PFS): 69% (CRT arm) vs. 24% (RT-alone arm). Three-year OS: 78% (CRT arm) vs. 47% (RT-alone arm).

Al-Sarraf M et al. (1998) Chemoradiotherapy versus radiotherapy in patients with advanced nasopharyngeal cancer: phase III randomized Intergroup study 0099. J Clin Oncol 16(4):1310–1317.

Al-Sarraf et al., 2001 → in the 5-year update, superiority of CRT continues. Five-year PFS: 78% (CRT arm) vs. 47% (RT-alone arm).

Al-Sarraf M et al. (2001) Superiority of 5-year survival with chemoradiotherapy vs. radiotherapy in patients with locally advanced nasopharyngeal cancer. Intergroup 0099 Phase III Study. Final report. Proc Am Soc Clin Oncol 20:2279

Meta-analysis 2002 → six trials. 1528 cases. Chemotherapy was used concurrently in only one trial (Al Sarraf et al.). Chemotherapy: 25–40% increase in disease-free survival (DFS) or PFS and 20% increase in OS. Significant survival increase was only observed in concomitant CRT.

Huncharek M et al. (2002) Combined Chemoradiation Versus Radiation Therapy Alone in Locally Advanced Nasopharyngeal Carcinoma: Results of a Meta-analysis of 1528 Patients from Six Randomized Trials. Am J Clin Oncol. 25(3):219–223

Lin et al., 2003 → Randomized Taiwan trial. 284 patients. Three-year DFS: 72% (concurrent arm) vs. 53% (RT alone arm).

Lin JC et al. (2003) Phase III study of concurrent chemoradiotherapy versus radiotherapy alone for advanced nasopharyngeal carcinoma: positive effect on overall and progression-free survival. J Clin Oncol 21(4):631–637

Wee et al., 2005 → Randomized Singapore trial. 221 patients. The randomization arm included concurrent cisplatin with adjuvant cisplatin/5-FU. Three-year DFS: 72% (concurrent arm) vs. 53% (RT alone arm). Three-year OS: 80% vs. 65

Wee J et al. (2005) Randomized trial of radiotherapy versus concurrent chemoradiotherapy followed by adjuvant chemotherapy in patients with American Joint Committee on Cancer/International Union against cancer stage III and IV nasopharyngeal cancer of the endemic variety. J Clin Oncol 23(27):6730–6738

Zhang et al., 2005 → Randomized Chinese trial. 115 patients. Concurrent chemotherapy was oxaliplatin. RT was 70–74 Gy + 10 Gy boost. Two-year DFS: 96% vs. 83% and two-year OS: 100% vs. 77% in favor of concurrent CRT.

Zhang L et al. (2005) Phase III study comparing standard radiotherapy with or without weekly oxaliplatin in treatment of locoregionally advanced nasopharyngeal carcinoma: preliminary results. J Clin Oncol 23(33):8461–8468

Meta-analysis, 2006 → Addition of chemotherapy increases the 5-year OS rate by 4–6% and decreases the risk of death by 18% and the risk of event by 10%. Concurrent CRT is superior to induction or adjuvant chemotherapy.

Baujat B et al. (2006) Chemotherapy in locally advanced nasopharyngeal carcinoma: an individual patient data meta-analysis of eight randomized trials and 1753 patients. Int J Radiat Oncol Biol Phys 64(1):47–56

Lee et al., 2010 → Randomized Hong Kong trial. 348 patients. RT dose was 66 Gy. The randomization arm included concurrent cisplatin with adjuvant cisplatin/5-FU. Five-year DFS: 72% (concurrent arm) vs. 62% (RT alone arm). OS rates were not statistically significant (78% vs. 54%)

Lee AW et al. (2010) Randomized trial of radiotherapy plus concurrent-adjuvant chemotherapy vs. radiotherapy alone for regionally advanced nasopharyngeal carcinoma. J Natl Cancer Inst 102(15):1188–1198

Chen et al., 2011 → Stage II disease (T1-2N1M0 or T2N0M0 with parapharyngeal space involvement). Five-year OS: 94.5% (RT + cisplatin arm) vs. 85.8% (RT-alone arm). DM is significantly lower in the RT + cisplatin arm but no difference in loco-regional relapse-free survival (LRRFS).

Chen QY et al. (2011) Concurrent chemoradiotherapy vs radiotherapy alone in stage II nasopharyngeal carcinoma: phase III randomized trial. J Natl Cancer Inst 103(23):1761–1770

Neoadjuvant chemotherapy

VUMCa I trial—International Nasopharynx Cancer Study Group, 1996 → randomized phase II. 339 cases. N2–3 and undifferentiated (WHO II–III) histology (70 Gy RT alone) vs. neoadjuvant three cycles chemo + concurrent CRT (BEC in every 3 weeks). Two-year DFS: 60% in CRT arm vs. 40% in RT alone arm. No difference in OS.

International Nasopharynx Cancer Study Group (1996) Vumca I. Trial. Preliminary results of a randomized trial comparing neoadjuvant chemotherapy (cisplatin, epirubicin, bleomycin) plus radiotherapy vs. radiotherapy alone in stage IV (≥N2, M0) undifferentiated nasopharyngeal carcinoma: A positive effect on progression-free survival. Int J Radiat Oncol Biol Phys. 35(3):463–9.

AOCOA trial, 1998 → 334 patients. 70 Gy RT alone vs. neoadjuvant 2–3 cycles of cisplatin+epirubicin followed by 70 Gy RT. Neoadjuvant chemotherapy has no benefit

Chua DT et al. (1998) Preliminary report of the Asian-Oceanian Clinical Oncology Association randomized trial comparing cisplatin and epirubicin followed by radiotherapy versus radiotherapy alone in the treatment of patients with locoregionally advanced nasopharyngeal carcinoma. Asian-Oceanian Clinical Oncology Association Nasopharynx Cancer Study Group. Cancer 83(11):2270–2283

Ma et al., 2001 → 456 patients. 70 Gy RT alone vs. neoadjuvant 2–3 cycles of cisplatin+bleomycin+5FU) followed by 70 Gy RT. No significant survival benefit with neoadjuvant chemotherapy.

Ma J et al. (2001) Results of a prospective randomized trial comparing neoadjuvant chemotherapy plus radiotherapy with radiotherapy alone in patients with locoregionally advanced nasopharyngeal carcinoma. J Clin Oncol 19(5):1350–1357

Hui et al., 2009 → 65 patients. Concurrent 70 Gy RT and cisplatin vs. 2 cycles of docetaxel+cisplatin followed by the same concurrent regimen. Three-year OS: 94.1% vs. 67.7% in favor of neoadjuvant chemotherapy without DFS difference.

Hui EP et al. (2009) Randomized phase II trial of concurrent cisplatin-radiotherapy with or without neoadjuvant docetaxel and cisplatin in advanced nasopharyngeal carcinoma. J Clin Oncol 27(2):242–249

Xu et al., 2012 → 338 patients. Concurrent 70 Gy RT and cisplatin/5-FU followed by adjuvant 4 cycles of cisplatin/5-FU vs. neoadjuvant 2 cycles of cisplatin/5-FU followed by the same concurrent and adjuvant regimens. No survival difference.

Xu T et al. (2012) Preliminary results of a phase III randomized study comparing chemotherapy neoadjuvantly or concurrently with radiotherapy for locoregionally advanced nasopharyngeal carcinoma. Med Oncol 29(1):272–278

IMRT

RTOG 0225, 2009 → phase II randomized. 68 patients staged I-IVB were treated with 70 Gy IMRT to the GTV and 59.4 Gy with or without concurrent cisplatin followed by adjuvant cisplatin and 5-FU. Two-year LRRFS rate was 90% with minimal severe toxicity.

Lee N et al. (2009) Intensity-modulated radiation therapy with or without chemotherapy for nasopharyngeal carcinoma: radiation therapy oncology group phase II trial 0225. J Clin Oncol. 27:3684–90.

EBV DNA

Lin et al., *2004* → 101 patients with locally advanced nasopharyngeal cancer. Higher stage disease was associated with higher concentrations of plasma EBV DNA. Plasma EBV concentration was significantly higher in patients that relapsed. EBV DNA concentration 1 week after completion of RT was the best predictor for risk of relapse. Also, a detectable EBV DNA level after treatment was completed was the most important prognostic favor for relapse-free survival (RFS) and OS. Two-year OS: 100% in patients with a pretreatment plasma EBV DNA level < 1500 copies/ mm vs. 83.4% in patients with >1500 copies/mm.

Lin J-C, et al. (2004) Quantification of plasma Epstein-Barr virus DNA in patients with advanced nasopharyngeal carcinoma. N Engl J Med. 350:2461–70.

5.3.2 Oropharyngeal Cancer

In contrast to the decreasing incidence of head and neck cancers, the incidence of oropharyngeal carcinoma is increasing, mainly based on the increase in HPV-related disease. Besides tobacco and alcohol use and chronic irritation, HPV infection is an important etiologic factor for oropharyngeal cancer, predominantly in non-smoker and non-drinker men [22]. HPV-related oropharyngeal cancer has a significantly more favorable prognosis compared to HPV-unrelated disease. Oropharyngeal cancer has been recommended to be classified as having a low (HPV+, ≤10 pack-year smoking or >10 pack-year and N0–N2a), intermediate (HPV+, >10 pack-year smoking and N2b–N3 or HPV−, ≤10 pack-year smoking, T2–3), or high risk (HPV−, ≤10 pack-year smoking, T4 or HPV−, >10 pack-year smoking) of death on the basis of four factors [23, 24]. Studies are ongoing whether treatment can be de-escalated in patients with HPV-related disease but until then, treatment recommendations for these two clinical entities currently remain the same.

The oropharynx starts from the soft palate and extends to the superior portion of the epiglottis (Fig. 5.12) [17].

- **Normal anatomical structures of the oropharynx** (Fig. 5.13): Anterior: oral cavity
- Anterior–superior: front side of soft palate Mid-anterior: isthmus faucium and oral cavity Lower-anterior: base of tongue and vallecula Lateral: palatine tonsils
- Posterior: posterior pharyngeal wall and prevertebral fascia of C2–3 vertebrae
- **Regions of the oropharynx in which malignancies may develop**:
- Posterior one-third of tongue (base of tongue)
- Lingual site of epiglottis, vallecula Soft palate and uvula
- Tonsils and tonsil plicae
- Mucosa of posterior oropharyngeal wall

5.3.2.1 Pathology

Most oropharyngeal tumors are SCC although other types like lymphoma and minor salivary glands can also be observed.

- SCC: 75%
- Lymphoma: 15%
- Lymphoepithelioma: 5%
- Others: 5% (minor salivary gland cancer, sarcoma)

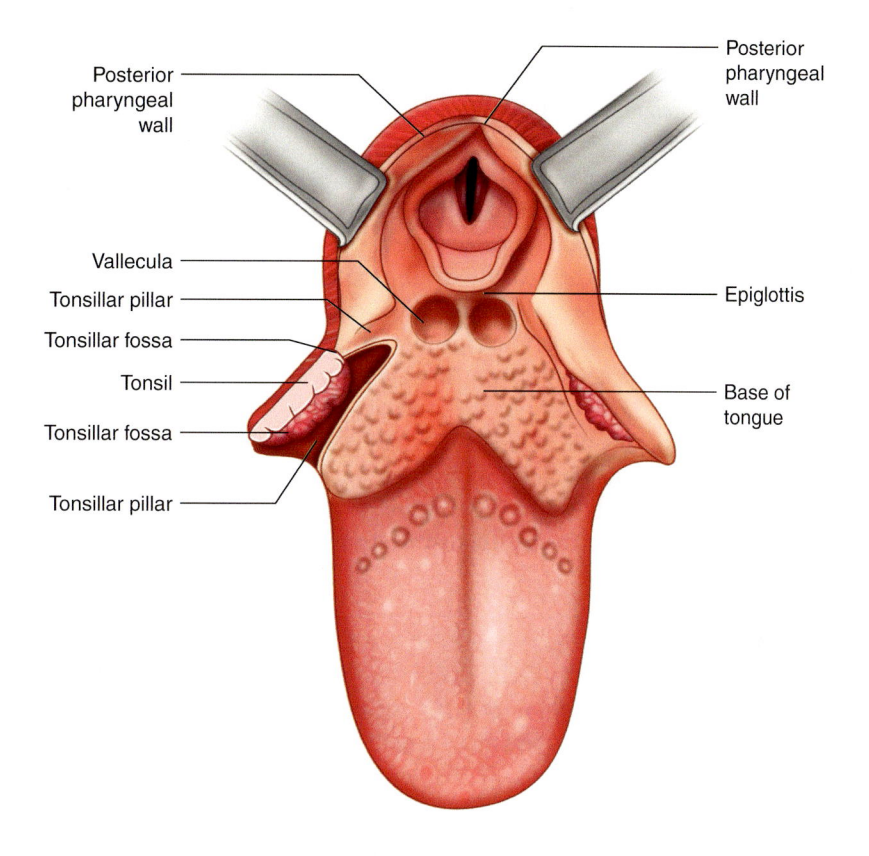

5.3 Pharyngeal Cancers

Fig. 5.12 Normal anatomical structures of the oropharynx (from [20], reproduced with the permission from the American Joint Committee on Cancer)

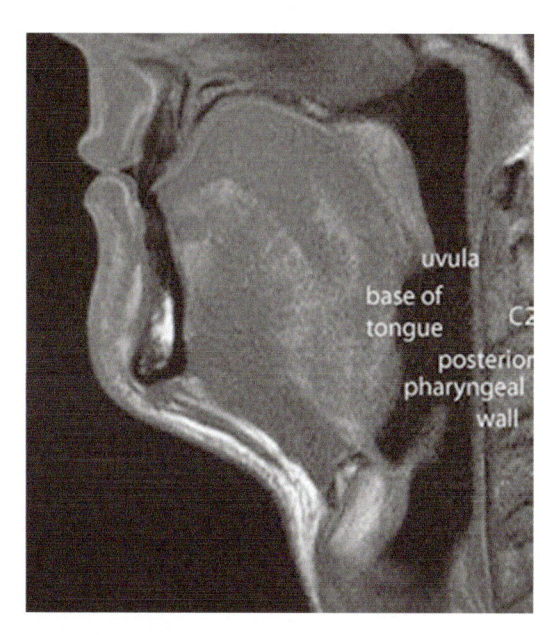

Fig. 5.13 Normal anatomical structures of the oropharynx

5.3.2.2 General Presentation

The first symptom is usually unilateral, progressive otalgia, and dysphagia with throat discomfort. Most patients have a feeling of a mass in the throat posteriorly. LN metastasis at presentation is frequent. Thus, a neck mass can be the first symptom at presentation. The risk of bilateral neck LN metastasis is very high, especially for the base of tongue cancers. In more advanced cases, dysfunctions in swallowing and speaking abilities, oral fetor, bleeding, hemoptysis, fixed tongue, trismus, and weight loss can be seen. The tumor is usually ulcerated in the routine physical exam, and unilateral tonsil enlargement can be observed. Therefore, unilateral tonsil hypertrophy in older patients should be examined for a possible malignancy.

A complete examination of all mucosal head and neck sites should be performed especially in smokers in order to evaluate a synchronous malignancy. Indirect laryngoscopy with a mirror should be performed to see the base of the tongue and vallecula. A tumor in this region can easily invade the surrounding soft tissues and present at more advanced stages. However, the actual size of the tumor is determined by bimanual palpation and advanced imaging modalities (MRI, CT). Mandibular invasion is also detected via radiological techniques. The risk of dDM is a bit higher than that for the oral cavity. Furthermore, the presence of a secondary malignancy or later development in the upper or lower airways is also higher. Thus, the patient should be evaluated and followed up for this possibility. A PET/CT scan should be performed in patients with advanced primary tumors and with N2 or greater nodal disease to assess DM. The final and exact diagnosis is made via biopsy.

5.3.2.3 Staging

The staging for oropharyngeal cancer has been revised in the eighth edition of AJCC based on HPV status. T staging is the same for both HPV-related and -unrelated disease [20].

> **Primary Tumor (T) (Fig. 5.14)**
> - T1: Tumor ≤2 cm in greatest dimension
> - T2: Tumor >2 cm but ≤4 cm in greatest dimension
> - T3: Tumor >4 cm in greatest dimension or extension to lingual surface of epiglottis
> - T4a: Moderately advanced local disease: Tumor invades the larynx, extrinsic muscles of tongue, medial pterygoid, hard palate, or mandible
> - T4b: Very advanced local disease: Tumor invades lateral pterygoid, pterygoid plates, lateral nasopharynx, skull base, or surrounding the carotid artery

N staging and stage grouping are different for HPV-related and -unrelated oropharyngeal cancers.

> **Nodal Stage (N) for HPV-Related Oropharyngeal Cancers**
> **Clinical staging**
> - cN1: 1 or more ipsilateral lymph node, all ≤6 cm
> - cN2: contralateral or bilateral lymph nodes, all ≤6 cm
> - cN3: Any lymph node >6 cm
>
> **Pathologic staging**
> - pN1: 4 or fewer lymph nodes
> - pN2: More than 4 lymph nodes

> **Nodal Stage (N) for HPV-Unrelated Oropharyngeal Cancers**
> **Clinical staging**
> - N1: A single ipsilateral lymph node, ≤3 cm, no ECE
> - N2a: A single ipsilateral lymph node, >3 cm but ≤6 cm, no ECE
> - N2b: Multiple ipsilateral lymph nodes all ≤6 cm, no ECE
> - N2c: Bilateral or contralateral lymph nodes, all ≤6 cm, no ECE
> - N3a: Any lymph node >6 cm, no ECE
> - N3b: Any lymph node with ECE
>
> **Pathologic staging**
> - N1: A single ipsilateral lymph node, ≤3 cm, no extracapsular extension (ECE)
> - N2a: A single ipsilateral or contralateral lymph node, ≤3 cm, with ECE or
> - A single ipsilateral lymph node, >3 cm but ≤6 cm, no ECE
> - N2b: Multiple ipsilateral lymph nodes all ≤6 cm, no ECE
> - N2c: Bilateral or contralateral lymph nodes, all ≤6 cm, no ECE
> - N3a: Any lymph node >6 cm, no ECE
> - N3b: Any lymph node with ECE

5.3 Pharyngeal Cancers

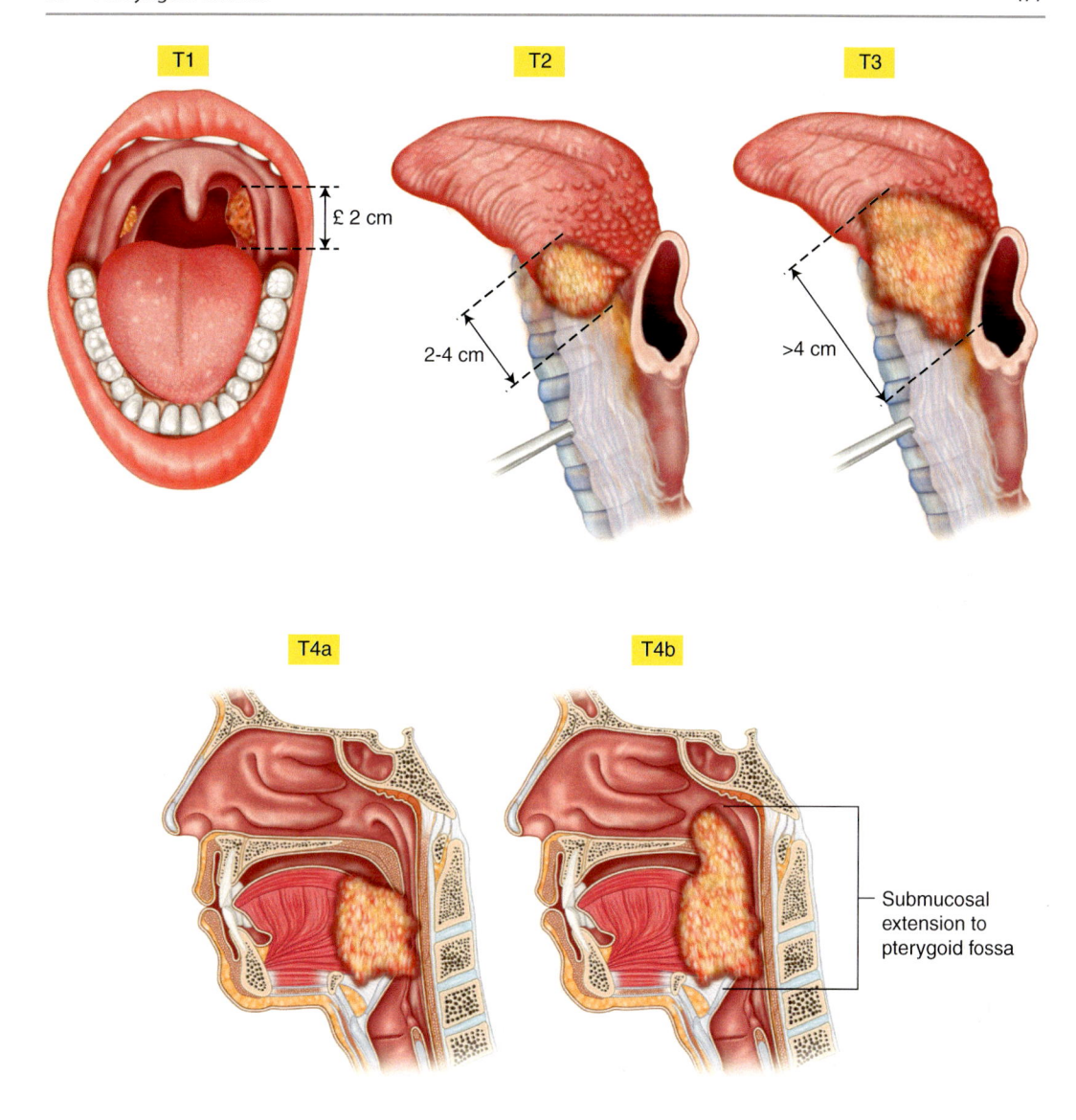

Fig. 5.14 T staging in oropharyngeal cancer (from [20], reproduced with the permission from the American Joint Committee on Cancer)

AJCC Stage Groups

- **Clinical stage for HPV-related cancer**
- Stage I: T1N0M0, T2N0M0, T1N1M0, T2N1M0
- Stage II: T1N2M0, T2N2M0, T3N0M0, T3N1M0, T3N2M0
- Stage III: T1N3M0, T2N3M0, T3N3M0, T4anyNM0
- Stage IV: Any T, any N, M1

Pathologic stage for HPV-related cancer
- Stage I: T1N0M0, T2N0M0, T1N1M0, T2N1M0
- Stage II: T1N2M0, T2N2M0, T3N0M0, T3N1M0, T4N0M0, T4N1M0
- Stage III: T3N2M0, T4N2M0
- Stage IV: Any T, any N, M1

Clinical and pathologic stage for HPV-unrelated cancer
- Stage I: T10M0
- Stage II: T2N0M0
- Stage III: T3N0M0, T1N1M0, T2N1M0, T3N1M0
- Stage IVA: T4aN0M0, T4aN1M0, T1N2M0, T2N2M0, T3N2M0, T4aN2M0
- Stage IVB: T4b, any N, M0; any T, N3M0
  Stage IVC: any T, any N, M1

5.3.2.4 Treatment Algorithm

Treatment Algorithm for p16-Negative Oropharyngeal Cancer [21]

T1-2, N0-1 → three alternative approaches:

- *Definitive RT* → consider surgery for residual disease
- *Primary tumor excision + uni/bilateral neck dissection*
  *At least one major risk factor (+) → CRT
  *At least one minor risk factor (+) → RT
- *Concurrent CRT (for N1 tumors)*

T3-4a, N0-1 → three alternative approaches
- *Concurrent CRT* → consider surgery for residual disease
- *Primary tumor excision + uni/bilateral neck dissection*
  *At least one major risk factor (+) → CRT
  *At least one minor risk factor (+) → RT
- *Induction chemotherapy followed by CRT*

T1-4, N2-3 → three alternative approaches
- *Concurrent CRT* → consider surgery for residual disease
- *Induction chemotherapy followed by CRT*

Consider surgery for residual disease
- *Primary tumor excision + uni/bilateral neck dissection*
  *At least one major risk factor (+) → CRT
  *At least one minor risk factor (+) → RT

T4b, N0-3 → alternative approaches based on performance status (PS)
- PS 0-1 → concurrent CRT or induction chemotherapy followed by RT or CRT
- PS 2 → RT or CRT
- PS 3 → palliative RT or
Single-agent chemotherapy or
Best supportive care

Treatment Algorithm for p16-Positive Oropharyngeal Cancer [21]

T1-2, N0 → two alternative approaches:
- *Definitive RT* → consider surgery for residual disease
- *Primary tumor excision + uni/bilateral neck dissection*
  *At least one major risk factor (+) → CRT
  *At least one minor risk factor (+) → RT

T0-2, N1 (single node ≤3 cm) → three alternative approaches:
- *Primary tumor excision + uni/bilateral neck dissection*
  *At least one major risk factor (+) → CRT
  *At least one minor risk factor (+) → RT
- *Definitive RT* → consider surgery for residual disease
- *Definitive CRT* → consider surgery for residual disease

T0-2, N1 (single node >3 cm or 2 or more ipsilateral nodes ≤6 cm) or
T1-4a, N2-3 → three alternative approaches:
- *Concurrent CRT* → consider surgery for residual disease
- *Primary tumor excision + uni/bilateral neck dissection*

> *At least one major risk factor (+) → CRT
> *At least one minor risk factor (+) → RT
> – *Induction chemotherapy followed by CRT*
> Consider surgery for residual disease
> T4b, N0-3 → alternative approaches based on PS
> – PS 0-1 → concurrent CRT or induction chemotherapy followed by RT or CRT
> – PS 2 → RT or CRT
> – PS 3 → palliative RT or
> Single-agent chemotherapy or
> Best supportive care

Poor prognostic factors

- *Major risk factors* (postoperative CRT indications):
  - Surgical margin positivity
  - ECE
- *Minor risk factors* (postoperative RT alone indications):
  - pT3, pT4
  - N2, N3
  - Nodal level IV or V involvement
  - Vascular tumor embolism

5.3.2.5 Radiotherapy

If tongue immobilization is desired, bite blocks can be used.

General GTV, CTV, and PTV definitions are given below.

GTV = Gross disease at the primary site or any gross lymph node
CTV1 = GTV + 5–10 mm
CTV2 = High-risk subclinical disease
CTV3 = Low-risk subclinical disease
PTV = CTV + 3–5 mm

For tonsil cancer (Table 5.6):

CTV1 should include the maxillary tuberosity, ipsilateral base of tongue minimally (even in no base of tongue extension), ipsilateral glossopharyngeal sulcus, and ipsilateral retromolar trigone. The lower margin should cover the superior tip of hyoid inferiorly.

CTV2 should include ipsilateral soft and hard palate to midline, ipsilateral glossotonsillar sulcus, ipsilateral base of tongue, pterygoid plate (if N+), lateral half of lateral pterygoid muscle (whole muscle in case of trismus or radiological involvement), ipsilateral lateral pharyngeal wall, and ipsilateral parapharyngeal space. Parotids should be covered as required in case of gross disease bordering.

For small (T1), well lateralized, node negative, or low bulk N1 disease without extension to soft palate or base of tongue, nodal treatment can be limited to ipsilateral neck alone.

If ipsilateral neck positive; ipsilateral levels II-IV should be covered in CTV2. Levels V and IB can be included based on tumor location and positive lymph node levels. Level IA needs coverage if there is extension to oral tongue or oral cavity. Retropharyngeal nodal coverage is required to start at jugular foramen if node positive, though is considered enough to start at the

Table 5.6 CTV doses and target definitions for tonsil cancer

Stage	CTV1 (70 Gy/33 fr)	CTV2 (59.4–63 Gy/33 fr)	CTV3 (54–57 Gy/33 fr)
Small T1, well lateralized, without extension to soft palate or base of tongue, N0	GTV + 5 mm + tonsillar region	İpsilateral Ib, II	İpsilateral III, IV, Va
T1–2 N0	GTV + 5 mm + tonsillar region	İpsilateral Ib, II, III	Contralateral II, III, IV, Va Bilateral RP ± bilateral IV–Vb (if lower neck uninvolved)
T1–4 N1-N2b	GTV + 5 mm + tonsillar region	İpsilateral Ib, II, III, IV, Va, RP (add Ia if oral tongue involved)	Contralateral II, III, IV, Va, RP ± bilateral IV–Vb (if lower neck uninvolved)
T1–4 N2c-3	GTV + 5 mm + tonsillar region	Bilateral Ib, II, III, IV, V, RP	Bilateral IV-Vb (if lower neck uninvolved)

CTV clinical target volume, *GTV* gross tumor volume, *RP* retropharyngeal

tip of atlas or transverse process of C1 if node negative (Figs. 5.15 and 5.16).

For base of tongue cancer:

CTV1 should include entire vallecula if involved.

CTV2 should include remaining base of tongue (for >T1 tumors with a circumferential 8–10 mm margin), ipsilateral glossotonsillar sulcus, preepiglottic space, ipsilateral posterior pharyngeal wall, and pterygoid plates and soft palate (if tonsil is involved) (Table 5.7).

5.3.2.6 Selected Publications
See also Sect. 5.10.

Ipsilateral RT

Princess Margaret, 2001 → retrospective. 228 cases. Ipsilateral RT only. T1–2 N0 tumors. Recurrence in contralateral neck was 3.5%. Conclusion: ipsilateral neck RT is feasible in selected cases with unilateral tonsil cancer.

O'Sullivan B et al. (2001) The benefits and pitfalls of ipsilateral radiotherapy in carcinoma of the tonsillar region. Int J Radiat Oncol Biol Phys 51(2):332–343.

Hyperfractionation

EORTC 22791, 1992 → 356 cases. T2–3, N0–1 oropharyngeal cancer except for base of tongue. Tumor size <3 cm. 70 Gy/7–8 weeks vs. 80.5 Gy HF 7 weeks (1.15 Gy BID). Increase in LC of T3 tumors receiving HF schedule. No increase in LC of T2 tumors. There was a trend for increased OS.

Horiot HJ et al. (1992) Hyperfractionation versus conventional fractionation in oropharyngeal carcinoma: final analysis of a randomized trial of the EORTC cooperative group of radiotherapy. Radiother Oncol. 25(4):231–241

MaRCH, 2006 → meta-analysis. 15 trials with 6515 patients were included. 44% patients had oropharyngeal cancer. The median follow-up was 6 years. Altered fractionated RT improves survival in patients with head and neck SCC (3.4% absolute improvement in 5-year OS). Comparison of the different types of altered RT suggests that HF has the greatest benefit with an absolute 8.2% improvement in OS at 5 years compared to a 2% absolute benefit with accelerated RT.

Bourhis J et al. (2006) Hyperfractionated or accelerated radiotherapy in head and neck cancer: a meta-analysis. Lancet 368:843–854

ORO 93–01, 2006 → Multicenter randomized phase III trial. 192 patients with stage III-IV oropharyngeal carcinoma (excluding T1N1 and T2N1) randomized to Arm A: CF RT (66–70 Gy in 33 to 35 fractions), Arm B: split-course accelerated HF RT (64 to 67.2 Gy in 1.6 Gy BID, with a planned two-week split at 38.4 Gy), and Arm C: CF RT with concurrent chemotherapy (three cycles of carboplatin and 5-FU). No difference in five-year OS (21% vs. 21% vs. 40%) or RFS (15% vs. 17% vs. 36%). Loco-regional control (LRC) was significantly better in arm C compared to arms A and B combined (20% vs. 48%). Toxicity rates were comparable.

Fallai C et al. (2006) Long-term results of conventional radiotherapy versus accelerated hyperfractionated radiotherapy versus concomitant radiotherapy and chemotherapy in locoregionally advanced carcinoma of the oropharynx. Tumori 92(1):41–54.

DAHANCA 6&7 trial re-evaluation for HPV status, 2011 → HPV-related disease was associated with a higher OS, disease-specific survival (DSS), and LRC rates. Accelerated RT significantly improved LRC compared to CF RT in both HPV-related and -unrelated oropharyngeal cancers. Also, DSS was significantly improved with accelerated RT in HPV-related disease.

Lassen P et al. (2011) The influence of HPV-associated p16-expression on accelerated fractionated radiotherapy in head and neck cancer: evaluation of the randomised DAHANCA 6&7 trial. Radiother Oncol 100:49–55.

IMRT

Washington University, 2004 → 74 oropharyngeal cancer patients treated with IMRT. 31 patients received definitive IMRT; 17 also received platinum-based chemotherapy. 43 patients received combined surgery and postoperative IMRT. Median follow-up: 33 months. Four-year OS: 87%, DFS: 81% (66% in the definitive vs. 92% in the postoperative RT group), LRC: 87% (78% in the definitive vs. 95% in the postoperative RT group), DM-free survival (DMFS): 90%

5.3 Pharyngeal Cancers

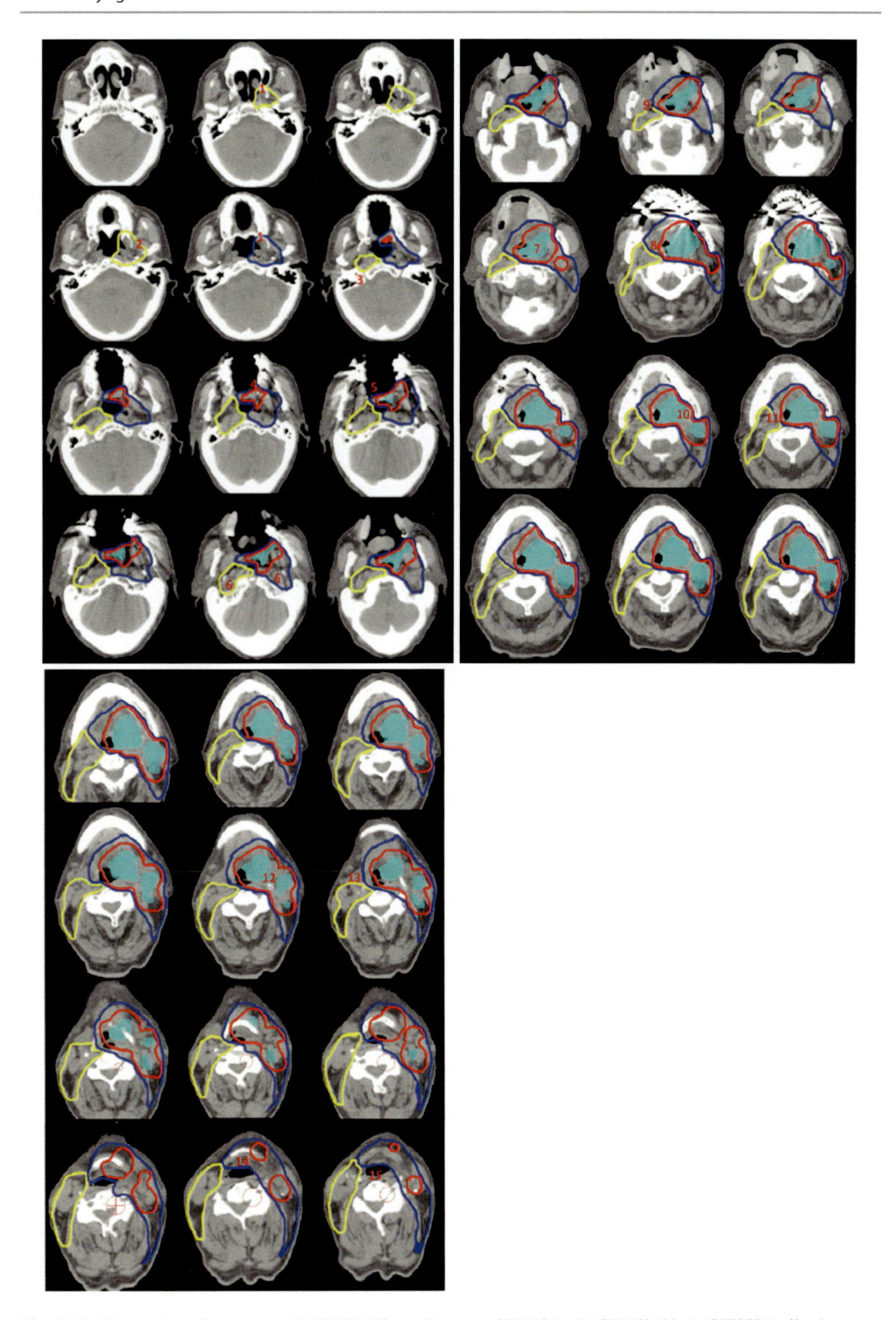

Fig. 5.15 Contouring of a patient with T3N2bM0 tonsil cancer. CTV70 (red), CTV63 (blue), CTV57 (yellow)

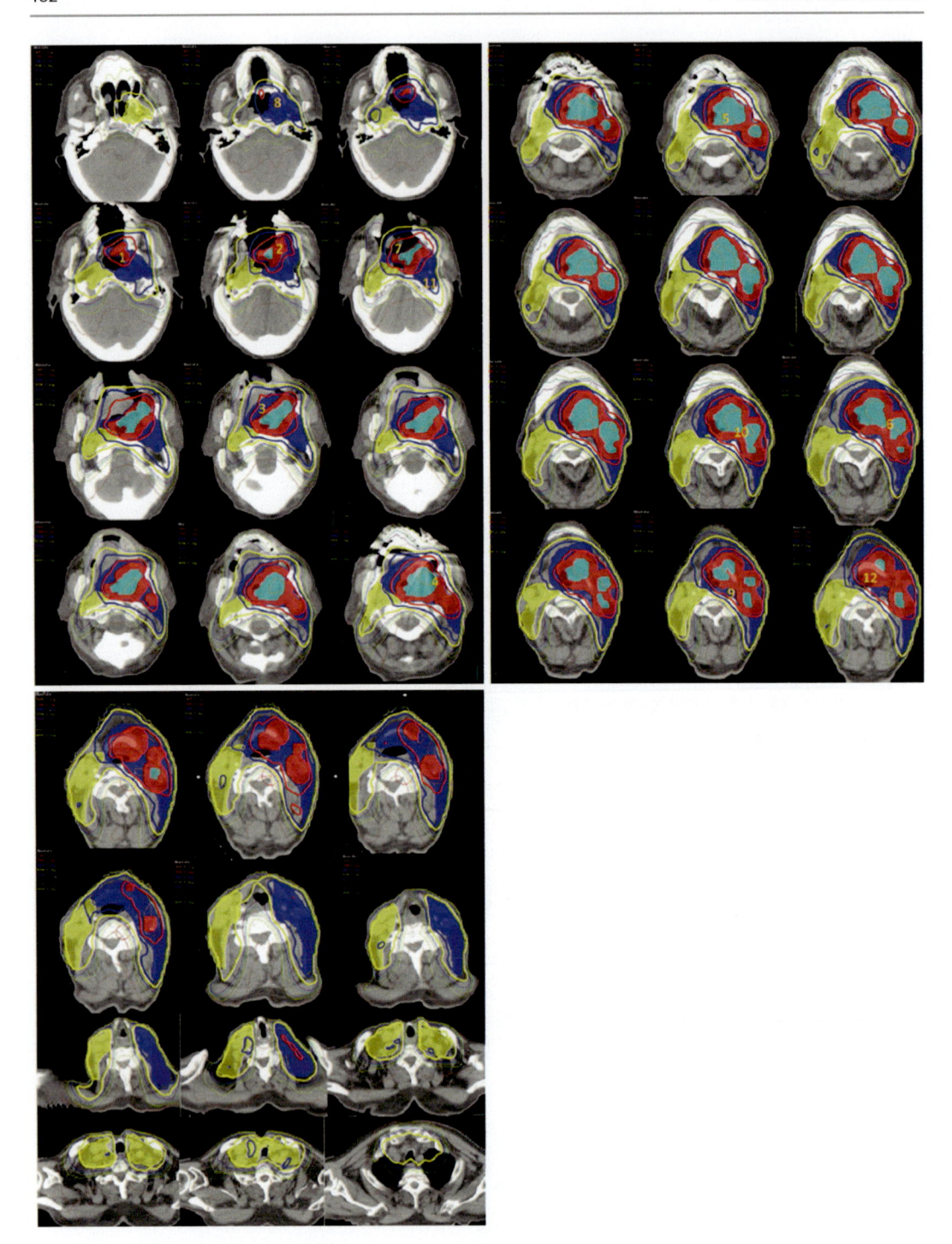

Fig. 5.16 IMRT plan for the same patient

(84% in the definitive vs. 94% in the postoperative group). GTV and nGTV were independent risk factors determining LRC and DFS for definitive oropharyngeal IMRT patients. IMRT was found to be an effective treatment modality for locally advanced oropharyngeal carcinoma.

Table 5.7 CTV doses and target definitions for base of tongue cancer

Stage	CTV1 (70 Gy/33 fr)	CTV2 (59.4–63 Gy/33 fr)	CTV3 (54–57 Gy/33 fr)
T1–2 N0	GTV + 5 mm	–	Bilateral II, III, IV, RP
T1–4 N1–2b	GTV + 5 mm	İpsilateral Ib, II, III, IV, Va, RP (add Ia if oral tongue involved)	Contralateral II, III, IV, Va, RP ± bilateral IV–Vb (if lower neck uninvolved)
T1–4 N2c-3	GTV + 5 mm	Ipsilateral Ib, II, III, IV, Va Bilateral RP	Bilateral IV–Vb (if lower neck uninvolved)

CTV clinical target volume, *GTV* gross tumor volume, *RP* retropharyngeal

Chao KS et al. (2004) Intensity-modulated radiation therapy for oropharyngeal carcinoma: impact of tumor volume. Int J Radiat Oncol Biol Phys 59(1):43–50

RTOG 0022, 2010 → 69 patients with early-stage (T1–2, N0–2) oropharyngeal cancer patients requiring bilateral neck RT were treated with 66 Gy at 2.2 Gy/fraction over 6 weeks to the primary tumor and involved nodes, and 54–60 Gy at 1.8–2.0 Gy/fraction simultaneously to the subclinical target volumes. Chemotherapy was not permitted. Median follow-up: 2.8 years. Two-year loco-regional failure (LRF) risk was 9% and higher in patients with underdosing of the tumor. No event was observed in non-smokers, possibly representing a surrogate for HPV-related disease. Two-year OS: 95%, DFS: 82%. No severe toxicity. Conclusion: Simultaneous integrated boost IMRT is a preferable option for patients with early-stage oropharyngeal cancer.

Eisbruch A et al (2010) Multi-institutional trial of accelerated hypofractionated intensity-modulated radiation therapy for early-stage oropharyngeal cancer (RTOG 00-22). Int J Radiat Oncol Biol Phys 76:1333–1338.

Concurrent chemoradiotherapy

Calais et al, 1999 → Phase III multicenter, randomized trial on 226 patients with stage III-IV oropharynx cancer. Arm A: RT alone (70 Gy CF), Arm B: Concurrent CRT (same RT with three cycles of carboplatin 70 mg/m^2/day and 5-FU 600 mg/m^2/day. Three-year OS: 31% vs. 51% in favor of CRT. Three-year DFS: 20% vs. 42% in favor of CRT. LRC rate: 66% in arm B vs. 42% in arm A. Grade 3–4 mucositis and hematologic toxicity significantly higher in arm B, skin toxicity not different between two arms.

Calais G et al. (1999) Randomized trial of radiation therapy versus concomitant chemotherapy and radiation therapy for advanced-stage oropharynx carcinoma. J Natl Cancer Inst 91(24):2081–6.

GORTEC 94–01, 2004 → Phase III multicenter randomized trial. 226 patients with stage III-IV oropharynx cancer randomized to standard RT (Arm A: 70 Gy in 35 fractions) and concurrent CRT (arm B: same RT with three cycles of carboplatin and 5-FU). Five-year OS: 16% vs. 22% DFS: 15% vs. 27%, and LRC: 25% vs. 48%, all significantly higher in arm B. Severe complication rate was similar.

Denis F et al. (2004) Final results of the 94–01 French Head and Neck Oncology and Radiotherapy Group randomized trial comparing radiotherapy alone with concomitant radiochemotherapy in advanced-stage oropharynx carcinoma. J Clin Oncol 22(1):69–76.

FNCLCC/GORTEC, 2006 → 163 patients with unresectable oropharyngeal and hypopharyngeal cancer were randomized to Arm A: HF RT (1.2 Gy BID, 80.4 Gy/46 day for oropharynx, 75.6 Gy/44 day for hypopharynx) and Arm B: same RT and concurrent chemotherapy (cisplatin and 5FU). Two-year OS: 20.1% vs. 37.8% and DFS: 25.2% vs. 48.2%, both in favor of arm B. Toxicity significantly higher in arm B.

Bensadoun RJ et al. (2006) French multicenter phase III randomized study testing concurrent twice-a-day radiotherapy and cisplatin/5-fluorouracil chemotherapy (BiRCF) in unresectable pharyngeal carcinoma: results at 2 years (FNCLCC-GORTEC). Int J Radiat Oncol Biol Phys 64(4):983–994.

Induction chemotherapy

GETTEC trial, 2000 → phase III trial. 318 patients with oropharyngeal cancer randomized to induction chemotherapy (three cycles of cisplatin and 5-FU) followed by RT or RT alone. RT was 70 Gy to the primary, 50 Gy to the neck when used definitively and 50–65 Gy when used postoperatively. Median follow-up: 5 years. OS was significantly improved with induction chemotherapy (5.1 years vs. 3.3 years). (caution: control arm was RT alone, not chemo-RT)

Domenge C et al. (2000). Randomized trial of neoadjuvant chemotherapy in oropharyngeal carcinoma. French Groupe d'Etude des Tumeurs de la Tete et du Cou (GETTEC). Br J Cancer 83:1594–1598.

Targeted therapy

Bonner et al, 2006 → randomized. 424 patients with stage III–IV head and neck cancers (60% had oropharyngeal cancer). Randomized to RT (70 Gy at 2 Gy/fractions qd, 72–76.8 Gy at 1.2 Gy BID or 72 Gy in 42 fractions concomitant boost 1.8 + 1.5 Gy) ± cetuximab (400 mg/m^2 loading dose IV 1 week before RT, followed by weekly infusions of 250 mg/m2 during RT). Median follow-up: 4.5 years. Three-year LRC: 47 vs. 34%; cetuximab gave a 32% reduction in LR progression. Three-year PFS: 42 vs. 31%; DM risks were similar. Three-year OS: 55 vs. 45%; cetuximab gave a 26% reduction in the risk of death. Improved survival and LRC were obtained with cetuximab (caution: control arm was RT alone, not chemo-RT). Patients with oropharyngeal cancer treated with cetuximab had higher 2-year LRC (50% vs. 41%), median LRRRF duration (49 months vs. 23 months), and median OS (>66 months vs. 30.3 months).

Bonner JA et al (2006) Radiotherapy plus cetuximab for squamous-cell carcinoma of the head and neck. N Engl J Med 354: 567–578

IMCL-9815, 2016 → Retrospective analysis of the Bonner study. p16-positive patients had longer OS when treated with RT alone or RT+cetuximab compared to p16-negative patients. Addition of cetuximab remained beneficial in terms of LRC, OS and PFS, regardless of p16 status in oropharyngeal cancer.

Rosenthal DI et al (2016). Association of human papillomavirus and p16 status with outcomes in the IMCL-9815 Phase III Registration Trial for patients with locoregionally advanced oropharyngeal squamous cell carcinoma of the head and neck treated with radiotherapy with or without cetuximab. J Clin Oncol 34:1300–1308.

Updated analysis of Bonner study, 2010 → Median OS: 49.0 months vs. 29.3 months with vs. without cetuximab. Five-year OS: 45.6% vs. 36.4% with vs. without cetuximab. OS was significantly improved in patients with an acneiform rash of at least grade 2 severity compared with patients with no rash or grade 1 rash after cetuximab.

Bonner JA et al (2010) Radiotherapy plus cetuximab for locoregionally advanced head and neck cancer: 5-year survival data from a phase 3 randomised trial, and relation between cetuximab-induced rash and survival. Lancet Oncol 11:21–28.

RTOG 0522, 2014 → Oropharyngeal cancer patients comprised 70%. Compared two concurrent cycles of 100 mg/m^2 cisplatin and accelerated concomitant boost to 72 Gy RT with or without cetuximab (loading dose of 400 mg/m^2 followed by weekly 250 mg/m^2). No improvement in LRC, DFS, or OS with the addition of cetuximab. Grade 3–4 mucositis and skin reactions increased with cetuximab. No differences in outcome as a function of HPV status.

Ang KK et al. (2014) Randomized phase III trial of concurrent accelerated radiation plus cisplatin with or without cetuximab for stage III to IV head and neck carcinoma: RTOG 0522. J Clin Oncol 32:2940–2950.

NRG Oncology RTOG 1016, 2019 → Noninferiority trial. 987 patients with T1-T2 N2a-N3 M0 or T3-T4 N0-N3 M0 HPV-positive oropharyngeal cancer were randomized to accelerated RT (70 Gy in 35 fractions over 6 weeks) and concurrent cetuximab or same RT and cisplatin. Cetuximab did not meet non-inferiority criteria for OS or PFS. Acute toxicity was higher with cisplatin but total toxicity rate was similar.

Conclusion: Concurrent cisplatin with RT is the standard treatment for patients with HPV-positive locoregionally advanced oropharyngeal carcinoma.

Gillison ML et al. (2019). Radiotherapy plus cetuximab or cisplatin in human papillomavirus-positive oropharyngeal cancer (NRG oncology RTOG 1016): a randomised, multicentre, non-inferiority trial. Lancet Oncol. 393(10166):40–50.

HPV status

Fakhry et al., 2010 → Prospective clinical trial among 96 patients with stage III-IV oropharynx or larynx cancer participated in an Eastern Cooperative Oncology Group (ECOG) phase II trial. Two cycles of induction chemotherapy (paclitaxel and carboplatin) followed by standard RT and concurrent weekly paclitaxel. HPV types 16, 33, or 35 was detected in 40% of patients. Patients with HPV-positive tumors had significantly higher response rates after induction chemotherapy (82% vs 55%) and after CRT (84% vs 57%). Two-year OS: 95% vs. 62% in favor of HPV+ tumors. Progression risk and death from any cause were also lower than HPV- tumors.

Fakhry C et al. (2010) Improved survival of patients with human papillomavirus-positive head and neck squamous cell carcinoma in a prospective clinical trial. J Natl Cancer Inst 100(4):261–9.

TROG 02.02 trial substudy, 2010 → Phase III trial. 465 patients with stage III-IV oropharyngeal cancer were randomly assigned to concurrent RT and cisplatin with or without tirapazamine. Two-year OS and failure-free survival (FFS) was significantly higher in patients with p16-positive tumors (91% vs. 74% and 87% vs. 72%). p16 was a significant prognostic factor on multivariable analysis. p16-positive patients had lower rates of LRF. Also, p16-negative patients had a trend for improved LRC with tirapazamine.

Rischin D et al. (2010) Prognostic significance of p16INK4A and human papillomavirus in patients with oropharyngeal cancer treated on TROG 02.02 phase III trial. J Clin Oncol 28(27):4142–8.

RTOG 0129, 2014 → 721 patients with stage III to IV carcinoma of the oral cavity, oropharynx, hypopharynx, or larynx randomized to 70 Gy in 35 fractions over 7 weeks (SFX) or 72 Gy in 42 fractions over 6 weeks (AFX-C). Both arms contained concurrent cisplatin (100 mg/m2). Median follow-up: 7.9 years. No difference in OS, PFS, LRF, or DM. For oropharyngeal cancer, p16-positive patients had significantly improved rates of OS, PFS, and LRF than p16-negative patients but not DM, even after adjustment for age, tumor stage, nodal stage, treatment assignment, and tobacco use. No difference in severe toxicity between the two arms and p-16 status.

Nguyen-Tan PF et al. (2014) Randomized Phase III Trial to Test Accelerated Versus Standard Fractionation in Combination with Concurrent Cisplatin for Head and Neck Carcinomas in the Radiation Therapy Oncology Group 0129 Trial: Long-Term Report of Efficacy and Toxicity. J Clin Oncol 32:3858–3867

De-escalated Radiotherapy for HPV Positive Tumors

ECOG 1308, 2016 → Prospective phase II trial. 90 patients with resectable HPV-related stage III-IV oropharyngeal cancer received induction chemotherapy (cisplatin, paclitaxel, cetuximab) followed by RT with concurrent cetuximab. RT dose was 54 Gy in 27 fractions in patients with a complete response to induction chemo and 69.3 Gy in 33 fractions in patients with less than a complete response. All patients received 51.3 in 27 fractions to uninvolved cervical LNs. Two-year OS and PFS were 91% and 78% for all patients, and 94% and 80% for complete responders. Among complete responders, patients with <10-pack-year tobacco use had significantly higher OS and PFS rates.

Marur S et al. (2016) E1308: phase II trial of induction chemotherapy followed by reduced-dose radiation and weekly cetuximab in patients with HPV-associated resectable squamous cell carcinoma of the oropharynx— ECOG-ACRIN Cancer Research Group. J Clin Oncol. 2016;35:490–7.

De-ESCALaTE HPV trial, 2018 → 334 patients with HPV-positive low-risk oropharyngeal cancer (non-smokers or lifetime smokers with a smoking history of <10 pack-years) were randomized to RT (70 Gy in 35 fractions) with concurrent cisplatin or RT with cetuximab. Two-year OS: 97.5% vs. 89.4%, 2-year recurrence: 6% vs. 16.1%, both in favor of cisplatin. Severe toxicity was similar.

Mehanna H et al. (2019) Radiotherapy plus cisplatin or cetuximab in low-risk human papillomavirus-positive oropharyngeal cancer (De-ESCALaTE HPV): an open-label randomized controlled phase 3 trial. Lancet Oncol 393(10166):51–60.

NRG Oncology HN0002, 2021 → Randomized, phase II trial. 308 patients with p16-positive, T1-T2 N1-N2b M0, or T3 N0-N2b M0 oropharynx SCC with ≤10 pack-years of smoking were randomized to 60 Gy IMRT in 6 weeks with concurrent weekly cisplatin or 60 Gy IMRT in 5 weeks. Two-year PFS: 90.5% vs. ≤87.6%, significantly higher with concurrent cisplatin. Two-year OS (96.7% vs. 97.3%) and short-term quality of life (QoL) were similar. Grade 3–4 acute toxicity rate was significantly higher with concurrent chemo (79.6% vs. 52.4%) but late toxicity was similar.

YOM SS et al. (2021) Reduced-Dose Radiation Therapy for HPV-Associated Oropharyngeal Carcinoma (NRG Oncology HN002). J Clin Oncol 39:956–965.

5.3.3 Hypopharyngeal Cancer

The hypopharynx is located between the epiglottis and the bottom of the cricoid cartilage (cricopharyngeal sphincter) (Fig. 5.17) [17].

> **Normal Anatomical Structures of the Hypopharynx** (Fig. 5.18)
> *Anterior–superior*: laryngeal entrance (aditus ad laryngeum), aryepiglottic folds, and arytenoids*Anterior–inferior*: postcricoid region
> *Posterior*: prevertebral fascia of C3–6 vertebrae
> *Lateral*: pyriform sinuses (food is transferred into the esophagus from these sinuses)
> *Inferior*: cervical esophagus

5.3.3.1 Pathology
Almost all hypopharyngeal malignancies are SCC, and the most frequent is the poorly differentiated type.

> The frequencies of all hypopharyngeal cancers according to subsite:
>
> - Pyriform sinus: 60%
> - Postcricoid region: 30%
> - Posterior wall: 10%

5.3.3.2 General Presentation
Primary hypopharyngeal cancer, particularly pyriform sinus cancer, frequently metastasizes to neck LNs. Postcricoid region tumors frequently bilaterally invade neck nodes as well as mediastinal and paratracheal lymphatics.

The major clinical findings in hypopharyngeal neoplasms are otalgia, palpable neck nodes, dysphagia, dysphonia, and weight loss. Actual dysphagia is a late finding where the tumor has become very advanced until it obstructs the upper airway and digestive passages. The first signs are usually odynophagia (pain and discomfort during swallowing) and an irritating feeling during the movement of nutrition from the pharyngoesophagus. Dysphagia is usually severe at presentation and almost always associated with weight loss. Reflective pain (otalgia) in the ear is common. Dysphonia is due to either the direct invasion of the larynx or vocal cord paralysis as a result of recurrent laryngeal nerve involvement, and dyspnea may also be seen. Fetor oris (foul smelling breath) and hemoptysis may also be observed. Aspiration of nutrition (particularly liquids) and related aspiration pneumonia may be seen, resulting from an obstruction at the level of the hypopharynx.

5.3 Pharyngeal Cancers

Fig. 5.17 Hypopharynx, oropharynx, and nasopharynx (from [20], reproduced with the permission from American Joint Committee on Cancer)

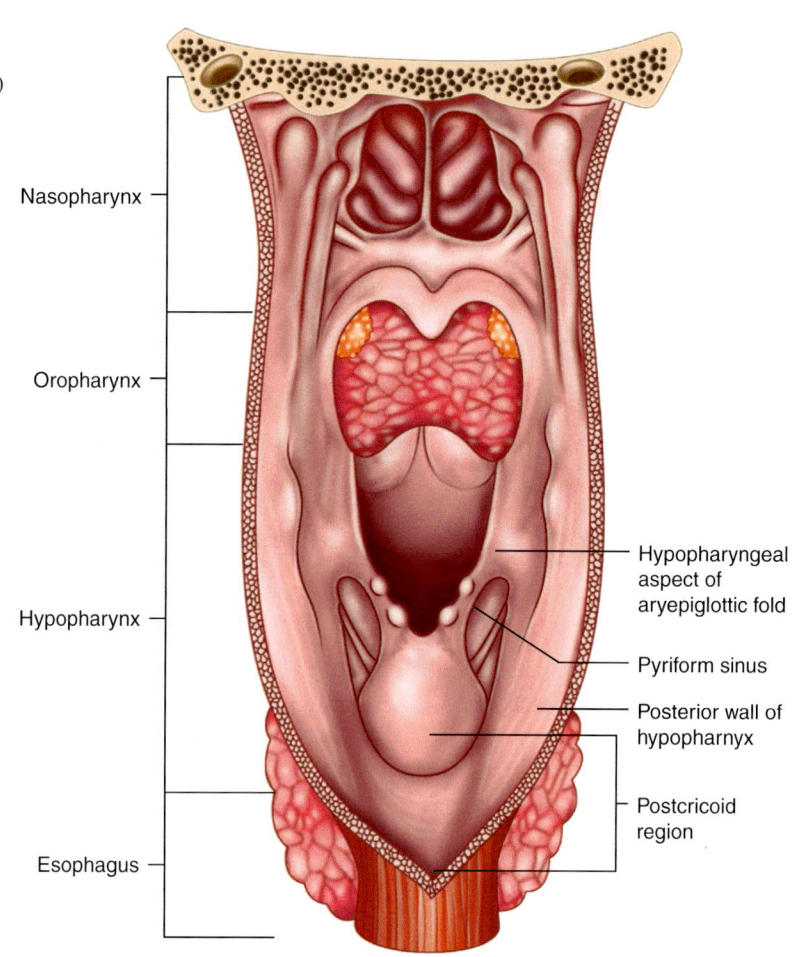

Unfortunately, most cases are at an advanced stage at presentation. The risk of DM is approximately 10%, and the most frequent sites are lung, bone, and liver, respectively.

- Posterior pharyngeal wall tumors usually drain into level II, III and IV neck nodes [9].
- Surgical findings also show that they frequently drain into the RP LNs.

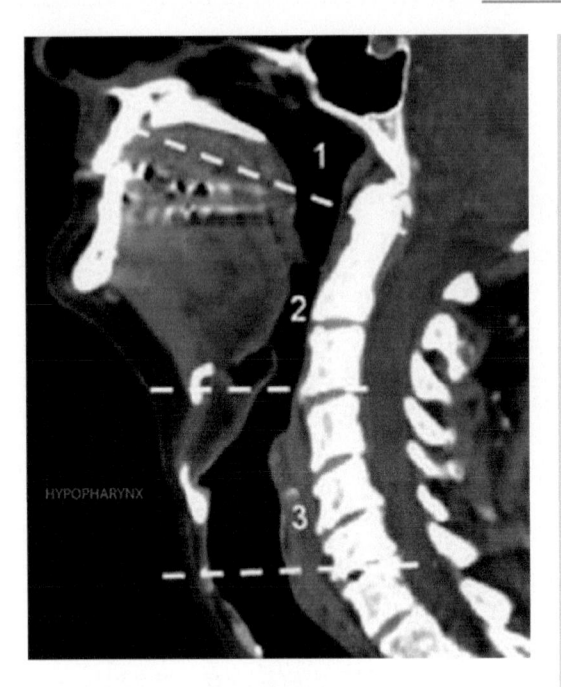

Fig. 5.18 Hypopharyngeal anatomy

The most important risk factor for hypopharyngeal cancer is tobacco and alcohol use [25]. Also, HPV-16 has been shown to be a risk factor in 20–25% of patients [26]. Plummer-Vinson syndrome is another risk factor [25].

5.3.3.3 Staging

Primary Tumor (T) (Fig. 5.19) [20]
- T1: Tumor limited to one subsite* of the hypopharynx and ≤2 cm in greatest dimension
- T2: Tumor invades more than one subsite* of the hypopharynx or an adjacent site, or measures >2 cm but ≤4 cm in greatest diameter without fixation of hemilarynx
- T3: Tumor measures >4 cm in greatest dimension or with fixation of hemilarynx or extension to esophagus
- T4a: Moderately advanced local disease: Tumor invades thyroid/cricoid cartilage, hyoid bone, thyroid gland, or central compartment soft tissue, which includes prelaryngeal strap muscles and subcutaneous fat
- T4b: Very advanced local disease: Tumor invades prevertebral fascia, encases carotid artery, or involves mediastinal structures
- *subsites of the hypopharynx are as follows*:
- *Pharyngoesophageal junction* (i.e., the postcricoid area), extending from the level of the arytenoid cartilages and connecting folds to the inferior border of the cricoid cartilage. *Pyriform sinus*, extending from the pharyngoepiglottic fold to the upper end of the esophagus, bounded laterally by the thyroid cartilage and medially by the surface of the aryepiglottic fold and the arytenoid and cricoid cartilages.
- *Posterior pharyngeal wall*, extending from the level of the floor of the vallecula to the level of the cricoarytenoid joints.

5.3.3.4 Treatment Algorithm for Hypopharyngeal Cancer [21]

- *Most T1, N0, and selected T2N0 (amenable to larynx-preservation)*
- **Definitive RT**
  – Residual disease after RT → Consider surgery
- **Surgery (partial laryngopharyngectomy + ipsilateral/bilateral neck dissection + hemithyroidectomy, and pretracheal and ipsilateral paratracheal LN dissection)**
  – At least one major risk factor → CRT
  – At least one minor risk factor → RT
- *T2-3, N0-3 (amenable to surgery) and T1N1-3* → three alternative approaches:
- **Induction chemotherapy**
  – Complete response in primary and neck → definitive RT or concurrent CRT
  – Partial response in primary → Concurrent CRT or surgery

5.3 Pharyngeal Cancers

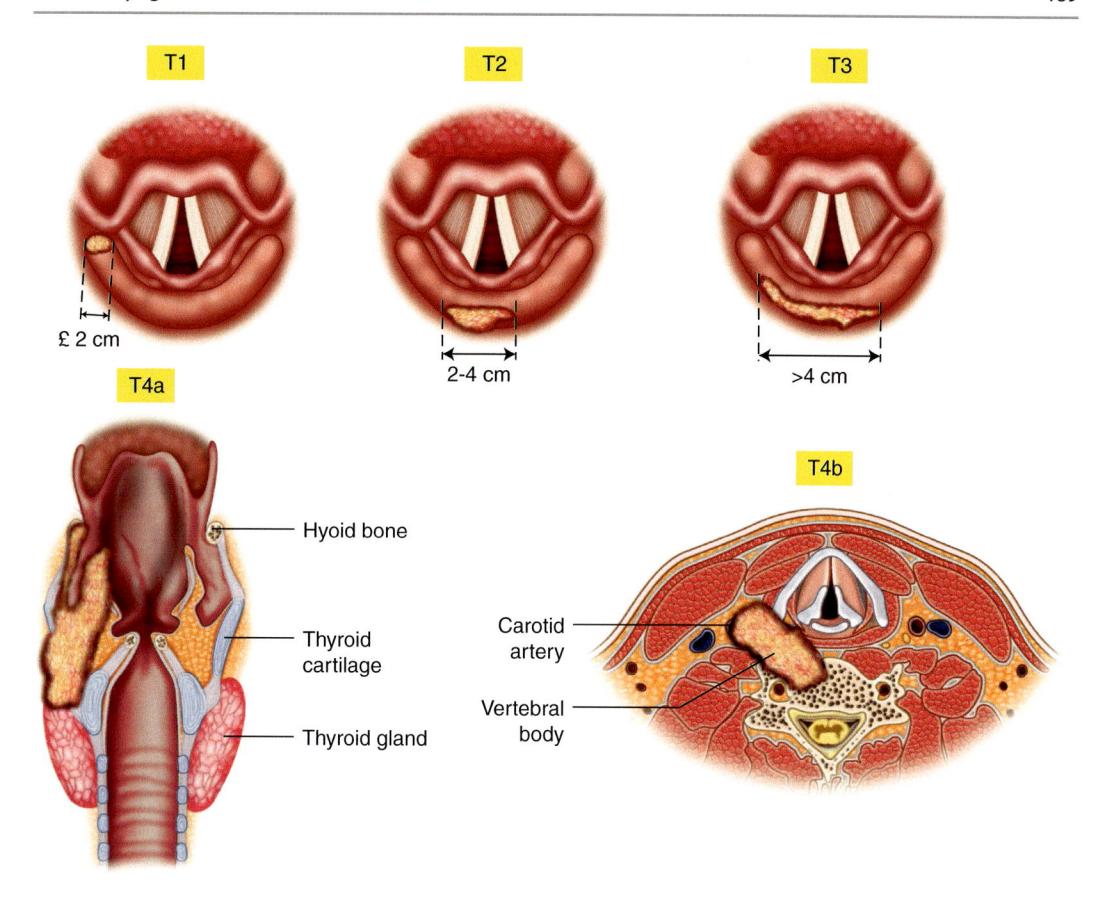

Fig. 5.19 T staging in hypopharyngeal cancer (from [20], reproduced with the permission from the American Joint Committee on Cancer)

- Less than partial response → Surgery
  Partial response in primary and/or neck after RT → consider surgery
    - After surgery:
        - At least one major risk factor → CRT
        - At least one minor risk factor → RT
- **Surgery (partial or total laryngopharyngectomy + ipsilateral/bilateral neck dissection +hemithyroidectomy, and pretracheal and ipsilateral paratracheal LN dissection)**
    - At least one major risk factor → CRT
    - At least one minor risk factor → RT
- **Concurrent CRT**

- Residual disease → consider surgery
  T4a, N0-3 → three alternative approaches:
- **Surgery (total laryngopharyngectomy + neck dissection +hemi- or total thyroidectomy, and pretracheal and ipsilateral or bilateral paratracheal LN dissection)**
    - At least one major risk factor → CRT
    - At least one minor risk factor → RT
- **Concurrent CRT**
    - Residual disease → consider surgery
- **Induction chemotherapy**
    - Complete response in primary and neck → definitive RT or concurrent CRT
    - Partial response in primary → Concurrent CRT or surgery

- Less than partial response → Surgery
  Partial response in primary and/or neck after RT → consider surgery
  After surgery:
  - At least one major risk factor → CRT
  - At least one minor risk factor → RT

Poor prognostic factors

- *Major risk factors:*
  - Positive surgical margin
  - ECE
- *Minor risk factors:*
  - pT3, pT4
  - N2, N3
  - Level IV and V LN involvement Vascular embolism

5.3.3.5 Radiotherapy

- Target volume definitions (Fig. 5.20)
- GTV = All gross disease
- CTV1 (66–70 Gy/2–2.12 Gy) = GTV + 5 mm
- CTV2 (59.4–66 Gy/1.8–2 Gy) = Entire CTV1 + at least 1-cm margin to include the entire hypopharynx, whole larynx, preepiglottic fat, prevertebral fascia
  - Positive LN levels
- CTV3 (46–54 Gy/1.63–2 Gy) = Elective node levels*
- PTV = 1 cm for the primary
- 3–5 mm for nodes

*Bilateral levels II-IV, RP, VI, +/−V (Fig. 5.21)

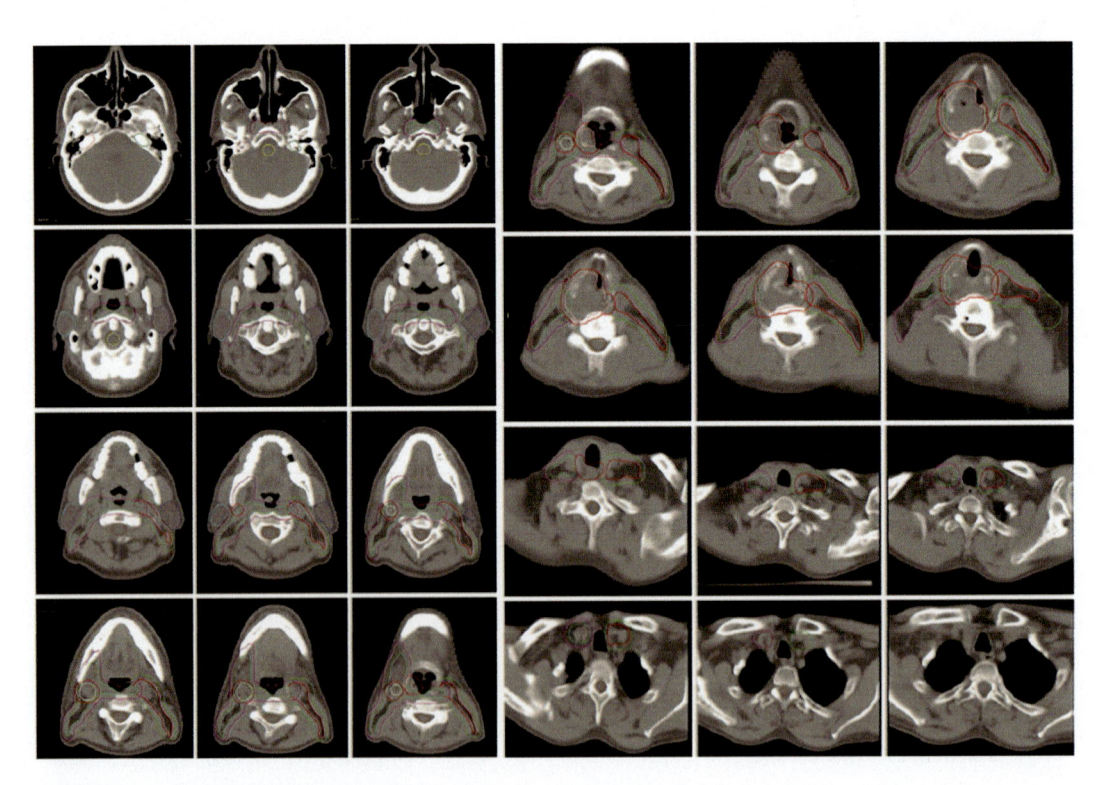

Fig. 5.20 contouring of a patient with T3N2bM0 hypopharyngeal cancer

5.3 Pharyngeal Cancers

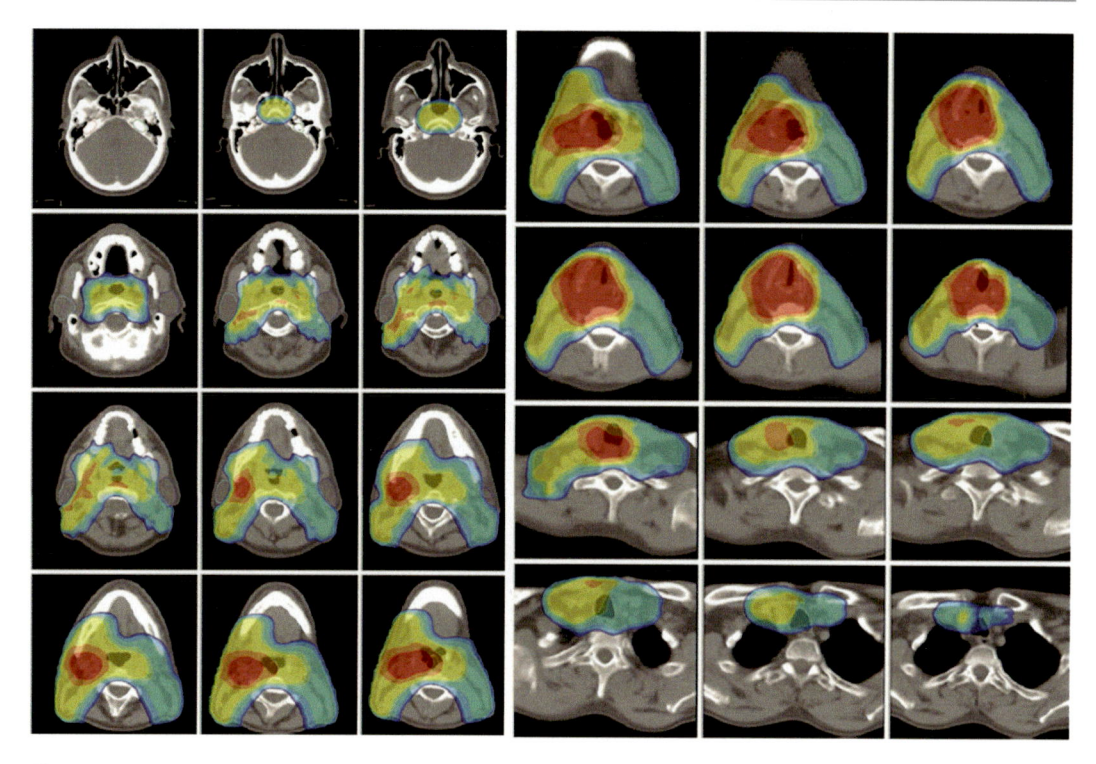

Fig. 5.21 Dose distribution of PTV70 (red), PTV59.4 (yellow), and PTV54 (blue) in the same patient

5.3.3.6 Selected Publications
Also see Sect. 5.4.6 and 5.10.

Surgery vs. RT

EORTC 24891, 1996 → 194 patients with stage II-IV cancer of pyriform sinus or aryepiglottic fold were randomized to either surgery with postoperative RT (50–70 Gy) or induction chemotherapy (cisplatin and 5-FU). In the second arm, patients with a complete response were treated by RT (70 Gy) and nonresponders underwent surgery with postoperative RT (50–70 Gy). Median survival: 25 vs. 44 months, 3- and 5-year OS: 43% and 35% vs. 57% and 30%, 3- and 5-year DFS: 32% and 27% vs. 43% and 25%, for surgery and induction chemo, respectively. Three- and 5-year larynx preservation in the induction-chemo arm: 42% and 35%.

Conclusion: Larynx preservation without jeopardizing survival is feasible in hypopharyngeal cancer. Induction chemotherapy followed by RT is the new standard.

Lefebvre JL et al. (1996) Larynx preservation in pyriform sinus cancer: preliminary results of a European Organization for Research and Treatment of Cancer phase III trial. EORTC Head and Neck Cancer Cooperative Group. J Natl Cancer Inst 88:890–899

EORTC 24891, update, 2012 → Ten-year OS: 13.8% vs. 13.1%, 10-year PFS: 8.5% vs. 10.8%, for surgery and chemo, respectively. Ten-year survival with a functional larynx: 8.7% in chemo arm.

Conclusion: This strategy does not compromise disease control or survival and allow more than half of the survivors to retain their larynx.

Lefebvre JL et al. (2012) Laryngeal preservation with induction chemotherapy for hypopharyngeal squamous cell carcinoma: 10-year results of EORTC trial 24,891. Ann Oncol 23:2708–2714.

Induction chemotherapy

EORTC 24954, 2009 → 450 patients with resectable advanced larynx (T3–T4) or hypopharynx (T2–T4) cancer with N0–N2 disease were randomized to sequential arm (two cycles of cisplatin and 5-FU; in responders another two cycles followed by 70 Gy RT) or the experimental arm (four cycles of cisplatin and 5-FU with split-course 60 Gy RT—20 Gy during the three 2-week intervals between chemo cycles). All nonresponders underwent salvage surgery and postoperative RT. Median follow-up: 6.5 years. Survival with a functional larynx, median OS (4.4 vs. 5.1 years), and median progression-free interval (3.0 vs. 3.1 years) were similar. Acute and late toxicity rates were also comparable.

Lefebvre JL et al. (2009) Phase 3 randomized trial on larynx preservation comparing sequential vs alternating chemotherapy and radiotherapy. J Natl Cancer Inst 101:142–152.

TREMPLIN, 2013 → 116 patients with stage III-IV larynx/hypopharynx cancer received three cycles of induction chemotherapy (docetaxel, cisplatin, 5-FU). Poor responders (<50% response) underwent salvage surgery, responders (≥50% response) were randomly assigned to conventional RT (70 Gy) with concurrent cisplatin (arm A) or concurrent cetuximab (arm B). No significant difference in OS, larynx preservation, larynx function preservation, or toxicity. There were fewer local treatment failures in arm A than in arm B; salvage surgery was feasible in arm B only.

Lefebvre JL et al. (2013) Induction chemotherapy followed by either chemoradiotherapy or bioradiotherapy for larynx preservation: the TREMPLIN randomized phase II study. J Clin Oncol 31(7):853–859.

5.4 Laryngeal Cancer

The Larynx is Divided into Three Compartments (Fig. 5.22) [8]

– *Supraglottic region*: Section above the vocal cords. This region includes epiglottis, aryepiglottic folds, arytenoids, ventricular bands (false vocal cords), and laryngeal ventricles.
– *Glottic region*: Region of the vocal cords. It includes the true vocal cords, interarytenoid area (posterior commissure), floor of the ventricle, and anterior commissure. The vocal cords consist of the vocal ligaments, vocalis muscle, and mucosal layers. The shell of the larynx is formed by the hyoid bone, thyroid cartilage, and cricoid cartilage.
– *Subglottic region*: This is the region inferior to the vocal cords until the first tracheal ring.
– The axial border between the glottic and supraglottic regions is the apex of the ventricle.
– The subglottis is considered to extend from 5 mm below the vocal cords to the inferior border of the cricoid cartilage or 10 mm below the apex of the ventricle.

The larynx is located between the tip of the thyroid (bottom of the C3 vertebral corpus) and the bottom of the cricoid cartilages (C6 vertebra) [8].

5.4.1 Pathology

Most (95–96%) malignant tumors of the larynx are SCC. Other malignancies, such as verrucous, basocellular and fusiform cell carcinomas, adenocarcinoma, adenoid cystic carcinoma, and mesenchymal malignancies such as sarcomas, are very rare.

The incidence of synchronous carcinoma in a patient with laryngeal cancer is approximately 1%. The probability of a metachronous primary tumor is 5–10%.

- Laryngeal cancer is the most frequent tumor seen together with another type of malignancy

5.4 Laryngeal Cancer

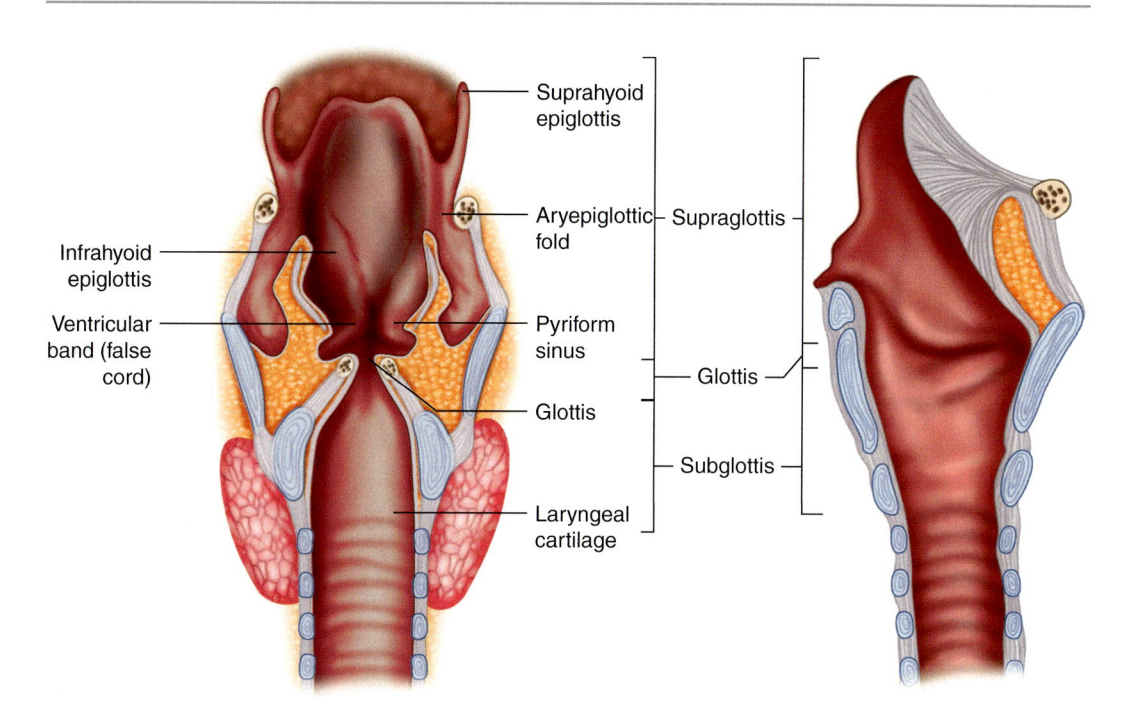

Fig. 5.22 Anatomy of the larynx (from [20], reproduced with the permission from the American Joint Committee on Cancer)

5.4.2 General Presentation

Hoarseness. This is the most common sign, particularly for glottic lesions. Laryngeal cancer should be ruled out if the hoarseness lasts for more than 2 weeks with a detailed ENT exam.

Dysphagia (difficulty swallowing). This is more common in supraglottic cancers.

Dyspnea and stridor. This symptom is seen when the tumor mass in the larynx obstructs the airway. Supraglottic tumors must be larger than glottic and subglottic tumors to narrow the airways.

Otalgia (ear pain). This is a referred pain that is particularly seen in supraglottic cancers via Arnold's nerve, a branch of the vagus nerve.

Cough. This is observed in cases with involvement of the superior laryngeal nerve (sensory nerve of the larynx), causing swallowing dysfunctions and sensory disorders. Hemoptysis associated with cough may be observed in ulcerated tumors.

Neck mass. Swelling or skin invasion seen in the laryngeal region or the bottom of the hyoid is a sign of advanced disease, since it shows that the tumor has invaded the thyroid cartilage and thyrohyoid membranous barrier. A mass combined with the larynx at the front of the neck may be palpated. Sometimes lymph node metastasis (Delphian node) in the front of the larynx at the level of the cricothyroid or thyrohyoid membranes may be observed as an anterior mass. Neck masses palpated at lateral sites are a sign of neck metastases.

Supraglottis (Fig. 5.23)
- T1: Tumor limited to one subsite* of supraglottis with normal vocal cord mobility
- T2: Tumor invades mucosa of more than one adjacent subsite* of supraglottis or glottis or region outside the supraglottis

(e.g., mucosa of base of tongue, vallecula, or medial wall of pyriform sinus) without fixation of the larynx
- T3: Tumor limited to the larynx with vocal cord fixation and/or invades any of the following: postcricoid space, pre-epiglottic space, paraglottic space, and/or inner cortex of the thyroid cartilage
- T4a: Moderately advanced local disease: Tumor invades through the thyroid cartilage and/or invades tissues beyond the larynx (e.g., trachea, soft tissues of the neck including deep extrinsic muscles of the tongue, strap muscles, thyroid, or esophagus)
- T4b: Very advanced local disease: Tumor invades prevertebral space, encases carotid artery, or invades mediastinal structures
- Subsites include the following: Ventricular bands (false cords), Arytenoids, Suprahyoid epiglottis, Infrahyoid epiglottis, Aryepiglottic folds (laryngeal aspect)

5.4.3 Staging

Glottis (Fig. 5.24)
- T1: Tumor limited to the vocal cord(s) which may involve anterior or posterior commissure, with normal mobility
  - T1a: Tumor limited to one vocal cord
  - T1b: Tumor involves both vocal cords
- T2: Tumor extends to the supraglottis and/or subglottis with impaired vocal cord mobility
- T3: Tumor limited to the larynx with vocal cord fixation and/or invades paraglottic space and/or inner cortex of the thyroid cartilage
- T4a: Moderately advanced local disease: Tumor invades through the outer cortex of the thyroid cartilage and/or invades tissues beyond the larynx (e.g., trachea, cricoid cartilage, soft tissues of neck including deep extrinsic muscles of the tongue, strap muscles, thyroid, or esophagus)
- T4b: Very advanced local disease: Tumor invades the prevertebral space, encases the carotid artery, or invades mediastinal structures

Subglottis (Fig. 5.25)
- T1: Tumor limited to the subglottis
- T2: Tumor extends to the vocal cord(s) with normal or impaired mobility
- T3: Tumor limited to the larynx with vocal cord fixation
- T4a: Moderately advanced local disease: Tumor invades the cricoid or thyroid cartilage and/or invades tissues beyond the larynx (e.g., trachea, soft tissues of the neck including deep extrinsic muscles of the tongue, strap muscles, thyroid, or esophagus)
- T4b: Very advanced local disease: Tumor invades the prevertebral space, encases the carotid artery, or invades mediastinal structures

- Supraglottic laryngeal cancer mainly metastasizes to level II cervical LN. Level Ib and V involvement is very rare.
  - Clinical LN positivity: 55%
  - Bilaterality: 16%
- The true vocal cords have no lymphatics, so lymphatic involvement is observed in the presence of supraglottic or subglottic tumor extension.
- Glottic spread is typically associated with metastasis to level II LNs.
- *LN involvement in glottic cancers:*
  - *T1: <2%*
  - *T2: 5%*
  - *T3: 15–20%*
  - *T4: 20-40%*
- Anterior commissure and anterior subglottic invasion is associated with pretracheal (level VI) LN metastasis.

5.4 Laryngeal Cancer

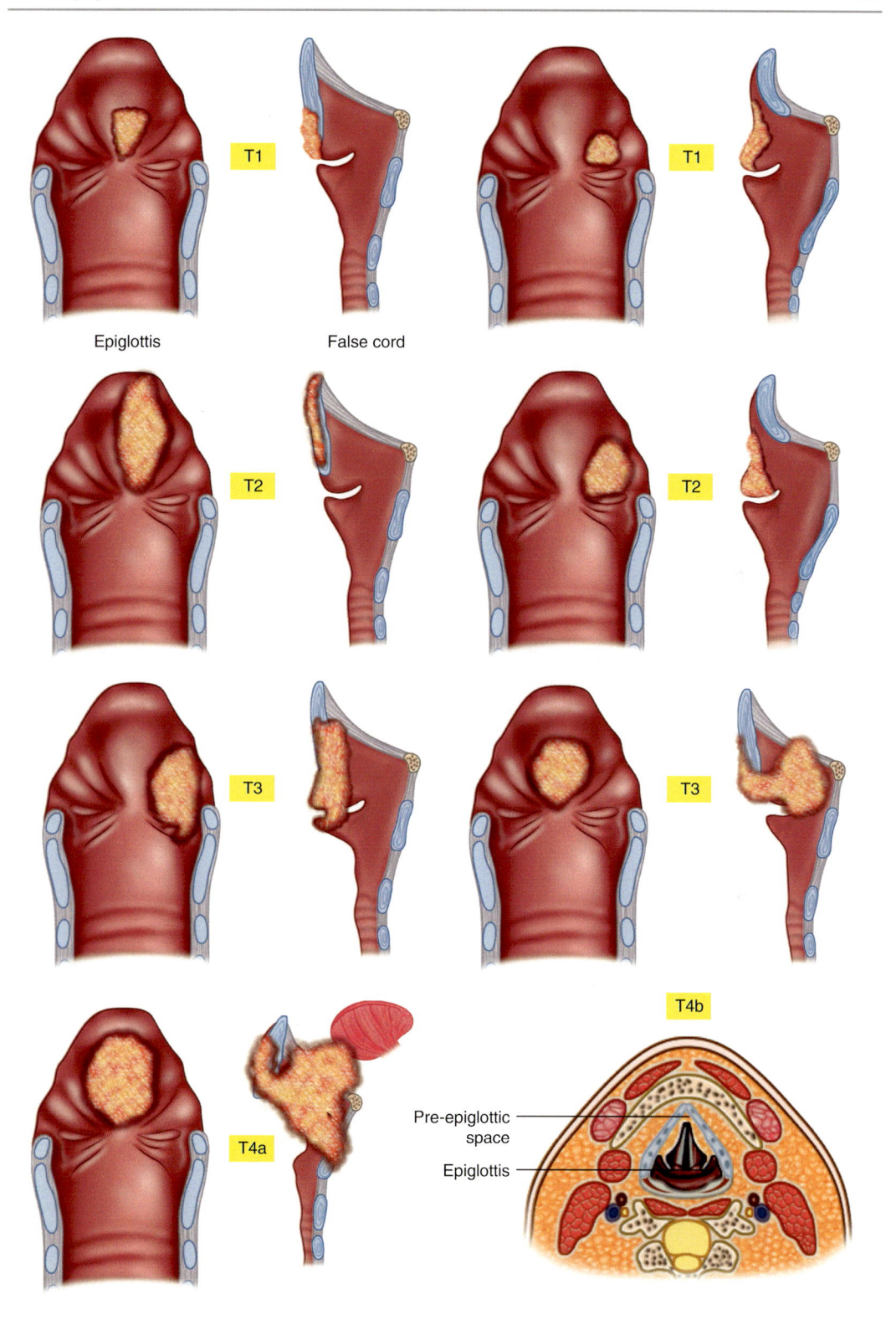

Epiglottis   False cord

Fig. 5.23 T staging for supraglottic laryngeal cancer (from [20], reproduced with the permission from the American Joint Committee on Cancer)

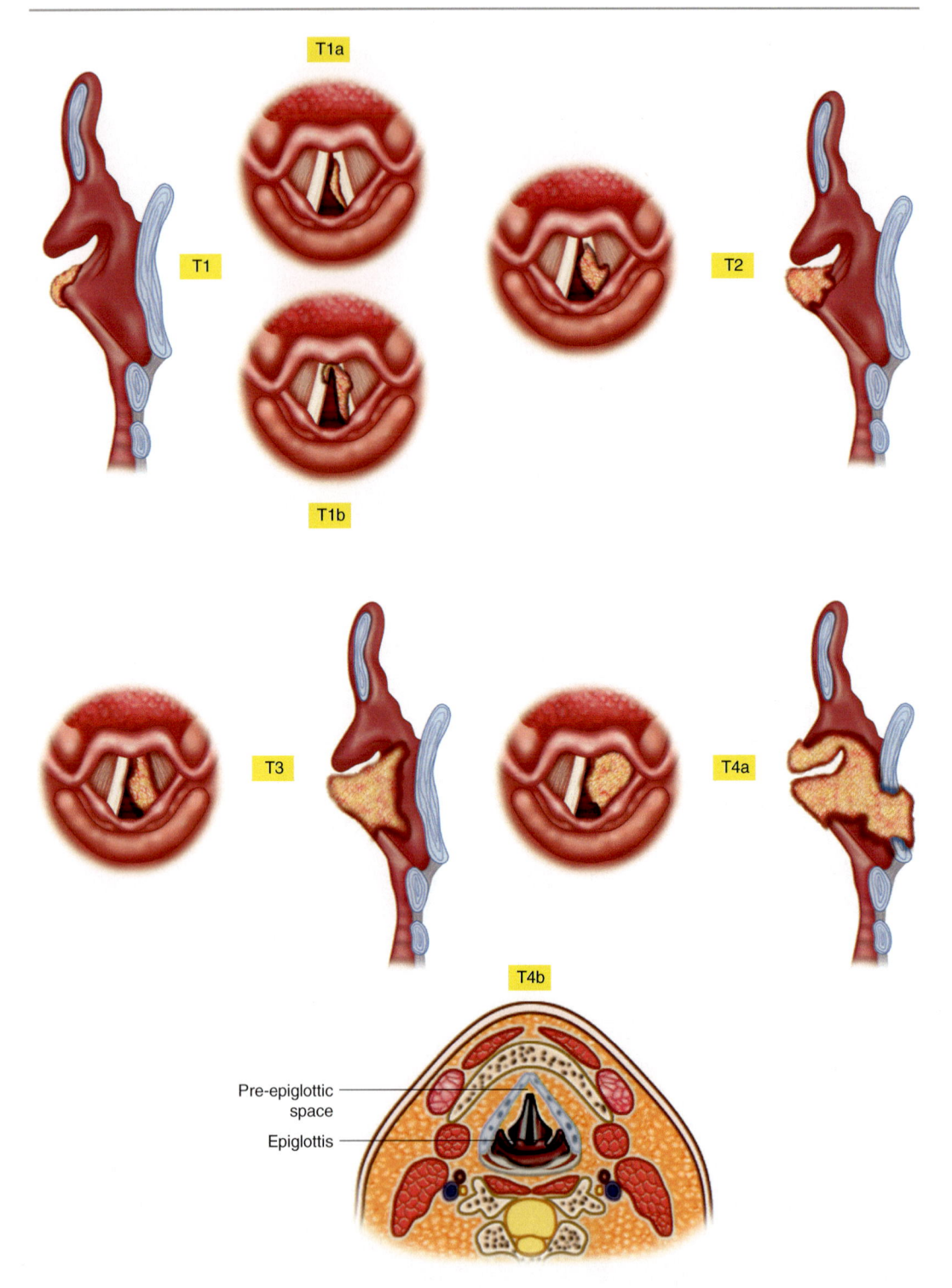

Fig. 5.24 T staging for glottic laryngeal cancer (from [20], reproduced with the permission from the American Joint Committee on Cancer)

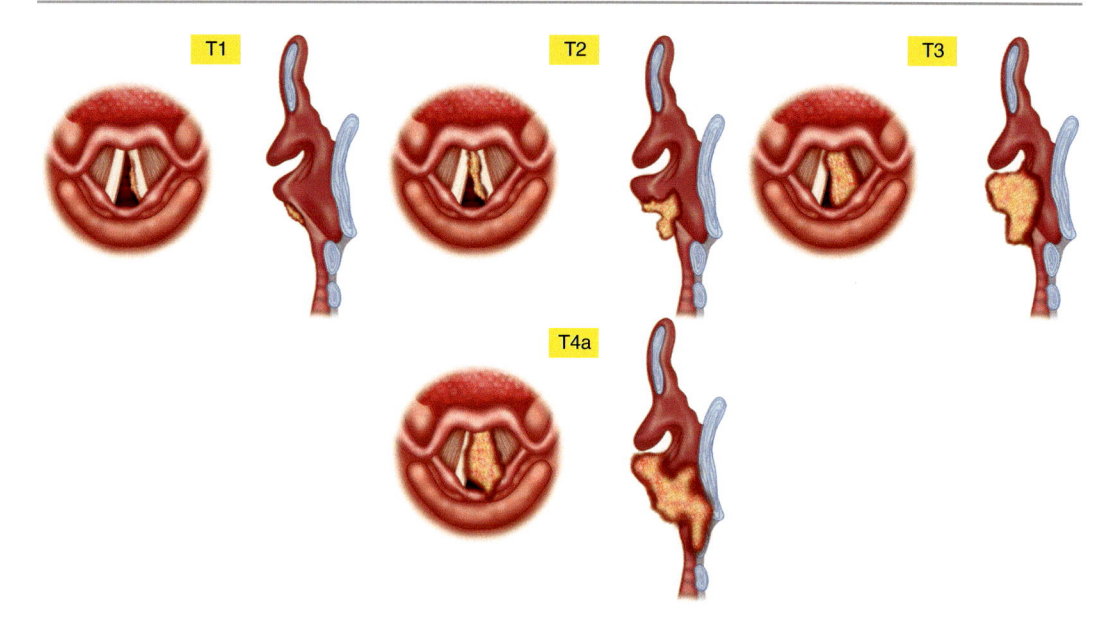

Fig. 5.25 T staging for subglottic laryngeal cancer (from [20], reproduced with the permission from the American Joint Committee on Cancer)

5.4.4 Treatment Algorithm

Treatment Algorithm for Laryngeal Cancer [21]

Glottic
- Tis
  - Endoscopic excision (stripping, laser)
  - Definitive RT
- T1–2, N0 or select T3N0 (amenable to larynx preserving)
  - Definitive RT
  - Surgery (Partial laryngectomy ± selective neck dissection)
    Consider adjuvant RT or CRT based on pathologic adverse factors*
- T3, N0-3 (requiring total laryngectomy)
  - Concurrent CRT or
  - Surgery (total laryngectomy + bi/ipsilateralthyroidectomy+pretracheal and ipsilateral paratracheal LN dissection + bi/ipsilateral neck dissection)
    Consider adjuvant RT or CRT based on pathologic adverse factors*
  - Induction chemotherapy
    Complete response: Definitive RT
    Partial response: RT or CRT
    Less than partial response: Laryngectomy
- T4a, N0-3
  - Surgery (total laryngectomy+thyroidectomy+pretracheal and ipsilateral paratracheal LN dissection+/-bi/ipsilateral neck dissection)
    Consider adjuvant RT or CRT based on pathologic adverse factors*
- In selected patients that decline surgery:
  - Concurrent CRT
  - Induction chemotherapy
    Complete response: Definitive RT
    Partial response: RT or CRT
    Less than partial response: Laryngectomy
- T4b, N0-3 or unresectable nodal disease or unfit for surgery
  - PS 0-1: Concurrent CRT or induction chemotherapy
  - PS 2: RT or concurrent CRT
  - PS 3: Palliative RT or single-agent chemo or best supportive care

Supraglottic
- *T1–2, N0-3 and selected T3N0-1 (amenable to larynx preserving)*
  - Endoscopic resection + neck dissection
    - Consider adjuvant RT or CRT based on pathologic adverse factors*
  - Open partial supraglottic laryngectomy + neck dissection
    - Consider adjuvant RT or CRT based on pathologic adverse factors*
  - Definitive RT
- *T3N0 (requiring total laryngectomy)*
  - Concurrent CRT
  - RT
  - Surgery (total laryngectomy + thyroidectomy + ipsilateral, central, or bilateral neck dissection)
    - Consider adjuvant RT or CRT based on pathologic adverse factors*
  - Induction chemotherapy
    - Complete response: Definitive RT
    - Partial response: RT or CRT
    - Less than partial response: Laryngectomy
- *Most T3, N1-3 (requiring total laryngectom)*
  - Concurrent CRT
  - Surgery (total laryngectomy+ipsilateral thyroidectomy+neck dissection)
    - Consider adjuvant RT or CRT based on pathologic adverse factors*
  - Induction chemotherapy
    - Complete response: Definitive RT
    - Partial response: RT or CRT
    - Less than partial response: Laryngectomy
- *T4a N0-3*
  - Surgery (total laryngectomy + thyroidectomy + pretracheal and ipsilateral paratracheal LN dissection ± bi/ipsilateral neck dissection)
    - Consider adjuvant RT or CRT based on pathologic adverse factors*
- In selected patients that decline surgery:
  - Concurrent CRT
  - Induction chemotherapy
    - Complete response: Definitive RT
    - Partial response: RT or CRT
    - Less than partial response: Laryngectomy
- *T4b, N0-3 or unresectable nodal disease or unfit for surgery*
  - PS 0-1: Concurrent CRT or induction chemotherapy
  - PS 2: RT or concurrent CRT
  - PS 3: Palliative RT or single-agent chemo or best supportive care

Postoperative CRT indications
- Surgical margin (+)
- ECE

Postoperative RT indications
- Perineural invasion (PNI)
- Lymphovascular space involvement (LVSI)
- Multiple LN involvement
- T3–4 tumor

5.4.5 Radiotherapy (Figs. 5.26 and 5.27)

- **T1-2N0 glottic cancer**
- GTV = Gross laryngeal lesion
- CTV = Entire larynx
- PTV = 1 cm
- CTV dose = 63 Gy in 28 fractions (2.25 Gy/fraction) in T1 tumors
  - 65.25 Gy in 29 fractions (2.25/fraction) in T2 tumors
- **Locally advanced glottic and all stages of supraglottic and subglottic cancers**
- GTV = All gross disease
- CTV1 (70 Gy) = GTV + 5–10 mm

5.4 Laryngeal Cancer

- CTV2 (59.4 Gy) = entire larynx (from the top of thyroid notch to the bottom of thyroid cartilage)
  - Surgical bed in adjuvant cases
  - All soft tissue involvement
  - Positive LN levels
- CTV3 (54 Gy) = Elective nodal regions
- *Levels II–IV should be covered.
- *Level Ib should be included if level II is positive.
- *Level V is not included unless levels II-IV are massively involved.
- *Level VI should be included in case of subglottic extension.
- PTV = 1 cm for CTV1 primary, 3–5 mm for other CTVs.
- **Indications for Treating Stoma**
  - Nodes: multiple LNs or positive LN close to stoma
  - Emergency tracheostomy
  - Subglottic extension >5 mm
  - Surgical scar crosses stoma
  - Anterior soft tissue/skin involvement
  - Positive/close margins

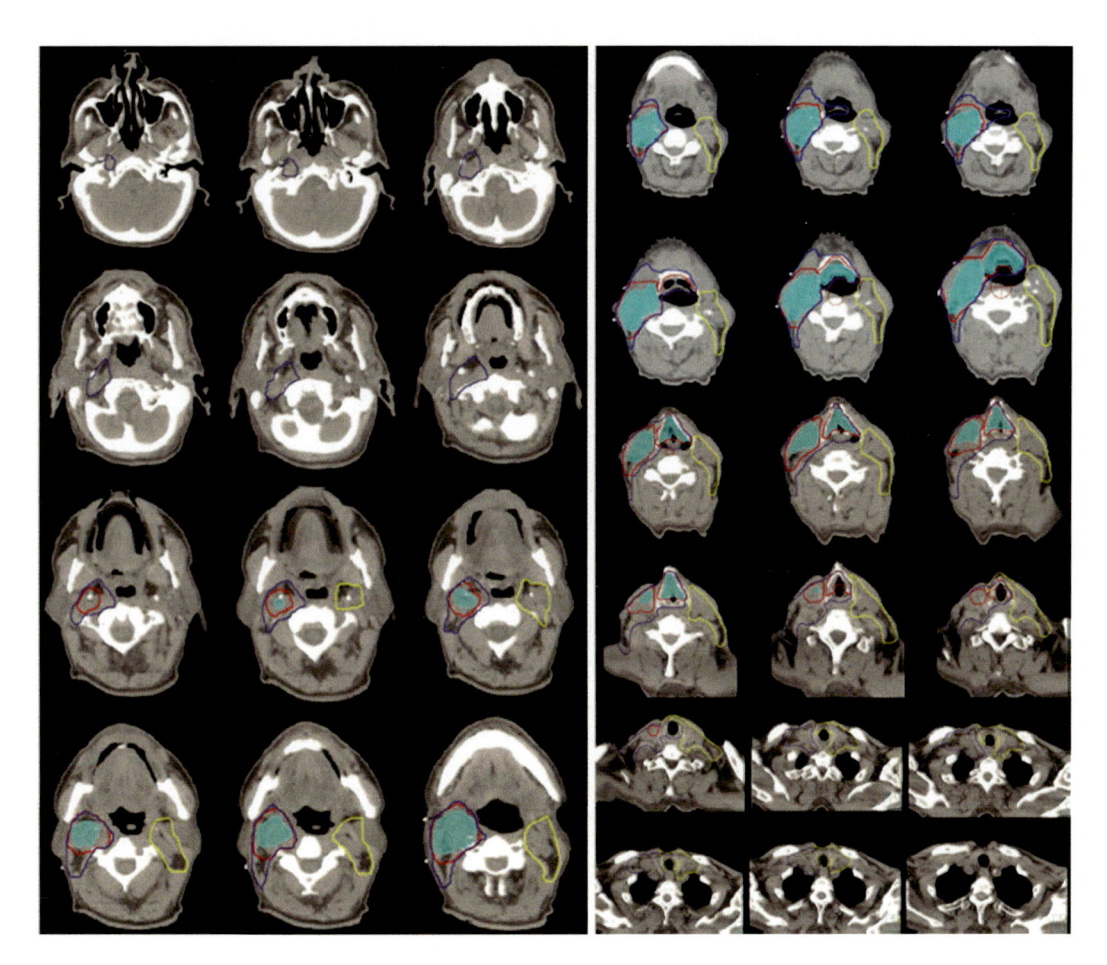

Fig. 5.26 Contouring of a patient with T2N2bM0 supraglottic laryngeal cancer

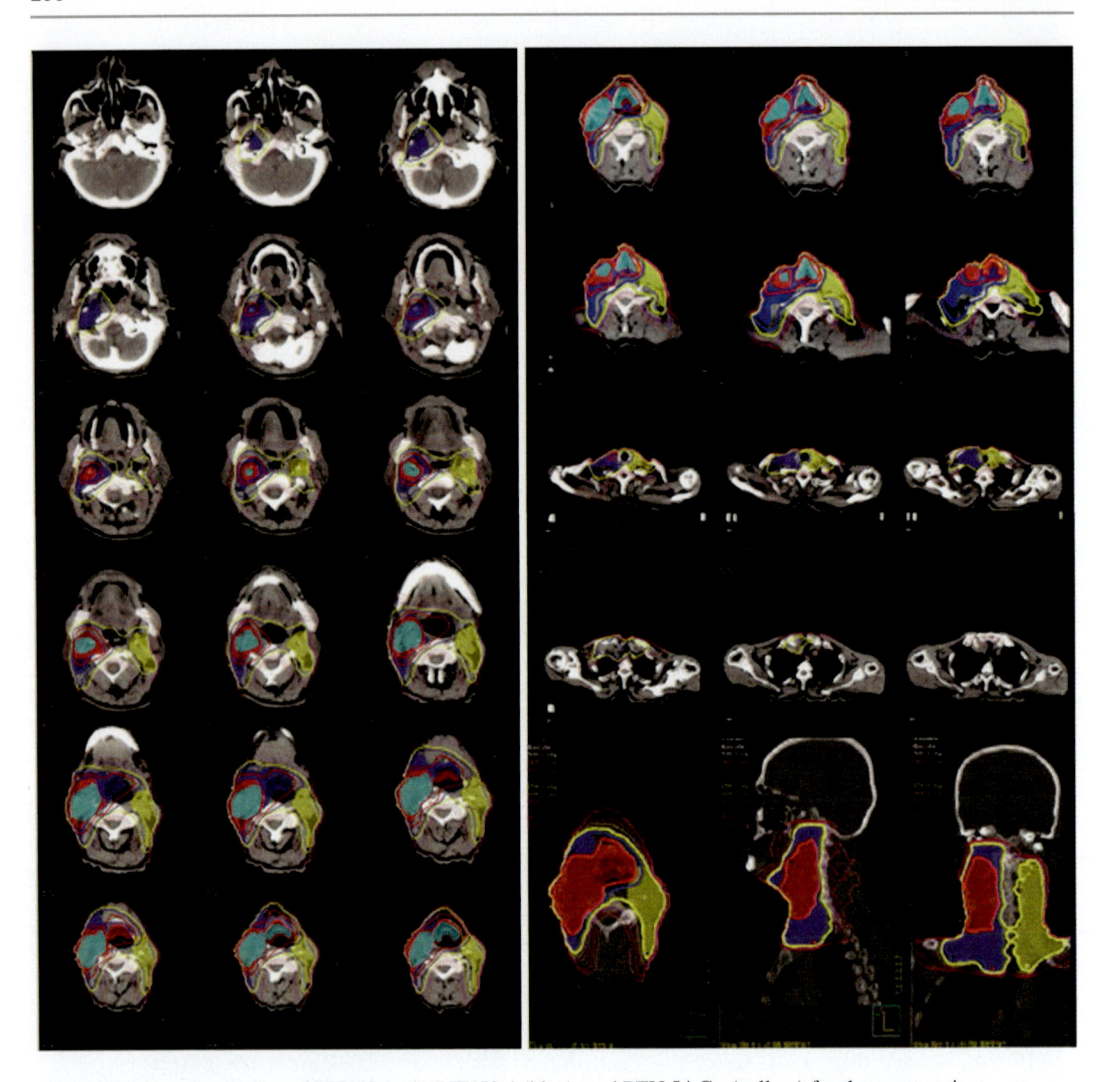

Fig. 5.27 Dose distribution of PTV70 (red), PTV59.4 (blue), and PTV 54 Gy (yellow) for the same patient

5.4.6 Selected Publications

See also Sect. 5.3.3.6 and 5.10.

Fractionation in early stage

Yamazaki et al, 2006 → 180 patients with T1N0M0 glottic cancer were randomized to Arm A (60-66 Gy in 2-Gy fractions) or Arm B (56.25-63 Gy in 2.25-Gy fractions). Five-year LC significantly higher in Arm B (77% vs. 92%). Five-year cause-specific survival (CSS) and toxicity rates were similar.

Yamazaki H et al (2006) Radiotherapy for early glottis carcinoma (T1N0M0): results of prospective randomized study of radiation fraction size and overall treatment time. Int J Radiat Oncol Biol Phys 64(1):77-82.

RTOG 95-12, 2014 → 250 patients with T2N0M0 glottic cancer were randomized to CF RT (70 Gy in 2 Gy/fractions) or HF RT (79.2 Gy in 1.2 Gy/fractions BID). Median follow-up: 7.9 years. Five-year LC: 70% vs. 78%, 5-year DFS: 40% vs. 49%, and 5-year OS: 63% vs. 72, all statistically insignificant. HF RT arm resulted in higher rates of acute toxicity. Late toxicity was comparable.

Trotti A et al (2014) Randomized trial of hyperfractionation versus conventional fractionation in T2 squamous cell carcinoma of the vocal cord (RTOG 9512). Int J Radiat Oncol Biol Phys 89(5):958-963.

Preoperative vs. postoperative Radiotherapy

RTOG 7303, 1991 → 277 patients with operable advanced supraglottic larynx and hypopharynx cancer were randomized to preoperative RT (50 Gy) or postoperative RT (60 Gy). Follow-up: 9-15 years. LRC was significantly better in the postoperative RT arm without absolute survival benefit. In patients with supraglottic larynx cancer, 78% of LRF occurred in the first 2 years. Local failure rate within 2 years: 31% (preop RT) vs. 18% (postop RT). After 2 years, DM and second primaries became the predominant failure pattern, especially in postop RT patients. Toxicity rates were comparable.

Conclusion: Because of an increased incidence of late DM and secondary primaries, additional therapeutic intervention is required beyond surgery and postoperative RT to impact significantly upon survival.

Tupchong L et al (1991) Randomized study of preoperative versus postoperative radiation therapy in advanced head and neck carcinoma: long-term follow-up of RTOG study 73-03. Int J Radiat Oncol Biol Phys 20(1):21-8.

Larynx Preservation

VA Larynx Trial, 1991 → 332 patients with stage III-IV laryngeal cancer were randomized to induction chemotherapy (three cycles of cisplatin and 5-FU) followed by RT or surgery followed by RT. In the chemo group, tumor response was assessed after two cycles; patients with a response received a third cycle followed by definitive RT (66-76 Gy), patients with no response or recurrence after chemo and RT underwent salvage laryngectomy. After two cycles of chemo, tumor response rate was 85% (31% complete and 54% partial). Median follow-up: 33 months. Two-year OS: 68% in both arms. Patterns of recurrence differed significantly; more LR but fewer DM in the chemo arm. 36% in the chemo arm required total laryngectomy. Overall larynx preservation: 64%.

The Department of Veterans Affairs Laryngeal Cancer Study Group (1991) Induction Chemotherapy plus Radiation Compared with Surgery plus Radiation in Patients with Advanced Laryngeal Cancer. N Engl J Med 324:1685-1690.

VA Larynx Trial, functional update, 1998 → Assessment of functional outcomes related to communication (including amount of speech therapy), swallowing and eating, and employment status. Patients with a preserved larynx were better in speech communication. Only 6% total laryngectomy patients developed usable esophageal speech, 8% remained nonvocal, 55% ended up as using artificial electrolarynx (55%), and 31% with tracheoesophageal speech.

Hillman RE et al (1998) Functional outcomes following treatment for advanced laryngeal cancer. Part I--Voice preservation in advanced laryngeal cancer. Part II--Laryngectomy rehabilitation: the state of the art in the VA System. Research Speech-Language Pathologists. Department of Veterans Affairs Laryngeal Cancer Study Group. Ann Otol Rhinol Laryngol Suppl 172:1-27.

GETTEC, 1998 → 68 patients with previously untreated T3 larynx cancer were randomized to total laryngectomy followed by RT or induction chemotherapy followed by RT in good responders, and by total laryngectomy plus RT in poor responders. The rate of laryngectomy in chemo arm was 58%. Two-year survival: 84% vs. 69% in favor of no chemo. DFS was also improved in the no-chemo arm.

Conclusion: Larynx preservation for patients, selected on the basis of having responded to induction chemotherapy, cannot be considered a standard treatment at the present time.

Richard JM et al (1998) Randomized trial of induction chemotherapy in larynx carcinoma. Oral Oncol 34:224–228

RTOG 9111, 2003 → 547 patients with stage III-IV glottis and supraglottic laryngeal cancer were randomized to Arm 1: RT alone vs Arm 2: induction chemo (two cycles of cisplatin and 5-FU) followed by RT (if partial-complete response) or surgery +/- adjuvant RT (if <partial response) vs Arm 3: concurrent CRT (cisplatin 100 mg/m^2). RT dose was 70 Gy in all arms. All cN2 patients got planned neck dissection after RT. Median follow-up: 3.8 years. Two-year larynx preservation rate: 70% vs. 75% vs. 88%, 2-year LRC rate: 56% vs. 61% vs. 78%, both in favor of Arm 3. DFS improved in both arms with chemo. OS rates were similar in all groups.

Severe toxicity rate lowest with RT alone: 61% vs. 82% vs. 81%. Mucosal toxicity of Arm 3 was significantly higher than the other two arms.

Forastiere A et al (2003) Concurrent Chemotherapy and Radiotherapy for Organ Preservation in Advanced Laryngeal Cancer. N Engl J Med 349:2091-2098

RTOG 9111, update, 2013 → 520 patients were analyzed. Median follow-up: 10.8 years. Ten-year larynx preservation rate significantly better in arm 3, and arm 2 was not better than arm 1. Both chemotherapy regimens significantly improved laryngectomy-free survival rate compared to arm 1. OS was similar in all arms. Late toxicity was similar.

Conclusion: Induction PF followed by RT and concurrent cisplatin/RT show similar efficacy for LFS. LRC and larynx preservation were significantly improved with concurrent cisplatin/RT.

Forastiere A et al (2013) Long-term results of RTOG 91-11: a comparison of three nonsurgical treatment strategies to preserve the larynx in patients with locally advanced larynx cancer. J Clin Oncol 31(7):845-52.

TAX 324, subgroup analysis, 2009 → 166 patients with locally advanced laryngeal and hypopharyngeal cancer were randomized to induction chemo with TPF or PF, both followed by chemoradiotherapy. Median OS: 59 vs. 24 months, median PFS: 21 vs. 11 months, both significantly higher with TPF. Among operable patients, 3-year LFS: 52% vs. 32%, significantly higher with TPF. Significantly fewer TPF patients had surgery (22% vs. 42%).

Posner MR et al (2009) Sequential therapy for the locally advanced larynx and hypopharynx cancer subgroup in TAX 324: survival, surgery, and organ preservation. Ann Oncol 20:921–927.

GORTEC 2000–01, 2015 → 213 patients with operable (requiring total laryngectomy) stage III-IV larynx or hypopharynx cancer were randomized to three cycles of induction chemotherapy with either TPF or PF, followed by RT for responders. Five- and 10-year larynx preservation rates: 74.0% vs. 58.1% and 70.3% vs. 46.5% in favor of TPF. Five- and 10-year larynx dysfunction-free survival rates: 67.2% vs. 46.5% and 63.7% vs. 37.2%. OS, DFS, and LCR were comparable. Grade 3-4 late toxicities of the larynx: 9.3% with TPF vs. 17.1% with PF.

Janoray G et al (2015) Long-term Results of a Multicenter Randomized Phase III Trial of Induction Chemotherapy With Cisplatin, 5-fluorouracil, ± Docetaxel for Larynx Preservation. J Natl Cancer Inst 108(4):djv368.

5.5 Oral Cavity Cancers

The oral cavity is located between the vermillion line (the mucosa–skin boundary of the upper and lower lips) anteriorly and the isthmus faucium posteriorly. It is an anatomical space that is bounded by the floor of the mouth inferiorly, the hard palate superiorly, and the buccal mucosa laterally. Although the oral cavity is a space, it is not a homogeneous region, and it includes various subsites [17].

Normal Anatomical Structures of the Oral Cavity (Fig. 5.28)
Upper and lower lips
 Anterior two-thirds of tongue
 Floor of the mouth
 Retromolar trigone
 Alveolar ridge (includes the alveolar processes of the maxilla and mandible and the overlying mucosa)
 Buccal mucosa (includes the mucosal surfaces of the cheek and lips)
 Hard palate

The oral cavity is divided into two parts via the arches of the upper and lower teeth when the mouth is closed [17]:

- Anterior part: vestibulum oris (entrance to the oral cavity)
- Posterior part: cavum oris proprium (true oral cavity)

5.5 Oral Cavity Cancers

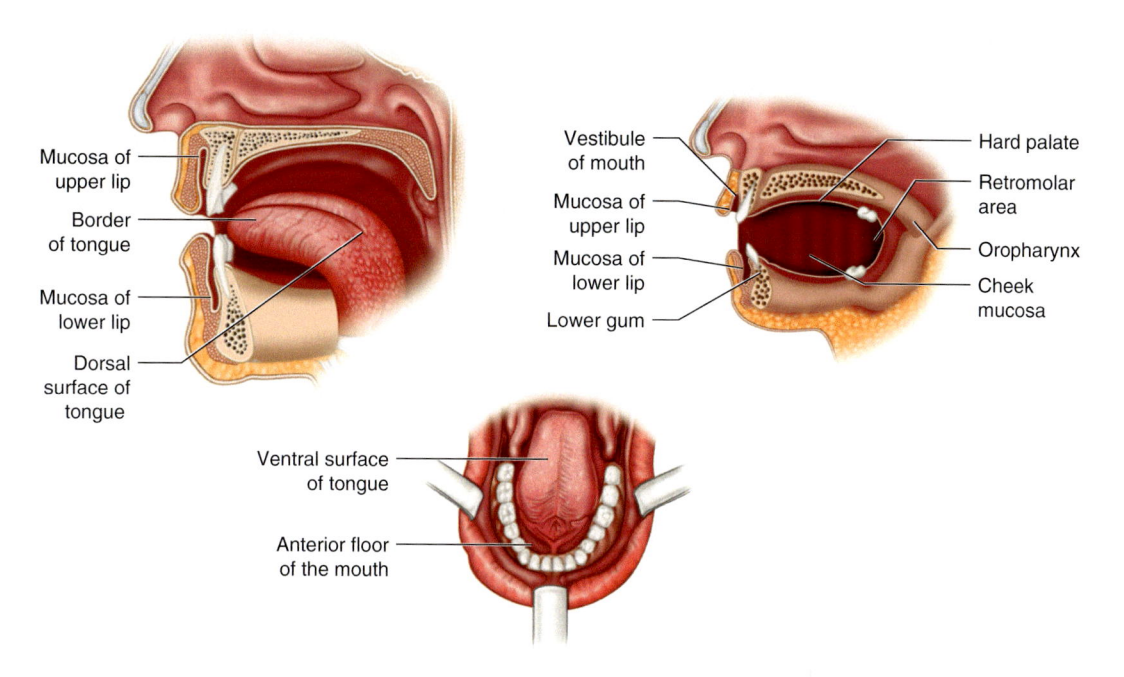

Fig. 5.28 Normal anatomical structures of the oral cavity (from [20], reproduced with the permission from the American Joint Committee on Cancer)

Retromolar trigone. This is located between the ramus of the mandible and the last molar tooth, connecting these two spaces when the mouth is closed. This region is important due to the fact that patients with maxillofacial trauma or maxillary fixation may feed via the retromolar trigone [17].

- Oral cavity cancers are the most common noncutaneous malignancy of the head and neck.
- Risk factors:
  - Tobacco
  - Alcohol
  - Ultraviolet radiation (for lip cancer)
  - HPV (for oral tongue cancer)
  - Herpes Simplex Virus
  - Genetic syndromes (Xeroderma pigmentosum, ataxia telangiectasia, Li-Fraumeni, bloom syndrome, Fanconi anemia)

5.5.1 Pathology

Premalignant lesions of the oral cavity:

Some epithelial lesions have the potential for malign transformation (Fig. 5.29). These are as follows:

- *Leukoplakia.* This is the most common premalignant lesion in the head–neck region. The white macules may have several histological features, from simple hyperkeratosis, dysplasia, and carcinoma in situ to invasive carcinoma. They are most commonly observed in the oral cavity and laryngeal mucosa. The risk of malignant transformation in these lesions is higher in smokers and in women. Leukoplakia may regress spontaneously without therapy.
- *Erythroplakia.* This is less common than leukoplakia. The oral cavity is the most frequent localization. The risk of malignant transformation is much higher, and biopsy is essential for histological evaluation. The treatment of choice is surgical excision.

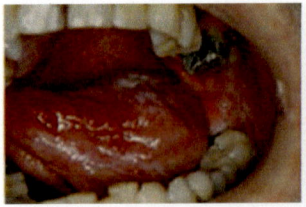

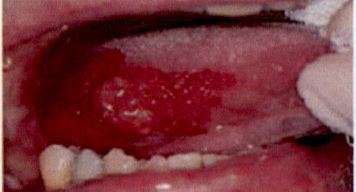

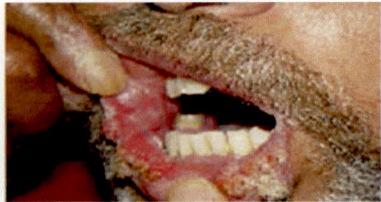

Fig. 5.29 Leukoplakia, erythroplakia, and lichen planus (from *left* to *right*)

- *Lichen planus.* Erosive lichen planus in the oral cavity may transform into malignancy. These are benign lesions and are frequently localized in the buccal mucosa and tongue symmetrically. The erosive type is seen in the floor of the mouth. Biopsy is essential to determine lichen type and for differential diagnosis from leukoplakia.
- In addition, oral submucous fibrosis, pemphigus, papilloma, chronic fungal infections, and Plummer–Vinson syndrome are also followed for possible malignant transformations.

- Most (95%) oral cavity malignancies are SCC.
- *Other malignant tumors of the oral cavity:*
- Epithelial origin: malignant melanoma
- Glandular origin: adenocarcinoma, adenoid cystic carcinoma Lymphoid origin: lymphoma
- Soft tissue origin: sarcoma, Kaposi's sarcoma
- Dental tissue origin: Ameloblastoma

5.5.2 General Presentation

The most prominent symptoms are a scar that does not heal and swelling or masses. The swelling or mass can be located in the primary region as well as the neck, since they have the potential to metastasize to cervical lymph nodes. In addition, pain (e.g., during chewing, swallowing, throat pain, ear pain), bleeding, speaking disorders, swallowing dysfunctions, and dyspnea can also be seen.

5.5.3 Staging (Table 5.8)

Primary Tumor (T) (Fig. 5.30) [20]
T1: Tumor ≤2 cm in greatest dimension and ≤5 mm depth of invasion (DOI)
  T2: Tumor ≤2 cm and DOI 5-10 mm or

- Tumor >2 cm but ≤4 cm and DOI ≤10 mm

  T3: Tumor >4 cm in greatest dimension

- Any tumor >10 mm DOI

  T4a: Moderately advanced local disease:

- Lip: Tumor invades through cortical bone, inferior alveolar nerve, floor of the mouth, or skin of the face (i.e., chin or nose)
- Oral cavity: Tumor invades adjacent structures (e.g., through cortical bone of the mandible or maxilla, maxillary sinus, and skin of the face)

  T4b: Very advanced local disease: Tumor invades masticator space, pterygoid plates, or skull base and/or encases internal carotid artery.

N staging is similar to those of other head–neck cancers except for Nasopharynx.

5.5 Oral Cavity Cancers

Table 5.8 Lymph node metastasis rate by oral cavity subsite (%) [27]

Subsite	Level Ia	Level Ib	Level II	Level III
Oral tongue	9	18	73	18
Floor of mouth	7	64	43	0
Retromolar trigone	0	25	63	12.5

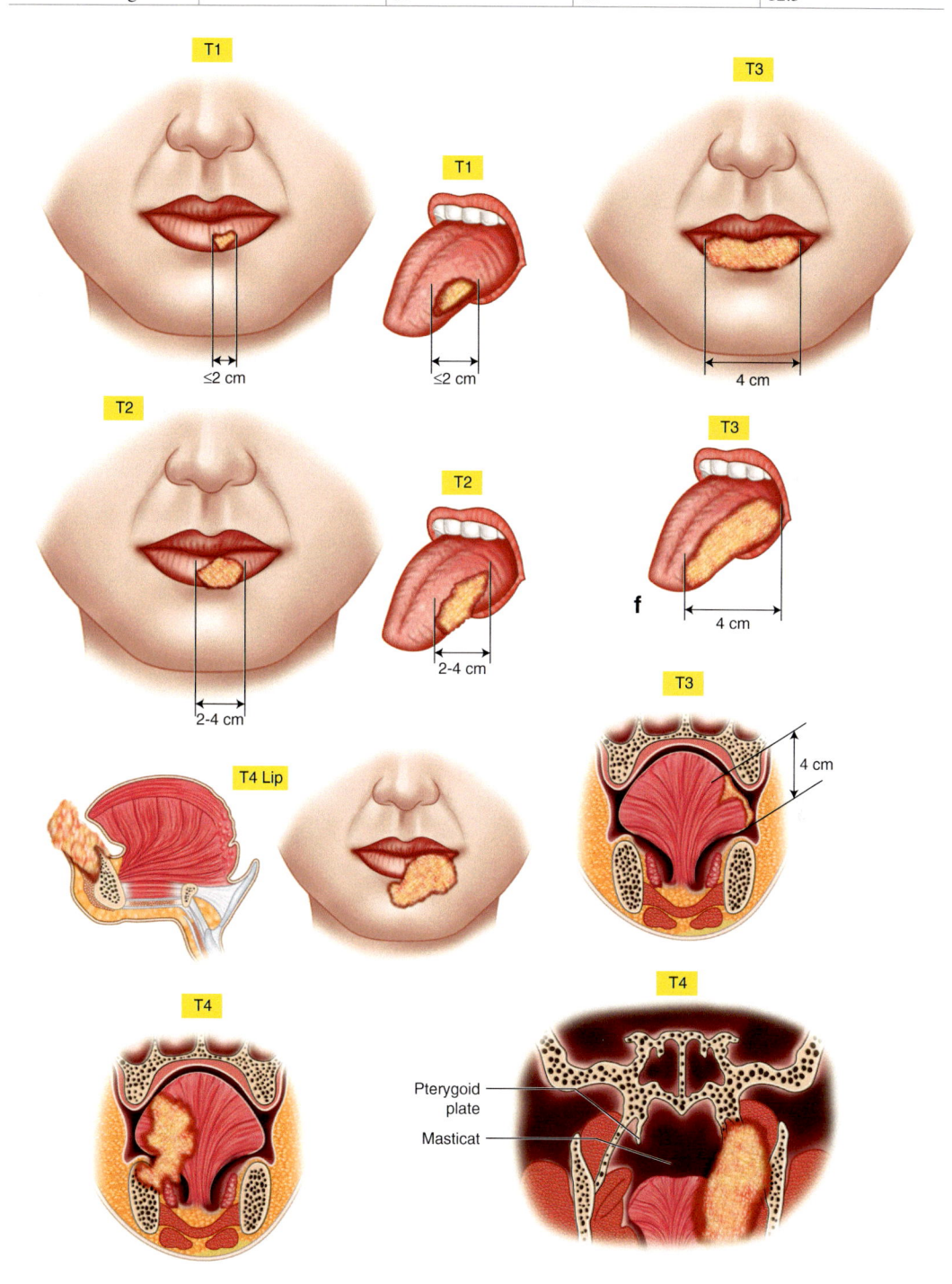

Fig. 5.30 T staging for oral cavity cancer (from [20], reproduced with the permission from the American Joint Committee on Cancer)

- Approximately 50% of all oral cavity malignancies are located in the lower lip or tongue. The second most common subsite is the floor of the mouth [27].
- Five-year estimated survival for all stages of all subsites is 64.5%, and for localized disease, it is 83.7% [28].
- Five-year survival is nearly 90% in lip cancers smaller than 2 cm in size, but decreases to 40% in advanced cases [29].
- Five-year survival is nearly 90% in stage I tongue cancers, but decreases to 10% in stage IV cases [30].

When clinical N (−) tongue cancers are examined histopathologically after neck dissection, pathological LN positivity is nearly 33% [31].

5.5.4 Treatment Algorithm [21]

T1-2 N0
- Surgery (preferred)
    Consider RT depending on pathologic risk factors*
- Definitive RT

T3N0, T1-3 N1-3, T4a N0-3
- Surgery
    Consider RT depending on pathologic risk factors*

T4b N0-3 or unresectable nodal disease or unfit for surgery
- PS 0-1: Concurrent CRT or induction chemotherapy
- PS 2: RT or concurrent CRT
- PS 3: Palliative RT or single-agent chemo or best supportive care

Postoperative CRT indications (major risk factors):

- ECE
- Surgical margin (+)

Postoperative RT indications (minor risk factors):

- Multiple LN (+)
- PNI
- LVSI

5.5.5 Radiotherapy (Figs. 5.31, 5.32, 5.33, and 5.34)

- GTV = All gross disease
- CTV1 (66–70 Gy) = GTV+0.5–2 cm (including entire surgical bed if surgical margin is +)
    Gross LN+0.5 cm
- CTV2 (59.4–60 Gy) = primary tumor bed and positive LN levels
- Includes sublingual gland, midline genioglossus and geniohyoid muscles, periosteum, and mandible for *floor of the mouth* tumors
- Includes floor of the mouth, base of tongue, muscles of tongue, anterior tonsillar pillar, and mandible for *oral tongue* tumors
- Includes underlying muscles, skin, infratemporal fossa, parotid gland, and facial nerve for *buccal mucosa* tumors
- Includes adjacent buccal mucosa, anterior tonsillar pillar, lower gum, maxilla, and mandible for *retromolar trigone* tumors.
- CTV2 for LNs = Ipsilateral levels Ib, II, and III
- *Small, DOI <2cm tumors with negative surgical margins do not require elective neck RT (if N0)
- *For *tip of the tongue* tumors, and tumors exceeding midline, upper neck should be included bilaterally.

5.5 Oral Cavity Cancers

- *For *upper lip* tumors, periauricular and intraparotid lymph nodes should be included.
- *Tumors with verrucous histology do not generally require elective neck irradiation as they rarely metastasize to LNs.
- CTV3 (54 Gy) = Elective neck node levels (contralateral upper neck and bilateral lower neck)

- *If level II is positive, ipsilateral posterior neck and RP LN should be included.
- *For T1-2 N0 *lip* tumors; no CTV3, for T3-4 N0; ipsilateral levels I-III
- *For N0 *oral tongue and floor of the mouth* tumors; bilateral levels I-IV, for N+; contralateral levels I-V
- *For T1-2 N0 *retromolar trigone and buccal mucosa* tumors; ipsilateral levels I–IV, for T3-4 N0 and N+; contralateral II-IV or I-V+RP
- PTV = 3–5 mm

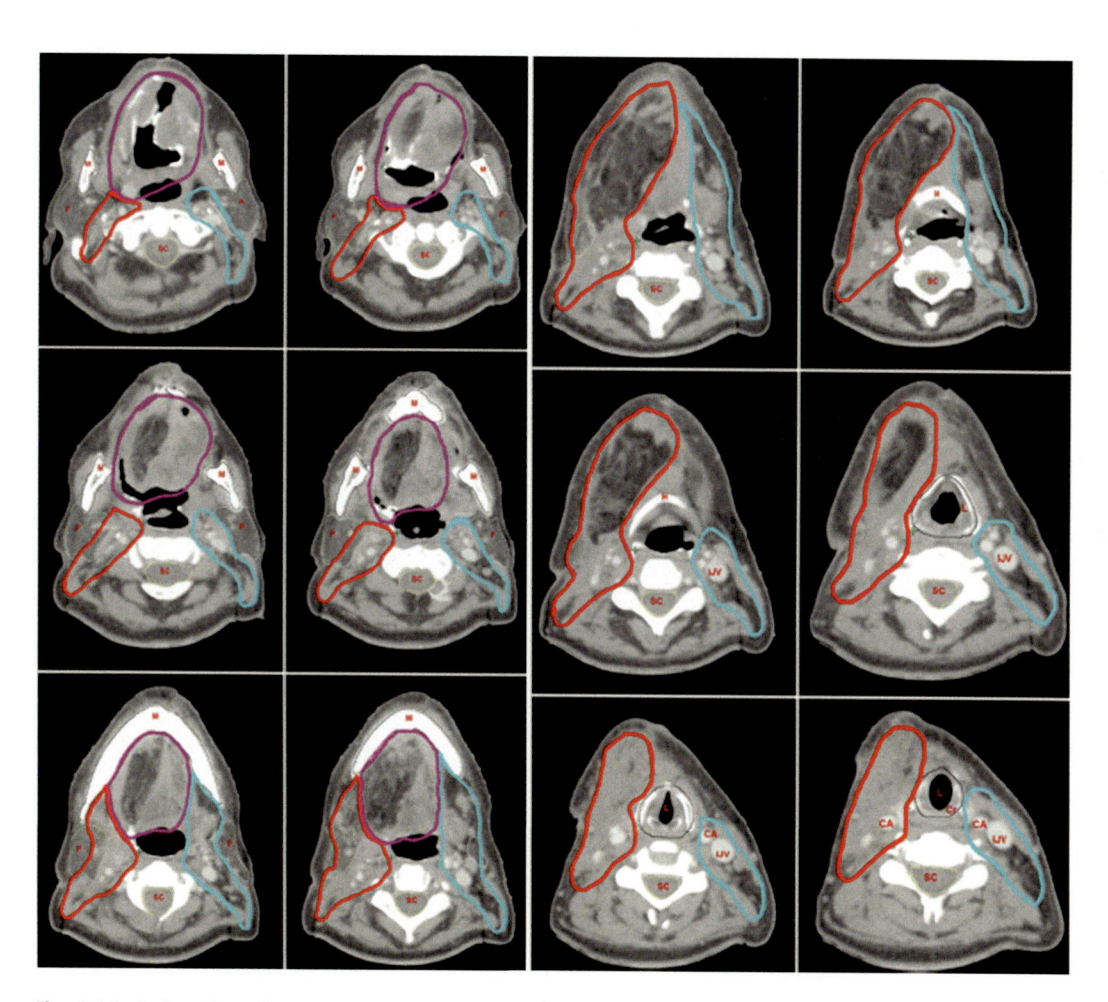

Fig. 5.31 Delineation of target volumes for T2N0M0 oral tongue cancer. CTV60 (magenta), CTV57 (red), CTV54 (blue). (*BS* brain stem, *OC* oral cavity, *M* mandible, *P* parotid gland, *SC* spinal cord, *E* epiglottis, *V* vallecula, *Ps* pyriform sinus, *H* hyoid bone, *IJV* internal jugular vein, *L* larynx, *CA* common carotid artery, *Cr* cricoid cartilage, *TG* thyroid gland, *T* trapezius muscle, *E* esophagus, *TCV* transverse cervical vessels)

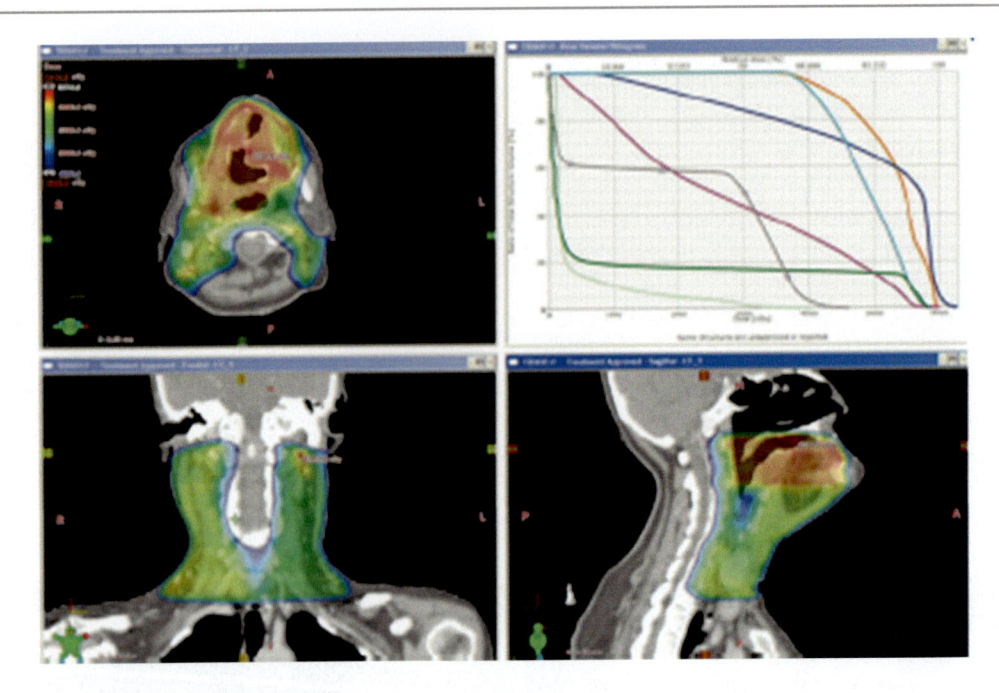

Fig. 5.32 IMRT plan for the same patient

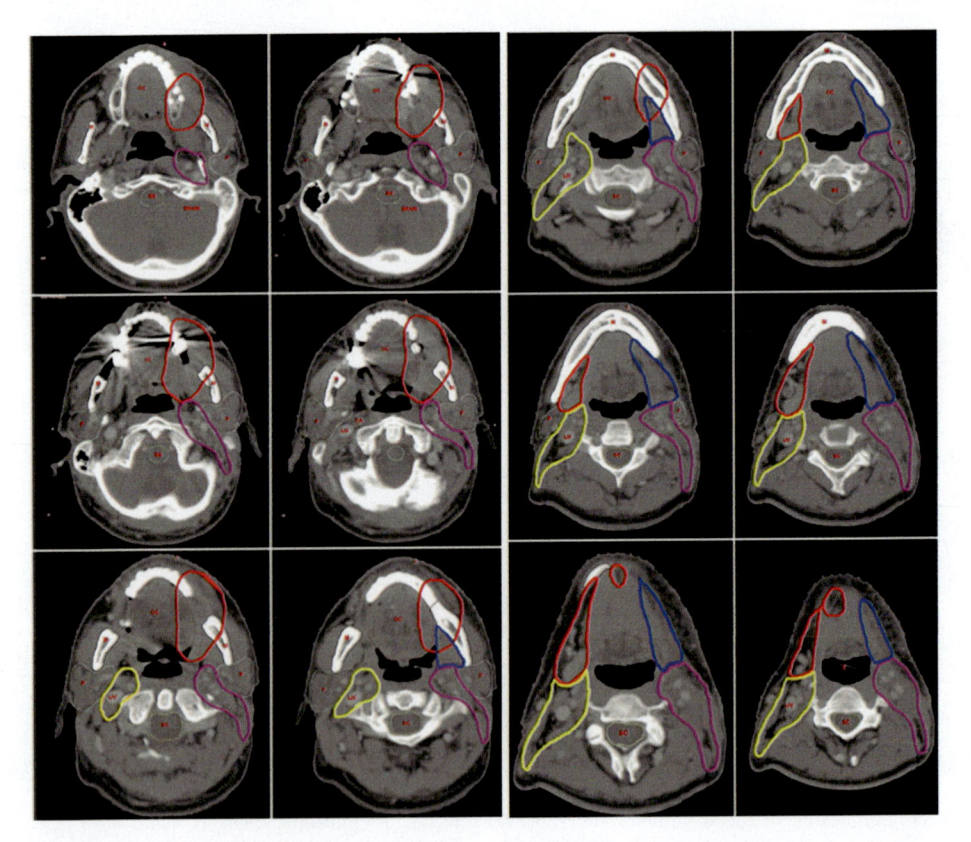

Fig. 5.33 Delineation of target volumes for T1N2bM0 buccal mucosa cancer. CTV64 (magenta), CTV60 (red), CTV57 (yellow), CTV54 (blue). (M mandible, P parotid gland, SC spinal cord, H hyoid bone, IJV internal jugular vein, L larynx, CA common carotid artery, Cr cricoid cartilage, OC oral cavity, BS Brain stem, V valleculae, E Epiglottis, Ps pyriform sinus, TG thyroid gland, T trapezius muscle)

5.5 Oral Cavity Cancers

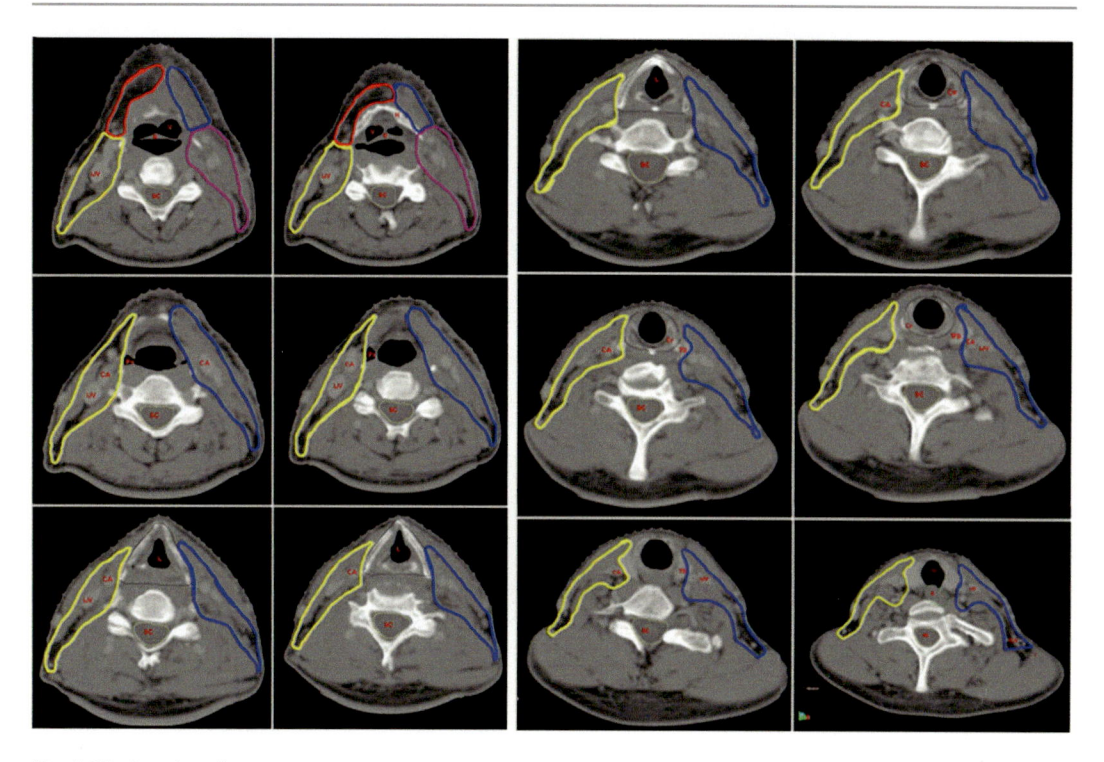

Fig. 5.33 (continued)

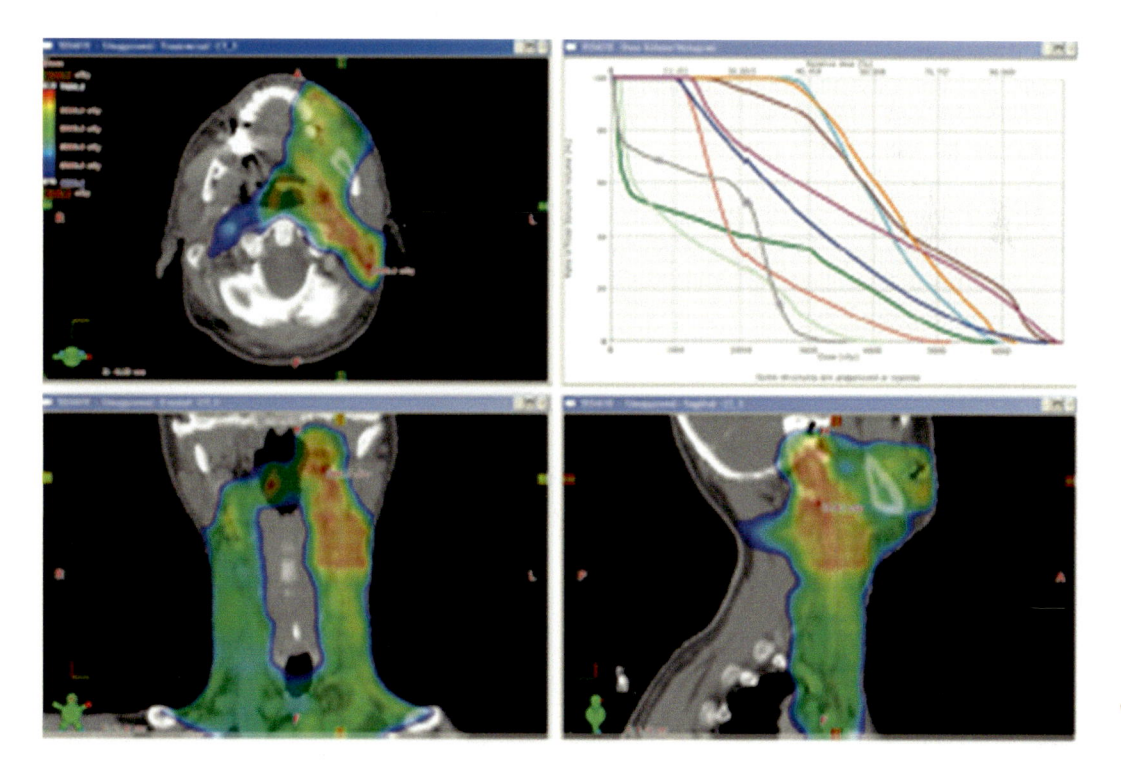

Fig. 5.34 IMRT plan for the same patient

- A tongue depressor or an be used to spare upper teeth, upper gingiva and soft palate and to stabilize the tongue in floor of mouth, lower lip, retromolar trigone, and tongue tumors.

5.5.6 Selected Publications

See also Sect. 5.3.2.6 and 5.10.

Tata Memorial trial, 2015 → 596 patients with T1–2 lateralized oral cavity cancer (85% oral tongue) with cN0 disease were randomized to elective neck dissection at the time of primary cancer surgery or therapeutic neck dissection at the time of nodal relapse. Elective nodal dissection included ipsilateral selective neck dissection of Ib, II, and III LNs. RT was applied to all patients with positive LNs, DOI ≥10 mm, or positive surgical margins. Three-year OS: 80% vs. 67.5%, and 3-year DFS: 69.5% vs. 45.9%, both significantly higher for elective neck dissection. Relapse pattern was different; mostly in the nodes in the therapeutic nodal dissection arm, and was non-nodal local or distant failure in the elective neck dissection arm. Adverse event rate: 6.6% vs. 3.6%.

D'Cruz A et al. (2015) Elective versus therapeutic neck dissection in node-negative oral cancer. N Engl J Med. 373:521–9.

Indian Randomized Trial, 1996 → patients with T3–4 N0–2b M0 buccal mucosa cancer underwent surgery and then were randomized to observation or postoperative RT (58–65 Gy in 1.8–2-Gy fractions). Three-year DFS: 38% vs. 68%; significantly higher in RT arm. OS was comparable (84% vs. 94%).

Mishra RC et al. (1996) Post-operative radiotherapy in carcinoma of buccal mucosa, a prospective randomized trial. Eur J Surg Oncol 22(5):502–504

5.6 Sinonasal Cancers

Sinonasal cancers include malignancies of the nasal cavity and paranasal sinuses (maxillary sinus, frontal sinus, ethmoid sinuses, and sphenoid sinus) (Fig. 5.36).

Paranasal Sinuses (Fig. 5.35) [17]

Frontal sinus. Not present at birth. Pneumatization becomes prominent at 12 years, and reaches adult size in 20 years.

Maxillary sinus. Filled with fluid at birth. The base reaches to the level of the base of the nose at 8 years and reaches adult size in adolescence.

Ethmoid sinus. Reaches adult size in 8–12 years. Consists of 3–15 cells. The border with sphenoid sinus is important due to their neighboring structures (optic nerves laterally and chiasm posteriorly).

Sphenoid sinus. Starts growing at the age of 3 years and reaches adult size at adolescence. The optic nerve, internal carotid artery, and maxillary sinus lie from superior to inferior in the lateral wall of the sphenoid sinus. The superior wall is separated from the dura by a bone 1 mm thick.

Nasal Cavity

Inferior–posterior: hard palate

Inferior–anterior: maxillary bone (os palatinum)

Superior–anterior: nasal bones (os nasale), superior lateral cartilage

Superior–posterior: lamina cribrosa (cribriform plate)

Superior–lateral: lamina papyracea, which separates bone from orbita

*Four subsites: nasal vestibule, lateral walls (corresponds with medial walls of maxillary sinuses), floor, and septum

5.6.1 Pathology

Sinonasal cancers are rare tumors that constitute 3% of all head–neck cancers and 1% of all cancers.

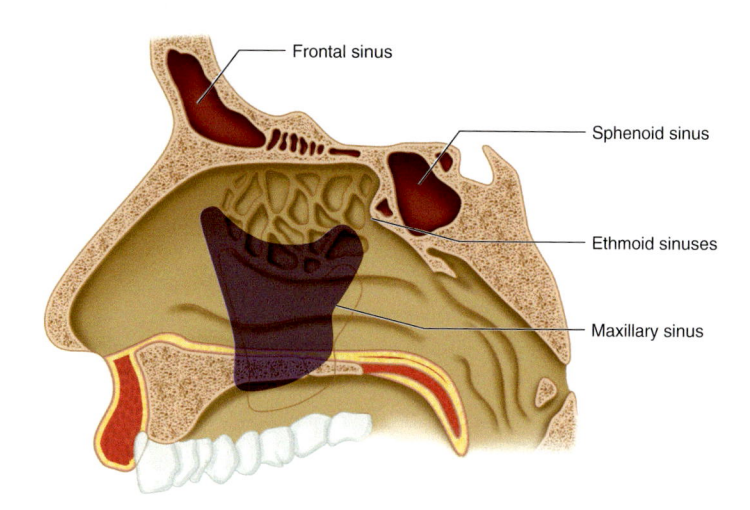

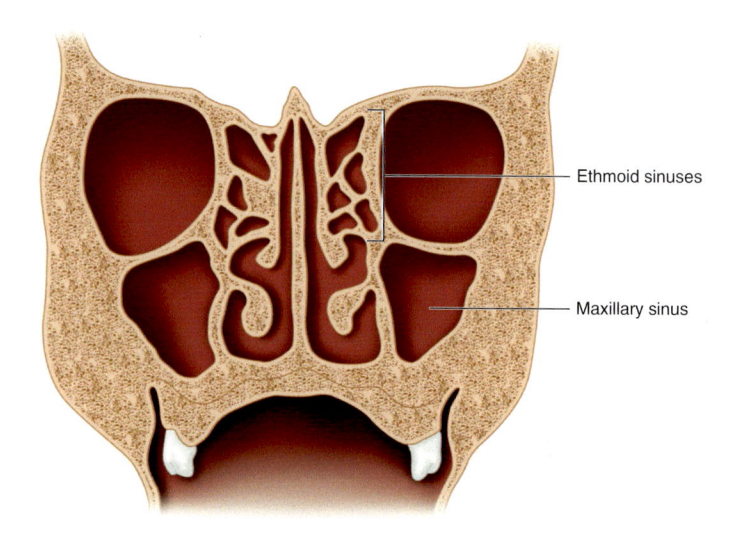

5.6 Sinonasal Cancers

Fig. 5.35 Nasal cavity and paranasal sinuses (anterior view)

5.6.2 General Presentation

The benign and malignant pathologies of the sinonasal region have similar clinical signs and symptoms at presentation. Therefore, malignant diseases are usually advanced at diagnosis.

- Most sinonasal cancers are SCC.
- Minor salivary gland neoplasms (adenocarcinoma, adenoid cystic carcinoma, mucoepidermoid carcinoma) 10–15%
- Melanoma 5–10% in nasal cavity
- Other rare histologies:
- Neuroendocrine carcinomas (small cell carcinoma, estesioneuroblastoma, and sinonasal undifferentiated carcinoma)
- Lymphoma
- Sarcoma
- Plasmacytoma

Symptoms in Maxillary Sinus Cancers
- Nasal symptoms (50%): nasal obstruction, rhinorrhea, epistaxis, and nasal mass

- Oral symptoms (25–35%): sensitivity in upper teeth, pain, trismus, ulceration, fullness in hard palate, and alveolar site
- Ocular symptoms (25%): propulsion of orbita upwards, unilateral increase in tears, diplopia, swelling in eye lids, and exophthalmos
- Facial symptoms: infraorbital nerve hypoesthesia, swelling in cheek, pain and asymmetry in face
- Otological symptoms: serous otitis media and hearing loss due to nasopharyngeal extension

The classical triad of sinonasal cancer is seen in advanced stages:

1. Facial asymmetry
2. Palpable or visible mass in oral cavity
3. Mass in nasal cavity observed during anterior rhinoscopy

- All three findings are seen in 40–60% of patients at diagnosis; at least one of them is observed in 90% of cases.

Ohngren's line. This is used in anatomical definitions of paranasal tumors. It is a virtual line passing from the medial chantus of the eye to the angle of the mandible, and it divides the maxillary antrum into anteroinferior and posterosuperior parts. A tumor in front of this line is more benign, while one behind it is more malignant in character.

- Tumors that pass the base of the maxillary sinus and reach the oral cavity are an exception.

5.6.3 Staging

Nasal Cavity and Ethmoid Sinus [20] (Fig. 5.36)
Primary tumor (T)
- TX: Primary tumor cannot be assessed Tis: Carcinoma *in situ*
- T1: Tumor restricted to any one subsite, with or without bony invasion
- T2: Tumor invading two subsites in a single region or extending to involve an adjacent region within the nasoethmoidal complex, with or without bony invasion
- T3: Tumor extends to invade the medial wall or floor of the orbit, maxillary sinus, palate, or cribriform plate
- T4a: Moderately advanced local disease: Tumor invades anterior orbital contents, skin of the nose or cheeks, minimal extension to anterior cranial fossa, pterygoid plates, sphenoid or frontal sinuses
- T4b: Very advanced local disease: Tumor invades orbital apex, dura, brain, middle cranial fossa, cranial nerves other than V_2, nasopharynx, or clivus

Maxillary Sinus (Fig. 5.37)
Primary tumor (T)
- TX: Primary tumor cannot be assessed
- T0: No evidence of primary tumor
- Tis: Carcinoma *in situ*
- T1: Tumor limited to maxillary sinus mucosa with no erosion or destruction of bone
- T2: Tumor causing bone erosion or destruction including extension into the

5.6 Sinonasal Cancers

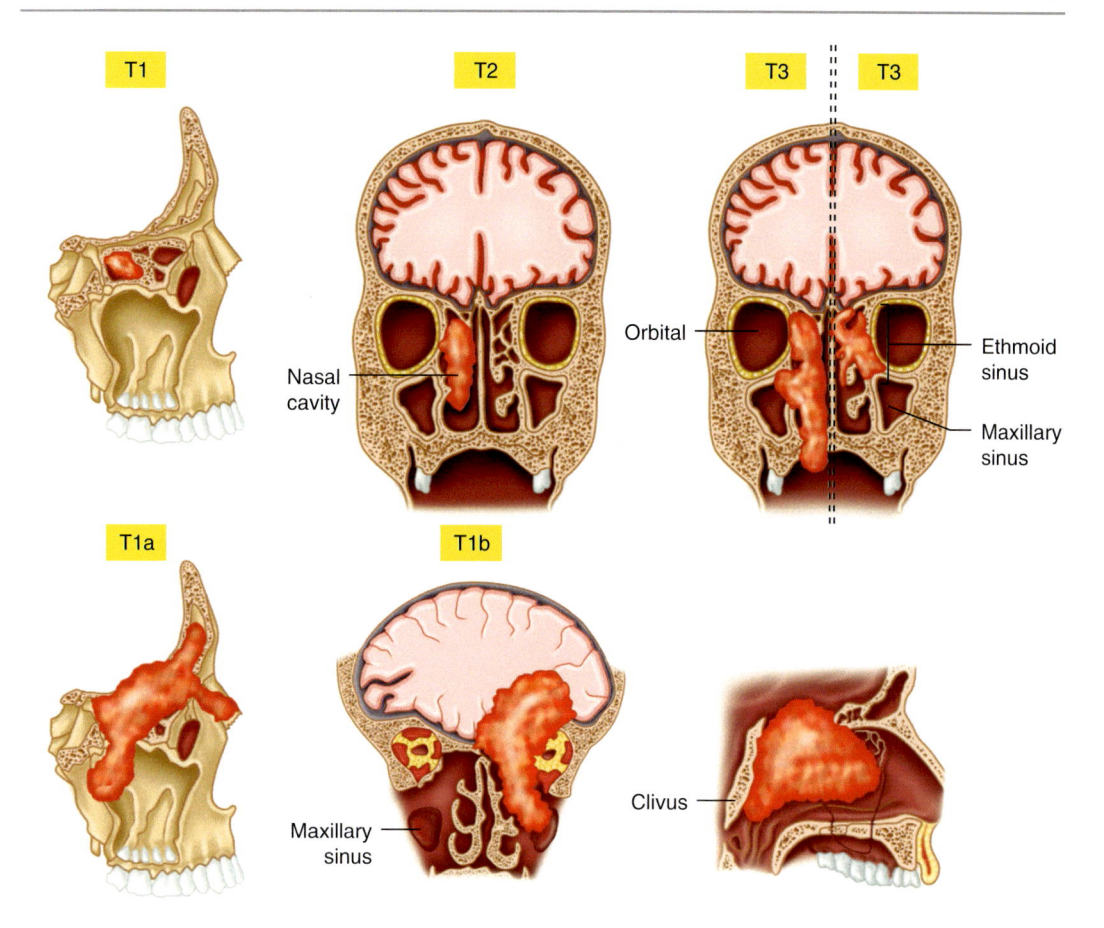

Fig. 5.36 T staging in nasal cavity and ethmoid sinus cancers (from [20], reproduced with the permission from the American Joint Committee on Cancer)

hard palate and/or the middle nasal meatus, except extension to the posterior wall of maxillary sinus and pterygoid plates
- T3: Tumor invades the bone of the posterior wall of maxillary sinus, subcutaneous tissues, floor or medial wall of the orbit, pterygoid fossa, ethmoid sinuses
- T4a: Moderately advanced local disease: Tumor invades anterior orbital contents, skin of the cheek, pterygoid plates, infratemporal fossa, cribriform plate, sphenoid, or frontal sinuses
- T4b: Very advanced local disease: Tumor invades orbital apex, dura, brain, middle cranial fossa, cranial nerves other than V_2, nasopharynx, or clivus

AJCC Stage Groupings
- Stage I: T1N0M0
- Stage II: T2N0M0
- Stage III: T3N0M0, T1N1M0, T2N1M0, T3N1M0
- Stage IVA: T4aN0M0, T4aN1M0, T1N2M0, T2N2M0, T3N2M0, T4aN2M0
- Stage IVB: T4b, any N, M0; any T, N3, M0 Stage IVC: Any T, any N, M1

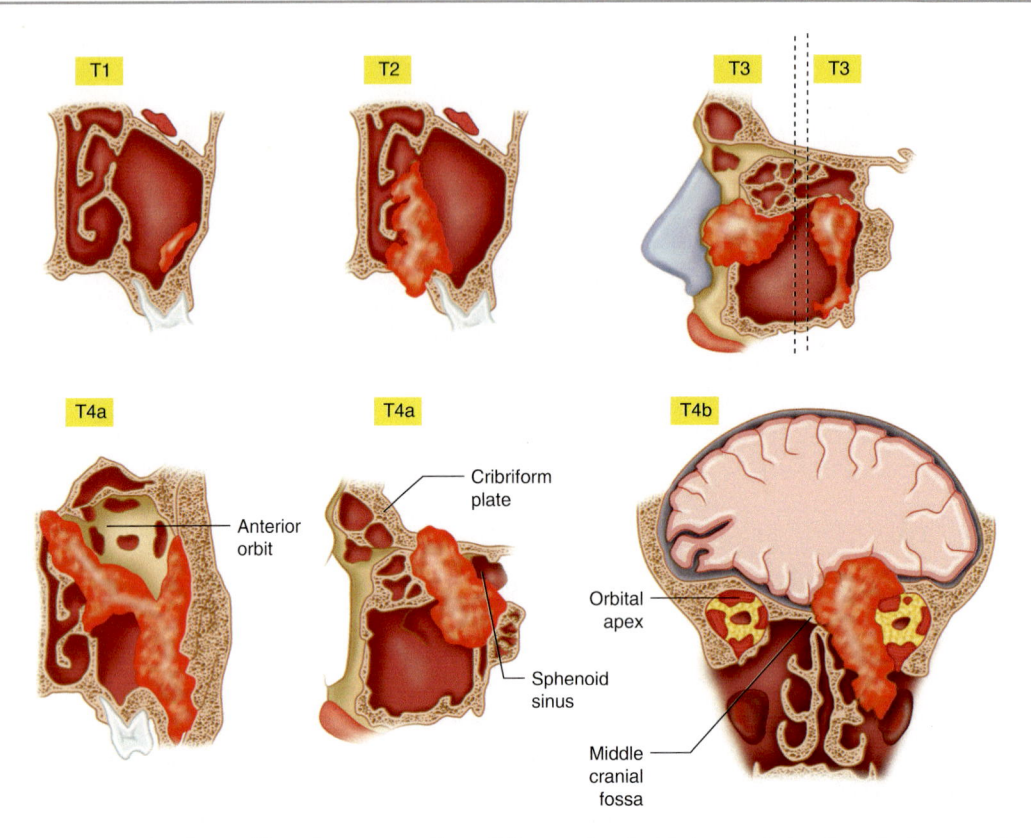

Fig. 5.37 T staging for maxillary sinus cancers (from [20], reproduced with the permission from the American Joint Committee on Cancer)

N staging is the same as for other head and neck cancers, except for nasopharyngeal cancers.

> Sinonasal cancers rarely metastasize to neck lymphatics [32].
>
> - 10% at presentation
> - Elective neck RT is not routinely recommended
> - Clinical, radiological, or pathological LN (+): neck RT is given
> - Elective neck RT is recommended for posterior pharyngeal wall extension (due to rich lymphatic supply to this region)
>
> Sinonasal cancers primarily drain into the RP, levels Ib, II, and periparotid LNs.
> Nasal vestibule tumors primarily drain into levels Ia, Ib), facial and preauricular LNs.

Esthesioneuroblastoma (Olfactory neuroblastoma) [33]

- This is a neurogenic tumor that originates from the olfactory epithelium which is limited to the upper conchae, the nasal septum, and the cribriform area.
- It can be seen at any age, but two-thirds of patients are between 10 and 34 years of age.
- Symptoms are nonspecific, like other intranasal tumors (epistaxis, anosmia, and nasal obstruction). Therefore, most patients are at advanced stages at presentation.
- It usually appears as a reddish mass.
- Although it grows slowly, local invasion is possible, and it has a metastasis potential of nearly 20%. The most frequently involved sites are lung and cervical lymph nodes.

Kadish Staging System for Estesioneuroblastoma
- Kadish A: Tumor limited to the nasal cavity
- Kadish B: Tumor in the nasal cavity with extension to paranasal sinuses
- Kadish C: Extension beyond the nasal cavity and paranasal sinuses involving the cribriform lamina, skull base, orbit, or intracranial cavity.
- Kadish D: Regional lymph node or distant metastasis

Adenoid cystic carcinoma [34]

These are also known as cylindromas due to their histological appearance.

- They have a tendency for early perineural invasion.
- PNI is generally along the maxillary and mandibular branches of the trigeminal nerve.
- Regional LN involvement is very rare, but the risk of DM (mostly to lung) is high.

Salivary Gland Tumors of the Sinonasal Region
- These constitute 6–17% of all sinonasal cancers
- In order of frequency: adenoid cystic carcinoma; adenocarcinoma; malignant pleomorphic adenoma, mucoepidermoid carcinoma, undifferentiated carcinoma
- All are generally very aggressive

5.6.4 Treatment Algorithm

Nasal Cavity and the Ethmoid Sinus [21]
Newly diagnosed T1–2
- Resection (preferred) → RT or observe (T1) or concurrent CRT
- Alternative → definitive RT

Newly diagnosed T3–4

- Resection (preferred) → RT or concurrent CRT
- Concurrent CRT
- Chemotherapy

Diagnosed after incomplete resection with gross residual disease

- Surgery (preferred) → RT or concurrent CRT
- RT
- Concurrent RT

Diagnosed after incomplete resection with no residual disease

- RT
- Surgery, if feasible → RT or observe (T1 only) or concurrent CRT

Maxillary Sinus [21]
T1–2 N0, all histologies except adenoid cystic

Resection
- Negative margin → observation
- PNI or LVSI → Consider RT or CRT
- Positive margin → re-resection
- Negative margin → Consider RT
- Positive margin → RT or CRT

T1-2 N0, adenoid cystic

Resection → RT (preferred)

- Consider observation if negative margin, no PNI

T3–4a N0, T1–4a N+, all histologies
- Resection ± neck dissection (for N+) → RT or CRT

T4b N0–3

- PS 0–1: Concurrent CRT or induction chemotherapy
- PS 2: RT or concurrent CRT
- PS 3: Palliative RT or single-agent chemo or best supportive care

5.6.5 Radiotherapy

Target Delineation Guidelines
(Figs. 5.38, 5.39, 5.40, and 5.41)
  GTV = All gross disease (if present)
  CTV1 = Primary tumor bed (GTV + 1–1.5 cm)
  CTV1ln = GTVln + 0.5–1 cm
  CTV2 = Entire surgical bed

- *Include cribriform plate if ethmoid sinus or olfactory region is involved
- *Include all other sinuses for paranasal sinus tumors if explored during surgery
- *Include bony orbit if orbital exenteration was performed

CTVln = Ipsilateral Ib and II

- *Include lower neck if palate, nasopharynx, skin of the cheek or anterior nose, or gingiva is involved.

CTV3 = Subclinical disease

- *For N0; ipsilateral I-II and RP
- *For N+; ipsilateral and contralateral negative levels and bilateral RP
- *Include whole tract of nerve to the skull base if PNI+
- *In all adenoid cystic cancers, local nerve pathway should be delineated to the skull base.
  - PTV1 = CTV + 3–5 cm

Dose Recommendations
CTV1 = 70 Gy
CTV2 = 59.4-63 Gy
CTV3 = 54-56 Gy

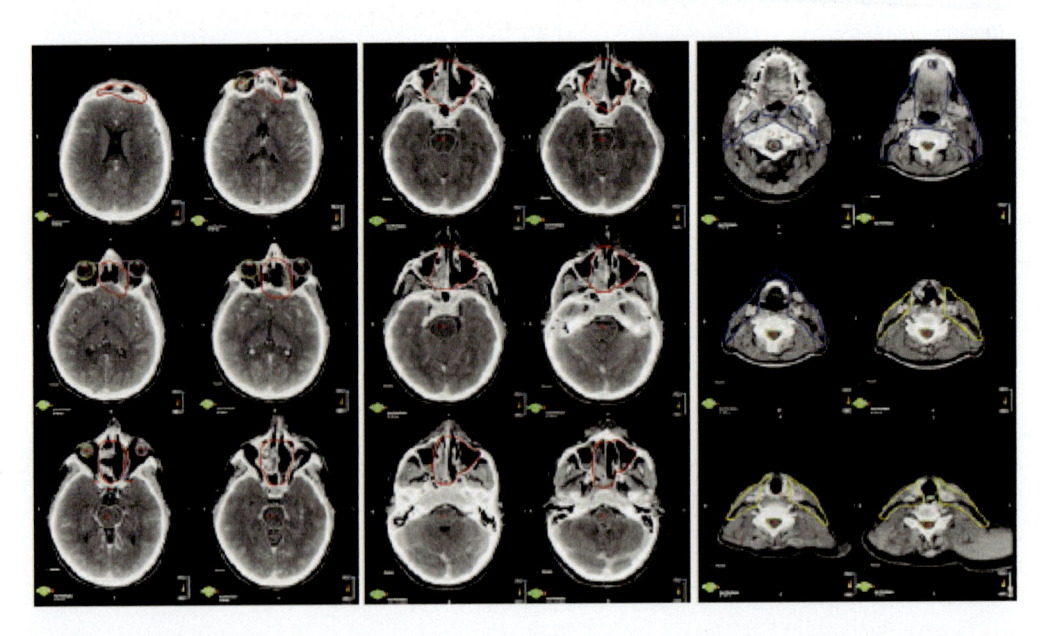

Fig. 5.38 Delineation of a patient with T4bN2bM0 maxillary sinus cancer. CTV60 for primary (red), CTV60 for lymphatics (blue), CTV54 (orange)

5.6 Sinonasal Cancers

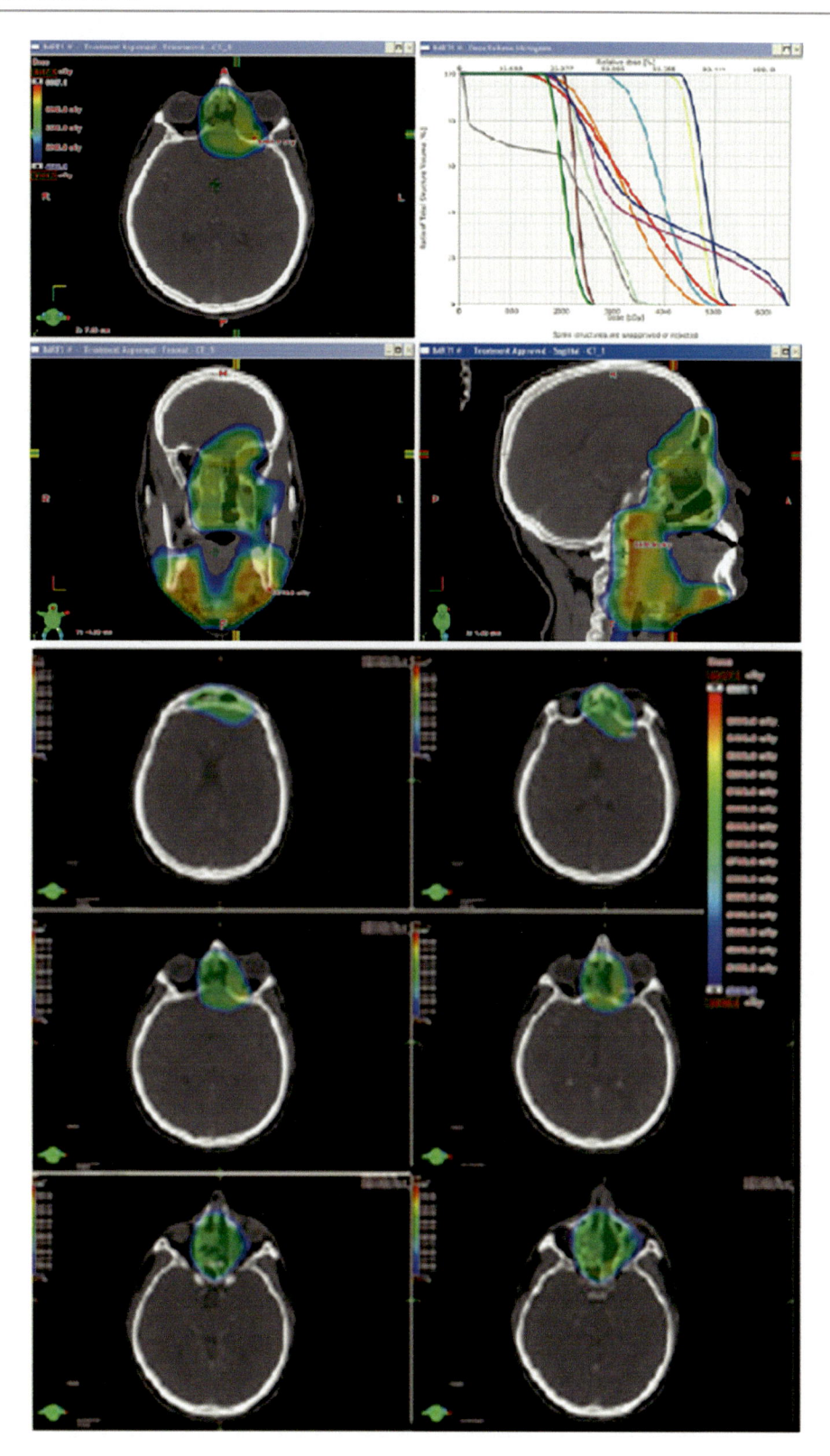

Fig. 5.39 IMRT plan for the same patient

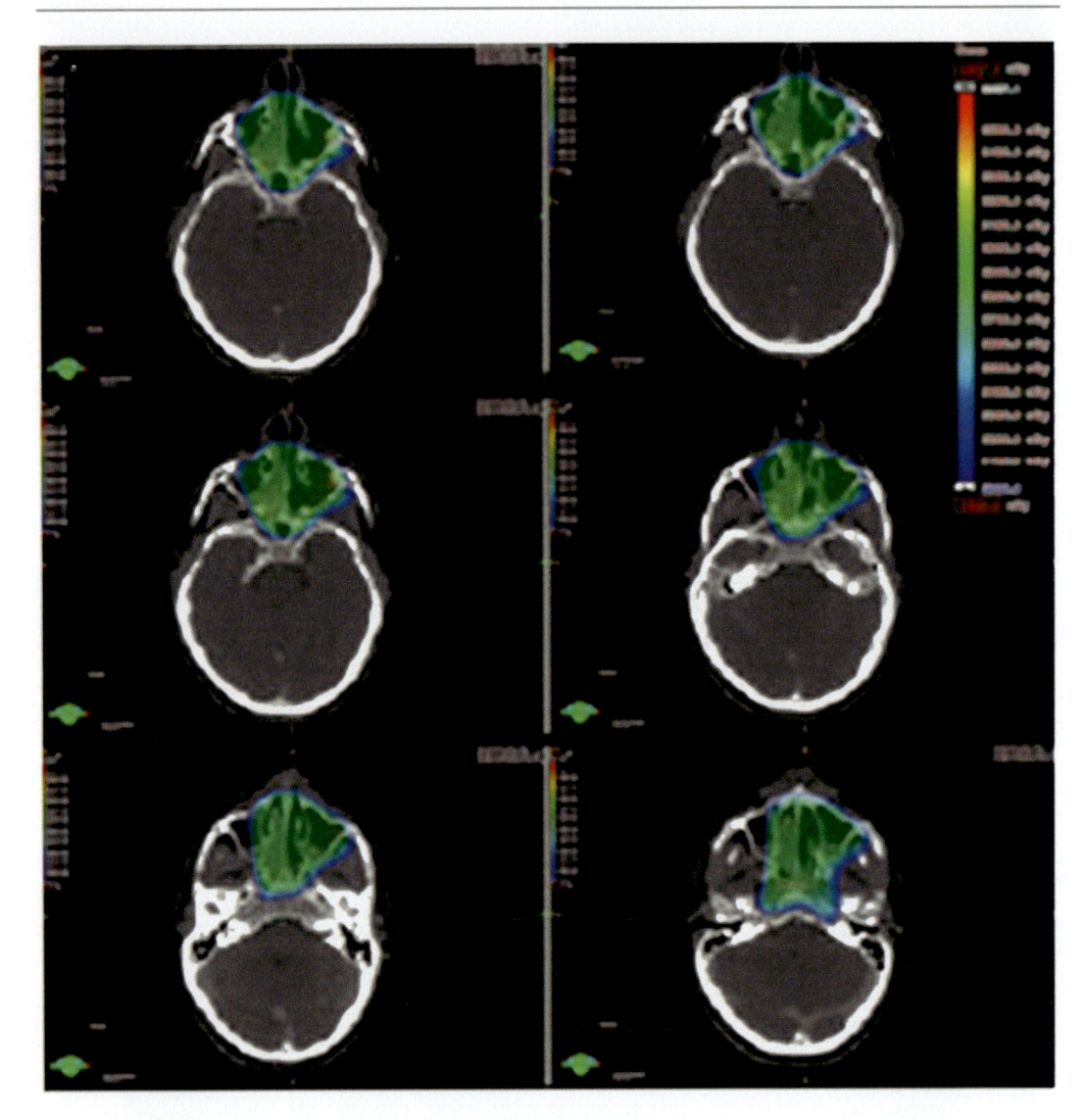

Fig. 5.39 (continued)

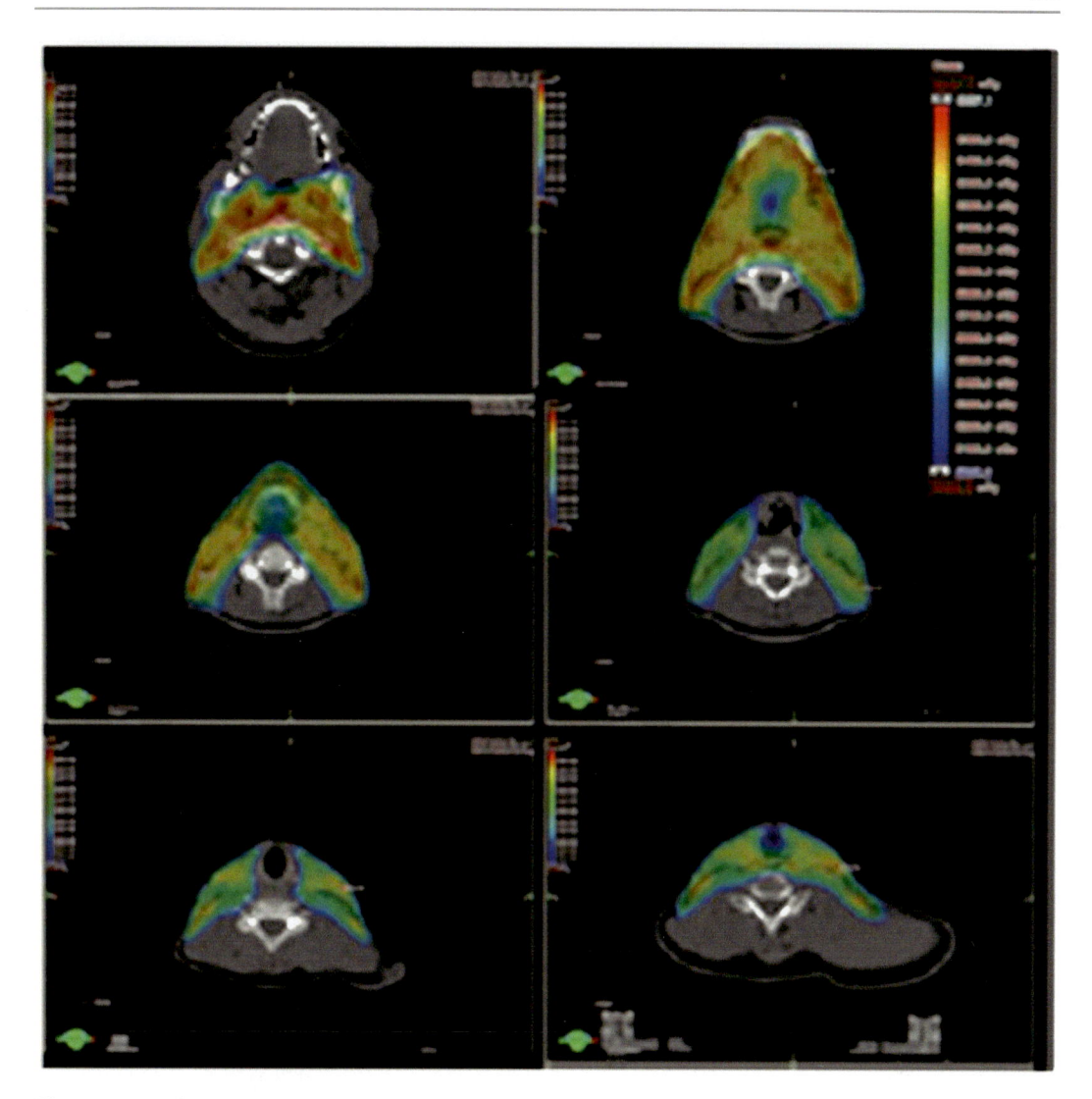

Fig. 5.39 (continued)

5.6.6 Selected Publications

No prospective randomized trials.

Stanford, 2000 → 97 patients with maxillary sinus cancer (61 surgery + RT, 36 RT alone). Five- and 10-year OS: 34% and 31%. Most nodal recurrences were level I–II. Five-year nodal relapse risk 0% vs. 20% with and without elective neck RT. Five-year DM risk: 81% vs. 29% with and without neck relapse.

Le QT (2000) Lymph node metastasis in maxillary sinus carcinoma. Int J Radiat Oncol Biol Phys 46(3):541–549

Review, 2001 → 220 patients with nasal and paranasal sinus carcinoma with a minimum 4-year follow-up were retrospectively reviewed. Five-year OS: 40%, LC: 59%. Prognostic factors were: histology (best for glandular carcinoma), T stage, localization (best for nasal cavity and worst for ethmoid sinus), and treatment (best for surgery alone, worst for RT alone).

Dulguerov P et al. (2001) Nasal and paranasal sinus carcinoma: are we making progress? A series of 220 patients and a systematic review. Cancer 92(12):3012–29.

Washington University, 2004 → 106 patients with paranasal sinus carcinoma. Most tumors in the maxillary (76%) or ethmoid (18%) sinus, and locally advanced. All patients underwent RT, combined with surgery in 65%; 2% received

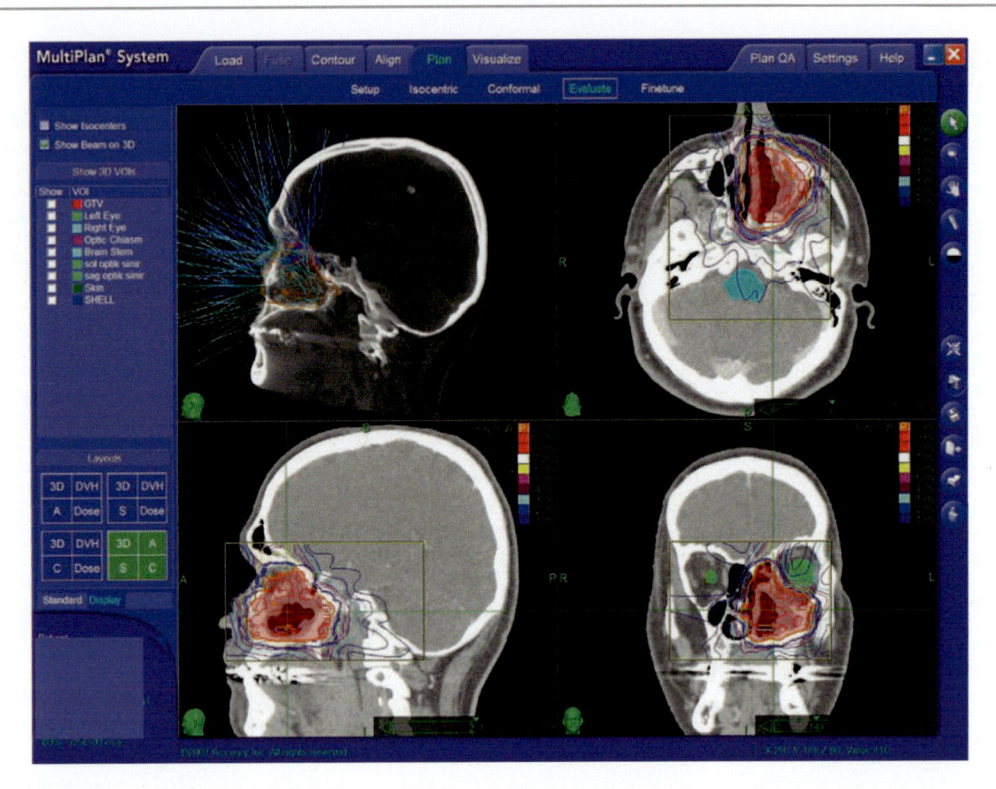

Fig. 5.40 Robotic radiosurgery planning in a patient with paranasal carcinoma (reproduced courtesy of Hacettepe University)

Fig. 5.41 Major salivary glands

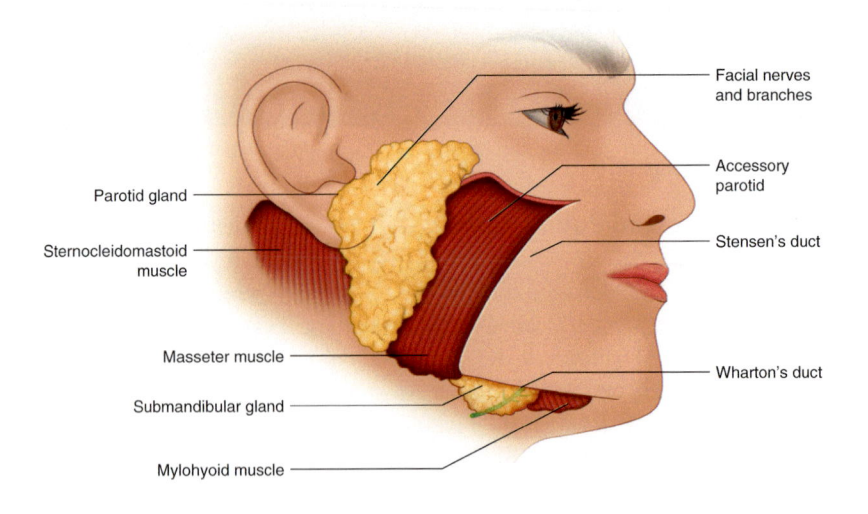

chemotherapy. Median follow-up: 5 years. Five-year LC: 58%, LRC: 39%, DFS: 33%, OS:27%. Significant improvement in DFS with surgery (35 vs. 29%). Nodal status is a statistically significant predictor for LRC and DFS. DFS improved slightly with combined modality treatment. OS rates remained suboptimal.

Blanco AI, Chao KS, Ozyigit G et al. (2004) Carcinoma of paranasal sinuses: long-term outcomes with radiotherapy. Int J Radiat Oncol Biol Phys 59(1):51–58

MGH Photon/Proton Regimen, 2006 → 36 advanced-stage sinonasal cancer cases. Median GTV dose: 69.6 Gy CGE, with accelerated hyperfractionation and concomitant boost with proton. Acceptable ophthalmological complications. No LC data

Weber DC (2006) Visual outcome of accelerated fractionated radiation for advanced sinonasal malignancies employing photons/protons. Radiother Oncol 81(3):243–249

Chen et al., 2007 → 127 patients with sinonasal carcinoma underwent RT between 1960 and 2005. Conventional RT: 59, 3D CRT: 45, IMRT: 23. 65% received adjuvant RT, 15% received chemotherapy. Five-year OS: 52%, DFS: 54%, LC: 62%. RT technique or decade of treatment did not affect the outcomes. But severe late toxicity rate declined with every proceeding decade.

Chen AM et al. (2007) Carcinomas of the paranasal sinuses and nasal cavity treated with radiotherapy at a single institution over five decades: are we making improvement? Int J Radiat Oncol Biol Phys 69(1):141–7.

ENB, Rare Cancer Network Study, 2010 → 77 patients with non-metastatic estesioneuroblastoma between 1971 and 2004. Kadish A: 11, B: 29, C: 37 patients. T1: 9, T2: 26, T3: 16, T4a: 15, T4b: 11. 88% with N0 disease. Total excision was possible in 44 patients. Median follow-up: 72 months. Five-year OS: 64%, DFS: 57%, LRC: 62%, and LC: 70%. Positive prognostic factors were T1–3 tumor, N0, R0/R1 resection, and total RT dose ≥54 Gy.

Ozsahin M et al. (2010) Outcome and prognostic factors in olfactory neuroblastoma: a rare cancer network study. Int J Radiat Oncol Biol Phys 78(4):992–7.

5.7 Major Salivary Gland Tumors

The salivary glands consist of three large, paired major glands, i.e., the parotid, submandibular, and sublingual glands and 6000–10,000 minor glands within the mucosal surface of the upper aerodigestive tract (Fig. 5.41).

- **Parotid gland** is an encapsulated gland just beneath the skin that is surrounded by a tight connective tissue. It is the largest salivary gland located in the retromandibular–preauricular region. The surrounding connective tissue at the bottom of the gland is less dense, and there is an aperture here from which infections or tumor may extend into the pterygopalatine fossa or the paraphrayngeal space.
  - The canal of the parotid (Stenon's duct) is approximately 6 cm in length; it exits from the anterior side of the gland and crosses the masseter muscle, before entering the oral cavity by passing between the buccinator muscle and the buccal mucosa.
  - The facial nerve leaves the temporal bone by exiting the stylomastoid foramen and entering the parotid parenchyma after a distance of 0.5–1.5 cm.
- **Submandibular salivary gland** is bounded anteriorly by the anterior belly of the digastrics, posteriorly by the stylomandibular ligament, and superiorly by the mandible. The majority of the gland is inferior to the mylohyoid muscle and is covered by the cervical fascia.
  - The canal of the gland (Wharton's duct) is approximately 5 cm in length. It extends anteriorly from beneath the floor of the mouth mucosa, and ends in the ostium at the sublingual caruncle.
- **Sublingual salivary gland** is the smallest major salivary gland. It is located in the floor of the mouth mucosa. Its posterior is in contact with the anterior part of the submandibular gland.
  - It has multiple canals (Bartholin's canals) that exit through multiple ostia located in the sublingual plicae of the mouth floor.

- **Minor salivary glands** are located in the upper aerodigestive tract, palate, buccal mucosa, base of tongue, pharynx, trachea, cheek, lip, gingiva, floor of mouth, tonsil, paranasal sinuses, nasal cavity, and nasopharynx.

5.7.1 Pathology

Salivary gland tumors are found in the:

- Parotid glands: 70%
- Submandibular glands: 8%
- Sublingual and minor glands: 22%

Parotid tumors are benign in nearly 75% of cases.
This ratio is 57% for the submandibular glands; malign tumors are more common in the sublingual and minor salivary glands.
Ninety-five percent of salivary gland tumors are seen in adults.

The most common benign salivary gland tumor is the pleomorphic adenoma, followed by Warthin's tumor.

Malignant tumors are varied, and those most commonly seen are adenoid cystic carcinoma, mucoepidermoid carcinoma, acinic cell carcinoma, carcinoma ex pleomorphic adenoma, undifferentiated carcinoma, salivary duct carcinoma, and adenocarcinoma.

Pleomorphic adenoma. Almost all are observed in the parotid gland, while a small proportion is in the submandibular gland.

- Almost all are ipsilateral, and are more frequent in females.
- The first sign is usually the presence of a slowly growing, painless retromandibular mass.
- Facial functions are normal, even in large benign tumors.
- Facial pain and paresis is a sign of malignancy.
- A painless, hard nodular mass is palpated.
- Pleomorphic adenoma has a pseudocapsule, so there is almost certainly residual tumor after simple excision, resulting in frequent relapses.
- They have a 5–10% risk of malignant transformation.

Warthin's tumor. Almost all are observed in older men, and bilaterally in 10–15%.

- This is only seen in parotids.
- An ovoid, mobile, fluctuating, elastic, and painless mass is usually located in the parotid tail.
- The tumor includes lymphoid tissue, so enlargement with associated pain may be seen with upper airway infections.

Adenoid cystic carcinoma. This is the most common malignant tumor of salivary glands other than the parotids.

- It has a tendency to localize in the submandibular, sublingual, and minor salivary glands, but seldom in the parotids.
- The facial nerve is almost always involved in such tumors originating from the parotids
- Nerve invasion is possibly due to the extension of the tumor into perineural lymphatics, which is a characteristic of adenoid cystic cancer.
- Local LN metastasis is rare, while lung metastasis is frequent.
- Recurrences can be seen even 10 years after therapy; 5-year OS is 60%.

Mucoepidermoid carcinoma. More than 90% of these cases are found in the parotids, followed by the hard palate and minor salivary glands.

- This is the salivary gland tumor most frequently seen in children.
- Well-differentiated lesions behave like benign tumors; undifferentiated types have a tendency for local invasion and DM, like high-grade malignancies.
- Well-differentiated low-grade tumors grow slowly, while poorly differentiated high-grade tumors grow rapidly and result in pain, facial paralysis, LN involvement, and DM.
- Five-year OS is 90% in well-differentiated and very low in poorly differentiated tumors.

Acinic cell tumor. This originates from the parotids in particular.

- It presents as a very slowly growing mass.
- It metastasizes in nearly 10% of cases, and is mostly bilateral.
- Risk of recurrence is low.
- Prognosis is better than for carcinomas, although it is malignant.

5.7.2 General Presentation

Salivary gland tumors generally start with a progressively growing painless mass. Benign salivary gland tumors usually present as a slowly growing mass with no discomfort.

Parotid neoplasms usually present as a mass in the tail, whereas submandibular gland tumors present as a swelling of the entire gland and those of minor salivary glands as a painless mass under healthy mucosa.

Signs of a malignant tumor: rapidly growing mass, pain, associated nerve paralysis (facial nerve, lingual nerve), fixation to the skin or deep tissues (a fixed mass), ipsilateral lymphadenopathy.

Most malignant salivary gland tumors are seen in patients 50–60 years old, 2% in children <10 years old, and 16% in patients <30 years old.

5.7.3 Staging

Primary Tumor (T) (Fig. 5.42) [20]
- T1: Tumor ≤2 cm in greatest dimension without extraparenchymal extension*
- T2: Tumor 2-4 cm in greatest dimension without extraparenchymal extension*
- T3: Tumor >4 cm and/or tumor having extraparenchymal extension*
- T4a: Moderately advanced local disease: Tumor invades the skin, mandible and/or ear canal and/or facial nerve
- T4b: Very advanced local disease: Tumor invades the skull base and/or pterygoid plates and/or carotid artery
- *Extraparenchymal extension is clinical or macroscopic evidence of invasion of soft tissues. Microscopic evidence alone does not constitute extraparenchymal extension for classification purposes.

N staging is the same as other head and neck cancers other than nasopharynx.

Stage Grouping

Stage I: T1N0M0
  Stage II: T2N0M0
  Stage III: T3N0M0, T1-3 N1M0
  Stage IVA: T1-3 N2M0, T4a N0-2 M0
  Stage IVB: T4b Any N M0
  Stage IVC: AnyT AnyN M1

LN Drainage
- *Parotid lymphatics*: preauricular, periparotid, and intraparotid, levels II-IV.
- *Submandibular/sublingual gland lymphatics*: levels I-II, less often III-IV.
- *Minor salivary*: depends on location, have highest propensity for LN spread

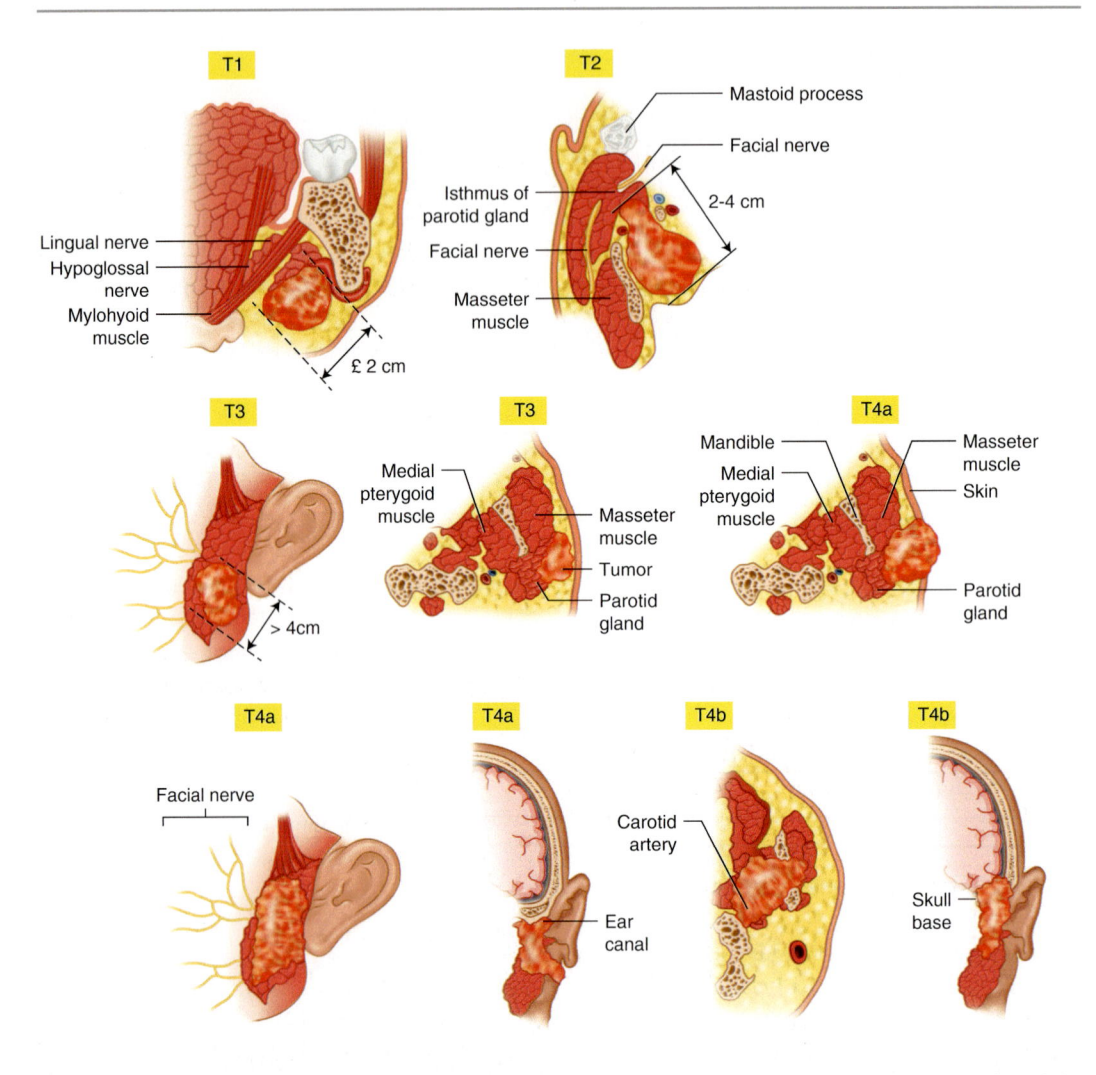

Fig. 5.42 T staging for major salivary glands (from [20], reproduced with the permission from the American Joint Committee on Cancer)

5.7.4 Treatment Algorithm

Treatment Algorithm for Salivary Gland Cancer [21]
Clinically benign or T 1-2 malignant

Complete resection
- Benign → Observation
- Low-grade → Consider RT if tumor spillage or PNI
- Intermediate or high-grade or adenoid cystic → RT

T3-4a
Surgery ± neck dissection (preferred)

- Complete resection → Consider RT based on pathologic risk factors

Definitive RT or CRT
T4b
- Definitive RT or CRT

5.7.5 Radiotherapy (Figs. 5.43 and 5.44)

Fig. 5.43 Contouring of a patient with T2N2bM0 parotid gland cancer. CTV64 (red)

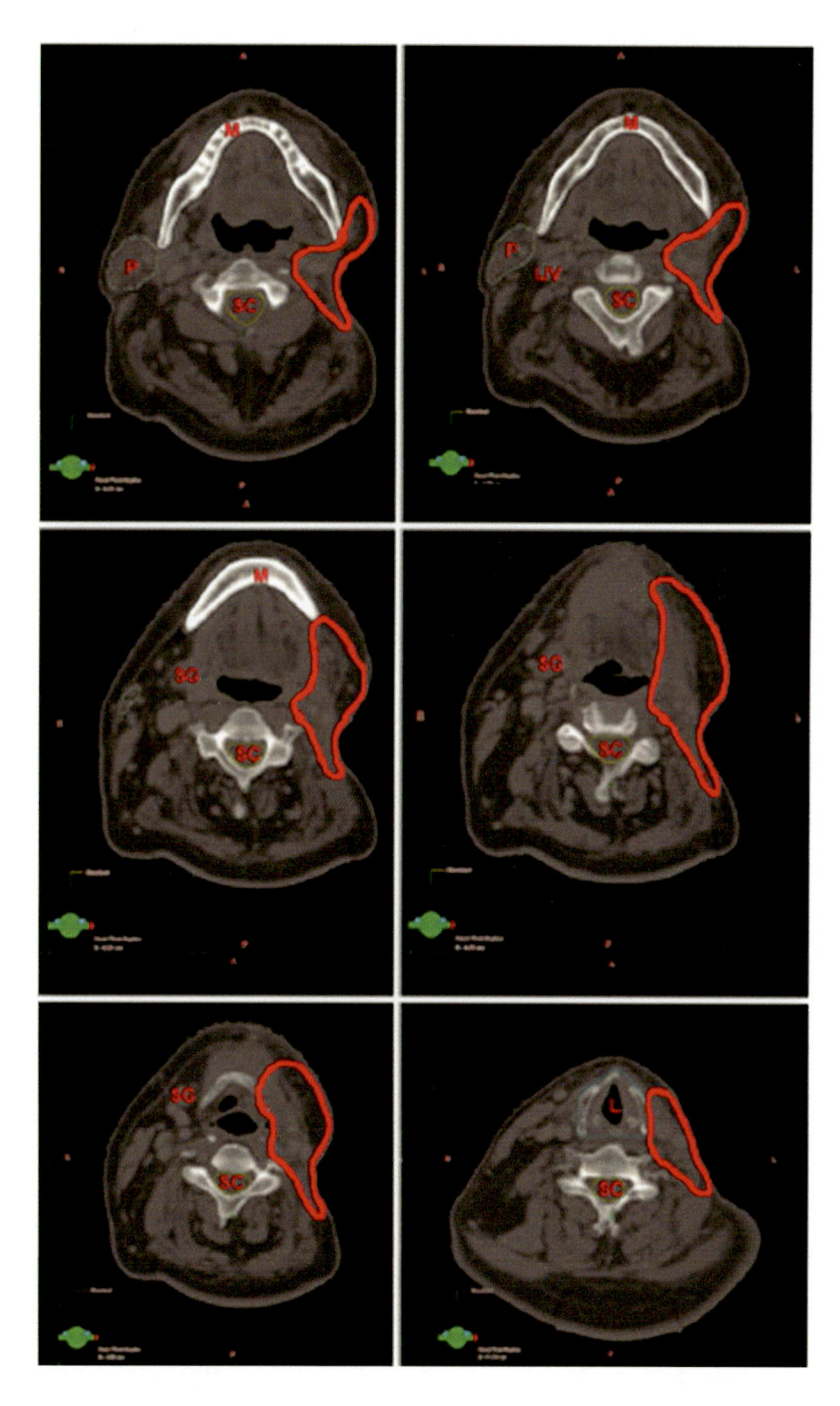

Postoperative RT indications in malignant salivary gland tumors
- Surgical margin (+) or close (1–5 mm)
- T3-4 tumor
- Lymph node (+) or ECE+
- High-grade tumor
- Recurrent tumor
- Bone involvement

RT indications in benign salivary gland tumors
- Inoperable or unresectable tumor
- Facial nerve involvement
- Recurrent tumor
- Subtotal excision

Target Volume Delineation
GTV = All gross disease
CTV1 (60 Gy) = Entire ipsilateral gland, parapharyngeal space, infratemporal fossa
CTV1ln = Ipsilateral subdigastric nodes
CTV2 (54 Gy) = Entire operative bed
CTV3 (50 Gy) = Low-risk regions of subclinical disease
CTV3ln (parotid) = Ipsilateral levels Ib, II, III if N0

- Ipsilateral levels I–IV if N+

*Elective neck irradiation is indicated in parotid tumors of locallay advanced disease, squamous and undifferentiated histologies, facial nerve involvement at diagnosis, and recurrence.
CTV3ln (submandibular) = Ipsilateral levels I–IV

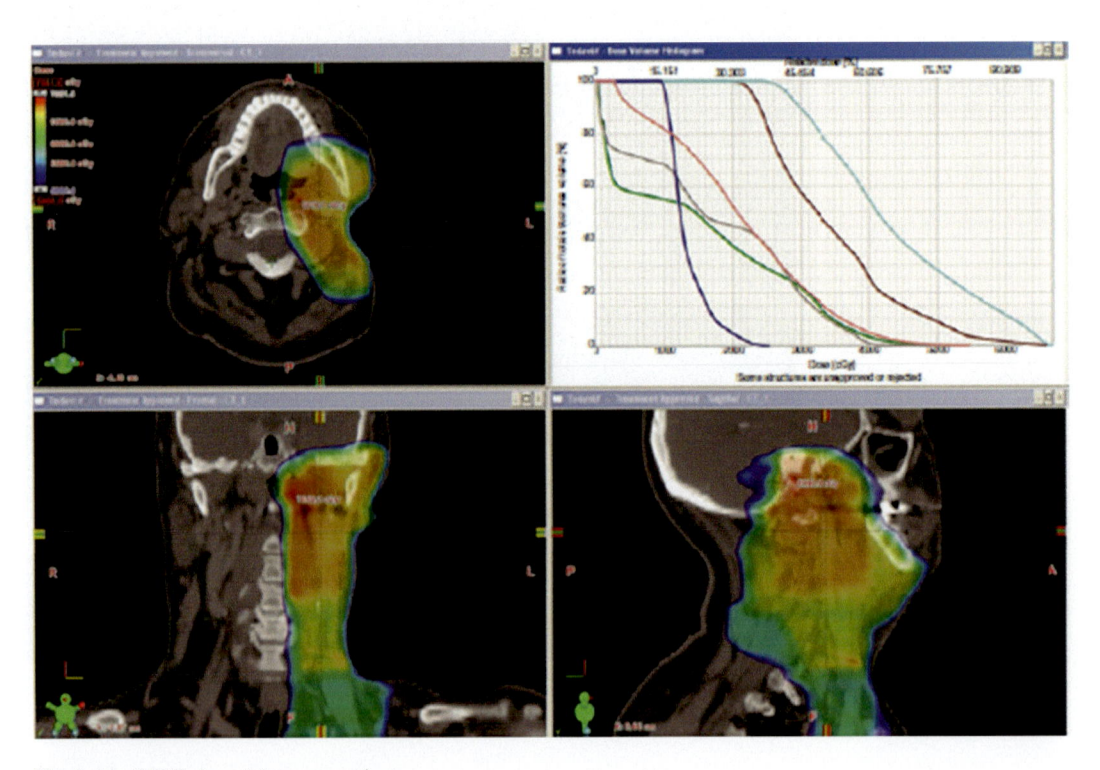

Fig. 5.44 IMRT plan of the same patient

5.7 Major Salivary Gland Tumors

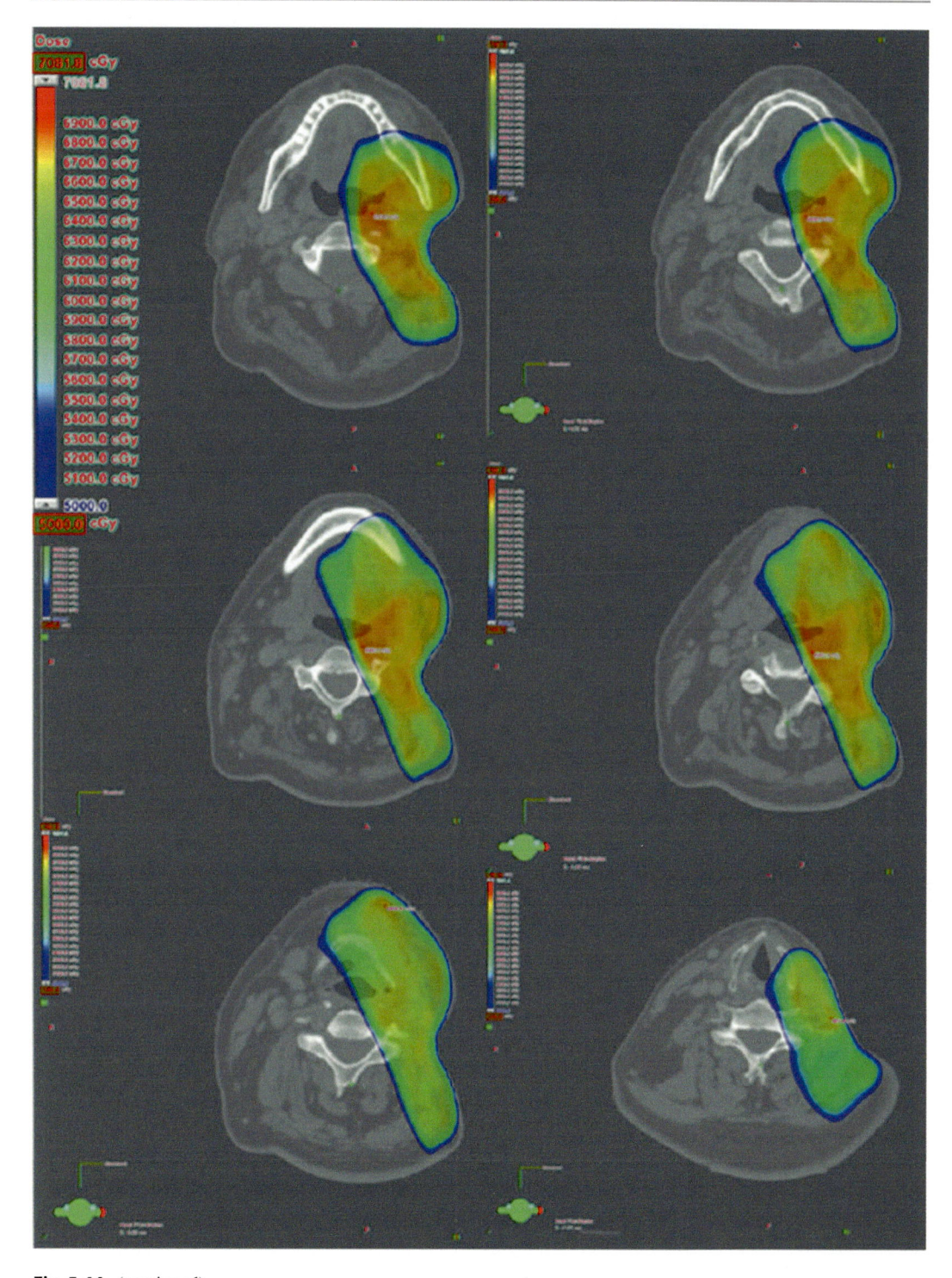

Fig. 5.44 (continued)

- Bilateral levels I–IV if tumor extends towards midline

*Elective neck irradiation is not necessary in small acinic cell and adenoid cystic cancers.

*Delineation of the cranial nerve pathway is not required in adenoid cystic carcinomas of submandibular gland with focal PNI as the recurrence is very low and morbidity is high

*For minor salivary gland adenoid cystic carcinomas, delineation of the cranial nerve pathway is required.

*Include the base of skull in tumors of the palate and paranasal sinuses

*Elective neck irradiation is indicated for tumors of the tongue, floor of the mouth, pharynx, and larynx.

*Include the pathway from the gland to the skull base if lingual and hypoglossal nerves are invaded in sublingual tumors.

PTV = 3–5 mm

5.7.6 Selected Publications

Dutch Head and Neck Oncology Cooperative Group, 2004 → 565 patients with malignant salivary gland tumors were retrospectively analyzed (332 parotid, 76 submandibular, 129 oral cavity, 28 pharynx/larynx). Median follow-up: 74 months. Ten-year OS: 50%, LC: 78%, regional control: 87%, DMFS: 67%. Postoperative RT decreased the risk of LR by 9.7 and regional recurrence by 2.3.

Terhaard CHJ et al. (2004) Salivary gland carcinoma: Independent prognostic factors for locoregional control, distant metastases, and overall survival: Results of the Dutch Head and Neck Oncology Cooperative Group. Head Neck 26:681–693.

Adjuvant radiotherapy

Dutch Head–Neck Oncology Cooperative Group, 2005 → 538 patients with malignant salivary gland tumors were retrospectively analyzed. 498 underwent surgery, 78% received postoperative RT with median 62 Gy. 40 underwent primary RT of 28–74 Gy. Adjuvant RT significantly increased 10-year LC rate in patients with T3–T4 tumors, close surgical margins, incomplete resection, bone invasion, and PNI. No dose-response relationship. Postoperative RT significantly improved regional control in the pN(+) neck (86% vs. 62%). A marginal dose-response was seen for elective neck treatment, in favor of ≥46 Gy. Also, a dose-response relationship was shown for patients treated with primary RT; 5-year LC improved with 66–70 Gy.

Terhaard CHJ et al. (2005) The role of radiotherapy in the treatment of malignant salivary gland tumors. Int J Radiat Oncol Biol Phys 61(1):103–111

Zeidan Y, 2013 → Retrospective analysis of 90 patients with minor salivary gland carcinoma (64% adenoid cystic carcinoma; 43% sinonasal, 39% oral cavity) after surgery and postoperative RT. Median follow-up was 71 months. Five- and 10-year LC: 90% and 88%, decreased with advanced T stage. Five- and 10-year DM: 24% and 28%; decreased with advanced tumor stage. Five- and 10-year OS: 76% and 63%; decreased with advanced T stage and N stage.

Zeidan Y et al. Long-term outcomes of surgery followed by radiation therapy for minor salivary gland carcinomas. Laryngoscope 123:2675–2680.

Holtzman et al, 2017 → 291 patients were retrospectively analyzed; 67 received primary RT, 224 received surgery followed by adjuvant RT. Five-, 10-, and 15-year LC: 82%, 77%, and 73%; LRC: 77%, 72%, and 67%; DMFS: 74%, 70%, and 70%; and OS: 63%, 47%, and 38%. Combined treatment increased LRC and OS. Severe toxicity rate: 5%.

Holtzman A et al. (2017) Outcomes after primary or adjuvant radiotherapy for salivary gland carcinoma. Acta Oncol 56(3):484–489.

Definitive radiotherapy

UCsF, 2006 → 45 malignant salivary gland tumors treated with RT alone. Median 66 Gy. Five-year LC: 70%; 10-year LC: 57%. LRs are frequent in T3-4 tumors and for RT doses <66 Gy.

Chen AM et al. (2006) Long-term outcome of patients treated by radiation therapy alone for salivary gland carcinomas. Int J Radiat Oncol Biol Phys 66(4):1044–1050

Elective nodal RT

UCsF; 2007 → 251 N0 malignant salivary gland tumors. Adenocystic 33%, mucoepidermoid 24%, adenocarcinoma 23%. Gross total resection R0 44%, R1 56%. No neck dissection. All with adjuvant RT. Median primary RT dose 63 Gy. Elective neck RT: ipsilateral 69%, bilateral 31%. Nodal relapse: T1 7%, T2 5%, T3 12%, T4 16%. Elective nodal RT: 10-year nodal relapse risk decreased from 26% to 0% (decrease in risk: SCC 67%, undifferentiated 50%, adenocarcinoma 34%). Whether or not elective nodal RT was given, no nodal relapse was observed in adenocystic (0/84) and acinic cell (0/21) tumors.

Conclusion: elective nodal RT is required for high-grade tumors, but not for adenoid cystic and acinic cell tumors.

Chen AM (2007) Patterns of nodal relapse after surgery and postoperative radiation therapy for carcinomas of the major and minor salivary glands: what is the role of elective neck irradiation? Int J Radiat Oncol Biol Phys 67(4):988–994

Adenoid cystic carcinoma

Garden et al, *1995* → 198 patients with adenoid cystic carcinoma received postoperative RT between 1962 and 1991 (122 in minor salivary glands). 42% had microscopic positive margins, 28% had close (≤5 mm) or uncertain margins, 69% had PNI (28% in major nerves). Median RT dose was 60 Gy (50–69 Gy). Median follow-up: 93 months. Five-, 10-, and 15-year LC: 95%, 86%, and 79%. LR rate: 18% vs. 9% vs. 5% in patients with positive margins vs. close or uncertain margins vs. negative margins (p = 0.02). Failure rate: 18% vs. 9% in patients with vs. without a major nerve PNI (p = 0.02). Trend toward better LC with increasing dose; significant in patients with positive margins (40 vs. 88% with <56 Gy vs. ≥56 Gy. Most common recurrence pattern: DM (37%).

Garden AS et (1995) The influence of positive margins and nerve invasion in adenoid cystic carcinoma of the head and neck treated with surgery and radiation. Int J Radiat Oncol Biol Phys 32(3):619–26.

Mendenhall et al, *2004* → 101 patients were treated with RT alone or combined with surgery between 1966 and 2001. Median follow-up: 6.6 years. In all patients, 5- and 10-year LC: 77% and 69%, OS: 68% and 49%, DMFS: 80% and 73%. Five- and 10-year LC: 56% and 43% vs. 94% and 91%; and OS: 57% and 42% vs. 77% and 55% with RT vs. combined treatment. Prognostic factors for LC: T stage and treatment, for OS: T stage and PNI.

Conclusion: Optimal treatment for patients with adenoid cystic carcinoma is surgery and adjuvant RT.

Mendenhall WM et al. (2004) Radiotherapy alone or combined with surgery for adenoid cystic carcinoma of the head and neck. Head Neck 26(2):154–62.

MsKCC, 2007 → 59 adenoid cystic carcinomas (oral cavity 28%, paranasal sinus 22%, parotid 14%, submandibular gland 14%). T1–4 tumors. Treated with surgery + RT. Included cranial base in 90% of cases. Median follow-up: 5.9 years. Five-year LC: 91%; OS: 87%. Ten-year LC: 81%; OS: 65%. Poor prognostic factors: T4 tumor, gross and/or clinical PNI, LN (+). Adjuvant RT after surgery had excellent LC rates.

Gomez DR (2008) Outcomes and prognostic variables in adenoid cystic carcinoma of the head and neck: a recent experience. Int J Radiat Oncol Biol Phys 70(5):1365–1372

DAHANCA, 2015 → Retrospective. 201 patients with adenoid cystic carcinoma treated with a curative intent between 1990 and 2005. Median follow-up: 7.5 years. Ten-year OS: 58%, DSS: 75%, LRC: 70%. Recurrence rate: 36%, DM: 18%. Tumor stage and margin status were independent prognostic factors for OS and LRC. RT did not improve survival, but the LRC rate.

Bjorndal K et al. (2015) Salivary adenoid cystic carcinoma in Denmark 1990–2005: Outcome and independent prognostic factors including the benefit of radiotherapy. Results of the Danish Head and Neck Cancer Group (DAHANCA). Oral Oncol 51(12):1138–42.

Fig. 5.45 Thyroid gland

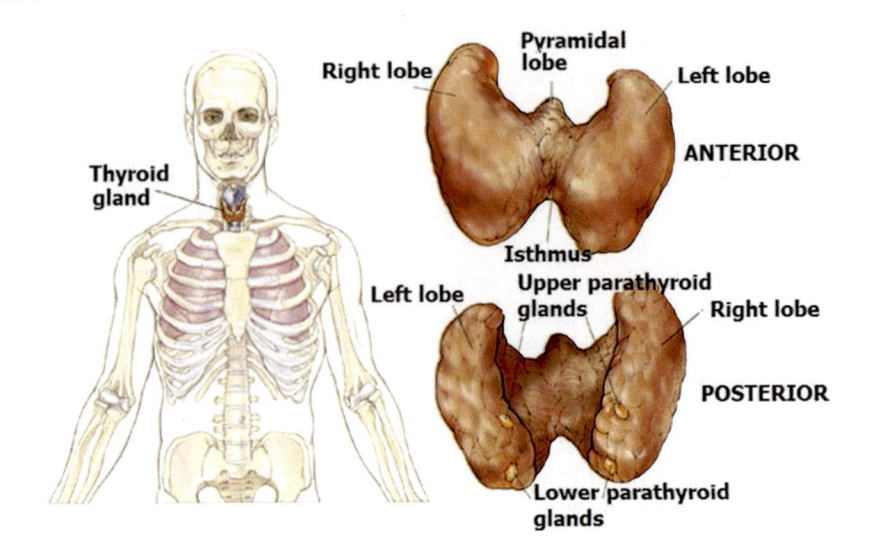

Neutron RT

RTOG-MRC Neutron Trial, 1993 → randomized. 32 inoperable or recurrent major/minor salivary gland tumors, Neutrons (17–22 nGy) vs. photons/electrons (55 Gy/4 weeks or 70 Gy/7 weeks). Ten-year LRC: 56% in neutron vs. 17% photon/electron arm ($p = 0.009$). Median survival: 3 years in neutron vs. 1.2 years in photon/electron arm. No difference in OS (25–15%).

Laramore G et al. (1993) Neutron versus photon irradiation for unresectable salivary gland tumors: final report of an RTOG-MRC randomized clinical trial. Int J Radiat Oncol Biol Phys 27(2):235–240

5.8 Thyroid Cancer

The thyroid gland originates from the pharyngeal pouches of the base of the tongue in the middle of the neck and move inferiorly within the thyroglossal canal (Fig. 5.45). It consists of two lobes connected to each other by an isthmus. The upper pole of the thyroid gland extends into the space formed by the sternothyroid muscle, inferior constrictor muscle, and posterior thyroid lamina. The lower pole extends to the level of the fifth or sixth tracheal rings. The lobes of the thyroid are in close proximity to the common carotid arteries posterolaterally, the prevertebral muscles, recurrent laryngeal nerves, sympathetic trunks, vagus, and phrenic nerves posteriorly, and the trachea and cricoid. The thyroid glands move together with the larynx and trachea during the process of swallowing.

Thyroid cancer is the most frequently seen endocrine tumor, but it constitutes only a very small proportion of other thyroid gland disorders. Thyroid cancers represent 1% of all new cancer cases each year. This cancer leads to 0.5% of all cancer-related mortality [35].

> Solitary thyroid nodules have a malignancy incidence of 10–30%. Therefore, all thyroid nodules should be carefully examined for malignancy risk [36].

Epidemiological studies have not demonstrated any prominent relation between thyroid cancer and iodine deficiency in the diet. A relationship between the presence of a benign goiter and well-differentiated thyroid cancer has not been demonstrated beyond doubt. Follicular carcinoma is frequently observed in geographical regions where goiter is endemic [35].

5.8 Thyroid Cancer

The risk of a palpable thyroid anomaly is high in patients receiving low-dose RT to the head–neck and upper mediastinal regions: 25% of patients receiving 0.02 Gy of external RT develop a goiter; 25% of these, or 7% of all cases receiving thyroid irradiation, develop thyroid cancer (most often papillary adenocarcinoma) [35].

- The latent period for the development of thyroid anomalies is 10–20 years [37].

Only a small percentage of all papillary and follicular thyroid cancers have a genetic basis. Gardner syndrome and Cowden's disease (familial goiter and skin hamartoma) have a well-established relationship to differentiated thyroid carcinomas.

Thyroid lymphomas, particularly the B cell type, are commonly seen in autoimmune thyroiditis.

The most commonly involved lymphatics are level VI, followed by the lower neck (levels III-IV), SCF, and level VII.

5.8.1 Pathology

Most thyroid cancers are derived from the follicular epithelium, except for medullary carcinoma which is derived from the parafollicular C cells [38].

Follicular Epithelial Cell
1. Well-differentiated thyroid cancer
   - Papillary carcinoma
     – Classic
     – Follicular variant
     – Oncocytic variant
     – Unfavorable variants (diffuse sclerosing, tall cell, columnar cell, hobnail)
   - Follicular carcinoma
     – Classic
     – Hurthle cell carcinoma
2. Poorly differentiated thyroid cancer: insular carcinoma
3. Undifferentiated thyroid cancer: Anaplastic carcinoma

Parafollicular C-cell
Medullary carcinoma

Papillary carcinoma

- This is the most common (80–90%) thyroid cancer. Ninety percent of radiation-related cancers are papillary thyroid cancers. They grow slowly, and are three times more common in females.
- They are characterized by papillary extension of the thyroidal epithelium and connective tissue. The tumor tissue contains concentric calcium deposits called psammoma bodies.
- In 80% of cases, they are multicentric within the gland.

Follicular carcinoma

- These comprise 5–15% of all malign thyroid cancers. They are commonly seen in older females.
- They are less multicentric than papillary thyroid cancers.
- They are grossly indistinguishable from follicular adenomas. Diagnosis is dependent on two features: tumor invasion through the entire tumor capsule or tumor invasion into a blood vessel located in the tumor capsule or immediately outside the tumor capsule.
- They have a 15% risk of LN metastasis, but more commonly show early hematogenous metastases to bone, lung, and liver.
- They have a good prognosis, but also a tendency for frequent recurrence.

Hurthle cell carcinoma (oncocytic or oxyphilic cell carcinoma)

- Two-three percent of all thyroid malignancies are of this type.
- Again, the presence of capsular or vascular invasion is required to classify it as a malignant neoplasm.
- They take up less radioactive iodine (RAI) than follicular cancers, and bilaterality and LN involvement are 25% higher than those of follicular cancer.
- They have a poorer prognosis than follicular cancer.

Insular carcinoma

- The prognosis is intermediate between differentiated and undifferentiated carcinomas.
- They tend to be widely invasive, and many will extend beyond the gland.

Anaplastic carcinoma

- These comprise 1% of all thyroid cancers.
- They are usually observed in females and in patients over 65 years of age.
- The tumor has no capsule, and extensively invades surrounding tissues.
- The disease progresses rapidly and metastasizes to the lungs. Masses can be identified using ultrasonography (USG, but no iodine uptake is observed in scintigraphy. Diagnosis is performed with fine needle aspiration.
- Thyroidectomy is impossible in most cases. Tracheostomia may be required to open the airway due to its rapid course.
- Survival is very short (between 3 weeks and 6 months).

Medullary carcinoma

- Two percent of all thyroid malignancies are medullary carcinomas.
- They originate from parafollicular C cells. Amyloid in stroma is diagnostic.
- They are more common in females and in older patients.
- Half of the patients present with cervical lymphadenopathy.
- Ninety percent of them are sporadic, while the remaining can be termed familial and are a component of multiple endocrine neoplasia (MEN IIa, MEN IIb) syndromes.
- Sporadic cases have a solitary nodule, while familial cases have bilateral multicentric tumors.
- Prognosis is somewhere between those for differentiated and anaplastic thyroid cancers.
- They do not usually respond to RAI treatment and external RT.

The ability to concentrate iodine base on histology:
Usually concentrate to a curative degree

- Papillary carcinoma (classic, follicular and oncocytic variants)
- Follicular carcinoma (classic)

Often do not concentrate to a curative degree

- Unfavorable papillary variants
- Hurthle cell carcinoma

Rarely concentrate to a curative degree

- Insular carcinoma

Never concentrate to a curative degree

- Anaplastic carcinoma
- Medullary carcinoma

5.8.2 General Presentation

- Benign thyroid nodule incidence is high in females aged between 20 and 40 years, and these cases have a cancer risk of approximately 5–10%.

5.8 Thyroid Cancer

- Males show a higher incidence of cancer below 20 years and over 40 years of age, soa age and gender are important risk factors.

- Swallowing difficulty, stridor, and hoarseness may be signs of malignancy, as well as a large goiter and retrosternal thyroid enlargement.
- Hoarseness is almost always a sign of malignancy, but is rarely seen in large benign thyroid adenomas.
- The characteristic sign of cancer is a firm and fixed mass larger than 2 cm in size.
- Thyroid mass along with substantial lymphadenopathy may also be a sign of malignancy.

5.8.3 Staging

Primary Tumor (T) (Fig. 5.46) [20]

- T0: No evidence of primary tumor
- T1: Tumor ≤2 cm in greatest dimension, limited to the thyroid
  - T1a: Tumor ≤ 1 cm
  - T1b: Tumor 1–2 cm
- T2: Tumor >2 cm but ≤4 cm in greatest dimension, limited to the thyroid
- T3: Tumor >4 cm in greatest dimension limited to the thyroid, or gross extrathyroidal extension invading only strap muscles
  - T3a: Tumor >4 cm limited to the thyroid
  - T3b: Gross extrathyroidal extension invading only strap muscles (sternohyoid, sternothyroid, thyrohyoid, or omohyoid) from a tumor of any size

- T4: Includes gross extrathyroidal extension
  - T4a: Invasion of subcutaneous soft tissues, larynx, trachea, esophagus, or recurrent laryngeal nerve from a tumor of any size
  - T4b: Invasion of prevertebral fascia or encasing the carotid artery or mediastinal vessels form a tumor of any size

Regional Lymph Nodes (N) (Fig. 5.47)

- NX: Regional lymph nodes cannot be assessed
- N0: No regional lymph node metastasis
  - N0a: ≥1 cytologically or histologically confirmed benign lymph nodes
  - N0b: No radiologic or clinical evidence of locoregional lymph node metastasis
- N1: Regional lymph node metastasis
  - N1a: Unilateral or bilateral metastasis to level VI or VII (pretracheal, paratracheal, or prelaryngeal/Delphian, or upper mediastinal lymph nodes)
  - N1b: Metastasis to unilateral, bilateral, or contralateral neck lymph nodes (levels I, II, III, IV, or V) or retropharyngeal lymph nodes

AJCC Stage Groupings
Differentiated
<55 years

- Stage I: any T, any N, M0
- Stage II: any T, any N, M1

≥55 years
- Stage I: T1-2 N0/NX M0
- Stage II: T1-2 N1 M0, T3a-b any N M0

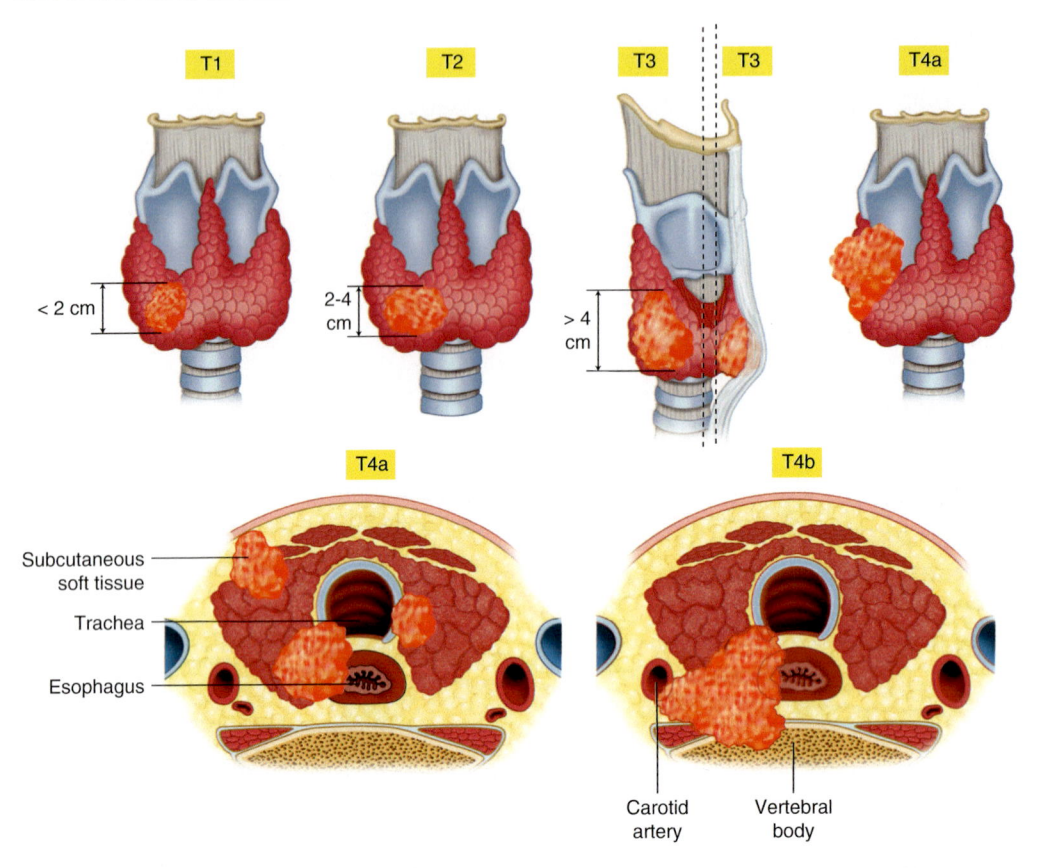

Fig. 5.46 T staging for thyroid cancer (from [20], reproduced with the permission from the American Joint Committee on Cancer)

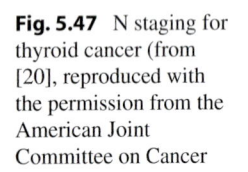

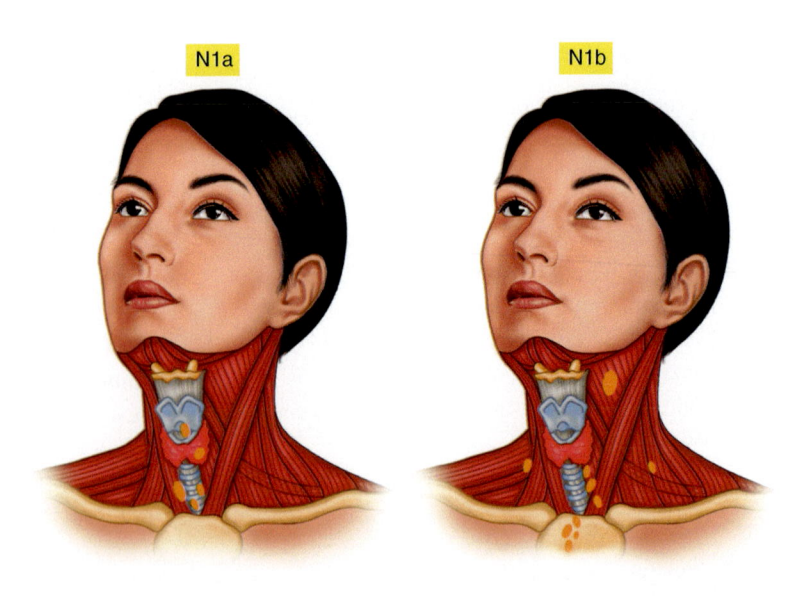

Fig. 5.47 N staging for thyroid cancer (from [20], reproduced with the permission from the American Joint Committee on Cancer

- Stage III: T4a any N M0
- Stage IVA: T4b any N M0
- Stage IVB: any T any N M1

Anaplastic
All anaplastic carcinomas are considered stage IV.
Stage IVA: T1-3a N0/NX M0
Stage IVB: T1-3a N1 M0, T3b-4 any N M0
Stage IVC: any T any N M1

5.8.3.1 Treatment Algorithm

Papillary carcinoma
- <1 cm → if LN (−) active surveillance with USG or lobectomy (high-risk)*
  - *posterior location, abutting the trachea, or apparent invasion, etc.
  - → if LN (+) manage as ≥1 cm
- ≥1 cm → total thyroidectomy or lobectomy ± neck dissection
  - Consider RAI or RT
- *Follicular carcinoma (including hurthle cell carcinoma)*
- Total thyroidectomy
  - Consider RAI or RT
- *Medullary carcinoma*
- Total thyroidectomy ± neck dissection
  - Consider RT for incomplete resection
- *Anaplastic carcinoma*
- Stage IVA or IVB → resectable → total thyroidectomy and lymph node dissection
  - Consider RT
  - → unresectable → RT and chemotherapy
- Stage IVC → total thyroidectomy and LN dissection with RT-chemo palliative care

5.8.4 Radiotherapy

External RT is used for definitive, palliative, or adjuvant purposes either alone or in combination with RAI. Indications for external RT in the management of thyroid cancers can be summarized as follows:

- Primary treatment for unresectable or RAI (−) tumors
- Large tumors that cannot be controlled with RAI (e.g., mediastinal involvement)
- Residual tumors in trachea, esophagus, or neck after surgery and RAI
- Tumors compressing vital organs
- Superior vena cava syndrome
- Recurrence (RAI uptake not important)
- Metastasis or relapse after maximal RAI

Target Volume and Dose Definitions

GTV = All gross disease
  CTV1 (66-70 Gy) = GTV + 3–5 mm
  CTV2 (59.5-66 Gy) = Thyroid bed (from hyoid bone to aortic arch) + positive LN levels

- Include involved-side tracheoesophageal groove
- Include tracheostomy stoma (if present) to the skin surface

  CTV3 (54 Gy) = Elective neck levels (level II-VII)
- *Include level II and V if level III positive
- *Include level VII if level VI positive
- *Include RP and/or level I if adjacent level II positive
PTV = 3–5 mm

RT in anaplastic carcinoma

- Concurrent CRT can be used.
- Accelerated HF RT (1.5–1.6 Gy bid) can be given (Fig. 5.48).

5.8.5 Selected Publications

Adjuvant RT

Princess Margaret Hospital, 2005 → 729 patients with differentiated thyroid cancer treated between 1958 and 1998 were retrospectively analyzed. Median follow-up: 11.3 years. Ten-year CSS: 87.3%, 10-year LRRFS: 84.9%. RAI resulted in a statistically significant improvement in LRRFS. No benefit from RAI in low-risk patients stage I ≤45 years. For patients >60 years with extrathyroidal extension but no gross residual disease, adjuvant RT significantly increased 10-year CSS (81% vs. 64.6%) and 10-year LRRFS (86.4% vs. 65.7%).

Brierley J et al. (2005) Prognostic factors and the effect of treatment with radioactive iodine and external beam radiation on patients with differentiated thyroid cancer seen at a single institu-

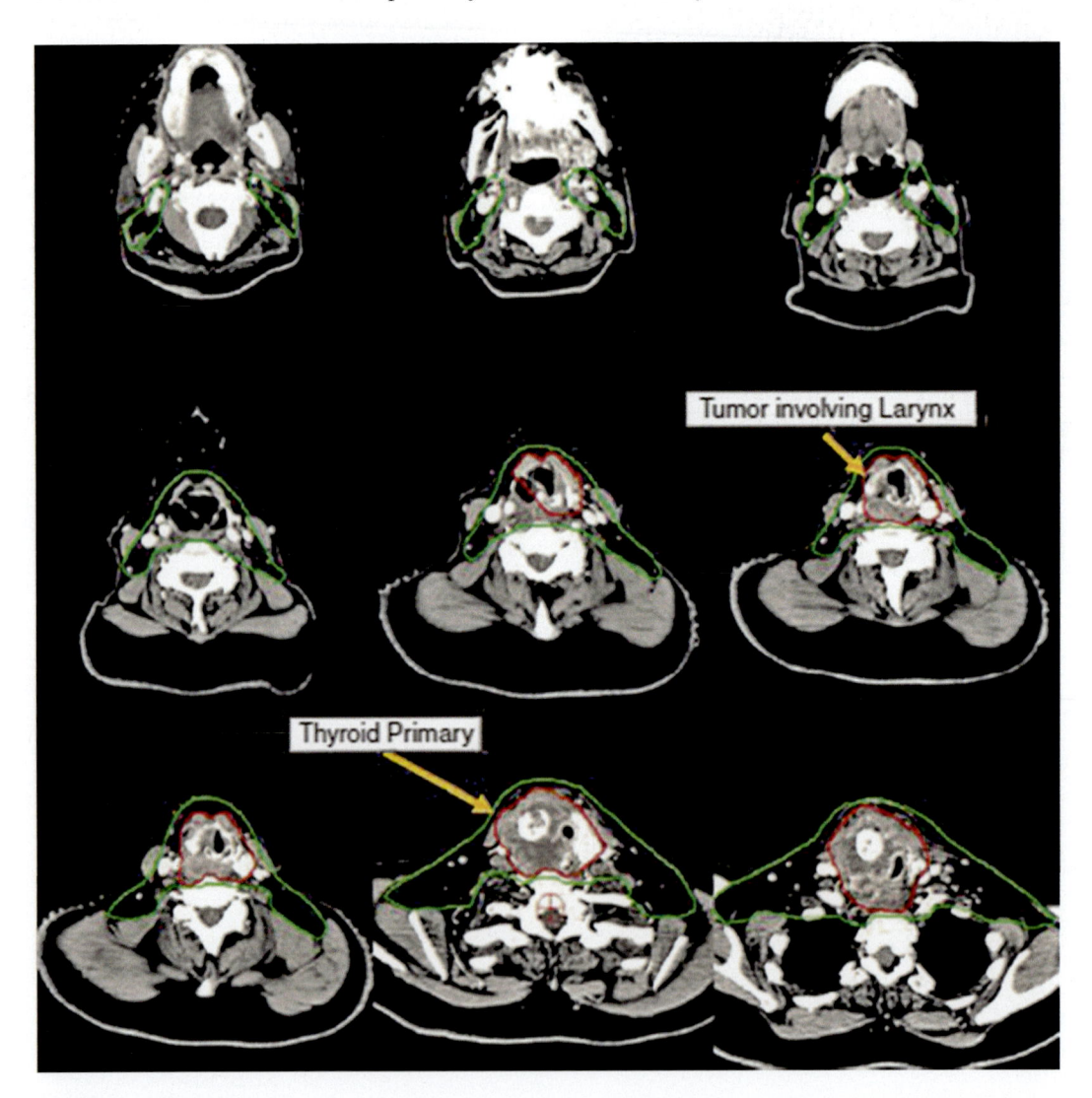

Fig. 5.48 Contouring of a patient with T4b N1b M0 anaplastic thyroid carcinoma. CTV70 (red), CTV60 (green)

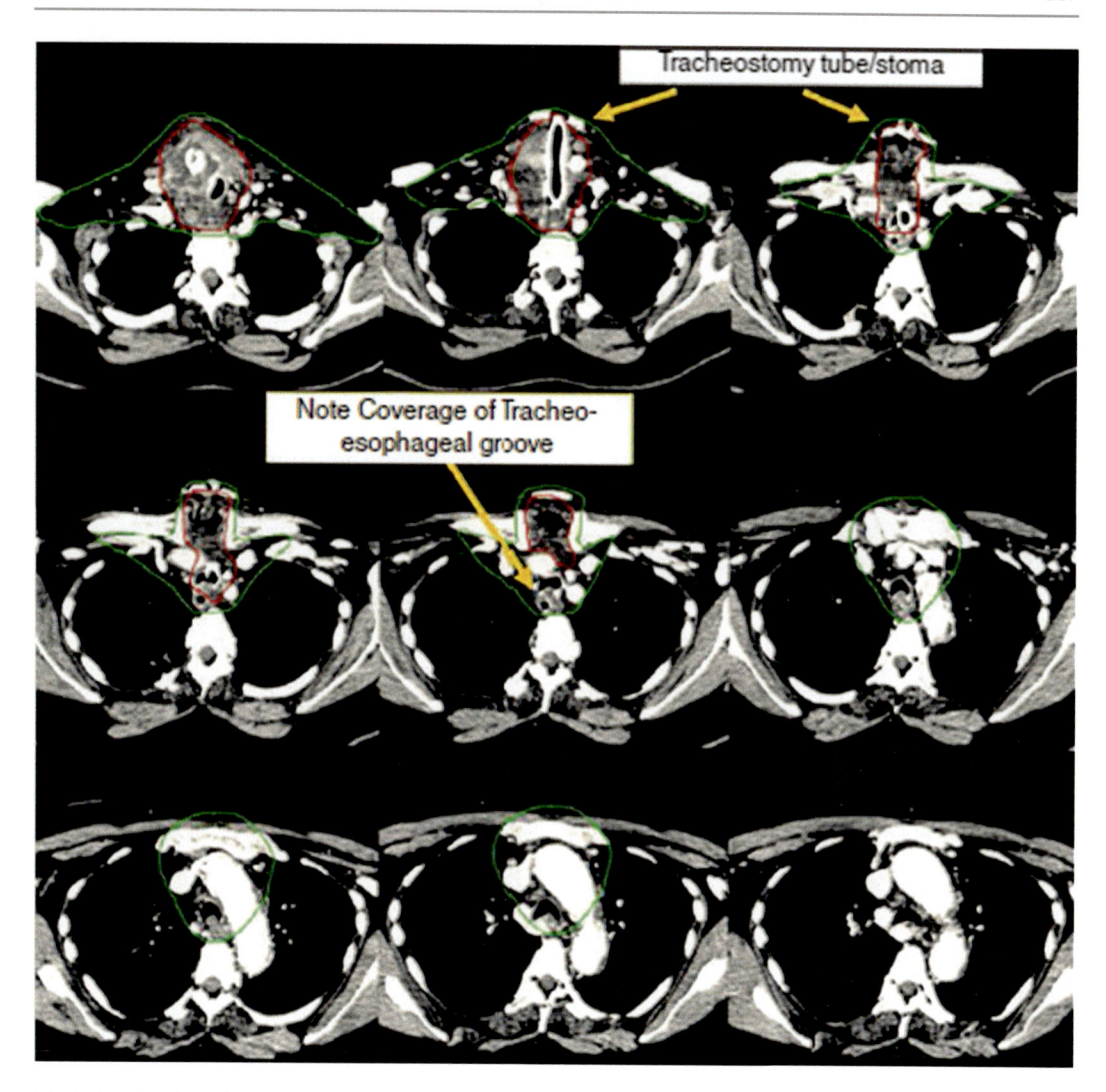

Fig. 5.48 (continued)

tion over 40 years. Clin Endocrinol (Oxf) 63(4):418–27

Chow et al, *2006* → 1297 patients with papillary thyroid carcinoma were retrospectively analyzed. After surgery alone, bilateral thyroidectomy significantly yielded lower rate of LR (vs. lobectomy). RT significantly improved outcomes in patients with gross residual disease, positive resection margins, T4 disease, N1b disease, and LN metastasis >2 cm. RAI was effective in T2-T4 disease.

Chow SM et al. (2006) Local and regional control in patients with papillary thyroid carcinoma: specific indications of external radiotherapy and radioactive iodine according to T and N categories in AJCC sixth edition. Endocr Relat Cancer 13(4):1159–72

Concurrent chemoradiotherapy

MSKCC, 2014 → 66 patients with gross residual/unresectable non-anaplastic non-medullary thyroid cancer were treated with RT between 1990 and 2012. Median OS: 42 months. Three-year LRRFS: 77.3%. LRRFS was significantly lower in poorly differentiated histology (89.4% vs. 66.1%), but DMFS was significantly worse (43.9% vs. 82.5%). Concurrent CRT only increased the rate of acute grade 3 hoarseness.

Romesser PB et al. (2014) External beam radiotherapy with or without concurrent chemotherapy in advanced or recurrent non-anaplastic non-medullary thyroid cancer. J Surg Oncol 110(4):375–82.

→

Medullary thyroid cancer

Princess Margaret Hospital, 1996 → 73 patients. Median 40 Gy postoperative RT in 40 of them. Prognostic factors for CSS: extraglandular extension and postoperative gross residual disease. RT was beneficial in cases with microscopic residue (+), extraglandular extension, LN (+).

Brierley J et al. (1996) Medullary thyroid cancer: analyses of survival and prognostic factors and the role of radiation therapy in local control. Thyroid 6(4):305–310

Hurthle cell thyroid cancer

Mayo Clinic, 2003 → retrospective. 18 patients. Five cases with adjuvant RT, seven unresectable cases with salvage RT, six metastatic cases with palliative RT. Five-year OS: 66.7%. Five-year CSS: 71.8%.

Foote RL et al. (2003) Is there a role for radiation therapy in the management of Hurthle cell carcinoma? Int J Radiat Oncol Biol Phys 56(4):1067–1072

Anaplastic thyroid cancer (altered fractionated RT)

Princess Margaret Hospital, 2006 → 47 cases between 1983 and 2004. 23 cases with radical RT (14 cases with once daily and 9 cases with twice daily; 1.5 Gy/45–66 Gy). 24 cases with palliative RT (<40 Gy). Six-month LC in palliative arm: 64%. Six-month LC in radical RT arm: 94%. Median OS was 3.3 months longer in the HF arm than the CF arm (13.6 months vs. 10.3 months).

Wang Y et al (2006) Clinical outcome of anaplastic thyroid carcinoma treated with radiotherapy of once- and twice-daily fractionation regimens. Cancer 107(8):1786–1092

National Cancer Database, 2013 → 2742 patients diagnosed with anaplastic thyroid cancer between 1998 and 2008 were reviewed. Older age and omission of treatment were associated with greater mortality; Thyroidectomy or chemotherapy increased median survival from 2 to 6 months, RT from 2 to 5 months. Median survival with surgery, RT, and chemotherapy was 11 months for stage IVA, 9 months for stage IVB, and 5 months for stage IVC.

Haymart MR et al. (2013) Marginal treatment benefit in anaplastic thyroid cancer. Cancer 1;119(17):3133-9.

5.9 Radiotherapy in Unknown Primary Head–Neck Cancers (Table 5.9)

Treatment Algorithm [21] (Table 5.10)

Table 5.9 Possible primaries based on involved neck node levels

Neck node level	Possible primaries
IA	Lower lip, anterior tongue, floor of mouth, skin
IB	Anterior alveolar ridge, oral cavity, anterior nasal cavity, submandibular glands, skin
II	Oral cavity, nasal cavity, oropharynx, nasopharynx, hypopharynx, larynx, major salivary glands
III	Oropharynx, nasopharynx, hypopharynx, larynx, oral cavity
IV	Hypopharynx, larynx, cervical esophagus, thyroid
V	Nasopharynx, oropharynx, cervical esophagus, thyroid, hypopharynx, subglottic larynx
VI	Hypopharynx, subglottic larynx, cervical esophagus, thyroid
Retropharyngeal	Nasopharynx, oropharynx, hypopharynx, soft palate

Table 5.10 Suggested neck node levels to delineate according to suspected primary site and N stage

Suspected primary site	N1–2a	N2b–3
Nasopharynx	I–IV, RP	II–V, RP
Oropharynx	II–IV (add RP if posterior pharyngeal wall suspected)	I–V, RP
Oral cavity	I–III (add IV if anterior tongue suspected)	I–V
Hypopharynx	II–IV	I–V, RP
Larynx	II–IV	II–V

RP retropharyngeal

5.9 Radiotherapy in Unknown Primary Head–Neck Cancers

Table 5.11 RTOG/EORTC guidelines for the delineation of elective nodal CTV

Level	Superior	Caudal	Anterior	Posterior	Lateral	Medial
Ia	Geniohyoid muscle	Plane tangent to body	Symphysis menti,	Body of hyoid	Medial edge of	–
	Plane tangent to	Of hyoid bone	Platysma muscle	Bone	Anterior belly of	
	Basilar edge of				Digastric muscle	
	Mandible					
Ib	Mylohyoid muscle cranial edge of submandibular gland	Plane through central part of hyoid bone	Symphysis menti, platysma muscle	Posterior edge of submandibular gland	Basilar edge/ inner side of mandible, platysma muscle skin	Lateral edge of anterior belly of digastric muscle
IIa	Caudal edge of lateral process of C1	Caudal edge of body of hyoid bone	Posterior edge of submandibular gland; anterior edge of internal carotid artery; posterior edge of posterior belly of digastric muscle	Posterior border of internal jugular vein	Medial edge of sternocleidomastoid	Medial edge of internal carotid artery, paraspinal (levator scapulae) muscle
IIb	Caudal edge of lateral process of C1	Caudal edge of body of hyoid bone	Posterior border of internal jugular vein	Posterior border of the sternocleidomastoid muscle	Medial edge of sternocleidomastoid	Medial edge of internal carotid artery, paraspinal (levator scapulae) muscle
III	Caudal edge of the body of hyoid bone	Caudal edge of cricoid cartilage	Posterolateral edge of the sternohyoid muscle anterior edge of sternocleidomastoid muscle	Posterior edge of the sternocleidomastoid muscle	Medial edge of sternocleidomastoid	Internal edge of carotid artery, paraspinal (scalenus) muscle
IV	Caudal edge of cricoid cartilage	2 cm cranial to sternoclavicular joint	Anteromedial edge of sternocleidomastoid muscle	Posterior edge of the sternocleidomastoid muscle	Medial edge of sternocleidomastoid	Medial edge of internal carotid artery, paraspinal (scalenus) muscle
V	Cranial edge of body of hyoid bone	CT slice encompassing the transverse cervical vessels	Posterior edge of the sternocleidomastoid muscle	Anterior border of the trapezius muscle	Platysma muscle skin	Paraspinal (levator scapulae, splenius capitis) muscle
VI	Caudal edge of body of thyroid cartilage	Sternal manubrium	Skin, platysma muscle	Separation between trachea and esophagus	Medial edges of thyroid gland, skin and anterior–medial edge of sternocleidomastoid muscle	–
Retro	Base of skull	Cranial edge of the	Fascia under the	Prevertebral	Medial edge of the	Midline
Pharyngeal		Body of hyoid bone	Pharyngeal mucosa	Muscle (longus Colli, longus Capitis)	Internal carotid Artery	

M muscle

When a head-neck cancer is determined histologically based on nodal excision but the primary site cannot be established, it is defined as an unknown primary head- neck cancer.

- These constitute 2–9% of all head–neck cancers.
- The most common primary is the tonsil (45%), base of tongue (40%) or pyriform sinus (10%).
- Biopsy from high-risk regions should be done by panendoscopy following physical examination and radiological tests; diagnostic tonsillectomy can also be performed.
- Patients should be evaluated for EBV and HPV. N stage should be decided based on EBV and HPV status.
  - If EBV (+) → stage like nasopharynx cancer
  - If HPV (+) → stage like p16+ oropharynx cancer
  - If both (−) → stage like p16− oropharynx cancer
- Histopathology is 50–75% SCC, followed by undifferentiated carcinoma (20%).

- Treatment is directed based on to neck node histopathology.
- *Adenocarcinoma (thyroglobulin and calcitonin negative)*
- Levels I–III → neck dissection ± parotidectomy → RT to neck ± parotid bed
- Levels IV, V → evaluate for infraclavicular primary → neck dissection
  - N1, ECE (−) → RT or observe
  - N2-3, ECE (−) → RT or CRT
  - ECE (+) → CRT or RT
- *Poorly differentiated/non-keratinizing SCC or NOS or non-thyroid anaplastic or SCC of neck node*
- N1 → neck dissection* or RT

- N2 → neck dissection* or CRT or induction chemo followed by CRT
  - N1, ECE (−) → RT or observe
  - *N2-3, ECE (−) → RT or CRT
  - *ECE (+) → CRT or RT

Target Volume Delineation (Figs. 5.49 and 5.50)
GTV = All gross LNs
CTV1 (60–70 Gy) = GTV + 5–10 mm (+ surgical bed if ECE+ or soft tissue invasion)
CTV2 (59.4–66 Gy) = Surgical bed and potential primary regions (entire nasopharynx, base of tongue, ipsilateral tonsillar fossae, pyriform sinus and highly suspected regions depending on the location and laterality of involved LN)

- *If LN is in level IA/IB, include oral cavity and nasopharynx can be excluded.
- *Nasopharynx can also be excluded in p16+/EBV- patients.
- *Larynx sparing is not appropriate if level 3 is positive without level 2 positivity.

CTV2ln = Positive neck node levels
CTV3 (46–54 Gy) = Elective neck node levels
PTV = 3–5 mm

5.10 Selected Publications for Head and Neck Cancers

Risk groups
Review, 2007 → definition of subgroups of head and neck SCC with different risks of treatment failure according to AJCC classification: low risk: T1–2N0; intermediate risk: T3N0, T1–3N1; high-risk: T4aN0–1, T1–4N2; very high risk: T4b any N, any T N3; and poor prognosis: M1.

5.10 Selected Publications for Head and Neck Cancers

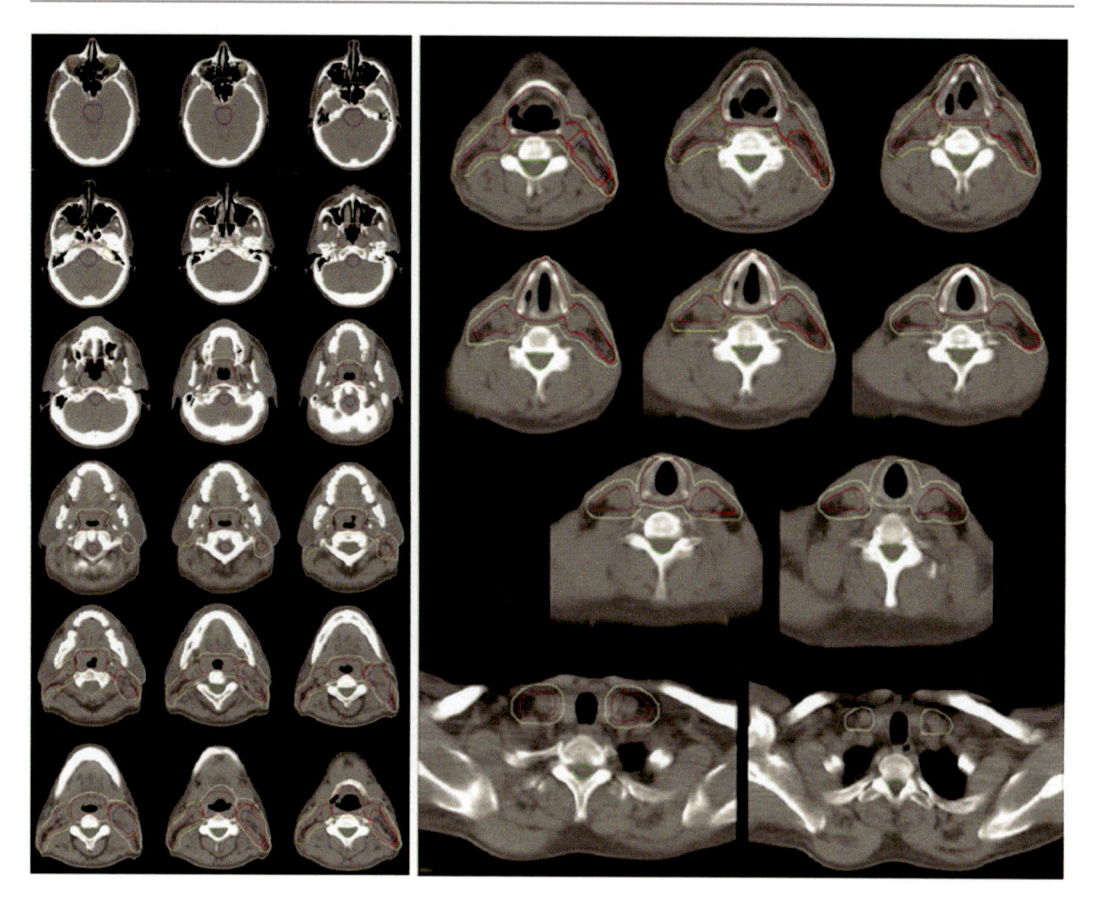

Fig. 5.49 Contouring of a patient with TxN2bM0 unknown primary head and neck cancer

Corvo R (2007) Evidence-based radiation oncology in head and neck squamous cell carcinoma. Radiother Oncol 85:156–170

Fractionation trials

Sanchiz et al., 1990 → 859 patients with locally advanced head and neck cancer randomized to once-daily RT (Group A) vs. BID RT (Group B) vs. once-daily RT and concurrent 5-FU (Group C). Median duration of response and OS significantly improved in Groups B and C compared to Group A. No significant difference between Groups B and C.

Sanchiz F, et al. (1990) Single fraction per day versus two fractions per day versus radiochemotherapy in the treatment of head and neck cancer. Int J Radiat Oncol Biol Phys 19(6):1347–1350.

CHaRT, 1997 → CHART: 54 Gy given in 36 fractions over 12 days, vs. CF RT: 66 Gy given in 33 fractions over 6.5 weeks. Similar LC was achieved with CHART, supporting the importance of repopulation as a cause of radiation failure.

Dische S et al. (1997) A randomized multicentre trial of CHART versus conventional radiotherapy in head and neck cancer. Radiother Oncol 44:123–136

RTOG 90–03, 2000 → randomization: (1) CF at 2 Gy/fraction/day, 5 days/week, to 70 Gy/35 fractions/7 weeks; (2) HF at 1.2 Gy/fraction, twice daily, 5 days/week to 81.6 Gy/68 fractions/7 weeks; (3) accelerated fractionation with a split at 1.6 Gy/fraction, twice daily, 5 days/week, to 67.2 Gy/42 fractions/6 weeks, including a 2-week rest after 38.4 Gy; or (4) accelerated fractionation with concomitant boost at 1.8 Gy/fraction/day, 5 days/week and 1.5 Gy/fraction/day to a boost field as a second daily treatment for the last 12 treatment days to 72 Gy/42 fractions/6 weeks. HF and accelerated fractionation with concomitant boost are more efficacious than

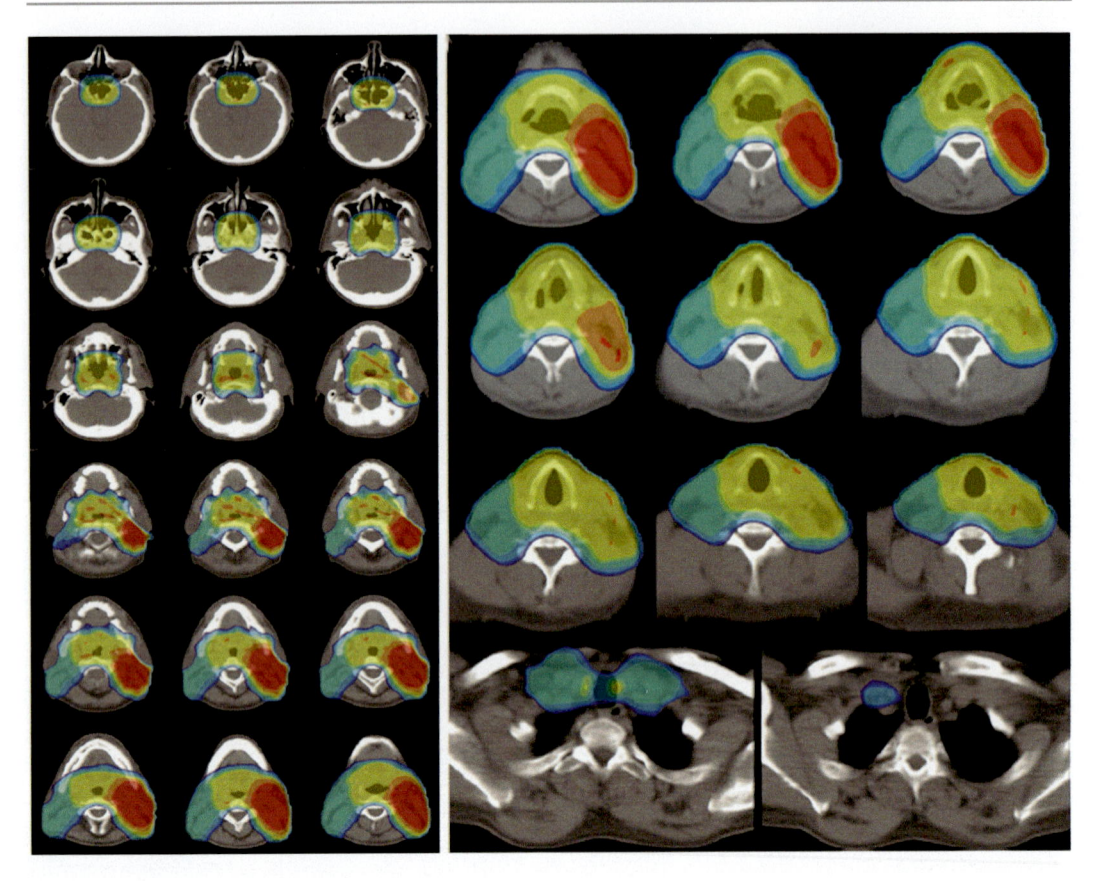

Fig. 5.50 IMRT plan of the same patient

CF for locally advanced head and neck cancer. Acute but not late effects are also increased.

Fu KK et al. (2000) A Radiation Therapy Oncology Group (RTOG) phase III randomized study to compare hyperfractionation and two variants of accelerated fractionation to standard fractionation radiotherapy for head and neck squamous cell carcinomas: first report of RTOG 9003. Int J Radiat Oncol Biol Phys 48:7–16

RTOG 90–03, 2014 → Five-year LRC and OS improved with HF RT without increasing late toxicity.

Beitler JJ et al. (2014) Final Results of Local-Regional Control and Late Toxicity of RTOG 90–03; A Randomized Trial of Altered Fractionation Radiation for Locally Advanced Head and Neck Cancer Int J Radiat Oncol Biol Phys. 89(1): 13–20.

Trans-Tasman Radiation Oncology Group Study, 2001 → 350 patients with stage III-IV oral cavity, oropharynx, larynx, and hypopharynx cancer randomized to CRT (70 Gy in 35 fractions in 49 days) or accelerated RT (ART: 1.8 Gy BID to 59.4 Gy in 33 fractions in 24 days). Median follow-up: 53 months. No significant difference in terms of DFS, DSS, or LRC. Acute mucosal toxicity was higher in the ART arm but not in the late term. Other late toxicities were lower in the ART arm.

Poulsen MG et al. (2001) A randomised trial of accelerated and conventional radiotherapy for stage III and IV squamous carcinoma of the head and neck: a Trans-Tasman Radiation Oncology Group Study. Radiotherapy and Oncology 60:113–122.

DAHANCA 6 and 7, 2003 → 1476 patients with head and neck cancer randomized to 2 Gy/fraction, 6 fractions vs. 5 fractions per week. Five-year LRRFS: 76% vs. 64% for 6 and 5 fractions. Five-year CSS:73% vs. 66% for 6 fractions and 5 fractions. No OS advantage with accelerated fractionation.

Overgaard J et al. (2003) Five compared with six fractions per week of conventional radiotherapy of squamous-cell carcinoma of head and neck: DAHANCA 6 and 7 randomised controlled trial. Lancet 362(9388):933–940.

GORTEC, 2006 → randomized: CF RT (70 Gy in 7 weeks to the primary tumor and 35 fractions of 2 Gy over 49 days) vs. very accelerated RT (62 to 64 Gy in 31–32 fractions of 2 Gy over 22–23 days; 2 Gy/fraction bid). The very accelerated RT regimen was feasible and provided a major benefit in LRC, but had a modest effect on survival.

Bourhis J et al. (2006) Phase III randomized trial of very accelerated radiation therapy compared with conventional radiation therapy in squamous cell head and neck cancer: a GORTEC trial. J Clin Oncol 24:2873–2878

German Meta-analysis, 2006 → 32 trials with a total of 10,225 patients were included in the meta-analysis. RT combined with simultaneous 5-FU, cisplatin, carboplatin, and mitomycin C as a single drug or combinations of 5-FU with one of the other drugs result in a large survival advantage irrespective of the RT schedule employed. If RT is used as a single modality, HF leads to a significant improvement in OS. Accelerated RT alone, especially when given as split course schedule or extremely accelerated treatments with a decreased total dose increases OS.

Budach W (2006) A meta-analysis of hyperfractionated and accelerated radiotherapy and combined chemotherapy and radiotherapy regimens in unresected locally advanced squamous cell carcinoma of the head and neck. BMC Cancer 6:28

IAEA-ACC study, 2010 → 900 patients with larynx, pharynx, or oral cavity cancer randomized to accelerated RT (ART: 6 fr/week) and CF RT (5 fr/week) to 66–70 Gy in 33–35 fractions. Five-year LRC: 42% vs. 30% in favor of ART. Acute toxicity was higher with ART but late toxicity was similar.

Overgaard J et al. (2010) Five versus six fractions of radiotherapy per week for squamous-cell carcinoma of the head and neck (IAEA-ACC study): a randomised, multicentre trial. Lancet Oncol 11(6):553–60.

GORTEC 99–02, 2012 → 840 patients with stage III-IV larynx, hypopharynx, oropharynx, and oral cavity cancer randomized to conventional 70 Gy RT (in 35 fractions) with concurrent carboplatin and 5-FU vs accelerated 70 Gy RT in 30 fractions (6 weeks) with concurrent carboplatin and 5-FU vs. very accelerated RT alone (64.8 Gy in 18 fractions). Best outcomes and toxicity achieved in conventional RT and chemo. Acute toxicity and feeding tube dependence significantly worse with very accelerated RT.

Bourhis J et al. (2012) Concomitant chemoradiotherapy versus acceleration of radiotherapy with or without concomitant chemotherapy in locally advanced head and neck carcinoma (GORTEC 99–02): an open-label phase 3 randomised trial. Lancet Oncol 13(2):145–53

Chemoradiotherapy trials

Pignon meta-analysis, 2000 → 63 trials (10,741 patients) of locoregional treatment with or without chemotherapy yielded a pooled hazard ratio for death of 0.90 ($p < 0.0001$), corresponding to an absolute survival benefit of 4% at 2 and 5 years in favor of chemotherapy. There was no significant benefit associated with adjuvant or neoadjuvant chemotherapy.

Pignon JP (2000) Chemotherapy added to locoregional treatment for head and neck squamous cell carcinoma: three meta-analyses of updated individual data. Lancet 355 949–955

Ang et al, 2001 → Multi-institutional, prospective, randomized trial. 213 patients with locally advanced oral cavity, oropharynx, larynx, or hypopharynx cancer underwent surgery, and then were randomized based on risk factors (e.g., primary disease site, surgical margin status, perineural invasion, number and location of positive LNs, and ECE) to no adjuvant RT for the low-risk group vs. 57.6 Gy in 6.5 weeks for the intermediate-risk group vs. 63 Gy in 5 weeks (with concomitant boost technique) or 7 weeks for the high-risk group. LRC and OS rates were significantly higher in low- and intermediate-risk compared to high-risk patients. Among high-risk patients, LRC and OS rates were higher with a 5-week RT compared to 7-week RT, although not statistically significant. Prolonged interval between surgery and RT in the 7-week schedule was associated with significantly lower LRC and OS rates and similar toxicity. Cumulative duration of combined therapy had a significant impact on LRC and OS.

Ang KK et al. (2001) Randomized trial addressing risk features and time factors of surgery plus radiotherapy in advanced head-and-neck cancer. Int J Radiat Oncol Biol Phys 51(3):571–8.

Adelstein, 2003 → 259 patients with unresectable head and neck cancer were randomized to arm A (70 Gy in 2 Gy/fr) or arm B (same RT and concurrent cisplatin) or arm C (concurrent cisplatin and 5-FU with split-course RT- 30 Gy given with the first and 30–40 Gy with the third cycle of chemo). Median follow-up: 41 months. Three-year OS: 23% vs. 37% vs. 27%, significantly higher for arm B. Severe toxicity significantly lower in arm A (52% vs. 89% vs. 77%).

Adelstein DJ (2003) An intergroup phase III comparison of standard radiation therapy and two schedules of concurrent chemoradiotherapy in patients with unresectable squamous cell head and neck cancer. J Clin Oncol 21:92–98

SAKK, 2004 → 224 patients with head and neck cancers (excluding nasopharynx and paranasal sinus) were randomized to HF RT (74.4 Gy/1.2 Gy BID) or same RT with concurrent cisplatin (20 mg/m2). LRC and DFS significantly improved with cisplatin. Acute and late toxicity was similar in both arms.

Huguenin P et al (2004) Concomitant cisplatin significantly improves locoregional control in advanced head and neck cancers treated with hyperfractionated radiotherapy. J Clin Oncol 22(23):4665–4673.

MACH-NC meta-analysis, 2011 → 87 randomized trials, 16,192 patients with cancers of oral cavity, oropharynx, hypopharynx, and larynx. Median follow-up: 5.6 years. Adding chemotherapy is beneficial in all tumor locations, and the benefit is higher for concurrent administration. Five-year absolute benefits associated with concurrent chemotherapy is 8.9% for oral cavity 8.1% for oropharynx, 5.4% for larynx, and 4% for hypopharynx.

Blanchard P et al. (2011) Meta-analysis of chemotherapy in head and neck cancer (MACH-NC): A comprehensive analysis by tumour site. Radiother Oncol 100:33–40

Postoperative radiotherapy

Peters et al, 1993 → 240 patients with stage III-IV oral cavity, oropharynx, hypopharynx, or larynx cancer were randomized to postoperative RT dose levels between 52.2 Gy and 68.4 Gy, all in 1.8 Gy-fractions. Failure was significantly higher with ≤ 54 Gy compared to >57.6 Gy. No significant dose response could be demonstrated above 57.6 Gy except for ECE; recurrence rate was significantly higher at 57.6 Gy than at ≥ 63 Gy. The only independent prognostic factor was ECE. Also, at least two of these risk factors increased risk of recurrence: oral cavity primary, positive/close margins, perineural invasion, ≥ 2 positive LNs, largest node >3 cm, treatment delay >6 weeks, and poor PS. Toxicity was higher with ≥63 Gy.

Peters LJ et al (1993) Evaluation of the dose for postoperative radiation therapy of head and neck cancer: First report of a prospective randomized trial. Int J Radiat Oncol Biol Phys 26(1):3-11

RTOG 9501, 2004 → 416 patients with stage III–IV head and neck cancer (43% oropharyngeal) were randomized to RT (60–66 Gy) or RT and concurrent cisplatin. Inclusion criteria were positive surgical margins, ECE, and at least 2 LN involvement. Concurrent postoperative CRT significantly improved LRC and DFS in high-risk patients. OS was comparable. Combined treatment increased adverse effects.

Cooper JS (2004) Postoperative concurrent radiotherapy and chemotherapy for high risk squamous cell carcinoma of the head and neck. N Engl J Med 350:1937–1944

EORTC 22931, 2004 → 334 patients with stage III-IV head and neck cancer (30% oropharyngeal) were randomized to RT (66 Gy) or RT and concurrent cisplatin. Inclusion criteria were positive surgical margins, ECE, PNI, vascular tumor embolism, and level IV-V LN positivity in oral cavity and oropharyngeal cancer. Median follow-up: 60 months. Five-year PFS: 36% vs. 47%, 5-year OS: 40% vs. 53%, 5-year LRR: 31%

vs. 18%, all in favor of combined treatment. Severe adverse effects were more frequent after combined therapy.

Bernier J (2004) Postoperative irradiation with or without concomitant chemotherapy for locally advanced head and neck cancer. N Engl J Med 350:1945–1952

EORTC 22931 and RTOG 9501 combined, 2005 → Patients with positive surgical margins and ECE significantly benefited from concurrent CRT. Patients with stage III-IV disease, PNI, vascular tumor embolism, and level IV-V LNs had a trend for improved outcomes with combined treatment. Concurrent CRT did not lead to an improvement in patients with at least 2 positive LNs.

Bernier J et al (2005) Defining risk levels in locally advanced head and neck cancers: a comparative analysis of concurrent postoperative radiation plus chemotherapy trials of the EORTC (#22931) and RTOG (# 9501). Head Neck 27(10):843–850.

RTOG 9501 update, 2012 → After ten-year follow-up, positive surgical margins, and ECE continues to benefit from combined treatment with higher LRC, DFS, and CSS rates. Trend for improved OS with concurrent chemotherapy. Long-term toxicity is comparable.

Cooper JS et al (2012) Long-term follow-up of the RTOG 9501/intergroup phase III trial: postoperative concurrent radiation therapy and chemotherapy in high-risk squamous cell carcinoma of the head and neck. Int J Radiat Oncol Biol Phys 84(5):1198–1205.

Induction chemotherapy

TAX 323, 2007 → 358 patients with unresectable stage III-IV oral cavity, oropharynx, hypopharynx and larynx cancer were randomized to TPF (docetaxel, cisplatin, 5-FU) vs. PF for four cycles. Patients without progression received RT. Median follow-up: 32.5 months. Median PFS: 11 vs. 8.2 months, median OS: 18.8 vs. 14.5 months, both in favor of TPF. TPF reduced the risk of death 27%. Toxicity was more common with TPF.

Vermorken JB et al. (2007) Cisplatin, Fluorouracil, and Docetaxel in Unresectable Head and Neck Cancer. N Engl J Med 357:1695–1704.

TAX 324, 2011 → 501 patients with unresectable stage III-IV oral cavity, oropharynx, hypopharynx, and larynx cancer were randomized to TPF vs. PF for three cycles. All patients received 70–74 Gy RT with concurrent weekly carboplatin. Median follow-up: 72.2 months. Five-year OS: 52% vs. 42%, median survival: 70.6 vs. 34.8 months, median PFS: 38.1 vs. 13.2 months, all in favor of TPF. No significant difference in toxicity.

Lorch J et al (2011) Long term results of TAX324, a randomized phase III trial of sequential therapy with TPF versus PF in locally advanced squamous cell cancer of the head and neck. Lancet Oncol 12(2):153–9.

PARADIGM, 2013 → 145 patients with stage III-IV oral cavity, oropharynx, hypopharynx, and larynx cancer were randomized to Arm A (induction chemo with docetaxel, cisplatin, 5-FU for three cycles. Then to Arm A1-poor responders received 72 Gy accelerated RT and concurrent weekly docetaxel or Arm A2-favorable responders received conventional 70 Gy RT and concurrent weekly carboplatin) vs. Arm B (72 Gy accelerated RT with concurrent three weekly cisplatin without induction). No difference in survival (73% vs. 78%) but more adverse events with induction chemo.

Haddad R et al (2013) Induction chemotherapy followed by concurrent chemoradiotherapy (sequential chemoradiotherapy) versus concurrent chemoradiotherapy alone in locally advanced head and neck cancer (PARADIGM): a randomized phase 3 trial. Lancet Oncol 14:257–64.

DeCIDE trial, 2014 → 285 patients with stage III-IV oral cavity, oropharynx, hypopharynx, and larynx cancer (all with N2-3 disease) were randomized to induction chemo with docetaxel, cisplatin, and 5-FU followed by 74-75 Gy RT in 1.5 Gy BID fractions vs. concurrent docetaxel, 5-FU, and hydroxyurea with same RT. No difference in DFS, RFS, and OS but more adverse events with induction chemo.

Cohen E et al (2014) Phase III randomized trial of induction chemotherapy in patients with N2 or N3 locally advanced head and neck cancer. J Clin Oncol. 32:2735–43.

5.11 Pearl Boxes

Regional Lymph Nodes (N) (Fig. 5.51)
- *N0*: no regional lymph node metastasis
- *N1*: metastasis in a single ipsilateral lymph node ≤3 cm in greatest dimension
- *N2a*: metastasis in a single ipsilateral lymph node >3 cm but ≤6 cm in greatest dimension
- *N2b*: metastasis in multiple ipsilateral lymph nodes ≤6 cm in greatest dimension
- *N2c*: metastasis in bilateral or contralateral lymph nodes ≤6 cm in greatest dimension
- *N3*: metastasis in a lymph node > 6 cm in greatest dimension

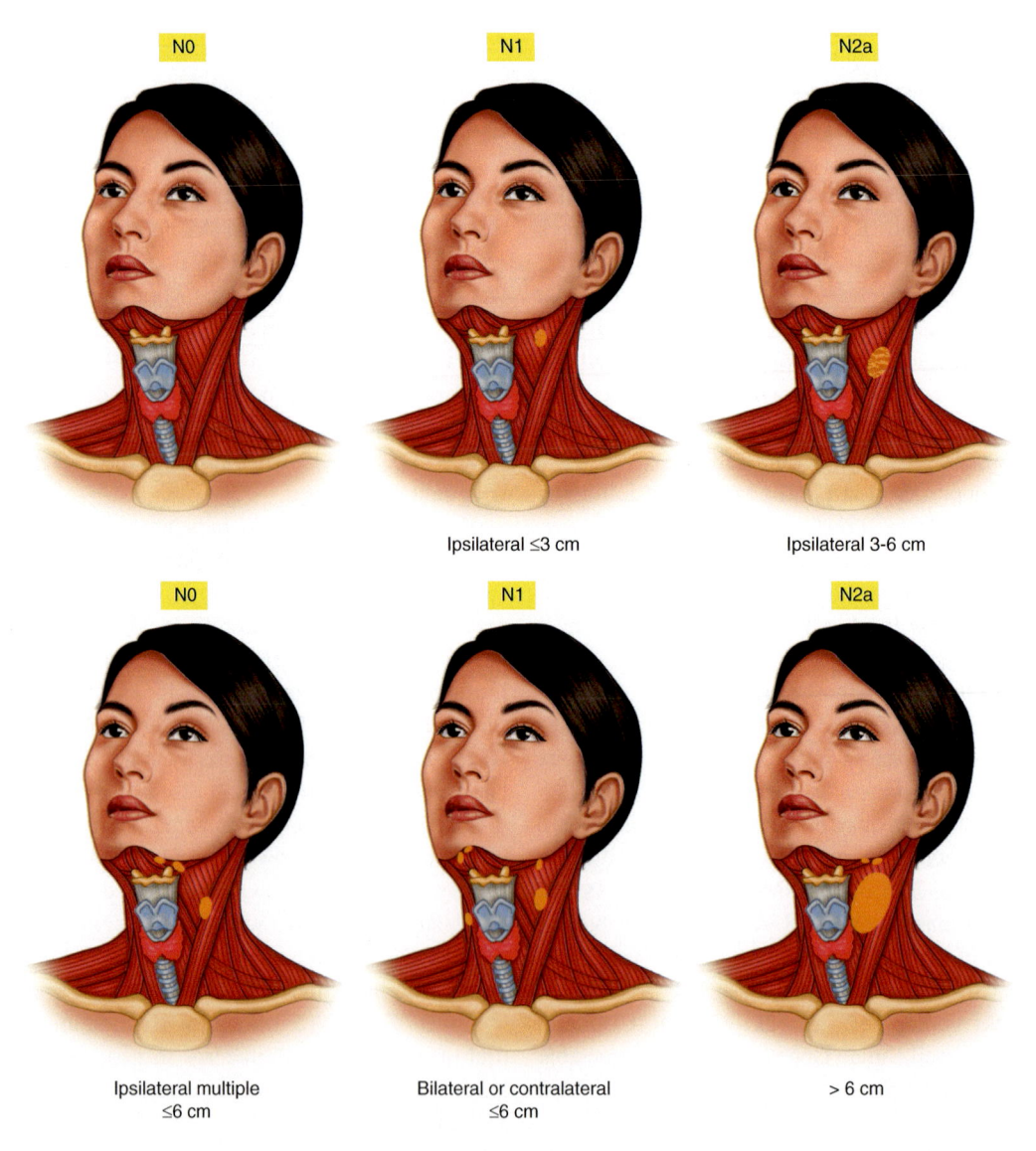

Fig. 5.51 N staging for head and neck cancers, excluding nasopharyngeal and thyroid cancers (from [20], reproduced with the permission from the American Joint Committee on Cancer)

5.11 Pearl Boxes

Recurrence Factors in Head–Neck Cancers After Surgery

- Close or positive surgical margins
- Perineural invasion
- Lymphovascular space invasion
- Involvement of two or more LNs
- LN >3 cm
- Extracapsular nodal extension
- 6 weeks after completion of surgery

Head–Neck Cancers Not Requiring Elective Neck RT

- T1–2 glottic laryngeal cancers
- Oral cavity cancers (T1 hard palate, gingiva, superficial oral tongue, buccal mucosa, T1–2 lip)
- Maxillary sinus cancers other than those with SCC
- Early-stage T1–2 nasal cavity cancers
- Low-grade adenoid cystic tumors
- N0 or N1 neck with no ECE after neck dissection

Level	Superior	Caudal	Anterior	Posterior	Lateral	Medial
Ia	Geniohyoid muscle plane tangent to basilar edge of mandible	Plane tangent to body of hyoid bone	Symphysis menti, platysma muscle	Body of hyoid bone	Medial edge of anterior belly of digastric muscle	–
Ib	Mylohyoid muscle cranial edge of submandibular gland	Plane through central part of hyoid bone	Symphysis menti, platysma muscle	Posterior edge of submandibular gland	Basilar edge/ innerside of mandible, platysma muscle skin	Lateral edge of anterior belly of digastric muscle
IIa	Caudal edge of lateral process of C1	Caudal edge of the body of hyoid bone	Posterior edge of submandibular gland; anterior edge of internal carotid artery; posterior edge of posterior belly of digastric muscle	Posterior border of internal jugular vein	Medial edge of sternocleidomatoid	Medial edge of internal carotid artery, paraspinal (levator scapulae) muscle
IIb	Caudal edge of lateral process of C1	Caudal edge of the body of hyoid bone	Posterior border of internal jugular vein	Posterior border of the sternocleidomastoid muscle	Medial edge of sternocleidomastoid	Medial edge of internal carotidartery, paraspinal (levator scapulae) muscle
III	Caudal edge of the body of hyoid bone	Caudal edge of cricoid cartilage	Posterolateral edge of the sternohyoid muscle anterioredge of sternocleidomastoid muscle	Posterior edge of the sternocleidomastoid muscle	Medial edge of sternocleidomastoid	Internal edge of carotid artery, paraspinal (scalenus) muscle
IV	Caudal edge of cricoid cartilage	2 cm cranial to sternoclavicular joint	Anteromedial edge of sternocleidomastoid muscle	Posterior edge of the sternocleidomastoid muscle	Medial edge of sternocleidomastoid	Medial edge of internal carotid artery, paraspinal (scalenus) muscle
V	Cranial edge of body of hyoid bone	CT slice encompassing the transverse cervical vessels	Posterior edge of the sternocleidomastoid muscle	Anterior border of the trapezius muscle	Platysma muscle skin	Paraspinal (levator scapulae, splenius capitis) muscle
VI	Caudal edge of body of thyroid cartilage	Sternal manubrium	Skin, platysma muscle	Separation between trachea and esophagus	Medial edges of thyroid gland, skin and anterior–medial edge of sternocleidomastoid muscle	–
Retro pharyngeal	Base of skull	Cranial edge of the body of hyoid bone	Fascia under the pharyngeal mucosa	Prevertebral muscle (longus colli, longus capitis)	Medial edge of the internal carotid artery	Midline

M muscle

Table 5.11

References

1. Head and neck cancer (2003). Springer, Berlin
2. Wannenmacher MDJ, Wenz F (2006) Strahlentherapie. Springer, Berlin
3. Schwartz LH, Ozsahin M, Zhang GN, Touboul E, De Vataire F, Andolenko P, Lacau-Saint-Guily J, Laugier A, Schlienger M (1994) Synchronous and metachronous head and neck carcinomas. Cancer 74(7):1933–1938. https://doi.org/10.1002/1097-0142(19941001)74:7<1933::aid-cncr2820740718>3.0.co;2-x
4. Albanes D, Heinonen OP, Taylor PR, Virtamo J, Edwards BK, Rautalahti M, Hartman AM, Palmgren J, Freedman LS, Haapakoski J, Barrett MJ, Pietinen P, Malila N, Tala E, Liippo K, Salomaa ER, Tangrea JA, Teppo L, Askin FB, Taskinen E, Erozan Y, Greenwald P, Huttunen JK (1996) Alpha-tocopherol and beta-carotene supplements and lung cancer incidence in the alpha-tocopherol, beta-carotene cancer prevention study: effects of base-line characteristics and study compliance. J Natl Cancer Inst 88(21):1560–1570. https://doi.org/10.1093/jnci/88.21.1560
5. Mellin H, Friesland S, Lewensohn R, Dalianis T, Munck-Wikland E (2000) Human papillomavirus (HPV) DNA in tonsillar cancer: clinical correlates, risk of relapse, and survival. Int J Cancer 89(3):300–304
6. Robbins KT, Clayman G, Levine PA, Medina J, Sessions R, Shaha A, Som P, Wolf GT, American Head and Neck Society; American Academy of Otolaryngology—Head and Neck Surgery (2002) Neck dissection classification update: revisions proposed by the American Head and Neck Society and the American Academy of Otolaryngology-Head and Neck Surgery. Arch Otolaryngol Head Neck Surg 128(7):751–758. https://doi.org/10.1001/archotol.128.7.751
7. Gregoire V, Ang K, Budach W, Grau C, Hamoir M, Langendijk JA, Lee A, Le QT, Maingon P, Nutting C, O'Sullivan B, Porceddu SV, Lengele B (2014) Delineation of the neck node levels for head and neck tumors: a 2013 update. DAHANCA, EORTC, HKNPCSG, NCIC CTG, NCRI, RTOG, TROG consensus guidelines. Radiother Oncol 110(1):172–181. https://doi.org/10.1016/j.radonc.2013.10.010
8. Delaere P (2006) Clinical and endoscopic examination of the head and neck. In: Hermans R (ed) Head and neck cancer imaging. Springer, Berlin, p 18
9. Lindberg R (1972) Distribution of cervical lymph node metastases from squamous cell carcinoma of the upper respiratory and digestive tracts. Cancer 29(6):1446–1449. https://doi.org/10.1002/1097-0142(197206)29:6<1446::aid-cncr2820290604>3.0.co;2-c
10. Chao KS, Ozyigit G, Tran BN, Cengiz M, Dempsey JF, Low DA (2003) Patterns of failure in patients receiving definitive and postoperative IMRT for head-and-neck cancer. Int J Radiat Oncol Biol Phys 55(2):312–321. https://doi.org/10.1016/s0360-3016(02)03940-8
11. Butler EB, Teh BS, Grant WH 3rd, Uhl BM, Kuppersmith RB, Chiu JK, Donovan DT, Woo SY (1999) Smart (simultaneous modulated accelerated radiation therapy) boost: a new accelerated fractionation schedule for the treatment of head and neck cancer with intensity modulated radiotherapy. Int J Radiat Oncol Biol Phys 45(1):21–32. https://doi.org/10.1016/s0360-3016(99)00101-7
12. Eisbruch A, Harris J, Garden AS, Chao CK, Straube W, Harari PM, Sanguineti G, Jones CU, Bosch WR, Ang KK (2010) Multi-institutional trial of accelerated hypofractionated intensity-modulated radiation therapy for early-stage oropharyngeal cancer (RTOG 00-22). Int J Radiat Oncol Biol Phys 76(5):1333–1338. https://doi.org/10.1016/j.ijrobp.2009.04.011
13. Lee N, Xia P, Quivey JM, Sultanem K, Poon I, Akazawa C, Akazawa P, Weinberg V, Fu KK (2002) Intensity-modulated radiotherapy in the treatment of nasopharyngeal carcinoma: an update of the UCSF experience. Int J Radiat Oncol Biol Phys 53(1):12–22. https://doi.org/10.1016/s0360-3016(02)02724-4
14. Kam MK, Leung SF, Zee B, Chau RM, Suen JJ, Mo F, Lai M, Ho R, Cheung KY, Yu BK, Chiu SK, Choi PH, Teo PM, Kwan WH, Chan AT (2007) Prospective randomized study of intensity-modulated radiotherapy on salivary gland function in early-stage nasopharyngeal carcinoma patients. J Clin Oncol 25(31):4873–4879. https://doi.org/10.1200/JCO.2007.11.5501
15. Pow EH, Kwong DL, McMillan AS, Wong MC, Sham JS, Leung LH, Leung WK (2006) Xerostomia and quality of life after intensity-modulated radiotherapy vs. conventional radiotherapy for early-stage nasopharyngeal carcinoma: initial report on a randomized controlled clinical trial. Int J Radiat Oncol Biol Phys 66(4):981–991. https://doi.org/10.1016/j.ijrobp.2006.06.013
16. CT-based delineation of lymph node levels in the N0 neck: DAHANCA, EORTC, GORTEC, RTOG consensus guidelines. www.rtog.org. Accessed 16 Mar 2021
17. Moore KL (2006) Clinically oriented anatomy, 5th edn. Lippincott Williams & Wilkins, Philadelphia
18. Greene FL, Fleming ID et al (2002) American Joint Committee on Cancer. AJCC cancer staging manual, 6th edn. Springer, New York
19. El-Naggar ACJ, Grandis J et al (2017) WHO classification of head and neck tumours. IARC Press, Lyon
20. Compton CC, Byrd DR, Garcia-Aguilar J, Kurtzman SH, Olawaiye A, Washington MK (2012) Pharynx. In: Compton C, Byrd D, Garcia-Aguilar J, Kurtzman S, Olawaiye A, Washington M (eds) AJCC cancer staging atlas. Springer, New York. https://doi.org/10.1007/978-1-4614-2080-4_4
21. National Comprehensive Cancer Network (NCCN) www.nccn.org/professionals/physician_gls/PDF/head-and-neck.pdf. Accessed March 2021

22. Broglie MA, Stoeckli SJ, Sauter R, Pasche P, Reinhard A, de Leval L, Huber GF, Pezier TF, Soltermann A, Giger R, Arnold A, Dettmer M, Arnoux A, Muller M, Spreitzer S, Lang F, Lutchmaya M, Stauffer E, Espeli V, Martucci F, Bongiovanni M, Foerbs D, Jochum W (2017) Impact of human papillomavirus on outcome in patients with oropharyngeal cancer treated with primary surgery. Head Neck 39(10):2004–2015. https://doi.org/10.1002/hed.24865

23. Ang KK, Harris J, Wheeler R, Weber R, Rosenthal DI, Nguyen-Tan PF, Westra WH, Chung CH, Jordan RC, Lu C, Kim H, Axelrod R, Silverman CC, Redmond KP, Gillison ML (2010) Human papillomavirus and survival of patients with oropharyngeal cancer. N Engl J Med 363(1):24–35. https://doi.org/10.1056/NEJMoa0912217

24. O'Sullivan B, Huang SH, Siu LL, Waldron J, Zhao H, Perez-Ordonez B, Weinreb I, Kim J, Ringash J, Bayley A, Dawson LA, Hope A, Cho J, Irish J, Gilbert R, Gullane P, Hui A, Liu FF, Chen E, Xu W (2013) Deintensification candidate subgroups in human papillomavirus-related oropharyngeal cancer according to minimal risk of distant metastasis. J Clin Oncol 31(5):543–550. https://doi.org/10.1200/JCO.2012.44.0164

25. Popescu CR, Bertesteanu SV, Mirea D, Grigore R, Lonescu D, Popescu B (2010) The epidemiology of hypopharynx and cervical esophagus cancer. J Med Life 3(4):396–401

26. Ribeiro KB, Levi JE, Pawlita M, Koifman S, Matos E, Eluf-Neto J, Wunsch-Filho V, Curado MP, Shangina O, Zaridze D, Szeszenia-Dabrowska N, Lissowska J, Daudt A, Menezes A, Bencko V, Mates D, Fernandez L, Fabianova E, Gheit T, Tommasino M, Boffetta P, Brennan P, Waterboer T (2011) Low human papillomavirus prevalence in head and neck cancer: results from two large case-control studies in high-incidence regions. Int J Epidemiol 40(2):489–502. https://doi.org/10.1093/ije/dyq249

27. Byers RM, Wolf PF, Ballantyne AJ (1988) Rationale for elective modified neck dissection. Head Neck Surg 10(3):160–167. https://doi.org/10.1002/hed.2890100304

28. NCI (2017) SEER cancer stat facts. Oral cavity and pharynx cancer. http://www.seer.cancer.gov/statfacts/html/oralcav.html. Accessed 24 Mar 2021

29. Sessions DG, Spector GJ, Lenox J, Parriott S, Haughey B, Chao C, Marks J, Perez C (2000) Analysis of treatment results for floor-of-mouth cancer. Laryngoscope 110(10 Pt 1):1764–1772. https://doi.org/10.1097/00005537-200010000-00038

30. Sessions DG, Spector GJ, Lenox J, Haughey B, Chao C, Marks J (2002) Analysis of treatment results for oral tongue cancer. Laryngoscope 112(4):616–625. https://doi.org/10.1097/00005537-200204000-00005

31. Million RR (1994) Management of head and neck cancer: a multidisciplinary approach, 2nd edn. JB Lippincott, Philadelphia

32. Katz TS, Mendenhall WM, Morris CG, Amdur RJ, Hinerman RW, Villaret DB (2002) Malignant tumors of the nasal cavity and paranasal sinuses. Head Neck 24(9):821–829. https://doi.org/10.1002/hed.10143

33. Monroe AT, Hinerman RW, Amdur RJ, Morris CG, Mendenhall WM (2003) Radiation therapy for esthesioneuroblastoma: rationale for elective neck irradiation. Head Neck 25(7):529–534. https://doi.org/10.1002/hed.10247

34. Mendenhall WM, Morris CG, Amdur RJ, Werning JW, Hinerman RW, Villaret DB (2004) Radiotherapy alone or combined with surgery for adenoid cystic carcinoma of the head and neck. Head Neck 26(2):154–162. https://doi.org/10.1002/hed.10380

35. Cancer: principles and practice of oncology (2005). 7th edn. Lippincott Williams & Wilkins, Philadelphia

36. Burch HB (1995) Evaluation and management of the solid thyroid nodule. Endocrinol Metab Clin North Am 24(4):663–710

37. Ron E, Lubin JH, Shore RE, Mabuchi K, Modan B, Pottern LM, Schneider AB, Tucker MA, Boice JD Jr (1995) Thyroid cancer after exposure to external radiation: a pooled analysis of seven studies. Radiat Res 141(3):259–277

38. Amdur RJ DR (2019) Thyroid cancer. In: Halperin EC WD, Perez CA, Brady LW (ed) Principles and practice of radiation oncology. 7th edn. Wolters Kluwer, China, pp 3504–3551

Lung Cancer

6.1 Introduction

Lung cancer is the malignant transformation of lung tissue. The term "lung cancer" or "bronchogenic carcinoma" refers to tumors that develop from bronchi, bronchioles, lung, or tracheal parenchyma. Approximately 1.8 million people are diagnosed with lung cancer and approximately 1.6 million people die due to lung cancer annually [1]. Lung cancer is the most frequent cancer in males, and its incidence is rapidly increasing in females.

The trachea starts just after the larynx, and is 12 cm in length. It extends from the level of the C6 vertebra to the T4 vertebra, where it divides into two main bronchi (carina level). Its posterior wall is flat and has a membranous structure. Its other walls consist of serial half-ring-shaped cartilages with muscular and membranous structure. Tracheal mucosa is lined with ciliary epithelial cells to clear away small foreign bodies that come in with the air.

The right and left bronchi have similar structures with the trachea, but their cartilages are more irregular. The right main bronchus is shorter, thicker, and more vertical, while the left main bronchus is thinner, longer, and more horizontal. The bronchi divide into thin bronchioles.

There are two lungs in the thorax cavity, the right and left lungs (Fig. 6.1a–c). The right lung has three and the left lung has two lobes. Each lung has a half conical base at the diaphragm. They have outer and inner faces, one base, and one apex. Structures entering and exiting the lungs pass from the middle part of the inner face (hilar region), which includes the main arteries, veins, lymphatics, and the nerves of the lungs.

The bronchioles end with small saccules known as alveoli, which are filled with air. Small capillaries spread across the walls of these saccules.

The pleura is a serous membrane that covers the outer part of the lung together with the upper part of the diaphragm. There is a region called the pleural space between the inner and outer sheets that is filled with fluid. This helps the lungs to inflate and deflate during respiration.

- Nonsmall cell lung cancer (NSCLC) and small cell lung cancer (SCLC) account for 95% of all lung cancers.
- Smoking is the primary risk factor for development of lung cancer and 90% of patients with lung cancer have history of smoking [2].
- For NSCLC, brain metastasis is more frequent in adenocarcinoma and less frequent for squamous cell carcinoma. Risk of brain metastasis increases as the tumor size and lymph node metastases increase [3].
- Approximately 10% of patients with SCLC suffer from syndrome of inappropriate antidiuretic hormone secretion (SIADH) [4]. These patients frequently have symptoms of nausea

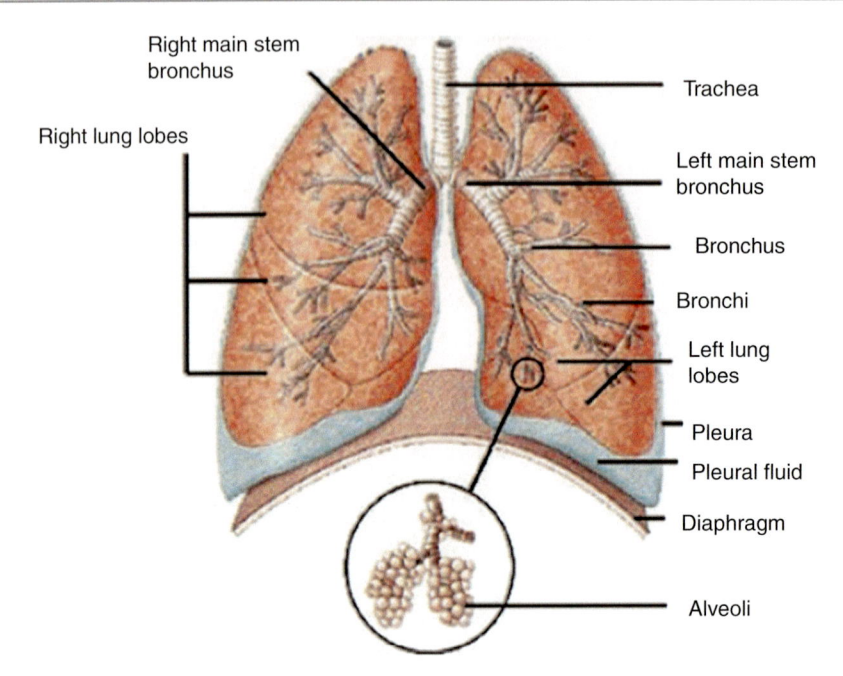

Fig. 6.1 Anatomy of the trachea and lobar bronchi; also the segmental bronchi are represented in the picture. T trachea, RUL segmental bronchi of the right upper lobe, ML segmental bronchi of the middle lobe, RLL segmental bronchi of the right lower lobe, LUL segmental bronchi of the left upper lobe, LLL segmental bronchi of the left lower lobe. Anatomy of the lung segments, frontal and lateral view. S1 apical segment of the right upper lobe, S1+2 apical-posterior segment of the left upper lobe, S2 posterior segment of the right upper lobe, S3 anterior segment of the (right or left) upper lobe, S4 on the right, lateral segment of the middle lobe and, on the left, lingular upper segment, S5 medial segment of the middle lobe and, on the left, inferior ligular segment, S6 apical segment of the (right or left) lower lobe, S 7 , 8 , 9 , 10 on the right, respectively, medial basal segment, anterior basal, lateral basal, and posterior basal segment of the lower lobe, S 7 + 8 , 9 , 10 on the left, anteromedial basal segment, lateral basal, and posterior basal segment of the lower lobe. IVC inferior vena cava, Ao aorta, T trachea. The arrow shows the right major fissure and the left one; the arrowheads show the small left fissure (Atlas of Imaging Anatomy, Lucio Olivetti (2015), p 90–92)

and vomiting. Other symptoms may also develop due to cerebral edema.
- Approximately 20% of patients with NSCLC have bone metastases at initial diagnosis [5].
- Adrenal masses are detected in approximately 10% of patients with NSCLC; however, one-fourth of these adrenal masses are due to metastasis from NSCLC [6].
- Hypercalcemia in lung cancer patients may occur more frequently due to bone metastases and less frequently due to release of parathyroid hormone-related protein (PTH-rP), calcitriol, or other osteoclast activating cytokines. Symptoms of hypercalcemia include anorexia, nausea, vomiting, constipation, lethargy, polyuria, polydipsia, and dehydration. Confusion and coma may be observed at later stages due to renal failure and nephrocalcinosis.
- Lambert-Eaton myasthenic syndrome (LEMS) is the most frequent neurological paraneoplastic syndrome and may be seen in approximately 1–3% of patients with SCLC [7]. Majority of patients with LEMS suffer from slowly progressing proximal muscle weakness.
- Horner syndrome may be seen in lung cancer patients as a neurological paraneoplastic syndrome with symptoms of ipsilateral ptosis, miosis, and anhidrosis due to involvement of inferior cervical sympathetic chain or the stellate ganglion.
- Pancoast syndrome is a neurological paraneoplastic syndrome seen typically in superior sulcus tumors and squamous cell subtype of NSCLC; it is rare in SCLC. Symptoms include

pain in shoulder and arms, weakness and atrophy of hand muscles, and Horner syndrome.
- Superior vena cava syndrome (SVCS) may occur more frequently in SCLC due to obstruction of vena cava superior. Patients typically present with dyspnea, facial, and neck swelling. Cough, pain, and dysphagia are less frequent symptoms. Physical findings may include distended neck veins, significant venous pattern over the chest, facial edema, and plethoric appearance. Chest X-ray typically shows right hilar and mediastinal widening due to mass.

Principles of Diagnosis and Treatment for Lung Cancer
- The initial step includes verification of malignancy diagnosis of SCLC, NSCLC, or other subtypes along with thorough evaluation of disease stage and patient's performance status.
- Staging is based on the TNM classification of AJCC for NSCLC; however, Veterans Administration Lung Study Group (VALSG) staging with limited stage (confined to one hemithorax) and extensive stage (extending beyond one hemithorax) classification may also be utilized for staging of SCLC [8].
- Treatment of lung cancer may be associated with significant toxicity. In this context, patient's performance status may affect treatment method and purpose.
- Performance status may be assessed by Karnofsky Performance Status (KPS) and Eastern Cooperative Oncology Group Performance Scale (ECOG-PS) [9].
- Evaluation of a patient with suspected lung cancer should typically include assessment of paraneoplastic syndromes, lymphadenopathies in the neck and axillary region, and weight loss. Neurological examination should be performed due to the risk of brain metastasis.
- Patients with advanced-stage disease at initial imaging workup who are deemed ineligible for surgical resection should undergo biopsy before the commencement of therapy. Selection of biopsy technique is dependent on the patient's performance status, disease extent, accessibility of tumor, and experience of the surgeon.
- Acquisition of adequate tissue is crucial for testing molecular biomarkers which may be utilized in systemic treatment.
- Thoracentesis is indicated for the assessment of malignant dissemination in the presence of pleural effusion.
- Baseline evaluation of pulmonary function tests should be performed for the planning of surgery and radiotherapy.
- Smoking cessation counseling and treatment should be performed for active smokers.

Imaging
- Chest radiograph may fail to diagnose subcentimetric lung tumors.
- Contrast-enhanced computed tomography (CT) scanning which is capable of detecting subcentimetric lesions is the most frequently used imaging method for lung cancer diagnosis. Specific criteria may improve identification of tumors with high malignancy potential.
  - Progression of tumor size is an important factor. Masses with doubling times shorter than 400 days are considered suspectful [10].
  - Spiculated lesions are usually adenocarcinomas and may behave more aggressively.
  - Pulmonary nodules smaller than 3 centimeters with >50% ground-glass opacity have low risk of vessel invasion and mediastinal lymph node metastasis. These nodules are less metabolic on positron emission tomography (PET) imaging and have a more favorable diagnosis compared to pulmonary nodules with solid component of >50%.
- Abdominal CT, brain magnetic resonance imaging (MRI), bone scanning, and PET-CT may be used for the detection of metastases.

Mediastinal Staging
- Mediastinal staging is indispensable for early-stage lung cancer patients considered for surgical resection.
- Endoscopic ultrasound is sensitive for the assessment of nodal disease. Nodal stations

including 2R/2L, 4R/4L, 7, and 10 may be assessed by endobronchial ultrasound (EBUS). Esophageal ultrasound may assist in the evaluation and sampling of fourth, seventh, eighth, and ninth nodal stations [11].
- Bronchoscopic deep biopsy is usually required for SCLC due to typical submucosal disease location.

PET/CT
- PET/CT is a relatively newer method for staging of lung cancer. The most commonly used tracer is 18F-fluoro-2-deoxyglucose (FDG) and it accumulates in tumors with increase in metabolic activity.
    - PET may detect the disease independently from the anatomy as it provides functional measurement results; however, false-negative findings may be a concern since FDG is not only a cancer-specific imaging agent.

Video-Assisted Thoracic Surgery (VATS)
- VATS offers an alternative to anterior mediastinotomy for assessment of fifth and sixth nodal stations.
- It also provides access to eighth and ninth nodal stations.
- Right subcarinal and paratracheal lymph nodes may be assessed by VATS.
- VATS may be used for the evaluation of pleural cavity and primary tumor (Table 6.1).

6.1.1 Nonsmall Cell Lung Cancer (NSCLC)

NSCLC account for 75–80% of all lung cancers.

6.1.1.1 Pathology
Three Major Subtypes of NSCLC [12]
- Adenocarcinoma
- Squamous cell carcinoma
- Large cell carcinoma

Adenocarcinoma

- Forty percent of all lung cancers in the USA are adenocarcinomas [12].

Table 6.1 Staging workup

Test	comment
Thorax CT	• Provides information about lesion size, lymphadenopathies, and pleural effusion • Should encompass the liver and adrenals
PET–CT	• Has a sensitivity of 97% and specificity of 78% for assessment of pulmonary nodules • Enables evaluation of mediastinal lymphadenopathies • Has poor utility for lesion size <10 mm
Brain MRI	• Indicated for patients with neurological symptoms • Suggested for presurgical evaluation in patients with >T3 tumors or nodal involvement
Thorax	• May be used for anatomical characterization of certain tumors at presurgical assessment • May assist in evaluating invasion of the aorta and brachial plexus
Bone scan	• Indicated for patients with symptoms of bone metastases • Should be omitted if PET-CT is performed
Pulmonary function tests	• Aids in assessment of pulmonary reserve for patients planned to undergo surgical resection

- This is the most frequent lung cancer in females, and its incidence is increasing.
- Generally starts from the peripheral parts of the lung.

Squamous cell carcinoma

- Also known as epidermoid cancer.
- Constitutes 30–35% of all lung cancers in the USA [12].
- Commonly seen in males and older people.
- Generally starts at main bronchi.
- Has a tendency to grow relatively slowly.
- Has no tendency for early metastasis.
- Highly correlated with cigarette smoking.

Large cell carcinoma

- Constitutes 5–15% of all lung cancers in the USA [12].

6.1 Introduction

Table 6.2 Prevalence of NSCLC and SCLC

Histology	Prevalence	Comments
NSCLC		
Adenocarcinoma	40–50%	– Positive immunohistochemistry staining on thyroid transcription factor 1 (TTF-1), Napsin-A – Most common NSCLC
Squamous cell carcinoma	20–30%	– Positive immunohistochemistry staining on p63, p40
Large cell carcinoma	5–10%	– Neuroendocrine morphology and differentiation
NSCLC-Other	5–10%	
SCLC	10–15%	– Positive immunohistochemistry staining on neuroendocrine markers

- Generally starts in the small bronchioles.
- May appear in any part of the lungs.
- Is generally large in size at diagnosis.
- Has a tendency to metastasize to the mediastinum and the brain.
- Is an undifferentiated carcinoma without squamous, glandular, or small cell differentiation histologically (Table 6.2).
- World Health Organization (WHO) has made several changes in the classification of lung cancer in 2015. In contrast with prior classification systems, 2015 WHO classification is more dependent on immunohistochemical characterization in addition to light microscopy [13].

6.1.1.2 Major Changes in WHO 2015 Lung Cancer Classification:

- Adenocarcinoma
  - Bronchoalveolar carcinoma description has been abandoned.
  - Mixed adenocarcinoma description has been abandoned.
  - Mucinous cystadenocarcinoma description has been abandoned and these tumors have been redescribed as colloid adenocarcinoma.
  - Adenocarcinoma in situ and minimally invasive adenocarcinoma descriptions have been added.
- Squamous cell carcinoma
  - Changes have been made in papillary, clear cell, small cell, and basaloid carcinoma.
  - Clear cell has been regarded as a cytologic feature rather than a subtype.
- Large cell carcinoma
  - Basaloid carcinoma has been moved and became a subtype of squamous cell carcinoma.
  - Large cell neuroendocrine carcinoma has been moved to neuroendocrine carcinoma.
  - Lymphoepithelioma like carcinoma has been moved to other and undefined carcinomas.
  - Clear cell and rhabdoid cell have been excluded from subtypes and regarded as descriptive features.
- Sarcomatoid carcinoma
  - Molecular testing has been suggested for diagnosis.
- Neuroendocrine carcinoma
  - Small and large cell carcinomas which were formerly described as discrete groups have been united in the neuroendocrine carcinomas group.

6.1.1.3 General Presentation

Most patients have nonspecific symptoms or present no signs until the disease has progressed significantly. Consequently, only a small portion of lung cancers are detected early, when curative treatment has the greatest chance of success.

Symptoms and Signs
- Cough
- Dyspnea
- Fatigue
- Chest, shoulder, arm, or back pain
- Recurrent pneumonia
- Hemoptysis
- Anorexia and weight loss
- General pain
- Hoarseness
- Wheezing
- Swelling in face or neck

Sometimes the symptoms of lung cancer may not be related to the lungs or respiration. Since lung cancer is usually diagnosed at an advanced stage, the primary cancer may have spread to other parts of the body. Therefore, symptoms like headache, bone fractures, thrombosis, and fatigue, related to the affected organs, may be seen.

Paraneoplastic syndromes
- Gynecomastia (generally in large cell carcinoma).
- Hypercalcemia (generally in squamous cell carcinoma).
- Hypertrophic pulmonary osteoarthropathy (generally in adenocarcinoma).
- Inappropriate ADH secretion is most frequently observed in SCLC.
- *Pancoast syndrome.* This is observed in superior sulcus tumors. It is characterized by lower brachial plexopathy, Horner syndrome, and pain in shoulder and ulnar site of arm.

Horner syndrome. This characterized by enophthalmos, ptosis, miosis, ipsilateral sweating loss, and hoarseness due to laryngeal nerve invasion (Figs. 6.2, 6.3, and 6.4).

6.1.1.4 Staging

AJCC/UICC Eighth Edition [14]
AJCC Staging
- Tx: Primary tumor cannot be assessed or tumor proven by the presence of malignant cells in sputum or bronchial washings but not visualized by imaging or bronchoscopy
- T0: No evidence of primary tumor
- Tis: Carcinoma in situ
- T1: A tumor that is 3 cm or smaller in greatest dimension is surrounded by lung or visceral pleura, and is without bronchoscopic evidence of invasion more proximal than the lobar bronchus (i.e., not in the main bronchus)
  - T1a (mi): Minimally invasive adenocarcinoma
  - T1a: Tumor ≤1 cm in greatest dimension
  - T1b: Tumor >1 cm but ≤2 cm in greatest dimension
  - T1c: Tumor >2 cm but ≤3 cm in greatest dimension
- T2: A tumor > 3 cm but ≤ 5 cm or tumor with any of the following features:
  - Involves main bronchus regardless of distance from the carina but without the involvement of the carina
  - Invades the visceral pleura
  - Associated with atelectasis or obstructive pneumonitis that extends to the hilar region, involving part or all of the lung
  - T2a: Tumor >3 cm but ≤4 cm in greatest dimension
  - T2b: Tumor >4 cm but ≤5 cm in greatest dimension
- T3: A tumor > 5 cm but ≤ 7 cm in greatest dimension or associated with separate tumor nodule(s) in the same lobe as the primary tumor or directly invades any of the following structures: chest wall (including the parietal

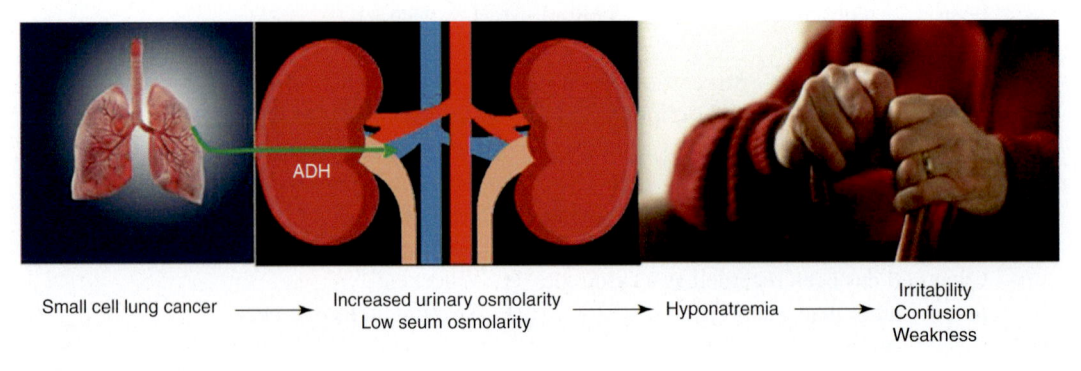

Small cell lung cancer → Increased urinary osmolarity Low seum osmolarity → Hyponatremia → Irritability Confusion Weakness

Fig. 6.2 Pathophysiology of SIADH

pleura and superior sulcus tumors), phrenic nerve, and parietal pericardium
- T4: A tumor > 7 cm in greatest dimension or associated with separate tumor nodule(s) in a different ipsilateral lobe than that of the primary tumor or invades any of the following structures: diaphragm, mediastinum, heart, great vessels, trachea, recurrent laryngeal nerve, esophagus, vertebral body, and carina
- Nx: Regional lymph nodes cannot be assessed
- N0: No regional lymph node metastasis
- N1: Metastasis in ipsilateral peribronchial and/or ipsilateral hilar lymph nodes and intrapulmonary nodes, including involvement by direct extension
- N2: Metastasis in ipsilateral mediastinal and/or subcarinal lymph node(s)
- N3: Metastasis in contralateral mediastinal, contralateral hilar, ipsilateral or contralateral scalene, or supraclavicular lymph node(s)
- M0: No distant metastasis
- M1: Distant metastasis present
  – M1a: Separate tumor nodule(s) in a contralateral lobe; tumor with pleural or pericardial nodule(s) or malignant pleural or pericardial effusions
  – M1b: Single extrathoracic metastasis
  – M1c: Multiple extrathoracic metastases in one or more organs

N2 lymph nodes

1. upper mediastinal
2. upper paratracheal
3. prevascular (3a) and retrotracheal (3p)
4. lower paratracheal
5. subaortic
6. paraaortic
7. subcarinal
8. paraesophageal
9. pulmonary ligament

N1 lymph nodes

10. hilar
11. interlobar
12. lobar
13. segmental
14. subsegmental (Figs. 6.5 and 6.6, Table 6.3)

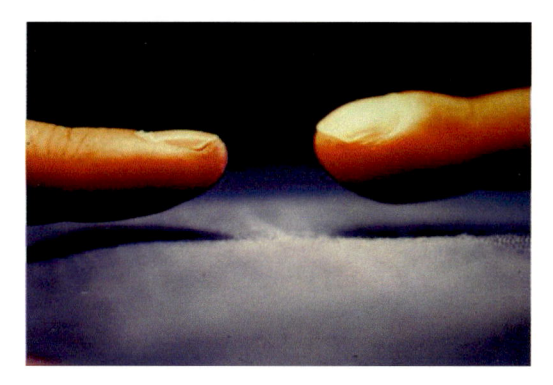

Fig. 6.3 Hypertrophic osteoarthropathy

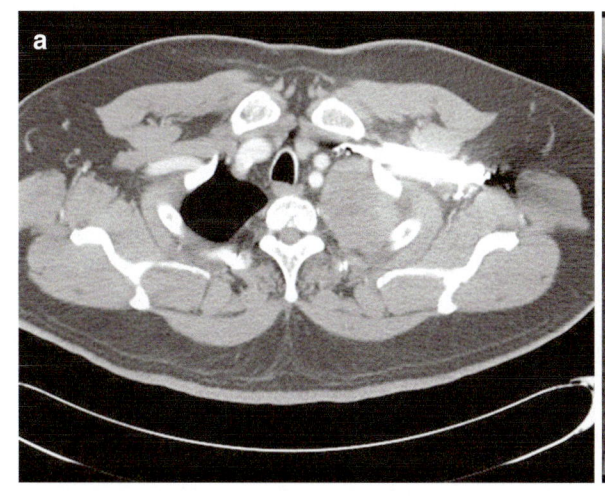

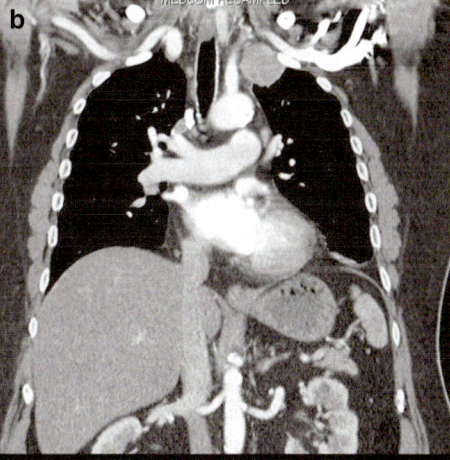

Fig. 6.4 (a, b) Chest radiograph and thorax CT images of a patient with Pancoast syndrome

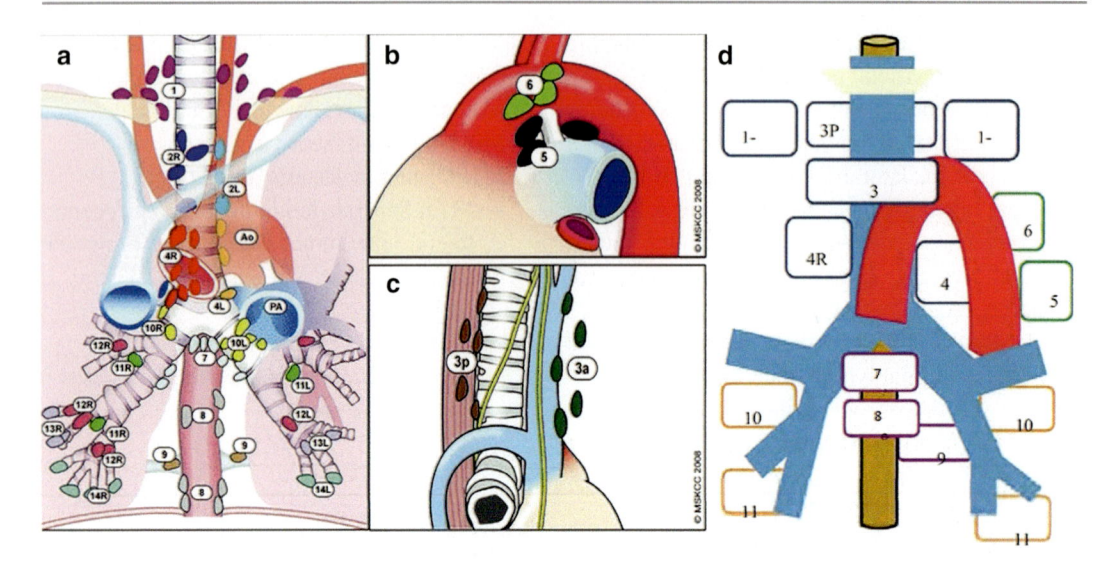

Fig. 6.5 (a) Distribution of lymph nodes by stations, except stations 3, 5, and 6. (b) The position of stations 5 and 6 lymph nodes. (c) The position of stations 3a and 3p lymph nodes. (d) Lymph node mapping in lung cancers [Greene FL., Page DL., Fleming, ID., et al. American Joint Committee on Cancer. AJCC cancer staging manual, 6th ed. New York: Springer, 2002. p.169 Fig.19.2, reproduced with the permission from the American Joint Committee on Cancer (AJCC), Chicago, Illinois)]

Lymphatic Drainage of Lung Lobes
- Right upper lobe → ipsilateral mediastinum
- Left upper lobe → ipsilateral and contralateral mediastinum
- Right lower lobe → subcarinal lymph nodes > right upper mediastinum > right lower mediastinum
- Left lower lobe → subcarinal lymph nodes > right or left upper mediastinum > right or left lower mediastinum

6.1.1.5 Treatment Algorithm

Surgery is the standard treatment for early-stage NSCLC which allows for pathologic verification of primary tumor along with assessment of nodal status. Indications for subsequent chemotherapy and/or radiotherapy are based on surgical assessment (Table 6.4).

Stereotactic Body Radiation Therapy (SBRT)

Medically inoperable patients with early-stage NSCLC have been previously treated with conventional radiotherapy which resulted in a 3-year survival rate of 30% [15]. Low rates of local control and survival for these patients have been attributed to inadequate tumor doses. However, SBRT has achieved improved local control rates comparable with that of surgical resection. Motion limiting immobilization equipment, four-dimensional (4D) CT simulation for breathing adapted radiotherapy, image guidance techniques for verification of treatment accuracy, sophisticated multileaf collimators, and intensity-modulated radiation therapy (IMRT) techniques have contributed to acquired treatment results with SBRT.

- SBRT may be performed by several treatment platforms including Cyberknife, Novalis, Synergy, Trilogy, Tomotherapy, TruBeam, and VitalBeam.
- Accuracy of treatment with SBRT may be improved by the use of onboard MRI for 4D volumetric tumor motion assessment.
- Proper use of technology and specialization of the physician are critical aspects of management with SBRT.
- AAPM Task Group 101 has published the requirements and principles for SBRT [16] (Figs. 6.7 and 6.8, Table 6.5).

6.1 Introduction

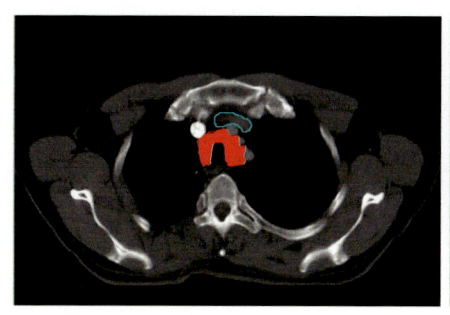

Red: Station 1-2 (left-right) **Turquoise:** Brachiocephalic vein

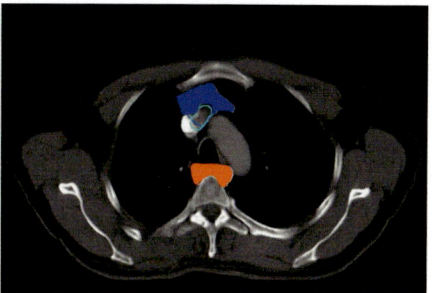

Blue: Station 3 (anterior) **Orange:** Station 3 (posterior) **Turquoise:** Brachiocephalic vein

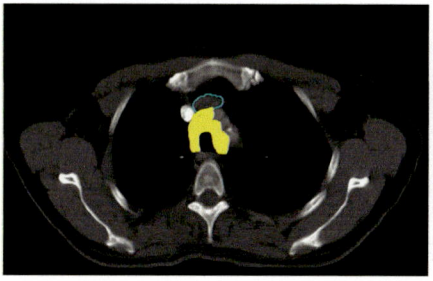

Yellow: Station 4 (left-right) **Turquoise:** Brachiocephalic vein

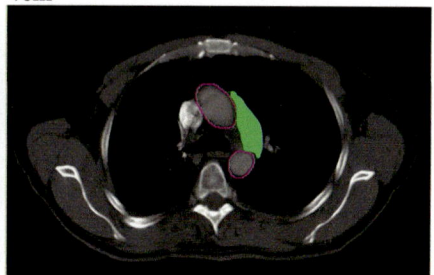

Green: Station 5 **Purple:** Aorta

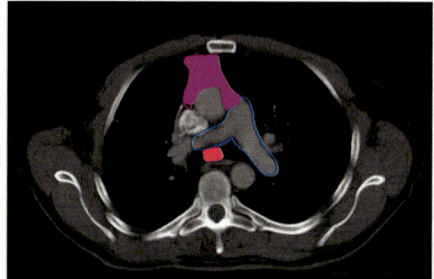

Pink: Station 6
Red: Station 7
Blue: Pulmonary artery

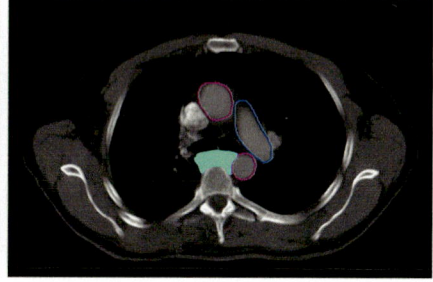

Pink: Aorta
Blue: Pulmonary artery
Turquoise: Station 8

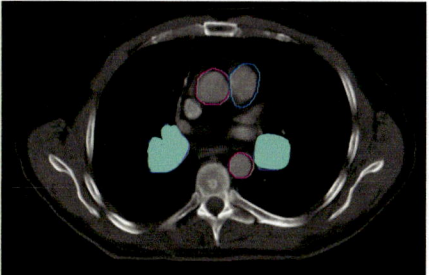

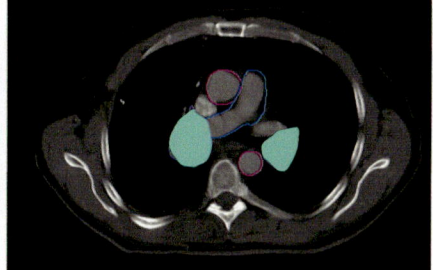

Pink: Aorta
Blue: Pulmonary artery
Turquoise: Station 10 (left-right)

Pink: Aorta
Blue: Pulmonary artery
Turquoise: Station 10 (left-right)

Fig. 6.6 Intrathoracic lymph nodes in serial axial CT slices

Table 6.3 Lung cancer stage grouping as per the AJCC eighth edition

Stage	T	N	M
Occult carcinoma	TX	N0	M0
Stage 0	Tis	N0	M0
Stage IA1	T1a(mi)	N0	M0
	T1a	N0	M0
Stage IA2	T1b	N0	M0
Stage IA3	T1c	N0	M0
Stage IB	T2c	N0	M0
Stage IIA	T2b	N0	M0
Stage IIB	T1a-T1c	N1	M0
	T2a	N1	M0
	T2b	N1	M0
	T3	N0	M0
Stage IIIA	T1a-T1c	N2	M0
	T2a-T2b	N2	M0
	T3	N1	M0
	T4	N0	M0
	T4	N1	M0
Stage IIIB	T1a-T1c	N3	M0
	T2a-T2b	N3	M0
	T3	N2	M0
	T4	N2	M0
	T1a–T1c	N3	M0
	T2a–T2b	N3	M0
	T3	N2	M0
	T4	N2	M0
Stage IVA	Any T	Any N	M1a
	Any T	Any N	M1b
Stage IVB	Any T	Any N	M1c

Table 6.4 Treatment algorithm for NSCLC

Stage	Treatment
Stage I–II	• Surgery (adjuvant chemotherapy if tumor size >4 cm or N+) • Stereotactic body radiation therapy (SBRT) for eligible patients
Stage IIA	• Preoperative chemotherapy followed by surgery or definitive chemoradiotherapy
Stage IIIB	• Chemoradiotherapy
Stage IV	• Platin based chemotherapy for patients with no mutations • Targeted therapies for patients with EGFR mutation or ALK rearrangement

Systemic Treatment for Unresectable Locally Advanced NSCLC

- Concurrent chemoradiotherapy is the principal treatment for patients with good performance status.
  - Cisplatin + etoposide and carboplatin + paclitaxel are common regimens
    Cisplatin (50 mg m^2 first, eighth, 29th, 36th days);
    Etoposide (50 mg m^2, days 1–5 and 29–33).
- Induction chemotherapy before RT is not recommended.
- Sequential chemotherapy and radiotherapy is better tolerated by patients with poor performance status.
- Risk of esophagitis is higher with concurrent chemoradiotherapy rather than sequential chemotherapy and radiotherapy.
- Median survival is 17–20 months and 5 year survival is approximately 20% with concurrent chemoradiotherapy (Table 6.6).

6.1.1.6 Radiotherapy

Conventional (2D) Radiotherapy
Radiotherapy for NSCLC is delivered either alone or in combination with chemotherapy for curative intent, or palliatively to relieve symptoms, prophylactically to prevent the occurrence of disease, as a preoperative neoadjuvant treatment to decrease tumor size in order to make it operable, or postoperatively to eradicate microscopic disease.

Simulation is performed in the supine position with arms up over the head.

Developmental Steps in Radiotherapy Techniques for NSCLC

- *Conventional radiotherapy.* Very large fields were used; dose escalation was not possible due to large volumes of lung, heart, and spinal cord. Elective nodal irradiation was used.
- *Conformal radiotherapy.* Limited fields are used, including only the primary tumor and the involved lymphatic region. Therefore, dose escalation is possible for curative approaches. Elective nodal radiotherapy is not used. However, incidental elective nodal RT was demonstrated in some trials by evaluating isodoses.
- *Stereotactic radiotherapy.* This is used for small, peripheral T1–2N0M0 tumors. It is a

6.1 Introduction

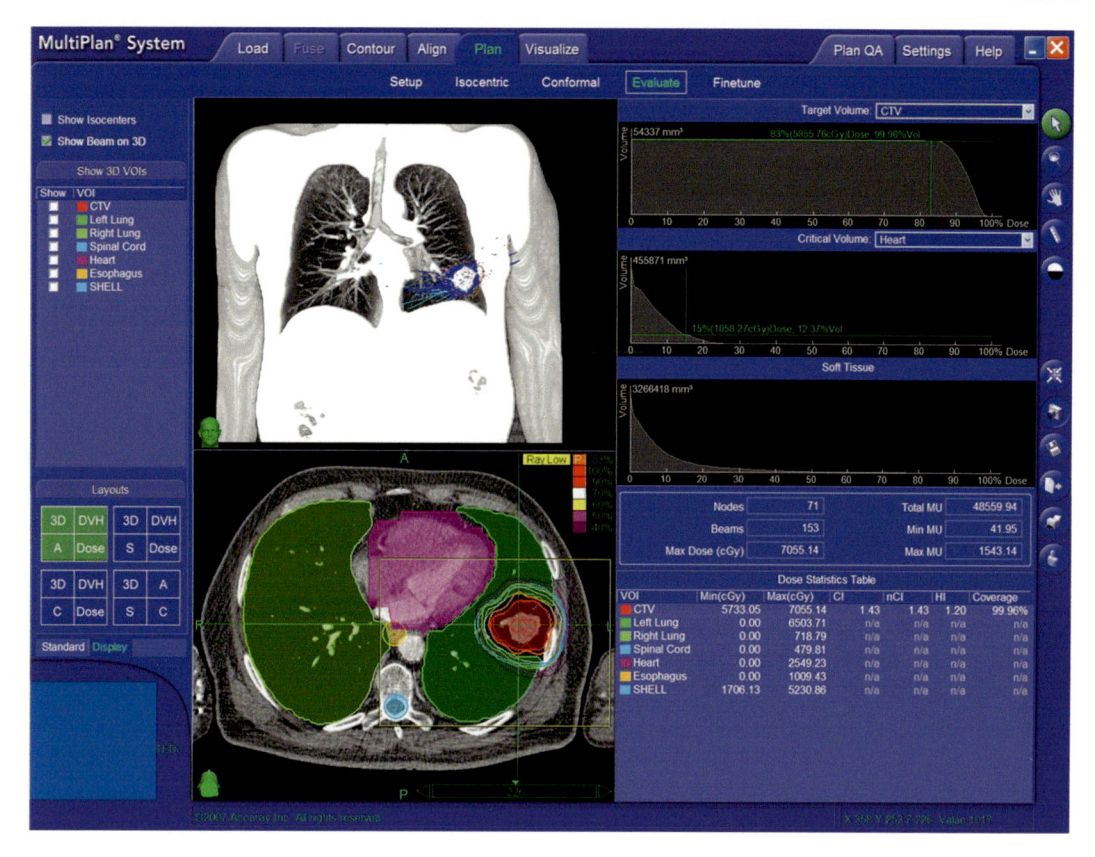

Fig. 6.7 Robotic radiosurgery for a T1N0M0 peripheral nonsmall cell lung cancer (Courtesy of Hacettepe University)

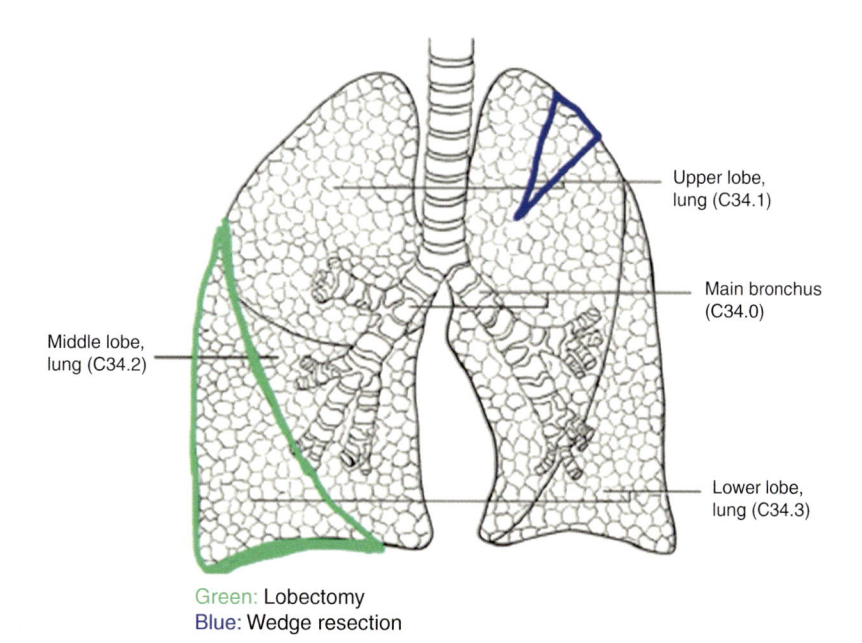

Fig. 6.8 Blue: Wedge resection and Green: lobectomy

Table 6.5 Stage-based surgical approach in NSCLC

Surgery	Comments
Stage IA	• Lobectomy (VATS/Torachotomy) [including mediastinal sampling] • VATS lobectomy is safe, with comparable rates of complications and shorter hospitalization duration • 5-year survival 73–92% with surgery • No benefit with complete mediastinal dissection
Small peripheral tumors (<2 cm)	• Role of sublobar resection may be considered for selected stage IA tumors • Anatomical segmentectomy may be preferred rather than nonanatomical wedge resection
Stage IB (T2aN0)	• Lobectomy is standard treatment • Pneumonectomy or sleeve resection may be required for central tumors or for tumors passing the fissure and extending to adjacent lobe • 5-year survival is 55–60%
Stage II	• Lobectomy standard treatment • Pneumonectomy or sleeve resection may be required • No benefit for incomplete resection • En bloc resection for T3 tumors if R0 resection is possible • 5-year survival is 35–45%
Superior sulcus tumors (Pancoast)	• Surgery after chemotherapy and RT • Standard RT dose 45 Gy • Surgery is contraindicated in the presence of N2-N3 disease and for patients with brachial plexus involvement extending beyond C8 and T1

Surgery	Comments
Stage IIIA	• Surgery followed by chemotherapy is recommended for T3N1 or T4N0–1 • Neoadjuvant (upfront) surgery is not recommended in the presence of N2 disease • Induction chemotherapy or chemoradiotherapy followed by surgery is recommended for patients with N2 disease • Trimodal treatment (surgery + KT + RT) is not optimal for patients requiring pneumonectomy • Surgery is not recommended for bulky N2 disease or in the presence of multiple mediastinal lymph node station involvement
Stage IIIB/IV	• Limited role for surgery • May be considered for patients with peripheral tumors and solitary brain metastasis • Surgery after systemic treatment may be considered for oligometastatic disease with indolent biology

highly effective and curative approach with minimal toxicity. More than 100 Gy can be given to the GTV. No elective nodal RT is used.

Typical conventional RT fields include the primary tumor +2 cm and the adjacent lymph node region +1 cm.

Conventional RT field in upper lobe tumors (Fig. 6.9)

- Bilateral supraclavicular field
- Upper mediastinum
- Subcarinal field (Two vertebrae below carina) or (Five to six centimeters below carina)
- Primary tumor +2 cm (Two anterior–posterior parallel–opposed fields)

Conventional RT fields in hilar tumors (Fig. 6.10)

- Superior: thoracic inlet
- Inferior: 8–9 cm below carina
- Includes mediastinum
- Primary tumor +2 cm (Two anterior–posterior parallel–opposed fields)

Conventional RT field in lower lobe tumors (Fig. 6.11).

- Superior: thoracic inlet.
- Inferior: 8–9 cm below carina (may be more inferior depending on whether the tumor includes the diaphragm).
- Includes mediastinum.

6.1 Introduction

Table 6.6 Radiotherapy indications for NSCLC

Radiotherapy	Comments
Stage I (inoperable due to medical comorbidity[a])	• SBRT recommended for peripheral lesions • Local control with SBRT is approximately 90% for T1-2N0 tumors
Stage I–II	• RT has no role if surgical margin is negative • 60 Gy external RT is recommended if surgical margin is microscopically positive
Stage IIIA, N2 (+) Surgery planned	• Induction chemoradiotherapy may be considered for selected patients • Standard preoperative RT dose 45 Gy delivered in 1.8 Gy/fx
Stage IIIA, N2 (+) Surgically resected	• 50–54 Gy external RT focused on the ipsilateral hilus, bronchial stump and high-risk nodal stations may be considered after adjuvant chemotherapy in the presence of multistation N2 disease
Stage IIIA/B, Unresectable	• 60–66 Gy cisplatin-based concurrent chemoradiotherapy is standard • Sequential chemotherapy and RT may be considered for patients with poor performance status • Induction chemotherapy has no role • N(−): CTV = GTV + 5 mm (up to 8 mm) • N(+): CTV = GTV + 5 mm (up to 8 mm) and involved lymph nodes + 3–5 mm
Stage IV	• Palliative intent RT for airway obstruction or hemoptysis • Palliative intent RT for bone metastasis to relieve pain • Palliative intent RT for brain metastasis • Palliative intent RT for tumor-related spinal cord compression

[a]Criteria for inoperability: baseline $FEV_1 < 40\%$, predicted postoperative $FEV_1 < 30\%$ or significantly limited diffusion capacity

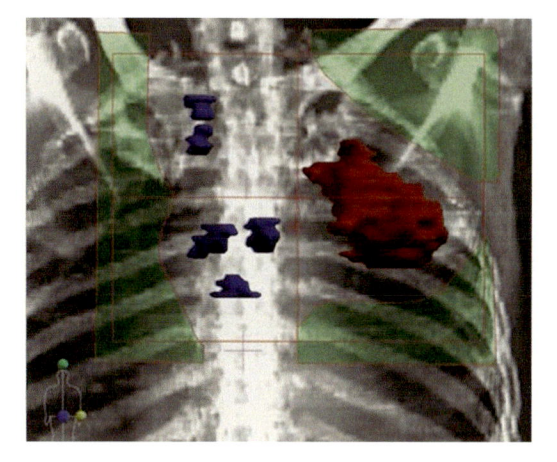

Fig. 6.9 Conformal RT field for upper lobe tumor

- Primary tumor +2 cm (Two anterior–posterior parallel–opposed fields).

Conventional RT field in lower lobe tumor + gross mediastinal lymph node involvement

- Superior: thoracic inlet
- Inferior: 8–9 cm below carina (may be more inferior depending on the tumor localization)
- Includes supraclavicular field
- Includes mediastinum
- Primary tumor +2 cm (Two anterior–posterior parallel–opposed fields) (Fig. 6.12)

Conventional Definitive RT Doses

Anterior–posterior parallel–opposed fields can be used until 46 Gy, then the spinal cord is spared and primary + involved lymphatics are given more than 60 Gy.

- Six to ten megavolt photons are recommended.

Planning in postoperative cases is similar. It is indicated for a close or positive surgical margin, extracapsular nodal extension, or multiple N2 lymph nodes.

- Microscopic residual (−): 46–54 Gy (1.8–2 Gy/day)
- Microscopic residual (+): field is localized after 46 Gy, total dose 60–66 Gy (1.8–2 Gy/day)

Possible borders for postoperative RT

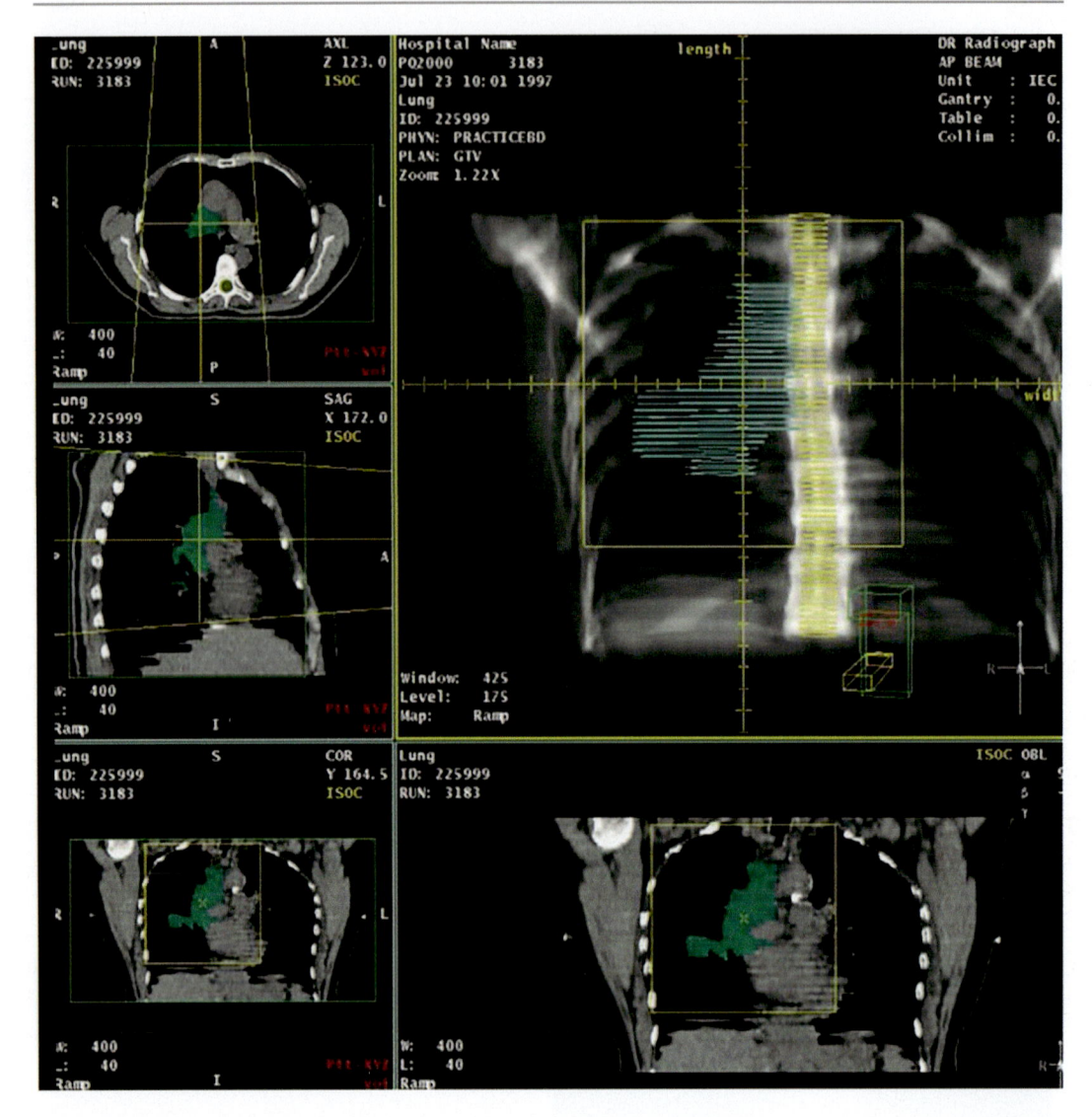

Fig. 6.10 Conformal RT field for hilar tumor

Mediastinum ± (microscopic residual + 2 cm) Mediastinum → superior: thoracic inlet (superior to sternoclavicular joint); inferior: T9–10 intervertebral space; lateral: costochondral joints (ipsilateral lateral field is enlarged if that site is to be irradiated) (Fig. 6.13).

3 Dimensional Conformal Radiotherapy/IMRT Simulation

- Patient positioning on CT simulator should be the stable treatment position.
- Supine positioning with arms up over the head and supported by arm support.
- Ideal simulation is performed by 4D CT simulator with the mid-inspiration approach. Internal target volume (ITV) with individualized margins may be used.
- Breath holding, respiratory gating/tracking techniques may be utilized.
- If 4D CT is not available, planning CT should be acquired at superficial breathing (Figs. 6.14 and 6.15).

Contouring/Dose

- Available clinical and imaging data including endoscopy/bronchoscopy reports, invasive

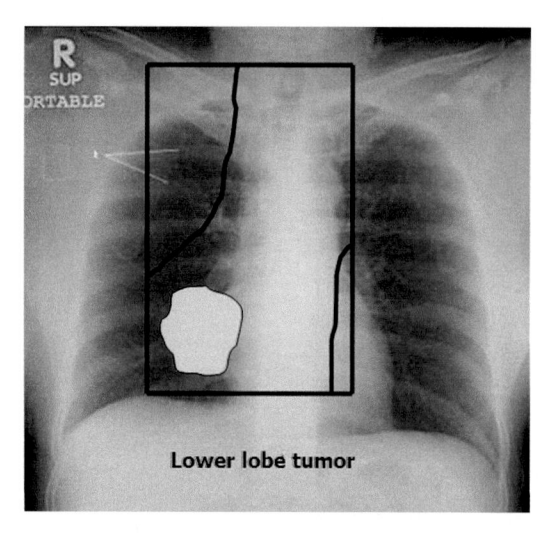

Fig. 6.11 Conventional RT field for lower lobe tumor

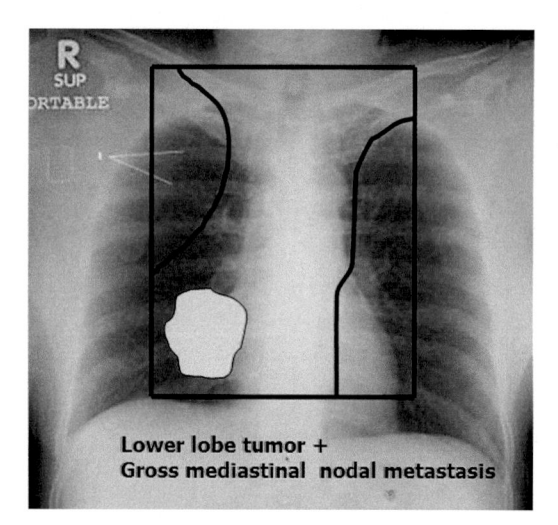

Fig. 6.12 Conventional RT field for lower lobe tumor with gross mediastinal LN involvement

mediastinal staging data, CT, MRI, and PET findings should be used in GTV definition.
- Safety margins should not be included in GTV.
- Visualized tumor on CT should be delineated as GTV.
  - Parenchyma window settings: W = 1600 and L = −600
  - Mediastinum window settings: W = 400 and L = 20).
- Primary tumor and lymph nodes with FDG uptake on PET should be included in GTV.
- A CTV margin of at least 6 mm should typically be added to generate the primary tumor CTV.
  - However, tighter margins may also be considered in selected cases [17].
- Adjacent tissues such as bone or pleura should not be included in CTV unless involved by tumor.
- Elective nodal RT including the bilateral supraclavicular, mediastinal, and hilar lymph nodes was previously common; however, selective nodal RT directed at the involved lymph node regions is currently favored due to the absence of data on the benefit of elective nodal irradiation.
- PTV is generated by expansion of the CTV with 8–15 mm margin to account for tumor motion, geometric uncertainties, and patient positioning errors.
  - However, individualized tighter margins may be used if 4D CT simulation, gating, and/or tracking is utilized.
  - Manual editing of PTV is not recommended.
- RT doses
  - 66 Gy for definitive RT
  - 60 Gy for concurrent chemoradiotherapy
- Margins of 1 cm circumferentially and 1.5 cm craniocaudally may be added to the CTV to generate the PTV if 4D CT simulation is not available (Fig. 6.16).

Dose constraints
- Spinal cord $D_{max} \leq 50.5$ Gy
- Total lung $V_{20Gy} \leq 37\%$
  - Mean lung dose: ≤ 20 Gy
- Brachial plexus $D_{max} < 66$ Gy
- Esophagus mean dose < 34 Gy
- Heart $V_{60Gy} < 1/3$, $V_{45Gy} < 2/3$, $V_{40Gy} < 100\%$

Brachial Plexus Contouring
Brachial plexus is rarely included in critical structure delineation typically due to poor visualization. However, brachial plexopathy with resul-

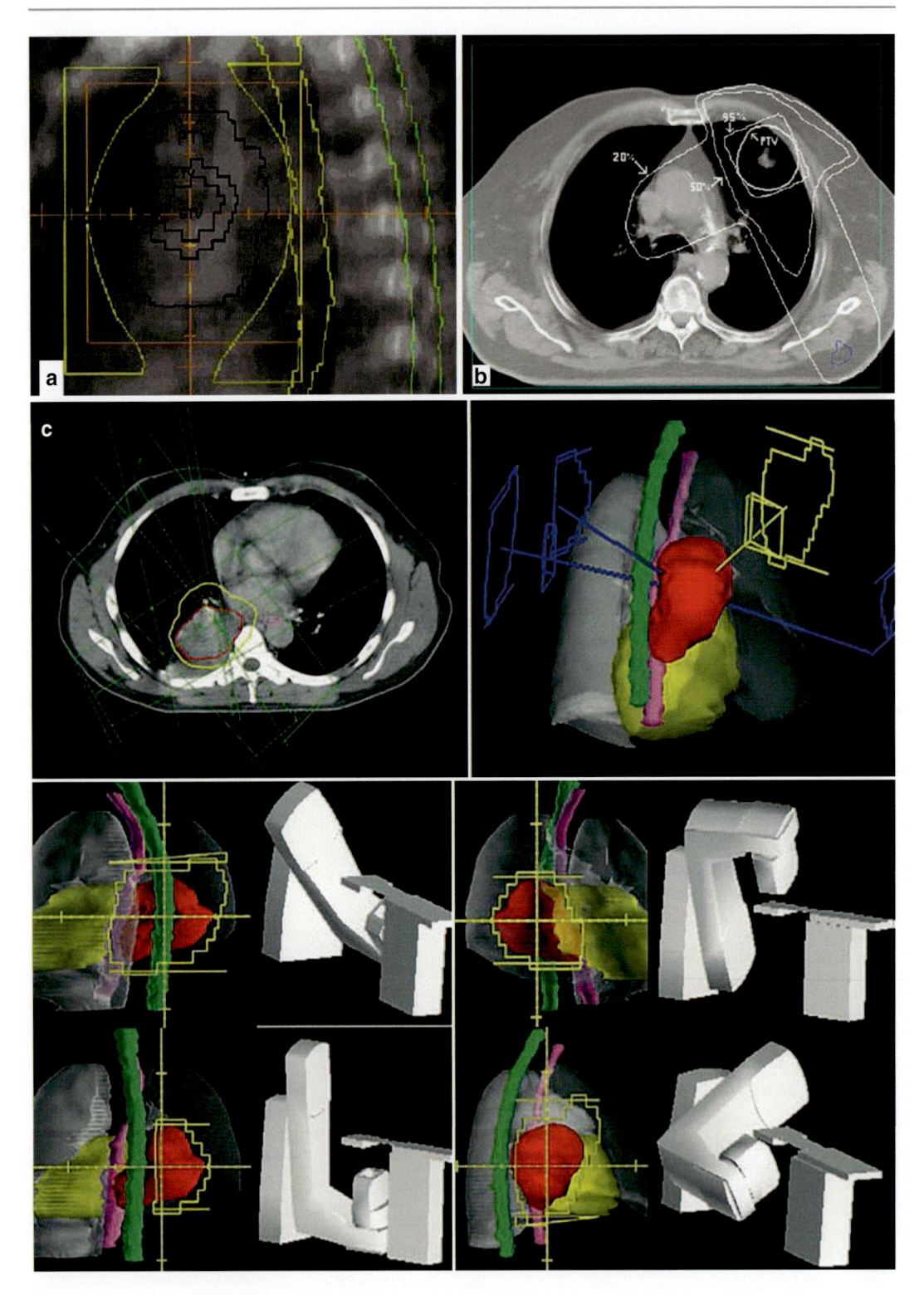

Fig. 6.13 (a–c) Conformal RT fields in lung cancer

Fig. 6.14 CT simulation with arm support

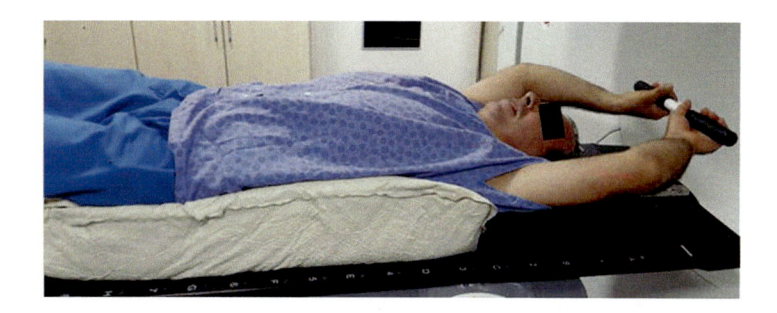

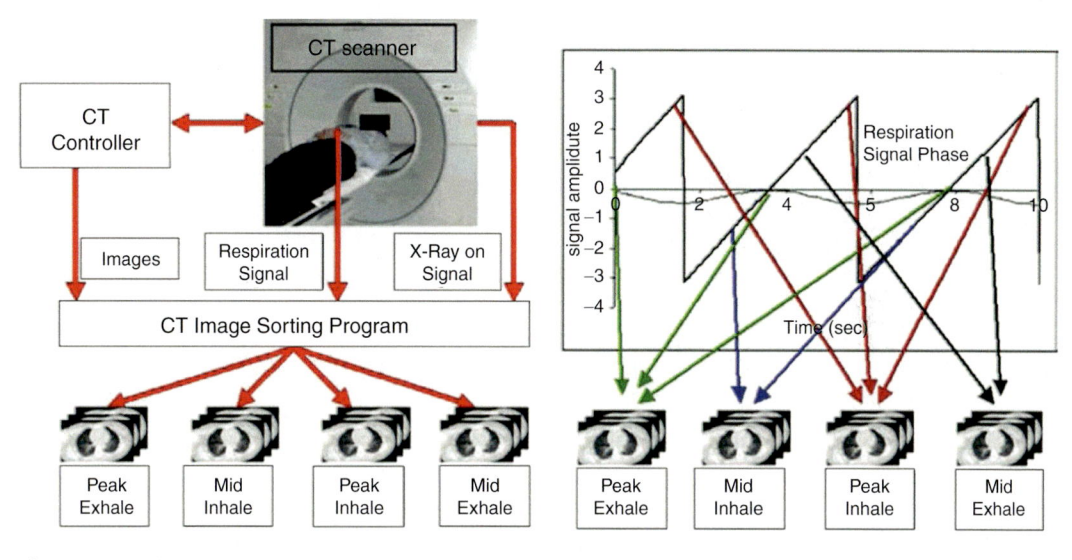

Fig. 6.15 4D simulation with tracking

tant morbidity may develop when violation of tolerance dose occurs. In this context, brachial plexus should be considered as a critical structure particularly for Pancoast tumors. Position of the arm is important for contouring.

1. C5, T1 and T2 vertebrae are identified.
2. Subclavian and axillary neurovascular bundle is identified.
3. Anterior and middle scalene muscles are identified and delineation by use of a 5-mm diameter paint tool is started from the neural foramina from C5 to T1. The space between anterior and middle scalene muscles is delineated on CT slices where no neural foramen is present.
4. Contouring is continued to neurovascular bundle inferiorly and one to two CT slices below clavicular head laterally.
5. Clavicula and first rib are landmarks (Fig. 6.17).

6.1.1.7 Selected Publications

Limited resection vs. *lobectomy*

North America Lung Cancer Study Group 821 (LCSG 821) 1995 (1982–88) → 247 pT1N0M0 peripheral NSCLC. Randomized to limited resection vs. lobectomy.

- Locoregional relapse: 17% in limited resection arm vs. 6% in lobectomy.
- Fatality: 39% in limited resection arm vs. 30% in lobectomy.
- Conclusion: lobectomy is recommended rather than limited excision.

Ginsberg RJ et al. (1995) Randomized trial of lobectomy vs. limited resection for T1 N0 nonsmall cell lung cancer. Ann Thorac Surg 60(3):615–622

Surgery ± external RT.

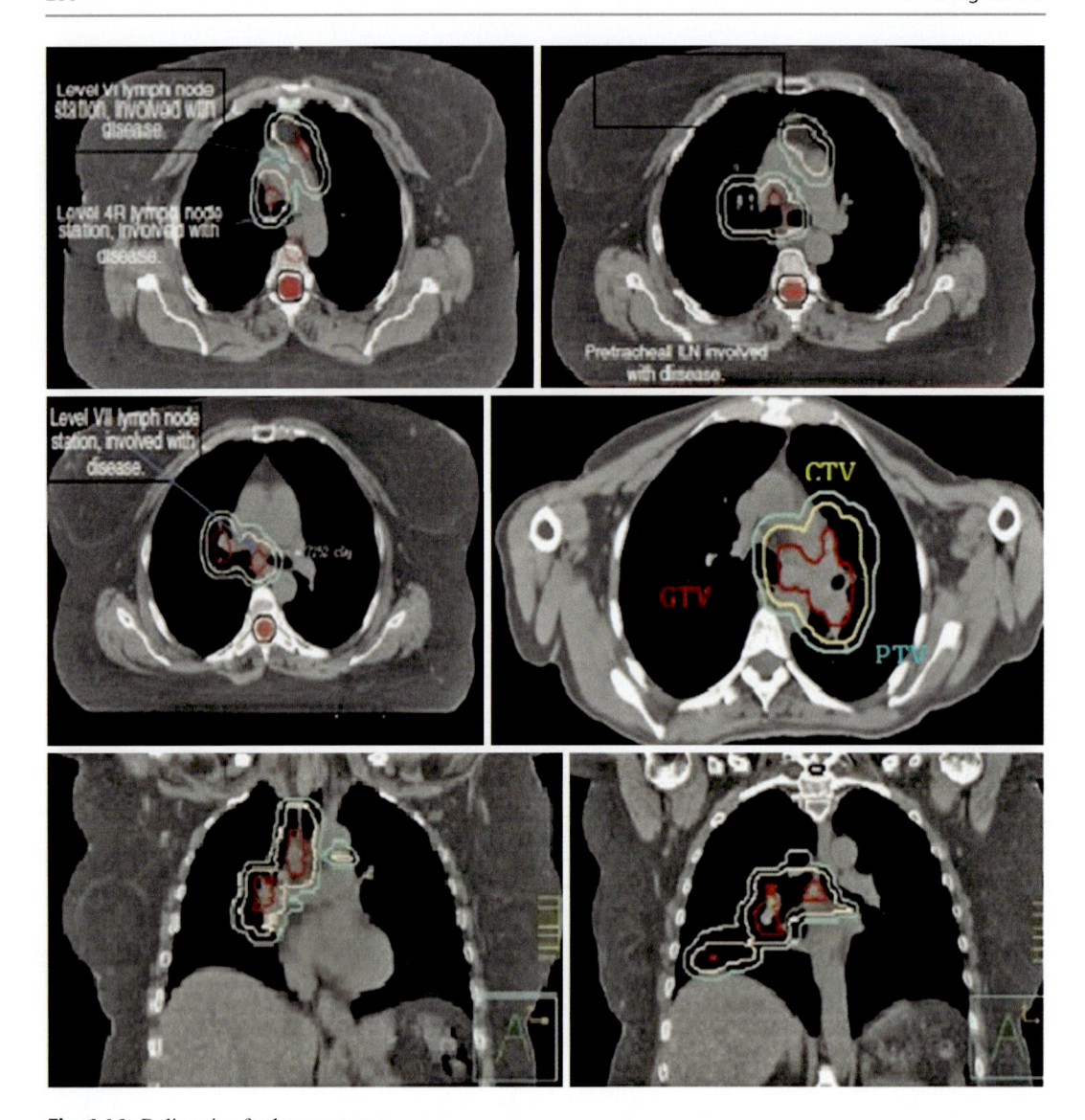

Fig. 6.16 Delineation for lung cancer

Rome, 2002 (1989–1997) → 104 stage I NSCLC. Randomized to surgery vs. surgery + RT. RT dose: 54 Gy (51 cases).

- Five-year OS: 67% in combined arm vs. 58% in surgery-alone arm ($p = 0.048$).
- Five-year locoregional recurrence: 2% in combined arm vs. 23% in surgery-alone arm.
- 11% toxicity in RT arm, but no long-term toxicity.
- 37% percent radiological lung fibrosis in RT arm.
- Conclusion: adjuvant RT increases local control after surgery, and is promising in regard to increasing survival.

Trodella L (2002) Adjuvant radiotherapy in nonsmall cell lung cancer with pathological stage I: definitive results of a phase III randomized trial. Radiother Oncol 62(1):11–19

GETCB, France. 1999 (1986–1994) → 728 stage I–III NSCLC cases. Randomized after surgery to surveillance vs. 60 Gy RT.

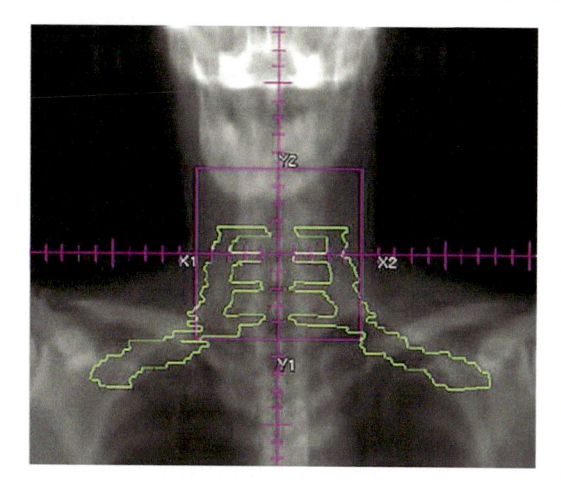

Fig. 6.17 Brachial plexus contouring

- Five-year OS: 30% in combined arm vs. 43% in surveillance.
- Death other than NSCLC: 31% in combined arm vs. 8% in surveillance (most deaths were due to high fraction dose; subgroup analysis showed that these deaths were at stage I–II, not stage III).
- No difference in local relapse and distant metastasis.
- However, presence of N0 cases and use of Co-60, high fraction dose (>2.5 Gy) are criticized.

Dautzenberg B et al (1999) A controlled study of postoperative radiotherapy for patients with completely resected nonsmall cell lung carcinoma. Groupe d'Etude et de Traitement des Cancers Bronchiques. Cancer 86(2):265–273*Lung Cancer Study Group 773 (LCSG 773), 1986* → 230 stage II–III NSCLC cases.
Randomized after surgery to RT (50 Gy, mediastinum) vs. surveillance.

- Local relapse: 1% in RT vs. 20% in surveillance
- No difference in overall survival

The Lung Cancer Study Group (1986) Effects of postoperative mediastinal radiation on completely resected stage II and stage III epidermoid cancer of the lung. N Engl J Med 315(22):1377–1381*Adjuvant chemotherapy vs. adjuvant chemoradiotherapy*

CALGB 9734, 2007→ randomized. 44 stage IIIA (N2) NSCLC cases. 37 patients: adjuvant chemotherapy after surgery (carboplatin + paclitaxel) vs. RT (2–4 weeks after chemotherapy, 50 Gy RT).

- No difference in median survival.
- No difference in overall survival.
- Study was closed early.

Perry MC (2007) A phase III study of surgical resection and paclitaxel/carboplatin chemotherapy with or without adjuvant radiation therapy for resected stage III non-small-cell lung cancer: Cancer and Leukemia Group B 9734. Clin Lung Cancer 8(4):268–272*INT 0115, ECOG EST 3590/RTOG 91–05, 2000* → Randomized. 488 stage II–IIIA
NSCLC cases, four cycles of cisplatin/etoposide + concurrent RT vs. RT (50.4 Gy).

- No difference in OS
- No difference in in-field relapses
- No difference in treatment-related toxicity

Keller SM et al (2000) A randomized trial of postoperative adjuvant therapy in patients with completely resected stage II or IIIA nonsmall-cell lung cancer. N Engl J Med 343(17):1217–1222Prophylactic cranial RT (PCI) in NSCLC
Germany, 2007 (1994–2001) → Randomized. 112 operable stage IIIA (diagnosed with mediastinoscopy). Surgery + adjuvant 50–60 Gy RT vs. preoperative CT (three cycles cisplatin/etoposide) + concurrent chemoradiotherapy (cisplatin/etoposide, 2 daily fractions of 1.5 Gy, total dose 45 Gy) + surgery + PCI (30 Gy in 15 fractions).

- Study closed early due to benefit of the chemotherapy arm.
- Brain metastasis as a first site: 8% in PCI arm vs. 35% in non-PCI arm (statistically significant).
- Brain metastasis rate: 9% in PCI arm vs. 27% in non-PCI arm (statistically significant).

Pottgen C (2007) Prophylactic cranial irradiation in operable stage IIIA nonsmall-cell lung cancer treated with neoadjuvant chemoradiotherapy: results from a German multicenter randomized trial. J Clin Oncol 25(31):4987–4992

CHART, 1997 → 563 patients. CHART (continuous hyperfractionated accelerated radiotherapy) regimen, which uses 36 small fractions of 1.5 Gy given three times per day, to give 54 Gy in only 12 consecutive days. NSCLC localized to the chest with a performance status of 0 or 1. Patients were randomized to CHART or conventional radiotherapy (60/2 Gy in 6 weeks). Overall there was a 24% reduction in the relative risk of death, which is equivalent to an absolute improvement in 2-year survival of 9%, from 20 to 29% ($p = 0.004$). The largest benefit occurred in patients with squamous cell carcinomas (34% reduction in the relative risk of death; an absolute improvement at 2 years of 14% from 19 to 33%). Severe dysphagia occurred more often in the CHART group (19 vs. 3%). There were no important differences in short-term or long-term morbidity. CHART compared with conventional radiotherapy gave a significant improvement in survival of patients with NSCLC.

Saunders M et al (1997) Continuous hyperfractionated accelerated radiotherapy (CHART) vs. conventional radiotherapy in nonsmall-cell lung cancer: a randomized multicentre trial. CHART Steering Committee. Lancet 350(9072):161–165

UCSD, 1990 → Patients randomly assigned to group one received cisplatin and vinblastine and then began radiation therapy on day 50 (60 Gy over a 6-week period). Patients assigned to group two received the same radiation therapy but began it immediately and received no chemotherapy. In patients with stage III NSCLC, induction chemotherapy significantly improves median survival (by about 4 months) and doubles the number of long-term survivors as compared with radiation therapy alone.

Dillman RO et al (1990) A randomized trial of induction chemotherapy plus high-dose radiation vs. radiation alone in stage III nonsmall-cell lung cancer. N Engl J Med 323(14):940–945

Institute Gustave Roussy, 1991 → Randomized study comparing radiotherapy alone with combination of radiotherapy and chemotherapy in nonresectable squamous cell and large-cell lung carcinoma. The radiation dose was 65 Gy in each group, and chemotherapy included vindesine, cyclophosphamide, cisplatin, and lomustine. The 2-year survival rate was 14% in RT alone and 21% in the combined arm ($p = 0.08$). The distant metastasis rate was significantly lower in the combined arm ($p < 0.001$). Local control was poor in both groups (17 and 15%, respectively) and remained the major problem.

Le Chevalier T et al (1991) Radiotherapy alone vs. combined chemotherapy and radiotherapy in nonresectable nonsmall-cell lung cancer: first analysis of a randomized trial in 353 patients. J Natl Cancer Inst 83(6):417–423

The Netherlands, 1992 → 331 patients with nonmetastatic inoperable NSCLC randomized to one of three treatments: RT for 2 weeks (3 Gy given ten times, in five fractions per week), followed by a 3-week rest period and then radiotherapy for 2 more weeks (2.5 Gy given ten times, five fractions per week); RT on the same schedule, combined with cisplatin given on the first day of each treatment week; or RT on the same schedule, combined with cisplatin given daily before radiotherapy. Survival was significantly improved in the radiotherapy and daily cisplatin group as compared with the radiotherapy group ($p = 0.009$).

Schaake-Koning C et al (1992) Effects of concomitant cisplatin and radiotherapy on inoperable nonsmall-cell lung cancer. N Engl J Med 326(8):524–530

Sause, 1995 → Patients were randomly assigned to receive either 60 Gy of RT delivered at 2 Gy per fraction, 5 days a week, over a 6-week period; induction chemotherapy consisting of cisplatin (100 mg/m^2) on days 1 and 29 and 5 mg/m^2 vinblastine per week for five consecutive weeks beginning on day 1 with cisplatin, followed by standard RT starting on day 50; or 69.6 Gy delivered at 1.2 Gy per fraction twice daily (hyperfractionated radiation therapy). Induction chemotherapy followed by RT in unresectable NSCLC was superior to hyperfractionated radiation therapy or standard radiation therapy alone, yielding a statistically significant short-term survival advantage.

Sause WT et al (1995) Radiation Therapy Oncology Group (RTOG) 88–08 and Eastern Cooperative Oncology Group (ECOG) 4588: preliminary results of a phase III trial in regionally advanced, unresectable nonsmall-cell lung cancer. J Natl Cancer Inst 87(3):198–205

PORT meta-analysis → 2,128 patients from nine randomized trials (published and unpublished) were analyzed by intention to treat. There were 707 deaths among 1,056 patients assigned to postoperative radiotherapy and 661 among 1,072 assigned to surgery alone. Median follow-up was 3.9 years for surviving patients. The results show a significant adverse effect of postoperative radiotherapy on survival (hazard ratio 1.21 [95% CI 1.08–1.34]). This 21% relative increase in the risk of death is equivalent to an absolute detriment of 7% (3–11) at 2 years, reducing overall survival from 55 to 48%. Subgroup analyses suggest that this adverse effect was greatest for patients with stage I/II, N0–N1 disease, whereas for those with stage III, N2 disease there was no clear evidence of an adverse effect. Postoperative radiotherapy was detrimental to patients with early-stage completely resected NSCLC and should not be used routinely for such patients. The role of postoperative radiotherapy in the treatment of N2 tumors is not clear and may warrant further research.

PORT Meta-analysis Trialists Group (1998) Postoperative radiotherapy in nonsmall-cell lung cancer: systematic review and meta-analysis of individual patient data from nine randomized controlled trials. Lancet 352(9124):257–263

6.2 Small Cell Lung Cancer (SCLC)

Small cell lung cancer (SCLC) constitutes nearly 20% of all lung cancers. Its incidence is increasing rapidly compared to other lung cancers. SCLC has the most prominent association with cigarette smoking among all lung cancers. The probability of female cigarette smokers developing SCLC is higher than that of male cigarette smokers.

SCLC is rapidly fatal if not treated.

- Mean survival with a combination of chemotherapy and radiotherapy for limited-stage disease is 20 months. Two-year survival is 45%, and 5-year survival is 15–20%.
- Survival is 1–3 months in extensive-stage disease if not treated. Chemotherapy may prolong survival.

6.2.1 Pathology

Tumor cells of small cell lung cancer are found in small, dense packages with limited cytoplasm. Their nuclei have finely granular chromatin, and a nucleolus cannot be seen (Fig. 6.18).

6.2.2 General Presentation

Symptoms and findings of SCLC are similar to those for NSCLC. Since SCLC has a high tendency for distant metastasis (~60% at presentation), metastatic organ-specific signs and symptoms may be prominent at diagnosis.

Most cases die due to relapses, secondary malignancies, or other reasons during late periods.

- Most recurrences are observed in the first year of therapy. Although relapses in the following years are rare, most patients die due to the disease.

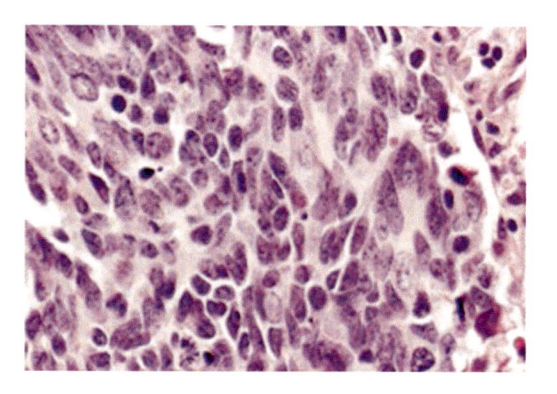

Fig. 6.18 SCLC histopathology

- Secondary malignancies are usually NSCLC, upper airway cancers, or gastrointestinal system tumors.
- Patients surviving more than 5 years have a sixfold greater risk of dying from noncancer-related causes.

6.2.3 Staging

SCLC is divided into limited- and extensive-stage SCLC as per the Veterans Affairs Administration Lung Cancer Study Group staging.

Limited-stage SCLC
- One-third of all SCLCs.
- Limited to one hemithorax, ipsilateral hilus, and mediastinum (can be treated with a reasonable radiotherapy portal).
- Ipsilateral pleural effusion: extensive stage.
- Ipsilateral supraclavicular lymph node does not affect limited-stage disease.

Extensive-stage SCLC
- Beyond the limits of its origin at the hemithorax, distant metastasis, pleural effusion (Any SCLC that cannot be treated with a reasonable radiotherapy portal.)

(Veterans Affairs Administration Lung Study Group 1957)

The *International Lung Cancer Study Group (ILCSG)* modified this staging system as follows. Contralateral mediastinal lymph nodes that can be treated with a reasonable RT field and benign pleural effusions also are included in limited-stage disease. All circumstances other than these are accepted as extensive-stage SCLC (1997) [18].

The ILCSG has defined limited-stage SCLC as stage I–IIIb according to the TNM staging system [18]. TNM staging is the same as that of NSCLC.

6.2.4 Treatment Algorithm

SCLC is highly chemosensitive and radiosensitive, but is not chemocurable or radiocurable (Table 6.7).

Table 6.7 Treatment algorithm for SCLC

Stage	Treatment
Limited-stage SCLC	Concurrent chemoradiotherapy (Chemotherapy → cisplatin + etoposide in every 3 weeks for four cycles) (RT → 1.5 Gy × 2/day, total 45 Gy, hyperfractionated course) or RT → 1.8–2 Gy/day, total dose 60–70 Gy if hyperfractionation is not possible
	Complete response → Prophylactic Cranial Irradiation (PCI) (25 Gy/2.5 Gy/fx or 30 Gy/2 Gy/fx)
Extensive-stage SCLC	Chemotherapy ± Palliative RT If response (+) → PCI Thoracic consolidative RT with 3 Gy/fx to 30 Gy or 1.5 Gy/fx to 54 Gy may be considered for patients with complete response to chemotherapy

6.2.5 Radiotherapy

The radiotherapy techniques for SCLC are similar to those for NSCLC.

Postchemotherapy tumor volume may be used for contouring of primary tumor; however, inclusion of prechemotherapy nodal volumes at initial diagnosis is recommended for RT planning.

RT should be started with the first or second cycle of chemotherapy.

Prophylactic Cranial RT
- Helmet-type cranial RT is applied.
- Anterior border includes orbital apex beyond 2 cm of lens. Inferior border is below the C2 vertebra including the temporal lobe, and 0.5 cm below the cribriform plate.
- For 3D conformal RT, contouring should start from the vertex and extend to 0.5 cm inferior of foramen magnum. PTV margin is 3–5 mm.

6.2.6 Selected Publications

NCI Canada, 1993 → patients randomized to early RT received 40 Gy in 15 fractions over 3 weeks to the primary site concurrent with the first cycle of EP (week 3), and late RT patients received the same radiation concurrent with the last cycle of EP (week 15). After completion of all chemotherapy and TI, patients without progressive disease received prophylactic cranial irradiation (25 Gy in 10 fractions over 2 weeks). Progression-free survival and overall survival were superior in the early RT arm. Patients in the late RT arm had a higher risk of brain metastases. The early administration of RT in the combined modality therapy of limited-stage SCLC is superior to late or consolidative TI.

Murray N et al (1993) Importance of timing for thoracic irradiation in the combined modality treatment of limited-stage small-cell lung cancer. The National Cancer Institute of Canada Clinical Trials Group. J Clin Oncol 11(2):336–344

Pignon, 1992 → the meta-analysis included 13 trials and 2140 patients with limited disease. The relative risk of death in the combined therapy group as compared with the chemotherapy group was 0.86 ($p = 0.001$), corresponding to a 14% reduction in the mortality rate. The benefit in terms of overall survival at 3 years was 5.4%. Thoracic radiotherapy moderately improves survival in patients with limited small-cell lung cancer who are treated with combination chemotherapy.

Pignon JP et al (1992) A meta-analysis of thoracic radiotherapy for small-cell lung cancer. N Engl J Med 327(23):1618–1624

Warde, 1992 → Meta-analysis of 11 randomized trials. The risk difference method showed that radiation therapy improved 2-year survival by 5.4%. Intrathoracic tumor control was improved by 25.3% in RT arms. The OR for excess treatment-related deaths in the thoracic radiation-treated patients was 2.54 ($p < 0.01$). This meta-analysis demonstrated a small but significant improvement in survival and a major improvement in tumor control in the thorax in patients receiving thoracic radiation therapy. However, this was achieved at the cost of a small increase in treatment-related mortality.

Warde P, Payne D (1992) Does thoracic irradiation improve survival and local control in limited-stage small-cell carcinoma of the lung? A meta-analysis. J Clin Oncol 10(6):890–895

Takada, 2002 → 231 patients with LS-SCLC. RT consisted of 45 Gy over 3 weeks (1.5 Gy twice daily), and the patients were randomly assigned to receive either sequential or concurrent TRT. All patients received four cycles of cisplatin plus etoposide every 3 weeks (sequential arm) or 4 weeks (concurrent arm). TRT was begun on day 2 of the first cycle of chemotherapy in the concurrent arm and after the fourth cycle in the sequential arm. Concurrent radiotherapy yielded better survival than sequential radiotherapy. The 2-, 3-, and 5-year survival rates for patients who received sequential radiotherapy were 35.1, 20.2, and 18.3%, respectively, as opposed to 54.4, 29.8, and 23.7%, respectively, for the patients who received concurrent radiotherapy. Hematologic toxicity was more severe in the concurrent arm.

Cisplatin plus etoposide and concurrent radiotherapy was more effective for the treatment of LS-SCLC than cisplatin plus etoposide and sequential radiotherapy.

Takada M et al (2002) Phase III study of concurrent vs. sequential thoracic radiotherapy in combination with cisplatin and etoposide for limited-stage small-cell lung cancer: results of the Japan Clinical Oncology Group Study 9104. J Clin Oncol 20(14):3054–3060

EORTC 08993/22993, 2007 (2001–2006) → 286 extensive-stage SCLC. Randomized after 4–6 cycles of CT in the case of complete response to PCI (20 Gy/5 fractions or 30 Gy/12 fractions) vs. surveillance.

- One-year symptomatic brain metastasis: 15% in PCI arm vs. 40% in surveillance arm.
- One-year OS: 27% in PCI arm, 13% in surveillance arm.
- Clinically insignificant neurological side effects in PCI arm.
- PCI dose: 20 Gy/5 fractions in 60% of cases, and efficient (but long-term sequelae unknown).
- PCI is beneficial in patients with extensive-stage disease who completely respond to 4–6 cycles of chemotherapy.

Slotman B et al (2007) Prophylactic cranial irradiation in extensive small-cell lung cancer. N Engl J Med 357(7):664–672

Gustave-Roussy PCI-88, France, 1998 → 211 SCLC cases with complete response. Randomized

PCI vs. surveillance. No standard RT schedule (RT dose: 24–30 Gy, fraction dose: £3 Gy and duration: <3 weeks). Median follow-up was 5 years.

- Trial closed early due to highly significant benefit of PCI.

Laplanche A (1998) Controlled clinical trial of prophylactic cranial irradiation for patients with small-cell lung cancer in complete remission. Lung Cancer 21(3):193–201

(Turrisi), Medical University of South Carolina, USA, 1999 → 417 SCLC cases. All patients had four cycles of cisplatin/etoposide and received concurrent 45 Gy thoracic RT, but were randomized into two groups, one receiving one fraction daily and the other two fractions daily.

- Five-year survival in hyperfractionated schedule was 26 vs. 16% in the conventional arm.
- Grade III esophagitis was significantly higher in the hyperfractionation arm.
- It was recommended that a 45 Gy hyperfractionation schedule or higher doses with conventional RT should be used.

Turrisi T et al. (1999) Twice-daily compared with once-daily thoracic radiotherapy in limited small-cell lung cancer treated concurrently with cisplatin and etoposide. N Engl J Med 340(4):265–271

(Auperin) Gustave-Roussy, France, 1999 → 7 randomized trials of PCI, including 987 cases with complete response to chemotherapy.

- Three-year survival: 20.7% in PCI arm vs. 15.3% in non-PCI arm (absolute benefit of 5%).
- PCI dose increase decreased the brain metastasis risk but not survival (8, 24–25, 30, 36–40 Gy).
- PCI should be done as early as possible after chemotherapy.
- Criticisms: number of patients in four trials <100, ~14% of cases with extensive-stage SCLC, and heterogeneous dose fractionation.

Auperin A (1999) Prophylactic cranial irradiation for patients with small-cell lung cancer in complete remission. Prophylactic Cranial Irradiation Overview Collaborative Group. N Engl J Med 341(7):476–484

Brussels, Belgium, 2001→ 12 randomized trials with 1,547 cases (PCI (+) and PCI (−)). PCI given with induction CT in five trials, or given after complete response to CT in five trials.

- PCI significantly decreased brain metastasis incidence (hazard ratio = HR = 0.48).
- PCI significantly increased OS (HR = 0.82).
- No long-term neurotoxicity.
- PCI should be given to patients with complete response to chemotherapy.

Meert AP (2001) Prophylactic cranial irradiation in small cell lung cancer: a systematic review of the literature with meta-analysis. BMC Cancer 1:5

References

1. Brambilla E, Travis WD (2014) Lung cancer. In: Stewart BW, Wild CP (eds) World cancer report. World Health Organization, Lyon
2. Alberg AJ, Samet JM (2003) Epidemiology of lung cancer. Chest 123:21S
3. Mujoomdar A, Austin JH, Malhotra R et al (2007) Clinical predictors of metastatic disease to the brain from non-small cell lung carcinoma: primary tumor size, cell type, and lymph node metastases. Radiology 242:882
4. Hansen O, Sørensen P, Hansen KH (2010) The occurrence of hyponatremia in SCLC and the influence on prognosis: a retrospective study of 453 patients treated in a single institution in a 10-year period. Lung Cancer 68:111
5. Toloza EM, Harpole L, McCrory DC (2003) Noninvasive staging of non-small cell lung cancer: a review of the current evidence. Chest 123:137S
6. Oliver TW Jr, Bernardino ME, Miller JI et al (1984) Isolated adrenal masses in nonsmall-cell bronchogenic carcinoma. Radiology 153:217
7. Elrington GM, Murray NM, Spiro SG, Newsom-Davis J (1991) Neurological paraneoplastic syndromes in patients with small cell lung cancer. A prospective survey of 150patients. J Neurol Neurosurg Psychiatry 54:764
8. Havemann K, Hirsch FR, Ihde DC et al (1989) Staging and prognostic factors in small cell lung cancer: a consensus report. Lung Cancer 5:119

References

9. Buccheri G, Ferrigno D, Tamburini M (1996) Karnofsky and ECOG performance status scoring in lung cancer: a prospective, longitudinal study of 536 patients from a single institution. Eur J Cancer 32A:1135
10. van Klaveren RJ, Oudkerk M, Prokop M et al (2009) Management of lung nodules detected by volume CT scanning. N Engl J Med 361(23):2221–2229
11. Gu P, Zhao YZ, Jiang LY, Zhang W, Xin Y, Han BH (2009) Endobronchial ultrasound-guided transbronchial needle aspiration for staging of lung cancer: a systematic review and meta-analysis. Eur J Cancer 45(8):1389–1396
12. Syrigos KN, Nutting CM, Roussos CI (eds.) (2006) Tumors of the chest, pp 4–13
13. Travis WD, Brambilla EW, Burke AP et al (2015) WHO classification of tumours of the lung, pleura, thymus, and heart. IARC Press, Lyon
14. Rami-Porta R, Asamura H, Travis WD, Rusch VW (2017) Lung. In: Amin MB (ed) AJCC cancer staging manual, 8th edn. AJCC, Chicago, p 431
15. Wisnivesky JP, Bonomi M, Henschke C, Iannuzzi M, McGinn T (2005) Radiation therapy for the treatment of unresected stage I-II non-small cell lung cancer. Chest 128(3):1461–1467
16. Benedict SH, Yenice KM, Followill D et al (2010) Stereotactic body radiation therapy: the report of AAPM task group 101. Med Phys 37(8):4078–4101
17. De Ruysscher D, Wanders R, van Haren E et al (2008) HI-CHART: a phase I/II study on the feasibility of high-dose continuous hyperfractionated accelerated radiotherapy in patients with inoperable non-small- cell lung cancer. Int J Radiat Oncol Biol Phys 71:132–138
18. Mountain CF (1997) Revisions in the international system for staging lung cancer. Chest 111(6):1710–1717

Breast Cancer

Breast cancer is the most common cancer, and the second most common cause of cancer-related death in females (Fig. 7.1). Breast cancer mortality rates have been decreasing due to improved breast cancer screening and improvements in adjuvant treatment [1].

The female breast lies between the second and sixth costae, and consists of 15–20 sections called lobes (Fig. 7.2). Each lobe ends in lobules, which are smaller than the lobes. In turn, these lobules end in milk saccules that secrete milk. All of these structures are connected to each other by canals.

> Thin bands extending from the chest wall muscles to the breast skin divide the lobes from each other. Each lobe excretes milk through one main canal. These main canals converge and exit from the nipple. All of these breast structures are surrounded by connective fatty and fibrous tissue.

7.1 Pathology

Most breast cancers originate from the interface between the ductal system and the lobules, called as the terminal ductal lobular unit (Fig. 7.3). Other tumors, such as carcinoma with neuroendocrine features, cystosarcoma phyllodes, sarcoma and lymphoma, and Paget's disease, constitute less than 5% of all breast malignancies [3].

> Breast cancers are histopathologically divided into two main groups: carcinoma in situ (noninvasive breast cancer) and invasive carcinoma.
>
> - *Carcinoma in situ*: malignant epithelial cells are limited to the basal membrane of ductus and acinus.
> - *Invasive carcinoma (infiltrative cancer)*: neoplastic cells invade the basal membrane and show stromal invasion; therefore, invasive cancers may invade lymphovascular spaces, and they have the ability to metastasize to regional lymph nodes and distant organs (Table 7.1).

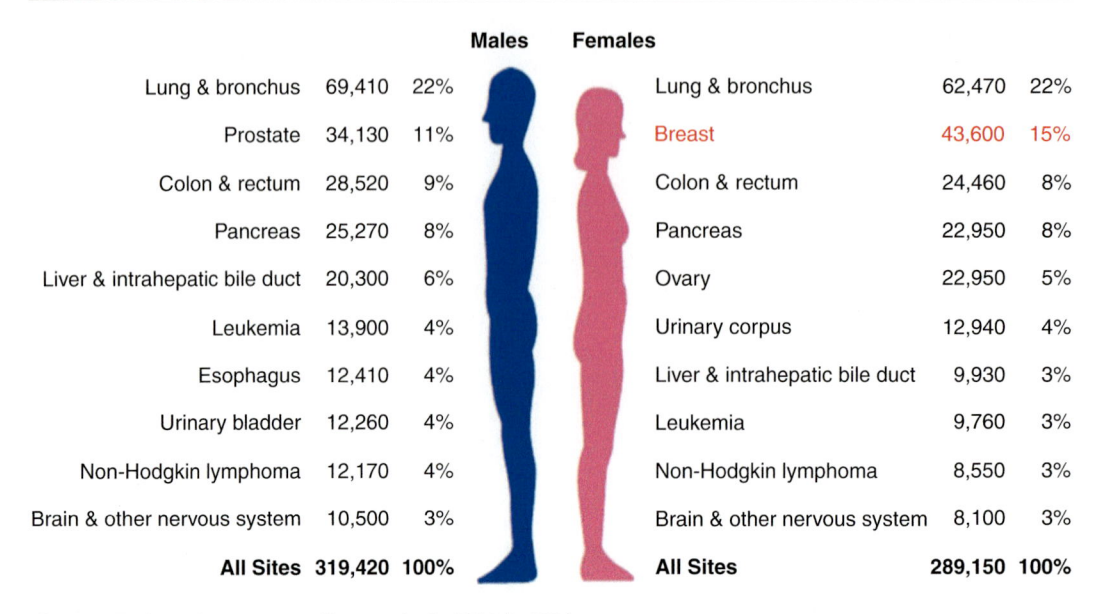

	Males			Females		
Lung & bronchus	69,410	22%		Lung & bronchus	62,470	22%
Prostate	34,130	11%		Breast	43,600	15%
Colon & rectum	28,520	9%		Colon & rectum	24,460	8%
Pancreas	25,270	8%		Pancreas	22,950	8%
Liver & intrahepatic bile duct	20,300	6%		Ovary	22,950	5%
Leukemia	13,900	4%		Urinary corpus	12,940	4%
Esophagus	12,410	4%		Liver & intrahepatic bile duct	9,930	3%
Urinary bladder	12,260	4%		Leukemia	9,760	3%
Non-Hodgkin lymphoma	12,170	4%		Non-Hodgkin lymphoma	8,550	3%
Brain & other nervous system	10,500	3%		Brain & other nervous system	8,100	3%
All Sites	**319,420**	**100%**		**All Sites**	**289,150**	**100%**

Fig. 7.1 Estimated cancer mortality rates in the USA in 2021

Fig. 7.2 Anatomy of the breast. (From [2], p 10, Fig. 2.1)

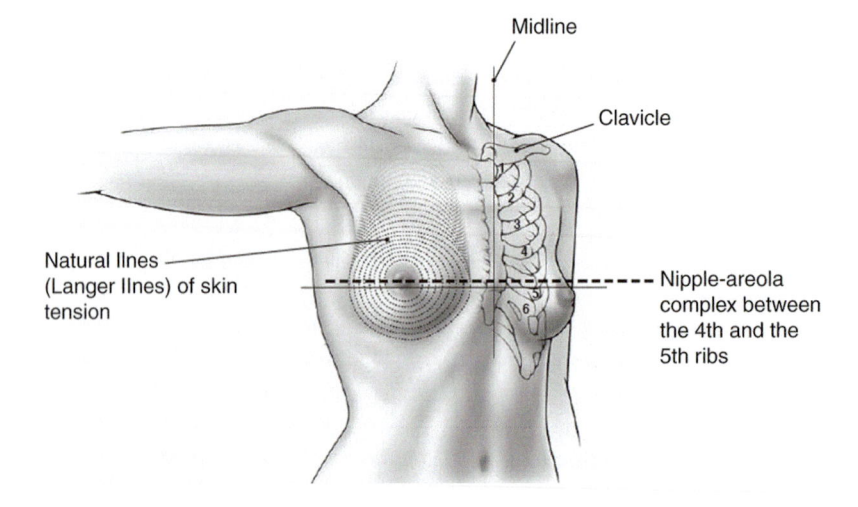

7.1 Pathology 279

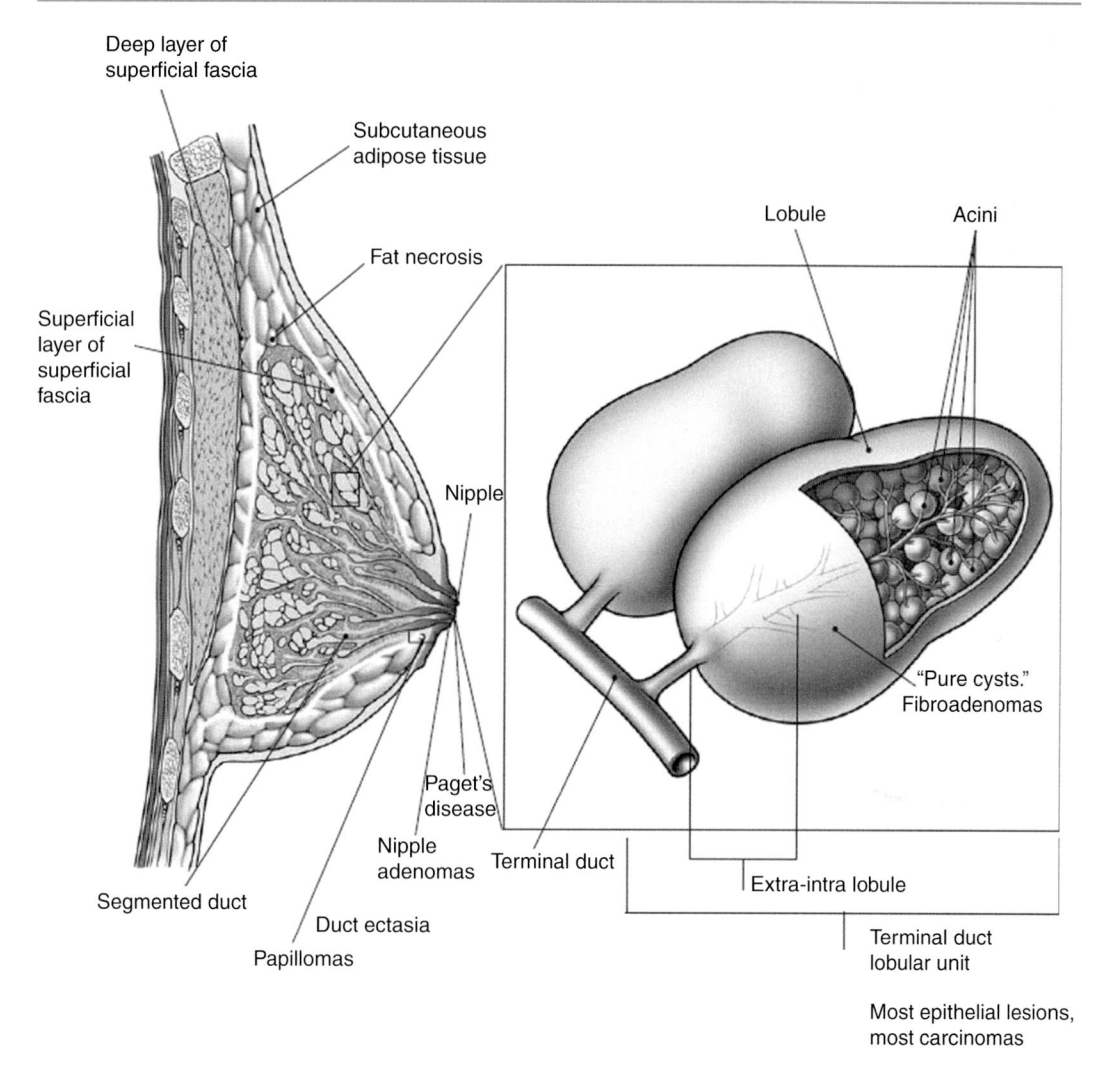

Fig. 7.3 Ductal lobular system in breast and their pathologies. (From [2], p 10, Fig. 2.3, reproduced with the permission from Springer Science and Business Media)

Table 7.1 American Joint Committee on Cancer histopathological classification of breast tumors [3]

Type

Precursor lesions
Ductal carcinoma in situ
Lobular neoplasia
    Lobular carcinoma in situ
        Classic lobular carcinoma in situ
        Pleomorphic lobular carcinoma in situ
    Atypical lobular hyperplasia
Intraductal proliferative lesions
    Usual ductal hyperplasia
    Columnar cell lesions including flat epithelial atypia
    Atypical ductal hyperplasia
Papillary lesions
    Intraductal papilloma
        Intraductal papilloma with atypical hyperplasia
        Intraductal papilloma with ductal carcinoma in situ
        Intraductal papilloma with lobular carcinoma in situ
    Intraductal papillary carcinoma
    Encapsulated papillary carcinoma
        Encapsulated papillary carcinoma with invasion
    Solid papillary carcinoma
        In situ
        Invasive
Invasive carcinoma of no special type (NST)
    Pleomorphic carcinoma
    Carcinoma with osteoclast-like stromal giant cells
    Carcinoma with choriocarcinomatous features
    Carcinoma with melanotic features
Invasive lobular carcinoma
    Classic lobular carcinoma
    Solid lobular carcinoma
    Alveolar lobular carcinoma
    Pleomorphic lobular carcinoma
    Tubulolobular carcinoma
    Mixed lobular carcinoma
Tubular carcinoma
Cribriform carcinoma
Mucinous carcinoma
Carcinoma with medullary features
    Medullary carcinoma
    Atypical medullary carcinoma
    Invasive carcinoma NST with medullary features
Carcinoma with apocrine differentiation
Carcinoma with signet-ring-cell differentiation
Invasive micropapillary carcinoma
Metaplastic carcinoma of no special type
    Low-grade adenosquamous carcinoma
    Fibromatosis-like metaplastic carcinoma
    Squamous cell carcinoma
    Spindle cell carcinoma
    Metaplastic carcinoma with mesenchymal differentiation
        Chondroid differentiation
        Osseous differentiation
        Other types of mesenchymal differentiation
    Mixed metaplastic carcinoma
    Myoepithelial carcinoma
Epithelial–myoepithelial tumors
Adenomyoepithelioma with carcinoma
Adenoid cystic carcinoma

Table 7.1 (continued)

Rare types
Carcinoma with neuroendocrine features
    Neuroendocrine tumor, well differentiated
    Neuroendocrine carcinoma poorly differentiated (small cell carcinoma)
    Carcinoma with neuroendocrine differentiation
Secretory carcinoma
Invasive papillary carcinoma
Acinic cell carcinoma
Mucoepidermoid carcinoma
Polymorphous carcinoma
Oncocytic carcinoma
Lipid-rich carcinoma
Glycogen-rich clear cell carcinoma
Sebaceous carcinoma

Invasive ductal carcinoma (infiltrating carcinoma of no special type or infiltrating carcinoma not otherwise specified (NOS) [3]. This is the most commonly seen invasive cancer (70–80%) and constitutes a wide range of cancers that do not have the specific features observed in other subtypes. It appears as solid cords of ductal tumor cells varying in size and degree of differentiation. Necrosis is rare but lymphatic invasion may be present. An associated in situ component is frequently seen. This situation may be important in relation to local relapses in patients with breast-conserving surgery.

Invasive ductal cancer with extensive in situ component [4]. Here there is a >25% in situ component within the tumor or its surroundings.

- This situation may be important in relation to local relapses in patients with breast conserving surgery.

Invasive lobular carcinoma [3]. This comprises about 8% of all invasive breast cancers, and is more commonly seen in females receiving hormone replacement therapy. This is characterized by small cells that infiltrate the stroma and adipose tissue individually and in a single-file pattern.

- Invasive lobular carcinomas are also much more commonly estrogren receptor (ER) positive than is invasive ductal carcinoma.
- It has a higher tendency for multifocality and bilaterality compared to other breast cancers.
- It also has a different metastasis pattern. Peritoneal, retroperitoneal, bone marrow, leptomeningeal, gastrointestinal, ovarian and uterine metastases are more frequent.
- Lung and pleural metastases are rare.

Inflammatory carcinoma [5]. This is a special clinical presentation of invasive breast cancer. It accounts for 0.5-2.5 % of all invasive breast cancers. As compared with locally advanced breast cancer, inflammatory carcinoma is diagnosed at an earlier age.

- Here, lymphatic drainage is damaged due to extensive dermal lymphatic invasion. Clinical findings are characterized by skin edema, erythema, induration, sensitivity and a "peau d'orange" appearance.
- Its name derives from the fact that it resembles inflammation, although there is no inflammation microscopically.
- The underlying invasive cancer is usually a high-grade infiltrative ductal carcinoma.

Ductal carcinoma in situ (DCIS) [6, 7]. This is a noninvasive malignant epithelial cell proliferation limited to the duct system that exhibits no basal membrane invasion.

- It may be limited to just a few terminal duct tubules or may include several lobules or segments in a very extensive form.
- DCIS constituted only 0.8–5% of all breast cancers before the introduction of routine screening with mammography. It now constitutes 15–20% of all breast cancers.
- 6.2–22% of all DCIS cases are bilateral, while 64–80% are multicentric.
- The risk of axillary metastasis is only 1–2%.
- DCIS is seen synchronously with invasive cancer in 2–46% of cases.
- The risk of invasive cancer development in a woman with DCIS is 10–16 times higher than in the normal female population.

One-third of patients with DCIS develop invasive disease within 10 years.

The primary goal of the treatment for DCIS is to prevent the development of invasive breast cancer. There are several prognostic indices for DCIS that can be used to standardize therapeutic approaches, the most common of which is the Van Nuys prognostic index (VPNI). Breast-conserving surgery (BCS) is recommended for patients with a low VNPI score (Table 7.2).

Table 7.2 Updated Van Nuys prognostic index

Parameter	1 point	2 points	3 points
Size (mm)	≤15	16–40	>40
Grade and necrosis	Grade I–II without necrosis	Grade I–II with necrosis	Grade III with/without necrosis
Surgical margin (mm)	≥10	1–9	<1
Age (years)	>60	40–60	<40

Total score of VNPI	Recommended treatment	10-year recurrence-free survival
4–6	BCS	97%

Important reminder: RT is currently recommended for all DCIS patients with BCS, regardless of VNPI. Clinical trial data show a reduction in ipsilateral breast recurrences with addition of RT to surgery [6].

Lobular carcinoma in situ (LCIS) [8, 9]. This is a very rare lesion found in 1% of all breast biopsies. It consists of uniform proliferated cells that fill and distort terminal duct lobular units.

- They are generally not macroscopic lesions and are diagnosed incidentally in an excised breast biopsy due to other reasons.
- They do not have a special appearance in mammography.
- LCIS is almost always observed in premenopausal women 35–55 years of age or postmenopausal women receiving hormone replacement therapy.
- LCIS cells are strongly ER positive.
- Approximately 25% of LCIS is associated with DCIS or invasive disease. It should be treated according to DCIS or invasive disease indications.
- The histological features differ between classic and nonclassic forms of LCIS. This difference impacts management.
- The relative risk of developing an invasive cancer in women with LCIS is approximately 7- to 11-fold higher than normal population.
- There is no risk of metastasis.

Treatment of LCIS. Surgical excision is recommended for any nonclassic LCIS diagnosed with biopsy or any LCIS with imaging-pathological discordance. Classical LCIS without high-risk factors can be observed with clinical and imaging follow-up since upgrade to invasive cancer rates are <5% in this setting.

- Historically, women with LCIS were treated with prophylactic bilateral mastectomy. Since most experts now consider this method too drastic for the moderate risk of invasive cancer, the decision of prophylactic surgery should be individualized.

7.2 General Presentation

Most patients in countries with established breast cancer screening programs present with an abnormal mammogram. However, approximately 15% of women present with a breast mass that is not detected on mammogram. Signs and symptoms are represented in Tables 7.3, 7.4, and 7.5.

7.3 Staging

Since there has been increasing knowledge on tumor biological characteristics that have both predictive and prognostic value, these characteristics were incorporated into the eighth edition of the AJCC staging for breast cancer in 2017 [3]. It was revised in January 1, 2018, and includes

1. Anatomical stage for use in global regions where biomarker tests (grade, ER, PR, and HER2) are not routinely available.
2. Clinical prognostic stage based on history, physical examination, imaging, biopsies, and biomarkers.
3. Pathological prognostic stage based on pathological findings at surgery and biomarkers. It can be used for all patients treated with surgery as initial treatment, but not for those who received neoadjuvant treatment.
4. AJCC uses "y" to designate stage after neoadjuvant therapy. The anatomical staging system is recommended for patients after neoadjuvant therapy. However, clinicians may find it useful to refer to the clinical prognostic staging system.

Note: AJCC eighth Edition Breast Cancer anatomical and prognostic staging systems were validated and the prognostic stage provided more

7.3 Staging

Table 7.3 Signs and symptoms in breast cancers

Signs and symptoms	Comment
Mass	Immovable
	Single dominant lesion
	Painless
	Unilateral and continuous
	No clear border
	Irregular and difficult to palpate
Pain	Ninety percent painless at presentation; pain occurs at later periods
Nipple discharge	Not frequent
	Unilateral generally bloody
Forgue sign	This is the fullness, verticality and upward position of the involved breast; upper quadrant cancers in the breast pull the nipple toward the tumor
Skin edema over breast	Tumor cells reach superficial dermal lymphatics by passing through lymph vessels within Cooper's ligaments; lymphatics are obstructed, lymph drainage is damaged, and a limited edema occurs on the skin
Nipple retraction or	This occurs as a result of tumor growth and its invasion of the nipple depression
Skin ulceration and muscle erythema	Tumor cells first invade the deep fascia, and then the pectoral and chest wall during the advanced stages of cancer
Axillary lymphadenopathies	–
Abnormal swelling in arm	Lymphatic drainage is damaged due to obstruction with tumor cells, resulting in lymphedema in arm

accurate stratification with regard to disease-specific survival than the anatomical stage [10]. The staging workup should include physical examination, diagnostic mammography, ultrasonography, and routine serum studies for patients clinical stage I or II disease. For patients with locally advanced disease, screening tests for metastatic disease should include a bone scan and computed tomography (CT) or PET-CT.

T Staging (Fig. 7.4) [3]

TX: Primary tumor cannot be assessed

T0: No evidence of primary tumor

Tis (DCIS)*: Ductal carcinoma in situ

Tis (Paget): Paget disease o f the nipple *not* associated with invasive carcinoma and/or carcinoma in situ (DCIS) in the underlying breast parenchyma. Carcinomas in the breast parenchyma associated with Paget disease are categorized based on the size and characteristics of the parenchymal disease, although the presence of Paget disease should still be noted.

T1: Tumor ≤ 20 mm in greatest dimension

- T1mi: Tumor ≤ 1 mm in greatest dimension
- T1a: Tumor > 1 mm but ≤5 mm in greatest dimension (round any measurement 1.0–1.9 to 2 mm).
- T1b: Tumor > 5 mm but ≤10 mm in greatest dimension
- T1c: Tumor > 10 mm but ≤20 mm in greatest dimension

T2: Tumor > 20 mm but ≤50 mm in greatest dimension

T3: Tumor > 50 mm in greatest dimension

T4: Tumor of any size with direct extension to the chest wall and/or to the skin (ulceration or macroscopic nodules); invasion of the dermis alone does not qualify as T4

T4a: Extension to the chest wall; invasion or adherence to pectoralis muscle in the absence of invasion of chest wall structures does not qualify as T4

T4b: Ulceration and/or ipsilateral macroscopic satellite nodules and/or edema (including peau d'orange) of the skin that does not meet the criteria for inflammatory carcinoma

Table 7.4 AJCC eighth edition clinical prognostic stage groups [3]

TNM	Grade	HER2 status	ER status	PR status	Clinical prognostic stage group
Tis N0 M0	Any	Any	Any	Any	0
T1N0 M0 T0 N1mi M0 T1N1mi M0	G1	Positive	Positive	Positive	IA
				Negative	IA
			Negative	Positive	IA
				Negative	IA
		Negative	Positive	Positive	IA
				Negative	IA
			Negative	Positive	IA
				Negative	IB
T1N0 M0 T0N1mi M0 T1N1miM0	G2	Positive	Positive	Positive	IA
				Negative	IA
			Negative	Positive	IA
				Negative	IA
		Negative	Positive	Positive	IA
				Negative	IA
			Negative	Positive	IA
				Negative	IB
T1N0 M0 T0N1mi M0 T1N1miM0	G3	Positive	Positive	Positive	IA
				Negative	IA
			Negative	Positive	IA
				Negative	IA
		Negative	Positive	Positive	IA
				Negative	IB
			Negative	Positive	IB
				Negative	IB
T0N1M0 T1N1M0 T2N0M0	G1	Positive	Positive	Positive	IB
				Negative	IIA
			Negative	Positive	IIA
				Negative	IIA
		Negative	Positive	Positive	IB
				Negative	IIA
			Negative	Positive	IIA
				Negative	IIA
T0N1M0 T1N1M0 T2N0M0	G2	Positive	Positive	Positive	IB
				Negative	IIA
			Negative	Positive	IIA
				Negative	IIA
		Negative	Positive	Positive	IB
				Negative	IIA
			Negative	Positive	IIA
				Negative	IIB

7.3 Staging

Table 7.4 (continued)

TNM	Grade	HER2 status	ER status	PR status	Clinical prognostic stage group
T0N1M0 T1N1M0 T2N0M0	G3	Positive	Positive	Positive	IB
				Negative	IIA
			Negative	Positive	IIA
				Negative	IIA
		Negative	Positive	Positive	IIA
				Negative	IIB
			Negative	Positive	IIB
				Negative	IIB
T2N1M0 T3N0M0	G1	Positive	Positive	Positive	IB
				Negative	IIA
			Negative	Positive	IIA
				Negative	IIB
		Negative	Positive	Positive	IIA
				Negative	IIB
			Negative	Positive	IIB
				Negative	IIB
T2N1M0 T3N0M0	G2	Positive	Positive	Positive	IB
				Negative	IIA
			Negative	Positive	IIA
				Negative	IIB
		Negative	Positive	Positive	IIA
				Negative	IIB
			Negative	Positive	IIB
				Negative	IIIB
T2N1M0 T3N0M0	G3	Positive	Positive	Positive	IB
				Negative	IIB
			Negative	Positive	IIB
				Negative	IIB
		Negative	Positive	Positive	IIB
				Negative	IIIA
			Negative	Positive	IIIA
				Negative	IIIB
T0N2M0 T1N2M0 T2N2M0 T3N1M T3N2M0	G1	Positive	Positive	Positive	IIA
				Negative	IIIA
			Negative	Positive	IIIA
				Negative	IIIA
		Negative	Positive	Positive	IIA
				Negative	IIIA
			Negative	Positive	IIIA
				Negative	IIIB
T0N2M0 T1N2M0 T2N2M0 T3N1M T3N2M0	G2	Positive	Positive	Positive	IIA
				Negative	IIIA
			Negative	Positive	IIIA
				Negative	IIIA
		Negative	Positive	Positive	IIA
				Negative	IIIA
			Negative	Positive	IIIA
				Negative	IIIB

(continued)

Table 7.4 (continued)

TNM	Grade	HER2 status	ER status	PR status	Clinical prognostic stage group
T0N2M0 T1N2M0 T2N2M0 T3N1M T3N2M0	G3	Positive	Positive	Positive	IIA
				Negative	IIIA
			Negative	Positive	IIIA
				Negative	IIIA
		Negative	Positive	Positive	IIA
				Negative	IIIA
			Negative	Positive	IIIA
				Negative	IIIB
T4N0M0 T4N1M0 T4N2M0 AnyTN3M0	G1	Positive	Positive	Positive	IIIA
				Negative	IIIB
			Negative	Positive	IIIB
				Negative	IIIB
		Negative	Positive	Positive	IIIB
				Negative	IIIB
			Negative	Positive	IIIB
				Negative	IIIC
T4N0M0 T4N1M0 T4N2M0 AnyTN3M0	G2	Positive	Positive	Positive	IIIA
				Negative	IIIB
			Negative	Positive	IIIB
				Negative	IIIB
		Negative	Positive	Positive	IIIB
				Negative	IIIB
			Negative	Positive	IIIB
				Negative	IIIC
T4N0M0 T4N1M0 T4N2M0 AnyTN3M0	G3	Positive	Positive	Positive	IIIB
				Negative	IIIB
			Negative	Positive	IIIB
				Negative	IIIB
		Negative	Positive	Positive	IIIB
				Negative	IIIC
			Negative	Positive	IIIC
				Negative	IIIC
Any T any N M1	Any	Any	Any	Any	IV

7.3 Staging

Table 7.5 AJCC eighth edition pathological prognostic stage groups [3]

TNM	Grade	HER2 status	ER status	PR status	Pathological prognostic stage group
Tis N0 M0	Any	Any	Any	Any	0
T1N0 M0 T0 N1mi M0 T1N1mi M0	G1	Positive	Positive	Positive	IA
				Negative	IA
			Negative	Positive	IA
				Negative	IA
		Negative	Positive	Positive	IA
				Negative	IA
			Negative	Positive	IA
				Negative	IA
T1N0 M0 T0N1mi M0 T1N1miM0	G2	Positive	Positive	Positive	IA
				Negative	IA
			Negative	Positive	IA
				Negative	IA
		Negative	Positive	Positive	IA
				Negative	IA
			Negative	Positive	IA
				Negative	IB
T1N0 M0 T0N1mi M0 T1N1miM0	G3	Positive	Positive	Positive	IA
				Negative	IA
			Negative	Positive	IA
				Negative	IA
		Negative	Positive	Positive	IA
				Negative	IA
			Negative	Positive	IA
				Negative	IB
T0N1M0 T1N1M0 T2N0M0	G1	Positive	Positive	Positive	IA
				Negative	IB
			Negative	Positive	IB
				Negative	IIA
		Negative	Positive	Positive	IA
				Negative	IB
			Negative	Positive	IB
				Negative	IIA
T0N1M0 T1N1M0 T2N0M0	G2	Positive	Positive	Positive	IA
				Negative	IB
			Negative	Positive	IB
				Negative	IIA
		Negative	Positive	Positive	IA
				Negative	IIA
			Negative	Positive	IIA
				Negative	IIA

(continued)

Table 7.5 (continued)

TNM	Grade	HER2 status	ER status	PR status	Pathological prognostic stage group
T0N1M0 T1N1M0 T2N0M0	G3	Positive	Positive	Positive	IA
				Negative	IIA
			Negative	Positive	IIA
				Negative	IIA
		Negative	Positive	Positive	IB
				Negative	IIA
			Negative	Positive	IIA
				Negative	IIA
T2N1M0 T3N0M0	G1	Positive	Positive	Positive	IA
				Negative	IIB
			Negative	Positive	IIB
				Negative	IIB
		Negative	Positive	Positive	IA
				Negative	IIB
			Negative	Positive	IIB
				Negative	IIB
T2N1M0 T3N0M0	G2	Positive	Positive	Positive	IB
				Negative	IIB
			Negative	Positive	IIB
				Negative	IIB
		Negative	Positive	Positive	IB
				Negative	IIB
			Negative	Positive	IIB
				Negative	IIB
T2N1M0 T3N0M0	G3	Positive	Positive	Positive	IB
				Negative	IIB
			Negative	Positive	IIB
				Negative	IIB
		Negative	Positive	Positive	IIA
				Negative	IIB
			Negative	Positive	IIB
				Negative	IIIA
T0N2M0 T1N2M0 T2N2M0 T3N1M T3N2M0	G1	Positive	Positive	Positive	IB
				Negative	IIIA
			Negative	Positive	IIIA
				Negative	IIIA
		Negative	Positive	Positive	IB
				Negative	IIIA
			Negative	Positive	IIIA
				Negative	IIIA

7.3 Staging

Table 7.5 (continued)

TNM	Grade	HER2 status	ER status	PR status	Pathological prognostic stage group
T0N2M0 T1N2M0 T2N2M0 T3N1M T3N2M0	G2	Positive	Positive	Positive	IB
				Negative	IIIA
			Negative	Positive	IIIA
				Negative	IIIA
		Negative	Positive	Positive	IB
				Negative	IIIA
			Negative	Positive	IIIA
				Negative	IIIB
T0N2M0 T1N2M0 T2N2M0 T3N1M T3N2M0	G3	Positive	Positive	Positive	IIA
				Negative	IIIA
			Negative	Positive	IIIA
				Negative	IIIA
		Negative	Positive	Positive	IIB
				Negative	IIIA
			Negative	Positive	IIIA
				Negative	IIIC
T4N0M0 T4N1M0 T4N2M0 AnyTN3M0	G1	Positive	Positive	Positive	IIIA
				Negative	IIIB
			Negative	Positive	IIIB
				Negative	IIIB
		Negative	Positive	Positive	IIIA
				Negative	IIIB
			Negative	Positive	IIIB
				Negative	IIIB
T4N0M0 T4N1M0 T4N2M0 AnyTN3M0	G2	Positive	Positive	Positive	IIIA
				Negative	IIIB
			Negative	Positive	IIIB
				Negative	IIIB
		Negative	Positive	Positive	IIIA
				Negative	IIIB
			Negative	Positive	IIIB
				Negative	IIIC
T4N0M0 T4N1M0 T4N2M0 AnyTN3M0	G3	Positive	Positive	Positive	IIIB
				Negative	IIIB
			Negative	Positive	IIIB
				Negative	IIIB
		Negative	Positive	Positive	IIIB
				Negative	IIIC
			Negative	Positive	IIIC
				Negative	IIIC
Any T any N M1	Any	Any	Any	Any	IV

T4c: Both T4a and T4b are present
T4d: Inflammatory carcinoma
*Note: Lobular carcinoma in situ (LCIS) is a benign entity and is removed from TNM staging in the AJCC Cancer Staging Manual, 8th Edition.

- cN3a: Metastases in ipsilateral infraclavicular lymph node(s)
- cN3b: Metastases in ipsilateral internal mammary lymph node(s) and axillary lymph node(s)
- cN3c: Metastases in ipsilateral supraclavicular lymph node(s)

Regional Lymph Nodes—Clinical (cN) (Fig. 7.5) [3]
cNX: Regional lymph nodes cannot be assessed (e.g., previously removed)
cN0: No regional lymph node metastasis (by imaging or clinical examination)
cN1: Metastasis to movable ipsilateral level I–II axillary lymph node(s)

- cN1mi: Micrometastases (approximately 200 cells, larger than 0.2 mm, but none larger than 2.0 mm)

cN2: Metastases in ipsilateral level I, II axillary lymph nodes that are clinically fixed or matted; or in ipsilateral internal mammary nodes in the absence of axillary lymph node metastases

- cN2a: Metastases in ipsilateral level I, II axillary lymph nodes fixed to one another (matted) or to other structures
- cN2b: Metastases only in ipsilateral internal mammary nodes in the absence of axillary lymph node metastases

cN3: Metastases in ipsilateral infraclavicular (level III axillary) lymph node(s) with or without level 1, II axillary lymph node involvement; or in ipsilateral internal mammary lymph node(s) with level I, II axillary lymph node metastases; or metastases in ipsilateral supraclavicular lymph node(s) with or without axillary or internal mammary lymph node involvement

Regional Lymph Nodes—Pathological (pN) (Fig. 7.6) [3]
pNX: Regional lymph nodes cannot be assessed (e.g., not removed for pathological study or previously removed)
pN0: No regional lymph node metastasis identified or isolated tumor cells (ITC) only

- pN0 (i+): ITCs only (malignant cell clusters no larger than 0.2 mm) in regional lymph node(s)
- pN0 (mol+): Positive molecular findings by reverse transcriptase polymerase chain reaction (RT-PCR); no ITCs detected

pN1: Micrometastases; or metastases in 1–3 axillary lymph nodes; and/or clinically negative internal mammary nodes with micrometastases or macrometastases by sentinel lymph node biopsy

- pN1mi: Micrometastases (approximately 200 cells, larger than 0.2 mm, but none larger than 2.0 mm)
- pN1a: Metastases in 1–3 axillary lymph nodes; at least one metastasis larger than 2.0 mm
- pN1b: Metastases in ipsilateral internal mammary sentinel nodes excluding ITCs
- pN1c: pN Ia and pN Ib combined

7.3 Staging

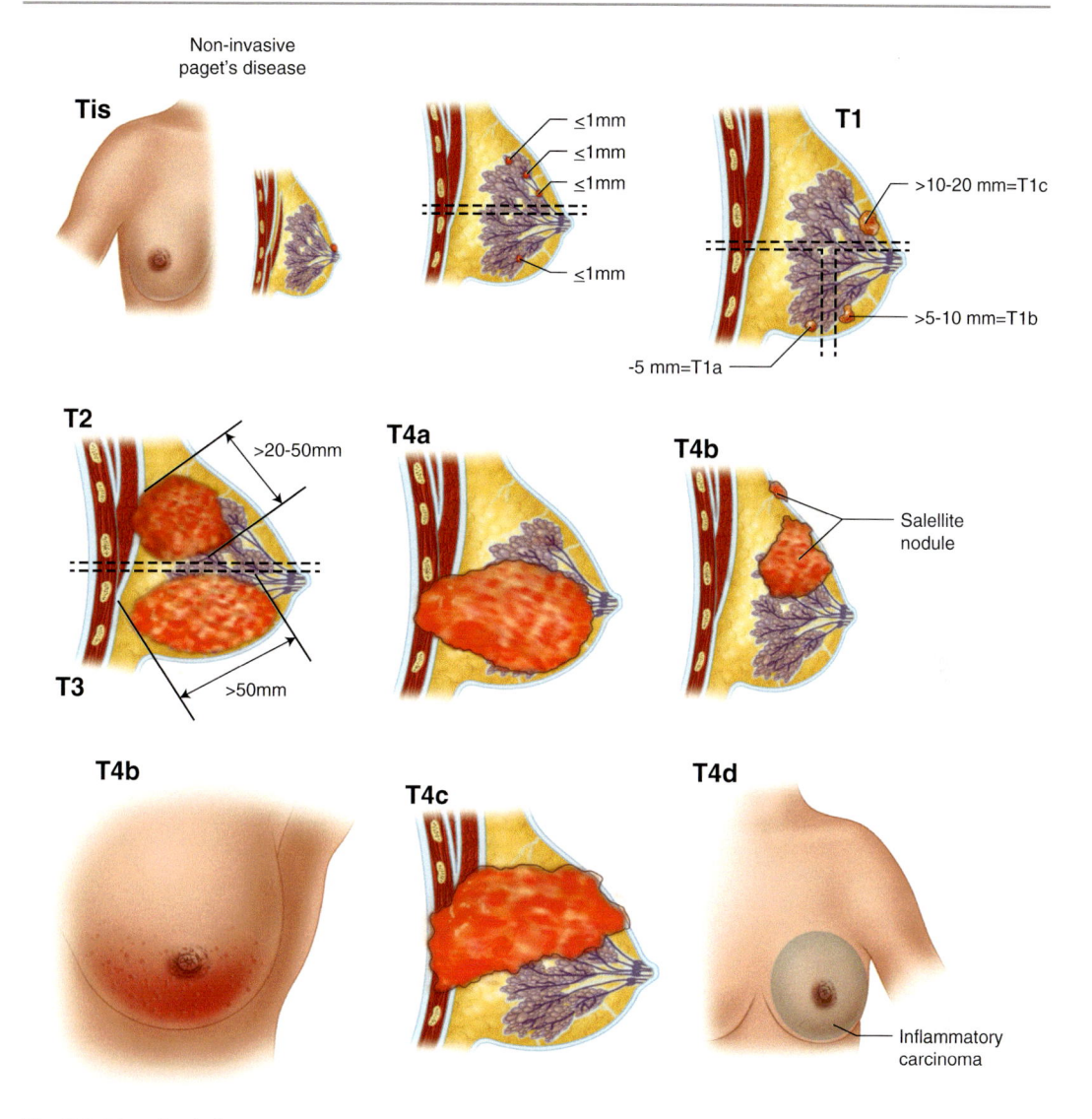

Fig. 7.4 T staging in breast cancer. (From [3] pp 597–598, Figs. 48.3–48.5, reproduced with the permission from the American Joint Committee on Cancer)

pN2: Metastases in 4–9 axillary lymph nodes; or positive ipsilateral internal mammary lymph nodes by imaging in the absence of axillary lymph node metastases

- pN2a: Metastases in 4–9 axillary lymph nodes (at least one tumor deposit larger than 2.0 mm)
- pN2b: Metastases in clinically detected internal mammary lymph nodes with or without microscopic confirmation; with pathologically negative axillary nodes

pN3: Metastases in 10 or more axillary lymph nodes; or in infraclavicular (level III axillary) lymph nodes; or positive ipsilateral internal mammary lymph nodes by imaging in the presence of one or more positive level I, II axillary lymph nodes; or in more than three axillary lymph nodes and micrometastases or macrometastases by sentinel lymph node biopsy in clinically negative ipsilateral internal mammary lymph nodes; or in ipsilateral supraclavicular lymph nodes

- pN3a: Metastases in 10 or more axillary lymph nodes (at least one tumor deposit larger than 2.0 mm); or metastases to the infraclavicular (level III axillary lymph) nodes
- pN3b: pN1a or pN2a in the presence of cN2b (positive internal mammary nodes by imaging); or pN2a in the presence of pN1b
- pN3c: Metastases in ipsilateral supraclavicular lymph nodes

Definition of Distant Metastasis (M) [3]
M0: No clinical or radiographic evidence of distant metastases*

cM0(i+): No clinical or radiographic evidence of distant metastases in the presence of tumor cells or deposits no larger than 0.2 mm detected microscopically or by molecular techniques in circulating blood, bone marrow, or other nonregional nodal tissue in a patient without symptoms or signs of metastases

M1: Distant metastases detected by clinical and radiographic means (cM) and/or histologically proven metastases larger than 0.2 mm (pM)

*Note that imaging studies are not required to assign the cM0 category.

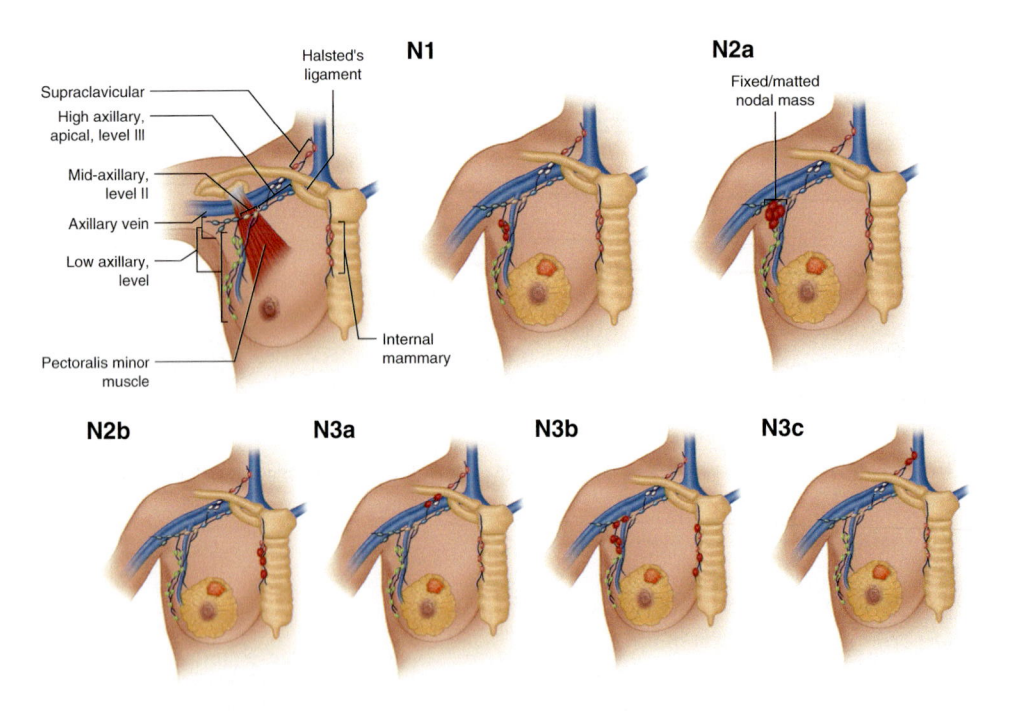

Fig. 7.5 Clinical N staging in breast cancer. (From [3] pp 596, 600, Figs. 48.2, 48.6, reproduced with the permission from the American Joint Committee on Cancer)

7.3 Staging

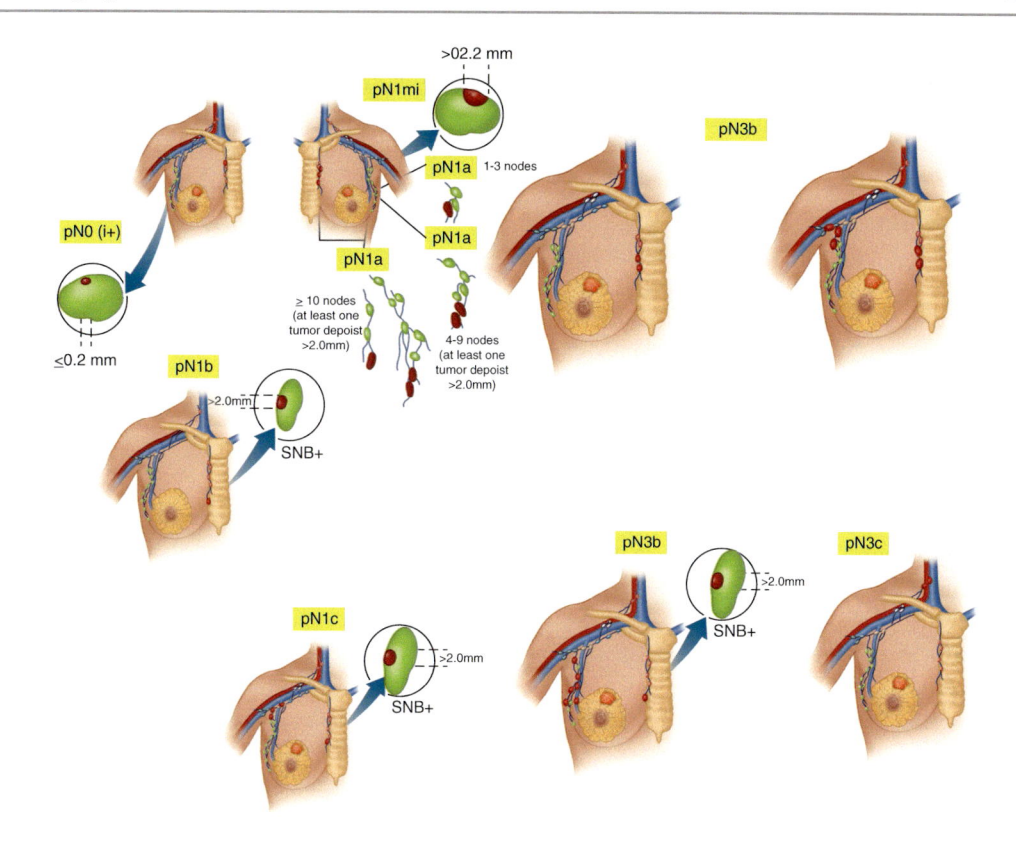

Fig. 7.6 Pathological N staging in breast cancer. (From [3] pp. 607–608, Figs. 48.10, 48.11 reproduced with the permission from the American Joint Committee on Cancer)

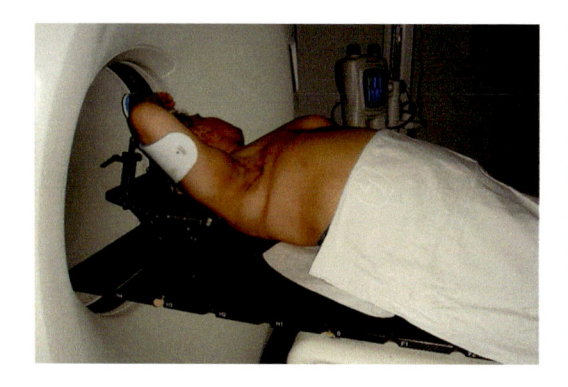

Fig. 7.7 Patient is positioned supine, ipsilateral arm above head, on an angled breast board, face moved toward the contralateral side. In order to keep the upper part of the supraclavicular field flat, arm is abducted almost with 90°. (Courtesy of Hacettepe University)

AJCC Anatomic Stage Groups
Stage 0: TisN0M0 Stage IA: T1N0M0
  Stage IB: T0N1miM0; T1N1miM0
  Stage IIA: T0N1M0; T1N1M0; T2N0M0 Stage IIB: T2N1M0; T3N0M0
  Stage IIIA: T0N2M0; T1N2M0; T2N2M0; T3N1M0; T3N2M0 Stage IIIB: T4N0M0; T4N1M0; T4N2M0
  Stage IIIC: Any TN3M0 Stage IV: Any T, Any N, M1
  Notes for anatomic stage grouping:

- T1 includes T1mi.
- M0 includes M0 (i+).

- If a patients presents with M1 disease prior to neoadjuvant systemic therapy, the stage is stage IV and remains stage IV regardless of response to neoadjuvant therapy.
- Staging following neoadjuvant therapy is denoted with a "yc" or "yp" prefix to the T and N classification. No stage group is assigned if there is a complete pathological response (pCR) to neoadjuvant therapy, for example, ypT0ypN0cM0.

Molecular Subtypes
- Luminal A: ER/PR+ HER2neu−, Ki67 is low
- Luminal B: ER/PR+ HER2Neu +/− Ki67 is high
- HER/neu enriched: ER/PR − HER2Neu +, regardless of Ki67 level
- Basal-like: ER/PR − HER2Neu−, regardless of Ki67 level

7.4 Treatment Algorithm

DCIS [11]
1. BCS + RT.
   - RT, decreases local recurrences
2. Total mastectomy (no axillary dissection)
   - Indications:
     - Diffuse microcalcifications Multicentricity
     - If the use of RT is contraindicated
     - Permanent surgical margin (+) despite re-excisions Patient request

*Note: Adjuvant tamoxifen/aromatase inhibitor (postmenopausal) is recommended for 5 years if ER (+).

LCIS [12]
1. Surgical excision for any nonclassic LCIS or any LCIS with imaging-pathological discordance

2. Lifelong surveillance ± tamoxifen (presence of high risk)
3. Young age, family history, or genetic predisposition: prophylactic bilateral mastectomy may be considered

Early-Stage Breast Cancer [Stage I–IIA and a Subset of Stage IIB Disease (T2N1)]
- BCS + axillary sentinel lymph node dissection (SLND) + RT

*Note: If SLND is negative, no further ALND is required.

- RT may not be needed; age > 70 years and adjuvant hormonal therapy is given
- *Contraindications for BCS:*
  - Multicentricity
  - Tumor size/breast size is too large
  - Diffuse microcalcification
  - Permanent surgical margin (+) despite re-excisions
  - Previous history of breast cancer
  - Pregnancy
  - Collagen vascular disease
  - Previous history of RT to breast

Alternative → Modified radical mastectomy (MRM) ± RT ± CT ± hormonal treatment (HT)
*Note: CT should be recommended in patients with high recurrence score on 21-gene assay.
Alternative → Neoadjuvant CT may be considered in some patients (particularly those with HER2-positive or triple-negative)

Locally Advanced Breast Cancer [a Subset of Stage IIB (T3N0) and Stage IIIA to IIIC]
- Neoadjuvant CT + surgery (BCS + AD or MRM) + RT ± HT ± CT
- *Alternative* → MRM + RT ± HT ± CT

- Anti-Her2 therapy (trastuzumab) should be added to standard CT regimen in patients with HER2+ disease

Stage IV
- CT/HT
  - Biphosphonates in bone metastases
  - ± Palliative RT
  - ± Targeted therapies (e.g., herceptin, lapatinib)

Special Circumstances

Isolated axillary lymph node metastasis (TxN(+)M0)

- Neoadjuvant CT + axillary dissection + RT (RT; breast + axilla ± supraclavicular field)
- Alternative → MRM ± RT ± HT ± CT

Breast cancer in pregnancy
First trimester → discuss termination of pregnancy

- If not: MRM ± CT, and after birth or at the second trimester ± RT ± HT

Second to third trimester → surgery (BCS + AD or MRM) ± CT, and after birth ± RT ± HT

Post-mastectomy RT (PMRT) Indications [13]
- Involvement of axillary lymph nodes
- T3 (>5 cm tumor), T4 (skin/chest wall invasion) or stage III tumor
- Positive surgical margin
- Gross extracapsular invasion in axilla
- Gross residual disease
- Invasion of pectoral muscle fascia
- Inadequate axillary dissection (<6 or <10 lymph nodes excised)

Late side effects of PMRT: Lymphatic edema, brachial plexopathy, radiation pneumonitis, costal fractures, cardiac toxicity, and secondary malignancies due to radiation; modern RT should be recommended, regardless of its side effects.

Relative PMRT Indications [13]
- Close surgical margin (<1 mm)
- Multiple primary tumors (multicentricity)

- Patients with complete (I–III) or level I–II axillary dissection (adequate axillary dissection) are not routinely recommended for radiotherapy to all fields if no axillary metastasis is detected.
- Supraclavicular lymph node radiotherapy is recommended for patients with four or more axillary lymph node metastases since the risk of recurrence in this region is very high. Supraclavicular lymph node radiotherapy is generally recommended for 1–3 involved axillary lymph nodes per updated ASCO/ASTRO/SSO recommendations in patients particularly with lymphovascular invasion or large primary. No axillary staging or no ALND in the case of +SLNB are relative indications for supraclavicular lymph node radiotherapy.
- Internal mammary lymph node (IMN) radiotherapy is a highly controversial issue. One should consider the high risk of cardiotoxicity regardless of technique since coronary arteries are very close to the high-dose region, particularly in left-sided breast cancers. RT should be administered to IMN if clinically or pathologically positive; otherwise, it is at the discretion of the treating physician.

Adjuvant Chemotherapy Indications [14]
- Tumor >1 cm or lymph node (+) (adjuvant CT decreases distant metastases as well as local relapses after surgery ± RT)

Neoadjuvant Chemotherapy Indications [15]
- Locally advanced breast cancer
- Initial therapy in stage IIIB–IIIC
- Inoperable stage IIB–IIIA (neoadjuvant CT increases probability of BCS in 20–30% of cases)

*Note: No survival advantage was seen with the administration of neoadjuvant chemotherapy compared with postoperative chemotherapy; however, there are few potential advantages of this approach. First of all, it may provide breast conservation and it allows assessment of the primary tumor's response. This can lead to early change to a different regimen in patients who do not respond or progress. The response to systemic therapy also reflects prognosis.

- **Breast-conserving surgery** [16]. This is the excision of the tumor together with a spe cific amount of surrounding healthy breast tissue (wide local excision, lumpectomy, quadrantectomy) (± axillary SLND or axillary lymph node dissection).
  - It is frequently used in the management of early-stage breast cancer since it provides good cosmesis with the preservation of the breast as well as good psycho social effects on the patient's social life.
- **Modified radical mastectomy** [17]. The entire breast, areola, nipple, pectoral muscle fascia, and pectoralis minor muscle are excised, and axillary dissection (to at least level I–II) is performed. MRM alone already includes AD, so it is not correct terminology to use MRM + AD in the patient's chart.

7.5 Radiotherapy

Radiotherapy for breast cancer decreases the locoregional relapse rate, increases survival, and palliates symptoms according to stage. Radiotherapy after BCS is an essential component of treatment for early-stage breast cancer [18]. Adjuvant radiotherapy during more advanced stages increases locoregional control and increases survival, particularly in patients with axillary lymph node metastases [19].

A breast board should be used in breast cancer radiotherapy, and the patient should put her arms up in a comfortable position at a fixed location (Figs. 7.6 and 7.7).

Breast RT After BCS and Chest Wall RT After MRM (Figs. 7.8 and 7.9) (RTOG Consensus Guidelines for Target Delineation in Breast Cancer) [20]
Breast clinical target volume
   Includes the apparent CT glandular breast tissue
   Cranial: Clinical reference and second rib insertion
   Caudal: Clinical reference and loss of CT apparent breast
   Anterior: Skin (5 mm inside to skin)
   Posterior: Excludes pectoralis muscles, chest wall muscles, ribs
   Lateral: Clinical reference and mid axillary line typically, excludes latissimus dorsi muscle
   Medial: Sternal-rib junction
   Lumpectomy GTV: includes seroma and surgical clips when present.
   Lumpectomy CTV: Lumpectomy GTV + 1–2 cm margins respecting the CTV breast anteriorly and posteriorly; excluding pectoralis muscle and chest wall
   PTV = CTV + 0.5 cm (depending on the institution's protocole)
Chest wall clinical target volume

Cranial: Caudal border of the clavicle head
Caudal: Clinical reference and loss of CT apparent contralateral breast
Anterior: Skin (3 mm inside to skin)
Posterior: Rib–pleural interface (includes pectoralis muscles, chest wall muscles, ribs)
Lateral: Clinical reference and mid axillary line typically, excludes latissimus dorsi muscle
Medial: Sternal-rib junction

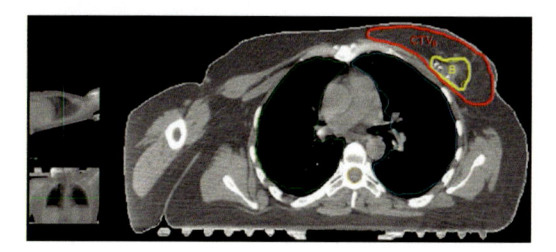

Fig. 7.8 Contouring of clinical target volumes of breast and tumor bed in a female with left-side breast cancer after breast-conserving surgery. CTV_B CTV for intact breast (red), B tumor bed for boost CTV (yellow)

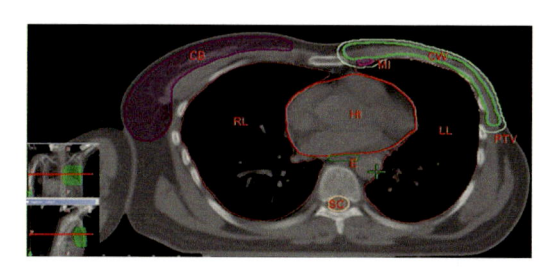

Fig. 7.9 Contouring of clinical target volumes of chest wall in a female with left-side breastcancer after mastectomy. PTV planning target volume for chest wall (light yellow), CW clinical target volume for chest wall (green)

- BCS or MRM scars, drain sites, and the borders of breast should be marked with wires to make them visible in the computed tomography. CT scans with 3 mm slice thickness are obtained starting form chin to at least 3 cm lower to the inframammary fold of breast.

Regional Lymphatics (Fig 7.10) RTOG Consensus Guidelines for Target Delineation in Breast Cancer) [20]

Supraclavicular field

Cranial: Caudal to the cricoid cartilage
Caudal: Junction of brachiocephalic-axillary veins or caudal edge of clavicle head
Anterior: Sternocleidomastoid muscle (SCM muscle)
Posterior: Anterior aspect of the scalene muscle
Lateral: Lateral edge of SCM muscle (cranial); junction of first rib and clavicle (caudal)
Medial: Excludes thyroid and trachea

Axilla level I

Cranial: Axillary vessels cross lateral edge of pectoralis minor muscle
Caudal: Pectoralis major muscle insert into ribs
Anterior: Plane defined by anterior surface of pectoralis major muscle and latissimus dorsi muscle
Posterior: Anterior surface of subscapularis muscle
Lateral: Medial border of latissimus dorsi muscle
Medial: Lateral border of pectoralis minor muscle

Axilla level II

Cranial: Axillary vessels cross medial edge of pectoralis minor muscle
Caudal: Axillary vessels cross lateral edge of pectoralis minor muscle
Anterior: Anterior surface of pectoralis minor muscle
Posterior: Ribs and intercostal muscles
Lateral: Lateral border of pectoralis minor muscle
Medial: Medial border of pectoralis minor muscle

Axilla level III

Cranial: Insertion of pectoralis minor muscle on cricoid

Caudal: Axillary vessels cross medial edge of pectoralis minor muscle
Anterior: Posterior surface of pectoralis major muscle
Posterior: Ribs and intercostal muscles
Lateral: Medial border of pectoralis minor muscle
Medial: Thoracic inlet
Internal mammary lymph nodes
Cranial: Superior aspect of the medial first rib
Caudal: Cranial aspect of the fourth rib

- Nodal regions contoured as target volume depend on the specific clinical case. The irradiation of regional lymphatics reduces the risk of nodal recurrence [21]. If the axilla is dissected adequately and negative, axillary RT is not indicated. However, if the patient has an inadequate axillary dissection or if there is evidence of extracapsular invasion, the full axillary region should be treated.
- When tangent field is placed encompassing all of breast, posterior of the borders must not exceed >2–3 cm in lung (Fig. 7.10). Minimum 2 cm extension is usually designed to encompass the breathing motion and movement of breast anteriorly. Wedges or field-in-field technique are used to improve dose homogeneity and to keep hot spots <10%.
- Most local relapses occur within or close to scar tissue. Thus, the entire mastectomy or incision scar should be irradiated. Additional electron fields can be used. For electron boost, energy is elected based on the depth of the tumor bed, with 90% isodose line covering the target.
- A breast board is a simple sloped plane that makes the chest wall parallel to the ground plane. It therefore prevents the lungs from receiving higher doses, simplifies setup, and enables dose homogeneity (Fig. 7.11).

Dose Recommendations

- Conventional doses: 1.8–2 Gy × 25 fractions to 45–50 Gy
- Hypofractionation doses: 2.67 Gy × 15 fraction to 40 Gy
- Boost dose: 10 Gy in 4–5 fractions (4–6 MV photon or 9–18 MeV electron energy may be used depending on the depth of the tumor bed)
- Surgical margin (+)/young age/close margin and re-excision is not possible: 14–16 Gy in 7–8 fractions or 12.5 Gy in 5 fractions may be used
- Normalization is done to 80–85% isodose so that a steep dose gradient can be used after these isodoses to spare underlying lung tissue
- The boost can be delivered by tangential fields when there is a deeply seated tumor bed

7.6 Selected Publications

Breast cancer prevention

NSABP P-1 Fisher B et al (1998) Tamoxifen for prevention of breast cancer: report of the National Surgical Adjuvant Breast and Bowel Project P-1 Study. J Natl Cancer Inst 90(18):1371–1388) → Randomized. More than 13,000 women with 5-year predicted risk of 1.66 for developing breast cancer per the Gail Model or women with lobular carcinoma in situ.

Placebo vs. tamoxifen (TMX) 20 mg for 5 years

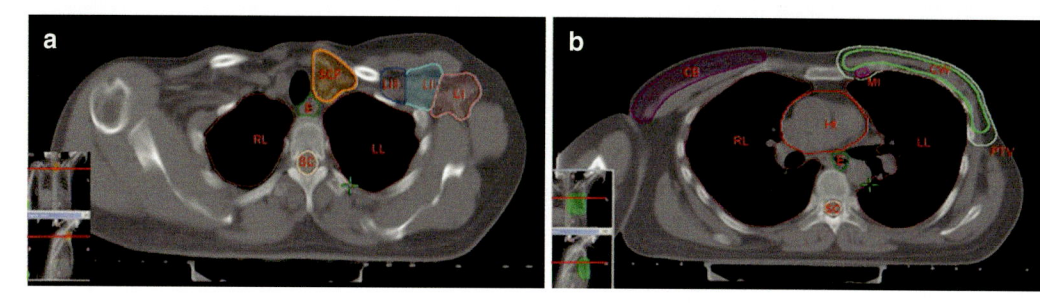

Fig. 7.10 Contouring of clinical target volumes of regional lymphatics in a female with left-side breast cancer after mastectomy. (**a**) Lymphatics, (**b**) chest wall. *SCF* supraclavicular fossa lymphatics (orange), *LI* level I axillary lymphatics (pink), *LII* level II axillary lymphatics (turquoise), *LIII* level III axillary lymphatics (dark blue), *MI* mammaria interna lymphatics (magenta)

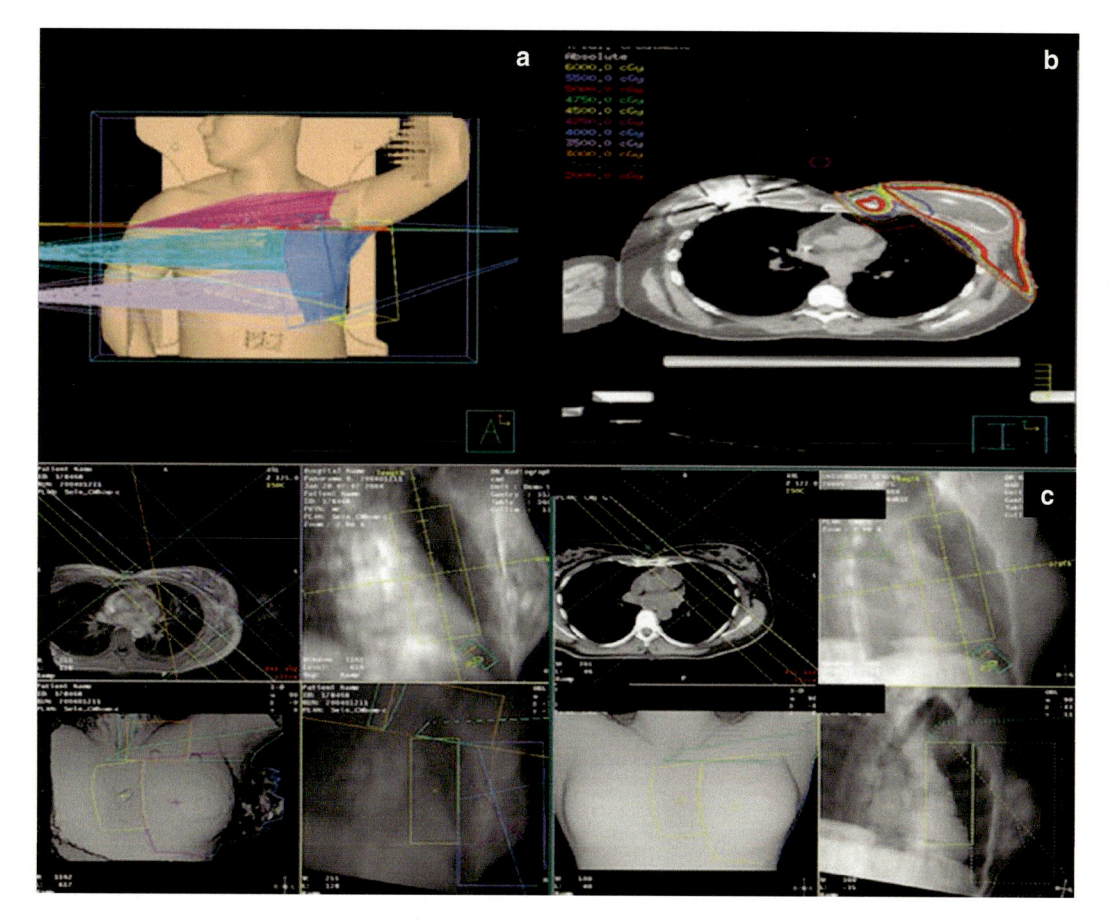

Fig. 7.11 (**a**) Setup of fields on patient body; (**b**) axial view of dose distribution; (**c**) conformal RT fields in breast cancer

TMX
- 49% decrease in invasive cancer risk
- 56% decrease in LCIS
- 86% decrease in atypical ductal hyperplasia
- 56% decrease in noninvasive breast cancer
- RR = 2.53 for endometrial cancer
- No increase in ischemic heart disease
- Decrease in fracture risk due to osteoporosis

Updated in 2005 with 7 years of median follow-up Fisher B et al (2005) Tamoxifen for the prevention of breast cancer: current status of the National Surgical Adjuvant Breast and Bowel Project P-1 study. J Natl Cancer Inst 97(22):1652–1662):

- 43% decrease in invasive breast cancer risk
- 37% decrease in noninvasive breast cancer risk
- 32% decrease in osteoporotic fracture risk
- No effect on the risk of developing ER-negative breast cancer

NSABP P-2 STAR: Study of Tamoxifen and Raloxifen) (Vogel VG et al (2006) Effects of tamoxifen vs raloxifene on the risk of developing invasive breast cancer and other disease outcomes: the NSABP Study of Tamoxifen and Raloxifene (STAR) P-2 Trial. JAMA 295:2727–2741) → Randomized. 19,747 cases.
No premenopausal women. TMX 20 mg vs. raloxifene (RLX) 60 mg for 5 years

Raloxifene Arm
- No difference in lowering the risk of invasive breast cancer
- No difference in noninvasive cancer risk
- Thromboembolic events, uterine cancers, and cataracts are decreased
- Similar for risks of LCIS, atypical ductal hyperplasia, osteoporotic fracture

IBIS-I Cuzick J et al (2002) First results from the International Breast Cancer Intervention Study (IBIS-I): a randomised prevention trial. Lancet 360(9336):817–824) → randomized. 7152 cases. Placebo vs. TMX 20 mg for 5 years. Median follow-up 50 months.

TMX Arm
- 32% decrease in noninvasive and invasive breast cancer risk
- Nonsignificant increase in endometrial cancer risk
- Significant increase in thromboembolic events

Updated in 2007 with 96-months of follow-up Cuzick J et al (2007) Long-term results of tamoxifen prophylaxis for breast cancer–96-month follow-up of the randomized IBIS-I trial. J Natl Cancer Inst 99(4):272-82

TMX Arm
- 34% decrease in ER+ invasive breast cancer
- No reduction in ER- invasive breast cancer
- Lower side effects in TMX group after completion of the active treatment

IBIS-II Cuzick J et al (2020) Use of anastrozole for breast cancer prevention (IBIS-II): long-term results of a randomised controlled trial. Lancet 395(10218):117 → Randomized. 3864 postmenopausal high-risk cases. Placebo vs. anastrozole 1 mg/day for 5 years. Median follow-up 131 months.

Anastrozole Arm
- 49% reduction in breast cancer
- 54% reduction in ER+ breast cancer with a continued significant effect in the period after treatment
- 59% reduction in DCIS
- No significant difference in death
- No excess of fractures or cardiovascular disease

MAP 3 Goss PE et al (2011) Exemestane for breast-cancer prevention in postmenopausal women. NEJM 364(25):2381 → Randomized. 4560 postmenopausal high-risk cases. Placebo vs. exemestane 25 mg for 5 years. Median follow-up 35 months.

Exemestane Arm

- 65% reduction in all breast cancers at 3 years follow-up
- 75% reduction in ER+ breast cancer
- Increased risk of bone fractures, skeletal symptoms, and vasomotor symptoms

Note: The choice of agent (tamoxifen, raloxifene, or aromatase inhibitor) is challenging and may be selected based on the menopausal status and side effects expected with each agent.

DCIS

Wong JS et al (2006) Prospective study of wide excision alone for ductal carcinoma in situ of the breast. J Clin Oncol 24(7):1031–6 → Prospective nonrandomized. 158 DCIS cases. DCIS ≤2.5 cm, grade 1–2, necrosis (+). No RT or TMX. Surgical margin ≥1 cm. Median follow-up 40 months.

- Local relapse: 2.4%/year and 12% at 5 years
- 69% DCIS relapse, 31% invasive breast cancer development

Updated in 2014:

- 15.6% local failure at 10 years
- Conclusion: radiotherapy should be given to prevent locoregional relapses, even in the presence of good prognostic factors

NSABP B-17 → Randomized. 818 DCIS cases. Lumpectomy vs. lumpectomy + RT. Surgical margin (±). RT: 50 Gy, boost (−).

1993 Results with 43 Months of Follow-up

- Five-year event-free survival: 74 vs. 84% in RT arm ($p < 0.05$)
- Five-year ipsilateral invasive breast cancer development: 10.5 vs. 2.9% in RT arm
- Five-year ipsilateral noninvasive breast cancer development: 10.4 vs. 7.5% in RT arm

Updated in 1998:

- Eight-year ipsilateral breast cancer development: 13.4 vs. 3.9% in RT arm
- Five-year ipsilateral noninvasive breast cancer development: 13.4 vs. 8.2% in RT arm

Updated in 2000:

- Ten-year ipsilateral breast cancer development: 16.4 vs. 7.1% in RT arm (62% decrease in breast cancer risk)
- Ten-year ipsilateral breast cancer development: 16.4 vs. 7.1% in RT arm (62% decrease in breast cancer risk)
- Ten-year ipsilateral noninvasive breast cancer development: 30.8 vs. 14.9% in RT arm (57% decrease in breast cancer risk)
- Similar rates of distant metastases and contralateral breast cancer

Predictive factors for ipsilateral recurrences: comedonecrosis and unknown or positive surgical margin.

NSABP B-24 Fisher B et al (1999) Tamoxifen in treatment of intraductal breast cancer: National Surgical Adjuvant Breast and Bowel Project B-24 randomised controlled trial. Lancet 353(9169):1993–2000) → Randomized. 1804 DCIS cases.

Lumpectomy + RT ± placebo vs. lumpectomy + RT ± TMX for 5 years

TMX Arm

- Five-year breast cancer events: 13.4% vs. 8.2% in TMX arm
- Lower risk of ipsilateral breast cancer in TMX arm even in patient with positive surgical margin and comedonecrosis

Results of NSABP B-17 and NSABP B-24 trials were updated in 2011 Wapnir IL et al (2011) Long-term outcomes of invasive ipsilateral breast tumor recurrences after lumpectomy in NSABP B-17 and B-24 randomized clinical trials for DCIS. J Natl Cancer Inst 16;103(6):478–88

- Addition of radiation reduced ipsilateral breast tumor recurrence by 52% compared to those with lumpectomy only
- Addition of TMX to lumpectomy and radiation reduced ipsilateral breast tumor recurrence by 32% compared to those without TMX
- Fifteen-year cumulative incidence of breast cancer death was 3.1% for lumpectomy only,

4.7% for lumpectomy + RT (B-17), 2.7% for lumpectomy + RT + placebo (B-24), and 2.3% for lumpectomy + RT + TMX

NSABP B-35 Margolese RG et al (2016) Anastrozole versus tamoxifen in postmenopausal women with ductal carcinoma in situ undergoing lumpectomy plus radiotherapy (NSABP B-35): a randomised, double-blind, phase 3 clinical trial) → Randomized. 3104 postmenopausal women with DCIS. Lumpectomy + RT + TMX vs. lumpectomy + RT + anastrozole. Median 9 years of follow-up.

- Anastrozole is superior only in women younger than 60 years of age in terms of breast cancer-free interval
- No difference in adverse events between the groups

EORTC 10853 Bijker N (2006) Breast-conserving treatment with or without radiotherapy in ductal carcinoma in situ: 10-year results of the European Organisation for Research and Treatment of Cancer Randomized Phase III Trial 10853—a study by the EORTC Breast Cancer Cooperative Group and EORTC Radiotherapy Group. J Clin Oncol 24(21):3381–3387) → Randomized. 1010 DCIS cases. DCIS <5 cm. Lumpectomy vs. lumpectomy + RT. Surgical margin (−); re-excision if surgical margin (+) intraoperatively. RT dose: 50 Gy, no boost dose.

- Ten-year local relapse: 26% in no RT arm vs. 15% in RT arm
- 47% decrease in risk, valid for all subgroups
- 42% decrease in invasive breast cancer risk
- 48% decrease in DCIS risk
- Local relapse risk increase: age < 40 years, grade 2–3, cribriform or solid growth pattern, suspicious surgical margin or wide local excision only
- No prognostic significance of DCIS size
- No difference in overall survival and distant metastasis

Updated in 2013 with a median 15.8 years of follow-up (Donker M et al (2013) Breast-conserving treatment with or without radiotherapy in ductal carcinoma In Situ: 15-year recurrence rates and outcome after a recurrence, from the EORTC 10853 randomized phase III trial. J Clin Oncol 31(32):4054–9

RT Arm
- Reduction in the risk of any local recurrence by 48%
- Fifteen-year local recurrence-free rate: 69% vs. 82% in RT arm
- No significant difference in breast cancer-specific survival or overall survival

Houghton J et al (2003) Radiotherapy and tamoxifen in women with completely excised ductal carcinoma in situ of the breast in the UK, Australia, and New Zealand: randomised controlled trial. Lancet 362(9378):95–102 → Randomized. 1701 DCIS cases. 2 × 2 trial; after BCS: (1) arm surveillance, (2) arm RT, (3) arm TMX, and (4) arm RT + TMX. RT dose: 50 Gy with no boost, and TMX: 20 mg for 5 years.

- No effect of TMX on ipsilateral invasive breast cancer risk but prevents ipsilateral DCIS risk (HR = 0.68)
- RT is beneficial for decreasing both ipsilateral invasive and noninvasive breast cancer risk (new event in ipsilateral breast risk: 16 vs. 7% in RT arms)
- No synergistic effect of RT + TMX

SweDCIS Warnberg F et al (2014) Effect of radiotherapy after breast-conserving surgery for ductal carcinoma in situ: 20 years follow-up in the randomized SweDCIS trial. J Clin Oncol 32:32, 3613–3618 → Randomized. 1046 women with DCIS. Breast-conserving surgery (BCS) + RT vs. BCS alone

- 12% of absolute risk reduction in the RT arm at 20 years
- No difference in breast cancer-specific death and overall survival

- Higher risk of invasive ipsilateral breast events and lower effect of RT in younger patients

Mastectomy + RT in early-stage breast cancer

NSABP B-04 Fisher B (2002) Twenty-five-year follow-up of a randomized trial comparing radical mastectomy, total mastectomy, and total mastectomy followed by irradiation. N Engl J Med 347(8):567–575) → A total of 1079 women with clinically negative axillary nodes underwent radical mastectomy, total mastectomy without axillary dissection but with postoperative irradiation, or total mastectomy plus axillary dissection only if their nodes became positive. A total of 586 women with clinically positive axillary nodes underwent either radical mastectomy or total mastectomy without axillary dissection but with postoperative irradiation.

- Ten-year results (1985): no survival difference between locoregional therapies
- Twenty-five-year results (2000): no differences in DFS, RFS, distant metastasis, and OS between treatment arms
- Conclusion: radical mastectomy had no advantage; surgery and RT for occult (+) axillary lymph nodes also had no survival advantage

Blichert-Toft M (1992) Danish randomized trial comparing breast conservation therapy with mastectomy: six years of life-table analysis. Danish Breast Cancer Cooperative Group. J Natl Cancer Inst Monogr (11):19–25 → Randomized. 908 early-stage breast cancer patients. Mastectomy vs. BCS + RT. Axillary dissection performed in all cases. Median follow-up: 3.3 years.

Six-Year Results
- DFS: 66 vs. 70% in RT arm
- OS: 82 vs. 79% in RT arm
- Conclusion: CRT + BCS is as effective as mastectomy in the management of early-stage breast cancer

Veronesi U et al (2002) Twenty-year follow-up of a randomized study comparing breast-conserving surgery with radical mastectomy for early breast cancer. N Engl J Med 347(16):1227–1232 → Randomized. 701 cases of breast cancer with $T <2$ cm. 25% axillary LN (+). Radical mastectomy vs. quadrantectomy + RT. RT: 50 + 10 Gy boost. Adjuvant CMF chemotherapy after 1976 in cases of axillary LN (+).

- Ipsilateral relapse: 2 vs. 9% in RT arm ($p < 0.05$)
- No difference in secondary malignancy and distant metastasis
- No difference in 20-year OS, cancer-specific survival, and cumulative incidence of contralateral breast carcinomas
- Conclusion: BCS + RT is an effective modality for the management of early-stage breast cancer that does not result in any significant increase in contralateral breast cancer

NSABP B-06 Fisher B et al (2002) Twenty-year follow-up of a randomized trial comparing total mastectomy, lumpectomy, and lumpectomy plus irradiation for the treatment of invasive breast cancer. N Engl J Med 347(16):1233–1241) → Randomized. 1851 stage I–II, $T <4$ cm, axillary LN (±). Lumpectomy vs. lumpectomy + RT vs. total mastectomy. Axillary dissection in all cases. RT: 50 Gy ± boost with no axillary RT. Melphalan + 5-FU chemotherapy for axillary LN (+). 10% surgical margin (+) in lumpectomy arm.

- Ipsilateral breast cancer recurrence: 39.3% in lumpectomy vs. 14.3% in lumpectomy + RT
- Relapses after lumpectomy: 73% in 5 years, 18% in 5–10 years, 8% after 10 years
- Relapses after lumpectomy + RT: 39% in 5 years, 29% in 5–10 years, 30% after 10 years
- All other events other than local recurrences were similar between the three arms
- Twenty-year DFS: 35% in lumpectomy vs. 35% lumpectomy + RT vs. 36% in mastectomy
- Twenty-year DMFS: 45% in lumpectomy vs. 46% lumpectomy + RT vs. 49% in mastectomy
- Twenty-year OS: 46% in lumpectomy vs. 46% lumpectomy + RT vs. 47% in mastectomy

- Conclusion: all three modalities were similar in terms of DFS, DMFS, and OS; late relapses were common in the RT arm, particularly after 5 years of therapy

EORTC 10801 Van Dongen JA et al (2000) Long-term results of a randomized trial comparing breast-conserving therapy with mastectomy: European Organization for Research and Treatment of Cancer 10,801 trial. J Natl Cancer Inst. 92(14):1143–50 → Randomized. 868 early-stage breast cancer patients. BCS + 50 Gy RT with a boost dose vs. modified radical mastectomy.

- No difference in 10-year overall survival and distant metastasis-free rates
- Lower locoregional recurrence rate at 10 years in mastectomy arm (12% vs. 20%)

Other randomized trials in early stage breast cancer → NCI (1979–1987), Gustave-Roussy (1972–1980).

General conclusion: There is no survival difference between mastectomy and BCS + RT (Table 7.6).

General conclusion. Ipsilateral breast relapse risk is 75% less in patients receiving RT in addition to BCS. This effect is more prominent in young women and node (+) patients.

Meta-analyses in early-stage breast cancer (BCS + RT vs. BCS)
Early Breast Cancer Trialists' Collaborative Group Meta-analysis → **1995 results:**

- Local recurrences and OS rates are similar
- Ten-year breast cancer-related mortality was 5% less in the RT arm
- Nonbreast cancer-related mortality was 25% more in the RT arm
- Most mortalities occurred after 60 years of age

Anon (1995) Effects of radiotherapy and surgery in early breast cancer. An overview of the randomized trials. N Eng J Med 333(22):1444–1455

2000 Results
- Included 40 randomized trials
- 50% of deaths were in LN (+) patients
- Local recurrence risk decreased from 27 to 8.8% in patients receiving RT
- Breast cancer-related mortality decreased
- Nonbreast cancer-related mortality increased
- Twenty-year OS: 37.1% in RT vs. 35.9% in surgery ($p = 0.06$)

Table 7.6 Selected randomized trials of BCS ± RT in early-stage breast cancer

Study	Surgery	Chemotherapy	RT dose	LR, RT (−) (%)	LR, (+)	RT (%)
NSABP B-06	Lumpectomy	CT for N (+)	50 Gy	39	14	
Milan	Quadrantectomy	N (+) high risk: CT	50 Gy + 10 Gy	23	6	
		N (+) low risk: TMX				
NSABP B-21	Lumpectomy	TMX	50 Gy ± boost	16	3	
Finland	Lumpectomy	–	50 Gy	18	7	
Sweden	Sector resection	9% (+)	48–54 Gy	14	4	
Canada	BCS	TMX	40 Gy/16 + 12.5 Gy/5	8	1	
CALGB 9343	Lumpectomy	TMX	45 Gy + 14 Gy	4	1	

RT radiotherapy, *CT* chemotherapy, *TMX* tamoxifen, *N* node

- Cardiac dose was higher in older studies with old RT techniques

Early Breast Cancer Trialists' Collaborative Group (2000) Favorable and unfavorable effects on long-term survival of radiotherapy for early breast cancer: an overview of the randomized trials. Lancet 355(9217):1757–1770

2005 Results
- Included 78 randomized trials with 42,000 cases
- Five-year local recurrence after BCS: 7% in RT (+) vs. 26% in RT (−) ($p < 0.05$)
- Five-year breast cancer-related mortality 30.5% in RT (+) vs. 35.9% in RT (−) ($p < 0.05$)

Clark M et al. (2005) Effects of radiotherapy and of differences in the extent of surgery for early breast cancer on local recurrence and 15-year survival: an overview of the randomized trials. Lancet 366(9503):2087–2106

2011 Results
- Included 10,801 women in 17 randomized trials with pN0 and pN+ disease
- RT decreased 10-year LR from 25 to 7.7% (pN0: 23% to 7.3%; pN+: 43% to 12.4%)
- RT reduced annual breast cancer death rate by 1/6

Early Breast Cancer Trialists' Collaborative Group (2011) Effect of radiotherapy after breast-conserving surgery on 10-year recurrence and 15-year breast cancer death: meta-analysis of individual patient data for 10,801 women in 17 randomised trials. Lancet 378:771–84

Other meta-analyses
BCS Project, 2004 → 15 randomized BCS ± RT trials

- Conclusion: RT decreases the risk of ipsilateral breast cancer development with minimal effect on mortality

Vinh-Hung V (2004) Breast-conserving surgery with or without radiotherapy: pooled-analysis for risks of ipsilateral breast tumor recurrence and mortality. J Natl Cancer Inst 96(2):115–121

McMaster University, 2000 → 18 randomized trials of BCS + RT or MRM. Conclusion: RT decreased local recurrence (OR = 0.25) and mortality (OR = 0.83).

Whelan TJ et al (2000) Does locoregional radiation therapy improve survival in breast cancer? A meta-analysis. J Clin Oncol 18(6):1220–1229

BCS + TMX ± RT in early-stage breast cancer
Austrian ABCSG 8A, 2007 → Randomized. 869 early-stage breast cancers (tm < 3 cm, G1–2, N0, ER or PR [+]). Anastrozole or TMX after lumpectomy: RT (50 ± 10 Gy boost) vs. RT (−)
2005 results with median 3.5 years of follow-up:

- Local recurrence: 0.2% in RT arm vs. 3.1% in no-RT arm
- Locoregional recurrences: 4.4% in RT arm vs. 6.7% in no-RT arm

2007 results Potter R (2007) Lumpectomy plus tamoxifen or anastrozole with or without whole breast irradiation in women with favorable early breast cancer. Int J Radiat Oncol Biol Phys 68(2):334–340)

- Five-year local recurrence: 0.4% in RT arm vs. 5.1% in no-RT arm
- Five-year locoregional recurrence: 2.1% in RT arm vs. 6.1% in no-RT arm
- No difference in terms of distant metastasis and overall survival
- Conclusion: radiotherapy is highly effective for decreasing local and regional recurrences

Intergroup (CALGB 9343, RTOG 97-02, ECOG), (Hughes KS et al (2004) Lumpectomy plus tamoxifen with or without irradiation in women 70 years of age or older with early breast cancer. N Engl J Med 351(10):971–977) → Randomized: >70 years 638 T1N0, ER (+) cases. TMX after BCS: randomized to RT vs. no-RT. Axilla dissection in 37% of cases. RT dose: 45 + 14 Gy boost.

Five-year results:

- Local recurrence 4% in TMX arm vs. 1% in TMX + RT arm ($p < 0.05$)
- No difference in salvage mastectomy, distant metastasis, and overall survival
- No difference in side effects and cosmesis after 4 years
- Conclusion: lumpectomy plus adjuvant therapy with tamoxifen alone can be an alternative approach for the treatment of women 70 years of age or older who have early, estrogen receptor-positive breast cancer

Updated in 2013 Hughes KS et al (2013) Lumpectomy plus tamoxifen with or without irradiation in women age 70 years or older with early breast cancer: long-term follow-up of CALGB 9343. J Clin Oncol 31(19):2382–7

Median follow-up 12.6 years

- Local recurrence 10% in TMX arm vs. 2% in TMX + RT arm
- No significant differences in time to mastectomy, time to distant metastasis, breast cancer-specific survival, or OS between the two groups

NSABP-B21 Fisher ER et al (2007) Pathobiology of small invasive breast cancers without metastases (T1a/b, N0, M0): National Surgical Adjuvant Breast and Bowel Project (NSABP) protocol B-21. Cancer 110(9):1929–36 → Randomized. 1009 pN0 patients with tumors ≤1 cm (both ER/PR ±). TMX vs. RT + placebo vs. RT + TMX

- Fourteen-year ipsilateral breast tumor recurrence: 19.5% in TMX arm, 10.8% in RT arm, 10.1% in TMX + RT arm
- TMX decreased contralateral breast primaries by 3.2%
- No difference in OS and DM
- TMX + RT should be considered for a tumor size of <1 cm

PRIME II (Kunkler IH et al (2015) 16: 266–73 → Randomized. 1326 pN0, >65 years of age. BCS + adjuvant endocrine therapy + RT vs. BCS + adjuvant endocrine therapy. Median follow-up 5 years.

- Lower ipsilateral breast tumor recurrence in RT arm (1.3% vs. 4.1%)
- No differences in regional recurrence, distant metastases, contralateral breast cancers, or new breast cancers

Radiotherapy boost trials in early-stage breast cancer
EORTC 22881/10882 (boost vs. no-boost) → Stage I–II. 5318 cases with BCS + axilla dissection + RT (50 Gy). 21% LN (+). Surgical margin (−). Randomized: 16 Gy boost vs. no-boost. Boost field: scar + 1.5 cm. Adjuvant chemotherapy in 28% of cases.

Cosmesis Results in 1999
- Excellent/good cosmesis: 71% in boost arm vs. 86% in no boost arm
- Conclusion: boost caused bad cosmesis

Updated in 2001:
- Five-year locoregional control: 4.3 vs. 7.3% in no-boost
- Decrease in recurrence risk: 40%; age < 40 years (10 vs. 20%), age > 60 years (2.5 vs. 4%)

Updated in 2007 Bartelink H (2007) Impact of a higher radiation dose on local control and survival in breast-conserving therapy of early breast cancer: 10-year results of the Randomized Boost versus No Boost EORTC 22881–10882 Trial. J Clin Oncol 25(22):3259–3265

- Ten-year local recurrence: 6 vs. 10% in no-boost ($p < 0.05$)
- Local relapses were significantly decreased with the boost in all age groups
- Conclusion: a boost should be given to all age groups to increase local control

Other randomized trials:
Lyon, 1997, (1986–1992)

Romestaing P et al (1997) Role of a 10-Gy boost in the conservative treatment of early breast cancer: results of a randomized clinical trial in Lyon, France. J Clin Oncol 15(3):963–968.

Netherlands, 2006,

Hurkmans CW (2006) High-dose simultaneously integrated breast boost using intensity-modulated radiotherapy and inverse optimization. Int J Radiat Oncol Biol Phys 66(3):923–930.

RTOG 1005 Trial (closed to accrual, results are awaited) → Randomized: 50 Gy/25 fr or 42.7 Gy/16 fr followed by a sequential boost of 12–14 Gy/6–7 fr vs. 40 Gy/15 fr with a concomitant boost of 3.2 Gy to the tumor bed (up to 48 Gy/15 fr).

IMPORT HIGH Trial (CRUK/06/003) (early results were presented at 2018 San Antonio Breast Cancer Symposium) → Randomized. 840 patients with BCS for pT1–3 pN03a M0 breast cancer. 40 Gy/15F to whole breast (WB) + 16 Gy/8fr sequential photon boost to tumor bed (40 + 16 Gy) vs. 36 Gy/15fr to WB, 40 Gy to partial breast +48 Gy (48 Gy) or + 53 Gy (53 Gy) in 15F SIB to tumor bed.

- Similar moderate/marked adverse events between the groups

Hypofractionation trials in breast cancer Canada, 2002 and 2010

Whelan T et al (2010) Long-Term Results of Hypofractionated Radiation Therapy for Breast Cancer. N Engl J Med 362:513-520 → Randomized: 50 Gy/25 fr vs. 42.5 Gy/16 fr. No boost was given. Median follow-up 12 years.

- Similar 10-year local recurrence rates between the groups (6.7% vs. 6.2%)
- Excellent or good global cosmetic outcome in 70% of patients in each arm

MRC START A&B (England), 2008 (1999–2002) → Two parallel randomized trials.

START-A: **START Trialists' Group (2008) The UK Standardisation of Breast Radiotherapy (START) Trial A of radiotherapy hypofractionation for treatment of early breast cancer: a randomised trial. Lancet Oncol 9(4):331–341** → 2236 women with early breast cancer (pT1–3a pN0–1 M0) were randomly assigned after primary surgery to receive 50 Gy in 25 fractions of 2.0 vs. 41.6 or 39 Gy in 13 fractions of 3.2 or 3.0 Gy over 5 weeks. After a median follow-up of 5.1 years, the rate of locoregional tumor relapse at 5 years was 3.6% after 50 Gy, 3.5% after 41.6 Gy, and 5.2% after 39 Gy. Photographic and patient self-assessments suggested lower rates of late adverse effects after 39 Gy than with 50 Gy. A lower total dose in a smaller number of fractions could offer similar rates of tumor control and normal tissue damage to the international standard fractionation schedule of 50 Gy in 25 fractions.

START-B: **START Trialists' Group (2008) The UK Standardisation of Breast Radiotherapy (START) Trial B of radiotherapy hypofractionation for treatment of early breast cancer: a randomised trial. Lancet 371(9618):1098–1107** → 2215 women with early breast cancer (pT1–3a pN0–1 M0) were randomly assigned after primary surgery to receive 50 Gy in 25 fractions of 2.0 Gy over 5 weeks or 40 Gy in 15 fractions of 2.67 Gy over 3 weeks. After a median follow-up of 6.0 years, the rate of locoregional tumor relapse at 5 years was 2.2% in the 40 Gy group and 3.3% in the 50 Gy group. Photographic and patient self-assessments indicated lower rates of late adverse effects after 40 Gy than after 50 Gy. A radiation schedule delivering 40 Gy in 15 fractions seems to offer rates of locoregional tumor relapse and late adverse effects that are at least as favorable as the standard schedule of 50 Gy in 25 fractions.

Results of both trials were updated in 2013 Havilan JS et al (2013) The UK Standardisation of Breast Radiotherapy (START) trials of radiotherapy hypofractionation for treatment of early breast cancer: 10-year follow-up results of two randomised controlled trials. Lancet Oncol 14(11):1086–1094.

START A, median follow-up 9.3 years

- Similar 10-year rates of local–regional relapse between the groups
- Less moderate or marked breast induration, telangiectasia, and breast edema in the 39 Gy regimen patients group than in the 50 Gy regimen group
- Similar moderate or marked normal tissue effects between 41·6 Gy and 50 Gy groups

START B, median follow-up 9.9 years:

- Similar 10-year rates of local–regional relapse between the groups
- Less breast shrinkage, telangiectasia, and breast edema in the 40 Gy group than in the 50 Gy group

Other randomized fractionation trials:
Royal Marsden, 2006
Owen JR (2006) Effect of radiotherapy fraction size on tumour control in patients with early-stage breast cancer after local tumour excision: long-term results of a randomised trial. Lancet Oncol 7(6):467–471*Egypt NCI, 2004*
Taher AN (2004) Hypofractionation versus conventional fractionation radiotherapy after conservative treatment of breast cancer: early skin reactions and cosmetic results. J Egypt Natl Canc Inst 16(3):178–187

Accelerated partial breast irradiation
Budapest (2013) Polgar C et al (2013) Breast-conserving therapy with partial or whole breast irradiation: ten-year results of the Budapest randomized trial. Radiother Oncol 108(2):197–202.) → Randomized: 258 women with pT1 pN0-1mi M0, grade 1–2, non-lobular breast cancer without the presence of extensive intraductal component and resected with negative margins were randomized after BCS to receive 50 Gy whole breast irradiation (WBI) or partial breast irradiation (PBI) [7 × 5.2 Gy high-dose-rate (HDR) multicatheter brachytherapy or 50 Gy electron beam (EB) irradiation]. Median follow-up 10.2 years.

- Similar 10-year local recurrence, overall survival, cancer-specific survival, and disease-free survival rates between WBI and PBI arms
- Higher rate of excellent–good cosmetic results in PBI arms (81% vs. 63%)

ELIOT (2013) (Veronesi U et al (2013) Intraoperative radiotherapy versus external radiotherapy for early breast cancer (ELIOT): a randomised controlled equivalence trial. Lancet Oncol 14(13):1269–77) → Randomized: 1305 early-stage breast cancer patients were randomized after BCS to receive either WBI (50 Gy + 10 Gy boost) or intraoperative radiotherapy with electrons (21 Gy).

- Higher 5-year event rate in the intraoperative radiotherapy arm (4.4% vs. 0.4%)
- Higher ipsilateral breast tumor recurrence in the intraoperative radiotherapy arm
- No difference in overall survival
- Fewer skin side effects in the intraoperative radiotherapy arm

TARGIT-A (2014) Vaidya JS et al (2014) Risk-adapted targeted intraoperative radiotherapy versus whole-breast radiotherapy for breast cancer: 5-year results for local control and overall survival from the TARGIT-A randomised trial. Lancet 383(9917):603–13.) → Randomized: 3451 patients >45 years. Targeted intraoperative radiotherapy versus whole-breast EBRT.

- Higher 5-year risk for local recurrence in the TARGIT arm (3.3% vs. 1.3%)
- Similar breast cancer mortality rates between the groups
- Decreased grade 3 or 4 skin complications in TARGIT arm

Lili L et al (2015) Accelerated partial breast irradiation using intensity-modulated radiotherapy versus whole breast irradiation: 5-year survival analysis of a phase 3 randomised controlled trial. Eur J Cancer 51(4):451–463 → Randomized: 520 patients >40 years with tumors 2.5 cm or smaller, no EIC, and margins 5 mm or wider to WBI (50 Gy in 25 fractions) or APBI (30 Gy in 5 fractions using IMRT).

- No difference in terms of ipsilateral breast tumor recurrence or overall survival
- Fewer toxicity and better cosmetic results in the APBI arm

UK IMPORT LOW (2017) (Coles CE et al (2017) 390: 1048–60.) → Randomized: Women aged >50 years who had undergone BCS for unifocal invasive ductal adenocarcinoma of grade 1–3, with a tumor size of 3 cm or less (pT1–2), none to three positive axillary nodes (pN0–1), and minimum microscopic margins of noncancerous tissue of 2 mm or more, were randomized to receive 40 Gy WBI (control) vs. 36 Gy WBI and 40 Gy to the partial breast (reduced-dose group) vs. 40 Gy to the partial breast only (partial-breast group). Median follow-up 72.2 months.

- Five-year estimates of local relapse cumulative incidence were 1.1% in the control group, 0.2% in the reduced-dose group, and 0.5% in the partial-breast group
- Similar adverse effects between the reduced-dose group and the partial-breast group
- Fewer adverse effects in breast appearance and breast harder or firmer in the reduced-dose group and the partial-breast group compared with whole-breast radiotherapy

Note*: APBI may be used in patients' age of >50 years with early-stage breast cancer without an EIC or LVI with microscopic margins 2 mm or wider, with negative axillary nodes. However, the optimal selection criteria, technical parameters, and choice between different modalities for APBI are not still known.

Axilla dissection vs. RT in early-stage breast cancer

NSABP B-04 Fisher B et al (2002) Twenty-five-year follow-up of a randomized trial comparing radical mastectomy, total mastectomy, and total mastectomy followed by irradiation. N Engl J Med 347(8):567–575. → Randomized. 1,665 operable cases. **Clinical axilla (+): radical mastectomy vs. total mastectomy + RT with no axilla dissection. Clinical axilla (−): radical mastectomy vs. total mastectomy vs. total mastectomy + RT. RT: chest wall and lymphatics (50 Gy/25 fractions) and 10–20 Gy** boost for LN (+). Lymphatics: axilla/SCF/internal mammary 45 Gy. No chemotherapy.

Twenty-five year results:

- 82% of events were due to breast cancer: mostly recurrences and metastases
- 57% of the events were in LN (+) vs. 33% in LN (−) patients; mostly distant metastases
- Distant metastases: 42% in LN (+) vs. 30% in LN (−)
- Recurrences in LN (+): 5% local and 4% regional
- Other most common events: non-breast cancer-related mortality (25%) and contralateral breast cancer (6%)
- No difference between disease and relapse-free survival between arms
- No benefit of axilla dissection clinically in LN (−) cases

The 10-year results of this trial in 1985 were the basis for the Fisher hypothesis: that breast cancer is a systemic not a locoregional disease.

Institute Curie, 2004 **Louis-Sylvestre C et al (2004) Axillary treatment in conservative management of operable breast cancer: dissection or radiotherapy? Results of a randomized study with 15 years of follow-up. J Clin Oncol 22(1):97–101)** → Clinical N0, $T < 3cm$, 658 cases; randomized to BCS + axilla dissection vs. BCS + axilla RT. RT: 50 + 10–15 Gy boost. Chemotherapy and hormone therapy in some LN (+) cases.

Fifteen-year results:

- No difference in OS and DFS
- Five-year distant metastasis: 10% in surgery vs. 12% in RT
- Fifteen-year distant metastasis: 25% in both arms
- Ipsilateral breast recurrences: 7% in 5 years, 12% in 10 years, and 17% in 15 years
- Fifteen-year isolated axilla recurrences: 1% in surgery vs. 3% in RT ($p = 0.04$)
- Conclusion: axilla dissection significantly decreased isolated axilla relapses compared to RT

EORTC 10981–22023 AMAROS Donker M et al (2014) Radiotherapy or surgery of the axilla after a positive sentinel node in breast cancer (EORTC 10981–22023 AMAROS): a randomised, multicentre, open-label, phase 3 non-inferiority trial. Lancet Oncol 15(12):1303–10) → Randomized: 4823 SLN bx (+) breast cancer patients with T1–2 tumor were randomized to receive ALND versus axillary radiotherapy. Median follow-up 6.1 years.

- No significant difference in 5-year axillary recurrence (0.43% in ALND arm vs. 1.19% in axillary radiotherapy arm)
- Lymphedema in the ipsilateral arm was higher in ALND arm at 1 year, 3 years, and 5 years

10-Year follow-up results were presented at 2018 San Antonio Breast Cancer Symposium

- No significant difference in 10-year axillary recurrence (0.93% in ALND arm vs. 1.82% in axillary radiotherapy arm)
- No difference in overall survival and distant metastasis-free survival between the groups
- No significant difference in 10-year local–regional recurrence between the groups (3.59% in ALND arm vs. 4.07% in axillary radiotherapy arm)
- More second primaries including contralateral breast in axillary radiotherapy arm ($p = 0.035$)

MA-20 Whelan TJ et al (2015) Regional nodal irradiation in early-stage breast cancer NEJM 373:307–316) → Randomized. 1832 women with node-positive or high-risk node-negative were randomized to BCS + WBI and regional nodal irradiation vs. BCS + WBI without regional nodal irradiation. A level I or II axillary dissection was required for patients with SLN (+). Median follow-up 9.5 years.

- No difference in overall survival (82.8% in the nodal irradiation group vs. 81.8% in the control group)
- Disease-free survival was better in the nodal irradiation group (82% vs. 77%)
- Higher rates of grade 2 or greater acute pneumonitis and lymphedema in the nodal irradiation group

EORTC 22922/10925 Poortmans PM et al (2015) Internal mammary and medial supraclavicular irradiation in breast cancer. N Engl J Med 373:317–327) → Randomized. 4004 patients were randomized to receive BCS + WBI with regional nodal irradiation versus BCS + WBI without regional nodal irradiation. Median follow-up 10.9 years.

- Ten-year overall survival was 82.3% in the regional nodal irradiation group and 80.7% in the control group ($p = 0.06$)
- Ten-year disease-free survival was 72.1% in the regional nodal irradiation and 69.1% in the control group ($p = 0.04$)
- Ten-year distant disease-free survival was 78% in the regional nodal irradiation and 75% in the control group ($p = 0.02$)
- Ten-year breast cancer mortality was 12.5% in the regional nodal irradiation and 14.4% in the control group ($p = 0.02$)

Axillary dissection vs. no axillary dissection
NSABP B-32 Krag DN et al (2010) Sentinel-lymph-node resection compared with conventional axillary-lymph-node dissection in clinically node-negative patients with breast cancer: overall survival findings from the NSABP B-32 randomised phase 3 trial Lancet Oncol 11(10):927–33. → Randomized. SLN bx (ALND if +) vs. upfront ALND.

- SLNbx had an overall accuracy of 97.1%, false-negative rate of 9.8%, and negative predictive value of 96.1%
- No difference in 8-year overall survival, disease-free survival, or sites of first treatment failure between upfront ALND vs. SLNBx

**ACOSOG Z0011 (Giuliano AE et al (2017) Effect of axillary dissection vs no axillary dissection on 10-year overall survival among women with invasive breast cancer and sentinel node metastasis. JAMA 318(10):918–926 → Randomized. 891 women with clinical T1 or T2 invasive breast cancer, no palpable axillary adenopathy, and one or two sentinel lymph nodes containing metastases were randomized to BCS with ALND versus BCS + SLND without ALND. Median follow-up 9.3 years.

- Ten-year overall survival was 86.3% in the SLND-alone group vs. 83.6% in the ALND group ($p = 0.02$)
- Similar 10-year regional recurrence between the groups

Conventional RT vs. IMRT
Canada, 2008
Pignol JP (2008) A multicenter randomized trial of breast intensity-modulated radiation therapy to reduce acute radiation dermatitis. J Clin Oncol 26:2085–209 → Randomized. 331 early-stage breast cancer cases with BCS. Standard conventional RT vs. IMRT (optional 16 Gy electron boost).

- Less wet desquamation, particularly on the inframammary fold in the IMRT arm: 26 vs. 43%
- Less wet desquamation on the entire breast in the IMRT arm: 31 vs. 48%
- V95 (dose received by 95% of the breast volume) was related to acute skin toxicity
- IMRT had a better dose distribution
- No differences in breast pain and quality of life

Royal Marsden, 2007
Donovan E (2007) Randomized trial of standard 2D radiotherapy (RT) versus intensity modulated radiotherapy (IMRT) in patients prescribed breast radiotherapy. Radiother Oncol 82(3):254–264 → Randomized. 306 early-stage breast cancer. RT: 50 + 11.1 Gy. IMRT vs. conventional RT. Median follow-up: 5 years.

- Photographic change in breast cosmesis: 40% in IMRT vs. 58% in non-IMRT
- Less palpable breast induration in IMRT arm
- No difference in breast pain and quality of life

- Late side effects decreased with minimal inhomogeneity

3D Conformal RT vs. IMRT
Cho BCJ et al (2002) Intensity modulated versus non-intensity modulated radiotherapy in the treatment of the left breast and upper internal mammary lymph node chain: a comparative planning study. Radiother Oncol 62(2):127–36 → 12 patients. Three different treatment plans: tangential photon fields with oblique IMC electron–photon fields with manually optimized beam weights and wedges, (2) wide split tangential photon fields with a heart block and computer-optimized wedge angles, and (3) IMRT tangential photon fields.

- Similar average NTCP for the ORs (heart and lung) between treatment plans
- The wide split tangent technique resulted in higher NTCP values (> or =2%) for the ORs
- The IMRT technique had the best breast and IMC target coverage

Beckham WA et al (2007) Is multibeam IMRT better than standard treatment for patients with left-sided breast cancer? Int J Radiat Oncol Biol Phys 69(3):918–24 → 30 patients with left-sided breast cancer. Conformal RT vs. IMRT.

- IM RT significantly improved homogeneity index (0.95 vs. 0.74), CI (0.91 vs. 0.48), volume of the heart receiving more than 30 Gy (V30-heart) (1.7% vs. 12.5%), and volume of lung receiving more than 20 Gy (V20-left lung) (17.1% vs. 26.6%) (all $p < 0.001$)
- IMRT increased the volume of normal tissues receiving low-dose RT

Rongsriyam K et al (2008) Dosimetric study of inverse-planed intensity modulated, forward-planned intensity modulated and conventional tangential techniques in breast conserving radiotherapy. J Med Assoc Thai 91(10):1571–82. →28 patients. Inverse IMRT vs. forward IMRT vs. conventional tangential technique.

- The iIMRT technique provides significantly improved PTV Dmax, PTV V105%, PTV V110%, target volume coverage, dose homogeneity, and dose conformity throughout the target volume of breast and reduced doses to all critical structures compared to the fIMRT and conventional techniques
- fIMRT technique significantly improved the dose distribution and reduced dose to OARs compared to conventional technique, although not better than iIMRT technique

References

1. de Gelder R, Heijnsdijk EA, Fracheboud J, Draisma G, de Koning HJ (2015) The effects of population-based mammography screening starting between age 40 and 50 in the presence of adjuvant systemic therapy. Int J Cancer 137:165–172
2. Jatoi IKM, Petit JY (2006) Atlas of breast surgery. Springer, New York
3. Compton CC, Byrd DR, Garcia-Aguilar J, Kurtzman SH, Olawaiye A, Washington MK (2012) Breast. In: Compton C, Byrd D, Garcia-Aguilar J, Kurtzman S, Olawaiye A, Washington M (eds) AJCC cancer staging atlas. Springer, New York. https://doi.org/10.1007/978-1-4614-2080-4_32
4. Kitchen PR, Cawson JN, Moore SE, Hill PA, Barbetti TM, Wilkins PA, Power AM, Henderson MA (2006) Margins and outcome of screen-detected breast cancer with extensive in situ component. ANZ J Surg 76:591–595
5. Hance KW, Anderson WF, Devesa SS, Young HA, Levine PH (2005) Trends in inflammatory breast carcinoma incidence and survival: the surveillance, epidemiology, and end results program at the National Cancer Institute. J Natl Cancer Inst 97:966–975
6. Gorringe KL, Fox SB (2017) Ductal carcinoma in situ biology, biomarkers, and diagnosis. Front Oncol 7:248
7. Valenzuela M, Julian TB (2007) Ductal carcinoma in situ: biology, diagnosis, and new therapies. Clin Breast Cancer 7:676–681
8. Hanby AM, Hughes TA (2008) In situ and invasive lobular neoplasia of the breast. Histopathology 52:58–66
9. Sokolova A, Lakhani SR (2021) Lobular carcinoma in situ: diagnostic criteria and molecular correlates. Mod Pathol 34:8–14
10. Weiss A, Chavez-MacGregor M, Lichtensztajn DY, Yi M, Tadros A, Hortobagyi GN, Giordano SH, Hunt KK, Mittendorf EA (2018) Validation study of the American Joint Committee on Cancer eighth edition prognostic stage compared with the anatomic stage in breast cancer. JAMA Oncol 4:203–209
11. Wapnir IL, Dignam JJ, Fisher B, Mamounas EP, Anderson SJ, Julian TB, Land SR, Margolese RG, Swain SM, Costantino JP, Wolmark N (2011) Long-term outcomes of invasive ipsilateral breast tumor recurrences after lumpectomy in NSABP B-17 and B-24 randomized clinical trials for DCIS. J Natl Cancer Inst 103:478–488
12. Masannat YA, Bains SK, Pinder SE, Purushotham AD (2013) Challenges in the management of pleomorphic lobular carcinoma in situ of the breast. Breast 22:194–196
13. Lee MC, Jagsi R (2007) Postmastectomy radiation therapy: indications and controversies. Surg Clin North Am 87:511–526, xi
14. Pritchard KI, Shepherd LE, O'Malley FP, Andrulis IL, Tu D, Bramwell VH, Levine MN, National Cancer Institute of Canada Clinical Trials Group (2006) HER2 and responsiveness of breast cancer to adjuvant chemotherapy. N Engl J Med 354:2103–2111
15. Symmans WF, Peintinger F, Hatzis C, Rajan R, Kuerer H, Valero V, Assad L, Poniecka A, Hennessy B, Green M, Buzdar AU, Singletary SE, Hortobagyi GN, Pusztai L (2007) Measurement of residual breast cancer burden to predict survival after neoadjuvant chemotherapy. J Clin Oncol 25:4414–4422
16. Weiss MC, Fowble BL, Solin LJ, Yeh IT, Schultz DJ (1992) Outcome of conservative therapy for invasive breast cancer by histologic subtype. Int J Radiat Oncol Biol Phys 23:941–947
17. Fisher B, Jeong JH, Anderson S, Bryant J, Fisher ER, Wolmark N (2002) Twenty-five-year follow-up of a randomized trial comparing radical mastectomy, total mastectomy, and total mastectomy followed by irradiation. N Engl J Med 347:567–575
18. Fisher B, Anderson S, Bryant J, Margolese RG, Deutsch M, Fisher ER, Jeong JH, Wolmark N (2002) Twenty-year follow-up of a randomized trial comparing total mastectomy, lumpectomy, and lumpectomy plus irradiation for the treatment of invasive breast cancer. N Engl J Med 347:1233–1241
19. EBCTCG, McGale P, Taylor C, Correa C, Cutter D, Duane F, Ewertz M, Gray R, Mannu G, Peto R, Whelan T, Wang Y, Wang Z, Darby S (2014) Effect of radiotherapy after mastectomy and axillary surgery on 10-year recurrence and 20-year breast cancer mortality: meta-analysis of individual patient data for 8135 women in 22 randomised trials. Lancet 383:2127–2135
20. Julia White AT, Arthur D, Buchholz T, MacDonald S, Marks L, Pierce L, Recht A, Rabinovitch R, Taghian A, Vicini F, Woodward W, X. Allen Li breast cancer atlas for radiation therapy planning: consensus definitions. In: Book breast cancer atlas for radiation therapy planning: consensus definitions
21. Freedman GM, Fowble BL, Nicolaou N, Sigurdson ER, Torosian MH, Boraas MC, Hoffman JP (2000) Should internal mammary lymph nodes in breast cancer be a target for the radiation oncologist? Int J Radiat Oncol Biol Phys 46:805–814

Genitourinary System Cancers

8.1 Prostate Cancer

Prostate cancer is among the most common cancers in men worldwide and the second most common cause of cancer-related mortality in males with an estimated 1,600,000 cases and 366,000 deaths annually [1]. Incidence rates for prostate cancer spiked dramatically in the late 1980s and early 1990s, in large part because of a surge in screening with the prostate-specific antigen (PSA) blood test. Risk factors for prostate cancer are increasing age, African ancestry, a family history of the disease, and certain inherited genetic conditions (e.g., Lynch syndrome and BRCA1 and BRCA2 mutations). Black men in the United States and the Caribbean have the highest documented prostate cancer incidence rates in the world. The 5-year relative survival rate for the 89% of men diagnosed with local or regional stage prostate cancer approaches 100% but drops to 30% for those diagnosed with metastatic disease [2]. The risk of prostate cancer in a man over the course of his lifetime is nearly 10%.

Prostate is the largest accessory gland of the male genital system. It is located in the pelvis, with its base superior and its apex inferior (Fig. 8.1). It is a conical organ that surrounds the prostatic urethra.

- The base of the prostate base is continuous with the bladder neck, while its apex is continuous with the membranous urethra.
- The prostate has four surfaces: anterior, posterior, and two inferolateral surfaces.
  - The anterior is convex and narrow and 2 cm from the pubic symphysis, with the space between being filled with a rich venous plexus (Santorini's plexus) and loose connective tissue. The anterior surface is connected to the pubic symphysis via two pubo-prostatic ligaments, while the inferolateral surfaces neighbor the levator and muscle. Between them is another rich venous plexus surrounded by prostatic sheaths (the lateral plexus).
  - The posterior surface is separated from the rectal ampulla by the prostatic capsule and Denonvilliers' fascia, and it neighbors both of the seminal vesicles and the ampulla of ductus deferens.
  - The cavernous sinus originating from the pelvic plexus (PP) is located within the periprostatic sheath, but outside of the prostate capsule, and is posterolateral to the prostate at the 5

and 7 o'clock positions. The membranous urethra perforates the urogenital diaphragm at the 3 and 9 o'clock positions and enters the corpus cavernosum. These nerves exhibit branching along their paths.

Nerve Entries into the Prostate Have a Low Resistance to Tumor Extension.

Vascular structures are found with the nerves, resulting in a neurovascular bundle. The distance between the capsule and the neurovascular bundle is shortest at the prostatic apex.

- The prostate is also surrounded by a thin sheath that includes rich venous plexus. This prostatic sheath is also continuous with the puboprostatic ligament anteriorly and the deep fascia of the transverse perineal muscle inferiorly.

Neurovascular bundle [4]. The relation of the cavernous nerves to the prostate is important for surgery. The cavernous nerves pass from the prostatic plexus and advance within the neurovascular bundle along with the prostate vessels. This bundle is located within retroperitoneal connective tissue posterolateral to the prostate and medial to the endopelvic fascia (Fig. 8.2).

Pelvic plexus (PP) (Fig. 8.3) [5]. The pelvic nerves originate at the S4–S3 and S2 levels as several anterior roots. These pelvic parasympathetic nerves combine with the sympathetic nerves originating from the hypogastric nerve.

- The upper part of the PP is called the vesical, and the lower part is the prostatic plexus

PP Branches
1. *Anterior branch*: advances within the inferolateral surface of bladder and the lateral surface of the seminal vesicles
2. *Anterior–inferior branch*: advances transversely with the lateral surface of the prostate and provides branches to the prostatovesical anastomosis
3. *Inferior branch*: advances between the posterolateral prostate surface and the rectum, and this branch is known as the neurovascular bundle of the prostate

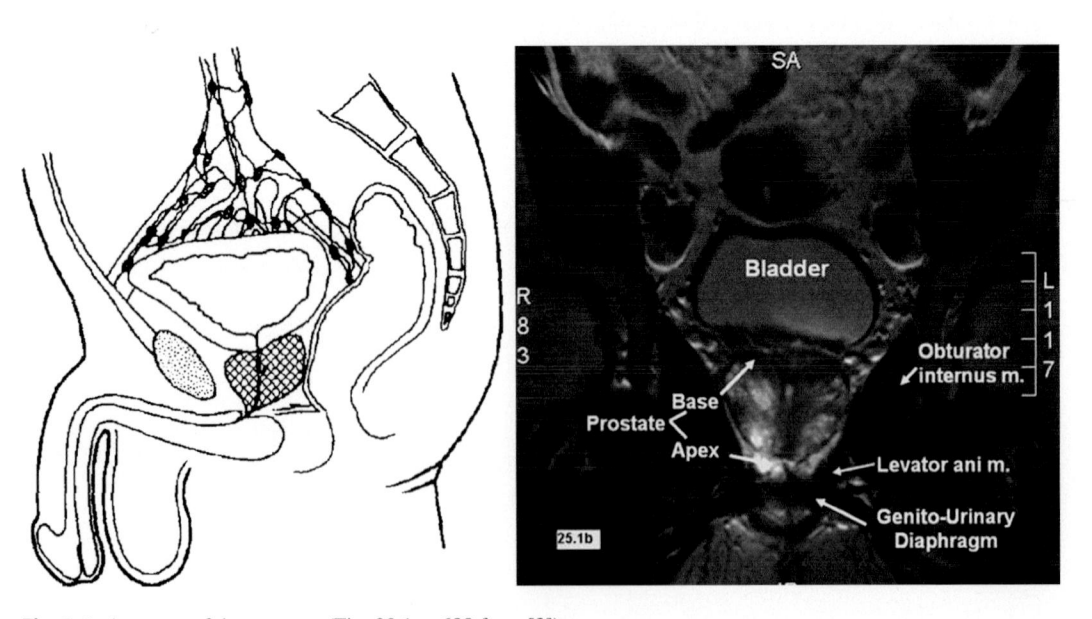

Fig. 8.1 Anatomy of the prostate. (Fig. 28.1, p 688 from [3])

Fig. 8.2 Prostate zonal anatomy

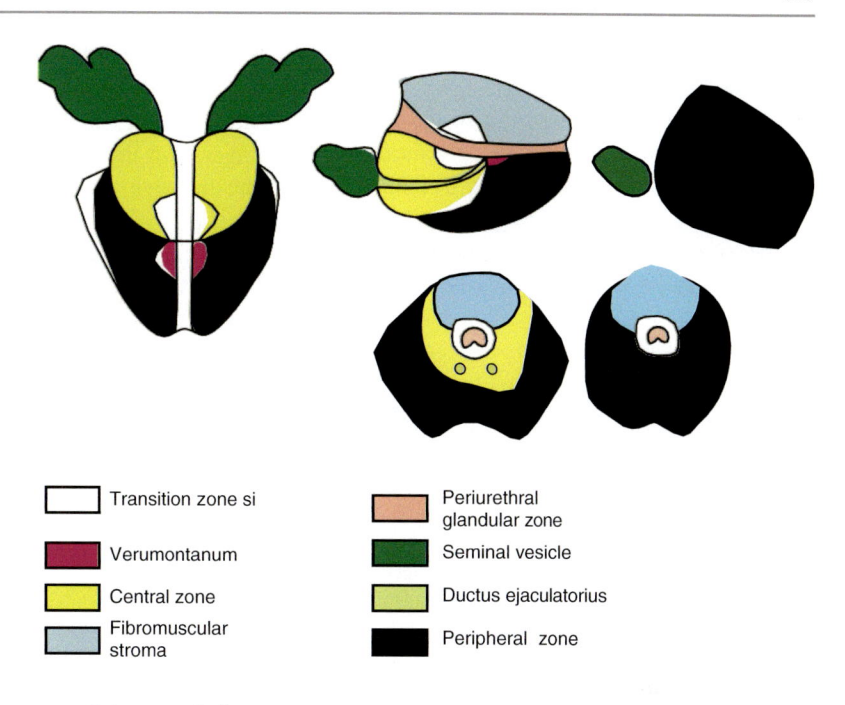

- Transition zone si
- Verumontanum
- Central zone
- Fibromuscular stroma
- Periurethral glandular zone
- Seminal vesicle
- Ductus ejaculatorius
- Peripheral zone

Fig. 8.3 Neurovascular bundle

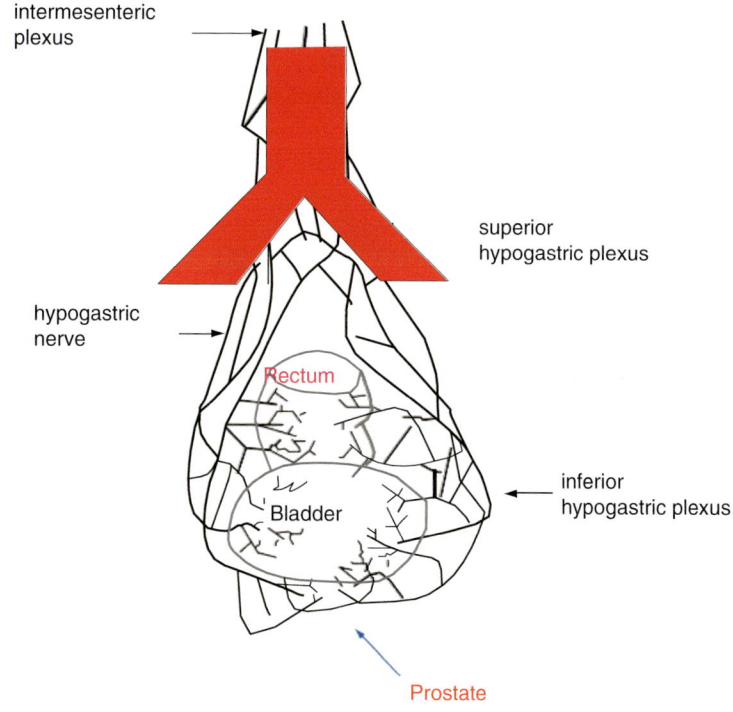

8.1.1 Pathology

More than 95% of prostate cancers are adenocarcinomas, and ~4% are transitional cell cancers [6]. Others are neuroendocrine carcinomas (small cell) and sarcomas. Prostatic intraepithelial neoplasia (PIN) is a precursor lesion.

- PIN is cytologically similar to prostate cancer but is differentiated from cancer by the presence of an intact basal membrane layer.
- PIN is generally classified into either high-grade PIN (HGPIN) or low-grade PIN (LGPIN) [7].
- The clinical importance of this distinction is that HGPIN is associated with cancer in

80% of cases, while this ratio is 20% for LGPIN.
- Prostate cancer develops from the peripheral zone in 70% of cases, from the central zone in 15–20% for central zone, and from the transitional zone in 5–10%.
- Most prostate cancers are multifocal and can be found in different zones of prostate with various grades.

Grading. The most commonly used histological grading system is that of Gleason, which evaluates major growth patterns and glandular differentiation [8].

- The Gleason grade varies between 1 and 5. The most commonly seen major pattern is combined with the secondary pattern, and the total is used to derive the Gleason score
- The Gleason score varies between 2 and 10. Grade 1 is similar to normal, while grade 5 corresponds to no glandular pattern.
- *Score:* 2–6 = well-differentiated; 7 = moderately differentiated; 8–10 = poorly differentiated.

The Gleason grades for the two most prevalent differentiation patterns are combined to create the Gleason score, and Gleason score is now incorporated into the newly adopted grade group system [9]. In the new grade group system, tumors are separated into five categories based upon the primary and secondary Gleason pattern as follows:

- Grade group 1 (Gleason score 3 + 3)
- Grade group 2 (Gleason score 3 + 4)
- Grade group 3 (Gleason score 4 + 3)
- Grade group 4 (Gleason score 4 + 4, 3 + 5, or 5 + 3)
- Grade group 5 (Gleason score 4 + 5, 5 + 4, or 5 + 5)

8.1.2 General Presentation

Most patients are asymptomatic in the early stages, and the presence of symptoms is generally the sign of locally advanced or metastatic disease.

- The extension of the tumor towards the urethra or neck of the bladder or direct invasion of the trigone causes irritative or obstructive symptoms and hematuria.
- Metastatic disease causes anorexia, weight loss, pathologic fractures, and bone pain as a result of metastatic bone disease. It may also cause spinal cordcompression and related symptoms of sensory loss, incontinence, and motor loss.
- Uremic symptoms may occur resulting from ureteral obstruction or retroperitoneal lymph node (LN) metastases.
- Hematospermia due to seminal vesicle invasion, lower extremity edema due to LN metastasis, and erectile dysfunction due to cavernous nerve invasion may also be seenin the advanced stages.

Findings. A firm and irregular prostate in a digital rectal exam (DRE) is typical of cancer, but cancer foci may be found in a normal prostate. Cachexia, globe vesicale, lower extremity lymphedema, deep venous thrombosis, spasticity, motor weakness, and supraclavicular lymphadenopathy may be observed during routine physical examination.

Laboratory findings. Anemia may be seen due to metastatic disease. Bilateral ureter obstruction causes azotemia. Alkaline phosphatase in bone metastasis and acid phosphatase in extraprostatic extension may increase.

Prostate-specific antigen [10]. PSA is 33 kD in weight and consists of a single glycoprotein chain of 237 amino acids and four carbohydrate side-chains, including sulfide bonds. It displays chymotrypsin-like features and is homologous with proteases in the family of human glandular kallikrein 3 (hK-3). It also exhibits 80% homology with the other prostate cancer marker, human glandular kallikrein 2 (hK-2). Human glandular kallikrein 1 (hK-1), mainly found in the pancreas and in renal tissue, has 73–84% homology with PSA.

- PSA is a neutral serine protease that liquifies seminal coagulate by hydrolyzing seminogelin I and II, which are seminal vesicle proteins.
- Only a small portion of the PSA is found in its free form, termed free-PSA (f-PSA), as most of it is bound to alpha 2-macroglobulin (AMG) and alpha 1-antichy- motrypsin (ACT).
- The half-life of PSA is around 2.2–3.2 days and reaches its lowest level 2–3 weeks after radical prostatectomy (RP).

The upper limit for serum prostate-specific antigen (PSA) is accepted to be 4.0 ng/mL. However, 20% of prostate cancer cases have PSA levels of below 4.0 ng/mL, and only 25% of cases with serum PSA levels of 4–10 ng/mL have prostate cancer in their biopsies. Therefore, new methods like age- and race-specific PSA, PSA density, PSA velocity, and free and complex PSA are being developed.

Age- and race-specific PSA. This may be useful for diagnosing organ-confined prostate cancer in young males that can be cured totally with radical therapy and may prevent unnecessary treatment of clinically insignificant small tumors in older patients. The reference intervals for age- and race-specific PSA values according to the best practice policy of the American Urology Society for PSA use and prostate biopsy indications are summarized in Table 8.1 [11].

Table 8.1 Age- and race-specific reference PSA levels

Age interval (years)	Asian	African	White American
40–49	0–2	0–2	0–2.5
50–59	0–3	0–4	0–3.5
60–69	0–4	0–4.5	0–0.4.5
70–79	0–5	0–5.5	0–6.5

An 8% increase in positive biopsy rates and diagnoses of organ-confined disease under 59 years can be obtained compared to when age-specific PSA values are used with the standard 4 ng/mL upper limit.

- **PSA density (PSAD).** More than 80% of men with high serum PSA levels have values of between 4 and 10 ng/mL. These high PSA levels are usually due to a high prevalence of benign prostate hyperplasia (BPH). The PSAD was introduced particularly for men with normal DRE and PSA levels of 4–10 ng/mL in order to differentiate between BPH and prostate cancer.
  - The PSAD is calculated by dividing the serum PSA level by the prostate volume, as measured by TRUS.
  - The threshold level of 0.15 or above indicates prostate cancer, while 0.15 or below indicates benign disease.
  - The main problems with the PSAD are errors in prostate volume measurement using TRUS, changes in the epithelial–stromal ratio of the prostate, and changes in PSA with age.
- **PSA velocity** (PSAV). This is the change in PSA level over time. This change may be due to either an increase in prostate volume or prostate cancer.
  - The concept of the PSAV was developed to aid the early diagnosis of organ-confined cancer, since such cases show high PSA changes in a short period of time (i.e., increased PSAV).

- The PSAV is calculated using a formula that incorporates at least three measured at 6-month (or more) intervals.

$$PSAV = 1/2 \times \left[\frac{(PSA2-PSA1)}{time} + \frac{(PSA3-PSA2)}{time} \right]$$

- The upper limit for the PSAV is taken to be 0.75 ng/mL/year; a value is considered to be a tumor-specific marker.
- The PSAV should be measured according to certain rules (measurement number of measurements).

Roach formulae. The risk of invasion is calculated using these formulae via the PSA level and the Gleason score:

$$\text{Capsule Invasion Risk} = \left(\frac{3}{2}\right) \times PSA + [(GS-3) \times 10]$$

$$\text{Seminal Vesicle Invasion Risk} = PSA + [(GS-6 \times 10]$$

$$\text{Lymph Node Metastasis Risk} = \left(\frac{2}{3}\right) \times PSA + [(GS-6) \times 10]$$

PSA Bounce Phenomenon

The PSA bounce is a temporary rise in PSA level after radiation therapy. Recent trials demonstrated that patients that show a PSA bounce are not at an increased risk of relapse compared to those who do not show a temporary rise. Between a third and a half of patients experience a PSA bounce regardless of the technique applied (i.e., external radiotherapy or brachytherapy), and it may occur anywhere between about 1 and 3 years after treatment. The magnitude of the bounce lies in the range 0.5–2 ng/mL and may last from a few months to around a year. The reason for the bounce is not known. Testosterone recovery after hormonal treatment may cause a PSA bounce in patients receiving androgen ablation therapy along with radiotherapy [12].

8.1.3 Staging

T Staging (Fig. 8.4) [14]
Clinical T staging
- T0: No evidence of primary tumor
- T1: Clinically inapparent tumor not palpable nor visible by imaging
  - T1a: Tumor incidental histologic finding in 5% or less of tissue resected
  - T1b: Tumor incidental histologic finding in more than 5% of tissue resected
  - T1c: Tumor identified by needle biopsy found in one or both sides, but not palpable
- T2: Tumor is palpable and confined within prostate
  - T2a: Tumor involves 50% or less of one lobe
  - T2b: Tumor involves more than 50% of one lobe but not both lobes
  - T2c: Tumor involves both lobes
- T3: tumor extends through the prostate capsule
  - T3a: Extracapsular extension (unilateral or bilateral) T3b: Tumor invades seminal vesicle(s)
- T4: Tumor is fixed or invades adjacent structures other than seminal vesicles: bladder neck, external sphincter, rectum, levator muscles, and/or pelvic wall

Pathological T Staging
- T2: Organ confined
- T3: Extraprostatic extension

- T3a: Extraprostatic extension (unilateral or bilateral) or microscopic invasion of bladder neck
- T3b: Tumor invades seminal vesicle(s)
- T4: Tumor is fixed or invades adjacent structures other than seminal vesicles such as external sphincter, rectum, bladder, levator muscles, and/or pelvic wall

   NOTE: There is no pathological T1 classification.

   NOTE: Positive surgical margin should be indicated by an R1 descriptor, indicating residual microscopic disease.

N Staging (Fig. 8.5)
- NX: Regional lymph nodes cannot be assessed
- N0: No regional lymph nodes metastasis
- N1: Metastasis in regional lymph nodes(s)
- **Regional lymph nodes (N)** (Fig. 8.5)
- Regional lymph nodes are the pelvic nodes below the bifurcation of the common iliac arteries.
   - Pelvic
   - Hypogastric
   - Obturator
   - Iliac (internal, external)
   - Sacral (lateral, presacral, promontory)
- Distant lymph nodes are outside the confines of the true pelvis.

   - Aortic (para-aortic, periaortic, or lumbar)
   - Common iliac
   - Inguinal (deep)
   - Superficial inguinal (femoral)
   - Supraclavicular
   - Cervical
   - Scalene
   - Retroperitoneal
- Involvement of distant LNs is classified M1a.

Distant metastasis (M)
- M0: No distant metastasis
- M1: Distant metastasis
   - M1a: Nonregional lymph node(s)
     M1b: Bone(s)
   - M1c: Other site(s) with or without bone disease

   NOTE: When more than one site of metastasis is present, the most advanced category is used. M1c is most advanced.

PSA Values
- <10
- ≥10–<20
- <20
- ≥20
- Any value

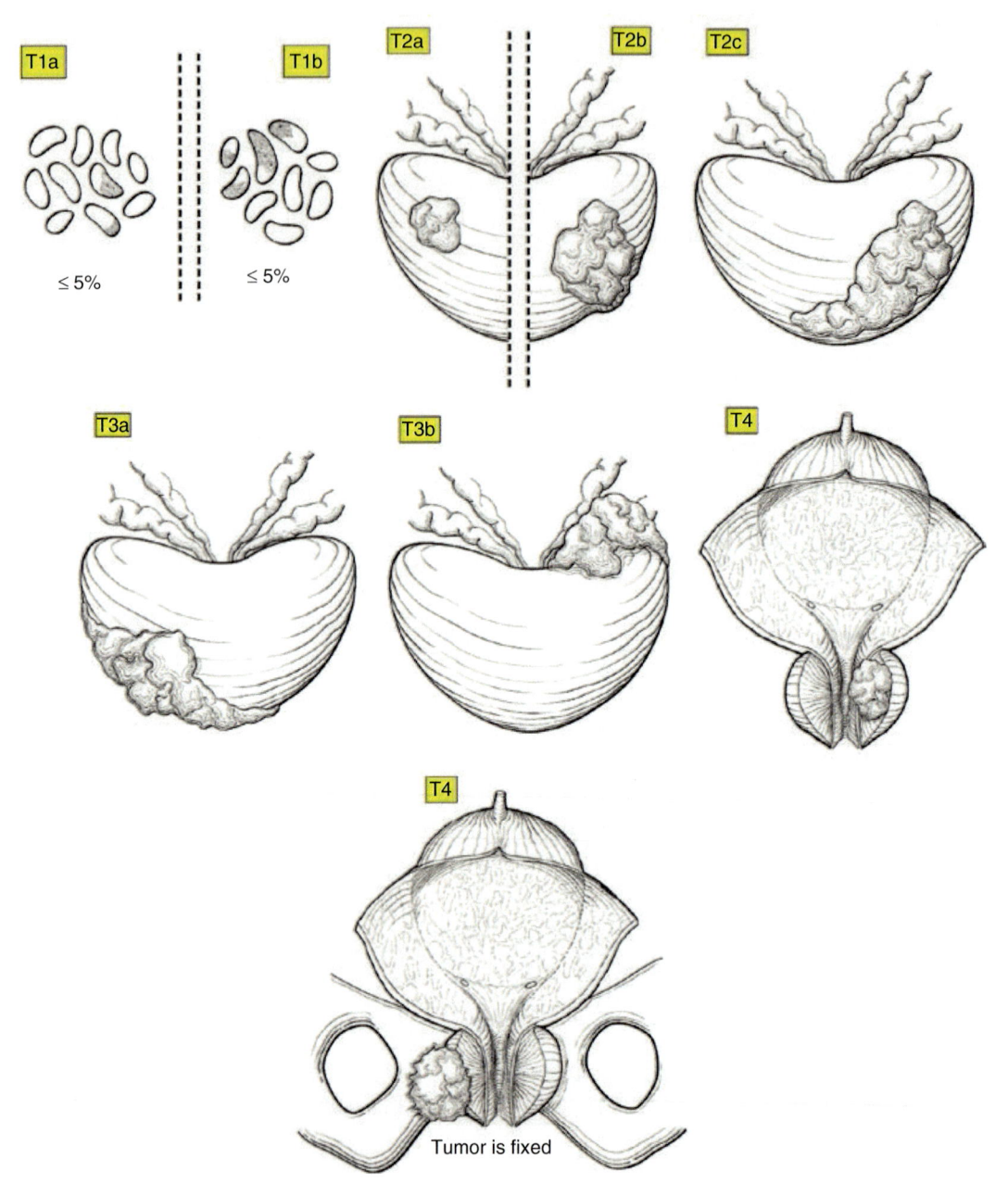

Fig. 8.4 T staging in prostate cancer. (from [13], p 295–298, Figs. 34.3–34.8, reproduced with the permission from the American Joint Committee on Cancer)

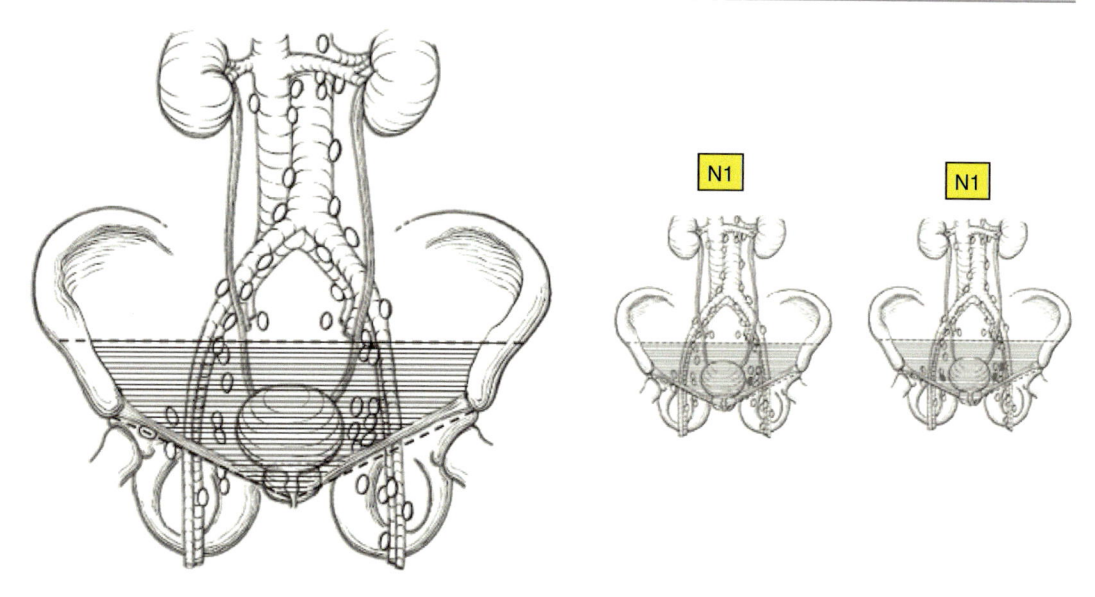

Fig. 8.5 Prostate lymphatics and N staging. (from [14], Compton C.C., Byrd D.R., Garcia-Aguilar J., Kurtzman S.H., Olawaiye A., Washington M.K. (2012) Prostate. In: Compton C., Byrd D., Garcia-Aguilar J., Kurtzman S., Olawaiye A., Washington M. (eds) AJCC Cancer Staging Atlas. Springer, New York, NY. https://doi.org/10.1007/978-1-4614-2080-4_41)

8.1.3.1 Histopathologic Grade (G)

Recently, the Gleason system has been compressed into so-called grade groups.

Grade group	Gleason score	Gleason pattern
1	≤6	≤3 + 3
2	7	3 + 4
3	7	4 + 3
4	8	4 + 4, 3 + 5, or 5 + 3
5	9–10	4 + 5, 5 + 4, or 5 + 5

AJCC stage groups					
T	N	M	PSA	Grade Group	Stage Group
cT1a-c, cT2a	N0	M0	<10	1	I
pT2	N0	M0	<10	1	I
cT1a-c, cT2a, pT2	N0	M0	≥10 < 20	1	IIA
cT2b-c	N0	M0	<20	1	IIA
T1–2	N0	M0	<20	2	IIB
T1–2	N0	M0	<20	3	IIC
T1–2	N0	M0	<20	4	IIC
T1–2	N0	M0	≥20	1–4	IIIA
T3–4	N0	M0	Any	1–4	IIIB
Any T	N0	M0	Any	5	IIIC
Any T	N1	M0	Any	Any	IVA
Any T	Any N	M1	Any	Any	IVB

Extension and invasion [15, 16]. This is related to extraprostatic extension, seminal vesicle invasion, distant metastasis, and Gleason score.

- Small and well-differentiated tumors (grade 1 and 2) are generally organ-confined, while large (>4 cm^3) and undifferentiated tumors (grade 4 and 5) are generally locally advanced or metastatic (LN/bone).
- Capsule penetration usually occurs from the perineural space.
- The trigone may be invaded in locally advanced prostate cancer, causing ureter obstruction.
- Rectal invasion is very rare, due to a barrier called Denonvilliers' fascia, located
- between the prostate and rectum.
- Lymphatic metastasis usually occurs in obturator LNs, as well as in presacral and periaortic LNs.
- Distant metastasis via the valveless Batson's venous plexus is commonly seen in the axial skeleton, particularly the lumbar vertebrae, followed by proximal femur, pelvis, thoracic vertebrae, costae, sternum, skull, and humerus.
- Bone metastasis of prostate cancer is typically osteoblastic. The involvement of long bones may cause fractures, and vertebral metastases may cause spinal cord compression.
- Visceral metastasis is usually to lungs, liver, and adrenal glands. Brain metastases are very rare and generally occur by direct metastatic extension of skull bones.

8.1.3.2 Imaging
Technetium-99 bone scan: if PSA >20, T2 and PSA >10, GS ≥8, T3–T4, or presence of symptoms

Pelvic CT/ MRI: if T3–T4 or T1–T2 and risk of LN involvement is >10% [17].

MR spectroscopy: Role in routine management is controversial

Prostate imaging reporting and data system (PI-RADS) — The International Prostate MRI Working Group developed PI-RADS to standardize prostate MRI examination performance and reporting. An updated version, PI-RADS v2.1, which was published in 2019 [18]. This system categorizes prostate lesions based on the likelihood of cancer according to a five-point scale, defined as the following:

- **PI-RADS 1**—Clinically significant cancer is highly unlikely to be present.
- **PI-RADS 2**—Clinically significant cancer is unlikely to be present.
- **PI-RADS 3**—The presence of clinically significant cancer is equivocal.
- **PI-RADS 4**—Clinically significant cancer is likely to be present.
- **PI-RADS 5**—Clinically significant cancer is highly likely to be present.

Next-Generation Imaging: F-18 NaF PET/CT, C-11 choline PET/CT, F-18 fluciclovine PET/CT, Ga-68 PSMA PET, or PET/MRI or whole-body MRI might be used in selective cases.

8.1.4 Treatment Algorithm

Risk stratification schema for localized prostate cancer, according to the National Comprehensive Cancer Network (NCCN) [17]

Risk group	Clinical/pathologic features
Very low	T1c AND Grade group 1 AND PSA <10 ng/mL AND Fewer than 3 prostate biopsy fragments/cores positive, ≤50% cancer in each fragment/core AND PSA density < 0.15 ng/mL/g
Low	T1 to T2a AND Grade group 1 AND PSA <10 ng/mL AND Does not qualify for very low risk
Favorable intermediate	No high or very high risk features No more than one intermediate risk factor: T2b to T2c OR Grade group 2 or 3 PSA 10 to 20 ng/mL AND Grade group 1 or 2 AND Percentage of positive biopsy cores <50%

Risk group	Clinical/pathologic features
Unfavorable intermediate	No high or very high risk features Two or three of the intermediate risk factors: T2b to T2c Grade group 2 or 3 PSA 10–20 ng/mL AND/OR Grade group 3 AND/OR ≥50% of positive biopsy cores
High	No very high risk features AND T3a OR Grade group 4 or 5 OR PSA >20 ng/mL
Very high	T3b to T4 OR Primary Gleason pattern 5 OR Two or three high-risk features OR >4 cores with Grade group 4 or 5

D'Amico risk stratification of localized prostate cancer:

- Low-risk—T1-T2a and Gleason score ≤6 and PSA ≤10 ng/mL
- Intermediate-risk—T2b and Gleason score 7 and/or PSA 10 to 20 ng/mL
- High-risk—≥T2c or Gleason score 8 to 10 or PSA >20 ng/mL

Treatment Algorithm for Prostate Cancer [17]

Very Low Risk
- Survival estimate >20 years:
  – Active surveillance (preferred)
  – External beam radiotherapy (EBRT) or brachytherapy
  – Radical prostatectomy (RP)
- Survival estimate 10–20 years: Active surveillance
- Survival estimate <10 years: Observation

Low Risk
- Survival estimate ≥10 years:
  – Active surveillance (preferred)
  – EBRT or brachytherapy
  – RP

- Survival estimate <10 years: Observation

Favorable Intermediate Risk
- Survival estimate >10 years:
  – Active surveillance (preferred)
  – EBRT or brachytherapy
  – RP ± pelvic lymph node dissection (PLND)* if predicted LN metastases risk is ≥2%
- Survival estimate 5–10 years:
- EBRT or brachytherapy
- Observation

Unfavorable Intermediate Risk
- Survival estimate >10 years:
  – RP ± pelvic lymph node dissection (PLND)* if predicted LN metastases risk is ≥2%
  – EBRT + short term androgen deprivation treatment (ADT) (4–6 months)
  – EBRT + brachytherapy ± ADT (4–6 months)
- Survival estimate 5–10 years:
  – Observation
  – EBRT + short term androgen deprivation treatment (ADT) (4–6 months)
  – EBRT + brachytherapy ± ADT (4–6 months)

High-Very High Risk
- Survival estimate >5 years or symptomatic:
  – EBRT + ADT (1.5–3 years) ± (docetaxel for very high risk only)
  – EBRT + brachytherapy + ADT (1–3 years)
  – RP + PLND*
- Survival estimate ≤5 years and asymptomatic: observation or ADT or EBRT
- *After surgery:
- Surgical margin (+) after RP; adjuvant RT or surveillance
- LN (+) after RP; ADT ± RT or surveillance

Any T, N1, M0
- Survival estimate >5 years or symptomatic:

- EBRT + ADT ± abiraterone
- ADT ± abiraterone
- Survival estimate ≤5 years and asymptomatic: observation or ADT

Disseminated Metastatic Disease
- Androgen ablation ± palliative RT ± bisphosphonates
- Abiraterone + dexamethazone, enzalutamide, apalutamide
- Chemotherapy (docetaxel + cabazitaxel/carboplatin or estramustine
- Stronsium, radium-223
- Immunotherapy (sipuleucel, pembrolizumab)
- *Residual disease or recurrence after RP*
    - Persistent local disease or high-risk local residual disease: RT ± ADT
    - No evidence of persistent local disease or metastasis: ADT or observation ± RT
- *Residual disease or recurrence after radiotherapy*
    - Biopsy (+) and no metastasis or low-risk group: surgery or salvage brachytherapy/SBRT, cryotherapy, high intensity focused ultrasound
    - If systemic or local therapies are not suitable: ADT or observation

Androgen ablation [10]. Several prostate cancers are hormone dependent, and it is well known that 70–80% of men with metastatic prostate cancer may respond to androgen ablation. Androgen ablation may be performed at different levels of the hypophyseal–testis axis by a variety of methods and agents (Table 8.2).

Radical Prostatectomy. Performed mainly by two approaches.
- The first and more popular is *radical retropubic prostatectomy* [19]. Here, the surgeon makes an incision 2 cm above the penis and 2–3 cm below the umbilicus and excises the prostate with its surrounding tissues from this incision. Pelvic LNs are dissected first with this technique, and the prostate is dissected if the LNs are not involved. These LNs are examined by frozen section, and the prostate is not usually dissected if they are positive for cancer. Another advantage of this technique is that the neurovascular bundle can be spared for erectile functions by nerve-sparing prostate surgery.
- The other technique, the *transperineal approach* [20], is usually not preferred

Table 8.2 Drugs used for androgen ablation

Level	Agent	Administration route	Dose (mg)	Usage frequency
Hypophysis (LHRH agonist)	Diethylstilbestrol	PO	1–3	Daily
	Goserelin	SC	10.8	Every 3 months
	Goserelin	SC	3.6	Monthly
	Leuprolide	IM/SC	11.25	Every 3 months
	Leuprolide	IM/SC	22.5	Every 6 months
Hypophysis (LHRH antagonist)	Degarelix	SC	240	Monthly
Surrenal gland	Ketoconazole	PO	400	Daily
	Aminoglutethimide	PO	250	Four times daily
Testis	Orchiectomy	–	–	–
Prostate	Bicalutamide	PO	50/150	Daily
	Flutamide	PO	250	Three times daily
	Nilutamide	PO	150	Daily

for prostate cancer. The incision is performed between the anus and testicles. There is less bleeding in this approach, and it is easier to perform than radical retropubic prostatectomy in obese men. However, the nerves are not spared, and the LNs cannot be dissected with this surgical approach. Therefore, the perineal technique is not usually preferred for the management of prostate cancer.

8.1.5 Radiotherapy

8.1.5.1 Relative Indications

Whole-Pelvis Radiotherapy [21]

Relative indications:
  T4 tumor
  LN metastases (+)
  Seminal vesicle invasion (+)
  LN metastases risk >15% *according to Roach formula*
  High risk group

Localized Prostate Field [20]
Indications
- T1–T2 tumor
- LN (−)
- Seminal vesicle invasion (−)
- LN involvement risk <15% according to the Roach formula
- Low- to moderate-risk patients
- High-risk patients receiving long-term androgen ablation therapy
- Intensity modulated radiation therapy is used with localized prostate fields. (Fig. 8.6).

8.1.5.2 Simulation

- Simulation is usually performed in the supine position. Some centers prefer the prone position in order to decrease the volume of the small intestine. Knee support may be used in the supine position for the same purpose.
- The bladder is visualized with a small amount of IV contrast material (40–50 cm^3) to differentiate the base of the prostate from the neck of the bladder.

Intensity-modulated radiation therapy (IMRT) is preferred over three-dimensional conformal radiation therapy. Although there are no randomized trials comparing IMRT with 3D-CRT, IMRT appears to be less toxic at equivalent tumor doses of radiation by better coverage of target and delivering less doses to nearby normal structures. Image-guided radiation therapy (IGRT) is suggested for daily localization of the prostate prior to each treatment.

IMRT Volumes (Fig. 8.6) [22]
GTV: nodule if visible; otherwise, the GTV is not delineated; it is adequate to delineate the CTV as described below.
  CTV: entire prostate gland (usually includes one-third of SV) ± SV (depends on SV invasion risk).
  PTV: may change according to institutional protocols. Usually it is CTV+ 1–8 cm in all directions, except 5–6 mm at posterior border for conformal RT and IMRT. In SBRT and other IGRT techniques, 0.5 mm is recommended in all but 0.3 mm in the posterior direction.
  CTV postoperative: Begin from the level of the cut end of the vas deferens and end at >8–12 mm inferior to the vesicourethral anastomosis (VUA). All surgical clips in the prostate and SV bed should be included.

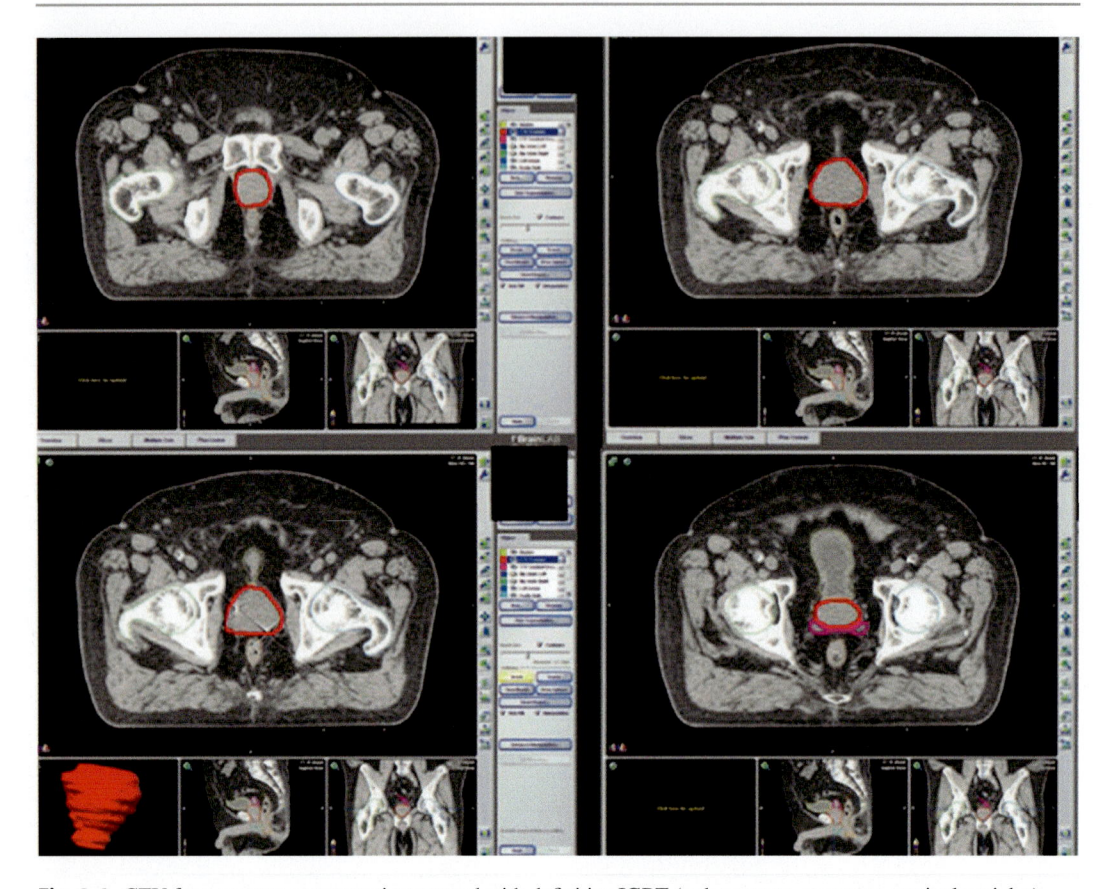

Fig. 8.6 CTV for a prostate cancer patient treated with definitive IGRT (red: prostate, magenta: seminal vesicles)

Critical organ dose for 3D-conformal RT and IMRT

Rectum

- V40 ≤45%, V65 ≤25%

Bladder

- V40 ≤ 60%, V65 ≤ 40%

Femoral heads

- V50 ≤ 10%,

Mean dose for femoral heads should be less than 50 Gy

Stereotactic body radiation therapy (SBRT) is an extreme hypofractionation in which the entire dose of radiation is administered in five or fewer fractions. Due to low alpha/beta ratio, SBRT might be quite effective in prostate cancer. It is an appropriate alternative to conventional fractionation RT for carefully selected men with low- or intermediate-risk prostate cancer who do not need nodal irradiation (Fig. 8.7). According to NRG-GU005 protocol for prostate cancer, 5 fractions of 7.25 Gy is delivered every other day.

CTV = prostate on T2-MRI fused with CT sim scan

PTV = CTV + 5 mm in all directions except for 3 mm posteriorly

8.1 Prostate Cancer

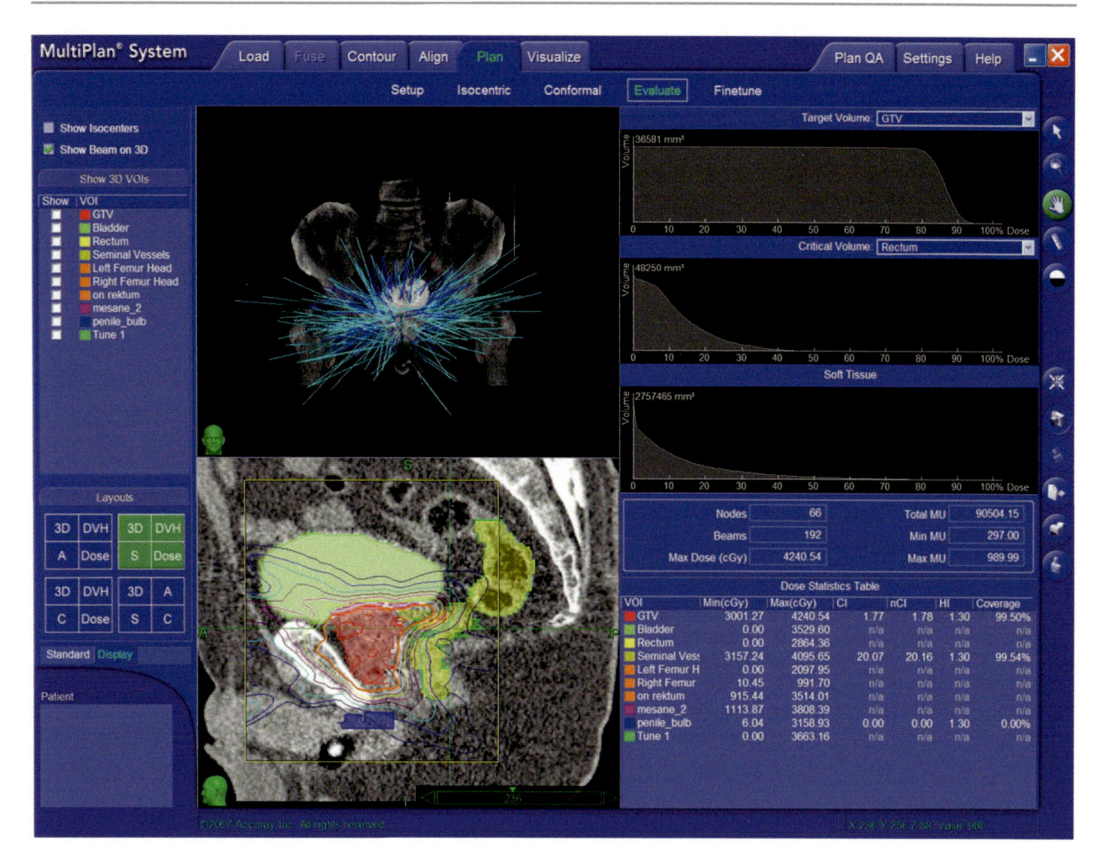

Fig. 8.7 Robotic radiosurgery planning for a patient with low-risk prostate cancer

Dose Constraints for SBRT:
Rectum: V18.12 Gy < 50% V29 Gy <20% V32.63 Gy <10% V34.4 Gy < 3 cc
Bladder: V18.12 < 10% V38.06 Gy < 0.03 cc
Urethra: V38.78 Gy < 0.03 cc
Femoral heads: V19.9 Gy < 10 cc V15.6 Gy < 1 cc
Penile bulb: V19.9 Gy < 3 cc

Dose/Energy
- 1.8–2 Gy daily fractions with a suitable X-ray energy according to patient thickness
  - (6–18 MV):
  - Whole pelvis (if indicated): 45–50 Gy
  - Prostate field: 78–79.2 Gy
  - Prophylactic dose to the seminal vesicles is 54 Gy to the proximal 1 cm. If there is seminal vesicle disease it receives full dose.
  - Prostate bed: 64–66 Gy for postoperative (at least 70 Gy for macroscopic disease)

8.1.6 Selected Publications

8.1.6.1 PSA Relapse Definition After RT

ASTRO consensus → three consecutive increases in PSA is the definition of biochemical failure after radiation therapy. The date of failure should be the midpoint between the post- RT nadir PSA and the first of the three consecutive rises.

- Biochemical failure is not a justification to initiate additional treatment. It is not equivalent to clinical failure. It is an appropriate early end point for clinical trials.

Anon (1997) Consensus statement: guidelines for PSA following radiation therapy. American Society for Therapeutic Radiology and Oncology Consensus Panel. Int J Radiat Oncol Biol Phys 37(5):1035–1041

Phoenix definition → (1) a rise by 2 ng/mL or more above the nadir PSA should be considered the standard definition for biochemical failure after external RT with or without hormonal treatment; (2) the date of failure should be determined "at call" (not backdated).

- Investigators are allowed to use the ASTRO Consensus Definition after RT alone (no hormonal therapy) with strict adherence to guidelines regarding "adequate follow-up".
- The reported date of control should be listed as 2 years short of the median follow-up.

Roach M III et al. (2006) Defining biochemical failure following radiotherapy with or without hormonal therapy in men with clinically localized prostate cancer: recommendations of the RTOG-ASTRO Phoenix Consensus Conference. Int J Radiat Oncol Biol Phys 65(4):965–974

8.1.6.2 Dose escalation

MRC RT01, 2007 → 843 men with localized prostate cancer who were randomly assigned to standard-dose CFRT or escalated-dose CFRT, both with neoadjuvant androgen suppression. Escalated-dose CFRT with neoadjuvant androgen suppression was clinically worthwhile in terms of biochemical control and decreased use of salvage androgen suppression. This additional efficacy was offset by an increased incidence of longer-term adverse events.

Dearnaley DP (2007) Escalated-dose versus standard-dose conformal radiotherapy in prostate cancer: first results from the MRC RT01 randomised controlled trial. Lancet Oncol 8(6):475–487

Dutch trial, 2006 → 664 stage T1b–4 prostate cancer patients were enrolled onto a randomized trial comparing 68 Gy with 78 Gy. The primary end point was freedom from failure (FFF). Median follow-up time was 51 months. Hormone therapy was prescribed for 143 patients. FFF was significantly better in the 78 Gy arm compared with the 68 Gy arm (64 vs. 54%, $p = 0.02$). There was no difference in late genitourinary toxicity of RTOG and EORTC grade 2 or more and a slightly higher nonsignificant incidence of late gastrointes- tinal toxicity of grade 2 or more.

Peeters ST (2006) Dose-response in radiotherapy for localized prostate cancer: results of the Dutch multicenter randomized phase III trial comparing 68 Gy of radiotherapy with 78 Gy. J Clin Oncol 24(13):1990–1996

8.1.6.3 Altered Fractionation

NCI Canada, 2005 → 66 Gy/33 fractions (2 Gy/fractions) vs. 52.5 Gy/20 fractions (2.62 Gy/fractions), 936 T1–T2, PSA <40 cases with no hormone therapy. Median follow-up: 5.7 years.

- Five-year DFS: 40% in hypofractionated arm vs. 47% in standard arm (NS).
- No difference in OS.
- Late grade III toxicity: 3.2% in both arms.

Lukka H (2005) Randomized trial comparing two fractionation schedules for patients with localized prostate cancer. J Clin Oncol 23(25):6132–6138

Moderate hypofractionation RT over conventional fractionation can be chosen if nodal irradiation is not needed, inconsistent with professional ASTRO, ASCO, and the AUA guidelines. (Morgan SC, et al. Hypofractionated Radiation Therapy for Localized Prostate Cancer: An ASTRO, ASCO, and AUA Evidence-Based Guideline. J Urol. 2018 Oct 9:S0022-5347(18)43963-8.) **Patients should be counseled about the small increased risk of gastrointestinal and genitourinary toxicity associated with this approach.** (Nossiter J, et al. Patient-Reported Functional Outcomes After Hypofractionated or Conventionally Fractionated Radiation for Prostate Cancer: A National Cohort Study in England. J Clin Oncol. 2020;38(7):744–752.

Multiple large randomized trials and a meta-analysis have evaluated the potential role of hypofractionated RT and the conclu-

sion of most was that efficacy is not inferior with moderate hypofractionation. CHHiP is a randomized, phase 3, non-inferiority trial that recruited men with localized prostate cancer (pT1b-T3aN0M0). Patients were randomly assigned (1:1:1) to conventional (74 Gy delivered in 37 fractions over 7·4 weeks) or one of two hypofractionated schedules (60 Gy in 20 fractions over 4 weeks or 57 Gy in 19 fractions over 3·8 weeks) all delivered with intensity-modulated techniques. With a median follow-up of 5.2 years, the study concluded that hypofractionated radiotherapy using 60 Gy in 20 fractions is non-inferior to conventional fractionation using 74 Gy in 37 fractions and is recommended as a new standard of care for external-beam radiotherapy of localized prostate cancer. There were no significant differences in either the proportion or cumulative incidence of side effects 5 years after treatment. (Dearnaley D, et al; CHHiP Investigators. Conventional versus hypofractionated high-dose intensity-modulated radiotherapy for prostate cancer: 5-year outcomes of the randomized, non-inferiority, phase 3 CHHiP trial. Lancet Oncol. 2016;17(8):1047–1060.)

In the Dutch HYPRO trial, 820 patients with intermediate- or high-risk localized prostate cancer were randomly assigned to RT with conventional fractionation (39 fractions of 2 Gy over eight weeks) or hypofractionation (19 fractions of 3.4 Gy in 6.5 weeks). At a median follow-up of 60 months, there was no statistically significant difference in the five-year relapse-free survival rate (77.1% vs. 80.5%, respectively, for conventional and hypofractionated treatment; HR 0.86, 95% CI 0.63–1.16). The rate of grade ≥2 gastrointestinal toxicity at three years was higher in the hypofractionation group (22% vs 18%), as was that of grade ≥3 genitourinary toxicity (19% vs 13%). (Incrocci L, et al. Hypofractionated versus conventionally fractionated radiotherapy for patients with localised prostate cancer (HYPRO): final efficacy results from a randomised, multicentre, open-label, phase 3 trial. Lancet Oncol. 2016;17(8):1061–1069.)

The PROFIT trial randomized 1206 men with intermediate-risk prostate cancer to either 7800 cGy in 39 fractions of 200 cGy over 7.8 weeks or 6000 cGy in 20 fractions of 300 cGy over 4 weeks. At a median follow-up of six years, the hypofractionated regimen had non-inferior biochemical-clinical failure. (Catton CN, et al. Randomized Trial of a Hypofractionated Radiation Regimen for the Treatment of Localized Prostate Cancer. J Clin Oncol. 2017:1884–1890.)

The RTOG 0415 trial randomly assigned 1115 patients with low-risk prostate cancer to RT with conventional fractionation (73.8 Gy in 41 fractions over 8.2 weeks) or a hypofractionated regimen (70.8 Gy in 28 fractions over 5.6 weeks). At a median follow-up of 5.9 years, hypofractionation met the predefined criteria for noninferiority, with a five-year disease-free survival rate of 86.3% versus 85.3% with conventional fractionation. However, hypofractionation was associated with a significantly increased rate of physician-reported, late, grade 2 and 3, gastrointestinal and genitourinary toxicity but not in patient-reported prostate cancer-specific (e.g., bowel, bladder, sexual) or general quality of life. (Lee WR, et al. Randomized Phase III Noninferiority Study Comparing Two Radiotherapy Fractionation Schedules in Patients With Low-Risk Prostate Cancer. J Clin Oncol. 2016;34(20):2325–32.) (Bruner DW, et al. Quality of Life in Patients With Low-Risk Prostate Cancer Treated With Hypofractionated vs Conventional Radiotherapy: A Phase 3 Randomized Clinical Trial. JAMA Oncol. 2019;5(5):664–670.)

A meta-analyses of 10 studies with 8278 men comparing hypofractionation with conventional fractionation revealed that moderate hypofractionation (up to a fraction size of 3.4Gy) results in similar oncologic outcomes in terms of disease-specific, metastasis-free, and overall survival. There appears to be little to no increase in both acute and late toxicity. (Hickey BE, et al. Hypofractionation for clinically localized prostate cancer. Cochrane Database Syst Rev. 2019;9(9):CD011462.)

8.1.6.4 Pelvic Field RT

RTOG 94-13, 2007 → this trial was designed to test the hypothesis that total androgen suppression and whole pelvic radiotherapy (WPRT) followed by a prostate boost improves progression-free survival (PFS) compared with total androgen suppression and prostate-only RT (PORT). This trial was also designed to test the hypothesis that neoadjuvant hormonal therapy (NHT) followed by concurrent total androgen suppression and RT improves PFS compared with RT followed by adjuvant hormonal therapy (AHT). It involved 1323 localized prostate cancer cases with an elevated PSA £100 ng/mL and an estimated risk of LN involvement of 15%. The difference in overall survival for the four arms was statistically significant ($p = 0.027$). However, no significant differences were found in PFS or overall survival between NHT vs. AHT and WPRT compared with PORT. A trend towards a difference was found in PFS ($p = 0.065$) in favor of the WPRT + NHT arm compared with the PORT + NHT and WPRT + AHT arms.

Lawton CA (2007) An update of the phase III trial comparing whole pelvic to prostate-only radiotherapy and neoadjuvant to adjuvant total androgen suppression: updated analy- sis of RTOG 94-13, with emphasis on unexpected hormone/radiation interactions. Int J Radiat Oncol Biol Phys 69(3):646–655

GETUG-01, France, 2007 → 444 patients with T1b–T3, N0 pNx, M0 prostate carcinoma were randomly assigned to either pelvic and prostate radiotherapy or prostate radiotherapy only. Short-term 6-month neoadjuvant and concomitant hormonal therapy was given to patients in the high-risk group. The pelvic dose was 46 Gy. The total dose recommended to the prostate was 66–70 Gy. With a 42.1-month median follow-up time, the 5-year PFS and overall survival were similar in the two treatment arms for the whole series and for each stratified group. Pelvic node irradiation was well tolerated but did not improve PFS.

Pommier P et al. (2007) Is there a role for pelvic irradiation in localized prostate adeno- carcinoma? Preliminary results of GETUG-01. J Clin Oncol 25(34):5366–5373

The POP-RT trial is a phase III trial comparing prophylactic whole-pelvic nodal radiotherapy to prostate-only radiotherapy (PORT) in high-risk prostate cancer. At a median follow-up of 68 months, 5-year BFFS was 95.0% with WPRT versus 81.2% with PORT, $p < 0.0001$). WPRT also showed higher 5-year DFS (89.5% vs 77.2%; $p = 0.002$), but 5-year OS was not different. Distant metastasis-free survival was also higher with WPRT (95.9% vs 89.2%; $p = 0.01$). Benefit in BFFS and DFS was maintained across prognostic subgroups. (Murthy V, et al. Prostate-Only Versus Whole-Pelvic Radiation Therapy in High-Risk and Very High-Risk Prostate Cancer (POP-RT): Outcomes From Phase III Randomized Controlled Trial. J Clin Oncol. 2021:JCO2003282. doi: https://doi.org/10.1200/JCO.20.03282. Epub ahead of print.)

8.1.6.5 Hormonal Therapy + Radiotherapy

RTOG 85-31, 2005 → 977 palpable primary tumor cases extending beyond the prostate (clinical stage T3) or those with regional lymphatic involvement. Patients who had undergone prostatectomy were eligible if penetration through the prostatic capsule to the margin of resection and/or seminal vesicle involvement was documented histologically. Patients were randomized to either RT and adjuvant goserelin (arm I) or RT alone followed by observation and the application of goserelin at relapse (arm II). In arm I, the drug was to be started during the last week of RT and was to be continued indefinitely or until signs of progression. Median follow-up was 7.6 years.

- Ten-year survival rate was significantly greater for the adjuvant arm than for the control arm: 49 vs. 39%, respectively ($p = 0.002$).
- Ten-year local failure rate for the adjuvant arm was 23 vs. 38% for the control arm ($p < 0.0001$).
- Ten-year rates for the incidence of distant metastases and disease-specific mortality were 24 vs. 39% ($p < 0.001$) and 16 vs. 22% ($p = 0.0052$), respectively.

- In patients with carcinoma of the prostate with an unfavorable prognosis, androgen suppression applied as an adjuvant after definitive RT was associated not only with a reduction indisease progression but with a statistically significant improvement in absolute survival. The improvement in survival appeared preferentially in patients with Gleason scores of 7–10.

Pilepich MV (2005) Androgen suppression adjuvant to definitive radiotherapy in prostate carcinoma—long-term results of Phase III RTOG 85-31. Int J Radiat Oncol Biol Phys 61(5):1285–1290

Harvard, 2008 → 206 men with localized but unfavorable-risk prostate cancer were randomized to receive RT alone or RT and hormone therapy. Median follow-up: 7.6 years.

- A significant increase in the risk of all-cause mortality (44 vs. 30 deaths; $p = 0.01$) was observed in men randomized to RT compared with RT and AST.
- The increased risk in all-cause mortality appeared to apply only to men randomized to RT with no or minimal comorbidity (31 vs. 11 deaths; $p < 0.001$).
- Among men with moderate or severe comorbidity, those randomized to RT alone vs. RT and AST did not have an increased risk of all-cause mortality (13 vs. 19 deaths; $p = 0.08$).
- The addition of 6 months of hormonotherapy to RT resulted in increased overall survival in men with localized but unfavorable-risk prostate cancer.

D'Amico AV (2008) Androgen suppression and radiation vs radiation alone for prostate cancer: a randomized trial. JAMA 299(3):289–295

RTOG 86-10, 2008 → 456 patients with bulky (5 × 5 cm) tumors (T2–4) with or without pelvic LN involvement. Patients received combined hormone therapy that consisted of goserelin 3.6 mg every 4 weeks and flutamide 250 mg t.i.d. for 2 months before and con- current with RT, or they received RT alone.

- Ten-year OS (43 vs. 34%) and median survival times (8.7 vs. 7.3 years) favored hormone therapy and RT, respectively ($p = 0.12$).
- There was a statistically significant improvement in 10-year disease-specific mortality (23 vs. 36%; $p = 0.01$), distant metastasis (35 vs. 47%; $p = 0.006$), DFS (11 vs. 3%; $p < 0.0001$), and biochemical failure (65 vs. 80%; $p < 0.0001$) with the addition of hormone therapy.
- No differences were observed in the risk of fatal cardiac events.
- The addition of 4 months of ADT to EBRT appears to have a dramatic impact on clinically meaningful end points in men with locally advanced disease, with no statistically significant impact on the risk of fatal cardiac events.

Roach M et al. (2008) Short-term neoadjuvant androgen deprivation therapy and external- beam radiotherapy for locally advanced prostate cancer: long-term results of RTOG 8610 J Clin Oncol 26(4):585–591

RTOG 92-02, 2008 → 1554 patients with T2c–T4 prostate cancer with no extra pelvic LN involvement and PSA less than 150 ng/mL. All patients received 4 months of goserelin and flutamide before and during RT. They were randomized to no further ADT (short-term ADT [STAD] + RT) or 24 months of goserelin (long-term ADT [LTAD] + RT). RT was 45 Gy to the pelvic nodes and 65–70 Gy to the prostate.

- At 10 years, the LTAD + RT group showed significant improvement over the STAD + RT group for all end points except overall survival: disease-free survival (13.2 vs. 22.5%; $p < 0.0001$), disease-specific survival (83.9 vs. 88.7%; $p = 0.0042$), local progression (22.2 vs. 12.3%; $p < 0.0001$), distant metastasis (22.8 vs. 14.8%; $p < 0.0001$), biochemical failure (68.1 vs. 51.9%; $p £ 0.0001$), and overall survival (51.6 vs. 53.9%, $p = 0.36$).

- A difference in overall survival was observed in patients with a Gleason score of 8–10 (31.9 vs. 45.1%; $p = 0.0061$).
- LTAD as delivered in this study for the treatment of locally advanced prostate cancer was superior to STAD for all end points except survival.
- The survival advantage for LTAD + RT in the treatment of locally advanced tumors with a Gleason score of 8–10 suggests that this should be the standard of treatment for these high-risk patients.

Horwitz EM et al. (2008) Ten-year follow-up of radiation therapy oncology group protocol 92-02: a phase III trial of the duration of elective androgen deprivation in locally advanced prostate cancer. J Clin Oncol 26(15):2497–2504

8.1.6.6 Stereotactic Body Radiation Therapy (Ultrahypofractionation)

Stereotactic body radiation therapy (SBRT) is an appropriate alternative to conventional fractionation RT for carefully selected patients with low- or intermediate-risk prostate cancer who do not need nodal irradiation. In HYPO-RT-PC trial, 1200 men with intermediate-risk or high-risk prostate cancer were randomly assigned to conventional fractionation external beam RT (78 Gy in daily 2 Gy fractions) or to SBRT (42.7 Gy in seven sessions of 6.1 Gy each, administered over 2.5 weeks). At a median follow-up of 5 years, the proportion of patients who were free of biochemical or clinical failure was similar (84% in both groups at 5 years). Patient reported early side effects were more pronounced with hypofractionation, and physician scored grade ≥ 2 urinary toxicity was higher in this group at the end of RT (28% vs 23%). No significant increases in late, grade ≥ 2 urinary or bowel toxicities were found, with small increase in urinary toxicity at 1 year with hypofractionation (6% vs 2%). (Widmark A, et al. Ultra-hypofractionated versus conventionally fractionated radiotherapy for prostate cancer: 5-year outcomes of the HYPO-RT-PC randomised, non-inferiority, phase 3 trial. Lancet. 2019;394(10196):385–395.)

PACE-B trial is a phase 3 non-inferiority trial that evaluates 874 men with low- or intermediate-risk prostate cancer. Patients were randomly assigned to fractionated IMRT or SBRT. In contrast to the HYPO-RT-PC trial, SBRT was not associated with greater acute genitourinary or gastrointestinal toxicity. (Brand DH, et al.; PACE Trial Investigators. Intensity-modulated fractionated radiotherapy versus stereotactic body radiotherapy for prostate cancer (PACE-B): acute toxicity findings from an international, randomised, open-label, phase 3, non-inferiority trial. Lancet Oncol. 2019;20(11):1531–1543.)

Pooled analysis from a multi-institutional consortium of prospective phase II trials evaluated the role of SBRT in 1100 patients. With a median follow-up of 36 months, the five-year actuarial BRFS rates for the low-, intermediate-, and high-risk groups were 95%, 84%, and 81%, respectively. (King CR, et al. Stereotactic body radiotherapy for localized prostate cancer: pooled analysis from a multi-institutional consortium of prospective phase II trials. Radiother Oncol. 2013;109(2):217–21.)

A systematic review of SBRT for localized prostate cancer included 38 trials and 6116 patients. At a median follow-up of 39 months, it was shown that increasing dose of SBRT was associated with improved biochemical control ($p = 0.018$) but worse late grade ≥ 3 GU toxicity ($p = 0.014$). (Jackson WC, et al. Stereotactic Body Radiation Therapy for Localized Prostate Cancer: A Systematic Review and Meta-Analysis of Over 6000 Patients Treated On Prospective Studies. Int J Radiat Oncol Biol Phys. 2019;104(4):778–789.)

In conclusion, SBRT seems to be a safe and effective treatment alternative to conventional fractionation RT for carefully selected patients with prostate cancer.

8.1.6.7 Prostatectomy ± RT

EORTC 22911, 2005 (1992–2001) → EORTC trial 22911 demonstrated that immediate postoperative irradiation significantly improved biochemical failure-free survival (BPFS) compared to wait-and-see (W and S) until relapse in patients with pT2–3 tumors and pathological risk factors (cancer extending beyond the capsule and/or involvement of seminal vesicles or positive surgical margins) after RP.

- Postoperative irradiation improved biochemical PFS in all patient groups.

Bolla M et al. (2005) Postoperative radiotherapy after radical prostatectomy: a randomised controlled trial (EORTC Trial 22911). Lancet 366(9485):572–578

Collette L et al. (2005) Patients at high risk of progression after radical prostatectomy: do they all benefit from immediate post-operative irradiation? (EORTC Trial 22911). Eur J Cancer 41(17):2662–2672

RTOG 90-19/SWOG-8794 Intergroup → 425 patients with stage pT3 N0 M0 prostate cancer. Men were randomly assigned to receive 60–64 Gy of external beam radiotherapy delivered to the prostatic fossa or observation. Median follow-up was 10.6 years.

- There were no significant between-group differences for overall survival.
- PSA relapse (median PSA relapse-free survival, 10.3 years for RT vs. 3.1 years for observation; $p < 0.001$) and disease recurrence (median recurrence-free survival, 13.8 years for RT vs. 9.9 years for observation; $p = 0.001$) were both significantly reduced with radiotherapy.
- Adverse effects were more common with radiotherapy vs. observation (23.8 vs. 11.9%), including rectal complications (3.3 vs. 0%), urethral strictures (17.8 vs. 9.5%), and total urinary incontinence (6.5 vs. 2.8%).

Thompson IM Jr et al. (2006) Adjuvant radiotherapy for pathologically advanced prostate cancer: a randomized clinical trial. JAMA 296(19):2329–2335

Swanson GP (2007) Predominant treatment failure in postprostatectomy patients is local: analysis of patterns of treatment failure in SWOG 8794. J Clin Oncol 25(16): 2225–2229

8.1.6.8 Adjuvant Versus Early Salvage Radiotherapy

Three randomized trials compared both approaches and found that treatment outcomes are similar with salvage as compared with adjuvant RT, but the risk of overtreatment and treatment related adverse effects are less with salvage therapy.

RADICALS-RT is a randomized, controlled phase 3 trial that compares adjuvant RT to salvage RT in 1396 patients with at least one risk factor (pathological T-stage 3 or 4, Gleason score of 7–10, positive margins, or preoperative PSA $\geq$10 ng/mL) for biochemical progression after RP. At a median follow-up of 4.9 years, five-year BRFS (PSA $\geq$0.1 ng/mL or three consecutive rises) was not significantly better with adjuvant RT (85% vs. 88%) and the number of men who were free from nonprotocol ADT was similar (93% vs 92%). Self-reported urinary incontinence was worse at 1 year for those in the adjuvant RT group (mean score 4.8 vs 4.0; $p = 0.0023$). Grade 3–4 urethral stricture within 2 years was reported in 6% of individuals in the adjuvant RT group versus 4% in the salvage RT group ($p = 0.020$). (Parker CC, et al. Timing of radiotherapy after radical prostatectomy (RADICALS-RT): a randomised, controlled phase 3 trial. Lancet. 2020;396(10260):1413–1421.)

Radiotherapy—Adjuvant versus Early Salvage (RAVES) compared adjuvant with early salvage RT in men with positive margins and/or pT3 disease in 333 patients. ADT was not allowed. With a median follow-up of 6.1 years, 50% patients in the salvage RT group had radiotherapy triggered by a PSA of $\geq$0.20 ng/mL. 5-year BRFS was 86% in the adjuvant RT group versus 87% in the salvage RT group ($p = 0.15$). The grade 2 or worse genitourinary toxicity rate was lower in the salvage RT group than in the adjuvant RT group. The grade 2 or worse gastrointestinal toxicity rate was similar between the groups. The study support the use of salvage RT as it

results in similar biochemical control to adjuvant RT, spares around half of men from pelvic radiation, and is associated with significantly lower genitourinary toxicity. (Kneebone A, et al. Adjuvant radiotherapy versus early salvage radiotherapy following radical prostatectomy (TROG 08.03/ANZUP RAVES): a randomised, controlled, phase 3, non-inferiority trial. Lancet Oncol. 2020;21(10):1331–1340.)

GETUG-AFU17 trial randomly assigned men 424 with pT3-T4 prostate cancer with positive surgical margins to immediate versus delayed RT, and all men received six months of triptorelin. At a median follow-up of 75 months, 54% of the men undergoing salvage RT had initiated RT after biochemical relapse. 5-year event-free survival was 92% in the adjuvant RT group and 90% in the salvage RT group (p = 0.42). Late genitourinary adverse events of grade 2 or worse were reported in 27 in the adjuvant RT group versus 7% in the salvage RT group (p < 0.0001). Late erectile dysfunction was grade 2 or worse in 28% in the adjuvant RT group and 8% in the salvage RT group (p < 0.0001). (Sargos P, et al. Adjuvant radiotherapy versus early salvage radiotherapy plus short-term androgen deprivation therapy in men with localised prostate cancer after radical prostatectomy (GETUG-AFU 17): a randomised, phase 3 trial. Lancet Oncol. 2020;21(10):1341–1352.)

Preplanned meta-analysis of all three trials (2153 patients), with a median follow-up ranging from 60 to 78 months, showed no evidence that event-free survival was improved with adjuvant RT compared with early salvage RT (p = 0.70). Until data on long-term outcomes are available, early salvage treatment would seem the preferable treatment policy as it offers the opportunity to spare many men radiotherapy and its associated side effects. (Vale CL, et al; ARTISTIC Meta-analysis Group. Adjuvant or early salvage radiotherapy for the treatment of localised and locally advanced prostate cancer: a prospectively planned systematic review and meta-analysis of aggregate data. Lancet. 2020;396(10260):1422–1431.)

8.1.6.9 Oligometastatic Disease

Radiotherapy to Primary
The phase III HORRAD trial randomly assigned 432 men with primary metastatic prostate cancer with bone metastases and a PSA >20 ng/mL to ADT with or without external beam RT (70 Gy in 35 fractions). Median follow-up was 47 months, and there was no difference in OS between patients receiving RT or not (45 vs. 43 months, p = 0.4). However, median time to PSA progression in the RT group was 15 months compared with 12 months in the control group (p = 0.02). An unplanned subgroup analysis suggested that OS might be favorably impacted in men with fewer than five metastases (HR 0.68, 95% CI 0.42–1.10). (Boevé LMS, et al. Effect on Survival of Androgen Deprivation Therapy Alone Compared to Androgen Deprivation Therapy Combined with Concurrent Radiation Therapy to the Prostate in Patients with Primary Bone Metastatic Prostate Cancer in a Prospective Randomised Clinical Trial: Data from the HORRAD Trial. Eur Urol. 2019;75(3):410–418.)

STOPCAP Systematic Review and Meta-analysis evaluated the pooled data of these two trials. There was an overall improvement in biochemical progression (p = 0.94 × 10^{-8}) and FFS (p = 0.64 × 10^{-7}). The effect of prostate RT varied by metastatic burden. There was 7% improvement in 3-year survival in men with fewer than five bone metastases. This analyses revealed that prostate RT should be considered for men with hormone sensitive metastatic prostate cancer with low metastatic burden. (Burdett S, et al; STOPCAP M1 Radiotherapy Collaborators. Prostate Radiotherapy for Metastatic Hormone-sensitive Prostate Cancer: A STOPCAP Systematic Review and Meta-analysis. Eur Urol. 2019;76(1):115–124.)

Radiotherapy to Metastases
After prior definitive therapy, patients may present with metachronous oligometastatic disease, which most of the time, is diagnosed using positron emission tomography (PET)/

computed tomography (CT). There are studies evaluating the role of metastases directed treatment (MDT) in this group of patients:

In a phase II trial (STOMP Trial), 62 asymptomatic patients with a biochemical recurrence after primary definitive treatment, one to three metastases on imaging, and a serum testosterone >50 ng/mL were randomly assigned to observation alone or to MDT. At a median follow-up for survival of 5.3 years (IQR 4.3–6.3), the 5-year overall survival was 85%, with 6 out of 14 deaths attributed to prostate cancer. The 5-year ADT-free survival was 8% for the surveillance group and 34% for the MDT group ($p = 0.06$). (Ost P. et al. Surveillance or metastasis-directed therapy for oligometastatic prostate cancer recurrence (STOMP): Five-year results of a randomized phase II trial. Journal of Clinical Oncology 2020 38:6_suppl, 10–10).

The phase II Observation versus stereotactic ablative RadiatIon for OLigometastatic prostate CancEr (ORIOLE) trial randomly assigned 54 men with recurrent, hormone-sensitive, oligometastatic prostate cancer (three or fewer lesions) to observation and no further treatment for six months or to SBRT to the metastatic sites outside of the prostate that were detected by conventional imaging. Progression at 6 months occurred in 19% receiving SBRT and 61% undergoing observation ($p = 0.005$). SBRT improved median PFS ($p = 0.002$). Total consolidation of PSMA avid disease decreased the risk of new lesions at 6 months (16% vs 63%; $p = 0.006$). No grade 3 or greater toxic events were observed. T cell receptor sequencing identified significant increased clonotypic expansion following SBRT and correlation between baseline clonality and progression with SBRT only (0.082085 vs 0.026051; $p = 0.03$). (Phillips R, et al. Outcomes of Observation vs Stereotactic Ablative Radiation for Oligometastatic Prostate Cancer: The ORIOLE Phase 2 Randomized Clinical Trial. JAMA Oncol. 2020;6(5):650–659.)

Treatment outcomes of MDT using 68 Ga-PSMA-PET/CT for oligometastatic or oligorecurrent prostate cancer was evaluated by Turkish Society for Radiation Oncology group study. In this retrospective study, 176 prostate cancer patients with 353 lesions received MDT. With a median follow-up of 22.9 months, 2-year local control rate at the treated oligometastatic site per patient was 93.2%. In multivariate analysis, an increased number of oligometastases and untreated primary prostate cancer were negative predictors for OS; advanced clinical tumor stage, untreated primary prostate cancer, BED3 value of ≤108 Gy, and MDT with conventional fractionation were negative predictors for PFS. (Hurmuz P, et al. Treatment outcomes of metastasis-directed treatment using 68Ga-PSMA-PET/CT for oligometastatic or oligorecurrent prostate cancer: Turkish Society for Radiation Oncology group study (TROD 09-002). Strahlenther Onkol. 2020;196(11):1034–1043.)

8.1.6.10 IMRT

MSKCC, 2008 → 478 patients were treated with 86.4 Gy using a five- to seven-field IMRT technique. The mean D95 and V100 for the planning target volume were 83 Gy and 87%, respectively. The median follow-up was 53 months. There was no acute grade 3 or 4 GI toxicity. Three patients (0.6%) had grade 3 GU toxicity. There was no acute grade 4 GU toxicity. Sixteen patients (3%) developed late grade 2 GI toxicity, and two patients (<1%) developed late grade 3 GI toxicity. Sixty patients (13%) had late grade 2 GU toxicity and 12 (<3%) experienced late grade 3 GU toxicity. The 5-year actuarial PSA relapse-free survivals according to the nadir plus 2 ng/mL definition were 98, 85, and 70% for the low-, intermediate-, and high-risk NCCN prognostic groups.

Cahlon O et al. (2008) Ultra-high dose (86.4 Gy) IMRT for localized prostate cancer: toxicity and biochemical outcomes. Int J Radiat Oncol Biol Phys 71(2):330–337

8.1.6.11 Surveillance vs. Prostatectomy

Scandinavian Prostate Cancer Group Study-4, 2005 (1989–1999) → 695 cases of early- stage T1–T2 prostate cancer. Surveillance vs. prostatectomy. Follow-up with PSA and DRE. Median follow-up: 8.2 years.

Mortality due to prostate cancer:
14.4% in surveillance vs. 8.6% in prostatectomy

After 10 years: 5% in survival difference (relative risk = 0.74)

Conclusion: prostate cancer mortality was less in treated patients.

Bill-Axelson A et al. (2005) Radical prostatectomy versus watchful waiting in early prostate cancer. N Engl J Med 352(19):1977–1984

Photon Therapy Versus Proton Therapy

A retrospective analysis of 421 men treated for prostate cancer with proton beam therapy with 842 matched controls treated with IMRT in the Medicare There was a statistically significant decrease in genitourinary toxicity at 6 months, but this difference had disappeared by 12 months. There were no other significant differences in toxicity between the two techniques at either 6 months or 1 year. The costs associated with proton beam therapy were approximately 75% higher compared with those associated with IMRT. (Yu JB, et al. Proton versus intensity-modulated radiotherapy for prostate cancer: patterns of care and early toxicity. J Natl Cancer Inst. 2013;105(1):25–32.)

A multi-institutional observational study (J-CROS1501PR) has been carried out to analyze outcomes of carbon-ion radiotherapy in prostate cancer. The five-year bRFS in low-risk, intermediate-risk, and high-risk patients was 92%, 89%, and 92%, respectively. The five-year CSS in low-risk, intermediate-risk, and high-risk patients was 100%, 100%, and 99%, respectively. The incidence of grade 2 late GU/GI toxicities was 4.6% and 0.4%, respectively, and the incidence of ≥G3 toxicities were 0%. (Nomiya T, et al. A multi-institutional analysis of prospective studies of carbon ion radiotherapy for prostate cancer: A report from the Japan Carbon ion Radiation Oncology Study Group (J-CROS). Radiother Oncol. 2016;121(2):288–293.)

8.2 Testicular Cancer

Testicular cancers constitute 1% of all cancers. A previous history of testicular cancer, cryptorchidism, Klinefelter's syndrome (47XXY), and other testicular feminizing syndromes are possible etiological factors for the development of these cancers [22].

Testis Anatomy (Fig. 8.8)

Scrotum. This includes the testicles, epididymis, and part of the spermatic cord. The tunica dartos layer sticks to the scrotal skin and consists of superficial and deep layers of superficial fascia. It contains smooth muscle fibers called the dartos muscle. This muscle is an important factor in the temperature regulation of the testes. This layer contains little fat tissue but is rich in elastic leaves.

Layers of the scrotum. These layers are located under the skin and tunica dartos:

1. *External spermatic fascia.* This is continuous with the fascia of the external abdominal oblique muscle.
2. *Cremasteric fascia and cremaster muscle.* This is formed by leaves of internal abdominal oblique muscle and (partially) transverse abdominal muscle. The innervation of this layer is provided by the right femoral branch of the genitofemoral nerve, and a clinically important reflex pathway is formed called the cremaster reflex (the excitation of medial part of the femoral region causes the cremaster muscle to contract and the testicles to be pulled up to the lower part of the abdomen).
3. *Internal spermatic fascia.* This is formed by the continuation of the transversalis fascia. This fascia is loosely attached to the parietal layer of the tunica vaginalis. It also surrounds the spermatic cord, testis, and epididymis.

8.2 Testicular Cancer

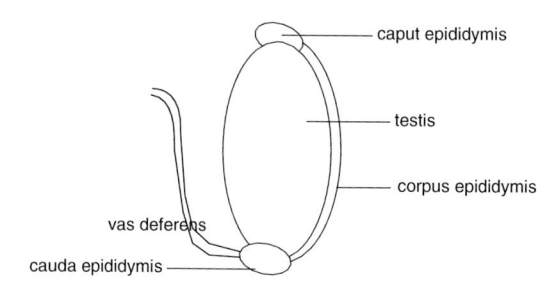

Fig. 8.8 Testis anatomy [38, p 42, Fig. 2]

Testis. These are two ovoid glands that produce sperm and testosterone and are located within the scrotum. They are 4–5 cm in height, 2.5–3 cm in width, 2–3 cm thick, and 10–14 g in weight.

- A strong membrane called the tunica albuginea, located in the inner visceral layer of the tunica vaginalis, surrounds the testicles.

8.2.1 Pathology

Malignant testicular germ cell cancers consist of a single cell line in 50% of cases, and 50% of those are seminomas [23]. Other malignant testicular germ cell cancers are formed by the combination of more than one clonogenic cell type. This histological structure is important for metastases and responds to chemotherapy. Polyembryoma is a rare histological pattern and can be classified as a single histological type, although it is also considered a mixed tumor.

1. Intratubular germ cell neoplasms, unclassified (old term: carcinoma in situ or CIS)
2. Malignant pure germ cell tumor (includes the single-cell type):
   Seminoma (34–55%)
   Embryonal carcinoma (23–4%)
   Teratoma (1–9%; locally invasive)
   Choriocarcinoma (1–4%; hematogenous spread)
   Yolk sac tumor (most frequent testicular tumor in childhood)
3. Malignant mixed germ cell tumor (includes more than one cell type): Embryonal carcinoma and teratoma (may additionally include seminoma or not)
   Embryonal carcinoma and yolk sac tumor (may additionally include seminoma or not) Embryonal carcinoma and seminoma
   Yolk sac tumor and teratoma (may additionally include seminoma or not) Choriocarcinoma and any of the other types
4. Polyembryoma

Testicular cancers are rare tumors. However, they are the most common cancers in males aged between 20 and 35 years [24].

8.2.1.1 General Presentation

The most common symptom is painless scrotal mass. There is hydrocele in 20% of cases. Bilateral gynecomastia, back pain due to retroperitoneal adenopathies, abdominal pain, nausea, vomiting, and constipation may also be seen. Signs and symptoms change according to the area of involvement in patients with extensive metastases.

Physical examination, ultrasonography, serum levels of beta human chorionic gonadotropin (b-hCG), a-fetoprotein (AFP), and lactate dehydrogenase (LDH) are used for the diagnosis of testicular cancers [25].

- AFP and b-hCG increase in 80–85% of extensive germ cell tumors.
- Pure seminomas have increased b-hCG but no increase in AFP in 15% of cases.
- b-hCG may cause false positivity in hypogonadism.
- AFP may increase in liver diseases and hepatitis.
- Testicular biopsy is not recommended in cases with suspected cancer.
  - Half-life of b-hCG is 24 h; that of AFP is 5 days.

8.2.1.2 Staging

Primary Tumor (T) (Fig. 8.9) [26]
Pathological T (pT)
- The extent of the primary tumor is classified after radical orchiectomy, so a pathologic stage is assigned.
- pTX: Primary tumor cannot be assessed
- pT0: No evidence of primary tumor
- pTis: Germ cell neoplasia in situ
- pT1: Tumor limited to testis (including rete testis invasion) without lymphovascular invasion
  - pT1a*: Tumor smaller than 3 cm in size
  - pT1b: Tumor 3 cm or larger in size
- pT2: Tumor limited to testis (including rete testis invasion) with lymphovascular invasion OR Tumor invading hilar soft tissue or epididymis or penetrating visceral mesothelial layer covering the external surface of tunica albuginea with or without lymphovascular invasion
- pT3: Tumor directly invades spermatic cord soft tissue with or without lymphovascular invasion
- pT4: Tumor invades scrotum with or without lymphovascular invasion
- *Subclassification of pT1 applies only to pure seminoma.

Regional Lymph Nodes (N) (Fig. 8.10)
- cNX: Regional lymph nodes cannot be assessed
- cN0: No regional lymph node metastasis
- cN1: Metastasis with a lymph node mass 2 cm or smaller in greatest dimension or multiple lymph nodes, none larger than 2 cm in greatest dimension
- cN2: Metastasis with a lymph node mass larger than 2 cm but not larger than 5 cm in greatest dimension or multiple lymph nodes, any one mass larger than 2 cm but not larger than 5 cm in greatest dimension
- cN3: Metastasis with a lymph node mass 5 cm or more in greatest dimension

- **Testicular lymphatics** (Fig. 8.11) [26]: interaortocaval, para-aortic (periaortic), paracaval, preaortic, precaval, retroaortic, retrocaval.
- Right testicular lymphatics generally drain into bilateral para-aortic lymph nodes, whereas left testicular lymphatics first drain into left para-aortic lymph nodes.

Distant Metastasis (M)
- M0: No distant metastasis
- M1: Distant metastasis
- M1a: Non-retroperitoneal nodal or pulmonary metastases
- M1b: Non-pulmonary visceral metastases

Serum Tumor Markers (S)
- SX: Marker studies not available or not performed
- S0: Marker study levels within normal limits
- S1: Lactate dehydrogenase (LDH) less than $1.5 \times N$*, and human chorionic gonadotropin (hCG) less than 5000 (mIU/mL), and Alpha-fetoprotein (AFP) less than 1000 (ng/mL)
- S2: LDH $1.5–10 \times N$* or hCG 5000–50,000 (mIU/mL), or AFP 1000–10,000 (ng/mL)
- S3: LDH more than $10 \times N$*, or hCG more than 50,000 (mIU/mL), or AFP more than 10,000 (ng/mL)
- *[Note: N indicates the upper limit of normal for the LDH assay.]

8.2 Testicular Cancer

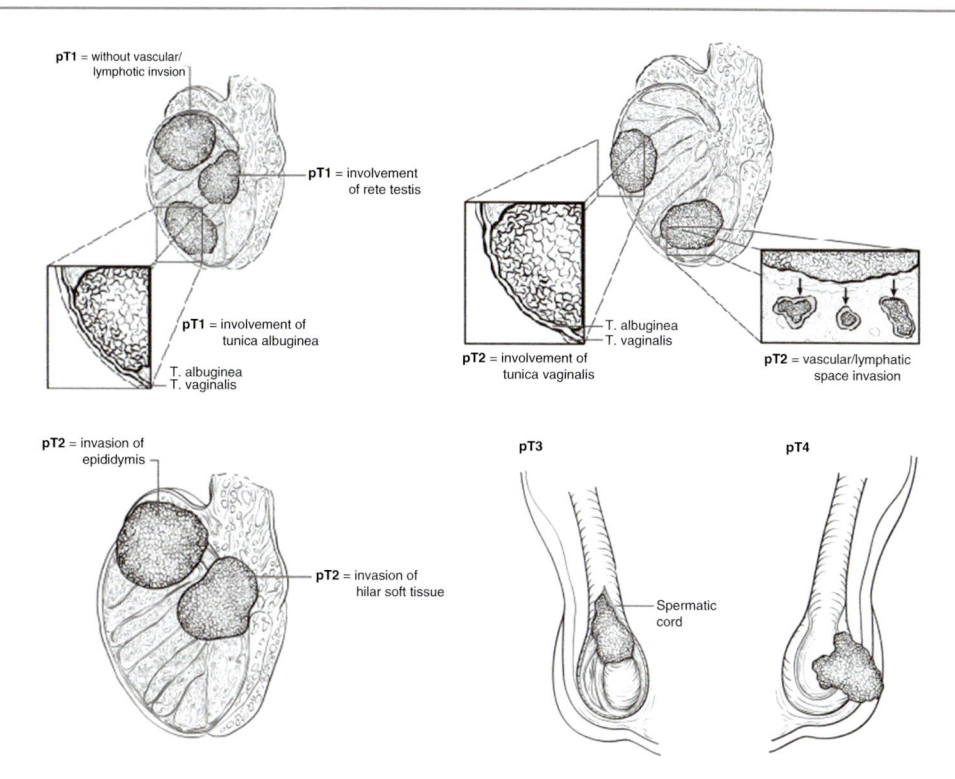

Fig. 8.9 T staging in testicular cancer. (from [26], Compton C.C., Byrd D.R., Garcia-Aguilar J., Kurtzman S.H., Olawaiye A., Washington M.K. (2012) Testis. In: Compton C., Byrd D., Garcia-Aguilar J., Kurtzman S., Olawaiye A., Washington M. (eds) AJCC Cancer Staging Atlas. Springer, New York, NY. https://doi.org/10.1007/978-1-4614-2080-4_42)

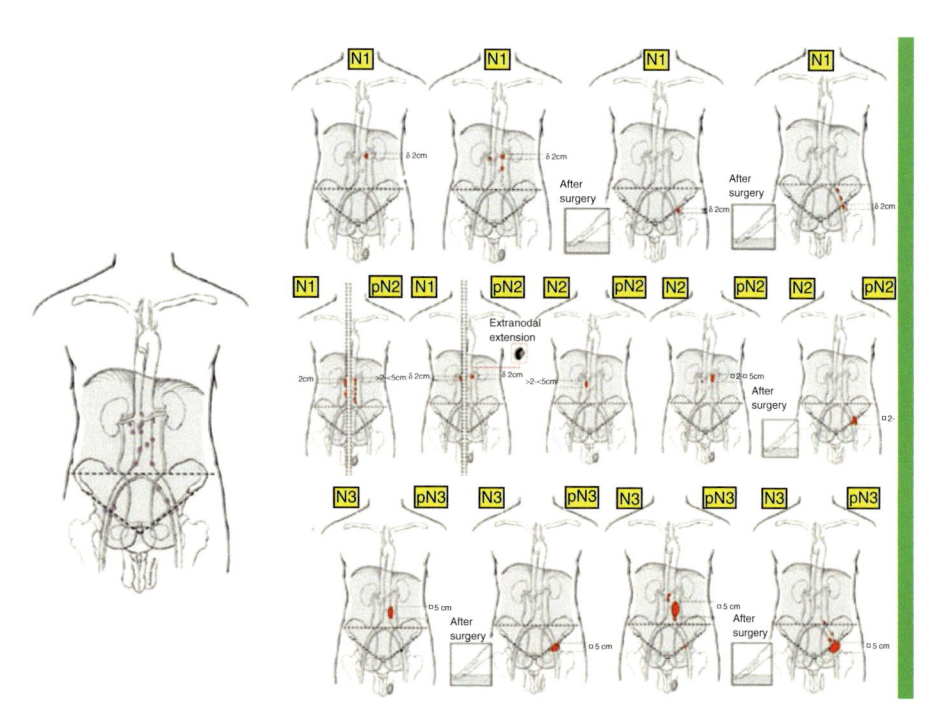

Fig. 8.10 N staging in testicular cancer. (from [26], Compton C.C., Byrd D.R., Garcia-Aguilar J., Kurtzman S.H., Olawaiye A., Washington M.K. (2012) Testis. In: Compton C., Byrd D., Garcia-Aguilar J., Kurtzman S., Olawaiye A., Washington M. (eds) AJCC Cancer Staging Atlas. Springer, New York, NY. https://doi.org/10.1007/978-1-4614-2080-4_42)

T	N	M	S	Stage group
Any pT/TX	Any N	M1a	S2	IIIB
Any pT/TX	N1-3	M0	S3	IIIC
Any pT/TX	Any N	M1a	S3	IIIC
Any pT/TX	Any N	M1b	Any S	IIIC

8.2.2 Treatment Algorithm

The treatment algorithm changes according to the histology of the testicular cancer (seminoma or nonseminoma) [26]. Seminoma is particularly sensitive to radiation. An international germ cell prognostic classification has been derived based on retrospective analyses. The most important property that all of these patients have in common is that all receive either a carboplatin- or a cisplatin-based chemotherapy regimen as the first line treatment.

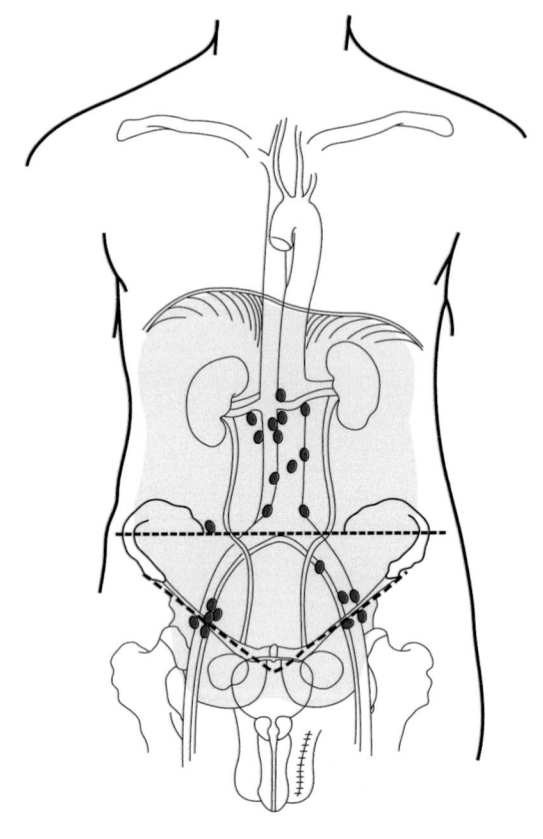

Fig. 8.11 Testicular lymphatics. (from [13], p 304, Fig. 35.2, reproduced with the permission of the American Joint Committee on Cancer)

AJCC Stage Groups

T	N	M	S	Stage group
pTis	N0	M0	S0	0
pT1-T4	N0	M0	SX	I
pT1	N0	M0	S0	IA
pT2	N0	M0	S0	IB
pT3	N0	M0	S0	IB
pT4	N0	M0	S0	IB
Any pT/TX	N0	M0	S1-3	IS
Any pT/TX	N1-3	M0	SX	II
Any pT/TX	N1	M0	S0	IIA
Any pT/TX	N1	M0	S1	IIA
Any pT/TX	N2	M0	S0	IIB
Any pT/TX	N2	M0	S1	IIB
Any pT/TX	N3	M0	S0	IIC
Any pT/TX	N3	M0	S1	IIC
Any pT/TX	Any N	M1	SX	III
Any pT/TX	Any N	M1a	S0	IIIA
Any pT/TX	Any N	M1a	S1	IIIA
Any pT/TX	N1-3	M0	S2	IIIB

Testis Cancer with Good Prognosis [27]
Nonseminoma
- Testicular or retroperitoneal origin Lung metastasis (−)
- Marker levels normal
  – LDH <1.5 times of upper limit
  – b-HCG (mIU/mL) <5000
  – AFP (ng/mL) <1000
- Fifty-six percent of nonseminomas Five-year PFS 89%, 5-year survival 92%

Seminoma
- Any origin
- Lung metastasis (−) Marker levels normal Ninety percent of seminomas
- Five-year PFS 82%, 5-year survival 86% *Testis cancer with moderate prognosis* [27]

Nonseminoma
- Testicular or retroperitoneal origin Lung metastasis (−)
- Marker levels moderately increased
  – LDH 1.5–10 times
  – b-HCG (mIU/mL): 5000–50,000
  – AFP (ng/mL): 1000–10,000

- Twenty-eight percent of nonseminomas
  Five-year PFS 75%, 5-year survival 80%

Seminoma
- Any origin
- Lung metastasis (+) Marker levels are normal Ten percent of seminomas
- Five-year PFS 67%, 5-year survival 72%

Testis cancers with poor prognosis [27]
Nonseminoma
- Mediastinal origin Lung metastasis (+)
- Marker levels are too high
  - LDH >10
  - b-HCG (mIU/mL) >50,000
  - AFP (ng/mL) >10,000
- Sixteen percent of nonseminomas
- Five-year PFS 41%, 5-year survival 48%

There is no poor prognostic group in seminoma.

The first step in the treatment of testicular tumors is surgery: orchiectomy (which should be performed by inguinal approach, not scrotal).

Treatment of Seminoma
Stage I
After orchiectomy
Surveillance (recurrence rate ~16%)
RT (pararaortic LN ± pelvic LN) (total 20 Gy) (fraction dose 1.8–2 Gy)
Chemotherapy (1–2 cycles of carboplatin)
Stage IIA–IIB
After orchiectomy
RT (pararaortic LN + pelvic LN)

- 20 Gy and boost to gross disease
- Stage IIA boost 30 Gy
- Stage IIB boost 36 Gy
- Or primary chemotherapy with etoposide/cisplatin (EP)x4cycles or bleomycin/etoposide/cisplatin (BEP)x3 cycles

Stage IIC–III
After orchiectomy Chemotherapy
4EP or 3 BEP

Treatment in nonseminomatous testis cancers [26]:

- After orchiectomy: chemotherapy ± LN dissection (selected cases).

8.2.3 Radiotherapy

Para-aortic Field (Fig.8.12)
Superior: below T10

- Renal hilar LNs
- Left renal hilar lymphatics should be included, particularly in left-sided testis cancers (left spermatic vein drains directly into left renal vein)
- Both kidneys should be spared, and maximum care should be taken to check for the presence of a horse-shoe kidney, which is a contraindication for radiotherapy

Pelvic + Iliac Field (Fig. 8.13)
Superior: lower border of para-aortic field (a gap is given) Inferior: middle obturator foramen
Medial: includes transverse processes of lumbar vertebrae Lateral: includes inguinal field

RT Field in Stage II Seminoma (Fig. 8.14)
- Paraaortic field + pelvic + inguinal field + contralateral inguinal field at the involved ipsilateral testicular site

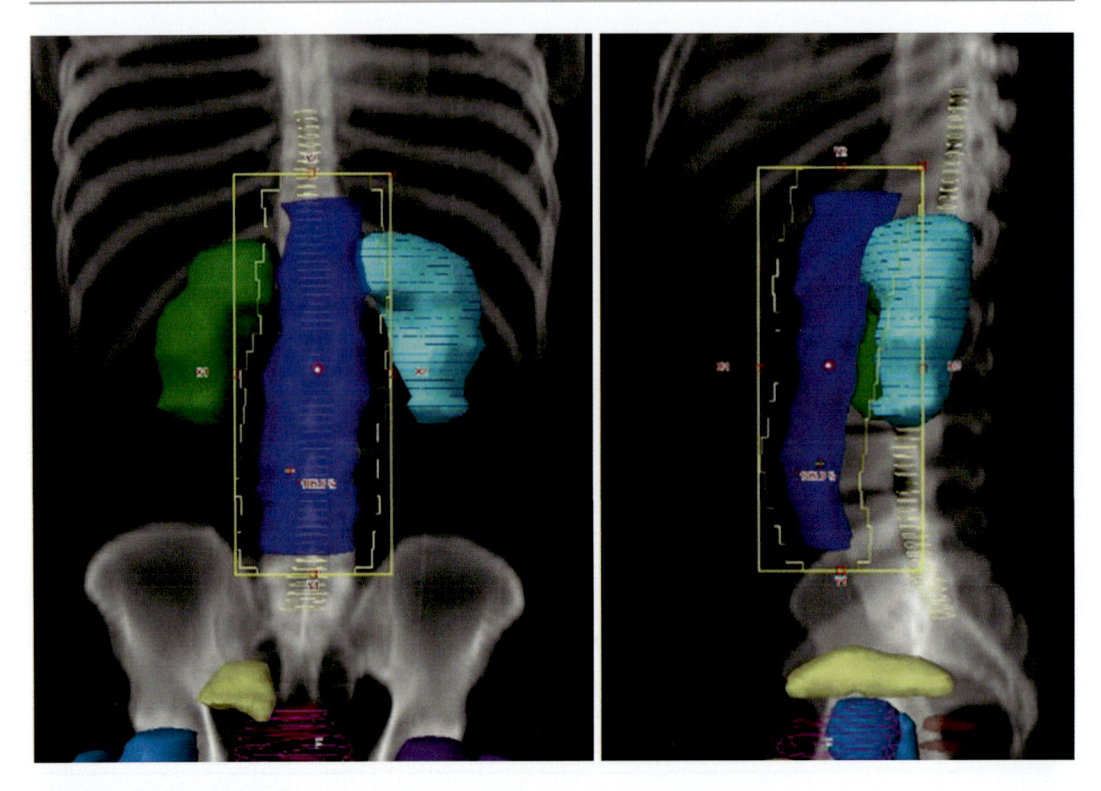

Fig. 8.12 Para-aortic region

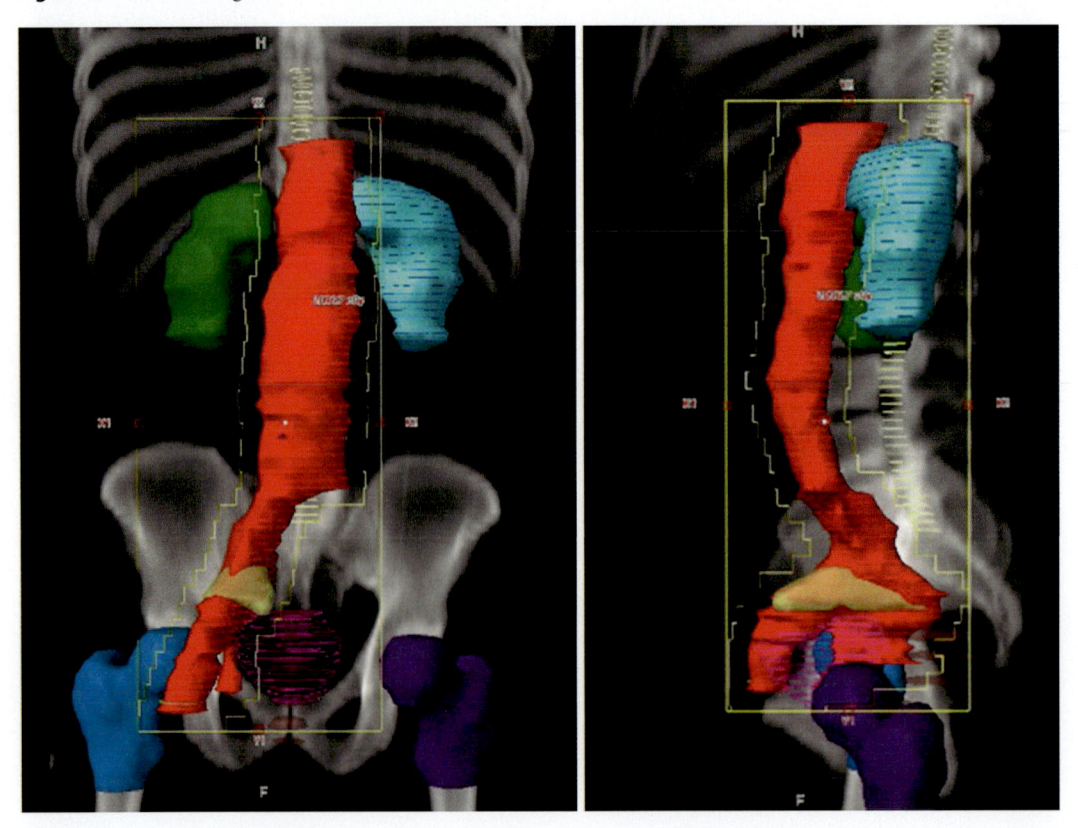

Fig. 8.13 Pelvic and iliac RT field

8.2 Testicular Cancer

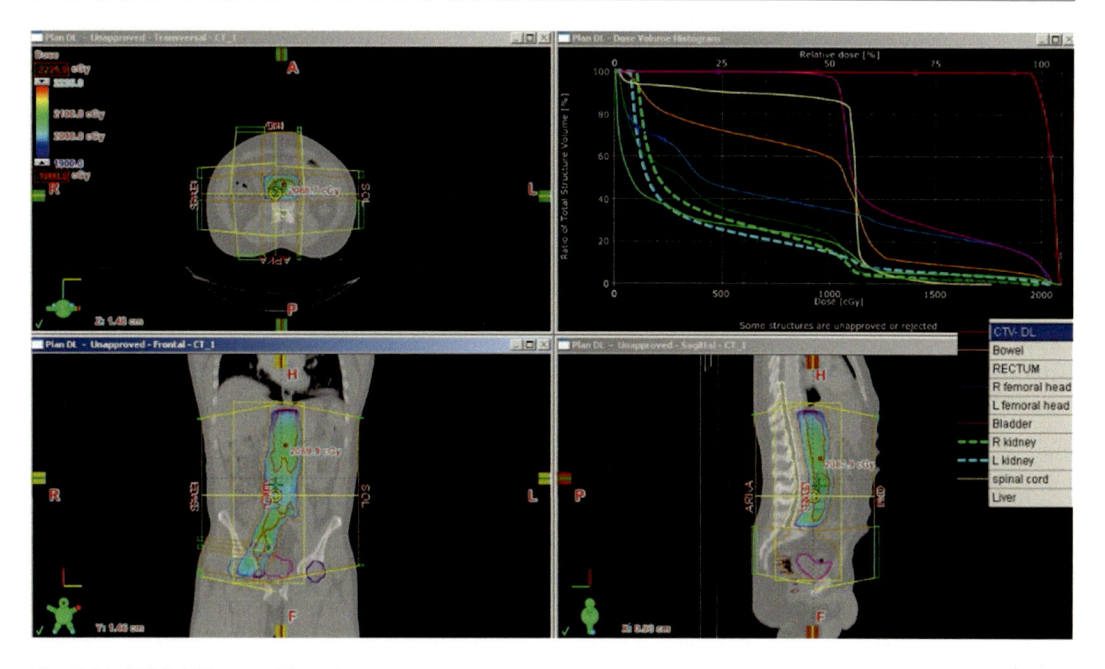

Fig. 8.14 RT field in stage II seminoma

- Stage I seminoma: only para-aortic RT can be used
- RT to ipsilateral iliac lymphatics is given in patients with a history of pelvic surgery
- Inguinal lymphatics and scrotum are irradiated in patients with scrotal orchiectomy

> The testis is a very radiosensitive organ [28]
> - 1 Gy oligospermia, 6 Gy sterility
> - Sperm count may decrease due to scattered radiation
> - Testicular shielding decreases these effects
> - Spermatids and spermatozoa are more radioresistant than spermatogonia

8.2.3.1 Selected Publications

Surveillance, 2000 → 303 men with stage I testicular germ cell tumors. Median follow-up was 5.1 years.

- Only three deaths (one from disease, one from neutropenic sepsis, and one from secondary leukemia).
- 52/183 (28%) patients with NSGCT and 18/120 (15%) patients with seminoma relapsed.
- The relapse-free survivals at 5 years were 82% for seminoma and 69% for NSGCT.
- Surveillance for stage I testicular germ cell tumors (both NSGCT and seminoma) was
- associated with a low mortality rate (3/303, 1%).

Francis R, Bower M, Brunström G et al. (2000) Surveillance for stage I testicular germ cell tumours: results and cost benefit analysis of management options. EurJCancer36(15):1925–1932*Surveillance, 2005* → prospective, single-arm study. 88 patients with stage I seminoma.

- The actuarial relapse-free rates were 83%, 80%, and 80% at 5, 10, and 15 years, respectively.
- Fifteen of seventeen relapses were below the diaphragm.
- Only one had a second relapse and was further salvaged by chemotherapy.
- All 17 relapsed patients remained free of recurrence after salvage treatment; none died of seminoma.

- Surveillance was found to be a safe alternative to postoperative adjuvant therapy for stage I testicular seminoma.

Choo R, Thomas G, Woo T et al. (2005) Long-term outcome of postorchiectomy surveillance for stage I testicular seminoma. Int J Radiat Oncol Biol Phys 61(3):736–740

Carboplatin vs. RT, 2005 → 1477 patients were randomly assigned to receive radiotherapy (para-aortic strip or dog-leg field) or one injection of carboplatin (dose based on the formula 7× [glomerular filtration rate + 25] mg.

- With a median follow-up of 4 years, relapse-free survival rates for radiotherapy and carboplatin were similar (96.7 vs. 97.7% at 2 years; 95.9 vs. 94.8% at 3 years, respectively).
- Patients given carboplatin were less lethargic and less likely to take time off work than those given radiotherapy.
- New, second primary testicular germ cell tumors were reported in ten patients allocated irradiation (all after para-aortic strip field) and in two allocated carboplatin (5-year event rate 1.96 vs. 0.54%, $p = 0.04$).
- One seminoma-related death occurred after radiotherapy, and none after carboplatin.
- This trial has shown the non-inferiority of carboplatin to radiotherapy in the treatment of stage I seminoma.

Oliver RT, Mason MD, Mead GM et al. (2005) Radiotherapy versus single-dose carboplatin in adjuvant treatment of stage I seminoma: a randomised trial. Lancet 366(9482):293–300

EORTC 30942, 2005 → 625 patients were randomly assigned 20 Gy/10 fractions over 2 weeks or 30 Gy/15 fractions during 3 weeks after orchiectomy.

- Four weeks after starting radiotherapy, significantly more patients receiving 30 Gy reported moderate or severe lethargy (20 vs. 5%) and an inability to carry out their normal work (46 vs. 28%). However, by 12 weeks, the levels in both groups were similar.
- With a median follow-up of 61 months, 10 and 11 relapses, respectively, were reported in the 30 and 20 Gy groups.
- Only one patient had died from seminoma (allocated to the 20 Gy treatment group).
- Treatment with 20 Gy in ten fractions was unlikely to produce relapse rates that were more than 3% higher than that for standard 30 Gy radiation therapy.

Jones WG, Fossa SD, Mead GM et al. (2005) Randomized trial of 30 versus 20 Gy in the adjuvant treatment of stage I testicular seminoma: a report on Medical Research Council Trial TE18, European Organisation for the Research and Treatment of Cancer Trial 30942. J Clin Oncol 23(6):1200–1208

MRC, PA vs. dog-leg field, 1999 → 478 men with testicular seminoma stage I (T1–T3; no ipsilateral inguinoscrotal operation before orchiectomy) were randomized to para-aortic (PA) vs. dog-leg RT fields. Median follow-up time was 4.5 years.

- Eighteen relapses, nine in each treatment group, had occurred 4–35 months after radiotherapy; among these, four were pelvic relapses, all occurring after PA radiotherapy.
- The 3-year relapse-free survivals were 96% after PA radiotherapy and 96.6% after DL.
- One patient (PA field) had died from seminoma.
- Survivals at 3 years were 99.3% for PA and 100% for DL radiotherapy.
- Acute toxicity (nausea, vomiting, leukopenia) was less frequent and less pronounced in patients in the PA arm.
- Within the first 18 months of follow-up, the sperm counts were significantly higher after PA than after DL irradiation.
- Patients with testicular seminoma stage I (T1–T3), undisturbed lymphatic drainage, and adjuvant radiotherapy confined to the PA LNs were associated with reduced hematologic, gastrointestinal, and gonadal toxicities, but they had a higher risk of pelvic recurrence compared with DL radiotherapy. The recur-

rence rate was low with either treatment. PA radiotherapy is recommended as the standard treatment in these patients.

Fossa SD et al. (1999) Optimal planning target volume for stage I testicular seminoma: A Medical Research Council randomized trial. Medical Research Council Testicular Tumor Working Group. J Clin Oncol 17(4):1146

8.3 Bladder Cancer

Bladder cancer is the fourth most common cancer in men, following prostate, lung, and colon cancers, and it is the eighth most common cancer in women. It constitutes 5–10% of all cancers in men [29]. The risk of developing bladder cancer up to 75 years of age is 2–5% in males and 0.5–1% in females. Risk factors are aniline dyes, cigarette smoking, pelvic radiotherapy, cyclophosphamide, phenacetin use, and *Schistosoma haematobium* infections (in African and Middle Eastern countries). Aromatic amines and 2-naphthylamine, benzidine, and 4-aminobiphenyl (used in leather production) are also well-known carcinogenic agents for the bladder.

- The bladder is a depot organ for urine collection with a capacity of 350–450 mL (Fig. 8.15). It is under the pubic symphysis when empty. It consists of the detrusor muscle (endodermal origin) and the trigone at its base (mesodermal origin).
  - The bladder is positioned next to the seminal vesicles and the ampulla of the vas deferens posteriorly, as well as the lower tips of the ureters and the rectum.
  - It is next to the uterus and vagina in females.
  - The superior portion of the bladder is covered with peritoneum and is close to the small intestines by this peritoneum.
  - The base of the bladder is next to the prostate in males.

- The inner surface of the bladder mucosa is covered with transient cells (urothelium). A submucosal consisting of elastic and connective tissue lies beneath a mucosal layer called the lamina propria. A smooth muscle layer composed of longitudinal, circular, and spiral leaves, as well as the serosa (adventitia), consisting of fibrous tissue, are found under the subserosa.

Medial umbilical ligament extending from the umbilicus to the top of the bladder is the remnant of the urachal canal, which has atrophied.

8.3.1 Pathology

Transitional cell carcinoma is responsible for 90% of all bladder cancers in developed countries. These type of cells can be papillary and superficial (70–75%) or solid and invasive (20–25%) [30].

- CIS (intraepithelial carcinoma) is another subtype seen in 10% of cases; this exhibits frequent mitoses.

Squamous cell cancers comprise 5–10% of all bladder cancers. They display a nodular and infiltrative pattern and constitute 70% of all bladder tumors in regions where schistosomiasis is endemic, like Egypt.

- It can also be related to suprapubic catheterization, chronic infection, and inflammation.

Adenocarcinoma is a rare form and comprises 2% of all bladder tumors [30]. Thirty to thirty-five percent of them originate from the urachal region, and the rest are related to bladder exstrophy and are of nonurachal origin.

Undifferentiated carcinomas have a high nucleus/cytoplasm ratio and generally form cell layers or clusters. They behave like small cell cancers and have a poor prognosis.

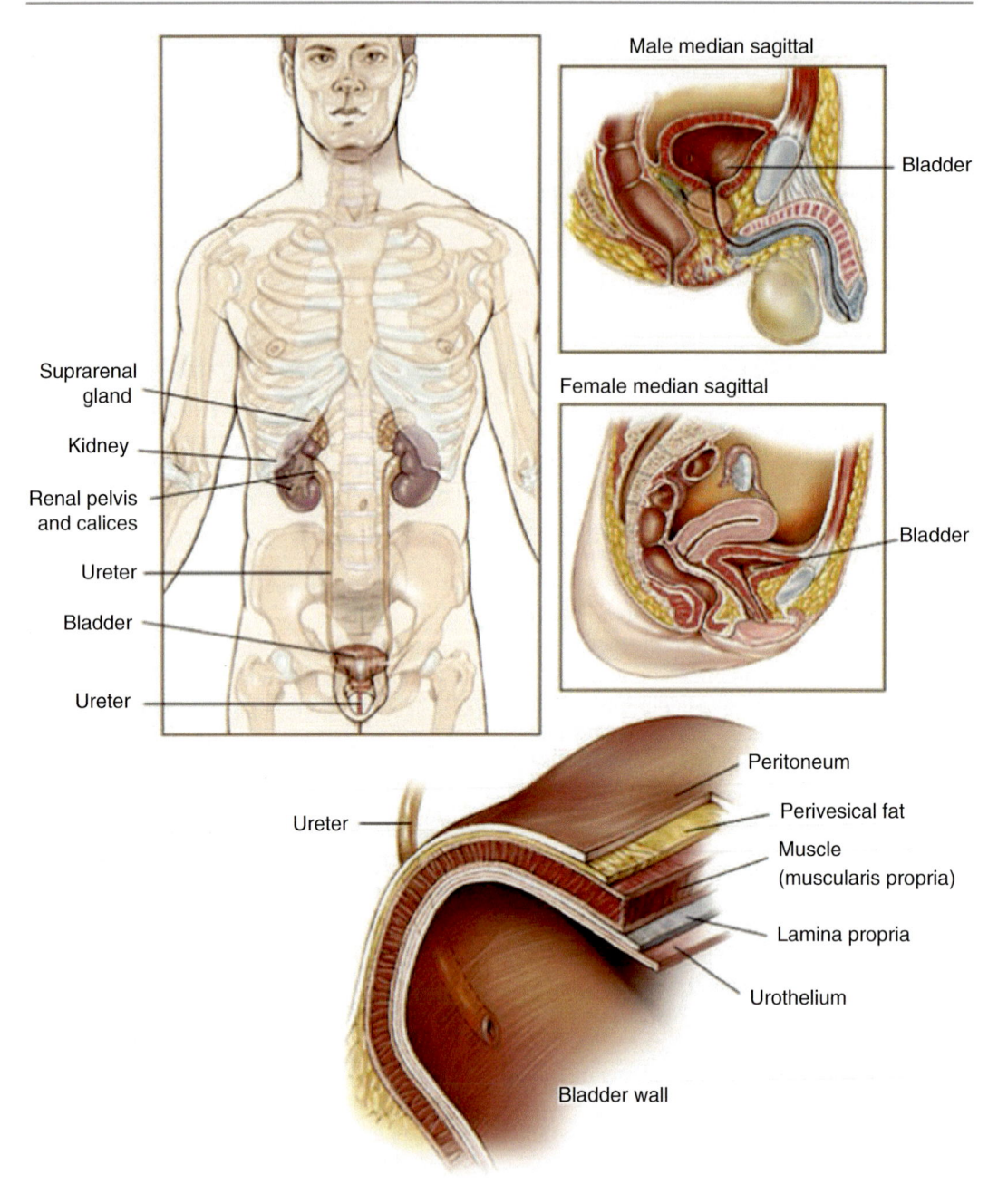

Fig. 8.15 Anatomy of the bladder

8.3.2 General Presentation

- Painless gross hematuria in 90% of cases.
- Polyuria, dysuria, and pollakiuria are commonly seen.
- Pain under umbilicus in bladder region.
- Pain and leg edema in advanced stages.

8.3.3 Staging

Primary Tumor (T) (Fig. 8.16) [31]
- TX: Primary tumor cannot be assessed
- T0: No evidence of primary tumor
- Ta: Noninvasive papillary carcinoma Tis: Carcinoma in situ (i.e., flat tumor)
- T1: Tumor invades subepithelial connective tissue T2: Tumor invades muscularis propria
  - pT2a: Tumor invades superficial muscularis propria (inner half)
  - pT2b: Tumor invades deep muscularis propria (outer half)
- T3: Tumor invades perivesical soft tissue
  - pT3a: Microscopically
  - pT3b: Macroscopically (extravesical mass)
- T4: Tumor invades any of the following: prostatic stroma, seminal vesicles, uterus, vagina, pelvic wall, or abdominal wall
- T4a: Tumor invades the prostatic stroma, uterus, vagina T4b: Tumor invades the pelvic wall, abdominal wall

Regional Lymph Nodes (N) (Fig. 8.17)
- NX: Lymph nodes cannot be assessed
- N0: No lymph node metastasis
- N1: Single regional lymph node metastasis in the true pelvis (perivesical, obturator, internal and external iliac, or sacral lymph node)
- N2: Multiple regional lymph node metastasis in the true pelvis (perivesical, obturator, internal and external iliac, or sacral lymph node)
- N3: Lymph node metastasis to the common iliac lymph nodes
- **Distant Metastasis (M)**
  - M0: No distant metastasis
  - M1: Distant metastasis
  - M1a: Distant metastasis limited to lymph nodes beyond the common iliacs
  - M1b: Non-lymph-node distant metastases

AJCC Stage Groups
- Stage Stage0a:TaN0M0
- Stage0is:TisN0M
- Stage I: T1N0M0
- Stage II: T2aN0M0; T2bN0M0
- Stage IIIA: T3aN0M0; T3bN0M0; T4aN0M0; T1-4aN1M0
- Stage IIIB: T1-4aN2-3M0
- Stage IVA: T4bAny NM0; Any T AnyNM1a;
- Stage IVB: Any T, Any NM1b

Bladder Lymphatics (Fig. 8.18) [31]
Iliac (external, internal), perivesical, pelvic, sacral (lateral, sacral promontorium (Gerota)).

Common iliac LNs are accepted as distant metastases (M1).

Bladder tumors are classified into three categories: superficial, invasive, and metastatic.

1. Superficial bladder tumors: limited to the epithelium (urothelium) → Ta, Tis, T1 (Tis is high grade and known as a flat tumor)
2. Invasive bladder tumors: invade the detrusor muscle and perivesical tissues beyond the uroepithelium (T2–T4)
3. Metastatic bladder tumors:

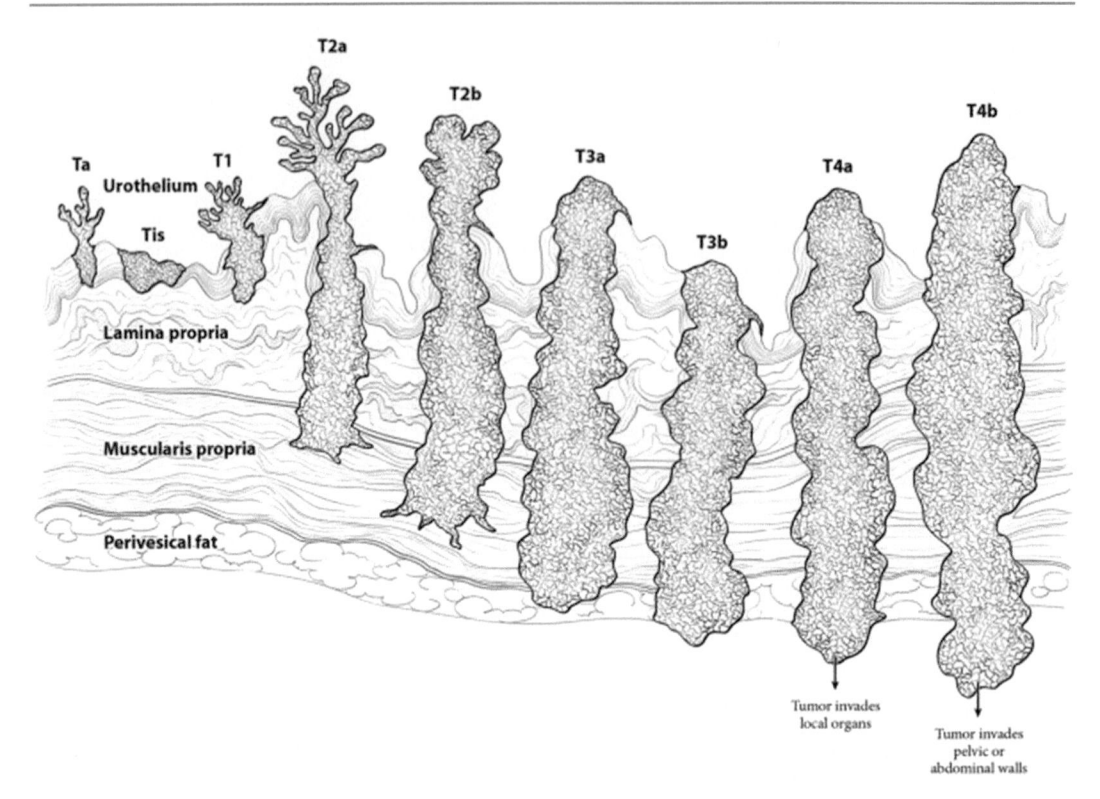

Fig. 8.16 T staging for bladder cancer. (from [31], Compton C.C., Byrd D.R., Garcia-Aguilar J., Kurtzman S.H., Olawaiye A., Washington M.K. (2012) Urinary Bladder. In: Compton C., Byrd D., Garcia-Aguilar J., Kurtzman S., Olawaiye A., Washington M. (eds) AJCC Cancer Staging Atlas. Springer, New York, NY. https://doi.org/10.1007/978-1-4614-2080-4_45)

- Seventy percent of all bladder tumors are superficial at diagnosis.
- However, 15–30% of all superficial tumors transform into invasive tumors
- Superficial tumors are seen singly in 70% and as multiple lesions in 30%.

There is a strong correlation between tumor grade and stage.

- Well-differentiated (grade I) and moderately differentiated tumors (grade II) are usually superficial, while undifferentiated (grade III) tumors are usually muscle invasive at diagnosis.

Papilloma [32]. This is a grade 0 bladder tumor. It is a papillary lesion with normal bladder mucosa around a thin fibrovascular nucleus.

- This lesion is very rarely seen, and almost never presents a relapse after total excision. If it is seen alone, it can be considered a benign neoplasm as a rule of thumb.
- However, it should be kept in mind that papillomas may be seen with high-grade urothelial cancers.

8.3.4 Treatment Algorithm

No Muscle Invasion: Ta, Tis, T1 [31, 33, 34]
TUR alone (relapse risk 30%), or TUR + 6-week intravesical BCG
Postoperative adjuvant treatment indications:

- Subtotal resection
- Residue (+) even after reoperations
- Urine cytology (+)
- Multifocality

8.3 Bladder Cancer

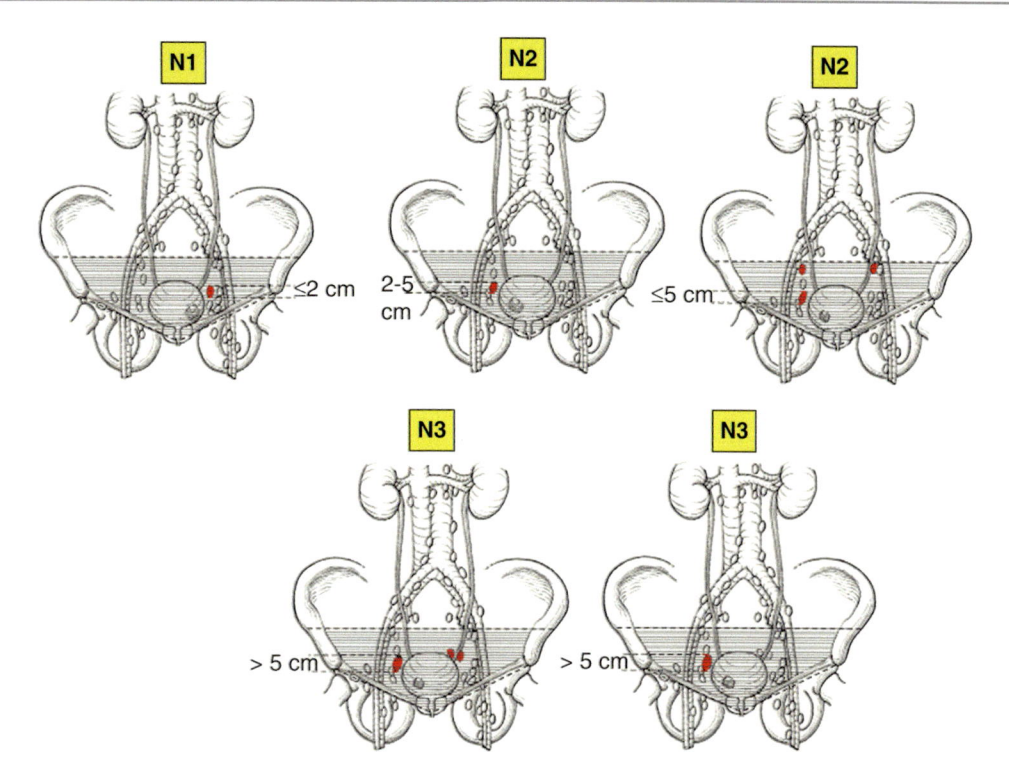

Fig. 8.17 N staging for bladder cancer. (from [31], Compton C.C., Byrd D.R., Garcia-Aguilar J., Kurtzman S.H., Olawaiye A., Washington M.K. (2012) Urinary Bladder. In: Compton C., Byrd D., Garcia-Aguilar J., Kurtzman S., Olawaiye A., Washington M. (eds) AJCC Cancer Staging Atlas. Springer, New York, NY. https://doi.org/10.1007/978-1-4614-2080-4_45)

- Grade II–III
- Tis, T1

(Following 6 months of BCG treatment, disease (+) → BCG every 6 months + mitomycin/doxorubicin every 3 weeks for 2 years)

Multiple recurrences or disease (+) following 1-year BCG → radical cystectomy/RT (Radiation tolerance may be worsened due to recurrent intravesical treatments.) *Muscle invasive* (+) [34]Treatment options:

- Radical cystectomy
- Partial cystectomy
- Bladder conserving surgery + chemoradiotherapy

Tumors with no CIS in the dome of the bladder may be treated with partial cystectomy. *Bladder-conserving surgery indications* → unifocal T2–3a, <5 cm, hydronephrosis (−), normal bladder functions, complete TUR

Multifocal T2–3a, T3b–4, hydroureter/hydronephrosis, subtotal resection → radical cystectomy

T3b–4 tumors: neoadjuvant chemotherapy + cystectomy + LN dissection + adjuvant chemotherapy*Local relapse*

Local relapse after cystectomy → cisplatin + RT (45 Gy pelvis, 60–64 Gy to relapse site) Relapse after chemoradiotherapy → cystectomy

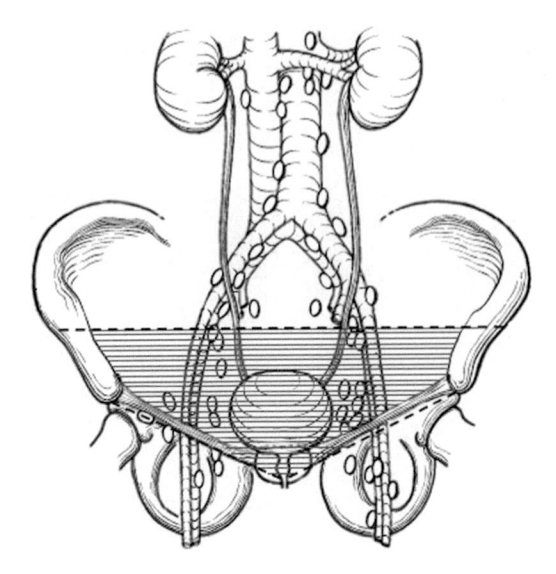

Fig. 8.18 Bladder lymphatics. (from [31], Compton C.C., Byrd D.R., Garcia-Aguilar J., Kurtzman S.H., Olawaiye A., Washington M.K. (2012) Urinary Bladder. In: Compton C., Byrd D., Garcia-Aguilar J., Kurtzman S., Olawaiye A., Washington M. (eds) AJCC Cancer Staging Atlas. Springer, New York, NY. https://doi.org/10.1007/978-1-4614-2080-4_45)

8.3.5 Radiotherapy

Radiation treatment should begin within 8 weeks after maximal TURBT. CT simulation using 3 mm or less slice thickness should be performed prior to planning. Planning is performed in either the supine or prone position with an empty bladder. IV contrast is given to visualize the bladder.

Conformal RT Volumes (Fig. 8.19) [35]

GTV_{tumor}: Any residual tumor seen on CT/MRI/cystoscopy.

$CTV_{bladder}$: CTV_{tumor} + whole bladder (The entire bladder, prostate, and prostatic urethra are included due to the multifocal nature of bladder cancer)

CTV_{nodal}: Regional lymph nodes: obturator, external iliac, and internal iliac lymph nodes

$CTV_{45Gy} = CTV_{bladder} + CTV_{nodal}$
$CTV_{54Gy} = CTV_{bladder}$, $PTV_{54Gy} = CTV_{54Gy} + 2$ cm
$CTV_{64Gy} = PTV_{64Gy} = GTV_{tumor} + 2$ cm

8.3.6 Selected Publications

Danish National Bladder Cancer Group, DAVECA protocol 8201, 1991 → 183 patients with T2–T4a entered a randomized study. Preoperative irradiation (40 Gy) followed by cystectomy vs. radical irradiation (60 Gy) followed by salvage cystectomy in cases of residual tumor.

- A trend for a higher survival rate following combined treatment with preoperative irradiation and cystectomy compared to radical irradiation followed by salvage cystectomy was observed.
- There was no difference in surgical complications between planned and salvage cystectomy, and there were no postoperative deaths among the cystectomized patients.
- All male patients experienced erective impotence after cystectomy.
- T-stage, response to radiotherapy and frequency of LN metastases were found to be of prognostic importance.

Sell A et al. (1991) Treatment of advanced bladder cancer category T2 T3 and T4a. A randomized multicenter study of preoperative irradiation and cystectomy versus radical irra- diation and early salvage cystectomy for residual tumor. DAVECA protocol 8201. Danish Vesical Cancer Group. Scand J Urol Nephrol Suppl 138:193–201

RTOG 85-12, 1993 (1986–1988) → T2–4, N0-2 or NX, M0 were treated with pelvic radiotherapy 40 Gy in 4 weeks and cisplatin 100 mg/m^2 on days 1 and 22. Complete responders were given an additional 24 Gy bladder boost plus a third dose of cisplatin; patients with residual tumor after 40 Gy were assigned radical cystectomy.

- The complete remission rate following cisplatin and 40 Gy for evaluable cases was 31/47 (66%).
- Acute toxicity was acceptable, with only two patients not completing induction therapy
- Patients with poorly differentiated tumors were more likely to achieve complete remission.

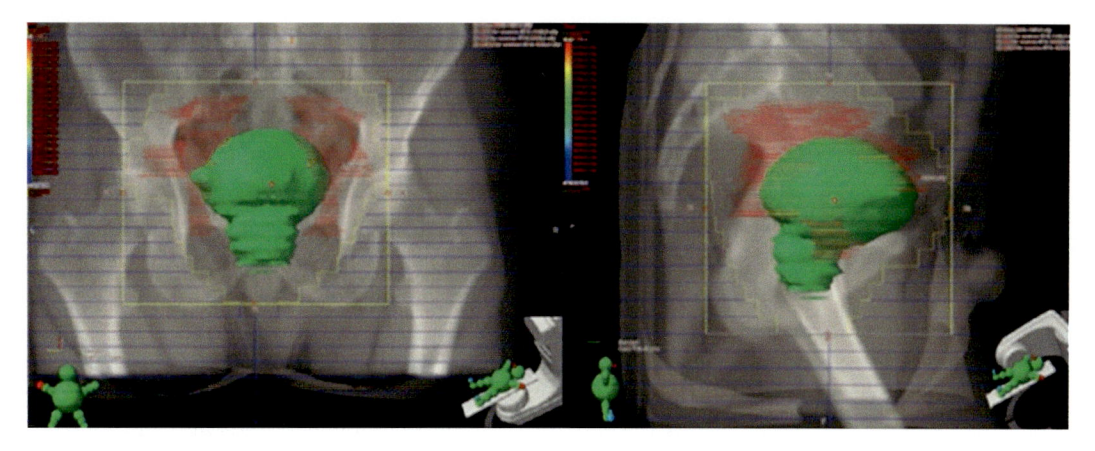

Fig. 8.19 Conformal RT fields in bladder cancer

- Actuarial survival was 64% at 3 years.
- The chemoradiotherapy regimen was moderately well tolerated and associated with tumor clearance in 66% of the patients treated.

Tester W et al. (1993) Combined modality program with possible organ preservation for invasive bladder carcinoma: results of RTOG protocol 85-12. Int J Radiat Oncol Biol Phys 25(5):783–790

RTOG 88-02, 1996 (1988–1990) → 91 patients with T2M0–T4AM0 suitable for radical cystectomy received two courses of methotrexate, cisplatin, and vinblastine (MCV regimen) followed by radiotherapy with 39.6 Gy and concurrent cisplatin. Operable patients who achieved complete response were selected for bladder preservation and treated with consolidation cisplatin–radiotherapy.

- 68 patients (75%) exhibited complete response.
- 14 patients with residual tumors underwent immediate cystectomy.
- 37 of 91 patients (40%) required cystectomy.
- Four-year cumulative risk of invasive local failure was 43%.
- Four-year actuarial risk of distant metastasis was 22%.
- Four-year actuarial survival rate of the entire group was 62%.
- Four-year actuarial rate of survival with bladder intact was 44%.
- Bladder preservation can be achieved in the majority of patients, and the overall survival was similar to that reported with aggressive surgical approaches.

Tester W et al. (1996) Neoadjuvant combined modality program with selective organ preservation for invasive bladder cancer: results of Radiation Therapy Oncology Group phase II trial 8802. J Clin Oncol 14(1):119–126

NCI-Kanada, 1996 → 99 eligible patients with T2–T4b. Patients and their physicians selected either definitive radiotherapy or precystectomy radiotherapy; patients were then randomly allocated to receive IV cisplatin 100 mg/m^2 at 2-week intervals for three cycles concurrent with pelvic radiation or to receive radiation without chemotherapy. Median follow-up was 6.5 years.

- Distant metastases were the same in both study arms.
- 25 of 48 control patients had a first recurrence in the pelvis, compared with 15 of 51 cisplatin-treated patients ($p = 0.036$).
- The pelvic relapse rates of the two groups were significantly reduced by concurrent cisplatin ($p = 0.038$).
- Concurrent cisplatin was found to improve pelvic control of locally advanced bladder cancer with preoperative or definitive radiation, but was not shown to improve overall survival.

- The use of concurrent cisplatin had no detectable effect on distant metastases.

Coppin CM et al. (1996) Improved local control of invasive bladder cancer by concurrent cisplatin and preoperative or definitive radiation. The National Cancer Institute of Canada Clinical Trials Group. J Clin Oncol 14(11):2901–2907

RTOG 89-03, 1998 (1990–1993) → 123 patients with T2–T4aNXM0. Randomized to two cycles of MCV before 39.6 Gy pelvic irradiation with concurrent cisplatin 100 mg/m² for two courses 3 weeks apart vs. chemoradiotherapy. The CR patients were treated with a boost dose of 25.2 Gy to a total of 64.8 Gy and one additional dose of cisplatin. Patients with residual disease underwent cystectomy. Median follow-up: 60 months.

- Five-year overall survival rate was 49; 48% in arm 1 and 49% in arm 2.
- Distant metastases at 5 years; 33% in arm 1 and 39% in arm 2.
- Five-year survival rate with a functioning bladder was 38, 36% in arm 1 and 40% in arm 2.
- Two cycles of MCV neoadjuvant chemotherapy had no impact on 5-year overall survival.

Shipley WU et al. (1998) Phase III trial of neoadjuvant chemotherapy in patients with invasive bladder cancer treated with selective bladder preservation by combined radiation therapy and chemotherapy: initial results of Radiation Therapy Oncology Group 89-03. J Clin Oncol 16(11):3576–3583

SWOG, 2001 (1993–1998) → 60 patients with T2–T4 with nodal involvement, medically or surgically inoperable, or refused cystectomy. 75 mg/m² cisplatin on day 1 and 1 g/m² daily, 5-fluorouracil on days 1–4 and definitive radiotherapy. Chemotherapy was repeated every 28 days, twice during and twice after radiation.

- The overall response rate was 51%.
- Overall 5-year survival was 32%.
- Five-year survival of the 25 patients who refused surgery was 45%.
- This combined modality may provide another alternative to cystectomy for patients refusing cystectomy.

Hussain MH et al. (2001) Combination cisplatin, 5-fluorouracil and radiation therapy for locally advanced unresectable or medically unfit bladder cancer cases: a Southwest Oncology Group Study. J Urol 165(1):56–60

RTOG 95-06, 2000, (1995–1997) → 34 patients with clinical stage T2–T4a, Nx M0 without hydronephrosis. After performing as complete a transurethral resection as possible, induction chemoradiotherapy (cisplatin and 5-fluorouracil, radiation was given twice a day, 3 Gy per fraction to the pelvis for a total of 24 Gy) was administered. Patients with a complete response received the same drugs combined with twice-daily radiation therapy to the bladder and a bladder tumor volume of 2.5 Gy per fraction for a total consolidation dose of 20 Gy. Median follow-up was 29 months.

- After induction treatment, 22 (67%) of the 33 patients had no tumor detectable on urine cytology or rebiopsy.
- Of the 11 patients who still had detectable tumors, six underwent radical cystectomy.
- No patient required a cystectomy for radiation toxicity.
- Three-year overall survival was 83%.
- Three-year survival with intact bladder was 66%.
- Both the complete response rate to induction therapy and the 3-year survival with an intact bladder were encouraging.

Kaufman DS et al. (2000) The initial results in muscle-invading bladder cancer of RTOG 95-06: phase I/II trial of transurethral surgery plus radiation therapy with concurrent cisplatin and 5-fluorouracil followed by selective bladder preservation or cystectomy depend- ing on the initial response. Oncologist 5(6):471–476

RTOG 97-06, 2003, (1998–2000) → 47 eligible patients with stage T2–T4aN0M0. Median follow-up was 26 months. TURBT within 6 weeks of the initiation of induction therapy (13

8.3 Bladder Cancer

days of concomitant boost RT, 1.8 Gy to the pelvis in the morning followed by 1.6 Gy to the tumor 4–6 h later and cisplatin (20 mg/m^2)). Three to four weeks after induction, the patients were evaluated for residual disease.

- CR rate was 74%.
- Three-year rates of locoregional failure, distant metastasis, overall survival, and bladder-intact survival were 27%, 29%, 61%, and 48%, respectively.
- Adjuvant MCV chemotherapy appears to be poorly tolerated.

Hagan MP et al. (2003) RTOG 97-06: Initial report of a phase I–II trial of selective bladder conservation using TURBT, twice-daily accelerated irradiation sensitized with cisplatin, and adjuvant MCV combination chemotherapy. Int J Radiat Oncol Biol Phys 57(3):665–672

RTOG 99-06, 2006 → 80 patients with T2–T4a; twice-daily radiotherapy with paclitaxel and cisplatin chemotherapy induction (TCI) was administered. Adjuvant gemcitabine and cisplatin were given to all patients.

- TCI resulted in 26% developing grade 3–4 acute toxicity, mainly gastrointestinal (25%).
- The postinduction complete response rate was 81% (65/80).
- Thirty-six patients died (22 of bladder cancer).
- Five-year overall and disease-specific survival rates were 56% and 71%, respectively.

Kaufman DS et al. (2008) Phase I–II RTOG Study (99-06) of patients with muscle-invasive bladder cancer undergoing transurethral surgery, paclitaxel, cisplatin, and twice-daily radiotherapy followed by selective bladder preservation or radical cystectomy and adjuvant chemotherapy. Urology 73:833–837

RTOG pooled analysis-2014: Four hundred sixty-eight patients with muscle invasive bladder cancer were enrolled onto six RTOG bladder-preservation studies, including five phase II studies (RTOG 8802, 9506, 9706, 9906, and 0233) and one phase III study (RTOG 8903). Complete response to combined modality treatment (CMT) was documented in 69% of patients. With a median follow-up of 4.3 years among all patients and 7.8 years among survivors (*n* = 205), the 5- and 10-year OS rates were 57% and 36%, respectively, and the 5- and 10-year DSS rates were 71% and 65%, respectively. Of the 205 patients alive at 5 years, 80% had an intact bladder. CMT protocols demonstrate long-term DSS comparable to modern immediate cystectomy studies, for patients with similarly staged MIBC. (Mak RH, et al. Long-term outcomes in patients with muscle-invasive bladder cancer after selective bladder-preserving combined-modality therapy: a pooled analysis of Radiation Therapy Oncology Group protocols 8802, 8903, 9506, 9706, 9906, and 0233. J Clin Oncol. 2014;32(34):3801–9.)

Massachusetts General Hospital evaluated treatment results in 348 patients with cT2 to T4a muscle-invasive bladder cancer treated in several RTOG studies. All patients underwent TURBT and concurrent chemoradiation to 64 to 65 Gy with concurrent cisplatin-based chemotherapy; some patients also received neoadjuvant chemotherapy and/or adjuvant chemotherapy. Clinical T stage and complete remission (CR) were significantly associated with improved DSS and OS. Use of neoadjuvant chemotherapy did not improve outcomes. No patient required cystectomy for treatment-related toxicity. CMT achieves a CR and preserves the native bladder in >70% of patients while offering long-term survival rates comparable to contemporary cystectomy series. These results support modern bladder-sparing therapy as a proven alternative for selected patients. (Efstathiou JA, et al. Long-term outcomes of selective bladder preservation by combined-modality therapy for invasive bladder cancer: the MGH experience. Eur Urol. 2012;61(4):705–11.)

A series of 415 patients of muscle invasive bladder cancer from Erlangen reported that early tumor stage and a complete TUR were the most important factors predicting CR and survival. Chemoradiotherapy was more effective than RT alone in terms of CR and survival. (Rödel C, et al. Combined-modality treatment and selective organ preservation in invasive bladder cancer: long-term results. J Clin Oncol. 2002;20(14):3061–71.)

The Bladder Cancer 2001 (BC2001) study was a multicenter randomized phase III trial that enrolled 360 patients with muscle-invasive bladder cancer. All patients underwent TURBT, followed by randomization to either RT alone or concurrent chemoradiotherapy. The chemotherapy regimen used was FU and mitomycin. At a median follow-up of 69.9 months, locoregional disease-free survival were 67% in the chemoradiotherapy group and 54% in the radiotherapy group ($p = 0.03$). Grade 3 or 4 adverse events were slightly more common in the chemoradiotherapy group than in the radiotherapy group during treatment (36.0% vs. 27.5%, $P = 0.07$) but not during follow-up (8.3% vs. 15.7%, $p = 0.07$). (James ND, et al.; BC2001 Investigators. Radiotherapy with or without chemotherapy in muscle-invasive bladder cancer. N Engl J Med. 2012;366(16):1477–88.).

References

1. Global Burden of Disease Cancer Collaboration (2017) Global, regional, and National Cancer Incidence, mortality, years of life lost, years lived with disability, and disability-adjusted life-years for 32 cancer groups, 1990 to 2015: a systematic analysis for the global burden of disease study. JAMA Oncol 3(4):524–548
2. American Cancer Society (2021) American Cancer Society: cancer facts and figures 2021. American Cancer Society, Atlanta
3. Schulsinger AR, Allison RR, Choi WH (2006) Bladder cancer. In: Levitt SH, Purdy JA, Perez CA, Vijayakumar S (eds) Technical basis of radiation therapy. Springer, New York, pp 574–575
4. Kurokawa K, Suzuki T, Suzuki K, Terada N, Ito K, Yoshikawa D, Arai Y, Yamanaka H (2003) Preliminary results of a monitoring system to confirm the preservation of cavernous nerves. Int J Urol 10(3):136–140
5. Ali M, Johnson IP, Hobson J, Mohammadi B, Khan F (2004) Anatomy of the pelvic plexus and innervation of the prostate gland. Clin Anat 17(2):123–129
6. Algaba F, Epstein JI, Aldape HC et al (1996) Assessment of prostate carcinoma in core needle biopsy–definition of minimal criteria for the diagnosis of cancer in biopsy material. Cancer 78(2):376–381
7. Konishi N, Shimada K, Ishida E, Nakamura M (2005) Molecular pathology of prostate cancer. Pathol Int 55(9):531–539. (review)
8. Gleason DF (1977) Histologic grading and clinical staging of prostatic carcinoma. In: Tannenbaum M (ed) Urologic pathology: the prostate. Lea & Febiger, Philadelphia, pp 171–197
9. Epstein JI, Zelefsky MJ, Sjoberg DD, Nelson JB, Egevad L, Magi-Galluzzi C, Vickers AJ, Parwani AV, Reuter VE, Fine SW, Eastham JA, Wiklund P, Han M, Reddy CA, Ciezki JP, Nyberg T, Klein EA (2016) A contemporary prostate cancer grading system: a validated alternative to the Gleason score. Eur Urol 69(3):428–435
10. Reynolds MA, Kastury K, Groskopf J, Schalken JA, Rittenhouse H (2007) Molecular markers for prostate cancer. Cancer Lett 249(1):5–13. (review)
11. Presti JR (2000) Neoplasm of the prostate gland. In: Tanagho EA, McAninch JW (eds) Smith's general urology, 15th edn. McGraw-Hill, San Francisco, pp 399–421
12. Akyol F, Ozyigit G, Selek U, Karabulut E (2005) PSA bouncing after short term androgen deprivation and 3D-conformal radiotherapy for localized prostate adenocarcinoma and the relationship with the kinetics of testosterone. Eur Urol 48:40–45
13. Levitt SH, Purdy JA, Perez CA (2006) Technical basis of radiation therapy, 4th revised edn. Springer, Berlin
14. Compton CC, Byrd DR, Garcia-Aguilar J, Kurtzman SH, Olawaiye A, Washington MK (2012) Prostate. In: Compton C, Byrd D, Garcia-Aguilar J, Kurtzman S, Olawaiye A, Washington M (eds) AJCC cancer staging atlas. Springer, New York. https://doi.org/10.1007/978-1-4614-2080-4_41
15. Kiyoshima K, Yokomizo A, Yoshida T, Tomita K, Yonemasu H, Nakamura M, Oda Y, Naito S, Hasegawa Y (2004) Anatomical features of periprostatic tissue and its surroundings: a histological analysis of 79 radical retropubic prostatectomy specimens. Jpn J Clin Oncol 34(8):463–468
16. Woods ME, Ouwenga M, Quek ML (2007) The role of pelvic lymphadenectomy in the man- agement of prostate and bladder cancer. ScientificWorldJournal 7:789–799. (review)
17. https://www.nccn.org/professionals/physician_gls/pdf/prostate.pdf
18. Turkbey B, Rosenkrantz AB, Haider MA et al (2019) Prostate imaging reporting and data system version 2.1: 2019 update of prostate imaging reporting and data system version 2. Eur Urol 76(3):340–351
19. Grubb RL III, Vardi IY, Bhayani SB, Kibel AS (2006) Minimally invasive approaches to localized prostate carcinoma. Hematol Oncol Clin North Am 20(4):879–895. (review)
20. Villavicencio H, Segarra J (2006) Perineal prostatectomy. Ann Urol (Paris) 40(5):317–327. (review)
21. Roach M III (2003) Hormonal therapy and radiotherapy for localized prostate cancer: who, where and

References

how long? J Urol 170(6 pt 2):S35–S40; discussion S40–S41 (review)
22. Ozyigit G, Selek U (eds.) (2017) Principles and practice of urooncology. Springer International Publishing AG
23. Fiks T, Sporny S (1985) Klinefelter's syndrome and neoplasms. Wiad Lek 38(17):1223–1226. (review)
24. Woodward PJ, Heidenreich A, Looijenga LHJ et al (2004) Germ cell tumours. In: Eble JN, Sauter G, Epstein JI, Sesterhenn IA (eds) Pathology and genetics of tumours of the urinary system and male genital organs. IARC Press, Lyon, pp 221–249
25. Bosl GJ, Bajorin DF, Sheinfeld J et al (2005) Cancer of the testis. In: DeVita VT Jr, Hellman S, Rosenberg SA (eds) Cancer: principles and practice of oncology, 7th edn. Lippincott Williams & Wilkins, Philadelphia, pp 1269–1290
26. Compton CC, Byrd DR, Garcia-Aguilar J, Kurtzman SH, Olawaiye A, Washington MK (2012) Testis. In: Compton C, Byrd D, Garcia-Aguilar J, Kurtzman S, Olawaiye A, Washington M (eds) AJCC cancer staging atlas. Springer, New York. https://doi.org/10.1007/978-1-4614-2080-4_42
27. Hansen EK, Roach M III (2018) Handbook of evidence-based radiation oncology. Springer, US, p 328
28. Shabbir M, Morgan RJ (2004) Testicular cancer. J R Soc Health 124(5):217–218. (review)
29. Naysmith TE, Blake DA, Harvey VJ, Johnson NP (1998) Do men undergoing sterilizing cancer treatments have a fertile future? Hum Reprod 13(11):3250–3255
30. Madeb R, Golijanin D, Knopf J, Messing EM (2007) Current state of screening for bladder cancer. Expert Rev Anticancer Ther 7(7):981–987. (review)
31. Compton CC, Byrd DR, Garcia-Aguilar J, Kurtzman SH, Olawaiye A, Washington MK (2012) Urinary bladder. In: Compton C, Byrd D, Garcia-Aguilar J, Kurtzman S, Olawaiye A, Washington M (eds) AJCC cancer staging atlas. Springer, New York. https://doi.org/10.1007/978-1-4614-2080-4_45
32. Mostofi FK, Davis CJ, Sesterhenn IA (1988) Pathology of tumors of the urinary tract. In: Skinner DG, Lieskovsky G (eds) Diagnosis and management of genitourinary cancer. WB Saunders, Philadelphia, pp 83–117
33. Soloway MS (1983) The management of superficial bladder cancer. In: Javadpour N (ed) Principles and management of urologic cancer, 2nd edn. Williams and Wilkins, Baltimore, pp 446–467
34. Epstein JI (2003) The new World Health Organization/International Society of Urological Pathology (WHO/ISUP) classification for TA, T1 bladder tumors: is it an improvement? Crit Rev Oncol Hematol 47(2):83–89. (review)
35. Bladder Cancer, Lee NY, et al. (eds.) (2014) Target volume delineation for conformal and intensity-modulated radiation therapy, medical radiology. Radiation oncology, Springer International Publishing Switzerland, p 377

Gynecological Cancers

9.1 Cervical Cancer

Worldwide, cervical cancer is the fourth most commonly diagnosed cancer and the fourth leading cause of cancer death in women, with an estimated 604,000 new cases and 342,000 deaths in 2020 [1]. It is the most common gynecological cancer in developing and underdeveloped countries. Over 90% of cervical cancer is caused by human papillomavirus (HPV) infection.

The uterus is located in the middle of the true pelvis between the rectum and bladder, and makes a right angle with the vagina. The upper two-third of the uterus is called the corpus, and the lower one-third is known as the cervix.

The cervix has a cylindrical shape and extends into the upper vagina (Fig. 9.1). It passes between the vagina and the upper uterine cavity via the endocervical canal.

- The endocervical canal is covered with glandular (columnar) cells, and the region around the cervical canal is called the endocervix.
- The region facing toward the vagina is termed the exocervix and is covered with squamous epithelial cells.
- Cervical cancers originating from the endocervix are adenocarcinomas, while those from the exocervix are squamous cell cancers.

9.1.1 Pathology

Cervical cancers are categorized into three major histological subtypes by the World Health Organization (WHO): squamous cell cancers, adenocarcinomas, and others. Approximately 70–80% of all cases are squamous cell cancers [2].

Squamous cell cancer almost always starts with abnormal metaplastic events in the transformation zone, resulting in sequential lesions from cervical intraepithelial neoplasias (CIN) of grades I, II, III, and microinvasive cancer.

9.1.1.1 The WHO Cervical Cancer Classification [2–4]

Squamous cell cancers → microinvasive squamous carcinoma, invasive squamous carcinoma, verrucous carcinoma, condylomatous carcinoma.

Adenocarcinoma → usual-type endocervical carcinoma, mucinous adenocarcinoma (gastric type and minimal deviation adenocarcinoma), endometrioid-type adenocarcinoma, clear cell carcinoma, serous adenocarcinoma, villoglandular adenocarcinoma, mesonephric carcinoma.

Other cancers → adenosquamous cancer, small cell cancer, adenoid cystic cancer, mucoepidermoid cancer, adenoid basal cancer, carcinoid cancer, undifferentiated cancer, lymphoma.

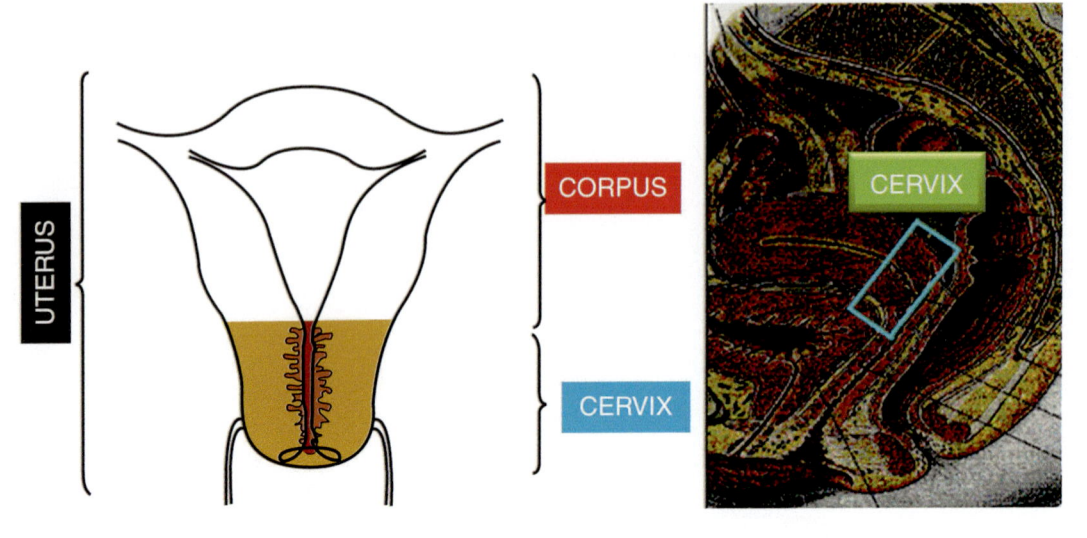

Fig. 9.1 Anatomy of the cervix

9.1.2 General Presentation

9.1.2.1 Early Period
(a) Postcoital bleeding
(b) Irregular menstruation
(c) Bloody discharge
(d) Discharge with foul odor

9.1.2.2 Late Period
(a) Leg and groin pain
(b) Fistulas (cervicovesical, vesicovaginal, cervicorectal, rectovaginal)
(c) Hydronephrosis and renal dysfunction due to ureter obstruction
(d) Edema in lower extremities
(e) Anemia

9.1.3 Staging

The first important step is a bimanual rectovaginal gynecological exam under general anesthesia. The International Federation of Gynecology and Obstetrics (FIGO) revised cervical staging system in 2018 [5]. This staging is performed by the use of clinical exam (inspection, palpation, colposcopy), endoscopic exam (cystoscopy and rectoscopy), direct radiological exams (intravenous pyelography, chest X-ray), findings from tumor biopsy (tumor depth, size, etc.), as well as advanced radiological imaging (PET/CT and/or MRI) and surgico-pathological findings [5]. The AJCC staging is consistent with FIGO staging [6].

9.1.3.1 FIGO Staging (Fig. 9.2) [6]
Stage I. Carcinoma strictly confined to the cervix; extension to the uterine corpus should be disregarded.

IA: Invasive cancer identified only microscopically, with maximum depth of invasion <5 mm*

IA1: Measured invasion of the stroma <3 mm in depth

IA2: Measured invasion of stroma ≥3 mm and <5 mm in depth

IB: Invasive carcinoma with measured deepest invasion ≥5 mm (greater than stage IA), lesion limited to the cervix uteri**

IB1: Invasive carcinoma ≥5 mm depth of stromal invasion and <2 cm in greatest dimension

IB2: Invasive carcinoma ≥2 cm and <4 cm in greatest dimension

IB3: Invasive carcinoma ≥4 cm in greatest dimension

Stage II. Carcinoma that extends beyond the cervix but has not extended onto the pelvic wall. The carcinoma involves the vagina but not extended as far as the lower third section.

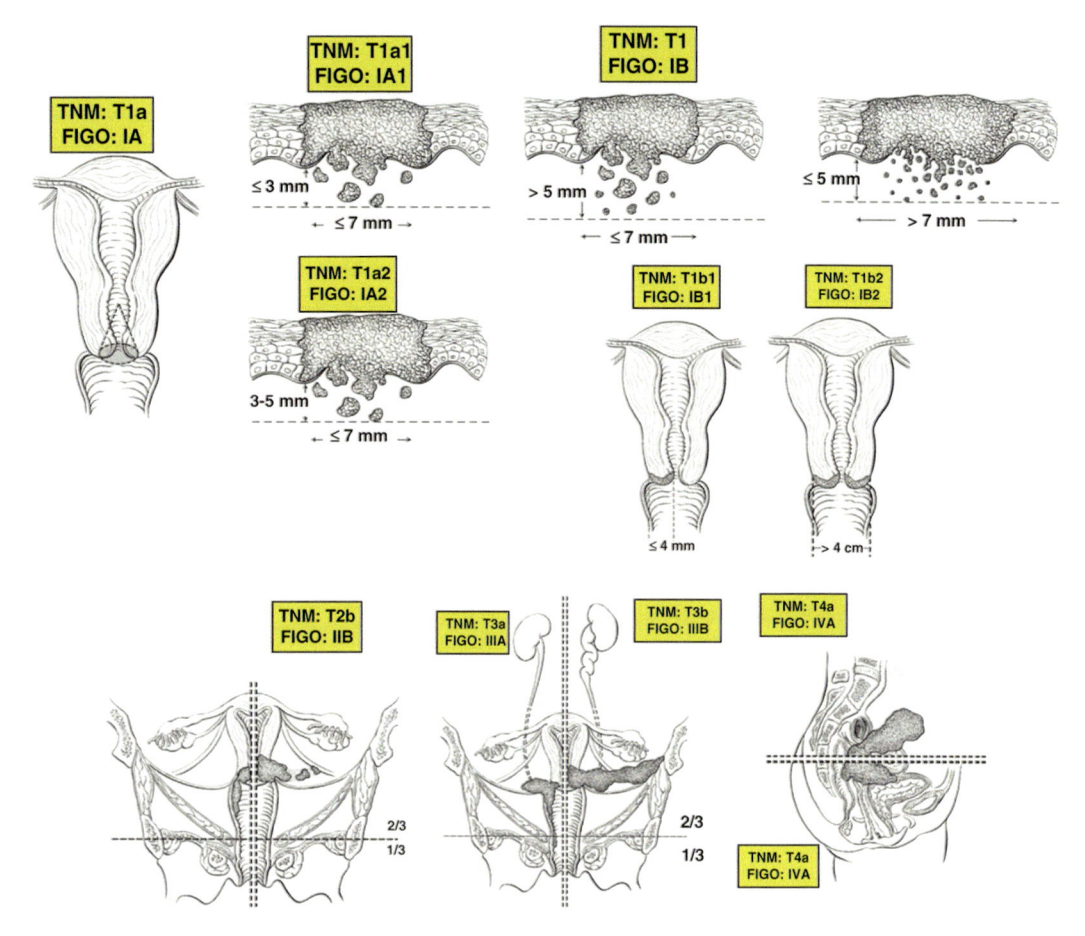

Fig. 9.2 Staging of cervical cancer. (From [6], Compton C.C., Byrd D.R., Garcia-Aguilar J., Kurtzman S.H., Olawaiye A., Washington M.K. (2012) Cervix Uteri. In: Compton C., Byrd D., Garcia-Aguilar J., Kurtzman S., Olawaiye A., Washington M. (eds) AJCC Cancer Staging Atlas. Springer, New York, NY. https://doi.org/10.1007/978-1-4614-2080-4_35)

IIA: Involvement limited to the upper two-thirds of the vagina without parametrial involvement

IIA1: Invasive carcinoma <4 cm in greatest dimension

IA2: Invasive carcinoma ≥4 cm in greatest dimension

IIB: Obvious parametrial involvement but not onto the pelvic sidewall

Stage III. Carcinoma that has extended onto the pelvic sidewall and/or involves the lower third of the vagina. On rectal examination, there is no cancer-free space between the tumor and the pelvic sidewall. All cases with a hydronephrosis or nonfunctioning kidney should be included unless they are known to be due to other causes. Involvement of pelvic and/or para-aortic lymph nodes.

IIIA: No extension onto the pelvic sidewall but involvement of the lower third of the vagina

IIIB: Extension onto the pelvic sidewall or hydronephrosis or nonfunctioning kidney

IIIC: Involvement of pelvic and/or para-aortic lymph nodes, irrespective of tumor size and extent (with r and p notations)***

IIIC1: Pelvic lymph node metastasis only

IIIC2: Para-aortic lymph node metastasis.

Stage IV. Carcinoma that has extended beyond the true pelvis or has involved (biopsy proven) the mucosa of the bladder and/or rectum. A bullous edema, as such, does not permit a case to be allotted to stage IV.

IVA: Spread of the tumor onto adjacent pelvic organs

IVB: Spread to distant organs (Including peritoneal spread, involvement of supraclavicular or mediastinal lymph nodes, para-aortic lymph nodes, lung, liver, or bone)

Note: FIGO no longer includes stage 0 (Tis).

* Imaging and pathology can be used, when available, to supplement clinical findings with respect to tumor size and extent, in all stages.

**The involvement of vascular/lymphatic spaces does not change the staging. The lateral extent of the lesion is no longer considered.

***Adding notation of r (imaging) and p (pathology) to indicate the findings that are used to allocate the case to stage IIIC.

9.1.3.2 Lymphatics of Cervix Cancer
(Fig. 9.3)

Paracervical.

Parametrial.

Hypogastric (internal iliac), including obturator.

External iliac.

Common iliac.

Presacral.

Para-aortic.

9.1.4 Treatment Algorithm

The local treatment decision should be made based on tumor size, stage, histology, lymph node (LN) involvement, possible complications of local therapies, requirement for adjuvant therapy, and patient choice. However, intraepithelial lesions are treated with superficial ablative modalities (cryosurgery or laser surgery); microinvasive cancers that are less than 3 mm in depth (stage IA1) with conservative surgery; early invasive cancers (stage IA2, IB1, IB2, and some small stage IIA tumors) with either radical surgery or radiotherapy (RT); and locally advanced cancers (stages IB3–IVA) with concurrent chemoradiotherapy (CRT) [7–10].

- *Modified radical hysterectomy (type II) and bilateral pelvic LN dissection.* This is the standard surgical approach for stage IA1 with lymphovascular space invasion (LVSI) and IA2 cervical cancer [11]. Uterus, cervix, upper 1–2 cm of the vagina, parametrial tissues, and medial half of the uterosacral and cardinal ligaments are removed.
- *Radical hysterectomy (type III) and bilateral pelvic LN dissection.* This is the standard surgical approach for stage IB1, IB2, and IIA1 cervical cancer [11]. Here, the uterus, surrounding tissues, and the upper half of the vagina are excised. The cardinal, sacrouterine, and vesicouterine ligaments are completely dissected together with the uterus. Pelvic lymphadenectomy is part of radical hysterectomy.
- *Vaginal hysterectomy (Schauta–Amreich operation) and radical trachelectomy* are alternative surgical approaches to spare fertility in patients with a tumor that is less than 2 cm and at stage IB1 [12]. In radical trachelectomy, the uterus is not dissected, and 2–3 cm of the upper vagina together with the cervix and the cardinal–sacrouterine ligaments are excised.

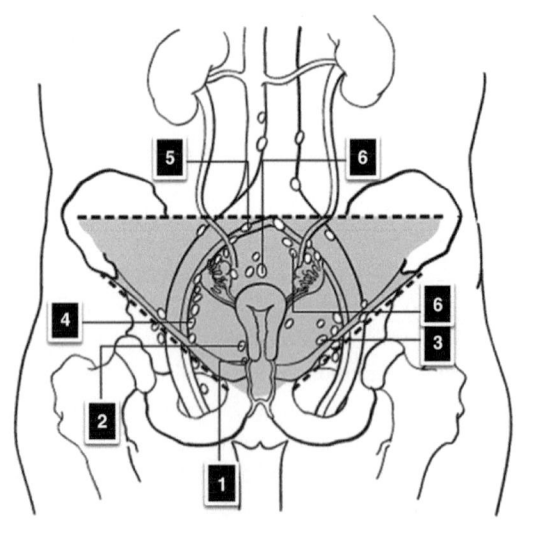

Fig. 9.3 Lymphatics of cervical cancer. (From [6], Compton C.C., Byrd D.R., Garcia-Aguilar J., Kurtzman S.H., Olawaiye A., Washington M.K. (2012) Cervix Uteri. In: Compton C., Byrd D., Garcia-Aguilar J., Kurtzman S., Olawaiye A., Washington M. (eds) AJCC Cancer Staging Atlas. Springer, New York, NY. https://doi.org/10.1007/978-1-4614-2080-4_35)

RT and surgery are equally effective in patients with stage IB cervical cancer, and survival is nearly 80–90% whether RT or surgery is used (Table 9.1) [7–10].

9.1.4.1 Treatment Algorithm for Cervix Cancer [7–10]

Preinvasive

Conization, loop electrosurgical excision procedure (LEEP), laser, cryotherapy, simple hysterectomy.

Stage IA

Total abdominal hysterectomy or cone biopsy. Alternative → Brachytherapy (BT) alone.

- (LDR → 65–75 Gy)
- (HDR → 5–6 × 7 Gy)

Stage IB1 and IB2

Radical hysterectomy + pelvic LN dissection, or External RT (ERT) + BT (ERT → 45 Gy, whole pelvis) (BT → HDR: 5 × 6 Gy, 4 × 7 Gy, LDR: 2 × 15–20 Gy).

Stages IB3 and IIA

Concurrent CRT (cisplatin-based).

- [ERT + BT]
- (ERT → 45 Gy, whole pelvis)
- (BT → HDR: 5 × 6 Gy, 4 × 7 Gy, LDR: 2 × 15–20 Gy)

Stage IIB

Concurrent CRT (cisplatin-based).

- [ERT + BT]
- (ERT → 45–50.4 Gy, whole pelvis)
- (BT → HDR: 5 × 6 Gy, 4 × 7 Gy, LDR: 2 × 15–20 Gy)

Stage IIIA

Concurrent CRT (cisplatin-based)

- [ERT + BT]
- (ERT → 45–50.4 Gy, whole pelvis)
- (BT → HDR: 5 × 6 Gy, 4 × 7 Gy, LDR: 2 × 17–20 Gy)

Stages IIIB and IVA

Concurrent CRT (cisplatin-based).

- [ERT + BT].
- (ERT → 50–54 Gy, whole pelvis ± para-aortic LN)
- (Gross LN → 10–15 Gy boost) (cumulative EQD2: 60–66 Gy)
- (BT → HDR: 5 × 6 Gy, 4 × 7 Gy, LDR: 2 × 20 Gy)

Stage IVB

Chemotherapy.

Postoperative radiotherapy indications [13]: Intermediate-risk factors (any two of these factors)

- LVSI.
- Deep cervical stromal invasion (greater than one-third of stromal depth).
- Tumor size >4 cm.

Table 9.1 Five-year survival in FIGO stage IB cervical cancer according to surgery or radiotherapy

References	Radiotherapy n	Five-year survival (%)	Radical hysterectomy n	Five-year survival (%)
Volterrani et al. (1983)	127	91	123	89
Inoue et al. (1984)	59	80	362	91
Lee et al. (1989)	–	–	237	86
Alvarez et al. (1991)	–	–	401	85
Burghardt et al. (1992)	–	–	443	83
Coia et al. (1990)	168	74	–	–
Lowrey et al. (1992)	130	81	–	–
Perez et al. (1992)	394	85	–	–
Eifel et al. (1994)	1494	81	–	–
Landoni et al. (1997)	167	74	170	74

N indicates patient number

Postoperative chemoradiotherapy indications [13]: High-risk factors (any of these factors)
- Surgical margin (+).
- LN (+).
- Parametrial involvement.

9.1.4.2 Prognostic Factors [14]

- Tumor size and volume (stage).
  Directly related to pelvic and para-aortic LN involvement
- Invasion depth.
  Related to parametrial extension
- Histology.
  RT response is better in squamous histology
  Adenocarcinomas are more fatal (controversial issue)
- LN involvement/size/number/location.
  As involvement (+) and/or LN size increases, survival decreases
- LVSI.
  Poor prognostic sign, frequent metastases
- Status of distant metastases.
- Uterine corpus extension, low hemoglobin level.

9.1.5 Radiotherapy

9.1.5.1 External Radiotherapy

External Radiotherapy Fields [15–17]
Conformal RT Volumes (Fig. 9.4).
  CTV = tumor/tumor bed, entire uterus, upper one-third of vagina, parametrium, iliac LNs.
  Para-aortic/common iliac LN (+); para-aortic LN.
  PTV = CTV + 0.5–1 cm.

- Three-dimensional conformal radiotherapy (3DCRT) or intensity modulated radiotherapy (IMRT) is recommended to optimize dose distribution while minimizing organs at risk doses [13]. However, routine daily online image guidance [i.e., cone-beam computed tomography (CT)] is necessary for IMRT.
- CT simulation is essential with standard bladder and bowel preparation.
- External RT fields are defined according to the surgical or radiological LN status:
  – Para-aortic LN (−): Pelvic RT (Fig. 9.5).

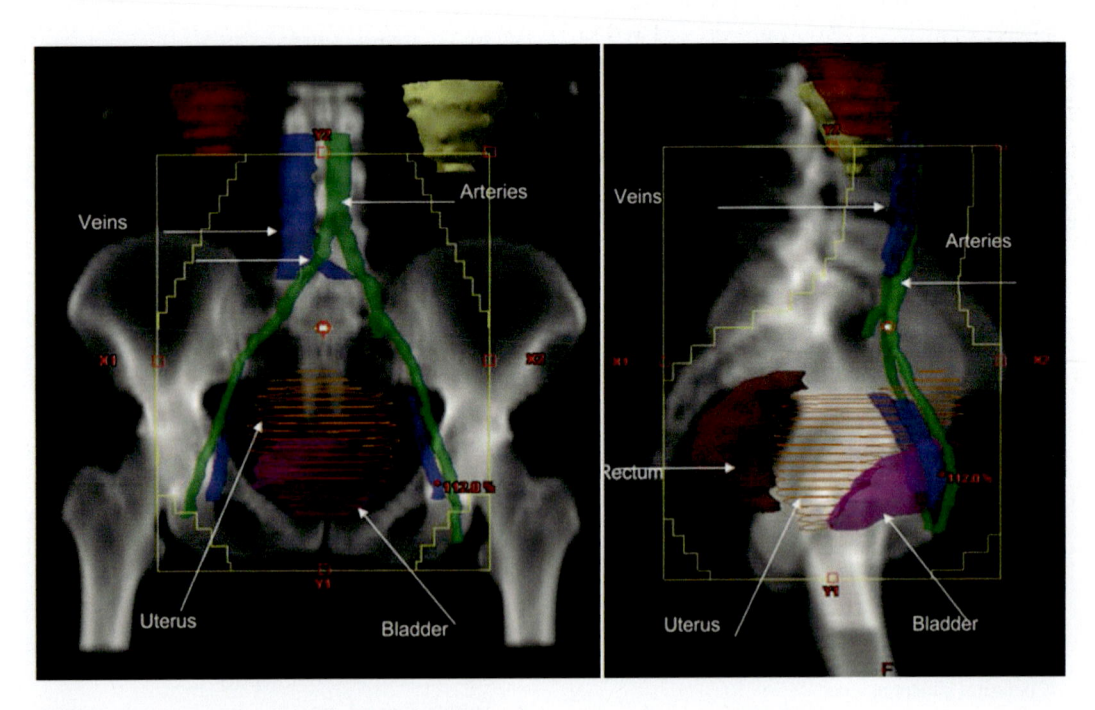

Fig. 9.4 Conformal RT volumes in cervical cancer

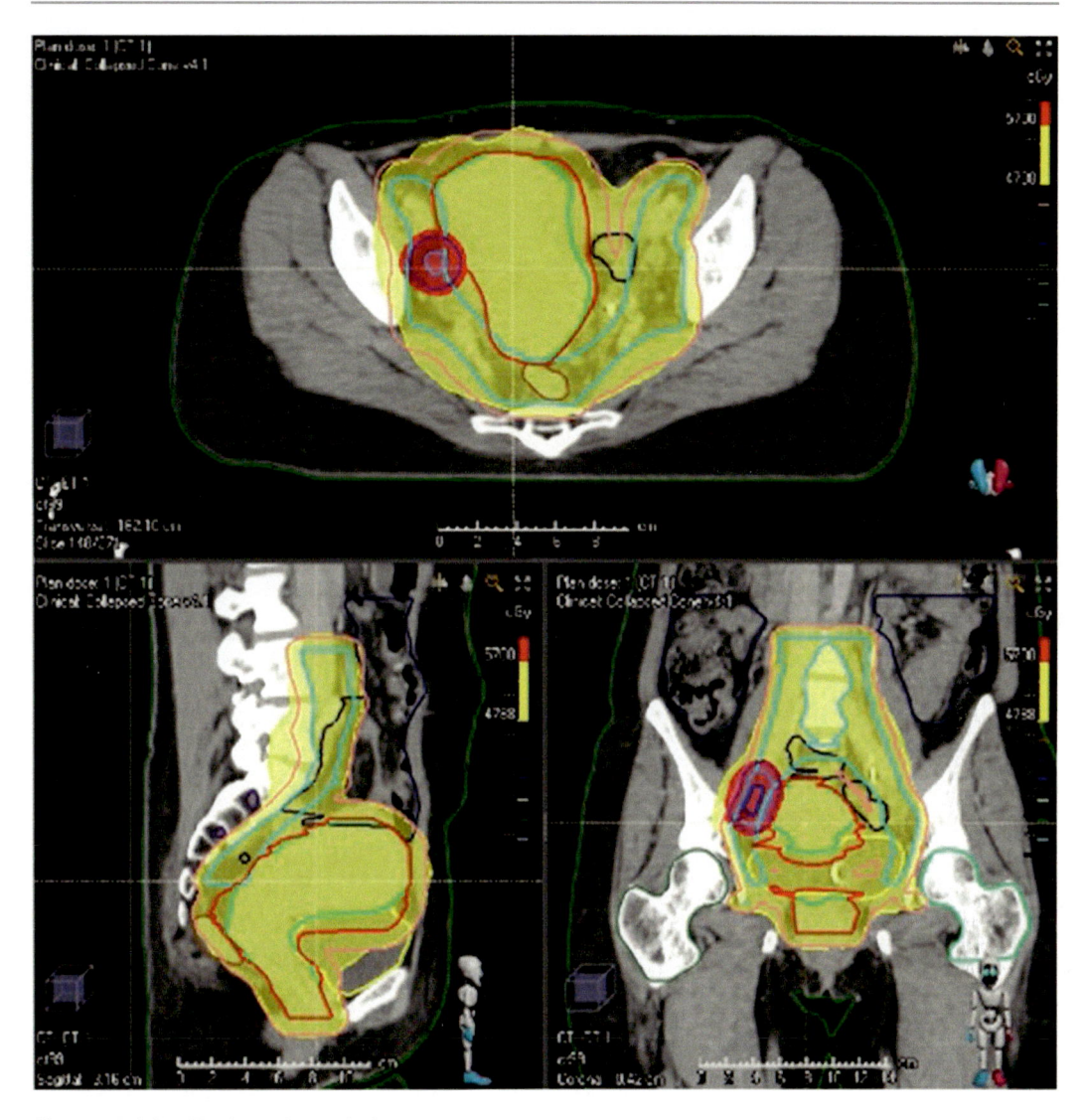

Fig. 9.5 Pelvic RT volumes in cervical cancer

- Common iliac/para-aortic LN (+): Pelvic and para-aortic radiotherapy (extended field radiotherapy = EFRT) (Fig. 9.6).
- Gross LN: + sequential or simultaneous integrated boost.
- RTOG published a consensus guideline for delineation of clinical target volume (CTV) in the postoperative pelvic RT of endometrial and cervical cancer [15, 16].
- Nodal CTV = obturator, presacral, external iliac, internal iliac, and common iliac LNs.
- Common iliac nodal CTV: Right and left common iliac artery/vein +7 mm margin → bifurcation of the common iliac artery.
- Presacral nodal CTV: Between the right and left common iliac nodal CTV, anterior to the S1–S2, 10–15 mm in diameter.
- External iliac nodal CTV: Bifurcation of the common iliac vessels → deep circumflex artery branches or beginning of the inguinofemoral vessels, + 7 mm margin (anterior: 10 mm).

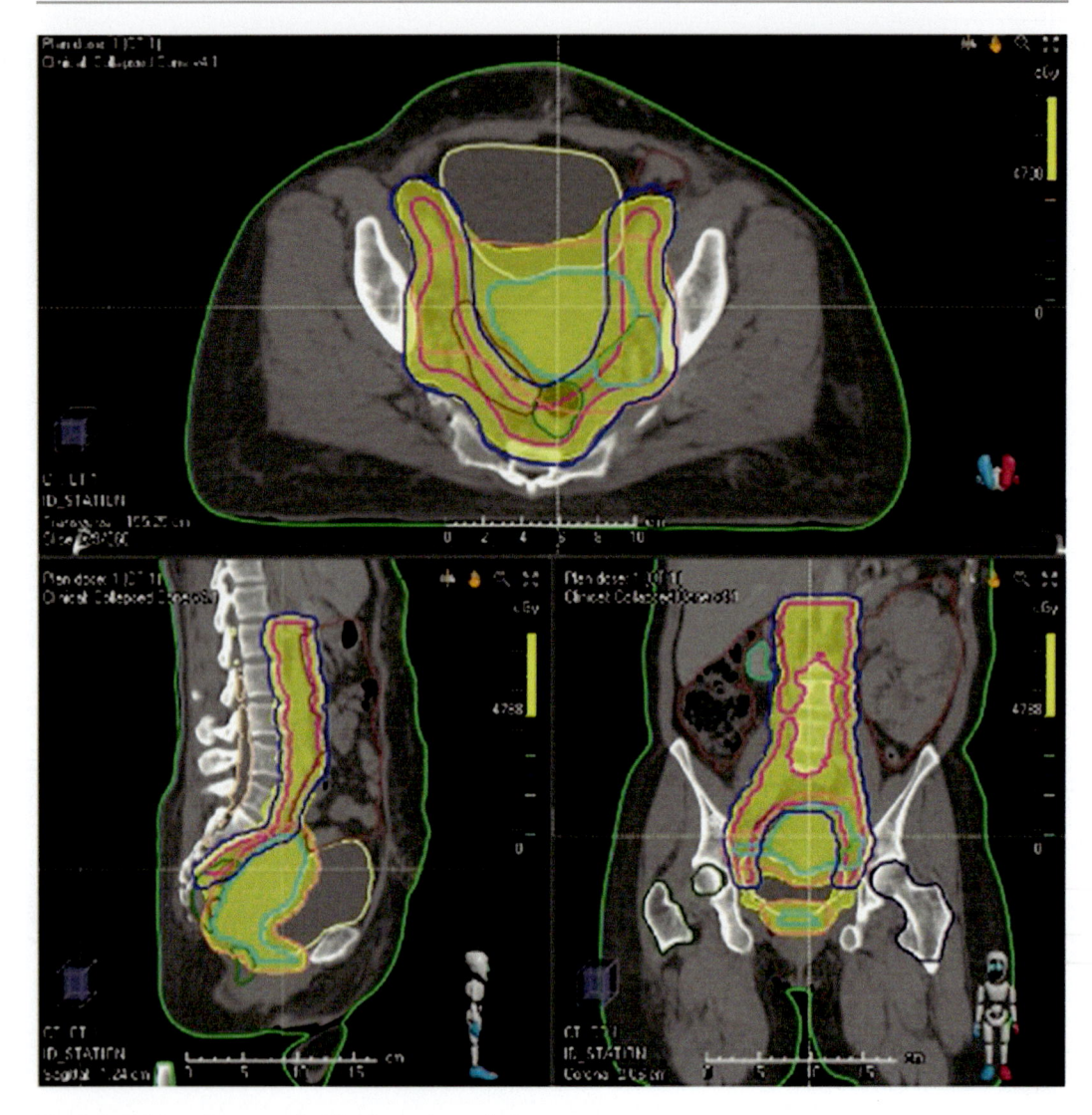

Fig. 9.6 Pelvic and para-aortic RT volumes in cervical cancer

- Internal iliac nodal CTV: Bifurcation of the common iliac vessels → vessels turn laterally prior to leaving the pelvis, + 7 mm margin.
- Obturator nodal CTV: Between the external and internal iliac nodal CTV, 15–18 mm in diameter, bifurcation of the internal and external iliac vessels → where the obturator vessels leave the pelvis through the obturator foramen.
- Para-aortic CTV = Paracaval, precaval, retrocaval, deep and superficial intercavoaortic, para-aortic, pre-aortic, and retro-aortic LNs.
  Superior: Left renal vein.
  Left lateral border: + (1–2 cm) lateral to the aorta.
  On the right: + (3–5 mm) around the vena cava inferior.
- All gross or suspicious LNs, lymphoceles, and surgical clips should be included in nodal CTVs.
- Nodal PTV = nodal CTV + 7 mm circumferential margin.

- Vaginal CTV = upper 1/3 vagina, parametrial and paravaginal tissues.
  - Includes the gross disease (if present), proximal vagina and parametrial/paravaginal tissues.
    Inferior: 3 cm below the upper border of the vagina/vaginal marker or 1 cm above the inferior extent of the obturator foramen.
    Anterior: Posterior bladder wall.
    Posterior: Anterior 1/3 mesorectum.
    Lateral: Medial border of the urogenital diaphragm.
  - Integrated target volume (ITV) = full bladder + empty bladder vaginal/obturator CTVs.
  - Vaginal PTV = vaginal ITV + 7 mm.
  - The rectum, bladder, bone, and muscles should be excluded from all CTVs.
  - Definitive treatment:
    Primary CTV = GTV, cervix, uterus, upper 1/2 vagina (upper 2/3 vagina if upper vagina involved or entire vagina if extensive vaginal involvement), bilateral parametria (including bilateral ovaries).
    Nodal CTV = common iliac, external iliac, internal iliac, obturator, and presacral nodes.
    Common iliac/para-aortic LN (+): Pericaval, interaortacaval, and para-aortic LNs.
    Distal third of the vaginal involvement: + bilateral inguinofemoral LNs.
    ITV is created as in postoperative cases.
    PTV = nodal CTV + 7 mm, primary CTV + (1.5–2 cm).
  - Small bowel, rectum, bladder, and bilateral femoral heads are defined as organs at risk (OARs) in all cases. Bilateral kidneys and spinal cord should also be contoured in patients treated with EFRT.

Prophylactic Para-Aortic Irradiation
- Prophylactic para-aortic field historically is not recommended since it has no effect on survival. It also decreases the tolerance of RT by increasing toxicity [18–21].
- However, it gains popularity again in recent years because para-aortic recurrence is the most important failure pattern after distant metastases in the image-guided BT era.
- The most controversial issue is cases with pelvic LN metastasis and no para-aortic LN metastasis on imaging.
- Today, prophylactic para-aortic RT (up to the left renal vein level) is recommended for patients with common iliac LN metastasis or ≥3 pelvic LN metastases.

9.1.5.2 Brachytherapy
BT plays a very important role in the treatment of gynecological cancers, particularly in cervical cancers [22]. Intracavitary, interstitial, or hybrid (intracavitary and interstitial) BT techniques can be used in cervical cancers. Today, three-dimensional (3D) image-guided adaptive BT (3D-IGBT) is the new standard.

Radioactive Sources Used in Brachytherapy
Sources used in BT usually emit gamma rays. Ra^{226} was commonly used until the 1960s in the form of tubes or needles. However, it was replaced with artificial radioactive sources due to protection problems. These new sources are also small and easily bendable for comfortable BT application. Several radioisotopes are currently used in BT (Table 9.2).

Source Loading Types in Brachytherapy
Manual afterloading: Radioactive sources are applied by long forceps from a certain distance behind a protective barrier. The patient treated with the radioactive sources should be isolated in hospital.

Remote afterloading: The radioactive source is stored within a shielded computer-controlled machine. The machine has a remote controller, and the radioactive source passes through special tubes and applicators placed inside the patient. Thus, such machines are called remote afterloading BT machines.

Two-Dimensional Point-Based Brachytherapy
ICRU 38 [23].
Since techniques, applicators, sources, and dose reference points of intracavitary applications may vary for gynecological malignancies

Table 9.2 Radionuclides used for brachytherapy applications

Radionuclide	Half-life	Radiation type(s)	Energy
Ra-226	1600 years	α, β, γ	0.78 MeV
Co-60	5.27 years	β, γ	1.25 MeV
Sr-90	28.7 years	β, γ	0.546 β en
Cs-137	30 years	β, γ	0.662 MeV
Au-198	64.7 h	β, γ	0.42 MeV
Ir-192	73.8 days	γ	0.38 MeV
I-125	60.1 days	α, γ, n	0.028 MeV
Y-90	64 h	α, β, γ	2.27 β en
Am-241	432 years	γ	0.06 MeV
Pd-103	17 days	α, β, γ	0.021 MeV
Sm-145	340 days	γ	0.041 MeV
Yb-169	32 days	γ	0.093 MeV

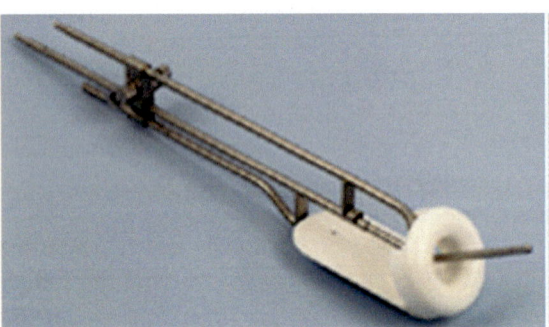

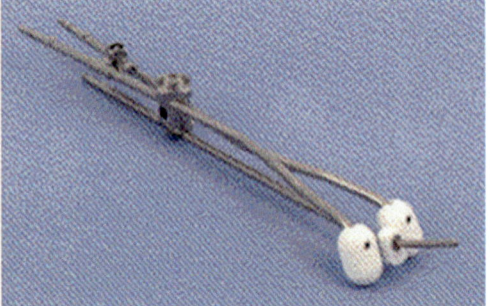

Fig. 9.7 Ring–tandem and ovoid–tandem applicators

between institutions, details should be given according to the historical ICRU-38 report (*Dose and Volume Specifications for Reporting Intracavitary Treatments in Gynecologic Malignancies*) published in 1985. This report standardizes application differences and provides a common language for gynecologic BT applications.

Every application of BT should report five major issues according to this publication: technique, total reference air KERMA (TRAK), reference volume, reference points and doses, and dose rate.

1. Technique

    The radioactive source, number and length of sources, and the type of applicator should be mentioned. Applicators are special devices that are placed into the organs to be treated, and the radioactive sources enter these applicators and exit out of them. These come in various shapes and types: metallic (Fletcher–Suit–Delclos); plastic (Delouche); individualized (Chassagne–Pierquin) (Fig. 9.7).

2. Reference air KERMA of source

    Kinetic energy released in the medium (KERMA): This is the combination of the initial kinetic energies of the charged ionizing particles that are liberated by an uncharged ionizing particle per unit mass of material. KERMA is measured in the same units as absorbed dose (Gy). The reference air KERMA is used to define visible activity. It is defined as the dose given at a distance of 1 m of air by a source with an activity of 1 MBq in 1 h. Its units are 1 μ/Gy m^2 = 1 cGy/h cm^2. Its value is 0.0342 for Ir-192.

3. Reference volume (Fig. 9.8)

    This is the volume surrounded by the reference isodose and is independent of technique. It is the combined volume in the 60 Gy isodose curve of external pelvic RT and intracavitary applications. The reference volume is

9.1 Cervical Cancer

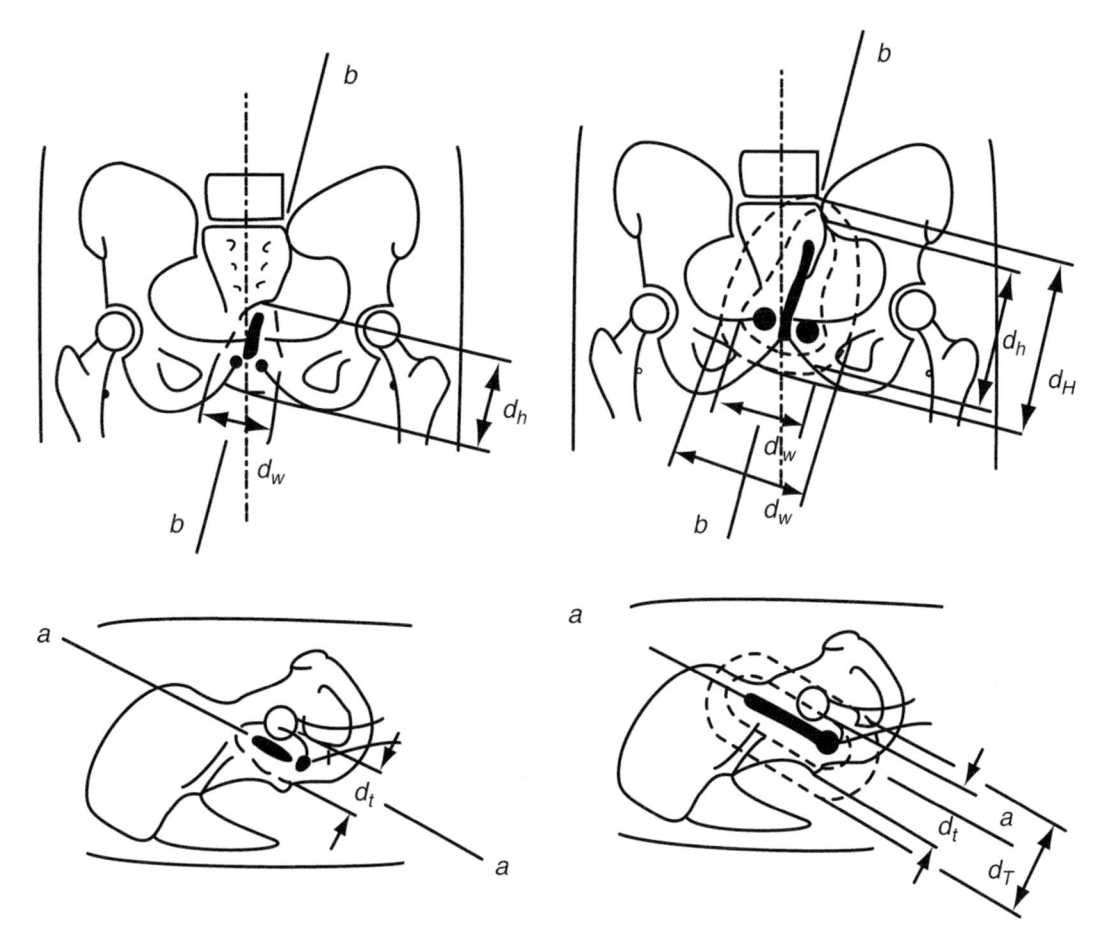

Fig. 9.8 Planes defining reference volumes

defined in three planes by combining reference isodoses:
- dh: (height) maximum size parallel to the intrauterine applicator in the oblique frontal plane, including the intrauterine applicator,
- dw: (width) maximum size vertical to the intrauterine applicator in the oblique frontal plane, including the intrauterine applicator,
- dt: (thickness) maximum size vertical to the intrauterine applicator in the oblique sagittal plane, including the intrauterine applicator.

4. Reference points (Fig. 9.9)
   Bladder reference point. A Foley catheter is placed into the bladder. The Foley balloon is filled with 7 cm³ of radio-opaque material.

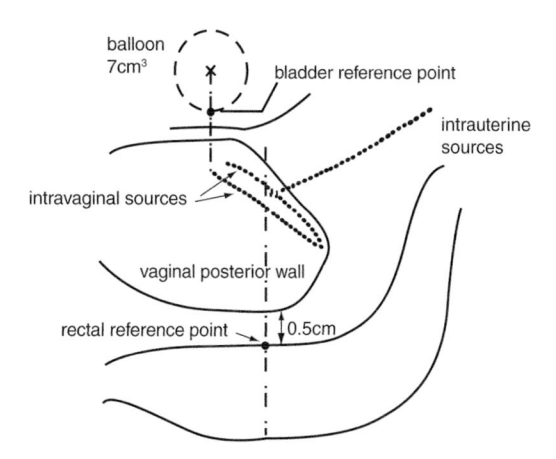

Fig. 9.9 Bladder and rectal reference points according to the ICRU-38 report

The catheter is pulled back and stabilized in the bladder neck. A straight line is marked in the anterior–posterior plane from the center of the balloon in a lateral graph. The reference point is the posterior point crossing the back of the balloon on this line. This is the center of the balloon in the anterior–posterior film.

Rectum reference point. An anterior–posterior straight line is marked from the bottom tip of the intrauterine tandem or the middle of the intravaginal ovoids. The point 5 mm back from the vaginal posterior wall is the rectum reference point.

Bony reference points; lymphatic trapezoid (Fletcher trapezoid) (Fig. 9.10). A line is marked 2 cm bilaterally from the middle of the L4 vertebra → points 1 and 2 are found. A line vertical from the middle of the S1 to S2 vertebrae to the top of the pubis symphysis is drawn, and two lines bilaterally from the middle of that line are drawn with a length of 6 cm → points 3 and 4 are found. Point 1 is combined with 3, and point 2 with 4. A trapezoid is formed by the combination of these lines; upper points → lower para-aortic lymphatics; lower points → external iliac lymphatics; middle points → common iliac lymphatics.

Pelvic wall reference points. These are used to calculate the doses for the distal parametrium and obturator lymphatics (Fig. 9.11). Two tangential lines are drawn from the upper and the most medial parts of the acetabulum vertical to each other, and the points where they cross are the pelvic wall reference points (Chassagne point) in the anterior–posterior film. The most upper points of right and left acetabulum are marked, and the lateral projection of the pelvic wall occurs midway between these two points.

5. Dose rate

This is the dose given per unit time. BT applications are categorized into three subgroups according to dose rate (Table 9.3): low dose

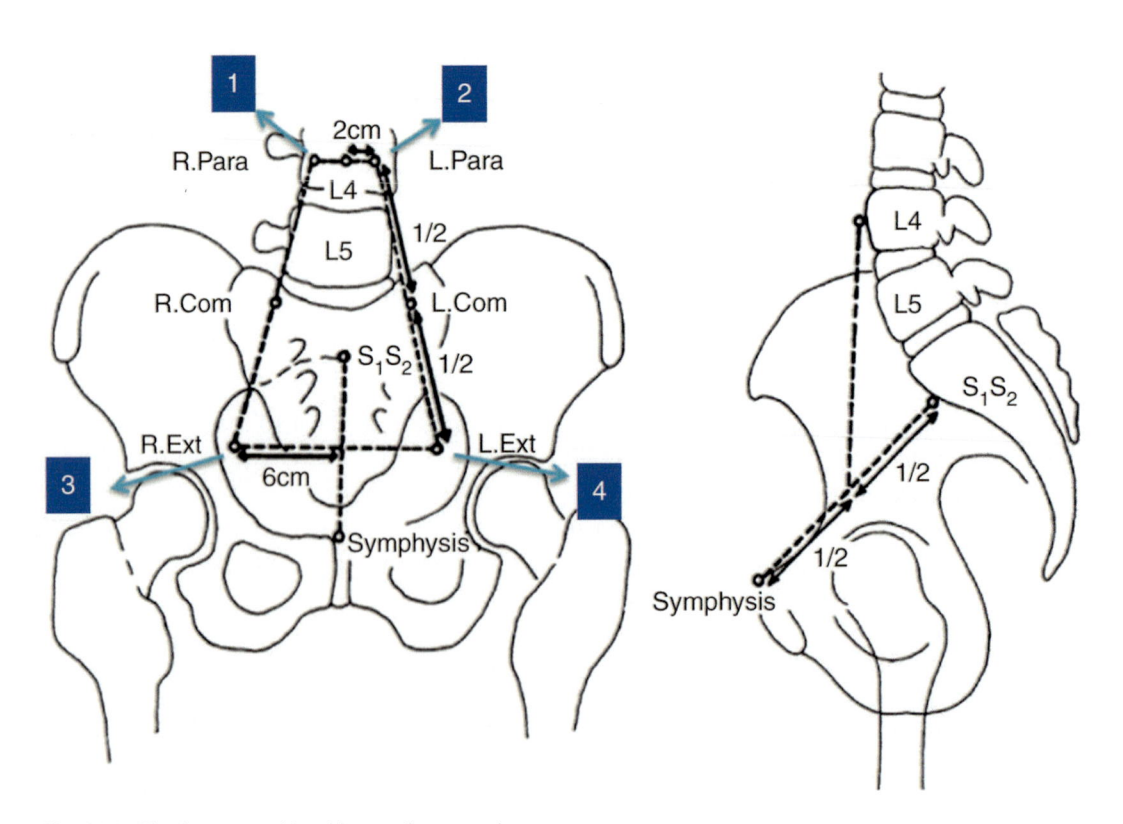

Fig. 9.10 Fletcher trapezoid and bony reference points

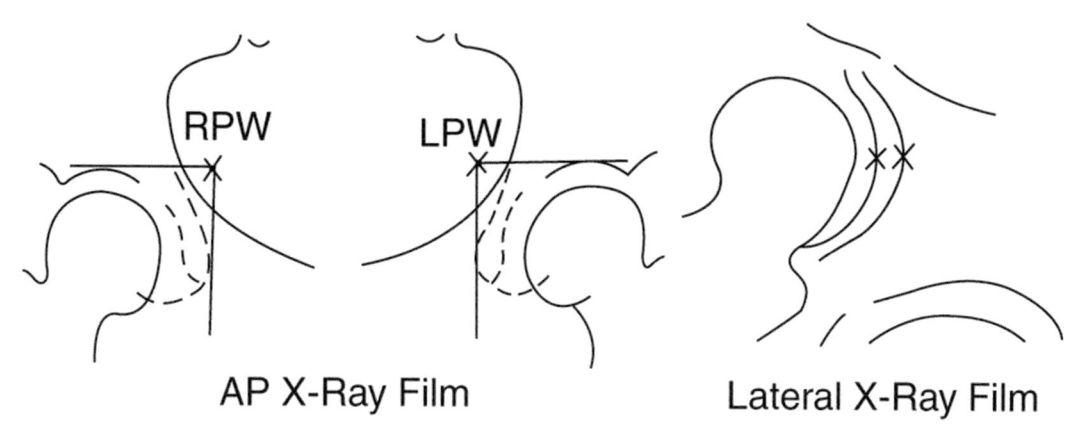

Fig. 9.11 Pelvic wall reference points

Table 9.3 The advantages and disadvantages of HDR brachytherapy

Advantages	Disadvantages
Better protection from radiation	Potential risks due to its radiobiological features
No requirement for hospitalization	Less experience compared to LDR
Short treatment time	Cost-effectiveness
Patient comfort	Expensive equipment
Low thromboembolism risk	Requires a special room
Less applicator movement	No possibility of correcting position during treatment
Dose optimization	Serious risk in case of machine failure

rate (LDR), 0.4–2 Gy/h; medium dose rate (MDR), 2–12 Gy/h; high dose rate (HDR), >12 Gy/h. Pulsed dose rate (PDR) BT is a new concept using modern afterloading technology and combines physical advantages of HDR BT with radiobiological advantages of LDR BT.

Dose Prescription Points (Fig. 9.12)

Dose prescription points are used for dose definitions. The most commonly used ones are points A and B.

Point A. This point is defined in the Manchester system as the point 2 cm superior to a horizontal line passing from the top of the lateral fornices at the midline, and 2 cm lateral from the midline. It is the point where radiation necrosis is first seen, and uterine artery crosses ureter. Point A relates to the maximum dose for healthy tissues and the minimum dose for the tumor.

Point B. This point is 3 cm lateral to point A (2 cm superior to the lateral fornices, and from that point 5 cm lateral from the midline), and it shows the dose taken by the obturator LNs.

Tolerance doses [22]. Rectum, 75 Gy (mean 68 Gy, <80% of the dose at point A); bladder 80 Gy (mean 70 Gy, 85% of the dose at point A); vagina 120–140 Gy (mean 125 Gy). The vaginal mucosa is a critical organ in cervical cancer RT, and 140% of the dose at point A should not be exceeded.

Point H [24]. This is recommended as an alternative to point A by the American Brachytherapy Society. The tips of the ovoids are connected with a horizontal line, and 2 cm superior and 2 cm lateral from the midpoint of this line are the H points on the right and left sides. They are practically 2 cm cranial to the top of the ovoids. They are not anatomical points like point A, and they may vary in position according to patient and application.

Point P. This is the most lateral point in the bony pelvis, and it defines the dose taken by external iliac LNs.

Three-Dimensional Volume-Based Brachytherapy [25–27]

- According to the recommendations of "The Groupe Européen de Curiethérapie and the European Society for Radiotherapy & Oncology (GEC-ESTRO)" working group

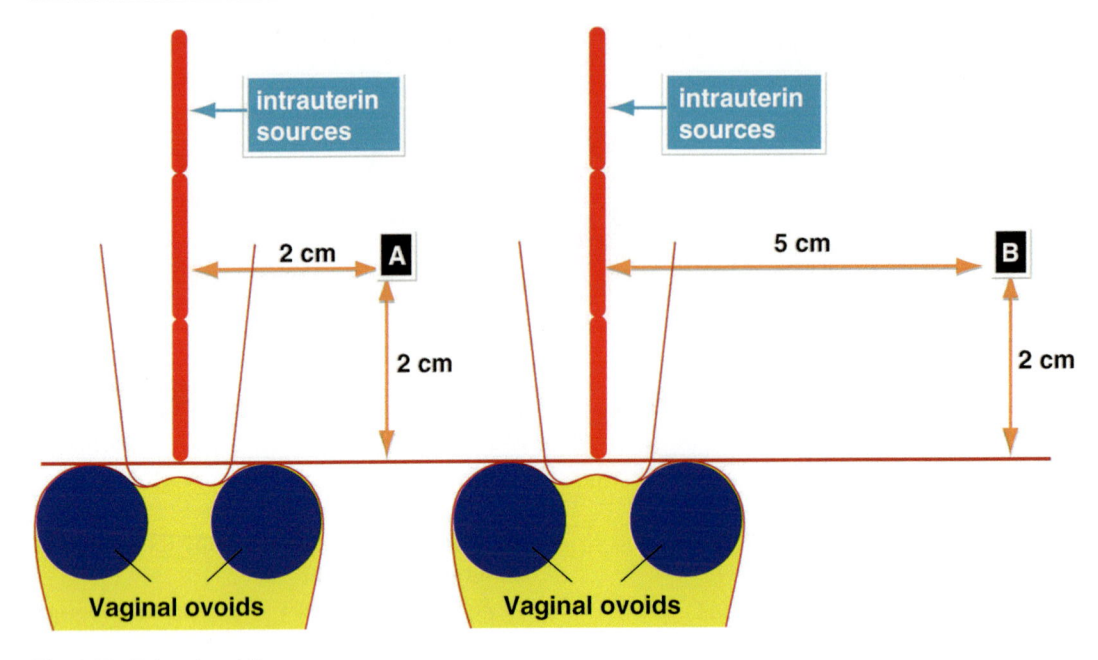

Fig. 9.12 Points A and B

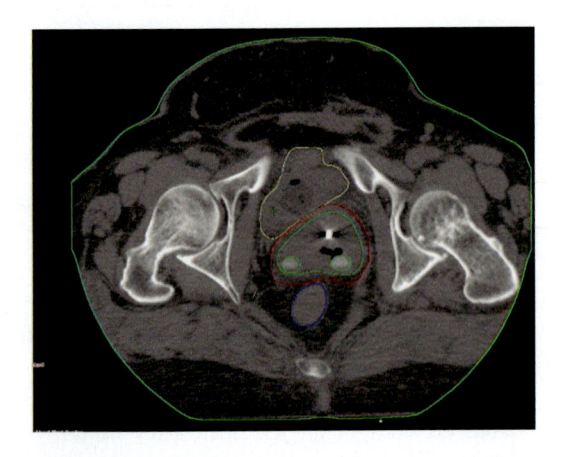

Fig. 9.13 Targets and organs at risk volumes in three-dimensional CT-guided brachytherapy

and International Commission on Radiation Units and Measurements (ICRU)-89, MRI is the gold standard imaging modality for 3D-IGBT for cervical cancer.
- CT-based BT can also be used by the guidance of gynecological examination and MRI immediately before BT.
- Accurate definition of both target volumes and organs at risk are critical (Fig. 9.13).
    - GTV = macroscopic tumor (if present) at time of BT + high signal intensity mass(es) (T2-weighted MRI) in cervix/corpus, parametria, vagina, bladder and rectum.
    - High-risk CTV (HRCTV) = GTV + cervix + gray zones in parametria, uterine corpus, vagina, or rectum, and bladder (T2-weighted MRI).
    - Intermediate-risk CTV (IRCTV) = HRCTV + (5–15 mm) (microscopic tumor at the time of BT including initial tumor load).
    - Organs at risk: Bladder, rectum, sigmoid, and vagina.
- Vaginal mucosal margin close/(+) after hysterectomy: external RT + vaginal BT.
    - 2 × 5.5 Gy at 5 mm
    - 3 × 6 Gy at vaginal surface
- According to the ICRU 89, the following parameters should be recorded in each fraction:
    - D98 and D90 for GTV.
    - D98 and D90 for HRCTV.
    - D98 for IRCTV.
    - D2cc of bladder, rectum, sigmoid, and rectovaginal point.
    - Dose to point A.
- Cumulative doses of external RT and BT should be calculated using the linear-quadratic

model [biologically equivalent dose in 2 Gy per fraction (EQD2)].
- EQD2 = D × [(d + α/β)/(2 + α/β).
- α/β = 10 Gy for tumor,
- α/β = 3 Gy for late responding tissue,
- GEC ESTRO dose recommendations for target and OARs (cumulative doses):
  - D98 GTV EQD2: >95 Gy.
  - D90 HRCTV EQD2: >90 Gy, <95 Gy.
  - D98 HRCTV EQD2: >75 Gy.
  - D98 IRCTV EQD2: >60 Gy.
  - Point A EQD2: >65 Gy.
  - D2cc rectum ˂65–75 Gy.
  - D2cc sigmoid ˂70–75 Gy.
  - D2cc bladder ˂80–90 Gy.
  - Rectovaginal point EQD2 ˂65–75 Gy.

The Application of Intracavitary Brachytherapy

The patient should be informed about the procedure before BT, and informed consent must be obtained. A detailed gynecological exam should be performed, and the dimensions of the vagina, the size and position of the uterus, and the localization, size, and extension of the tumor determined. The application of BT requires cervical dilatation, and thus general anesthesia, spinal anesthesia, paracervical blockage, or systemic analgesia with sedation are needed. Treatment is performed on an outpatient basis; the patient should have an open IV line and an empty rectum under sterile surgical conditions.

The patient is positioned in the lithotomy position on a gynecological table. The vulva, perineum, and pelvic region are cleaned and a Foley catheter is placed into the bladder. Its balloon is filled with 7 cc of radio-opaque material. A speculum is placed into the vagina, and the cervical os is visualized. The cervical canal is dilated sequentially with bougies of different thicknesses. The length of the uterine cavity is determined by hysterometry, and the tandem and ovoids or ring are lubricated with 1% viscous lidocaine. The tandem is first placed into the uterine cavity, and then the ovoids or ring are placed into the fornices and they are all stabilized. The angle between the longitudinal axis of the tandem and the diameter of the ring is always 90°.

Then, rectal retractor or rectal packing with a radio-opaque gauze is placed between the rectal wall and the applicators, and the same packing is placed between the bladder and the applicators. For conventional BT, dummy catheters are placed into the tandem and ring or ovoids for dose calculations in orthogonal films. An anterior–posterior and a lateral film are taken at the same position, which shows all of the applicators and the bony points of the Fletcher trapezoid. The dose is historically prescribed to point A. The dose distribution should be pear-shaped when the tandem is used. Its large side is located in the upper vagina, and its narrow edge is in the uterine fundus (Fig. 9.14). For 3D BT, CT or MRI imaging is obtained with the applicator in place. Then, target volumes and OARs are contoured and, finally, applicator reconstruction and treatment planning is done (Fig. 9.15). The source, place, and time are optimized in dosimetry; the plan that enables the maximum dose to be delivered to the target and the minimum dose to the organs at risk is selected, and the patient is taken to the treatment room. Applicators are connected to the treatment machine with special connecting cables. Health personnel move from the treatment room to the command room and start the BT session. One session takes between 5 and 15 min, depending

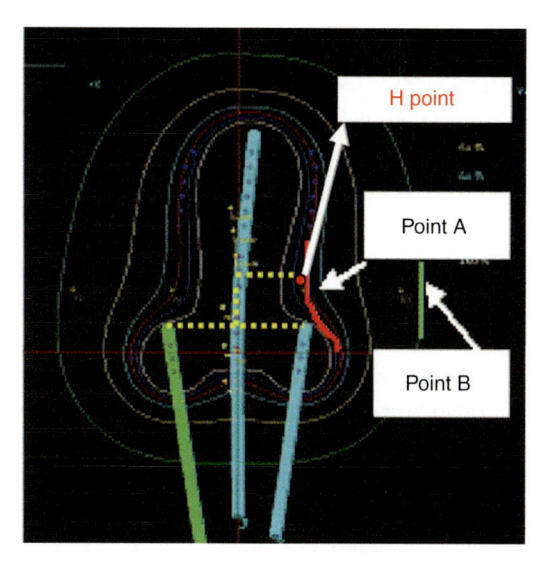

Fig. 9.14 Isodose distribution in point-based cervical brachytherapy

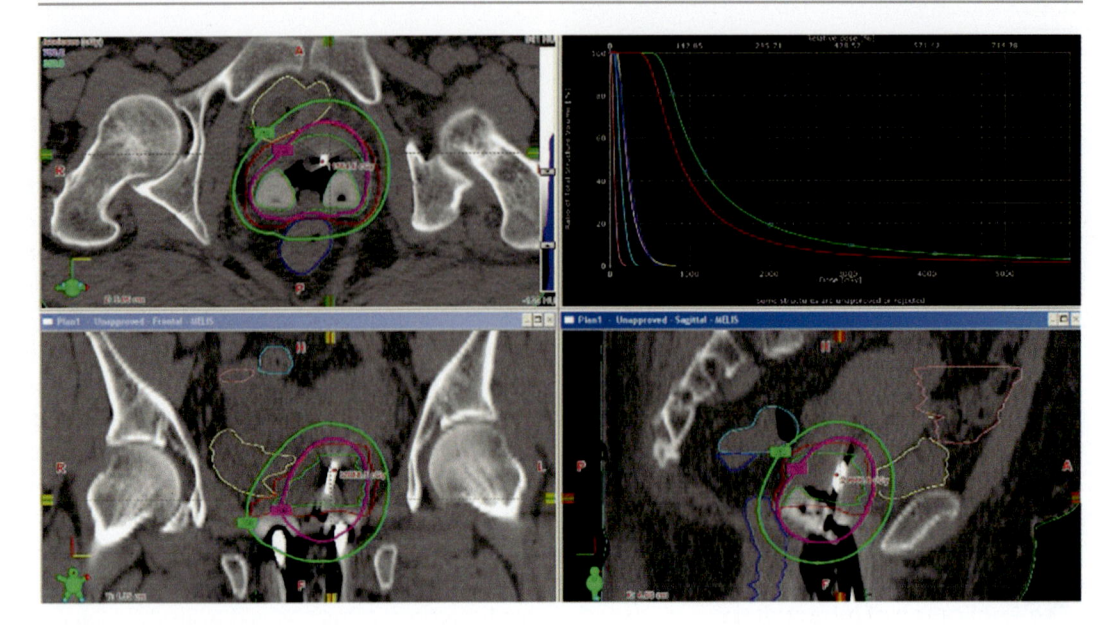

Fig. 9.15 Isodose distribution in image-guided cervical brachytherapy

on the source activity in that session of HDR BT. After the treatment has finished, the applicators are taken out. The same procedure is repeated during each session.

9.1.6 Selected Publications

Canada, 2002 → A systematic review of eight randomized trials of cisplatin administered concurrently with external-beam RT vs. RT without cisplatin for cervical cancer.

- A statistically significant effect in favor of cisplatin-based chemotherapy plus RT compared with RT without cisplatin [relative risk (RR) of death, 0.74].
- The pooled RR of death among the six trials that enrolled only women with locally advanced cervical cancer was 0.78. The pooled RR for the two trials in high-risk early-stage disease also demonstrated a statistically significant benefit of the addition of cisplatin-based chemotherapy to RT (RR = 0.56).
- Concurrent cisplatin-based chemotherapy plus RT improves overall survival over various controls in women with locally advanced cervical cancer, large stage IB tumors (prior to surgery), and high-risk early-stage disease (following surgery).

Lukka H et al. (2002) Concurrent cisplatin-based chemotherapy plus radiotherapy for cervical cancer: a meta-analysis. Clin Oncol (R Coll Radiol) 14(3):203–212.

England, 2001 → 17 published and two unpublished trials of CRT for cervical cancer.

- The absolute benefits in progression-free (PFS) and overall survival (OS) were 16 and 12%, respectively. Significant benefits of CRT on both local (odds ratio 0.61, $p < 0.0001$) and distant recurrence (0.57, $p < 0.0001$) were also recorded.
- Grade 3 or 4 hematological (odds ratio 1.49–8.60) and gastrointestinal (2.22) toxicities were significantly greater in the concomitant CRT group than the control group.
- Concomitant chemotherapy and RT improves OS and PFS and reduces local and distant recurrence in selected patients with cervical cancer, which may give a cytotoxic and sensitization effect.

Green JA et al. (2001) Survival and recurrence after concomitant chemotherapy and radiotherapy for cancer of the uterine cervix: a systematic review and meta-analysis. Lancet 358(9284):781–786.

Milan, 1997 → 343 patients with stage IB and IIA cervical carcinoma were randomized to surgery and to radical RT. Adjuvant RT (63%) was delivered after surgery for women with surgical stage pT2b or greater, less than 3 mm of safe cervical stroma, cut-through, or positive nodes.

- After a median follow-up of 87 (range 57–120) months, 5-year OS and disease-free survival (DFS) rates were identical in the surgery and RT groups (83% and 74%, respectively, for both groups).
- Significant factors for survival in univariate and multivariate analyses were cervical diameter, positive lymphangiography, and adenocarcinomatous histotype. Also, 48 (28%) surgery group patients had severe morbidity compared with 19 (12%) RT group patients ($p = 0.0004$).
- The combination of surgery and RT has the worst morbidity, especially in relation to urological complications. The optimum therapy for each patient should take into account clinical factors such as menopausal status, age, medical illness, histological type, and cervical diameter to yield the best cure with minimum complications.

Landoni F et al. (1997) Randomised study of radical surgery versus radiotherapy for stage Ib–IIa cervical cancer. Lancet 350(9077):535–540.

GOG 92, 2006 → 277 patients with stage IB cervical cancer with negative LNs but with two or more of the following features: more than one-third (deep) stromal invasion, capillary lymphatic space involvement, and tumor diameter of 4 cm or more. In total, 137 randomized to pelvic irradiation (RT: 46 Gy in 23 fractions to 50.4 Gy in 28 fractions) and 140 randomized to observation (OBS).

- The RT arm showed a statistically significant (46%) reduction in risk of recurrence [hazard ratio (HR) = 0.54, $p = 0.007$) and a statistically significant reduction in risk of progression or death (HR = 0.58, $p = 0.009$).
- The improvement in OS with RT did not reach statistical significance ($p = 0.074$).
- Pelvic RT after radical surgery significantly reduces the risk of recurrence and prolongs PFS in women with stage IB cervical cancer.

Sedlis A et al. (1999) A randomized trial of pelvic radiation therapy versus no further therapy in selected patients with stage IB carcinoma of the cervix after radical hysterectomy and pelvic lymphadenectomy: A Gynecologic Oncology Group Study. Gynecol Oncol 73(2):177–183.

Rotman M et al. (2006) A phase III randomized trial of postoperative pelvic irradiation in stage IB cervical carcinoma with poor prognostic features: follow-up of a gynecologic oncology group study. Int J Radiat Oncol Biol Phys 65(1):169–176.

GOG 109/Intergroup 0107/SWOG 8797/ RTOG 9112, 2000 → 268 patients with clinical stage IA2, IB, and IIA carcinoma of the cervix, initially treated with radical hysterectomy and pelvic lymphadenectomy, and who had positive pelvic LNs and/or positive margins and/or microscopic involvement of the parametrium, were randomized to receive whole pelvic RT (WPRT) or WPRT + hemotherapy.

- The addition of concurrent cisplatin-based chemotherapy to RT significantly improves PFS (80% vs. 63%, $p = 0.003$) and OS (81% vs. 71%, $p = 0.007$) for high-risk, early-stage patients who undergo radical hysterectomy and pelvic lymphadenectomy for carcinoma of the cervix.

Peters WA III et al. (2005) Concurrent chemotherapy and pelvic radiation therapy compared with pelvic radiation therapy alone as adjuvant therapy after radical surgery in high-risk early-stage cancer of the cervix. J Clin Oncol 18(8):1606–1613.

Monk BJ et al. (2005) Rethinking the use of radiation and chemotherapy after radical hysterectomy: a clinical-pathologic analysis of a Gynecologic Oncology Group/Southwest Oncology Group/Radiation Therapy Oncology Group trial. Gynecol Oncol 96(3):721–728.

GOG 71/RTOG 84–12, 2003 → 256 patients with exophytic or "barrel"-shaped tumors measuring ≥4 cm were randomized to either external RT and intracavitary irradiation or attenuated irradiation followed by extrafascial hysterectomy.

- There was no clinically important benefit of the use of extrafascial hysterectomy. However, there is good evidence to suggest that patients with 4, 5, and 6 cm tumors may have benefited from extrafascial hysterectomy.
- No difference in OS, but trend for higher local relapse without surgery.

Keys HM et al. (2003) Radiation therapy with and without extrafascial hysterectomy for bulky stage IB cervical carcinoma: a randomized trial of the Gynecologic Oncology Group. Gynecol Oncol 89(3):343–353.

GOG 123, 2007 → Women with bulky stage IB cervical cancers (tumor, ≥4 cm in diameter) were randomly assigned to receive RT alone or in combination with cisplatin (40 mg/m^2 of body surface area once a week for up to six doses, maximal weekly dose, 70 mg), followed in all patients by adjuvant hysterectomy. The cumulative dose of external pelvic and intracavitary radiation was 75 Gy to point A (cervical parametrium) and 55 Gy to point B (pelvic wall). Cisplatin was given during external RT, and adjuvant hysterectomy was performed 3–6 weeks later.

- Adding weekly infusions of cisplatin to pelvic RT followed by hysterectomy significantly reduced the risk of disease recurrence and death in women with bulky stage IB cervical cancers.
- Five-year PFS (71% vs. 60%) and OS (78% vs. 64%), without increasing serious late adverse effects.

Keys HM et al. (1999) Cisplatin, radiation, and adjuvant hysterectomy compared with radiation and adjuvant hysterectomy for bulky stage IB cervical carcinoma. N Engl J Med 340(15):1154–1161.

Stehman FB et al. (2007) Radiation therapy with or without weekly cisplatin for bulky stage 1B cervical carcinoma: follow-up of a Gynecologic Oncology Group trial. Am J Obstet Gynecol 197(5):503. e1–6.

GOG 141, 2007 → Patients with bulky FIGO stage IB cervical cancer, tumor diameter ≥4 cm. Prospective random allocation was to either neoadjuvant chemotherapy (NACT) (vincristine–cisplatin chemotherapy every 10 days for three cycles) before exploratory laparotomy and planned radical hysterectomy and pelvic/para-aortic lymphadenectomy (RHPPL), or RHPPL only.

- There is no evidence from this trial that NACT offered any additional objective benefit to patients undergoing RHPPL for suboptimal stage IB cervical cancer.

Eddy GL (2007) Treatment of ("bulky") stage IB cervical cancer with or without neoadjuvant vincristine and cisplatin prior to radical hysterectomy and pelvic/para-aortic lymphadenectomy: a phase III trial of the gynecologic oncology group. Gynecol Oncol 106(2):362–369.

NCI Canada, 2002 → 253 patients with stage IB to IVA squamous cell cervical cancer with central disease ≥5 cm or histologically confirmed pelvic LN involvement were randomized to receive RT (external-beam RT plus BT) plus weekly CDDP chemotherapy (40 mg/m^2) (arm 1) or the same RT without chemotherapy (arm 2).

- This study did not show a benefit to either pelvic control or survival upon the addition of concurrent weekly CDDP chemotherapy at a dose of 40 mg/m^2 to radical RT.

Pearcey R (2002) Phase III trial comparing radical radiotherapy with and without cisplatin chemotherapy in patients with advanced squamous cell cancer of the cervix. J Clin Oncol 20(4):966–972.

RTOG 90–01, 2004 → 403 women with stage IIB to IVA disease, stage IB to IIA disease with a tumor diameter ≥5 cm, or positive pelvic LNs were randomly assigned to receive either EFRT and BT or WPRT with concomitant cisplatin/5-FU and BT.

- The addition of fluorouracil and cisplatin to RT significantly improved the survival rate of women with locally advanced cervical cancer without increasing the rate of late treatment-related side effects [8-year OS (67% vs. 41%, $p < 0.0001$), DFS (61% vs. 36%, $p < 0.0001$)].

Morris M et al. (1999) Pelvic radiation with concurrent chemotherapy compared with pelvic and para-aortic radiation for high-risk cervical cancer. The New England journal of medicine 340(15):1137–1143.

Eifel PJ et al. (2004) Pelvic irradiation with concurrent chemotherapy versus pelvic and para-aortic irradiation for high-risk cervical cancer: an update of radiation therapy oncology group trial (RTOG) 90–01. J Clin Oncol 22(5):872–880.

EORTC, 1998 → 441 patients with stage I and IIB with proximal vaginal and/or parametrial involvement with positive pelvic LNs either on lymphangiogram or at surgery, stage IIB with distal vaginal and/or parametrial involvement, and III regardless of pelvic node status on lymphangiogram were randomized between pelvic irradiation and pelvic and para-aortic irradiation. Patients with clinically or surgically involved para-aortic nodes were not included. The external beam dose to the para-aortic area was 45 Gy.

- Routine para-aortic irradiation for all high-risk patients with cervical carcinoma is of limited value.

Haie C et al. (1988) Is prophylactic para-aortic irradiation worthwhile in the treatment of advanced cervical carcinoma? Results of a controlled clinical trial of the EORTC radiotherapy group. Radiother Oncol 11(2):101–112.

RTOG 79–20, 1995 → 367 patients with FIGO stage IB or IIA primary cervical cancers measuring 4 cm or greater in lateral diameter or with FIGO stage IIB cervical cancers (no para-aortic LN metastases) were randomized to RTOG protocol 79–20 to receive either standard pelvic only irradiation or pelvic plus para-aortic irradiation.

- There was a statistically significant difference in OS at 10 years for the pelvic plus para-aortic irradiation arm (55% vs. 44%, $p = 0.02$), without a difference in DFS.
- Grade 4–5 toxicity increased with EFRT (8% vs. 4%, $p = 0.06$).

Rotman M et al. (1995) Prophylactic extended-field irradiation of para-aortic lymph nodes in stages IIB and bulky IB and IIA cervical carcinomas. Ten-year treatment results of RTOG 79–20. JAMA 274(5):387–393.

GOG 191, 2008 → 114 patients with stage IIB–IVA cervical cancer and HGB <14.0 g/dL were randomly assigned to CRT ± recombinant human erythropoietin (R-HUEPO) (40,000 units s.c. weekly). R-HUEPO was stopped if HGB > 14.0 g/dL.

- The impact of maintaining HGB level >12.0 g/dL on PFS, OS, and local control remains undetermined.

Thomas G (2008) Phase III trial to evaluate the efficacy of maintaining hemoglobin levels above 12.0 g/dL with erythropoietin vs above 10.0 g/dL without erythropoietin in anemic patients receiving concurrent radiation and cisplatin for cervical cancer. Gynecol Oncol 108(2):317–325.

India, 1994 → Randomized trial of HDR vs. LDR BT.

- HDR intracavitary BT was found to be an equally good alternative to conventional LDR BT in the treatment of carcinoma of the uterine cervix.

Patel FD et al. (1994) Low dose rate vs. high dose rate brachytherapy in the treatment of carcinoma of the uterine cervix: a clinical trial. Int J Radiat Oncol Biol Phys 28(2):335–341.

Hacettepe, 2007 → 141 patients with stage I–II cervical cancer without para-aortic LN metastases and treated by surgery and postoperative RT. Indications for postoperative external RT were LN metastasis, positive surgical margins, parametrial involvement, pT2 tumor, and presence of any two minor risk factors like LVSI, deep stromal invasion, and tumor diameter between 2 and 4 cm. Median follow-up time was 55 months.

- Five-year OS, DFS, locoregional recurrence-free (LRFS), and distant metastases-free survival (DMFS) rates were 70, 68, 77, and 88%, respectively.
- Multivariate analyses revealed that the level and number of metastatic LNs and concomitant CT were unique significant prognostic factors for OS, DFS, and LRFS.
- Endometrial involvement was proven to be significant for DFS and DMFS.
- Patients with less than three LN metastases or those with only obturator LN involvement showed a similar prognosis to their counterparts with no LN metastases.
- Patients with either common iliac LN or more than three LN metastases had a significantly worse outcome.

Atahan IL et al. (2007) Radiotherapy in the adjuvant setting of cervical carcinoma: treatment, results, and prognostic factors. Int J Gynecol Cancer 17(4):813–820.
Study B9E-MC-JHQS, 2011 → 515 patients with FIGO stage IIB to IVA primary cervical cancer were randomized to arm A [cisplatin 40 mg/m² and gemcitabine 125 mg/m² weekly × 6 weeks with concurrent external RT (50.4 Gy), followed by BT (30–35 Gy in 96 h), and then two adjuvant 21-day cycles of cisplatin, 50 mg/m² on day 1, plus gemcitabine, 1000 mg/m² on days 1 and 8] or to arm B (cisplatin and concurrent external RT followed by BT).

- There was a statistically significant improvement in 3-year PFS (74.4% vs. 65%, $p = 0.029$) and OS for the arm A.
- Grade 3–4 toxicities increased in arm A than in arm B (86.5% vs. 46.3%, $p = 0.001$).
- Adjuvant gemcitabine and cisplatin chemotherapy following CRT may improve treatment outcomes.

Duenas-Gonzales A et al. (2011) Phase III, open-label, randomized study comparing concurrent gemcitabine plus cisplatin and radiation followed by adjuvant gemcitabine and cisplatin versus concurrent cisplatin and radiation in patients with stage IIB to IVA carcinoma of the cervix." J Clin Oncol 29:1678–1685.
Korea Cancer Center Hospital, 2014 → 2158 patients with stage IB–IIA cervical cancer with any intermediate-risk factor after radical hysterectomy were randomized to a development group or a validation group (1620 patients, 538 patients).

- Histology (adenocarcinoma/adenosquamous carcinoma), tumor size (≥3 cm), DSI, and LVSI were significantly associated with disease recurrence (four-factor model).
- The presence of any two of these factors defines the intermediate-risk group.
- According to the this study, histology was added as an intermediate-risk factor.

Ryu SY et al. (2014) Intermediate-risk grouping of cervical cancer patients treated with radical hysterectomy: a Korean Gynecologic Oncology Group study." Br J Cancer 110(2):278–285.
Pittsburgh University, 2014 → 61 patients with stage IB1 or IVA primary cervical cancers with PET-positive pelvic LNs were treated with extended field IMRT with concurrent cisplatin followed by HDR BT. External RT dose was 45 Gy in 25 fractions with concomitant boost to involved LNs (median 55 Gy in 25 fractions).

- Complete response was observed in 77% of patients.
- At a mean 29 months follow-up, while distant metastasis was the predominant failure pattern (23%), local or regional recurrences observed in 16.3% and 4.9% of patients, respectively.
- The rate of para-aortic recurrence in patients with pelvic-only LNs was 2.5%.
- The rate of late >grade 3 toxicity was 4%.
- Extended field IMRT was well tolerated with low regional recurrence in node-positive cervical cancer.

Vargo JA et al. (2014) Extended field intensity modulated radiation therapy with concomitant boost for lymph node-positive cervical cancer: analysis of regional control and recurrence patterns in the positron emission tomography/computed tomography era. Int J Radiat Oncol Biol Phys 90(5):1091–1098.

Saudi Arabia, 2014 → 102 patients with stage IIB to IVA locally advanced cervical cancer with radiological positive pelvic LN metastases without para-aortic LN metastases were randomized to extended-field concurrent chemoradiotherapy (EF-CCRT) or whole pelvic concurrent chemoradiotherapy (WP-CCRT) followed by HDR BT.

- At a median 60 months follow-up, EF-CCRT vs. WP-CCRT:
  – Overall para-aortic LN control: 97.1% vs. 82.1% ($p = 0.02$).
  – Distant-metastasis control: 86.9% vs. 74.7% ($p = 0.03$).
  – DFS: 80.3% vs. 69.1% ($p = 0.03$).
  – OS: 72.4% vs. 60.4% ($p = 0.04$).
- No difference in acute toxicity. Late toxicities were mild and minimal.
- Prophylactic EF-CCRT improves treatment outcomes compared to WP-CCRT in patients with radiological pelvic LN metastases without para-aortic LN metastases.

Asiri MA et al. (2014) Is extended-field concurrent chemoradiation an option for radiologic negative para-aortic lymph node, locally advanced cervical cancer? Cancer Management and Research 6:339–348.

RetroEMBRACE, 2016 → 731 patients treated with definitive external RT (mean 46 ± 2.5 Gy) followed by MRI-guided BT. 77.4% of patients received concomitant chemotherapy.

- Cumulative mean HRCTV D90 87 ± 15 Gy (EQD2$_{10}$), bladder D2cc 81 ± 22 Gy, rectum D2cc 64 ± 9 Gy, sigmoid D2cc 66 ± 10 Gy, and bowel D2cc 64 ± 9Gy (EQD23).
- The 5-year local control, pelvic control, cancer-specific survival, and OS were 89%, 84%, 73%, and 65%, respectively.
- The 5-year grade 3–5 morbidity was 5%, 7%, and 5% for bladder, gastrointestinal tract, and vagina, respectively.
- Compared to historical series, image-guided BT improved local control, pelvic control, OS, and cancer-specific survival of around 10% with limited toxicity.

Sturdza A et al. (2016) Image guided brachytherapy in locally advanced cervical cancer: Improved pelvic control and survival in RetroEMBRACE, a multicenter cohort study. Radiother Oncol 120(3):428–433.

INTERTECC-2, 2017 → 83 patients with clinical stage IB–IVA cervical cancer were definitively treated with concurrent weekly cisplatin and IMRT (45–50.4 Gy) followed by intracavitary BT boost. Patients with gross LNs were treated with a SIB (1.7 Gy/47.6 Gy to gross tumor and elective LNs, 1.93–2.12 Gy/54–59.4 Gy to gross LNs).

- At 26 months of follow-up, the incidence of grade ≥3 neutropenia and clinically significant gastrointestinal (GI) toxicity was 19.3% vs. 12%, respectively.
- Compared with patients treated without image-guided (IG)-IMRT (n = 48), those treated with bone marrow-sparing IG-IMRT (n = 35) had a significantly lower incidence of grade ≥3 neutropenia (8.6% vs. 27.1%, $p = 0.035$) and nonsignificantly lower incidence of grade ≥3 leukopenia (25.7% vs. 41.7%; $p = 0.13$) and any grade ≥3 hematologic toxicity (31.4% vs. 43.8%; $p = 0.25$).

IMRT reduces acute hematological and gastrointestinal (GI) toxicity for patients with locoregionally advanced cervical cancer.

Mell LK et al. (2017) Bone marrow-sparing intensity modulated radiation therapy with concurrent cisplatin for stage IB-IVA cervical cancer: an international multicenter phase II clinical trial (INTERTECC-2). Int J Radiat Oncol Biol Phys 97(3):536–545.

India, 2018 → 635 patients with clinical stages IB2, IIA, or IIB squamous cervical cancer were randomized to neoadjuvant chemotherapy (paclitaxel and carboplatin × 3 cycles) followed by surgery or concomitant CRT (pelvic RT with cisplatin × 5 cycles).

- This is the first phase III randomized trial comparing neoadjuvant chemotherapy followed by surgery to concomitant CRT.
- Concomitant CRT improved 5-year DFS (69.3% vs. 76.7%, $p = 0.038$), but no difference in 5-year OS (75.4% vs. 74.7%, $p = 0.87$).
- Cisplatin-based CRT is the standard treatment for patients with locally advanced cervical cancer.

Gupta S (2018) Neoadjuvant chemotherapy followed by radical surgery versus concomitant chemotherapy and radiotherapy in patients with stage IB2, IIA, or IIB squamous cervical cancer: A randomized controlled trial. J Clin Oncol 36(16):1548–1555.

NRG Oncology-RTOG 1203 (TIME-C), 2018 → 289 patients with cervical or endometrial cancer were postoperatively randomized to standard four-field pelvic radiotherapy or IMRT.

- Patient-reported acute gastrointestinal and urinary toxicity, and late diarrhea was significantly decreased with IMRT than standard RT.
- Postoperative pelvic IMRT in patients with cervical or endometrial cancer improves quality of life compared to 3DCRT.

Klopp AH et al. (2018) Patient-reported toxicity during pelvic intensity-modulated radiation therapy: NRG Oncology-RTOG 1203. J Clin Oncol 36:2538–2544.

Taiwan, 2018 → 198 patients with pelvic LN positive and para-aortic LN-negative cervical cancer (IB2-IVA) were treated with pelvic RT or subrenal vein radiotherapy (SRVRT). RT fields were selected based on pelvic LN location and number of pelvic LNs. RT dose was 45–50.4 Gy in 1.8 Gy per fraction.

- SRVRT group: More advanced stage disease based on FIGO stage, common iliac LNs, and number of pelvic LNs.

- At a median 63 months follow-up, para-aortic nodal failure: 23.7% pelvic RT vs. 1.3% SRVRT ($p < 0.001$).
- SRVRT resulted in significantly improved 5-year para-aortic LN recurrence-free survival (56.8% vs. 100%, $p < 0.001$) and cancer-specific survival (56.5% vs. 93.9%, $p < 0.001$) among patients with common iliac or ≥3 pelvic LNs.
- No increase in severe toxicities with SVRT.

Patients with common iliac LN metastasis and ≥3 pelvic LNs should be treated with SRVRT.

Lee J et al. (2018) Impact of para-aortic recurrence risk-guided intensity-modulated radiotherapy in locally advanced cervical cancer with positive pelvic lymph nodes. Gynecol Oncol 148(2):291–298.

Washington University, 2019 → 600 patients with stage IB1–IVA cervical cancer were treated with IMRT and 3D-IGBT or 2D-ERT and 2D-BT.

- 2D-ERT/BT vs. IMRT/3D-IGBT: Five-year freedom from relapse (FFR) 57% vs. 65% ($p = 0.04$), CSS 62% vs. 69% ($p = 0.04$), and OS 57% vs. 61% ($p = 0.04$).
- When stratified by LN status according to PET/CT, disease control was most improved with IMRT/3D-IGBT vs. 2D-ERT/BT in patients with positive pelvic LNs only ($p = 0.02$).
- Cumulatively, 15% patients had grade ≥3 late bowel/bladder toxicity (18% 2D ERT/BT vs. 11% IMRT/3D-IGBT, $p = 0.02$).

IMRT and 3D-IGBT were associated with improved survival and decreased toxicity.

Lin AJ et al. (2019) Intensity modulated radiation therapy and image-guided adapted brachytherapy for cervix cancer. Int J Radiat Oncol Biol Phys 103(5):1088–1097.

China Medical University, 2020 → 164 patients with stage Ib2–IIIb cervical cancer were prospectively randomized to concurrent weekly cisplatin chemotherapy and pelvic bone marrow-sparing intensity modulated radiotherapy (PBMS-IMRT) or control group (IMRT).

- ≥Grade 2 hematological toxicity: 50% PBMS-IMRT vs. 69.5% IMRT, $p = 0.02$.
- No difference in ≥grade 2 gastrointestinal toxicity ($p > 0.05$).
- Lumbosacral spine (LSS) V10 ≥87%, LSS mean ≥39 Gy, and pelvic bone V40 ≥28% were more likely to experience ≥grade 2 hematological toxicity.

Huang J et al. (2020) Pelvic bone marrow sparing intensity modulated radiotherapy reduces the incidence of the hematologic toxicity of patients with cervical cancer receiving concurrent chemoradiotherapy: a single-center prospective randomized controlled trial. Radiat Oncol 15:180.

Yale University, 2020 → 59 patients with LN-positive cervical cancer were treated with definitive CRT with a simultaneous integrated boost (SIB) to involved LNs. Median dose was 56.25 Gy in 25 fractions. While 56% of patients had pelvic LN metastases, 44% had pelvic and para-aortic LN metastases.

- At a median 30 months of follow-up, 3-year OS, PFS, and locoregional control rates were 67%, 60%, and 89%, respectively.
- Univariate analysis: Para-aortic LN metastasis and the presence of ≥4 grossly involved LNs were associated with a higher risk of disease progression.
- Acute and late ≥grade 3 gastrointestinal or genitourinary toxicity rates were 3% and 12%, respectively.

Definitive CRT with a SIB to grossly involved LNs for patients with LN-positive cervical cancer is well tolerated and is associated with excellent oncological outcomes.

Jethwa KR et al. (2020) Lymph node–directed simultaneous integrated boost in patients with clinically lymph node–positive cervical cancer treated with definitive chemoradiotherapy: clinical outcomes and toxicity. J Radiat Oncol 9:103–111.

Chinese, 2021 → 1048 patients with FIGO stage IB–IIA cervical cancer with adverse pathological factors after radical hysterectomy were randomized 1:1:1 to receive adjuvant RT, concurrent CRT (CCRT), or sequential CRT (SCRT). Adjuvant RT dose was 45–50 Gy. Chemotherapy consists of weekly cisplatin (30–40 mg/m^2) in CCRT arm and cisplatin (60–75 mg/m^2) plus paclitaxel (135–175 mg/m^2) in SCRT arm (in a 21-day cycle, given two cycles before and two cycles after RT).

- SCRT was associated with a higher rate of 3-year DFS than RT (90% vs. 82%) and CCRT (90% vs.85).
- SCRT also decreased cancer death risk compared with RT (5-year rate, 92% vs. 88%) after adjustment for LN involvement.
- No difference in DFS or cancer death risk in patients treated with CCRT or RT.

In this randomized clinical trial, SCRT, rather than CCRT, resulted in a higher DFS and lower risk of cancer death than RT among patients with early-stage cervical cancer in a postoperative setting,

Huang H et al. (2021) Effectiveness of sequential chemoradiation vs. concurrent chemoradiation or radiation alone in adjuvant treatment after hysterectomy for cervical cancer The STARS phase 3 randomized clinical trial. JAMA Oncol Jan 14:e207168.

9.2 Endometrial Cancer

Endometrial cancer is the fourth most common female cancer following breast, lung, and colon cancers. Such cancers have shown a prominent increase in incidence in developed countries, and have become the most common gynecological cancer. Although its incidence is high, it is only the seventh most common cause of cancer-related mortality due to early diagnosis and treatment [28].

The corpus is the major part of the uterus, and it extends into the fundus, where uterus combines with the Fallopian tubes. The isthmus is 0.5 cm long and located between the cervix and corpus (Fig. 9.16).

Fig. 9.16 Endometrial anatomy. (From [29], Compton C.C., Byrd D.R., Garcia-Aguilar J., Kurtzman S.H., Olawaiye A., Washington M.K. (2012) Corpus Uteri. In: Compton C., Byrd D., Garcia-Aguilar J., Kurtzman S., Olawaiye A., Washington M. (eds) AJCC Cancer Staging Atlas. Springer, New York, NY. https://doi.org/10.1007/978-1-4614-2080-4_36)

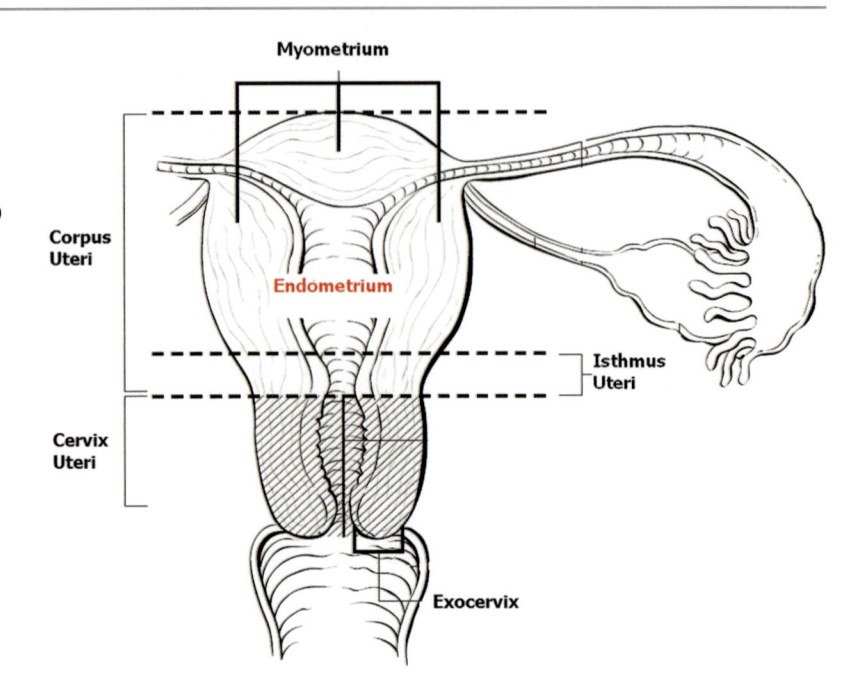

Histologically, the uterine corpus consists of three layers; from innermost to outermost, these are

1. **Endometrium:** Consists of a basal layer and a functional layer that includes endometrial glands.
2. **Myometrium:** Consists of smooth muscle and lymph vessels.
3. **Serosa:** The peritoneum that covers the uterine corpus from anterior and posterior, and the cervix only from posterior.

9.2.1 Pathology

Endometrial cancers are generally adenocarcinomas [30]. The basic developmental mechanism is a long period of unbalanced estrogenic excitation with progesterone. This excitation may be exogenous or endogenous. Estrogen-secreting ovarian neoplasms (granulosa cells or functional thecomas) and polycystic ovarian syndrome (Stein–Leventhal syndrome), resulting in the secretion of high levels of estrogen, may cause endometrial hyperplasia and subsequent carcinoma.

Endometrial adenocarcinomas are classified into two groups according to their histomorphological features, pathogenesis, and prognosis.

Type 1 endometrial adenocarcinomas. These generally develop from endometrial hyperplasia. There are hyperplasia foci around the carcinoma. Type 1 endometrial carcinomas are well-differentiated and are difficult to distinguish from normal endometrial glands. These types of carcinomas are strongly associated with estrogen stimulation and usually included "endometrioid-type" adenocarcinomas.

- They generally have low-grade histology, do not invade deep myometrium, and they have a good prognosis.
- They constitute 75–80% of all endometrial carcinomas.

Type 2 endometrial adenocarcinomas. These have no associated hyperplasia. Patients are older than those with type I cases. They are less differentiated neoplasias.

- Constitute 10–15% of endometrial cancers, and have a poor prognosis.
- No association with estrogen.

- High-grade and high malignant potential.
- Serous and clear cell carcinomas belong to this group of neoplasias.
- Types other than adenocarcinomas have high risks of recurrence and distant metastasis.

The Cancer Genome Atlas Research Network identified four genomic subgroups for endometrial cancer (2013) [31]:

- POLE-ultramutated: best prognosis.
- Mismatch repair deficient (MMRd) or microsatellite instable (MSI) (hypermutated): intermediate prognosis.
- Copy-number low (no specific molecular profile-NSMP): intermediate prognosis.
- Copy-number high (TP53 mutation): worst prognosis (serous-like).

Cellular Classification of Epithelial Tumors of the Uterus [30]

Endometrioid adenocarcinoma (75–80%)
Not otherwise specified
Villoglandular
Secretory
Ciliated adenocarcinoma
Adenocarcinoma with squamous differentiation
Serous adenocarcinoma (<10%)
Mucinous adenocarcinoma (1%)
Clear cell adenocarcinoma (4%)
Squamous cell carcinoma (<1%)
Mixed cell adenocarcinoma (10%)
Transitional carcinoma
Undifferentiated carcinoma

9.2.2 General Presentation

Particularly in older women, cervical stenosis causing hematometrium or pyometrium may not result in bleeding. This sign is a poor prognostic factor. More than 50% of cases with pyometrium and associated vaginal discharge have carcinoma in D&C (dilatation and curettage), and most have squamous carcinoma, which is a very rare form of endometrial carcinomas.

More than 90% of patients with endometrial cancer complain only of vaginal bleeding.

- Most such bleeding is postmenopausal in origin.
- Premenopausal cases have abnormal uterine bleeding.
- Rarely, patients have a feeling of pelvic compression or discomfort, which is a sign of extrauterine disease extension.

Also, 5–17% of patients are asymptomatic. In asymptomatic cases, the disease is frequently discovered through abnormal PAP smears, incidentally in hysterectomy specimens, or via abnormal radiological findings in the uterus (e.g., thickening of the endometrial wall in pelvic US).

9.2.3 Staging

Uterine cavity size, endocervical curettage findings, cystoscopy, and rectoscopy findings were used in the clinical staging of endometrial cancer until 1988. However, this system could not determine some important prognostic factors such as myometrial invasion depth and LN metastasis and had downstage the tumor (22% lower stages than surgical staging).

Surgical staging of endometrial cancer should include peritoneal lavage for cytological exam, biopsy of all suspicious lesions by abdominal and pelvic exploration, total hysterectomy (TH), bilateral salpingo-oophorectomy (BSO), and bilateral pelvic ± para-aortic LN dissection.

Uterus is thoroughly examined for tumor size, myometrial invasion depth, and cervical stromal invasion. All suspicious pelvic and para-aortic LNs should be examined pathologically.

Prognostic parameters like myometrial invasion depth, peritoneal cytology, LN involvement, and cervical and adnexal extension were added into the 1988 surgical–pathological staging system [29, 32–34]. FIGO revised the endometrial cancer staging system in 2009 [33].

9.2.3.1 FIGO Staging for Endometrial Cancer (Fig. 9.17) [29]

Stage I: Tumor limited to corpus uteri.
  Stage IA: Invasion to less than 50% of the myometrium.

Fig. 9.17 Endometrial cancer staging. (From [29], Compton C.C., Byrd D.R., Garcia-Aguilar J., Kurtzman S.H., Olawaiye A., Washington M.K. (2012) Corpus Uteri. In: Compton C., Byrd D., Garcia-Aguilar J., Kurtzman S., Olawaiye A., Washington M. (eds) AJCC Cancer Staging Atlas. Springer, New York, NY. https://doi.org/10.1007/978-1-4614-2080-4_36)

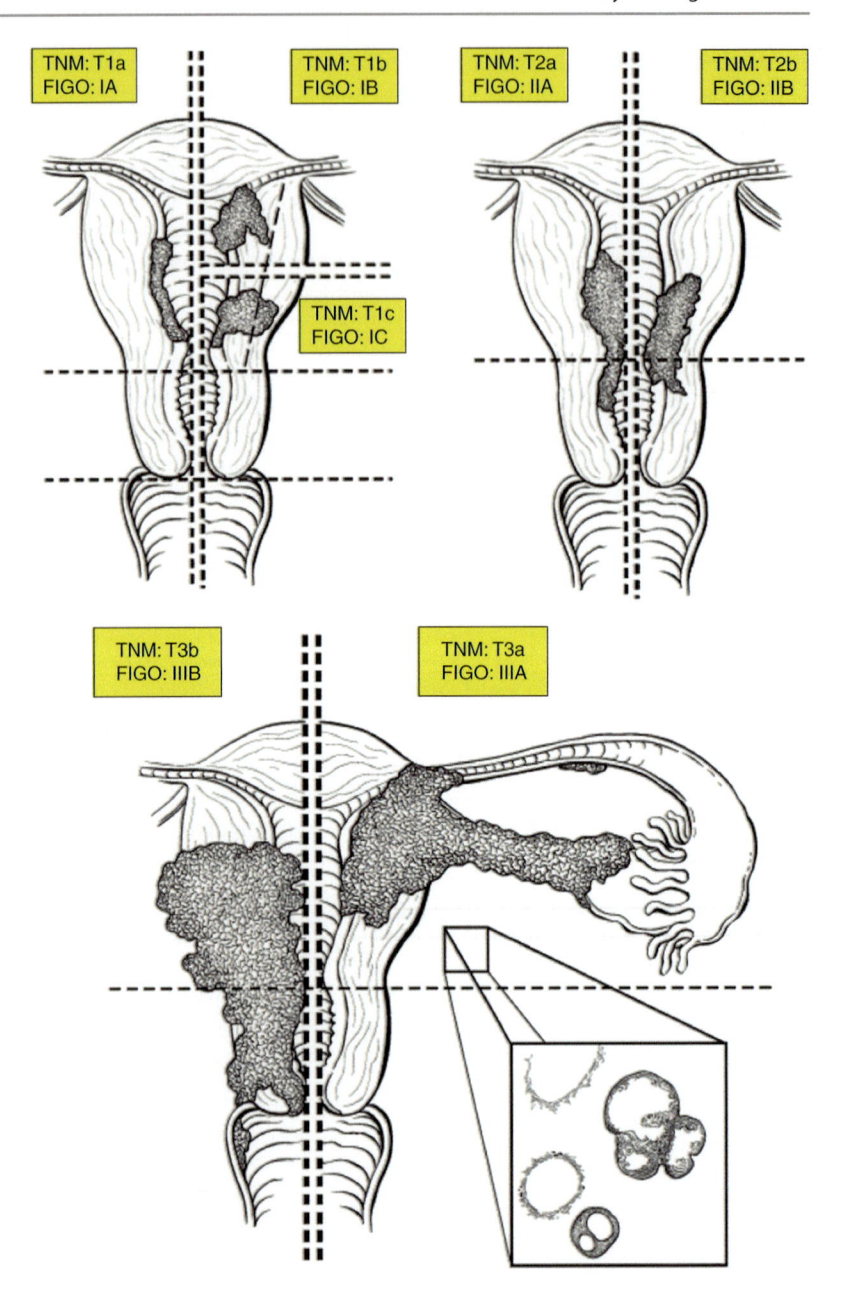

Stage IB: Invasion to greater than 50% of the myometrium.

Stage II: Tumor invades stromal connective tissue of the cervix but does not extend beyond uterus*.

Stage III: Local and/or regional spread of the tumor.

Stage IIIA: Tumor invades serosa and/or adnexa (direct extension or metastasis).

Stage IIIB: Vaginal or parametrial involvement.

Stage IIIC: Metastasis to pelvic and/or para-aortic lymph nodes.

Stage IIIC1: Metastases in pelvic lymph nodes.

Stage IIIC2: Metastases in para-aortic lymph nodes with or without pelvic lymph node involvement.

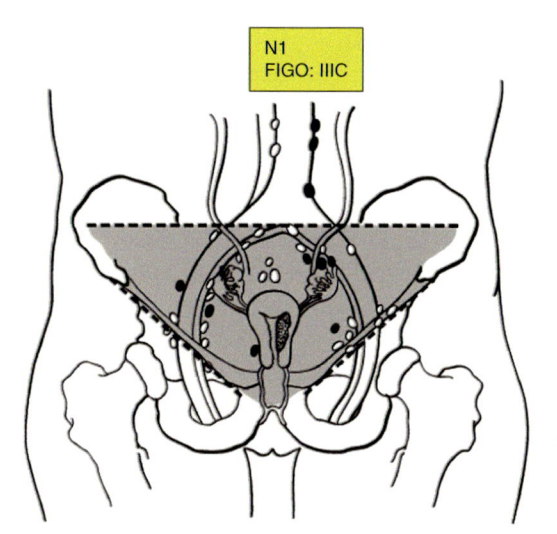

Fig. 9.18 Endometrial lymphatics. (From [29], Compton C.C., Byrd D.R., Garcia-Aguilar J., Kurtzman S.H., Olawaiye A., Washington M.K. (2012) Corpus Uteri. In: Compton C., Byrd D., Garcia-Aguilar J., Kurtzman S., Olawaiye A., Washington M. (eds) AJCC Cancer Staging Atlas. Springer, New York, NY. https://doi.org/10.1007/978-1-4614-2080-4_36)

Stage IVA: Tumor invasion of bladder and/or bowel mucosa.

Stage IVB: Distant metastases including intraperitoneal disease and/or inguinal lymph nodes (no pathological M0; use clinical M0).

*Endocervical glandular involvement only should be considered as stage I (not stage II).

9.2.3.2 Endometrial Cancer Lymphatics
(Fig. 9.18)
Parametrial.
　Obturator.
　Hypogastric (internal iliac).
　External iliac.
　Common iliac.
　Presacral.
　Para-aortic.

9.2.3.3 Prognostic Factors [31, 34]
- →*Myometrial invasion:*
  Deep myometrial invasion increases the risk of LN metastasis, extrauterine extension, and recurrence.
- →*Pathology:*
  Other than endometrioid types have increased recurrence and distant metastasis.
- →*Histological differentiation (grade):*
  It is almost always associated with increased recurrence risk.
- →*Lymphovascular space invasion:*
  It is approximately 15% in early endometrial cancers, but increases with an increased in myometrial invasion depth and tumor grade.
- →*Lymph node metastasis:*
  It is the most important prognostic factor.
- →*Peritoneal cytology:*
  While it is associated with poor prognosis in stage III disease, its impact on prognosis is unclear in early-stage disease.
- →*Adnexal metastasis:*
  It is a very high risk for recurrence.
- →*Lower uterine segment involvement:*
  It is associated with worse outcome.
- →*Tumor size:*
  Tumors >2 cm; LN metastasis is higher.
- →*Hormonal receptors:*
  Progesterone receptor (+) is more determinant for prognosis than estrogen receptor.
- →*Age:*
  Young age is a good prognostic factor since grade and poor histological types are more frequent in older ages.
- →*Molecular prognostic factors:*
  Patients with POLE mutations have the best prognosis, copy number high have the worst prognosis, and MMRd and NSMP have the intermediate prognosis. β catenin mutation (CTNNB1) is a poor prognostic factor.

9.2.4 Treatment Algorithm for Endometrial Cancer [35]

Surgery: TH + BSO + pelvic ± para-aortic LN dissection.

Adjuvant treatment of endometrial cancer is classically based on the risk of disease recurrence, which is defined by the clinicopathological risk factors and whether LN dissection is performed or not.

Clinicopathological risk factors: Stage, histopathology, grade, depth of myometrial invasion, LVSI, tumor size, age, and lower uterine segment involvement.

Recently, molecular risk grouping has been defined and added to the new European Society of Gynaecological Oncology (ESGO), the European Society for Radiotherapy and Oncology (ESTRO), and the European Society of Pathology (ESP) guideline [35].

- **Low-risk.**
  - Molecular classification unknown.
    – Stage IA Gr 1–2, endometrioid, LVSI (−)/focal.
  - Molecular classification known.
    – Stage I–II POLEmut endometrial carcinoma, no residual disease.
    – Stage IA MMRd/NSMP, Gr 1–2, endometrioid, LVSI (−)/focal.
  - *Surveillance after surgery.*
- **Intermediate-risk.**
  - Molecular classification unknown.
    – Stage IB Gr 1–2, endometrioid, LVSI (−)/focal.
    – Stage IA Gr 3, endometrioid, LVSI (−)/focal.
    – Stage IA, non-endometrioid carcinoma without myometrial invasion.
  - Molecular classification known.
    – Stage IB MMRd/NSMP, Gr 1–2, endometrioid, LVSI (−)/focal.
    – Stage IA MMRd/NSMP, Gr 3, endometrioid, LVSI (−)/focal.
    – Stage IA p53abn and/or non-endometrioid carcinoma without myometrial invasion.
  - *Adjuvant vaginal cuff BT is recommended.*
  - *Age <60 years, surveillance after surgery.*
- **High-intermediate (HIR) risk.**
  - Molecular classification unknown.
    – Stage I, endometrioid, substantial LVSI (regardless of grade and depth of invasion).
    – Stage IB Gr 3, endometrioid, regardless LVSI status.
    – Stage II.
  - Molecular classification known.
    – Stage I MMRd/NSMP endometrioid, substantial LVSI (regardless of grade and depth of invasion).
    – Stage IB MMRd/NSMP, Gr 3, endometrioid, regardless of LVSI status.
    – Stage II MMRd/NSMP endometrioid.
  - *Complete surgical staging (+).*
    – Vaginal cuff BT.
    – Substantial LVSI or stage II: pelvic RT.
    – High-grade and/or substantial LVSI: adjuvant chemotherapy.
  - *Complete surgical staging (−).*
    – Pelvic RT.
    – Substantial LVSI or stage II: adjuvant chemotherapy.
    – Gr 3, LVSI negative or stage II, grade 1 endometrioid: vaginal cuff BT.
- **High-risk.**
  - Molecular classification unknown.
    – Stage III–IVA, no residual disease.
    – Stage I–IVA, non-endometrioid with myometrial invasion, no residual disease.
  - Molecular classification known.
    – Stage III–IVA MMRd/NSMP, endometrioid, no residual disease.
    – Stage I–IVA p53abn endometrial carcinoma with myometrial invasion, no residual disease.
    – Stage I–IVA NSMP/MMRd non-endometrioid with myometrial invasion, no residual disease.
  - *Combined pelvic ± para-aortic RT and chemotherapy.*
  - *Chemotherapy: concurrent or sequential.*
- **Advanced stage or metastatic disease.**
  - Molecular classification unknown.
    – Stage III–IVA with residual disease.
    – Stage IVB.
  - Molecular classification known.
    – Stage III–IVA with residual disease (any molecular type).
    – Stage IVB (any molecular type).
  - *Systemic therapy is recommended.*
  - *Unresectable or symptomatic tumor: palliative RT.*

9.2.4.1 Special Situations
Medically inoperable.

Stage I, Gr 1–2 with minimal myometrial invasion (MRI): intracavitary BT alone.

Stage I, deep myometrial invasion with MRI or MRI not available: pelvic RT (45–50.4 Gy) + intracavitary BT.

Stage II: pelvic RT (45–50 Gy) + intracavitary BT.

Stage III: pelvic ± para-aortic RT (45–50 Gy) + intracavitary BT.

Recurrence.

No previous RT → pelvic ± para-aortic RT + BT boost (total dose 60–70 Gy).

Uterine sarcoma.

Surgery.

Stage II–IVA: adjuvant chemotherapy ± RT.

Stage IVB: systemic chemotherapy ± palliative RT.

9.2.5 Radiotherapy [15, 16]

- Patients are usually treated with RT in the postoperative setting.
- While a four-field box technique was generally used in the past, IMRT technique is the standard treatment approach today. With these modern techniques, more normal tissues are spared and less toxicity is observed. Routine daily online image guidance (i.e., cone-beam CT) is recommended for IMRT/VMAT.
- Patients are CT simulated with standard bladder and bowel preparation in the supine or prone (with belly board) position.
- Vaginal or rectal markers may be placed into the patient. Oral contrast material may be beneficial for the visualization of small bowel.
- Intravenous (IV) contrast may be helpful to define pelvic vascular structures.
- External RT fields are defined according to the surgical LN status:
  - Para-aortic LN (−): Pelvic RT (Fig. 9.19).
  - Para-aortic LN (+): Pelvic and para-aortic RT (extended field radiotherapy = EFRT) (Fig. 9.20).
- RTOG published a consensus guideline for delineation of CTV in the postoperative pelvic RT of endometrial and cervical cancer [15, 16].
- Nodal CTV = obturator, presacral, external iliac, internal iliac, and common iliac LNs.
  - Common iliac nodal CTV: right and left common iliac artery/vein +7 mm margin → bifurcation of the common iliac artery.
  - Presacral nodal CTV: Between the right and left common iliac nodal CTV, anterior to the S1–S2, 10–15 mm in diameter. If patients have gross cervical involvement, presacral LNs should be included.
  - External iliac nodal CTV: Bifurcation of the common iliac vessels → deep circumflex artery branches or beginning of the inguinofemoral vessels, + 7 mm margin (anterior: 10 mm).
  - İnternal iliac nodal CTV: Bifurcation of the common iliac vessels → vessels turn laterally prior to leaving the pelvis, +7 mm margin.
  - Obturator nodal CTV: Between the external and internal iliac nodal CTV, 15–18 mm in diameter, bifurcation of the internal and external iliac vessels → obturator vessels leave the pelvis through the obturator foramen.
  - Para-aortic CTV = Paracaval, precaval, retrocaval, deep and superficial intercavo-aortic, para-aortic, pre-aortic, and retro-aortic LNs.
    - Superior: Left renal vein.
    - Left lateral border: + (1–2 cm) lateral to the aorta.
    - On the right: + (3–5 mm) around the vena cava inferior.
  - All gross or suspicious LNs, lymphoceles, and surgical clips should be included in nodal CTVs.
  - Nodal PTV = nodal CTV + 7 mm circumferential margin.
- Vaginal CTV = upper 1/3 vagina, parametrial and paravaginal tissues.
  - Includes the gross disease (if present), proximal vagina, and parametrial/paravaginal tissues.

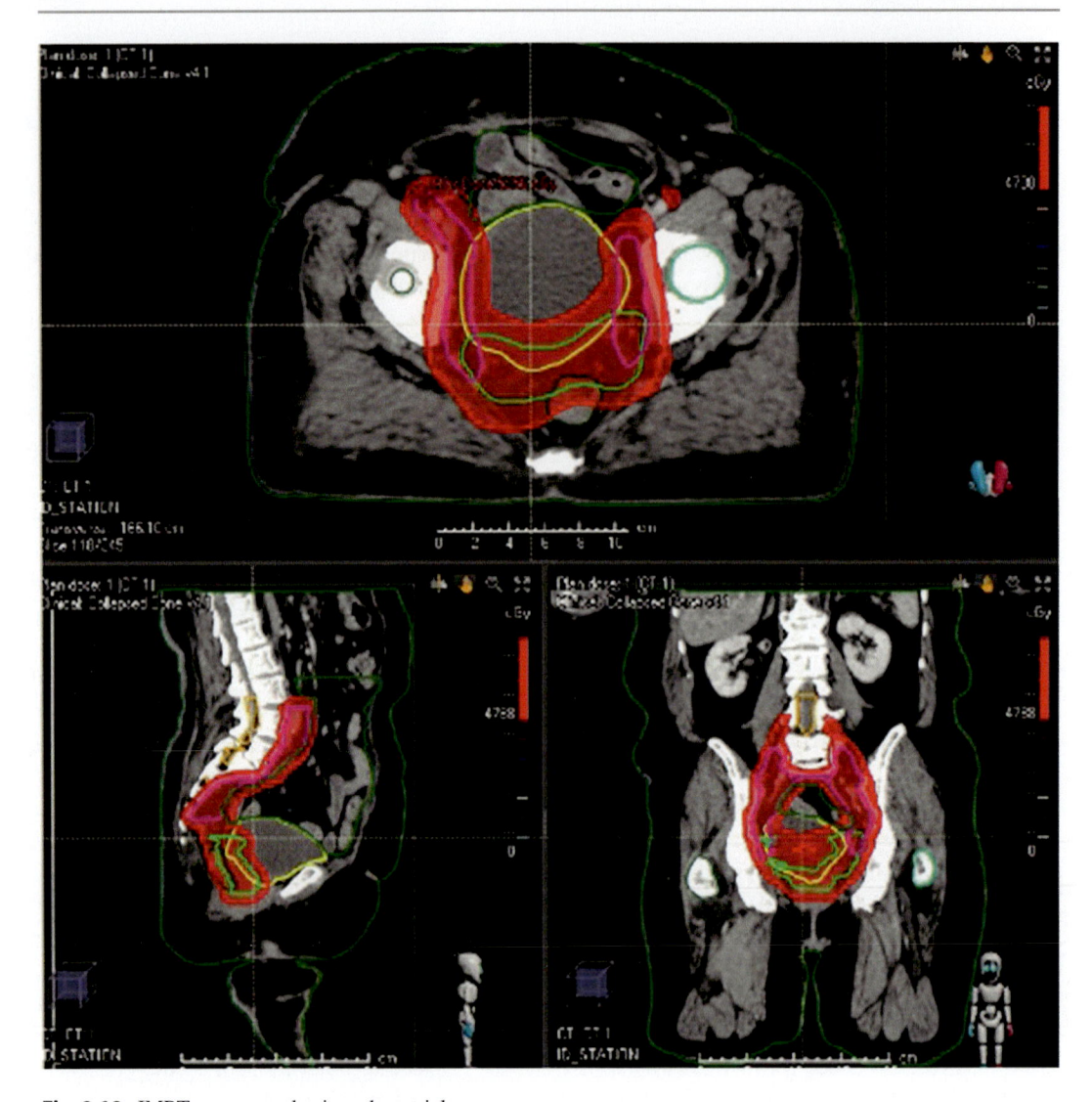

Fig. 9.19 IMRT treatment plan in endometrial cancer

- Inferior: 3 cm below the upper border of the vagina/vaginal marker or 1 cm above the inferior extent of the obturator foramen.
- Anterior: Posterior bladder wall.
- Posterior: Anterior 1/3 mesorectum.
- Lateral: Medial border of the urogenital diaphragm.
- Integrated target volume (ITV) = full bladder + empty bladder vaginal/obturator CTVs.
- PTV = vaginal ITV + 7 mm.

- Small bowel, rectum, bladder, and bilateral femoral heads are defined as organs at risk (OARs) in all cases (EFRT: bilateral kidneys and spinal cord).
- The OARs should be excluded from all CTVs.
- Recommended OARs doses are detailed below [36]:
  - Small bowel (peritoneal cavity) dose: V45 Gy <195 cc, V40 Gy <30%.
  - Bladder dose: V45 Gy <35%.
  - Rectum dose: V30 Gy <60%, V50 Gy <50%.

9.2 Endometrial Cancer

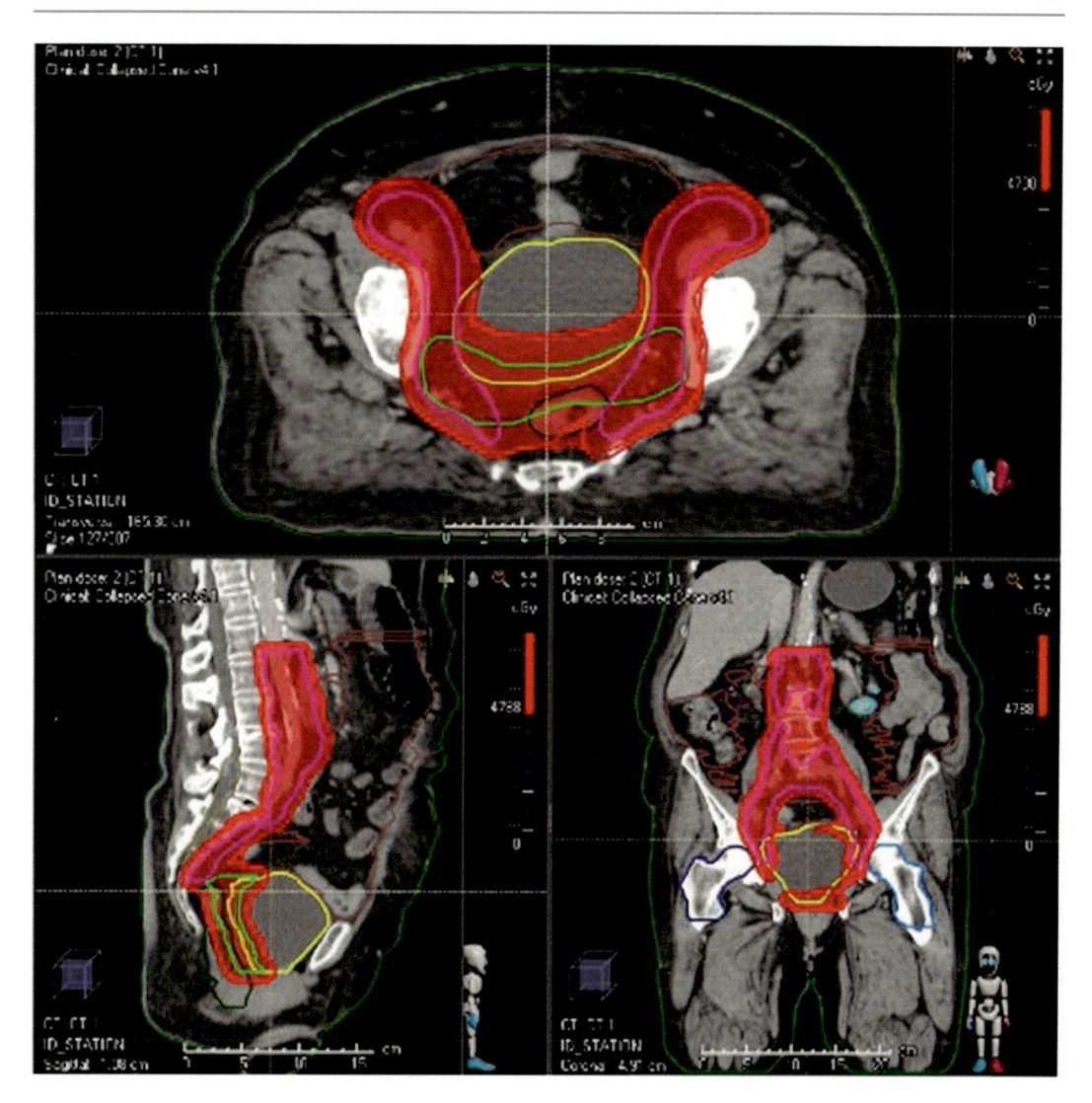

Fig. 9.20 Extended field RT in endometrial cancer

- Vaginal surface dose: 100 Gy.
- Femoral head: V30 Gy <15%, V50 Gy <5%.
- Bone marrow: Up to 90% receives 10 Gy, up to 37% receives 40 Gy.
- Kidney dose <18 Gy.
- Spinal cord dose <45 Gy.

9.2.5.1 Brachytherapy [37, 38]

Postoperative vaginal BT. Adjuvant vaginal BT alone is used in intermediate-risk patients. In patients with positive/close vaginal mucosal margin, grade 3 disease, extensive LVSI, or cervical stromal invasion, vaginal cuff BT boost can be applied following external RT. Vaginal cylinders (single channel/multichannel) are usually used in this technique, and target volume includes proximal 3–5 cm of the vagina (Fig. 9.21). Today, 3D-IGBT is usually used in endometrial cancer (Fig. 9.22). If the tumor has serous or clear cell histology, grade 3 disease, or extensive LVSI, the entire length of the vagina should be treated.

It is important to use appropriately sized cylinders that suit the vaginal cuff. Dose is usually prescribed to 0.5 cm from the cylinder surface or vaginal mucosa.

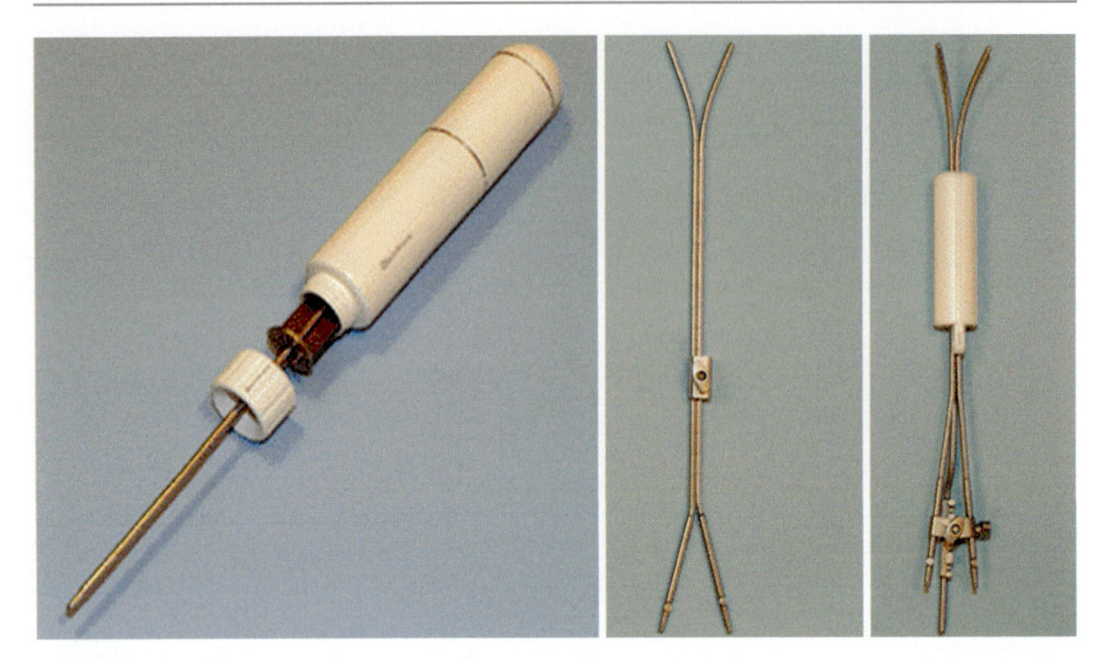

Fig. 9.21 Vaginal cylinders and Y applicator for endometrial brachytherapy applications

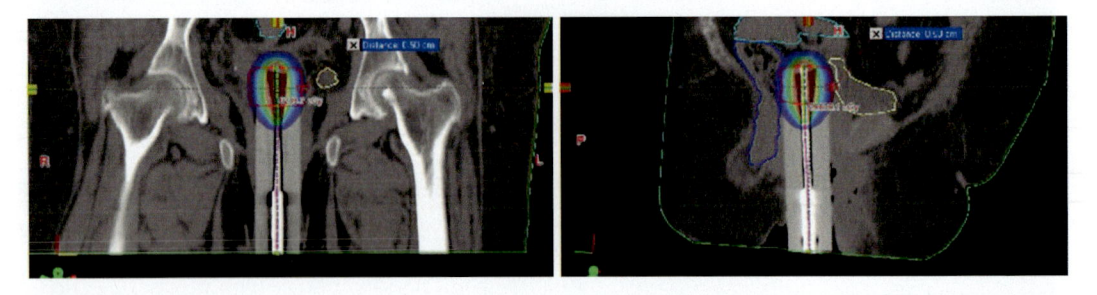

Fig. 9.22 Computed tomography-based vaginal cuff brachytherapy planning

Definitive BT. BT alone or combined with external RT can be used in medically inoperable patients with endometrial cancer [38]. CT- or MRI-guided BT is recommended. Y applicators can be used in the treatment of intact endometrial cancer (Fig. 9.21). Dose is prescribed to cover CTV (entire uterus, cervix, and upper 1–2 cm of the vagina). Cervical dilation is performed by Hegar dilators.

9.2.5.2 Dose–Fractionation
- Postoperative RT:
  - Vaginal BT alone:
    – Vaginal mucosa: 5 × 6 Gy, 6 × 4 Gy.
    – 5 mm below the vaginal mucosa: 3 × 7 Gy, 5 × 5 Gy
- External RT ± vaginal BT:
  – External RT (45 Gy) + BT (vaginal mucosa: 3 × 6 Gy, 5 mm below the vaginal mucosa: 3 × 5 Gy).
  – External RT (50.4 Gy) + BT (vaginal mucosa: 2 × 6 Gy, 5 mm below the vaginal mucosa: 2 × 5 Gy).
- For medically inoperable patients:
  - BT alone:
    – Stage I, Gr 1–2, minimal myometrial invasion.
    GTV D90 80–90 Gy, CTV D90 48–62.5 Gy.
    6 × 6 Gy/6 × 6.4 Gy/5 × 7.3 Gy/4 × 8.5 Gy/9–10 × 5 Gy
- External RT + BT:

- Stage I, Gr 1–2, deep myometrial invasion or stage II or stage III:
    GTV D90 80–90 Gy, CTV D90 65–75 Gy.
    45 Gy external RT + 3 × 6.5 Gy/3 × 6.3 Gy/4 × 5.2 Gy/5 × 5 Gy/2 × 8.5 Gy BRT
    50 Gy external RT + 6 × 3.75 Gy BT
- Recommended OARs doses:
    - D2cc rectum ˂70–75 Gy.
    - D2cc sigmoid ˂70–75 Gy.
    - D2cc bladder ˂80–100 Gy.

9.2.6 Selected Publications

Norwegian Study, 1980 → 540 patients with stage I endometrial cancer after TAH + BSO (no LN dissection) were randomized to intravaginal BT or intravaginal BT and external pelvic RT (ERT) (40 Gy) following surgery. 60 Gy BT was delivered to the surface of the vaginal mucosa.

- Vaginal and pelvic recurrences were significantly reduced in the BT and ERT arm (1.9 vs. 6.9%, $p < 0.01$).
- More patients in BT and ERT group developed distant metastases (9.9 vs. 5.4%).
- No significant difference in OS.
- Subgroup analysis: Stage IB, grade 3 tumors might benefit from additional ERT.
- Update: Median follow-up 20.5 years. No significant difference in OS. ERT significantly decreased survival and increased the risk of secondary cancer (39.7% vs. 25.6%, $p = 0.014$) in patients <60 years with low-risk endometrial cancer.

Aalders J et al. (1980) Postoperative external irradiation and prognostic parameters in stage I endometrial carcinoma: clinical and histopathologic study of 540 patients. Obstet Gynecol 56(4):419–427.

Onsrud M et al. (2013) Long-term outcomes after pelvic radiation for early-stage endometrial cancer. J Clin Oncol 31:3951–3956.

PORTEC-1, 2003 → 714 patients with FIGO stage I, either Gr 1 or 2 with ≥50% myometrial invasion or Gr 2–3 with <50% after TAH/BSO (no lymphadenectomy) were randomized to observation or adjuvant WPRT (46 Gy).

- High-intermediate risk (HIR) disease was defined: Age >60 years, Gr 3, >50% myometrial invasion; two of those three factors present. IB Gr 3 disease was not included.
- At a median follow-up of 73 months, 8-year locoregional recurrence (LRR) rates 15% vs. 4% ($p < 0.0001$).
- No significant difference in OS.
- Update: Median follow-up was 13.3 years. The addition of WPRT significantly decreased 15-year locoregional recurrence (LRR) rates (15.5% vs. 6%, $p < 0.0001$). 74% of the LRRs were isolated vaginal recurrences. No difference in OS or FFS.

Creutzberg CL et al. (2003) Survival after relapse in patients with endometrial cancer: Results from a randomized trial. Gynecol Oncol 89(2):201–209.

Creutzberg CL et al. (2011) Fifteen-year radiotherapy outcomes of the randomized PORTEC-1 trial for endometrial carcinoma. Int J Radiat Oncol Biol Phys 81(4):e631–638.

GOG 99, 2004 → 448 patients with "intermediate risk" endometrial adenocarcinoma were randomized after surgery to either no additional therapy (NAT) or WPRT. A HIR subgroup of patients was defined as those with (1) moderate to poorly differentiated tumor, presence of LVSI, and outer third myometrial invasion; (2) age 50 or greater with any two risk factors listed above; or (3) age of at least 70 with any risk factor listed above. All other eligible participants were considered to be in a low-intermediate risk (LIR) subgroup.

- Adjunctive RT in early-stage intermediate-risk endometrial carcinoma decreases the risk of recurrence, but should be limited to patients whose risk factors fit the definition of HIR.

Keys HM et al. (2004) A phase III trial of surgery with or without adjunctive external pelvic radiation therapy in intermediate risk endometrial adenocarcinoma: a Gynecologic Oncology Group study. Gynecol Oncol 92(3):744–751.

PORTEC, 2005 → 714 stage I endometrial carcinoma patients randomly assigned to postoperative pelvic RT or no further treatment, excluding those with stage IC, grade 3, or stage IB, grade 1 lesions.

- Ten-year locoregional relapse rates were 5% (RT) and 14% (controls; $p < 0.0001$), and 10-year OS rates were 66 and 73%, respectively ($p = 0.09$).
- In view of the significant locoregional control benefit, RT remains indicated in stage I endometrial carcinoma patients with high-risk features for locoregional relapse.

Scholten AN (2005) Postoperative radiotherapy for Stage 1 endometrial carcinoma: long-term outcome of the randomized PORTEC trial with central pathology review. Int J Radiat Oncol Biol Phys 63(3):834–838.

Sorbe et al. *2005* → 290 low-risk endometrial carcinomas (stages IA–IB and grades 1–2). The HDR MicroSelectron afterloading equipment (iridium-192) was used. Perspex vaginal applicators with diameters of 20–30 mm were used, and the dose was specified at 5 mm from the surface of the applicator. Six fractions were given, and the overall treatment time was 8 days. The size of the dose per fraction was randomly set to 2.5 Gy (total dose of 15.0 Gy) or 5.0 Gy (total dose of 30.0 Gy).

- The overall locoregional recurrence rate of the complete series was 1.4%, and the rate of vaginal recurrence was 0.7%. There was no difference between the two randomized groups.
- The vaginal shortening (measured by colpometry) was not significant ($p = 0.159$) in the 2.5 Gy group (mean, 0.3 cm) but was highly significant ($p < 0.000001$) in the 5.0 Gy group (mean 2.1 cm) after 5 years.
- Mucosal atrophy and bleeding were significantly more frequent in the 5.0 Gy group. Symptoms noted in the 2.5 Gy group were no different from those that could be expected in a normal group of postmenopausal women.
- The fractionation schedule recommended for postoperative vaginal irradiation in low-risk endometrial carcinoma is six fractions of 2.5 Gy when the HDR technique is used.

Sorbe B (2005) Intravaginal high-dose-rate brachytherapy for stage I endometrial cancer: a randomized study of two dose-per-fraction levels. Int J Radiat Oncol Biol Phys 62(5):1385–1389.

GOG 122, 2006 → 422 patients with stage III or IV endometrial carcinoma having a maximum of 2 cm of postoperative residual disease were randomized to whole-abdominal irradiation (WAI) and doxorubicin–cisplatin (AP) chemotherapy. RT dose was 30 Gy in 20 fractions, with a 15 Gy boost. Chemotherapy consisted of doxorubicin 60 mg/m^2 and cisplatin 50 mg/m^2 every 3 weeks for seven cycles, followed by one cycle of cisplatin.

- Chemotherapy with AP significantly improved PFS and OS compared with WAI.

Randall ME et al. (2006) Randomized phase III trial of whole-abdominal irradiation versus doxorubicin and cisplatin chemotherapy in advanced endometrial carcinoma: a Gynecologic Oncology Group study. J Clin Oncol 24(1):36–44.

ASTEC/EN.5 Study Group, 2009 → 905 patients with early-stage endometrial cancer (intermediate or high risk) after surgery were randomized to observation or immediate external pelvic RT (40–46 Gy). Vaginal BT could be used regardless of ERT allocation.

- At a median follow-up of 58 months, no significant differences were observed in OS (83.9% vs. 83.5%, $p = 0.31$) and DSS (89.9% vs. 88.5%, $p = 0.28$) rates.
- Acute and late toxicity were greater in the external pelvic RT arm than observation arm.
- Meta-analysis of GOG 99, PORTEC-1, and ASTEC/EN.5 study showed no benefit of

external pelvic RT on OS in early-stage endometrial cancer. For these patients, vaginal BT may be sufficient.

ASTEC/EN.5 Study Group (2009) Adjuvant external beam radiotherapy in the treatment of endometrial cancer (MRC ASTEC and NCIC CTG EN.5 randomised trials): pooled trial results, systematic review, and meta-analysis. Lancet 373(9658):137–146.

JGOG 2033, 2007 → Adjuvant WPRT vs. cyclophosphamide–doxorubicin–cisplatin (CAP) chemotherapy in women with endometrioid adenocarcinoma with deeper than 50% myometrial invasion.

- No statistically significant differences in PFS and OS were observed.
- Among 120 patients in a high- to intermediate-risk group defined as (1) stage IC in patients over 70 years old or with grade 3 endometrioid adenocarcinoma or (2) stage II or IIIA (positive cytology), the CAP group had a significantly higher PFS rate (83.8 vs. 66.2%, $p = 0.024$) and higher OS rate (89.7 vs. 73.6%, $p = 0.006$).
- Adverse effects were not significantly increased in the CAP group vs. the WPRT group.
- Adjuvant chemotherapy may be a useful alternative to RT for intermediate-risk endometrial cancer.

Susumu N (2008) Randomized phase III trial of pelvic radiotherapy versus cisplatin-based combined chemotherapy in patients with intermediate-and high-risk endometrial cancer: A Japanese Gynecologic Oncology Group study. Gynecol Oncol 108(1):226–233.

PORTEC-2, 2008 → 427 patients with HIR endometrial cancer (age >60 years and stage 1C grade 1–2 or stage 1B grade 3; any age and stage 2A grade 1–2 or grade 3 with <50% myometrial invasion) after surgery were randomized to external pelvic RT (46 Gy in 23 fractions) or vaginal BT (21 Gy HDR in 3 fractions, or 30 Gy LDR).

- At a median follow-up of 34 months, no significant differences in 3-year OS and RFS.
- Vaginal relapse: 2% vs. 0.9, $p = 0.97$.
- Pelvic relapses were significantly decreased with pelvic RT (0.7% vs. 3.6%, $p = 0.03$).
- Update: At a median follow-up of 45 months, no significant differences were observed in 5-year DFS, OS, and vaginal relapse rates. However, pelvic relapses were significantly decreased with pelvic RT (0.5% vs. 3.8%, $p = 0.02$). Acute grade 1–2 gastrointestinal toxicity was higher in external pelvic RT arm 53.8% vs. 12.6% ($p < 0.05$).
- Central pathology review and molecular analysis (2019):
  – No significant differences in 10-year vaginal recurrence, isolated pelvic recurrence, or distant metastases rates.
  – Ten-year pelvic recurrence was more common in the vaginal BT group (6.3% vs. 0.9%, $p = 0.004$), mostly combined with distant metastases.
  – Ten-year OS was 69.5% for vaginal BT vs. 67.6% for external pelvic RT ($p = 0.72$).
  – L1CAM and p53-mutant expression and substantial LVSI were risk factors for pelvic recurrence and distant metastases. If patients have these unfavorable risk factors, external pelvic RT should be recommended to reduce pelvic recurrences.

Nout RA et al. (2008) Vaginal brachytherapy versus external beam pelvic radiotherapy for high-intermediate risk endometrial cancer: Results of the randomized PORTEC-2 trial. J Clin Oncol 26:2008 (May 20 suppl; abstr LBA5503).

Nout RA et al. (2010) Vaginal brachytherapy versus pelvic external beam radiotherapy for patients with endometrial cancer of high-intermediate risk (PORTEC-2): an open-label, non-inferiority, randomised trial. Lancet 375:816–823.

Wortman BG et al. (2019) Ten-year results of the PORTEC-2 trial for high-intermediate risk endometrial carcinoma: improving patient selection for adjuvant therapy. Br J Cancer 119(9):1067–1074.

Hacettepe, 2008 → 128 patients with intermediate- to high-risk stage I endometrial adenocarcinoma (any stage I with grade 3 histology or stage IB grade 2 or any stage IC disease) were treated with HDR BT alone after complete surgical staging. A total dose of 27.5 Gy with HDR BT, prescribed at 0.5 cm, was delivered in five fractions on five consecutive days. Median follow-up was 48 months.

- Six (4.7%) patients developed either local recurrence ($n = 2$) or distant metastases ($n = 4$).
- Five-year OS and DFS rates were 96 and 93%, respectively.
- Only age was found to be a significant prognostic factor for DFS. Patients younger than 60 years had significantly higher DFS ($p = 0.006$).
- None of the patients experienced grade 3/4 complications due to the vaginal HDR BT.
- Vaginal cuff BT alone is an adequate treatment modality in stage I endometrial adenocarcinoma patients with intermediate- to high-risk features after complete surgical staging with low complication rates.

Atahan IL, Ozyar E, Yildiz F, Ozyigit G et al. (2008) Vaginal high dose rate brachytherapy alone in patients with intermediate- to high-risk stage I endometrial carcinoma after radical surgery. Int J Gynecol Cancer 18(6):1294–1299.

Sweden, 2009 → 645 patients with low-risk, stage IA–IB, grade 1–2, endometrioid-type endometrial cancer were randomized to surgery (TAH, BSO, and LN sampling) or surgery followed by vaginal BT (3–6 fractions, 3–8 Gy). BT dose was prescribed to 5 mm from the surface of the applicator.

- There were no significant differences in OS or vaginal recurrence (3.1% vs. 1.2%, $p = 0.114$) rates.
- Among the entire cohort, locoregional recurrence and distant metastases rates were 2.6% and 1.4%, respectively.
- Genitourinary side effects were slightly more common after surgery and vaginal BT arm (0.6% vs. 2.8%, $p = 0.063$).
- There is no benefit of vaginal BT in low-risk patients.

Sorbe B et al. (2009) Intravaginal brachytherapy in FIGO stage I low-risk endometrial cancer a controlled randomized study. Int J Gynecol Cancer 19:873–878.

NSGO-EC-9501/EORTC-55991 and MaNGO ILIADE-III pooled analysis, 2010 → 540 patients with high-risk, stage I–III endometrial cancer were randomized to adjuvant RT or adjuvant RT with sequential chemotherapy (four cycles of platinum-based chemotherapy given either before or after RT). Eligible patients had no residual tumor.

- The addition of sequential chemotherapy to adjuvant external RT improved 5-year PFS (78% vs. 69%, $p = 0.009$) and CSS (87% vs. 78%, $p = 0.01$).
- No significant differences in OS (82% vs. 75%, $p = 0.07$).
- The benefit of adjuvant chemotherapy was restricted to endometrioid-type endometrial cancers.

Hogberg T et al. (2010) Sequential adjuvant chemotherapy and radiotherapy in endometrial cancer - results from two randomised studies. Eur J Cancer 46(13):2422–2431.

Sweden, 2012 → 527 patients with medium-risk (stage IA–IC) endometrial cancer were randomized to vaginal BT alone or adjuvant external pelvic RT and vaginal BT.

- Five-year locoregional recurrence rates were 5% for those treated with vaginal BT alone compared to 1.5% for those treated with vaginal BT and external pelvic RT ($p = 0.013$).
- External pelvic RT and vaginal BT was well tolerated. Gr 3 late side effects were only <2%. However, combined treatment significantly increased toxicity.
- No significant differences in OS (90% vs. 89%, $p = 0.548$).

Sorbe B et al. (2012) External pelvic and vaginal irradiation versus vaginal irradiation alone as postoperative therapy in medium-risk endometrial carcinoma: a prospective randomized study. Int J Radiat Oncol Biol Phys 82(3):1249–1255.

NRG Oncology-RTOG 1203 (TIME-C), 2018 → 289 patients with cervical or endometrial cancer were postoperatively randomized to standard four-field pelvic radiotherapy or IMRT.

- Patient-reported acute gastrointestinal and urinary toxicity, and late diarrhea was significantly decreased with IMRT than standard RT.
- Postoperative pelvic IMRT in patients with cervical or endometrial cancer improves quality of life compared to 3DCRT.

Klopp AH et al. (2018) Patient-reported toxicity during pelvic intensity-modulated radiation therapy: NRG Oncology-RTOG 1203. J Clin Oncol 36:2538–2544.

GOG 258 → 813 patients with endometrial cancer were randomized to adjuvant chemo-RT or chemotherapy alone. Eligible patients had stage III–IVA (<2 cm residual disease) endometrioid or stage I/II serous or clear cell endometrial cancer. Adjuvant chemo-RT arm included concurrent cisplatin and ERT followed by carboplatin and paclitaxel × 4 cycles. Chemotherapy-alone arm included carboplatin and paclitaxel × 6 cycles.

- At 60 months, no differences in RFS (HR 0.90, 95% CI 0.74–1.10) for those receiving chemo-RT vs. chemotherapy alone.
- Addition of RT decreased 5-year incidence of vaginal (2% vs. 7%; HR 0.36, 95% CI 0.16–0.82) and pelvic/para-aortic recurrences (11% vs. 20%; 0.43, 95% CI 0.28–0.66).
- Distant recurrences were more common with chemo-RT (27% vs. 21%; HR 1.36, 95% CI 1.0–1.86).
- The rates of ≥grade 3 toxicities were similar between the treatment arms (58% vs. 63%).

Matei D et al. (2019) Adjuvant chemotherapy plus radiation for locally advanced endometrial cancer. N Engl J Med 380:2317–2326.

GOG 249, 2019 → 601 patients with high-intermediate and high-risk early-stage endometrial cancer were randomized to pelvic external RT (45–50.4 Gy) or vaginal cuff BT plus chemotherapy. Patients with GOG 99-based HIR criteria (based on age, tumor Gr, depth of invasion, and presence of LVI), stage II, or stage I–II serous (S) or clear cell (CC) cancer were included. Chemotherapy consists of three cycles of carboplatin (AUC 6) and paclitaxel (175 mg/m^2) every 21 days. Median follow-up was 53 months.

- Five-year RFS was 76% in both arms.
- Five-year OS was 87% for RT and 85% for vaginal cuff BT + chemotherapy, respectively.
- No significant differences in vaginal or distant recurrences.
- Pelvic or para-aortic nodal recurrences were significantly more common in the vaginal cuff BT + chemotherapy arm (4% vs. 9%).
- While acute toxicity was more common and more severe with vaginal cuff BT + chemotherapy, late toxicity was similar.
- Adjuvant pelvic RT is the standard of care for high-risk, early-stage endometrial cancer. No benefit of the addition of chemotherapy was demonstrated in non-endometrioid histologies.

Randall ME et al. (2019) Phase III trial: Adjuvant pelvic radiation therapy versus vaginal brachytherapy plus paclitaxel/carboplatin in high-intermediate and high-risk early-Stage endometrial cancer. J Clin Oncol 37:1810–1818.

PORTEC-3, 2019 → 686 patients with high-risk endometrial cancer were randomized to combined adjuvant chemotherapy and RT or pelvic RT alone. Eligible patients had stage I (endometrioid Gr 3 with >50% myometrial invasion and/or LVSI), stage II or III disease, or tumors with serous or clear cell histologies (stage I–III). Chemo-RT included two cycles of cisplatin (50 mg/m^2) concurrently with pelvic RT, followed by four cycles of carboplatin (AUC 5) and paclitaxel (175 mg/m^2). RT was given 48.6 Gy in 1.8 Gy fractions.

- Median follow-up was 72.6 months.
- Chemo-RT improved 5-year OS (81.4% vs. 76.1%, *p* = 0.034) and 5-year FFS (76.5% vs. 69.1%, p = 0.016).

- Post-hoc analysis:
  - Stage III endometrial cancer: 5-year OS 78.5% vs. 68.5%, $p = 0.043$, 5-year FFS 70.9% vs. 58.4%, $p = 0.011$.
  - Serous cancer: 5-year OS 71.4% vs. 52.8%, $p = 0.037$, 5-year FFS 59.7% vs. 47.9%, $p = 0.008$.
- The major failure pattern was distant metastases in both arms (21.4% vs. 29.1%, $p = 0.047$).
- No differences in grade 3 toxicities between treatment arms (8% vs. 5%, $p = 0.24$). However, ≥grade 2 adverse events were more common in the chemo-RT arm (38% vs. 23%, $p = 0.002$).
- Adjuvant chemoRT should be recommended to high-risk patients with stage III disease or serous-type endometrial cancer.

de Boer SM et al. (2019) Adjuvant chemoradiotherapy versus radiotherapy alone in women with high-risk endometrial cancer (PORTEC-3): patterns of recurrence and post-hoc survival analysis of a randomized phase 3 trial. Lancet 20(9):1273–1285.

PORTEC-3 (molecular analysis), 2020 → Molecular subgroup analysis of high-risk patients included in the PORTEC-3 trial was performed ($n = 410$). Four subgroups were defined: p53 abnormal (p53abn) 23%, POLE-ultramutated (POLEmut) 12%, MMR-deficient (MMRd) 33%, and no specific molecular profile (NSMP) 32%.

- Five-year RFS: p53abn 48%, POLEmut 98%, MMRd 72%, and NSMP 74% ($p < 0.001$).
- Five-year RFS: Chemotherapy and RT vs. RT alone.
  - p53abn 59% vs. 36% ($p = 0.019$); POLEmut 100% vs. 97% ($p = 0.637$); MMRd 68% vs. 76% ($p = 0.428$); NSMP 80% vs. 68% ($p = 0.243$).
- Molecular classification has strong prognostic value in high-risk endometrial cancer. p53abn tumors most benefit from combined adjuvant chemotherapy and RT.

Leon-Castillo A et al. (2020) Molecular classification of the PORTEC-3 trial for high-risk endometrial cancer: Impact on prognosis and benefit from adjuvant therapy. J Clin Oncol 38(29):3388–3397.

9.3 Vaginal Cancer

Vaginal cancer is rarely seen and comprises 1–2% of all gynecological cancers [39]. They are classified into primary and secondary vaginal cancers. Primary vaginal cancer originates from the vagina, while secondary vaginal cancer usually originates from the metastases of cervical and vulvar cancers.

The vagina is a 7–9-cm-long cylindrical organ extending from the vestibule to the uterine cervix (Fig. 9.23). Its wall consists of glandular mucous membrane, a muscular layer rich in vascular structures, and adventitial connective tissues. Vaginal mucosa contains mucosal-type stratified nonkeratinized squamous cells that are sensitive to hormones and show cyclic changes. They also have multiple mucosal ridges called rugae.

Primary vaginal cancers are mostly seen in patients over 60 years of age [41].

Lower third of vagina develops from urogenital sinus, and upper two-third of vagina develops from Mullerian canal similar to inner genital organs such as uterus, Fallopian tubes, and ovaries during embryogenesis.

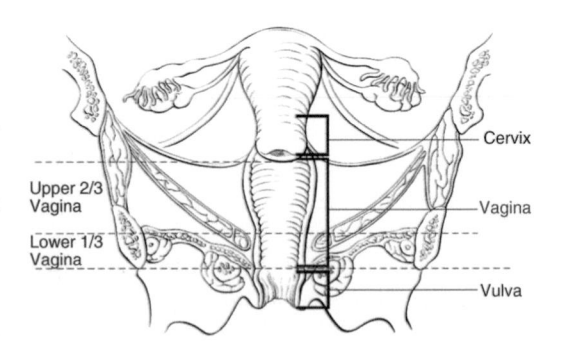

Fig. 9.23 Anatomy of the vagina. (From [40], Compton C.C., Byrd D.R., Garcia-Aguilar J., Kurtzman S.H., Olawaiye A., Washington M.K. (2012) Vagina. In: Compton C., Byrd D., Garcia-Aguilar J., Kurtzman S., Olawaiye A., Washington M. (eds) AJCC Cancer Staging Atlas. Springer, New York, NY. https://doi.org/10.1007/978-1-4614-2080-4_34)

9.3 Vaginal Cancer

9.3.1 Pathology

Mostly embryonal sarcomas and endodermal sinus tumors are observed in childhood, adenocarcinomas related to diethylstilbestrol (DES) in adolescence, and squamous cell cancer in adults. Although vaginal cancers are generally seen in older patients, their frequency among young females has increased in recent years due to HPV infections.

Primary vaginal cancers have several subtypes: pure epithelial, mesenchymal, mixed epithelial and mesenchymal, and germ cell tumors.

Nearly 80–90% of all pathological cases are squamous cell cancers, and most of these are grade II–III lesions [42].

Invasive squamous cell cancer. This is observed mostly in the posterior wall of the upper third of the vagina as an exophytic mass. Lesions may be 3–5 cm long and ulcerated. Advanced lesions are those that easily bleed. Endophytic tumors are less commonly seen. Most of them are microscopically moderately differentiated and show destructive infiltrative growth patterns in vaginal and paravaginal soft tissues.

Verrucous carcinoma. This is the least common variant of squamous cell cancer. They present with postmenopausal bleeding or vaginal discharge, and 0.8–10 cm in diameter, 2–3-cm-long verrucous lesions. These tumors grow slowly and rarely metastasize.

Malignant melanoma. These constitute 3% of all vaginal epithelial tumors. They originate from melanocytes found in various places in the vagina. Cases usually occur in the late reproductive or perimenopausal period. It behaves very aggressively and extends early.

Adenocarcinoma. This is observed in 10–15% of cases. It originates from the Bartholin and Skene glands. This is mostly observed in young women.

- Clear cell carcinoma develops due to DES exposure in utero. While patients with DES-associated clear cell carcinoma tend to have a good prognosis, patients with non-DES-associated adenocarcinoma have a worse prognosis.

Mesenchymal cancers. Pure primary tumors are very rare and follow an aggressive and lethal course. Sarcoma botryoides (embryonal rhabdomyosarcoma) and leiomyosarcomas are the most frequent. In addition, lymphomas, malignant fibrous histiocytomas, and postradiotherapy angiosarcomas may be seen.

Germ cell tumors. These are rarely seen in the vagina. Malignant ones are embryonic cell cancers and endodermal sinus tumors. They are observed in newborns.

9.3.2 General Presentation

Most patients present with abnormal vaginal bleeding and discharge. Pelvic pain and other symptoms change according to tumoral extension into surrounding tissues and organs. Anterior tumors produce bladder and urinary symptoms, while posterior ones yield tenesmus and defecation problems.

- Dyspareunia is observed in sexually active females.
- The disease is rarely diagnosed through abnormal PAP smear findings.
- The tumor is grossly exophytic in general, but it may be endophytic. Superficial ulcerations usually occur in advanced stages.
- Most of them localize in the upper one-third of the posterior vaginal wall.
- Chronic pelvic pain is observed in 5% of cases at diagnosis.

9.3.3 Staging

9.3.3.1 FIGO Staging (Fig. 9.24) [40, 43]

I: Tumor confined to the vagina.

II: Tumor invades paravaginal tissues but not pelvic sidewall*.

III: Tumor extends to pelvic sidewall* and/or causing hydronephrosis or nonfunctioning kidney.

Pelvic or inguinal lymph node metastasis.

IVA: Tumor invades mucosa of the bladder or rectum and/or extends beyond the true pelvis

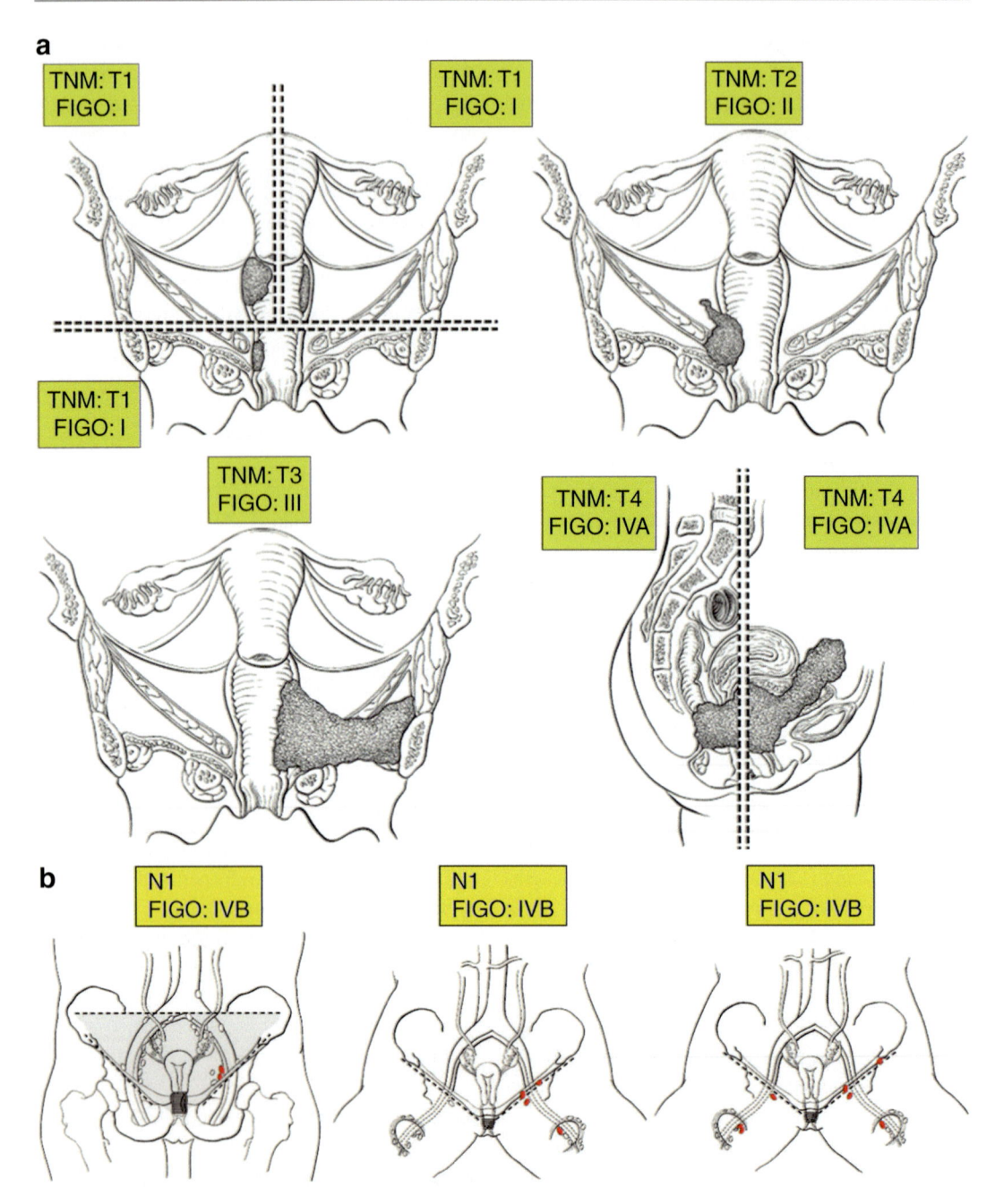

Fig. 9.24 Vaginal cancer staging. (**a**) Primary T staging, (**b**) N staging (From [40], Compton C.C., Byrd D.R., Garcia-Aguilar J., Kurtzman S.H., Olawaiye A., Washington M.K. (2012) Vagina. In: Compton C., Byrd D., Garcia-Aguilar J., Kurtzman S., Olawaiye A., Washington M. (eds) AJCC Cancer Staging Atlas. Springer, New York, NY. https://doi.org/10.1007/978-1-4614-2080-4_34)

9.3 Vaginal Cancer

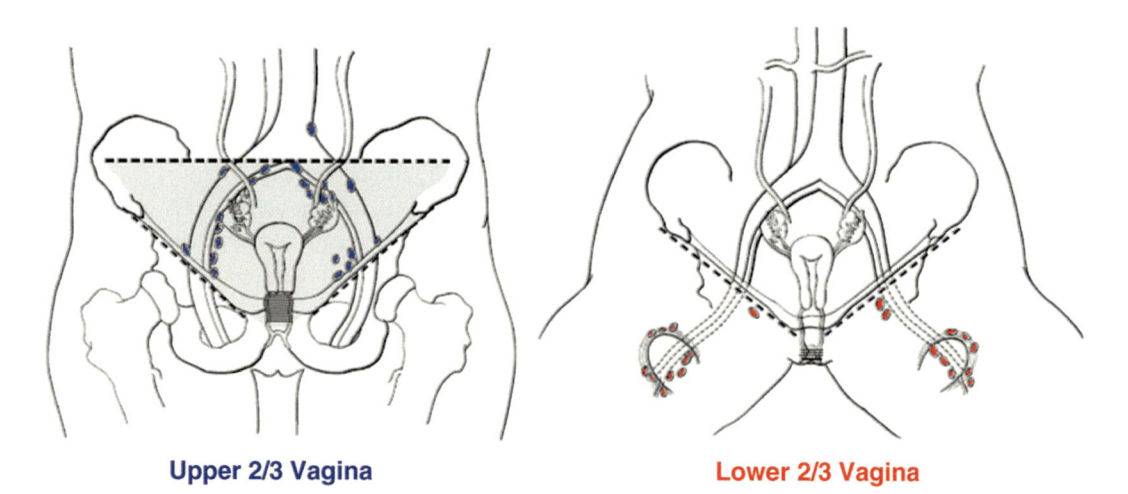

Upper 2/3 Vagina **Lower 2/3 Vagina**

Fig. 9.25 Vaginal lymphatics. (From [40], Compton C.C., Byrd D.R., Garcia-Aguilar J., Kurtzman S.H., Olawaiye A., Washington M.K. (2012) Vagina. In: Compton C., Byrd D., Garcia-Aguilar J., Kurtzman S., Olawaiye A., Washington M. (eds) AJCC Cancer Staging Atlas. Springer, New York, NY. https://doi.org/10.1007/978-1-4614-2080-4_34)

(bullous edema is not sufficient evidence to classify a tumor as T4).

IVB: Distant metastases.

*Note: Pelvic sidewall is defined as muscle, fascia, neurovascular structures, or skeletal portions of the bony pelvis.

9.3.3.2 Lymphatics of Vagina (Fig. 9.25)

Upper 2/3 of vagina: Obturator, internal iliac (hypogastric), external iliac, and common iliac LNs.

Lower 1/3 of vagina: First into the inguinal and femoral LNs, and subsequently to the pelvic LNs.

Posterior vaginal wall lesions can also drain to the presacral lymph nodes.

9.3.4 Treatment Algorithm

9.3.4.1 Treatment Algorithm for Vaginal Cancer [39, 44]

Stage 0 (VAIN 3 or CIS)

Cryotherapy, laser, topical 5-FU, 5% imiquimod or wide local excision.

[resistance to treatment; 60–70 Gy (EQD2) image-guided BT may be offered for full-thickness vaginal wall].

Stage I (tumor diameter <2 cm and upper one-third of vagina)

Surgery → partial vaginectomy + radical hysterectomy + pelvic LN dissection + vaginal reconstruction

- Surgical margin (+) or close; postoperative RT.

Alternative → BT (60–70 Gy).

Stage I (tumor diameter ≥2 cm and mid to lower third of vagina)

Pelvic RT + vaginal BT.

(45–50 Gy to the whole pelvis, cumulative EQD2 70–80 Gy)

(lower one-third of vagina (+), inguinal LN is included)

[concurrent weekly cisplatin (40 mg/m^2) should be added to RT].

Stage II

Pelvic RT + vaginal BT.

(45–50 Gy to the whole pelvis, cumulative EQD2 70–80 Gy)

(lower one-third of vagina (+), inguinal LN is included)

[concurrent weekly cisplatin (40 mg/m^2) should be added to RT].

Stage III/IV

Pelvic RT + vaginal BT.
(45–50 Gy to the whole pelvis, cumulative EQD2 70–80 Gy)
(lower one-third of vagina (+), inguinal LN is included)
(LN (+), 10–15 Gy to involved LNs)
Concurrent weekly cisplatin (40 mg/m^2) should be added to RT.
Pelvic exenteration in patients with vesicovaginal or rectovaginal fistula.
Special situations
Clear cell adenocarcinoma.
Surgery (radical hysterectomy, vaginectomy, pelvic + ara-aortic LN dissection).
Definitive RT in selected cases like stage II–IV
Metastatic disease.
Palliative RT ± chemotherapy
Recurrence.
Pelvic exenteration.

9.3.5 Radiotherapy

9.3.5.1 External RT Fields [45]
- Simulation is usually performed in the frog-leg position.
- Vaginal entrance and rectum can be marked by vaginal marker and rectal tube.
- CT-based treatment planning is recommended (with full and empty bladder CT).
- Gynecological examination is the most important part for local extent of disease assessment.
- MRI (T2-weighted images) and PET/CT can be useful for target delineation and treatment planning.
- The initial macroscopic gross tumor volume (GTV-T$_{init}$) at the time of diagnosis can only be delineated by T2-weighted MRI.
- The low-risk clinical target volume (CTV-T$_{LR}$) consists of the GTV-T$_{init}$, whole vagina, paravaginal space, paracolpia, parametria, and cervix.
- Nodal CTV (CTV-N) includes external iliac, internal iliac, and obturator nodes. If tumor involves the posterior vaginal wall or rectovaginal septum, presacral and mesorectal nodes should be included.
- Inguinal/femoral LN (+) or lower vaginal tumor; RT field should be include inguinal/femoral lymphatics (Fig. 9.26).
- Common iliac/para-aortic LNs should be included in the case of metastases at the level of common iliac or para-aortic veins (up to the renal vein).
- All pathological LNs should be defined (GTV-N$_x$/CTV-N$_x$) for external RT boost (SIB or sequential).
- Organs at risk for vaginal cancer include the bladder, rectum, sigmoid, bowel, urethra, anal canal, and noninvolved vagina, as well as vulva and clitoris for lower-third vaginal tumors.
- Full and empty bladder CT are fused to account for changes in bladder volume, then vaginal ITV will be defined.
- PTV margin is determined based on available treatment planning imaging, immobilization devices, and image-guidance technique (usually 1–1.5 cm).
- External RT should be delivered with 3DCRT or IMRT/VMAT (with daily cone beam CT) techniques to decrease toxicity.
- Total dose: 45–50 Gy in 1.8–2 Gy/day with high-energy photons; then BT if indicated.
- Gross LNs should receive 60–65 Gy.

9.3.5.2 Brachytherapy [45]
- A cumulative dose of 70–80 Gy (EQD2) has been recommended.
- Intracavitary (thickness ≤0.5 cm) or interstitial BT can be used according to the tumor thickness.
- Image-guided adaptive BT improves local control and decreases toxicity. Macroscopic gross residual tumor volume at the time of BT (GTV-T$_{res}$), high-risk clinical target volume (CTV-T$_{HR}$) including GTV-T$_{res}$ and all pathological tissues, and initial tumor extension at diagnosis that is called intermediate-risk clinical target volume (CTV-T$_{IR}$) should be delineated.
- A minimal 0.5 cm margin should be added around the CTV-T$_{HR}$ to define CTV-T$_{IR}$.

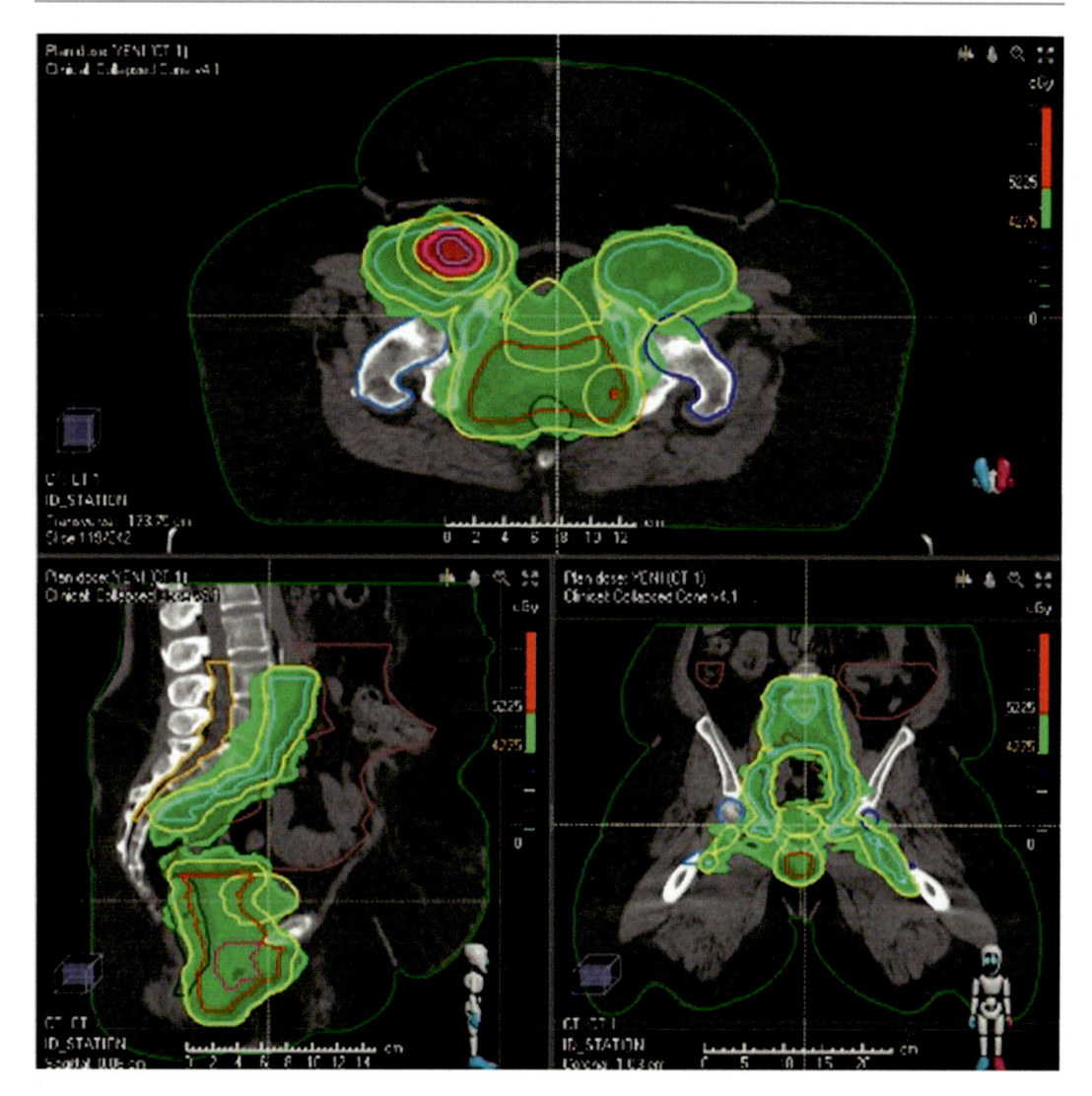

Fig. 9.26 VMAT field in lower one-third and inguinal LN (+) vaginal cancer

9.3.6 Selected Publications

M.D. Anderson, 2005 → 193 patients were treated with definitive RT for squamous cell carcinoma of the vagina.
- Five-year DSS rates were 85% for the 50 patients with stage I, 78% for the 97 patients with stage II, and 58% for the 46 patients with stage III–IVA disease ($p = 0.0013$).
- Five-year DSS rates were 82% and 60% for patients with tumors ≤4 cm or >4 cm, respectively ($p = 0.0001$).
- Five-year pelvic disease control rates were 86% for stage I, 84% for stage II, and 71% for stage III–IVA ($p = 0.027$).
- Excellent outcomes can be achieved with definitive RT for invasive squamous cell carcinoma of the vagina.
- Treatment must be individualized according to the site and size of the tumor at presentation and the response to initial external-beam RT.

Frank SJ (2005) Definitive radiation therapy for squamous cell carcinoma of the vagina. Int J Radiat Oncol Biol Phys 62(1):138–147.

MSKCC, 1992 → 36 patients were treated with combined external RT and BT, 11 patients were treated with external RT alone, and 2 patients were treated with BT alone.

- Five-year survival was 44% for stage I, 48% for stage II, 40% for stage III, and 0% for stages IVa and IVb.
- There was a significant increase in the 5-year actuarial survival for those patients who had BT as part of their treatment compared to those patients treated with external RT alone (50% vs. 9%) ($p < 0.001$). For stages II and III, there was a trend toward improved actuarial and crude DFS with the use of a temporary Ir-192 interstitial implant as part of the treatment compared to the use of intracavitary BT as part of the treatment (80% vs. 45%) ($p = 0.25$) and (75% vs. 44%) ($p = 0.08$), respectively.
- BT plays an important role in the management of primary vaginal cancer. A temporary interstitial implant should be used over an intracavitary form of therapy for more invasive disease.

Stock RG (1992) The importance of brachytherapy technique in the management of primary carcinoma of the vagina. Int J Radiat Oncol Biol Phys 24(4):747–753.

Washington University, 1988 → 165 patients with vaginal cancer.

- Ten-year DFS rates were stage 0, 94%; stage I, 75%; stage IIA, 55%; stage IIB, 43%; stage III, 32%; stage IV, 0%.
- RT dose delivered to the primary tumor or the parametrial extension was critical in achieving successful results.
- High incidence of distant metastases underscored the need for earlier diagnosis and effective systemic cytotoxic agents if survival is to be significantly improved in these patients.

Perez CA (1988) Definitive irradiation in carcinoma of the vagina: long-term evaluation of results. Int J Radiat Oncol Biol Phys 15(6):1283–1290.

Ontario, 2007 → 12 patients were treated with concurrent weekly CRT. Median follow-up was 50 months. All patients received pelvic external RT concurrently with weekly intravenous cis-platinum chemotherapy (40 mg/m^2) followed by BT. The median dose of external RT was 4500 cGy given in 25 fractions over 5 weeks.

- Five-year OS, PFS, and locoregional PFS rates were 66, 75, and 92%, respectively.
- It was found to be feasible to deliver concurrent weekly cis-platinum chemotherapy with high-dose radiation, leading to excellent local control and an acceptable toxicity profile.

Samant R (2007) Primary vaginal cancer treated with concurrent chemoradiation using cis-platinum. Int J Radiat Oncol Biol Phys 69(3):746–750.

UC Davis, 2004 → 14 patients were treated with primary therapy consisting of concurrent RT and chemotherapy.

- CRT was found to be an effective treatment for squamous carcinoma of the vagina.

Dalrymple JL (2004) Chemoradiation for primary invasive squamous carcinoma of the vagina. Int J Gynecol Cancer 14(1):110–117.

Pittsburgh University, 2012 → 30 patients with vaginal cancer were treated with 3-D HDR interstitial BT. Twenty-eight (93.3%) patients received external RT at a median 45 Gy followed by HDR BT at 3.75–5.0 Gy per fraction in five fractions.

- One- and two-year locoregional control rates were 84.4% and 78.8%, respectively.
- One- and two-year OS rates were 82.1% and 70.2%, respectively.
- No grade ≥3 gastrointestinal complications. Late grade 3 vaginal ulceration and grade 4 vaginal necrosis were seen in two patients.
- 3-D HDR BT was feasible and safe in the treatment of vaginal cancers.

Beriwal S (2012) Three-dimensional image-based high-dose-rate interstitial brachytherapy for vaginal cancer. Brachytherapy 11:176–180.

Dimopoulos, 2012 → 13 patients were treated with external RT (45–50.4 Gy) and MRI-guided BT ± chemotherapy.

- Three-year local control and OS rates were 92% and 85%, respectively.
- Grade 3–4 complications observed in three patients.
- MRI-guided BT was found promising.

Dimopoulos JC (2012) Treatment of locally advanced vaginal cancer with radiochemotherapy and magnetic resonance image-guided adaptive brachytherapy: dose-volume parameters and first clinical results. Int J Radiat Oncol Biol Phys 82:1880–1888.

Stanford University, 2013 → 91 patients were treated with definitive external RT ± BT. The mean total dose was 70.1 Gy.

- Total RT dose >70 Gy was associated with improved locoregional control and OS, but not statistically significant.
- Tumor size >4 cm was associated with poor outcomes.
- Grade 3–4 gastrointestinal and skin toxicity developed in 11% of patients. No grade 3–4 toxicities <72 Gy.
- No locoregional recurrence or grade 3–4 toxicities were reported in patients treated with IMRT.
- IMRT may result in excellent outcomes with reduced toxicity.

Hiniker SM (2013) Primary squamous cell carcinoma of the vagina: prognostic factors, treatment patterns, and outcomes. Gynecol Oncol 131:380–385.

Dana Farber Cancer Institute, 2013 → 71 patients with primary vaginal cancer were treated with definitive RT ($n = 51$) or CRT ($n = 20$). In the CRT group, all patients were treated with external RT and BT.

- CRT improved 3-year OS (56% vs. 79%, $p = 0.037$) and DFS (43% vs. 73%, $p = 0.011$).
- CRT was a significant predictor for DFS on multivariate analysis.

- Concurrent chemotherapy may improve treatment outcomes in patients with vaginal cancer.

Miyamoto DT (2013) Concurrent Chemoradiation for Vaginal Cancer. PLoS One 8(6):e65048.

SEER Database, 2016 → 2517 patients with vaginal cancer were treated with external RT alone or BT (alone or in combination with external RT).

- BT boost improved median OS (6.1 years vs. 3.6 years, $p < 0.001$) independent of other prognostic factors. Largest benefit was observed in patients with tumors >5 cm.
- An absolute 13% reduction in the risk of death with the addition of BT boost independent of stage.
- BT boost was associated with a longer OS than external RT alone.

Orton A (2016) Brachytherapy improves survival in primary vaginal cancer. Gynecol Oncol 141: 501–506.

Ontario, 2021 → 67 patients with vaginal cancer (stage I–IVA) were treated with external RT (45 Gy) and 3-D HDR interstitial BT (5–7.5 Gy/fraction, 2–4 fractions). Median total EQD2 for tumor was 74 Gy. Concurrent cisplatin was used in 70.2% of the patients.

- Three-year local control, PFS, and OS rates were 84.5%, 66.4%, and 81.5%, respectively.
- Concurrent cisplatin was associated with improved OS ($p = 0.014$) and PFS ($p = 0.032$).
- Grade 3–4 genitourinary/gastrointestinal toxicity was 10.4%.
- 3-D CT- or MRI-guided HDR interstitial BT was safe and effective.

Goodman CD (2021) 3D image-guided interstitial brachytherapy for primary vaginal cancer: A multi-institutional experience. Gynecol Oncol 160:134–139.

National Cancer Database, 2021 → 1094 patients treated with definitive chemotherapy and external RT ± BT were identified.

- BT boost was associated with improved 5-year OS (62.9% vs. 49.3%, $p = 0.0126$) on propensity score analyses.
- Overall treatment time ≤63 days was associated with improved 5-year OS (67.8% vs. 54.5%, $p = 0.0031$) in patients treated with external RT and BT.
- Other good prognostic factors for combined treatment: younger age, no comorbidity score, and negative lymph nodes.

Reshko LB (2021) The impact of brachytherapy boost and radiotherapy treatment duration on survival in patients with vaginal cancer treated with definitive chemoradiation. Gynecol Oncol 20:75–84.

SEER Database, 2021 → 1813 patients were treated with definitive RT ($n = 676$) or CRT (concurrent or adjuvant) ($n = 1137$).

- CRT was associated with improved 5-year OS and CSS.
- Age, histological type, tumor size, surgery, and FIGO stage were all independent prognostic factors for OS and CSS.
- CRT was improved survival in patients with vaginal cancer compared with RT alone, except for those with adenocarcinoma, tumor size <2 cm, or FIGO stage I.

Zhou W (2021) Radiotherapy plus chemotherapy is associated with improved survival compared to radiotherapy alone in patients with primary vaginal carcinoma: A retrospective SEER study. Front Oncol 10:570933.

9.4 Vulvar Cancer

Vulvar cancer is a rare tumor with an estimated 45,240 new cases and 17,427 deaths in 2020 [1].

The vulva consists of mons pubis, clitoris, labia majora, labia minora, vaginal vestibule, Bartholin's glands, prepuce over clitoris, posterior fourchette, and perineal body.

HPV infection, vulvar intraepithelial neoplasia (VIN), cigarette smoking, lichen sclerosus, and immunodeficiency syndromes are classically associated with the development of vulvar cancer. It has a bimodal age distribution, both young and elderly patients can be affected.

Prognostic factors: LN metastases (unilateral/bilateral, number, location, extracapsular extension, etc.), tumor size, grade, tumor localization, LVSI, HPV, surgical margin, and age [46].

9.4.1 Pathology

9.4.1.1 Vulvar Intraepithelial Neoplasia (VIN) [47]

- Low-grade squamous intraepithelial lesion (LSIL) (=uVIN1).
  - Benign.
  - >90% HPV 6 and 11
- High-grade squamous intraepithelial lesion (HSIL) (=uVIN2–3).
  - Young patients.
  - >90% HPV 16, 18, and 33
  - 20% of vulvar cancers
- VIN differentiated type (dVIN).
  - Postmenopausal patients.
  - Associated with lichen sclerosis.
  - 80% of vulvar cancers.

Mostly vulvar cancers are squamous cell cancers (80–90%) [48].

Spray pattern invasion. This is characterized by infiltration of the surrounding stroma by individual tumor cells or cords. They have an increased risk of local recurrence and LN metastases.

Verrucous carcinoma. This is the least common variant of squamous cell cancer. These tumors grow slowly and rarely metastasize to lymph nodes.

Other histological types: malignant melanoma, basal cell carcinoma, Merkel cell tumors, carcinoid, sarcomas, transitional cell carcinoma, and adenoid cystic carcinoma.

9.4.2 General Presentation

Most patients present with a vulvar mass or ulcer. Patients also present with pruritus, vulvar pain,

dysuria, difficulty with defecation, bleeding, and discharge according to the extent of the disease.
- Most tumors are located in labia majora and minora.

9.4.3 Staging

9.4.3.1 FIGO Staging [33, 49]
I: Tumor confined to the vulva

IA: ≤2 cm, confined to the vulva or perineum and with stromal invasion ≤1 mm

IB: >2 cm or with stromal invasion >1 mm, confined to the vulva or perineum

II: Tumor of any size with extension to adjacent perineal structures (1/3 lower urethra, 1/3 lower vagina, anal involvement)

III: Tumor of any size ± extension to any of the following: Upper/proximal 2/3 of urethra, upper/proximal 2/3 vagina, bladder mucosa, rectal mucosa, or fixed to pelvic bone

IIIA: −1 or 2 lymph node metastasis (<5 mm)
−1 lymph node metastasis (≥5 mm).

IIIB: −2 or more lymph node metastasis (≥5 mm)
−3 or more lymph node metastasis (<5 mm).

IIIC: Positive lymph node with extracapsular extension

IV: Tumor invades other regional (2/3 upper urethra, 2/3 upper vagina), or distant structures

IVA: Tumor invades any of the following:
- Upper urethral or vaginal mucosa, bladder mucosa, rectal mucosa, or fixed to pelvic bone, or.
- Fixed or ulcerated inguinofemoral lymph nodes.

IVB: Any distant metastasis including pelvic lymph nodes

* The depth of invasion is defined as the measurement of the tumor from the epithelial stromal junction of the adjacent-most superficial dermal papilla to the deepest point of invasion.

9.4.3.2 Lymphatics of Vulva
Superficial inguinofemoral lymph nodes.
Deep inguinofemoral lymph nodes.
Pelvic lymph nodes (obturator, external iliac, internal iliac).

9.4.4 Treatment Algorithm

9.4.4.1 Treatment Algorithm for Vulvar Cancer [50, 51]
Stage 0 (CIS)
Local excision or CO_2 laser.
Stage IA (≤1 mm depth of invasion)
Wide local excision
- Surgical margin (+) or <8 mm, LVSI, >5 mm depth of invasion, and/or diffuse/spray pattern of invasion; postoperative RT to vulva.

Stage IB/II (>1 mm invasion)
Wide local excision + LN dissection.
Lateralized lesion: Ipsilateral (superficial) LN dissection or sentinel LN biopsy.
Central lesion, >5 mm depth, LVSI, poor differentiation: Bilateral (superficial) lymph node dissection.
If LN (+): Add deep inguinal dissection
- Surgical margin (+) or <8 mm, LVSI, >5 mm depth of invasion, and/or diffuse/spray pattern of invasion; postoperative RT to vulva.
- >1 LN (+) or extracapsular extension; postoperative RT to inguinal and pelvic LNs.

Alternatively: Preoperative chemo-RT (for lesions close to urethra, clitoris, or rectum) ± planned LN dissection.

9.5 50 Gy for cN0, 54 Gy for cN+

Either elective chemo-RT to groins or planned LN

Complete response (CR) in primary lesion: If biopsy negative observation.

If <CR or persistent disease: Resection, radical vulvectomy, or boost to primary (65–70 Gy).

Stage III/IVA
cN0: Bilateral LN dissection followed by chemo-RT (vulva or vulva and inguinal/pelvic lymph nodes).

cN+ (fixed or ulcerated): Preoperative chemo-RT followed by LN dissection.

45–50 Gy with cisplatin (50 mg/m^2), 5-FU/cisplatin, and 5-FU/mitomycin C.

If extracapsular extension: 60 Gy.
If gross residual disease: 65–70 Gy.

Persistent or recurrent disease: salvage surgery.

Stage III/IVA

Combination chemotherapy (carboplatin and paclitaxel) ± palliative RT.

9.5.1 Radiotherapy

9.5.1.1 External RT Fields (Fig. 9.27) [52]

- To improve dose homogeneity and limit normal tissue toxicity, IMRT or VMAT techniques should be preferred (with image guidance).
- Simulation is traditionally performed in the "frog-leg" position.
- CT-based treatment planning is recommended with specific bladder and rectal filling protocols.
- IV contrast material is very helpful for contouring.
- Tumor, anus, urethra, clitoris, and all surgical scars should be marked by wires or radio-opaque markers.
- Contouring: vulva.
  - GTV = any visible and/or palpable vulvar disease.
  - CTV = GTV/tumor bed + entire vulva, adjacent skin, mucosa, and subcutaneous tissue (GTV to CTV margin: minimum 1 cm).
  - If satellite lesions, LVSI, dermal lymphatic invasion, or muscle invasion: Add extra margins.
  - Vaginal involvement:
    GTV + 3 cm.
    Uncertainty in the proximal extent of vaginal extension or LVSI: Entire vaginal length.
  - Involvement of clitoris, urethra, anus, anal canal, or bladder: GTV + at least 2 cm of these structures. If disease extends into the mid or proximal urethra, CTV should include the entire urethra and bladder neck.
  - Close or positive margins: Marked with wire and at least 2 cm margin.
  - Excluding bony tissue and the soft tissues of the thigh and buttock.
- ITV should be created to account for bladder and rectal filling in locally advanced vulvar cancer.
- PTV = CTV or ITV + (0.7–1 cm).
- Contouring: LNs.
  - Inguinofemoral, external iliac, internal iliac, and obturator regions should be contoured bilaterally.
  - Involvement of anus/anal canal: Add perirectal (including mesorectum) and presacral LNs (S1–S3).
  - Involvement of proximal half of the posterior vaginal wall: Add presacral LNs (S1–S3).
  - Pelvic nodal CTV = vessels +7 mm.
  - Inguinal nodal CTV = inguinofemoral vessels + anteromedial ≥35 mm, anterior ≥23 mm, anterolateral ≥25 mm, and medial ≥22 mm.
  - PTV = nodal CTV + (7–10 mm).
- If patients have LN metastases, treatment volume should encompass vulva in addition to lymphatic region.
- If there is a skin involvement, CTV should include the skin with bolus.
- Small bowel, sigmoid, bladder, rectum, and bilateral femoral heads should be defined as critical structures.
- Postoperative dose: 1.8–2 Gy/day.
  - Negative surgical margin: 45–50 Gy.
  - Close margin: 50–55 Gy.
  - Positive margin: 54–59.9 Gy.
  - Microscopic nodal disease: 50 Gy.
  - Extracapsular extension: 60 Gy.
  - Gross disease: 65–70 Gy.
- Dose constraints for IMRT [53]:
  - Rectum: V40 Gy ≤40%, maximum dose (Dmax) 50 Gy.
  - Bladder: V40 Gy ≤40%, Dmax 50 Gy.
  - Small bowel: V35 Gy ≤35%, Dmax 50 Gy.
  - 95% of the volume of each PTV should receive the prescription dose.

9.5.2 Selected Publications

GOG 37, 1986 → 114 patients with vulvar cancer after radical vulvectomy and bilateral inguinal

9.5 50 Gy for cN0, 54 Gy for cN+

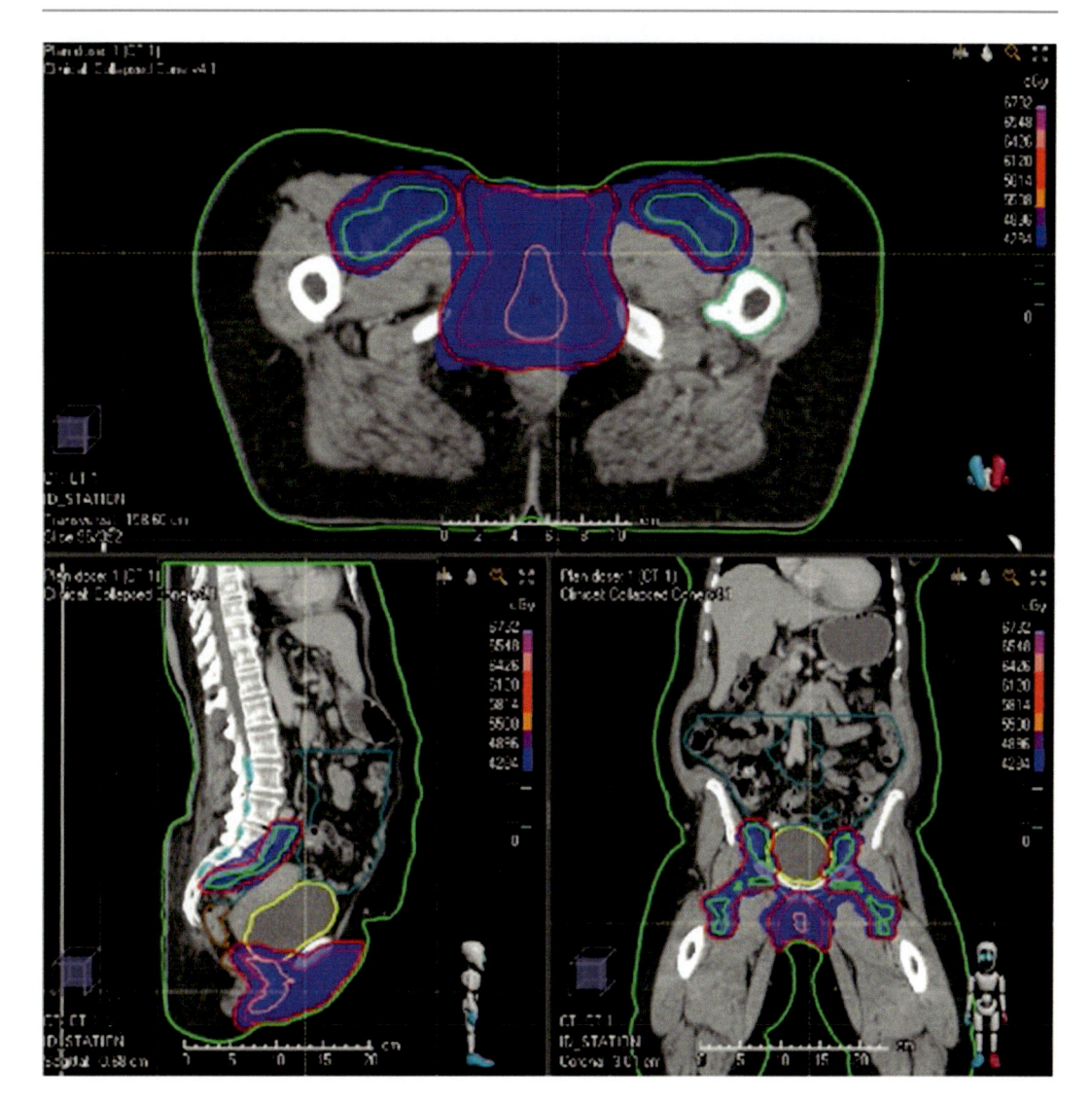

Fig. 9.27 VMAT field in inguinal LN (+) vulvar cancer

lymphadenectomy (inguinal LN+) were randomized to pelvic lymphadenectomy or pelvic RT (including groin but not to vulva, 45–50 Gy).

- Acute and chronic morbidity was similar.
- Major poor prognostic factors: Clinically suspicious or fixed ulcerated inguinal LN and ≥2 positive inguinal LNs.
- Adjuvant RT decreased groin recurrence (5% vs. 24%). However, no difference in pelvic recurrence.
- Two-year overall survival was improved in RT group (68% vs. 54%, $p = 0.03$), especially in patients with major poor prognostic factors.
- The addition of adjuvant RT to pelvic and inguinofemoral LNs after radical vulvectomy and bilateral inguinal lymphadenectomy was associated with superior outcomes.
- Update (2009): No difference in 6-year OS (51% vs. 41%, $p = 0.18$); however, 6-year cancer-related deaths were reduced in the RT group (51% vs. 29%, $p = 0.015$), especially in

patients with ≥2 positive inguinal LNs, ECE, and gross LN metastases.

Homesley HD et al. (1986) Radiation Therapy versus pelvic node resection for carcinoma of the vulva with positive groin nodes. Obstet Gynecol 68(6):733–740.

Kunos C et al. (2009) Radiation therapy compared with pelvic node resection for node-positive vulvar cancer: a randomized controlled trial. Obstet Gynecol 114:537–546.

GOG 88, 1992 → 58 patients with vulvar cancer (N0–1 inguinal nodes) were randomized to radical vulvectomy and inguinal lymphadenectomy or radical vulvectomy and inguinal RT (2 Gy/50 Gy to 3 cm depth below the skin). Vulva and pelvic LNs were not included in the RT portal.

- Closed prematurely due to excessive number of groin relapses on the inguinal RT arm (0% vs. 18.5%).
- Inguinal lymphadenectomy significantly improved treatment outcomes: 3-year DFS (90% vs. 65%, $p = 0.03$), 3-year OS (85% vs. 60%, $p = 0.04$).
- In patients with N0–1 vulvar carcinoma, RT to intact groin was significantly inferior to inguinal lymphadenectomy.

Stehman FB et al. (1992) Groin dissection versus groin radiation in carcinoma of the vulva: A Gynecologic Oncology Group study. Int J Radiat Oncol Biol Phys 24(2):389–396.

GOG 101, 1998 → 73 patients with unresectable vulvar cancer (stage III–IV) were treated with preoperative chemoradiotherapy (CRT). CRT scheme consists of split course and twice daily irradiation (1.7 Gy/47.6 Gy, 2-week break) with two courses of 5-FU and cisplatin. If patients had residual disease, surgery including bilateral inguinofemoral LN dissection was performed.

- Preoperative CRT resulted in 46.5% clinical CR and 53.5% gross residual disease. Only 2.8% patients had unresectable disease.
- Four-year OS was 55% in CRT group.
- Preoperative CRT in unresectable vulvar cancer may reduce the need for more radical surgery, including primary pelvic exenteration.

Moore DH et al. (1998) Preoperative chemoradiation for advanced vulvar cancer: a phase II study of the Gynecologic Oncology Group. Int J Radiat Oncol Biol Phys 42(1):79–85.

Oklahoma University, 2008 → 63 patients with advanced stage (III–IV) vulvar cancer were treated with primary surgery or primary CRT (weekly cisplatin or two cycles of cisplatin plus 5-FU with RT).

- Primary CRT was more commonly applied to younger patients (61 vs. 72 years old) that had larger tumors (6 vs. 3.5 cm) with less nodal metastasis (54% vs. 83%).
- No difference in OS, PFS, or recurrence rates.

Landrum LM et al. (2008) Comparison of outcome measures in patients with advanced squamous cell carcinoma of the vulva treated with surgery or primary chemoradiation. Gynecol Oncol 108(3):584–590.

Cochrane Systematic Review, 2011 → 141 patients with locally advanced vulvar cancer included in this review. The effectiveness and safety of CRT (neoadjuvant or primary) for women with locally advanced vulvar cancer compared to primary surgery were analyzed.

- No significant difference in OS.
- No significant difference in toxicity.

Shylasree TS et al. (2011) Chemoradiation for advanced primary vulval cancer. Cochrane Database Syst Rev.; (4): CD003752

GOG 173, 2012 → 452 patients with vulvar cancer underwent sentinel LN (SLN) biopsy followed by lymphadenectomy. Patients with squamous cell cancer, 2–6 cm tumor, ≥1 mm depth of invasion, tumor limited to vulva, clinically N0 were included.

- The sensitivity and specificity of SLN biopsy were 91.7% and 96%, respectively.

- False-negative rate: 8.3%.
- <4 cm tumor: false-negative rate: 2%.
- SLN biopsy may replace inguinofemoral lymphadenectomy in selected patients with vulvar cancer.

Levenback CF et al. (2012) Lymphatic mapping and sentinel lymph node biopsy in women with squamous cell carcinoma of the vulva: a gynecologic oncology group study. J Clin Oncol 30(31):3786–3791.

GOG 205, 2012 → 58 patients with unresectable (T3–T4) vulvar cancer were treated with CRT (1.8 Gy/57.6 Gy and weekly cisplatin) followed by surgery for residual tumor.

- Median follow-up time: 24.8 months.
- 64% of patients had clinical complete response (CR). Among these women, 78% had also pathological CR.
- Concomitant cisplatin-based CRT was effective and tolerable.

Moore DH et al. (2012) A phase II trial of radiation therapy and weekly cisplatin chemotherapy for the treatment of locally-advanced squamous cell carcinoma of the vulva: A gynecologic oncology group study. Gynecol Oncol 124:529–533.

NCDB, 2015 → 1797 patients with vulvar cancer who had pathological inguinal LN metastases were treated with adjuvant RT following extirpative surgery.

- 26.3% of patients received adjuvant chemotherapy and 76.6% had 1–3 LN metastases.
- The median survival significantly improved with adjuvant chemotherapy (29.7 months vs. 44 months, $p = 0.001$).
- Adjuvant chemotherapy resulted in 38% reduction in the risk of death (HR 0.62, 95% CI 0.48–0.79, $p < 0.001$).

Gill BS et al. (2015) Impact of adjuvant chemotherapy with radiation for node-positive vulvar cancer: A National Cancer Data Base (NCDB) analysis. Gynecol Oncol 137(3):365–372.

AGO-CaRE-1 Study, 2015 → 1249 patients had surgical groin staging were analyzed in this retrospective multicentric cohort study.

- 35.8% patients had LN metastases (1 LN+ 38.5%, 2 LN+ 22.8%).
- The 3-year PFS and OS rates for LN+ and negative patients were 35.2% and 56.2%, and 75.2% and 90.2%, respectively.
- LN+ patients who received adjuvant RT (54.6%) had better 3-year PFS (39.6% vs. 25.9%, $p = 0.004$) and OS (57.7% vs. 51.4%, $p = 0.17$).
- Adjuvant RT was associated with improved outcomes in LN+ patients.

Mahner S et al (2015) Adjuvant therapy in lymph node-positive vulvar cancer: the AGO-CaRE-1 study. J Natl Cancer Inst 107(3):dju426.

NCDB, 2017 → 3075 patients with positive surgical margins were analyzed. All patients underwent initial extirpative surgery.

- The 3-year OS significantly improved with adjuvant RT (58.5% vs. 67.4%, $p < 0.001$).
- Cumulative doses of ≥54 Gy were associated with improved OS.
- No survival benefit of ≥60 Gy.
- Optimal dose range in surgical margin-positive patients: 54–59.9 Gy.

Chapman BV et al. (2017) Adjuvant Radiation Therapy for Margin-Positive Vulvar Squamous Cell Carcinoma: Defining the Ideal Dose-Response Using the National Cancer Data Base. Int J Radiat Oncol Biol Phys 97(1):107-117

NCDB, 2017 → 1352 patients with vulvar cancer treated with definitive RT or definitive CRT were included. Median radiation dose was 59.4 (40–79.2) Gy.

- The 5-year OS was 49.9% with CRT vs. 27.4% with RT alone ($p < 0.001$).
- There was a 24% reduction of death risk with CRT (HR 0.76, %95 CI 0.63–0.91, $p = 0.003$).
- Definitive CRT was associated with higher OS compared to RT alone in patients with unre-

sectable or medically inoperable vulvar cancer.

Rao YJ et al. (2017) Improved survival with definitive chemoradiation compared to definitive radiation alone in squamous cell carcinoma of the vulva: A review of the National Cancer Database. Gynecol Oncol 146:572–579.

NCDB, 2017 → 2046 patients with locally advanced vulvar cancer treated with definitive RT or preoperative RT were included.

- The 3-year OS was 57.1% with preoperative RT/CRT followed by surgery vs. 41.7% with definitive RT/CRT ($p < 0.001$).
- In multivariate analysis, ≥55 Gy definitive RT/CRT had similar OS with preoperative RT/CRT followed by surgery.

Natesan D et al. (2017) Primary versus preoperative radiation for locally advanced vulvar cancer. Int J Gynecol Cancer 27(4):794–804.

FRANCOGYN Group, 2020 → 677 patients with vulvar cancer were treated with primary surgery (93.9%). In women with LN groin staging, 176 patients had at least one LN metastasis.

- No significant differences in 5-year OS between patients with one extracapsular LN metastasis and patients with one intracapsular LN metastasis, or with two LN metastases ($p = 0.62$ and $p = 0.63$, respectively).
- In women with a single intracapsular LN metastasis, LVSI and the absence of adjuvant inguinofemoral RT were negative predictive factors for RFS.

Serre E et al. (2020) Inguino-femoral radiotherapy in vulvar squamous cell carcinoma: clues to revised indications in patients with only one intracapsular lymph node metastasis. Acta Oncol 59(5):518–524.

van der Velden, 2021 → 96 patients with vulvar cancer who have single clinically occult intracapsular LN metastasis treated with inguinofemoral lymphadenectomy without adjuvant RT were analyzed.

- After a median follow-up of 64 months, 1% and 2.1% of patients had an isolated groin or groin and local recurrence, respectively.
- Size of the LN or LN ratio had no impact on outcomes.
- The 5-year disease-specific survival, OS, and groin recurrence-free survival rates were 79%, 62.5%, and 97%, respectively.
- Adjuvant RT to the groin and pelvis can be safely omitted in patients with a single occult intracapsular LN metastasis to prevent unnecessary toxicity and morbidity.

van der Velden J et al. (2021) Radiotherapy is not indicated in patients with vulvar squamous cell carcinoma and only one occult intracapsular groin node metastasis. Gynecologic Oncology 160(1):128–133.

References

1. Sung H, Ferlay J, Siegel RL, Laversanne M, Soerjomataram I, Jemal A, et al. (2021) Global cancer statistics 2020: GLOBOCAN estimates of incidence and mortality worldwide for 36 cancers in 185 countries. CA Cancer J Clin 0:1–41
2. Silverberg SG, Ioffe OB (2003) Pathology of cervical cancer. Cancer J 9(5):335–347. (review)
3. Kurman RJ, Carcangiu ML, Herrington CS, Young RH (2014) WHO classification of tumours of the female reproductive organs, 4th edn. IARC, Lyon
4. Stolnicu S, Barsan I, Hoang L, Patel P, Terinte C, Pesci A et al (2018) International endocervical adenocarcinoma criteria and classification (IECC): a new pathogenetic classification for invasive adenocarcinomas of the endocervix. Am J Surg Pathol 42:214–226
5. Bhatla N, Berek JS, Cuello Fredes M, Denny LA, Grenman S, Karunaratne K et al (2019) Revised FIGO staging for carcinoma of the cervix uteri. Int J Gynaecol Obstet 145(1):129–135
6. Compton CC, Byrd DR, Garcia-Aguilar J, Kurtzman SH, Olawaiye A, Washington MK (2012) Cervix uteri. In: Compton C, Byrd D, Garcia-Aguilar J, Kurtzman S, Olawaiye A, Washington M (eds) AJCC cancer staging atlas. Springer, New York, NY. https://doi.org/10.1007/978-1-4614-2080-4_35
7. Hansen EK, Mack Roach III (2018) Handbook of evidence-based radiation oncology, 3rd edn. Springer, US, pp 623–644
8. Thomas GM (1999) Improved treatment for cervical cancer–concurrent chemotherapy and radiotherapy. N Engl J Med 340(15):1198–1200

9. Pearcey R, Brundage M, Drouin P et al (2002) Phase III trial comparing radical radiotherapy with and without cisplatin chemotherapy in patients with advanced squamous cell cancer of the cervix. J Clin Oncol 20(4):966–972
10. Rose PG, Bundy BN (2002) Chemoradiation for locally advanced cervical cancer: does it help? J Clin Oncol 20(4):891–893
11. The NCCN guidelines for cervical cancer version 1.2021
12. Schlaerth AC, Abu-Rustum NR (2006) Role of minimally invasive surgery in gynecologic cancers. Oncologist 11(8):895–901. (review)
13. Chino J, Annunziata CM, Beriwal S, Bradfield L, Erickson BA, Fields EC et al (2020) Radiation therapy for cervical cancer: executive summary of an ASTRO clinical practice guideline. Pract Radiat Oncol 10(4):220–234
14. Cibula D, Potter R, Planchamp F et al (2018) The European Society of Gynaecological Oncology/European Society for Radiotherapy and Oncology/European Society of Pathology guidelines for the management of patients with cervical cancer. Radiother Oncol 127(3):404–416
15. Small W Jr, Mell LK, Anderson P, Creutzberg C, De Los SJ, Gaffney D et al (2008) Consensus guidelines for delineation of clinical target volume for intensity-modulated pelvic radiotherapy in postoperative treatment of endometrial and cervical cancer. Int J Radiat Oncol Biol Phys 71(2):428–434
16. Small W Jr, Bosch WR, Harkenrider MM, Strauss JB, Abu-Rustum N, Albuquerque KV et al (2021) NRG oncology/RTOG consensus guidelines for delineation of clinical target volume for intensity-modulated pelvic radiation therapy in postoperative treatment of endometrial and cervical cancer-an update. Int J Radiat Oncol Biol Phy 109(2):413–424
17. Lim K, Small W Jr, Portelance L, Creutzberg C, Jurgenliemk-Schulz IM, Mundt A et al (2011) Consensus guidelines for delineation of clinical target volume for intensity-modulated pelvic radiotherapy for the definitive treatment of cervix cancer. Int J Radiat Oncol Biol Phys 79(2):348–355
18. Eifel PJ, Burke TW, Delclos L et al (1991) Early stage I adenocarcinoma of the uterine cervix: treatment results in patients with tumors less than or equal to 4 cm in diameter. Gynecol Oncol 41(3):199–205
19. Peters WA III, Liu PY, Barrett RJ II et al (2000) Concurrent chemotherapy and pelvic radiation therapy compared with pelvic radiation therapy alone as adjuvant therapy after radical surgery in high-risk early-stage cancer of the cervix. J Clin Oncol 18(8):1606–1613
20. Keys HM, Bundy BN, Stehman FB et al (1999) Cisplatin, radiation, and adjuvant hysterectomy compared with radiation and adjuvant hysterectomy for bulky stage IB cervical carcinoma. N Engl J Med 340(15):1154–1161
21. Whitney CW, Sause W, Bundy BN et al (1999) Randomized comparison of fluorouracil plus cisplatin versus hydroxyurea as an adjunct to radiation therapy in stage IIB-IVA carcinoma of the cervix with negative Para-aortic lymph nodes: a Gynecologic oncology group and southwest oncology group study. J Clin Oncol 17(5):1339–1348
22. Viswanathan AN, Petereit DG (2007) Gynecologic brachytherapy. In: Devlin PM (ed) Brachytherapy: applications and technique, 1st edn. Lippincott Williams & Wilkins, Philadelphia, pp 225–254
23. International Commission on Radiation Units and Measurements (1985) Dose and volume specification for reporting intracavitary therapy in gynecology, vol 38. International Commission on Radiation Units and Measurements, Bethesda, MD
24. Wang KL, Yang YC, Chao KS, Wu MH, Tai HC, Chen TC, Huang MC, Chen JR, Su TH, Chen YJ (2007) Correlation of traditional point a with anatomic location of uterine artery and ureter in cancer of the uterine cervix. Int J Radiat Oncol Biol Phys 69(2):498–503
25. Potter R, Haie-Meder C, Van Limbergen E et al (2006) Recommendations from gynaecological (GYN) GEC ESTRO working group (II): concepts and terms in 3D image-based treatment planning in cervix cancer brachytherapy-3D dose volume parameters and aspects of 3D image-based anatomy, radiation physics, radiobiology. Radiother Oncol 78(1):67–77
26. Haie-Meder C, Potter R, Van Limbergen E et al (2005) Recommendations from gynaecological (GYN) GEC-ESTRO working group (I): concepts and terms in 3D image based 3D treatment planning in cervix cancer brachytherapy with emphasis on MRI assessment of GTV and CTV. Radiother Oncol 74(3):235–245
27. International Commission on Radiation Units and Measurements (2016) Prescribing R, and reporting brachytherapy for cancer of the cervix (ICRU report 89), Bethesda
28. Siegel RL, Miller KD, Jemal A (2020) Cancer statistics, 2020. CA Cancer J Clin 70(1):7–30
29. Compton CC, Byrd DR, Garcia-Aguilar J, Kurtzman SH, Olawaiye A, Washington MK (2012) Corpus uteri. In: Compton C, Byrd D, Garcia-Aguilar J, Kurtzman S, Olawaiye A, Washington M (eds) AJCC cancer staging atlas. Springer, New York, NY. https://doi.org/10.1007/978-1-4614-2080-4_36
30. WHO Classification of Tumours Editorial Board (2020) WHO classification of tumours: female genital tumours, 5th edn., vol 4.
31. Levine DA, The Cancer Genome Atlas Research Network (2013) Integrated genomic characterization of endometrial carcinoma. Nature 497:67–73
32. Photopulos GJ (1994) Surgicopathologic staging of endometrial adenocarcinoma. Curr Opin Obstet Gynecol 6(1):92–97. (review)
33. Pecorelli S (2009) Revised FIGO staging for carcinoma of the vulva, cervix, and endometrium. Int J Gynaecol Obstet 105(2):103–104. 26 Deeks E (2007) Local therapy in endometrial cancer: evidence based review. Curr Opin Oncol 19(5):512–515 (review)

34. Amant F, Moerman P, Neven P, Timmerman D, Van Limbergen E, Vergote I (2005) Endometrial cancer. Lancet 366(9484):491–505. (review)
35. Concin N, Matias-Guiu X, Vergote I et al (2021) ESGO/ESTRO/ESP guidelines for the management of patients with endometrial carcinoma. Int J Gynecol Cancer 31:12–39
36. Kloop AH, Moughan J, Portelance L, Miller BE, Salehpour MR, Hildebrandt E et al (2013) Hematologic toxicity in RTOG 0418: a phase 2 study of postoperative IMRT for gynecologic cancer. Int J Radiat Oncol Biol Phys 86(1):83–90
37. Small W Jr, Beriwal S, Demanes DJ et al (2012) American brachytherapy society consensus guidelines for adjuvant vaginal cuff brachytherapy after hysterectomy. Brachytherapy 11(1):58–67
38. Schwarz JK, Beriwal S, Esthappan J et al (2015) Consensus statement for brachytherapy for the treatment of medically inoperable endometrial cancer. Brachytherapy 14(5):587–599
39. Gadducci A, Fabrini MG, Lanfredini N, Sergiampietri C (2015) Squamous cell carcinoma of the vagina: natural history, treatment modalities and prognostic factors. Crit Rev Oncol Hematol 93:211–224
40. Compton CC, Byrd DR, Garcia-Aguilar J, Kurtzman SH, Olawaiye A, Washington MK (2012) Vagina. In: Compton C, Byrd D, Garcia-Aguilar J, Kurtzman S, Olawaiye A, Washington M (eds) AJCC cancer staging atlas. Springer, New York, NY. https://doi.org/10.1007/978-1-4614-2080-4_34
41. Shah CA, Goff BA, Lowe K, Peters WA 3rd, Li CI (2009) Factors affecting risk of mortality in women with vaginal cancer. Obstet Gynecol 113(5):1038–1045
42. Creasman WT, Phillips JL, Menck HR (1998) The National Cancer Data Base report on cancer of the vagina. Cancer 83:1033
43. FIGO Committee on Gynecologic Oncology (2009) Current FIGO staging for cancer of the vagina, fallopian tube, ovary, and gestational trophoblastic neoplasia. Int J Gynaecol Obstet 105:3–4
44. Westerveld H, Nesvacil N, Fokdal L, Chargari C, Schmid MP, Milosevic M, Mahantshetty UM, Nout RA (2020) Definitive radiotherapy with image-guided adaptive brachytherapy for primary vaginal cancer. Lancet Oncol 21:e157–e167. (review)
45. Shmid MP, Fokdal L, Westerveld H et al (2020) Recommendations from gynaecological (GYN) GEC-ESTRO working group – ACROP: target concept for image guided adaptive brachytherapy in primary vaginal cancer. Radiother Oncol 145:36–44
46. Te Grootenhuis NC, Pouwer AW, de Bock GH, Hollema H, Bulten J, van der Zee AGJ, de Hullu JA, Oonk MHM (2018) Prognostic factors for local recurrence of squamous cell carcinoma of the vulva: a systematic review. Gynecol Oncol 148(3):622–631
47. Bornstein J, Bogliatto F, Haefner HK, Stockdale CK, Preti M, Bohl TG, Reutter J, Terminology Committee ISSVD (2016) The 2015 international society for the study of vulvovaginal disease (ISSVD) terminology of vulvar squamous intraepithelial lesions. Obstet Gynecol 127(2):264–268
48. Saraiya M, Watson M, Wu X, King JB, Chen VW, Smith JS, Giuliano AR (2008) Incidence of in situ and invasive vulvar cancer in the US, 1998-2003. Cancer 113(10 Suppl):2865–2872
49. Compton CC, Byrd DR, Garcia-Aguilar J, Kurtzman SH, Olawaiye A, Washington MK (2012) Vulva. In: Compton C, Byrd D, Garcia-Aguilar J, Kurtzman S, Olawaiye A, Washington M (eds) AJCC cancer staging atlas. Springer, New York, NY. https://doi.org/10.1007/978-1-4614-2080-4_33
50. Oonk MHM, Planchamp F, Baldwin P et al (2017) European Society of Gynaecological Oncology Guidelines for the management of patients with vulvar cancer. Int J Gynecol Cancer 27(4):832–837
51. Hansen EK, Mack Roach III (2018) Handbook of evidence-based radiation oncology, 3rd edn. Springer, US, pp 699–712
52. Gaffney DK, King B, Viswanathan AN et al (2016) Consensus recommendations for radiation therapy contouring and treatment of vulvar carcinoma. Int J Radiat Oncol Biol Phys 95(4):1191–1200
53. Beriwal S, Heron DE, Kim H et al (2006) Intensity-modulated radiotherapy for the treatment of vulvar carcinoma: a comparative dosimetric study with early clinical outcome. Int J Radiat Oncol Biol Phys 64(5):1395–1400

Gastrointestinal System Cancers

10.1 Esophageal Cancer

Esophageal cancer is the fourth most common gastrointestinal cancer and constitutes 5% of all cancers. It is common in patients older than 60 years and in black populations [1].

The esophagus is a muscular tube that starts at the C6 vertebra after the hypopharynx and extends into the stomach at the T11 level (Fig. 10.1). The length of the esophagus changes according to age and gender (22–28 cm in adults; 23–30 cm in males; 20–26 cm in females). The distance from the teeth to the cricopharyngeal region is approximately 15 cm in males and 14 cm in females (that from the teeth to the esophagogastric junction is 38–42 cm in males and 2 cm less in females).

The surrounding organs, vessels, and muscles naturally compress the esophagus and cause three important narrowings that can be seen during endoscopy or fluoroscopy:

1. *Upper narrowing*: The narrowest place at the cricopharyngeal level; caused by the cricopharyngeus muscle.
2. *Middle narrowing*: Here, the aortic arch and the left main bronchus cross the esophagus anteriorly and left laterally.
3. *Lower narrowing*: This occurs at the level of the gastroesophageal sphincter and the diaphragmatic hiatus.

The esophagus is anatomically divided into three segments: cervical, thoracic, and abdominal:

- The cervical esophagus extends from the lower hypopharynx to the level of the sternal notch. It is between 15 and 20 cm from the front incisors.
- The thoracic esophagus extends from the sternal notch to the esophagogastric junction (EGJ) and is divided into three separate sections as upper, lower, and middle.
  - The upper thoracic esophagus is between the sternal notch and the vena azygos (between 20 and 25 cm).
  - The mid-thoracic esophagus is between the vena azygos and the inferior pulmonary vein (between 30 and 35 cm).
  - The lower thoracic esophagus is between the inferior pulmonary vein and the EGJ (between 35 and 40 cm).
- Abdominal esophagus includes GEJ and is located between 40 and 42 cm.

The esophagus has no serosal layer.

10.1.1 Pathology

Globally, the most common type of esophageal cancer is squamous cell carcinoma. However, in

Fig. 10.1 Anatomy of the esophagus (from Compton et al. [2])

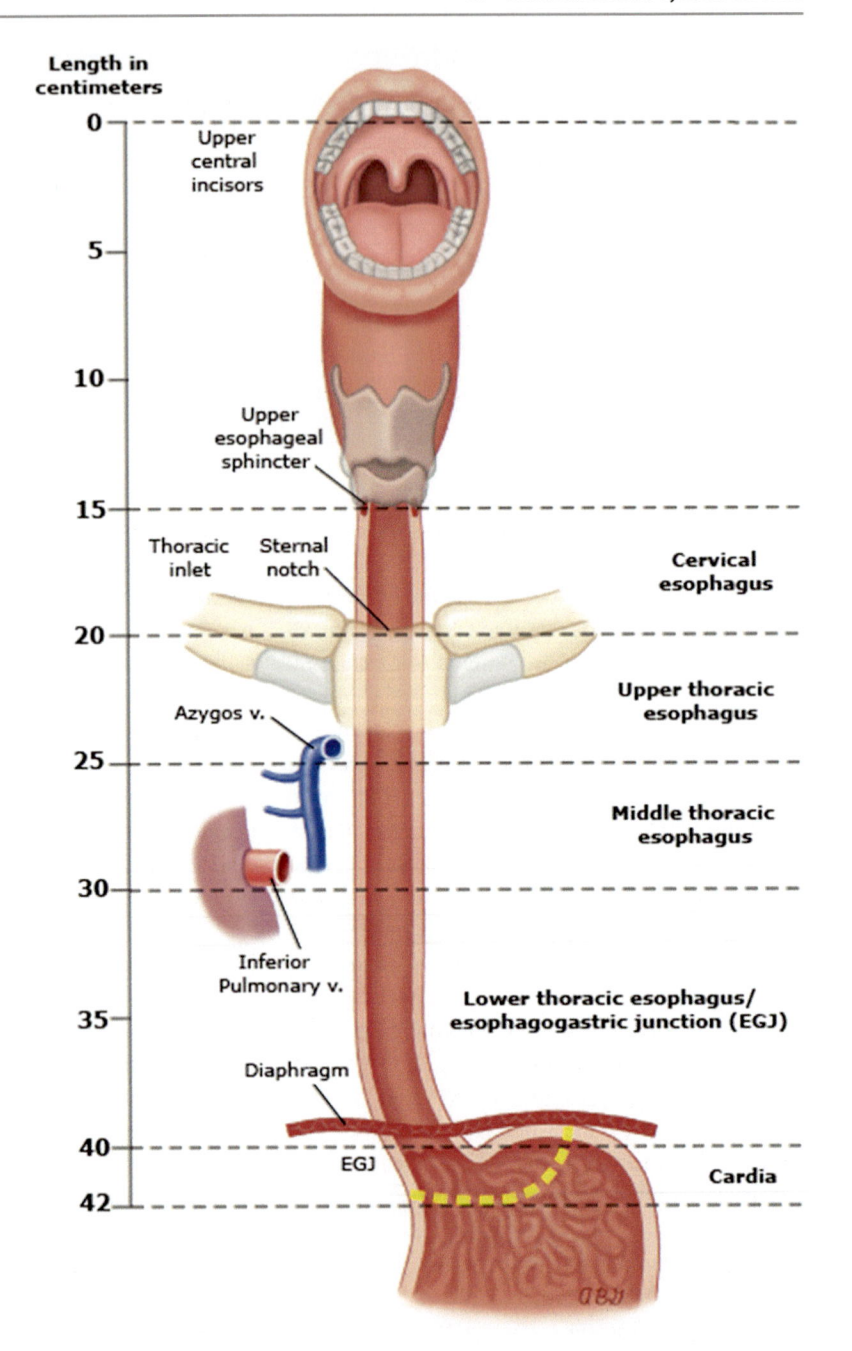

the USA, >60% of all esophageal cancers are adenocarcinoma [3].

- Small cell cancer shows neuroendocrine features and may secrete ADH, ACTH, and calcitonine. Prognosis is poor.

- Mucoepidermoid cancer is rarely seen. It is common in old age and mostly localized in the middle–lower parts of the esophagus.

Squamous Cell Cancers [3]. These constitute nearly 95% of all malignant esophageal tumors.

The major symptom is dysphagia. Symptoms start after the tumor fills two-thirds of the lumen, and diagnosis is usually late. Therefore, symptomatic patients commonly show a direct or lymphatic spread. The most common distant metastatic sites are lung, liver, pleura, bone, and kidneys. Histologically, most squamous cell cancers are poorly differentiated and show less keratinization. These tumors have no specific laboratory findings. Five-year survival is 5–30%.

Adenocarcinomas [3]. These are mostly localized in the lower parts and close to the esophagogastric junction, and compromise 3–5% of all malignant esophageal tumors. Adenocarcinoma generally develops in the stomach and extends directly into the esophagus. Primary esophageal adenocarcinoma is rare (2% in the upper half, 8–10% in the lower half). Primary adenocarcinoma originates from submucosal glands. However, adenocarcinoma has a better prognosis than squamous cell cancer.

Adenocarcinoma may develop in cases of Barrett's esophagitis.

Other Tumors. Small cell cancer, melanoma, adenoid cystic cancer (cylindroma), carcinosarcoma, pseudosarcoma, lymphoma, and metastatic tumors.

10.1.2 General Presentation

This disease is asymptomatic in its early stages. However, some signs (that can seem insignificant to patients) may be very important.

- The esophagus has a very rich lymphatic and nerve supply. Therefore, any tumor may cause segmental paralysis at that site and result in minimal swallowing difficulties. Retrosternal filling is then felt by the patient, who tries to tolerate this by drinking water. These findings form part of the daily routine and are insignificant to most patients.
- In case organic obstruction occurs, patients tend to chew the food more than before and will eat fluidy foods unaware of the obstruction. That is why most patients present with advanced metastatic stage.

Dysphagia is the most important finding. It is the first sign in 98% of cases. Dysphagia should not be mixed with odynophagia.

Retrosternal pain spreading to the neck, back, and epigastric region, especially with swallowing, shows the infiltration of the tumor into surrounding organs.

Weight loss, anorexia, vomiting, and pyrosis are common signs. Lower-region tumors may be mixed with signs of gastric ulcer. Tumor invasions or metastases cause respiratory system complications. Aspiration pneumonia and tracheoesophageal fistulas are dramatic clinical situations.

Hoarseness may be seen due to recurrent laryngeal nerve involvement. In addition, minor GIS bleeding may be observed. Symptoms related to metastases are common in advanced stages.

10.1.3 Staging

The tumor, node, metastasis (TNM) staging system of the American Joint Committee on Cancer (AJCC) for esophageal cancer is used universally. In the most recent (eighth edition, 2017) version, tumors involving the esophagogastric junction (EGJ) with the tumor epicenter no more than 2 cm into the proximal stomach are staged as esophageal cancers. In contrast, EGJ tumors with their epicenter located more than 2 cm into the proximal stomach are staged as stomach cancers, as are all cardia cancers not involving the EGJ, even if they are within 2 cm of the EGJ [4].

Esophageal cancer in AJCC 8th edition; they are grouped with the same T, N, M criteria regardless of histology (squamous/adenocarcinoma) and location (cervical, thoracic, esophagogastric) (Fig. 10.2).

- However, squamous cell carcinoma and adenocarcinomas are divided into different stage groups.
- These different stage groups also differ according to whether they received presurgical treatment or not; in this case, the "yp" prefix should be specified.

Fig. 10.2 T staging in esophageal cancer (from Compton et al. [2])

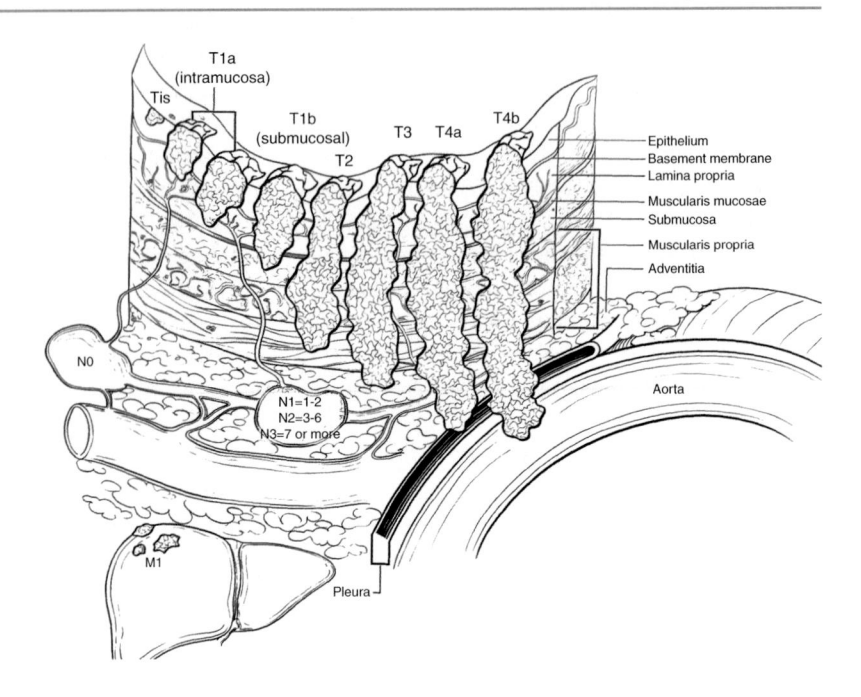

- In N0 cases, the prognosis depends not only on the T stage but also on the histology and degree of differentiation.
- The prognosis in squamous cell esophageal cancers also depends on the tumor location [4].
- AJCC 8th edition; emphasizes the importance of the involved lymph node number, not the involved nodal area.

Primary tumor (T), squamous cell carcinoma, and adenocarcinoma

T	
TX	Primary tumor cannot be assessed
T0	No evidence of primary tumor
Tis	High-grade dysplasia
T1	Tumor invades lamina propria, muscularis mucosae, or submucosa
T1a	Tumor invades lamina propria or muscularis mucosae
T1b	Tumor invades submucosa
T2	Tumor invades muscularis propria
T3	Tumor invades adventitia
T4	Tumor invades adjacent structures
T4a	Resectable tumor invading pleura, pericardium, or diaphragm
T4b	Unresectable tumor invading other adjacent structures, such as aorta, vertebral body, trachea, etc.

Primary tumor (T), squamous cell carcinoma, and adenocarcinoma

T	

Regional lymph nodes (N), squamous cell carcinoma, and adenocarcinoma

N	
NX	Regional lymph node(s) cannot be assessed
N0	No regional lymph node metastasis
N1	Metastasis in 1–2 regional lymph nodes
N2	Metastasis in 3–6 regional lymph nodes
N3	Metastasis in seven or more regional lymph nodes

Distant metastasis (M), squamous cell carcinoma, and adenocarcinoma

M	
M0	No distant metastasis
M1	Distant metastasis

10.1.3.1 Esophagus Lymphatics
(Fig. 10.3) [2]

Cervical esophagus: Scalen, internal jugular, upper and lower cervical, periesophageal, supraclavicular lymph nodes.

Upper thoracic esophagus: Upper periesophagiogastric LN at the level of azygos vein.

Middle thoracic esophagus: Subcarinal LN.

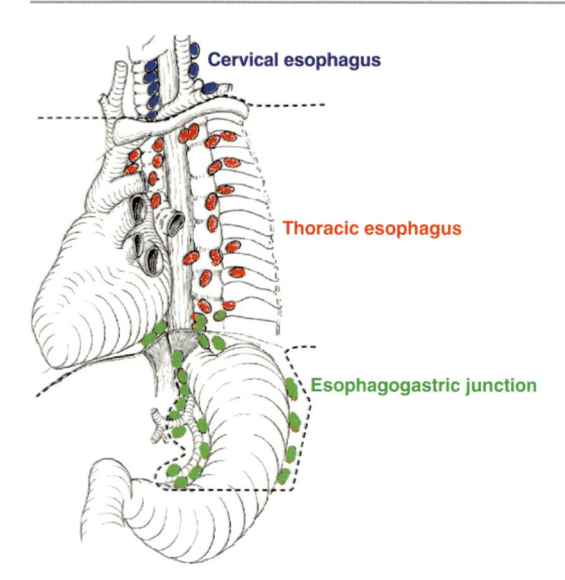

Fig. 10.3 Esophageal lymphatics (from Compton et al. [2])

Lower thoracic esophagus: Lower periesophagiogastric LN below the level of azygos vein.

Esophagogastric region: Lower periesophagiogastric LN below the level of azygos vein, diaphragmatic, pericardiac, left perigastric, celiac (Fig. 10.4).

10.1.3.2 Location of Primary

Cervical esophagus: Between hypopharynx and thoracic inlet (sternal notch).

Upper thoracic esophagus: Between thoracic inlet and lower border of azygos vein.

Middle thoracic esophagus: Between azygos vein and inferior pulmonary veins.

Lower thoracic esophagus/esophagogastric junction: Between inferior pulmonary veins and inferiorly by the stomach (Tables 10.1 and 10.2).

10.1.3.3 Importance of Pretreatment Staging in Esophageal Cancer

Pretreatment staging of newly diagnosed esophageal cancer cases is of vital importance. In cases limited to mucosa and submucosa, surgical/endoscopic resection alone may be sufficient.

- In cases with N+ or invasive esophageal and esophagogastric junction cancers, neoadjuvant treatment is indicated, even if R0 resection (microscopic surgical margin is negative). In cases with N (+) or esophageal and esophagogastric junction cancers that have invaded the esophageal wall, even if R0 resection (microscopic surgical margin negative), neoadjuvant therapy is indicated.
- The management of cT2N0 cases is controversial [5]. NCNN recommends surgery after neoadjuvant chemoradiotherapy in the distal esophagus and esophagogastric junction cancers (<2 cm, squamous cell carcinoma, well-differentiated).
  - Regardless of histology, 50–80% of patients with esophageal and esophagogastric junction cancers are locally advanced unresectable or metastatic and incurable [6]. The goal is palliative therapy, and very few have progression-free survival.

10.1.4 Treatment Algorithm

Pretreatment Diagnostic Studies
- Esophagography and upper GIS endoscopy (for imaging and biopsy) are performed initially.
- Endoscopic ultrasound (EUS) is the recommended method for locoregional staging [7].
- Endoscopic mucosal resection (EMR) or endoscopic submucosal dissection (ESD) can improve the accuracy of staging early superficial lesions [8].
  - Staging laparoscopy is controversial and is often recommended for suspected intraperitoneal metastatic disease with cT3-T4 adenocarcinomas located in the potentially resectable lower thoracic esophagus/EGJ [9].
- Bronchoscopy is recommended for subcarinal tumors and laryngoscopy for squamous cell carcinomas [10].
- PET-CT is recommended for lymphatic involvement and distant metastasis [11].
- USG or EUS-guided fine-needle aspiration biopsy is recommended for nodal staging [12].
- Bone scintigraphy and brain MRI are applied if there is any suspicion of metastasis.

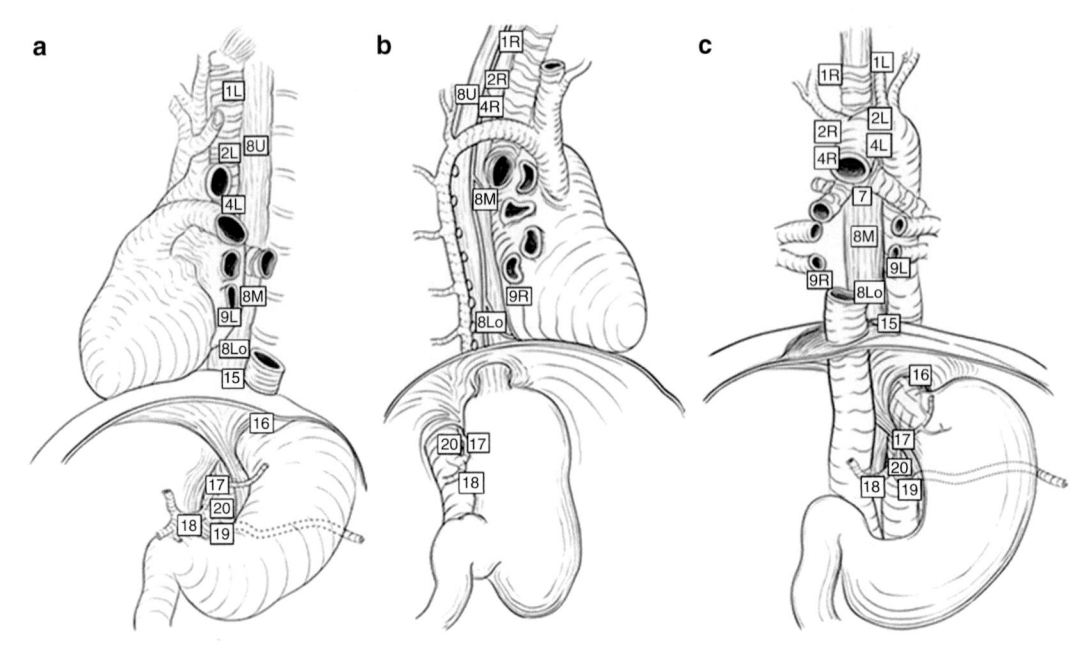

Fig. 10.4 (**a–c**) Lymph node maps for esophageal cancer. Regional lymph node stations for staging esophageal cancer from left (**a**), right (**b**), and anterior (**c**). **1R:** Right lower cervical paratracheal nodes, between the supraclavicular paratracheal space and apex of the lung. **1L:** Left lower cervical paratracheal nodes, between the supraclavicular paratracheal space and apex of the lung. **2R:** Right upper paratracheal nodes, between the intersection of the caudal margin of the brachiocephalic artery with the trachea and the apex of the lung. **2L:** Left upper paratracheal nodes, between the top of the aortic arch and the apex of the lung. **4R:** Right lower paratracheal nodes, between the intersection of the caudal margin of the brachiocephalic artery with the trachea and cephalic border of the azygos vein. **4L:** Left lower paratracheal nodes, between the top of the aortic arch and the carina. **7:** Subcarinal nodes, caudal to the carina of the trachea. **8U:** Upper thoracic paraesophageal lymph nodes, from the apex of the lung to the tracheal bifurcation. **8M:** Middle thoracic paraesophageal lymph nodes, from the tracheal bifurcation to the caudal margin of the inferior pulmonary vein. **8Lo:** Lower thoracic paraesophageal lymph nodes, from the caudal margin of the inferior pulmonary vein to the EGJ. **9R:** Pulmonary ligament nodes, within the right inferior pulmonary ligament. **9L:** Pulmonary ligament nodes, within the left inferior pulmonary ligament. **15:** Diaphragmatic nodes, lying on the dome of the diaphragm and adjacent or to behind its crura. **16:** Paracardial nodes, immediately adjacent to the gastroesophageal junction. **17:** Left gastric nodes, along the course of the left gastric artery. **18:** Common hepatic nodes, immediately on the proximal common hepatic artery. **19:** Splenic nodes, immediately on the proximal common hepatic artery. **20:** Celiac nodes, at the base of the celiac artery (from Compton et al. [2])

- Nutrition evaluation must be done.
- Tumor markers (e.g., CEA) are usually elevated in advanced disease; they have no role in diagnosis but may be important in prognosis and follow-up.
- P52, HER2, cyclin D1, and EGFR; are potential biomarkers in tumor growth/invasion and metastasis.

10.1.4.1 Surgical Treatment in Squamous Cell Esophageal Cancer

High-grade dysplasias

- Endoscopic treatments [5]
- Ablation (especially RF ablation)

10.1 Esophageal Cancer

Table 10.1 (A–C) Prognostic stage groups, squamous cell carcinoma

AJCC stage groups, squamous cell carcinoma

Clinical (cTNM)

T	N	M	Stage
(A)			
Tis	N0	M0	0
T1	N0-1	M0	I
T2	N0-1	M0	II
T3	N0	M0	II
T3	N1	M0	III
T1-3	N2	M0	III
T4	N0-2	M0	IVA
Any T	N3	M0	IVA
Any T	Any N	M1	IVB
(B)			

Post-neoadjuvant therapy (ypTNM)

T	N	M	Stage
T0-2	N0	M0	I
T3	N0	M0	II
T0-2	N1	M0	IIIA
T3	N1	M0	IIIB
T0-3	N2	M0	IIIB
T4a	N0	M0	IIIB
T4a	N1-2	M0	IVA
T4a	NX	M0	IVA
T4b	N0-2	M0	IVA
Any T	N3	M0	IVA
Any T	Any N	M1	IVB
T0-2	N0	M0	I

Pathological (pTNM)

pT	pN	M	Grade	Location	Staging
(C)					
Tis	N0	M0	N/A	Any	0
T1a	N0	M0	G1	Any	IA
T1a	N0	M0	G2-3	Any	IB
T1a	N0	M0	GX	Any	IA
T1b	N0	M0	G1-3	Any	IB
T1b	N0	M0	GX	Any	IB
T2	N0	M0	G1	Any	IB
T2	N0	M0	G2-3	Any	IIA
T2	N0	M0	GX	Any	IIA
T3	N0	M0	Any	Lower	IIA
T3	N0	M0	G1	Upper/middle	IIA
T3	N0	M0	G2-3	Upper/middle	IIB
T3	N0	M0	GX	Any	IIB
T3	N0	M0	Any	Location X	IIB
T1	N1	M0	Any	Any	IIB
T1	N2	M0	Any	Any	IIIA
T2	N1	M0	Any	Any	IIIA
T2	N2	M0	Any	Any	IIIB
T3	N1-2	M0	Any	Any	IIIB

Table 10.1 (continued)

Pathological (pTNM)

pT	pN	M	Grade	Location	Staging
T4a	N0-1	M0	Any	Any	IIIB
T4a	N2	M0	Any	Any	IVA
T4b	N0-2	M0	Any	Any	IVA
Any T	N3	M0	Any	Any	IVA
Any T	Any N	M1	Any	Any	IVB

TNM tumor, node, metastasis, *AJCC* American Joint Committee on Cancer, *UICC* Union for International Cancer Control, *N/A* not applicable. Used with permission of the American College of Surgeons, Chicago, Illinois. The original source for this information is the AJCC Cancer Staging Manual, Eighth Edition (2017) published by Springer International Publishing

Table 10.2 (A–C) Prognostic stage groups, adenocarcinoma

AJCC stage groups, adenocarcinoma

Clinical (cTNM)

T	N	M	Stage
(A)			
Tis	N0	M0	0
T1	N0	M0	I
T1	N1	M0	IIA
T2	N0	M0	IIB
T2	N1	M0	III
T3	N0-1	M0	III
T4a	N0-1	M0	III
T1-4a	N2	M0	IVA
T4b	N0-2	M0	IVA
Any T	N3	M0	IVA
Any T	Any N	M1	IVB

Post-neoadjuvant therapy (ypTNM)

ypT	ypN	M	Stage
(B)			
T0-2	N0	M0	I
T3	N0	M0	II
T0-2	N1	M0	IIIA
T3	N1	M0	IIIB
T0-3	N2	M0	IIIB
T4a	N0	M0	IIIB
T4a	N1-2	M0	IVA
T4a	NX	M0	IVA
T4b	N0-2	M0	IVA
Any T	N3	M0	IVA
Any T	Any N	M1	IVB

Pathological (pTNM)

pT	pN	M	G	Evre
(C)				
Tis	N0	M0	N/A	0
T1a	N0	M0	G1	IA
T1a	N0	M0	GX	IA
T1a	N0	M0	G2	IB
T1b	N0	M0	G1-2	IB
T1b	N0	M0	GX	IB
T1	N0	M0	G3	IC
T2	N0	M0	G1-2	IC
T2	N0	M0	G3	IIA
T2	N0	M0	GX	IIA
T1	N1	M0	Any	IIB
T3	N0	M0	Any	IIB
T1	N2	M0	Any	IIIA
T2	N1	M0	Any	IIIA
T2	N2	M0	Any	IIIB
T3	N1-2	M0	Any	IIIB
T4a	N0-1	M0	Any	IIIB
T4a	N2	M0	Any	IVA
T4b	N0-2	M0	Any	IVA
Any T	N3	M0	Any	IVA
Any T	Any N	M1	Any	IVB

TNM tumor, node, metastasis, *AJCC* American Joint Committee on Cancer, *UICC* Union for International Cancer Control, *N/A* not applicable. Used with permission of the American College of Surgeons, Chicago, Illinois. The original source for this information is the AJCC Cancer Staging Manual, Eighth Edition (2017) published by Springer International Publishing

T1aN0

- Endoscopic resection [13]
- EMR or ESD is preferred to surgery, especially in T1a and LVI (-) cases
- Esophagectomy

T1bN0

- Esophagectomy + systemic lymph node dissection [14] LND, thoracoabdominal is recommended.
- EMR or ESD ± RT
    - In cases who cannot tolerate surgery and/or chemoradiotherapy

T1b-3/N0-1

- Esophagectomy + systemic lymph node dissection [14]
    - As an adjuvant after induction therapies or as a neoadjuvant before.
    - The most common esophagectomy is transthoracic esophagectomy.
    - Lymph node dissection
        Extended two-site lymphadenectomy is recommended for all cases with squamous cell carcinoma located in the thoracic esophagus.
        For upper thoracic esophageal tumors, three-site dissection is recommended.
        In cases not receiving induction therapy (for adequate nodal staging), at least 12–15 lymph nodes should be dissected from the mediastinum and upper abdomen.

T4/N2-3

- Surgery is not recommended as a primary therapy [15].
- After induction, treatments or salvage can be tried as an option.

M1

- Surgery is generally contraindicated.

10.1.4.2 Radiotherapy in Squamous Cell Esophageal Cancer

Cervical esophagus

- Chemoradiotherapy is recommended as first-line therapy [16].
- RT alone can be used for palliation in patients who cannot tolerate chemoradiotherapy.

High-grade dysplasia and T1a

- RT has no role.

Local disease (T1b-3/N0-1)

- As induction therapy, concomitant chemoradiotherapy is appropriate and survival may be increased in responding patients [17].
- RT alone has not been shown to improve survival as induction therapy.
- Definitive chemoradiotherapy (sequential or concomitant) is an acceptable alternative for medically inoperable patients or those who refuse surgery.
- Concomitant chemoradiotherapy has been shown to increase local control and survival [18].
- RT alone can be used for palliation in patients who cannot tolerate chemoradiotherapy.
- In N+ cases who did not receive preoperative treatment and after complete surgery, chemoradiotherapy rather than chemotherapy alone is recommended intolerant cases [19].
- Basic indications for postoperative RT (PORT) [20].
- Presence of positive surgical margin (R1 resection) and residual tumor (R2 resection). In randomized studies, RT alone without chemotherapy in postoperative cases did not improve survival while increasing locoregional control [21].
- PORT can be considered in poorly differentiated ± LVI (+) pN0 cases with radically resected upper thoracic esophagus origin and deep invasion depth (above T3).
- PORT has been shown to increase local control in cases with insufficient lymph dissection [22].

Locally advanced disease (T4/N2–3)

- Induction chemoradiotherapy can increase the chances of cure with surgery by decreasing the stage [23].
- In cases with extensive nodal involvement [multiarea lymphatic (cervical, mediastinum, abdominal) or multilymphatic station], definitive chemoradiotherapy is an acceptable alternative.

M1 disease

- Concurrent chemoradiotherapy has a role in distant nodal involvement (supraclavicular and retroperitoneal). It can be applied for palliation purposes.

Recurrent disease

- Concurrent chemoradiotherapy is the most effective treatment for postoperative locoregional recurrence, particularly nodal recurrence [24].

10.1.5 Radiotherapy

The patient is simulated in the supine position. Oral barium contrast solution is used to visualize the esophagus and the site of the tumor (the narrowed part of the esophagus due to the tumor).

- Anterior–posterior field is used until 45 Gy, then 50.4 Gy by oblique or three fields (one anterior, two posterior–oblique fields) by sparing the spinal cord.

10.1.5.1 Conventional RT Fields
(Fig. 10.5)
Cervical Esophagus RT
- *Superior*: 5 cm proximal to tumor + supraclavicular LN + upper mediastinal LN

Inferior: 5 cm distal to tumor
Lateral: tumor + 2.5–3 cm + mediastinal LN + 2/3 of clavicle for SCF LN
Middle esophagus RT
Superior: 5 cm proximal to tumor + upper mediastinal LN
Inferior: 5 cm distal to tumor + mediastinal LN
Lateral: tumor + 2.5–3 cm + mediastinal LN
Lower esophagus RT
Superior: 5 cm proximal to tumor + mediastinal LN
Inferior: 5 cm distal to tumor + mediastinal LN + celiac LN (until L1–2 vertebrae)
Lateral: tumor + 2.5–3 cm + mediastinal LN

10.1.5.2 Conformal RT/IMRT Volumes
(Fig. 10.6) [7]
Simulation
- It is simulated when the hands are on the head.
- To account for intrathoracic organ movements if 4D simulation is not possible.
- CT acquisition rate can be reduced.
- The target volume and the mobility of critical structures are determined by shooting at maximum expiration and inspiration.
- If possible, 4D simulation is done.
  - If IMRT is being done, 4D simulation is important.
  - It is especially recommended for distal esophagus and EGJ tumors.
- Immobilization may vary depending on the location of the tumor.
  - Head–shoulder thermoplastic mask for cervical/upper thoracic esophagus.
  - Vacuum bed immobilization is used for the lower thoracic esophagus/EGJ tumors.
- It is recommended that the patient not eat or drink for 2–4 h before the simulation and each treatment to prevent intestinal gas from causing organ motility.
- "IV contrast for nodal contouring" can be used, especially if IMRT will be performed.

Contouring
- PET-CT is helpful in contouring.
- CTV:
  - Primary GTV +3–5 cm longitudinal margin; 0.5–1 cm circumferential margin is given.
  - 0.5–1 cm margin is given for the involved lymph node.

10.1 Esophageal Cancer

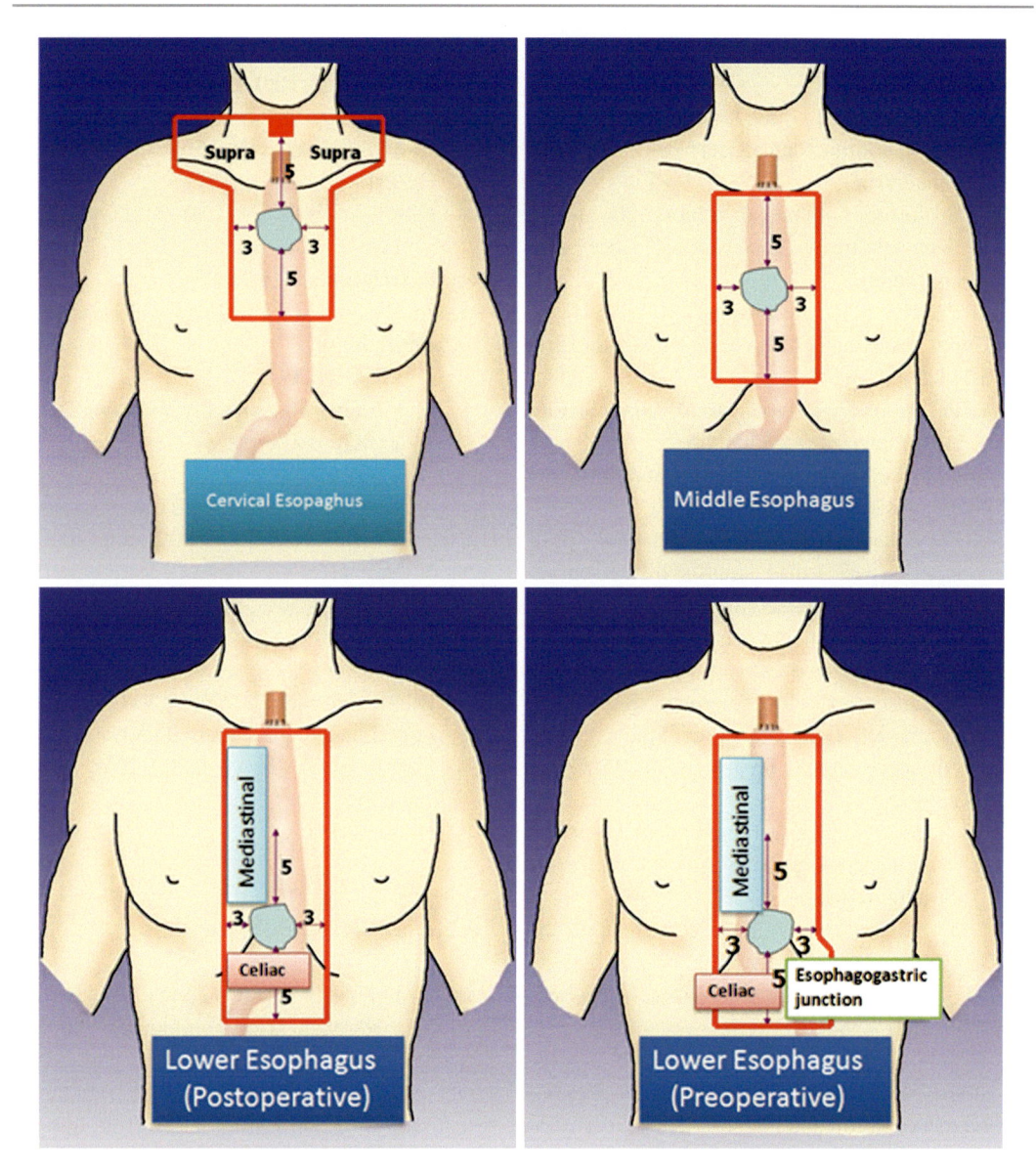

Fig. 10.5 RT fields in esophageal tumors

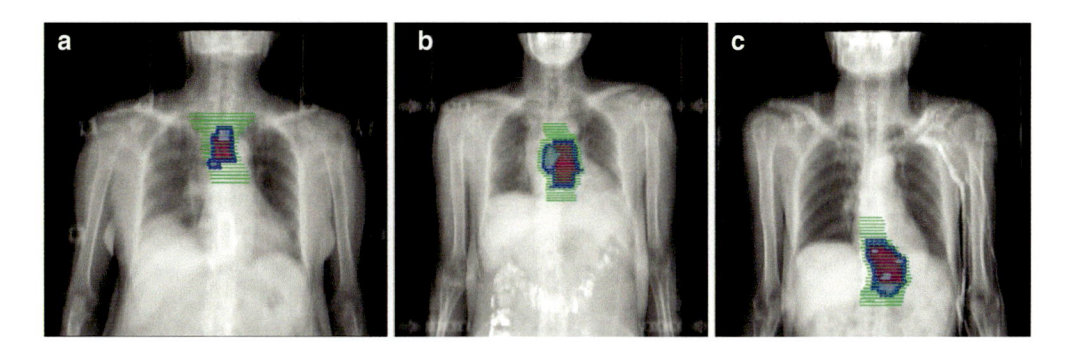

Fig. 10.6 Targets contouring of (**a**) upper thoracic esophagus; (**b**) middle thoracic esophagus; (**c**) lower thoracic esophagus. Red area indicates GTV-tumor, gray area includes GTV-nodal, and green area outlines PTV [25]

- The entire esophagus and lungs are contoured.
- Brachial plexus and larynx contouring is also recommended in cervical esophageal cancers.
- In lower esophageal tumors, the heart, liver, kidneys, and stomach are also contoured.
- Elective nodal contouring:
- Upper esophagus (cervical, upper thoracic esophagus)
  - Periesophageal, mediastinal, supraclavicular lymph nodes are included.
- Lower esophagus (mid and lower thoracic esophagus and EGJ tumors)
  - Periesophageal, lower mediastinal, perigastric (pericardiac, small and large curvature region), celiac (along the left gastric artery) lymph nodes are included.
- If it is adenocarcinoma:
  - Because of the low tolerance dose of the stomach, lymphatics of the greater curvature should be limited to 45 Gy in planning.
    Periesophageal lymph nodes; extend along the superior vena cava above the diaphragm.
    Paratracheal lymph nodes are with the left recurrent nerve.
    When CTV is contoured, normal tissues are excluded.
- ITV: If 4D simulation is not possible, it is the sum of the contoured CTV volumes at maximum expiration and inspiration.
  - PTV:
    If IGRT will not be done, CTV + 0.5–1 cm margin is given.
    If IGRT is to be made, CTV + 0.5 cm margin is given.

Dose/Fractionation
- Squamous cell carcinoma:
  - 40–50.4 Gy (1.80–2.00 Gy/fx) in neoadjuvant therapy
  - In definitive therapy: 50.4 Gy (1.80–2.00 Gy/fx)
    Up to 66 Gy in the cervical esophagus
  - PORT: 45–50 Gy (1.80–2.00 Gy/fx)

- Adenocarcinoma:
  - 40–50.4 Gy (1.80–2.00 Gy/fx) in neoadjuvant therapy
  - In definitive therapy: 50.4 Gy (1.80–2.00 Gy/fx)
  - PORT: 45–50.4 Gy (1.80–2.00 Gy/fx)

Dose Restrictions (OAR)
- Lung: [lung] – [PTV + 2 cm] ≤ 40 Gy
  - Entire lung:
    V30Gy <20%
    V20Gy <30% (ideally V20Gy <25%)
    V10Gy <40%
    V5Gy <60%
    Average lung dose <20 Gy
- Heart: D 100% <30 Gy, D 50% <40 Gy
  - V50Gy <1/3, V45Gy <2/3, V40Gy <100
- Liver: V30Gy ≤60%, liver mean dose ≤25 Gy
  - V35Gy <50%, V30Gy <100%
- Kidney:
  - Bilateral kidney: D 70% ≤ 20 Gy
  - Functional single kidney: D 80% ≤ 20 Gy
  - V50Gy <1/3, V30Gy <2/3, V23Gy <100
- Spinal cord:
  - Dmax is 45 Gy

10.1.6 Selected Publications

Hong Kong, 1993 → 60 patients with curative resection; 30 each were randomized into the radiotherapy group (CR + R) and the control group (CR). Seventy patients with palliative resection; 35 each were randomized into the radiotherapy group (PR + R) and the control group (PR).

- Patients receiving postoperative radiotherapy had lower survival due to irradiation-related death and the early appearance of metastatic diseases.

Fok M (1993) Postoperative radiotherapy for carcinoma of the esophagus: a prospective, randomized controlled study. Surgery 113(2):138–147

Scandinavian Trial, 1992 → 186 patients were randomized to four treatment groups: group 1, surgery alone; group 2, preoperative chemotherapy (cisplatin and bleomycin) and surgery; group 3, preoperative irradiation (35 Gy) and surgery; group 4, preoperative chemotherapy, radiotherapy, and surgery.

- Preoperative irradiation had a beneficial effect on intermediate-term survival, whereas the chemotherapy did not influence survival.

Nygaard K (1992) Pre-operative radiotherapy prolongs survival in operable esophageal carcinoma: a randomized, multicenter study of pre-operative radiotherapy and chemotherapy. The second Scandinavian trial in esophageal cancer. World J Surg 16(6):1104–1109; discussion 1110

Scotch, 1992 → 176 patients were randomly assigned to preoperative radiotherapy or surgery alone. Patients assigned to the radiotherapy arm received 20 Gy in ten fractions using parallel opposed 4 MV beams.

- Low-dose preoperative radiotherapy offered no advantage over surgery alone.

Arnott SJ (1992) Low dose preoperative radiotherapy for carcinoma of the oesophagus: results of a randomized clinical trial. Radiother Oncol 24(2):108–113

Meta-analysis, 2005 → 1147 patients from five randomized trials were used to assess whether preoperative radiotherapy improves overall survival. Median follow-up was 9 years.

- The hazard ratio (HR) of 0.89 (95% CI 0.78–1.01) suggests an overall reduction in the risk of death of 11% and an absolute survival benefit of 3% at 2 years and 4% at 5 years. This result is not conventionally statistically significant ($p = 0.062$).
- There was no clear evidence that preoperative radiotherapy improves the survival of patients with potentially resectable esophageal cancer.

Arnott SJ (2005) Preoperative radiotherapy for esophageal carcinoma. Cochrane Database Syst Rev (4):CD001799

CALGB 9781/RTOG 97-16, 2008 → Patients were randomized to treatment with either surgery alone or cisplatin (100 mg/m^2) and 5FU (1,000 mg/m^2/day × 4 days) on weeks 1 and 5 concurrent with radiation therapy (50.4 Gy to 1.8 Gy/fraction over 5.6 weeks) followed by esophagectomy with lymph node dissection.

- This trial demonstrated a survival advantage with the use of chemoradiation therapy followed by surgery in the treatment of esophageal cancer. The observed survival difference was statistically significant and suggested that trimodality therapy is an appropriate standard of care for patients with this disease.

Tepper J et al (2008) Phase III trial of trimodality therapy with cisplatin, fluorouracil, radiotherapy, and surgery compared with surgery alone for esophageal cancer: CALGB 9781. J Clin Oncol 26(7):1086–1092

TROG/AGITG, 2005 → 128 patients were randomly assigned to surgery alone and 128 patients to surgery after 80 mg/m^2 cisplatin on day 1800 mg/m^2 fluorouracil on days 1–4, with concurrent radiotherapy of 35 Gy given in 15 fractions.

- Preoperative chemoradiotherapy with cisplatin and fluorouracil did not improve progression-free or overall survival for patients with resectable esophageal cancer compared with surgery alone.

Burmeister BH et al (2005) Surgery alone versus chemoradiotherapy followed by surgery for resectable cancer of the oesophagus: a randomised controlled phase III trial. Lancet Oncol 6(9):659–668

Ireland, 1996 → Patients assigned to multimodal therapy received two courses of chemotherapy (fluorouracil and cisplatin) and radiotherapy (40 Gy, administered in 15 fractions beginning concurrently with the first course of chemotherapy), followed by surgery.

- Multimodal treatment was found to be superior to surgery alone for patients with resectable adenocarcinoma of the esophagus.

Walsh TN (1996) A comparison of multimodal therapy and surgery for esophageal adenocarcinoma. N Engl J Med 335(7):462–467
EORTC, 1997 → The preoperative combined therapy consisted of radiotherapy, in a dose of 18.5 Gy delivered in five fractions of 3.7 Gy for 2 weeks, and cisplatin administered before the first day of radiotherapy. The surgical plan included one-stage en bloc esophagectomy and proximal gastrectomy by the abdominal and right thoracic routes, to be performed immediately after randomization in the group assigned to surgery alone and 2–4 weeks after the completion of preoperative chemoradiotherapy in the group assigned to combined therapy.

- Preoperative chemoradiotherapy did not improve overall survival, but it prolonged disease-free survival and survival free of local disease.

Bosset JF et al (1997) Chemoradiotherapy followed by surgery compared with surgery alone in squamous-cell cancer of the esophagus. N Engl J Med 337(3):161–167
FFCD 9102, 2007 → 259 patients with operable T3N0-1M0 thoracic esophageal cancer. Patients received two cycles of fluorouracil (FU) and cisplatin and either conventional (46 Gy in 4.5 weeks) or split-course (15 Gy, days 1–5 and 22–26) concomitant radiotherapy. Patients with response were randomly assigned to surgery (arm A) or continuation of chemoradiation [arm B; three cycles of FU/cisplatin and either conventional (20 Gy) or split-course (15 Gy) radiotherapy].

- There was no benefit from the addition of surgery after chemoradiation compared with the continuation of additional chemoradiation in patients who responded to chemoradiation.

Bedenne L (2007) Chemoradiation followed by surgery compared with chemoradiation alone in squamous cancer of the esophagus: FFCD 9102. J Clin Oncol 25(10):1160–1168
Germany, 2005 → 172 patients with locally advanced squamous cell carcinoma of the esophagus were randomly allocated to either induction chemotherapy followed by chemoradiotherapy (40 Gy), followed by surgery, or the same induction chemotherapy followed by chemoradiotherapy (at least 65 Gy) without surgery.

- Adding surgery to chemoradiotherapy improved local tumor control but did not increase survival of patients with locally advanced esophageal SCC.

Stahl M et al (2005) Chemoradiation with and without surgery in patients with locally advanced squamous cell carcinoma of the esophagus. J Clin Oncol 23(10):2310–2317
Shandong Randomized Trial, 2006 → 269 patients with resectable esophageal cancer in the chest were randomized into two groups: 135 in the surgery group and 134 in the radiotherapy group.

- The results with late-course accelerated hyperfractionated conformal radiotherapy alone were found to be comparable to those achieved by radical surgery for patients with resectable esophageal cancer in the chest.

Sun XD (2006) Randomized clinical study of surgery versus radiotherapy alone in the treatment of resectable esophageal cancer in the chest. Zhonghua Zhong Liu Za Zhi 28(10):784–787
RTOG 85-01, 1992 → Patients with squamous cell or adenocarcinoma of the esophagus, T1–3 N0–1 M0. Combined modality therapy ($n = 134$): 50 Gy in 25 fractions over 5 weeks, plus cisplatin and fluorouracil. In the randomized study, combined therapy was compared with RT only ($n = 62$): 64 Gy in 32 fractions over 6.4 weeks.

- Combined therapy increased the survival of patients who had squamous cell or adenocarcinoma of the esophagus, T1–3 N0–1 M0, compared with RT alone.

Cooper JS et al (1999) Chemoradiotherapy of locally advanced esophageal cancer: long-term follow-up of a prospective randomized trial (RTOG 85-01). Radiation Therapy Oncology Group. JAMA 281(17):1623–1627

ECOG EST-1282, 1998 → Randomization: combined use of 5-FU, mitomycin C, and radiation therapy vs. radiation therapy alone.

- Two- and five-year survivals were 12 and 7% in the radiation alone arm and 27 and 9% in the chemoradiation arm.
- Patients treated with chemoradiation had a longer median survival (14.8 months) compared to patients receiving radiation therapy alone (9.2 months). This difference was statistically significant.

Smith TJ et al (1998) Combined chemoradiotherapy vs. radiotherapy alone for early stage squamous cell carcinoma of the esophagus: a study of the Eastern Cooperative Oncology Group. Int J Radiat Oncol Biol Phys 42(2):269–276

10.2 Gastric Cancer

Gastric cancer is the third most common cancer in the world and the second highest cause of cancer-related mortality [26]. Its incidence is high in Japan, China, and Russia, and it is less commonly seen in the USA and Canada. Low socioeconomic status, cigarette smoking, and high alcohol consumption are correlated with gastric cancer.

The stomach is histologically composed of five layers from the innermost to the outermost: mucosal, submucosal, muscularis propria, subserosal, and serosal.

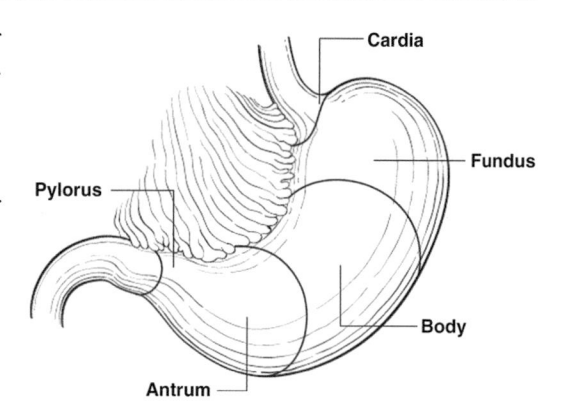

Fig. 10.7 Anatomy of the stomach (from Compton et al. [27])

- The gastric mucosa has three microscopic subtypes: cardiac, fundic, and antral mucosa.
- Submucosa is a loose connective tissue containing elastic leaves, arteries, veins, lymphatic veins, and Meissner's neural networks.
- The external muscular layer consists of three layers: outer longitudinal, inner circular, and oblique. The inner circular layer forms a pyloric sphincter at the gastroduodenal junction. Auerbach's (myenteric) plexus is located between the longitudinal and circular layers.

The stomach is an intraperitoneal organ that starts at the T11 vertebra and ends in the duodenum on the right side of the midline.

- It is divided into five anatomic regions (Fig. 10.7):
  - The *cardia* surrounds the gastroesophageal junction, and is 1–3 cm long.
  - The *fundus* is the proximal part of the stomach.
  - The *corpus* is next to the fundus.
  - The *antrum* is the angled part of the distal stomach.
  - The *pylorus* separates the antrum and duodenum.
- The superomedial border is the lesser curvature, and inferolateral border is the greater curvature of the stomach.

10.2.1 Pathology

There are various classification systems for gastric cancers [28]. Gastric cancers are commonly localized in the antrum/distal stomach (40%), the proximal stomach and the gastroesophageal junction (35%), and the corpus (25%).

1926 Bormann
- Type I (polypoid)
- Type II (fungating)
- Type III (ulcerated)
- Type IV (infiltrative)
- *1953 Stout*
- Fungating
- Penetrating
- Superficial
- Linitis plastica
- Nonspecific
- *1965 Lauren*
- Intestinal
- Diffuse
- *1977 Ming*
- Expansive
- Infiltrative
- *1981 Japanese Society for Gastric Cancer*
- Papillary
- Tubular
- Less differentiated
- Mucinous
- Signet-ring cell
- *2000 WHO classification*
- Adenocarcinoma
  - Intestinal type
  - Diffuse type
- Papillary adenocarcinoma
- Tubular adenocarcinoma
- Mucinous adenocarcinoma
- Signet-ring cell carcinoma
- Adenosquamous carcinoma
- Squamous cell carcinoma
- Small cell carcinoma
- Undifferentiated carcinoma
- Others

10.2.2 General Presentation

Early gastric cancers usually have no symptoms or signs. Epigastric fullness, pain, nausea, and vomiting may be seen, but none of them are specific for gastric cancers. Symptoms may be observed in the early period according to the localization of tumor. In addition, bleeding, obstructive findings, and masses can be found in physical exams.

- Lymph node metastasis ratio is 18% in serosa (−) cases, and 5-year survival in these cases after resection is over 50%.
- Lymph node metastasis ratio is 80% in serosa (+) cases, and prognosis is very poor in these patients.
- Implantation metastases from seeded tumors may develop when the tumor reaches the serosa. These tumors characteristically implant into the ovaries (*Krukenberg tumor*) and the cul-de-sac in the pelvis (*Blumer's shelf*).
- A left supraclavicular lymph node metastasis is called a *Virchow node*.
- A periumbilical lymph node metastasis is called a *Sister Mary Joseph nodule*.
- An axillary lymph node metastasis is known as an *Irish nodule*.
- Gastric/perigastric discomfort, pain, and fullness.
- Rapid weight loss.
- Nausea and vomiting.
- Fullness after meals.
- GIS bleeding or occult bleeding.
- Iron-deficiency anemia in two-thirds of patients.

10.2.3 Staging

Japanese Staging System for Gastric Cancer [29] (Table 10.3)

10.2 Gastric Cancer

Table 10.3 (A–C) Stomach cancer TNM, 8th edition

T category	T criteria
A: Primary tumor (T) (Fig. 10.8) [27]	
TX	Primary tumor cannot be assessed
T0	No evidence of primary tumor
Tis	Carcinoma in situ: Intraepithelial tumor without invasion of the lamina propria, high-grade dysplasia
T1	Tumor invades the lamina propria, muscularis mucosae, or submucosa
T1a	Tumor invades the lamina propria or muscularis mucosae
T1b	Tumor invades the submucosa
T2	Tumor invades the muscularis propria[a]
T3	Tumor penetrates the subserosal connective tissue without invasion of the visceral peritoneum or adjacent structures[b,c]
T4	Tumor invades the serosa (visceral peritoneum) or adjacent structures[b,c]
T4a	Tumor invades the serosa (visceral peritoneum)
T4b	Tumor invades adjacent structures/organs

[a] A tumor may penetrate the muscularis propria with extension into the gastrocolic or gastrohepatic ligaments, or into the greater or lesser omentum, without perforation of the visceral peritoneum covering these structures. In this case, the tumor is classified as T3. If there is perforation of the visceral peritoneum covering the gastric ligaments or the omentum, the tumor should be classified as T4

[b] The adjacent structures of the stomach include the spleen, transverse colon, liver, diaphragm, pancreas, abdominal wall, adrenal gland, kidney, small intestine, and retroperitoneum

[c] Intramural extension to the duodenum or esophagus is not considered invasion of an adjacent structure, but is classified using the depth of the greatest invasion in any of these sites

N category	N criteria
B: Regional lymph nodes (N)	
NX	Regional lymph node(s) cannot be assessed
N0	No regional lymph node metastasis
N1	Metastases in 1 or 2 regional lymph nodes
N2	Metastases in 3–6 regional lymph nodes
N3	Metastases in 7 or more regional lymph nodes
N3a	Metastases in 7–15 regional lymph nodes
N3b	Metastases in 16 or more regional lymph nodes

N category	N criteria
Distant metastasis (M)	
M category	M criteria
M0	No distant metastasis
M1	Distant metastasis

AJCC stage groups
Clinical (cTNM)

T	N	M	Stage
C: Prognostic stage groups			
Tis	N0	M0	0
T1	N0	M0	I
T2	N0	M0	I
T1	N1, N2, or N3	M0	IIA
T2	N1, N2, or N3	M0	IIA
T3	N0	M0	IIB
T4a	N0	M0	IIB
T3	N1, N2, or N3	M0	III
T4a	N1, N2, or N3	M0	III
T4b	Any N	M0	IVA
Any T	Any N	M1	IVB

Pathological (pTNM)

T	N	M	Stage
Tis	N0	M0	0
T1	N0	M0	IA
T1	N1	M0	IB
T2	N0	M0	IB
T1	N2	M0	IIA
T2	N1	M0	IIA
T3	N0	M0	IIA
T1	N3a	M0	IIB
T2	N2	M0	IIB
T3	N1	M0	IIB
T4a	N0	M0	IIB
T2	N3a	M0	IIIA
T3	N2	M0	IIIA
T4a	N1	M0	IIIA
T4a	N2	M0	IIIA
T4b	N0	M0	IIIA
T1	N3b	M0	IIIB
T2	N3b	M0	IIIB
T3	N3a	M0	IIIB
T4a	N3a	M0	IIIB
T4b	N1	M0	IIIB
T4b	N2	M0	IIIB
T3	N3b	M0	IIIC
T4a	N3b	M0	IIIC
T4b	N3a	M0	IIIC
T4b	N3b	M0	IIIC
Any T	Any N	M1	IV

(continued)

AJCC stage groups			
Clinical (cTNM)			
T	N	M	Stage
Post-neoadjuvant therapy (ypTNM)			
T	N	M	**Stage**
T1	N0	M0	I
T2	N0	M0	I
T1	N1	M0	I
T3	N0	M0	II
T2	N1	M0	II
T1	N2	M0	II
T4a	N0	M0	II
T3	N1	M0	II
T2	N2	M0	II
T1	N3	M0	II
T4a	N1	M0	III
T3	N2	M0	III
T2	N3	M0	III
T4b	N0	M0	III
T4b	N1	M0	III
T4a	N2	M0	III
T3	N3	M0	III
T4b	N2	M0	III
T4b	N3	M0	III
T4a	N3	M0	III
Any T	Any N	M1	IV

TNM tumor, node, metastasis, *AJCC* American Joint Committee on Cancer, *UICC* Union for International Cancer Control, *N/A* not applicable. Used with permission of the American College of Surgeons, Chicago, Illinois. The original source for this information is the AJCC Cancer Staging Manual, Eighth Edition (2017) published by Springer International Publishing

- *N1 (1. compartment)*: involvement of 1, 2, 3, 4, 5, 6 lymph nodes
- *N2 (2. compartment)*: involvement of 7, 8, 9, 10, 11 lymph nodes
- *N3 (3. compartment)*: involvement of 12, 13, 14, 15, 16 lymph nodes
- *P parameter*
- *P0:* peritoneal involvement (−)
- *P1:* peritoneal involvement (+)
- *PX:* peritoneal involvement is unknown
- *H parameter*
- *H0:* hepatic metastasis (−)
- *H1:* hepatic metastasis (+)
- *HX:* hepatic metastasis is unknown
- *N parameter*
- *N0:* lymphatic involvement (−)
- *N1:* 1. compartment involvement
- *N2:* 2. compartment involvement
- *N3:* 3. compartment involvement
- *CY (peritoneal cytology) parameter*
- *CY0:* peritoneal cytology is benign or indeterminate
- *CY1:* peritoneal cytology (+)
- *CYX:* peritoneal cytology is not performed
- *M parameter*
- *M0:* distant metastasis (−) (peritoneum, liver, cytological metastases)
- *M1:* distant metastasis (+) (other than peritoneum, liver, cytological metastases)
- *Japanese Staging System Groupings:*
- Stage Ia: T1N0
- Stage Ib: T1N1, T2N0
- Stage II: T1N2, T2N1, T3N0
- Stage IIIa: T4N0, T3N1, T2N2
- Stage IIIb: T4N1, T3N2
- Stage IV: H1, P1, CY1, M1, N3

10.2.3.1 Gastric Lymphatics (Fig. 10.9)

- 1, 2 Perigastric
- 3, 4 Lesser and greater curvature
- 5 Suprapyloric (right gastric)
- 6 Infrapyloric
- 7 Left gastric
- 8 Common hepatic
- 9 Celiac
- 10 Splenic hilum
- 11 Splenic
- 12 Hepaticoduodenal
- *Lymph nodes accepted as distant metastases*: posterior pancreatic (13), sup. mesenteric (14), middle colic (15), para-aortic (16), portal, retroperitoneal
- *Will Rogers Phenomenon*. The stage determined by the AJCC staging system may differ from the Japanese nodal–anatomic staging system due to extensive lymph node dissection in the latter. This may cause upstaging and is known as the Will Rogers phenomenon.

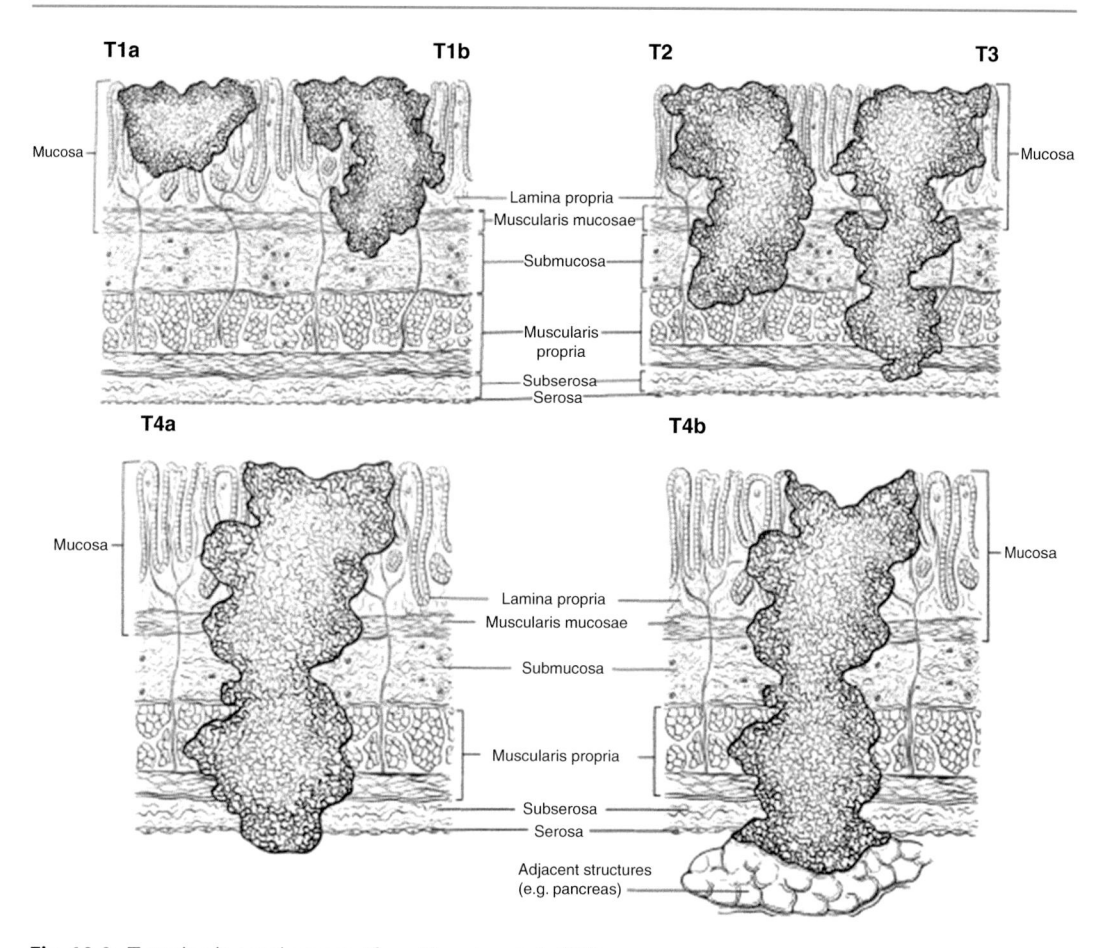

Fig. 10.8 T staging in gastric cancer (from Compton et al. [27])

10.2.4 Treatment Algorithm

Prognostic Factors

- Patient age: Poor prognosis in the young.
- Tumor stage: Invasion depth and lymph node metastasis.
- Localization: Prognosis is better in lower gastric tumors.
- Microscopic type and grade: Intestinal type is better than diffuse type according to the Lauren classification.
- Inflammatory reaction: Extensive inflammation around the tumor is a good prognostic sign.
- Perineural invasion.
- Lymphovascular space invasion.
- Surgical margin, lymph node involvement.
- Surgery: Radical LN dissection was reported to be better than standard lymph node dissection (controversial).

10.2.4.1 Diagnostic Studies Before Treatment

For primary tumor:

- Endoscopic biopsy is the first definitive diagnostic study [30].
- Double-contrast barium radiography is a complementary method.
  – It is useful in the diagnosis of linitis plastica.
- Endoscopic ultrasonography (EUS) is useful in investigating T staging and perigastric nodal lymphadenopathy by detecting the depth of invasion [31].

Fig. 10.9 Gastric lymphatics (from Compton et al. [27])

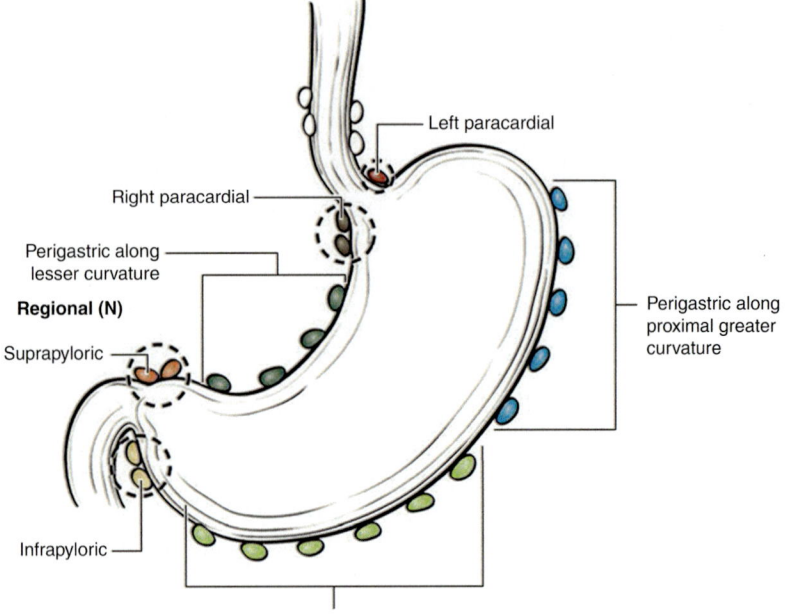

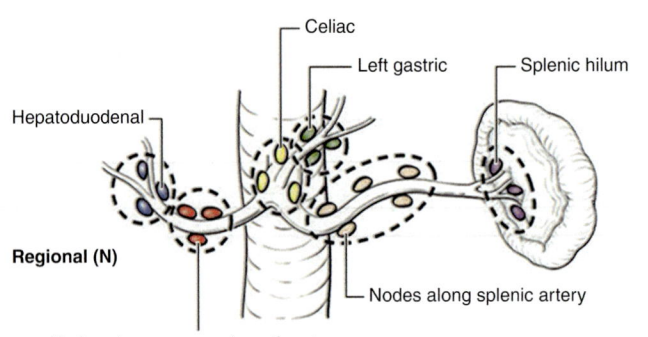

- In the evaluation of metastasis:
- CT, abdominal US, MRI.
- Staging laparoscopy with peritoneal cytology.
- Staging laparoscopy is recommended for all medically appropriate cases, except for T1 cases detected in EUS [32].
- PET-CT has a relatively limited value due to its low sensitivity in diffuse tumors histologically.
- It has 50% sensitivity in the diagnosis of peritoneal carcinomatosis [33].
- Tumor markers:
- CEA, CA19-9, CA72-4:
- Serum levels are associated with tumor stage and survival.
- They have no role in screening but are useful in diagnosing recurrence and distant metastasis [34].
- CA125, Sialyl Tn Antigens (STN):
- They often increase in the presence of peritoneal metastasis.
- Less than 10% of gastric cancers are hereditary [35] and 3–5% are associated with inherited predisposing syndromes:

Lynch syndrome, Peutz–Jeghers syndrome, familial adenomatous polyposis, Li–Fraumeni syndrome.

- A known entity is hereditary diffuse stomach cancer (HDGC).

- It is associated with young age and diffuse-type histology.
- The underlying genetic defect in 50% of HDGCs; CDH-1 (E-cadherin gene).
- The defect of this gene leads to a 40–60% risk of stomach cancer throughout life [2].
- HER2 receptor overexpression and amplification encoded by the ERBB2 gene; is detected in 7–34% of all gastric cancers [36].
- However, HER2 overexpression is not clinically correlated.
- HER2-targeted therapies provided a survival advantage in advanced gastric cancer (ToGA study) [37].
- However, in the phase III ARTIST study, HER2 overexpression in resected gastric cancer did not make a difference in disease-free survival between chemoradiotherapy vs. chemotherapy alone [38].
- Other molecular targets investigated [38]:
- Those for EGFR overexpression.
- Angiogenesis pathway targets, vascular endothelial growth factors (VEGFA, VEGFR), fibroblast growth factor receptors (FGFR), hepatocyte growth factor receptors (HGFR and tyrosine receptor kinase), and the mammalian target of rapamycin protein complex "mammalian target of rapamycin" [mTOR].

Surgery
- Surgery is the most important treatment method in potential curable gastric cancer.
- Surgical treatment should be in the form of gastrectomy with sufficient surgical margins and dissection of sufficient lymph nodes.
- At least 16 lymph nodes should be examined, 30 lymph nodes should be dissected [27].
- Preoperative or postoperative chemotherapy or chemoradiotherapy may be required depending on the tumor stage.
- In the presence of unresectable advanced disease and distant metastasis, the first treatment approach is chemotherapy. Palliative surgical interventions are indicated in the presence of obstruction or bleeding. There is no evidence of a survival benefit of noncurative reduction surgery and may reduce patient quality of life and compliance with chemotherapy.
    - R0 resection: There is no tumor microscopically at the surgical margin.
    - R1 resection: There is a tumor microscopically at the surgical margin.
    - R2 resection: There is a tumor macroscopically at the surgical margin.
    - D0 dissection: It is just the dissection of some of the perigastric lymph nodes.
    - D1 dissection: It is the dissection of perigastric lymph nodes.
    - D2 dissection: It is the dissection of perigastric + celiac axis lymph nodes. Celiac axis lymph nodes: The celiac trunk, left gastric artery, common hepatic artery, and lymph nodes around the splenic artery are dissected.
    - D3 dissection: Also, hepatoduodenal, peripancreatic, and mesentery root lymph nodes are also dissected.
    - D4 dissection: Also, para-aortic lymph nodes are also dissected [39].
- AJCC, as adequate lymph node dissection. It suggests pathological examination of 15 or more lymph nodes [27].

Surgery in Early-Stage Gastric Cancer
- Gastrectomy + D1/D2 lymphadenectomy [40]:
    - In distal 2/3 gastric cancers, distal gastrectomy is performed [41].
    - In proximal 1/3 gastric cancers, total gastrectomy is performed [42].
    - If the tumor is confined to the proximal 1/3 stomach only, a proximal gastrectomy is performed [42].
    - Optional pylorus-preserving gastrectomy can be applied in middle 1/3 gastric cancers.
    - Perigastric lymph nodes along the left gastric artery are dissected.
- Endoscopic resection:
- Indications for endoscopic mucosal resection (EMR) or endoscopic submucosal resection (ESD) [43]:

- Nodal positive probability should be extremely low. It should be differentiated adenocarcinoma without ulcerative findings. In the resected material (histologically), invasion should only be in the mucosa. There should be a lesion of Ø 2 cm or less.

Surgery in Advanced Stage (Non-metastatic) Gastric Cancer
- Gastrectomy + D2 lymphadenectomy:
- For distal or total gastrectomy, an adequate surgical margin should be >5 cm.
- If D2 lymphadenectomy cannot be performed, adjuvant chemoradiotherapy should be applied.

Radiotherapy in Advanced Stage (Non-Metastatic) Gastric Cancer

Adjuvant chemoradiotherapy; after curative gastrectomy, it is indicated if limited lymph node dissection (D0/D1) has been performed [44].

- 5 y; 1.8 Gy is administered at 25 fx.
- 5-FU; 425 mg/m2 + leucovorin 20 mg/m2, 5 days.
- Before RT; it is applied 5 days and 5 days after RT.
- After RT; two cycles are applied (every 28 days).

Capecitabine 825 mg/m^2 can be administered in two doses per day.

Preoperative chemoradiotherapy is still under investigation [45].

Chemotherapy in Advanced Stage (Non-Metastatic) Gastric Cancer

Perioperative chemotherapy

Three cures preoperative and Three cures postoperative CT:

- ECF (epirubicin, cisplatin, 5-FU)
- EOF (oxaliplatin instead of cisplatin)
- ECX (capecitabine instead of 5-FU)

Chemotherapy in Metastatic Stomach Cancer

Two-agent chemotherapy regimen is recommended.

In appropriate cases, three-agent chemotherapy is applied.
In HER2-positive cases, trastuzumab is added.

Treatment Algorithm for Gastric Cancer [5, 12]

- *Stage IA, stage IB (no invasion beyond muscularis propria (−) T2N0)*
- Surgery alone
- *Stage IB (T1N1 or invasion beyond muscularis propria (+) T2N0)*
- *Stage II–IV, M0*
- Surgery + one cycle CT + concurrent chemoradiotherapy with second cycle of CT
- (RT → 45 Gy) (CT → 5-FU + leucovorin)
- Stage I–IV, M0 (unresectable or inoperable)
- Concurrent chemoradiotherapy
- (RT → 45 Gy) (CT → 5-FU + leucovorin)
- Alternative → CT alone (if RT is not given)
- Low KPS → best supportive care
- RT alone for palliative purposes, but no effect on survival
- *M1*

Palliative CT ± RT or palliative surgery

- Total gastrectomy or subtotal gastrectomy is used for lower gastric cancers. D2 lymph node dissection may be added.
- Total gastrectomy + D2 lymph node dissection is performed for middle gastric tumors.
- Total gastrectomy + D2 lymph node dissection is standard procedure for upper gastric tumors.
- Para-aortic lymph node dissection is performed for tumors of the fundus or cardia due to extraperitoneal lymph node drainage.

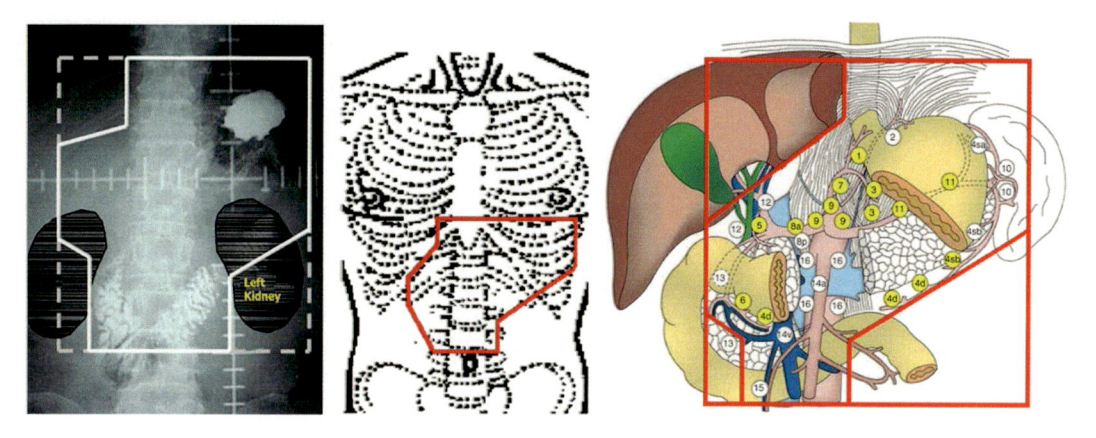

Fig. 10.10 Schematic illustration of radiotherapy fields for gastric cancer (from Neuhof and Wenz [46], Fig. 334b, p. 526, p. 526, Fig. 21.2, reproduced with the permission from Springer Science and Business Media)

10.2.5 Radiotherapy

10.2.5.1 2D Conventional Radiotherapy

Total RT dose: 45 Gy in 1.8 Gy/day.
Anterior–posterior (Fig. 10.10):

- Superior: Above the left diaphragm including the site of anastomosis with at least a 2 cm margin [nearly at the level of the T10–11 vertebrae (left hemidiaphragm)]
- Inferior: L3–4
- Lateral: Include porta hepatis and gastric bed

RT fields in gastric cancer:

- Celiac lymph nodes are at the level of the T12–L1 vertebrae.
- Porta hepatis lymphatics are included by extending the field 2 cm right lateral to the para-aortic field at the level of the T11–L1 vertebrae.
- Para-aortic lymphatics are included by extending the field below the L3 vertebra.

The superior border may be extended to include the site of anastomosis and the paraesophageal region in proximal tumors located in the cardia.

- Simulation is performed in the supine position. Hands should be at the sides if the anterior–posterior fields are used, or above the head if multiple fields are used, such as conformal therapies.
- Oral barium contrast is used to visualize the site of anastomosis, the esophagus, and the gastric remnant, as well as the small intestine.
- Kidneys should be visualized by IV contrast material.

10.2.5.2 3D Conformal RT/IMRT
(Fig. 10.11)
Simulation:

The patient is simulated in the supine position with the hands on the head. Wing board or vacuum mattress can be used for immobilization. CT imaging area is between the diaphragm/carina and L4/L5 vertebra. If possible, 4D simulation should be done. Patients should have fasted for 2–3 hours before simulation and treatment. IV and oral contrast; It is applied for imaging the tumor bed with nodal contouring. Preoperative imaging is required to determine the size of the primary tumor.

Contouring:
Contouring in Operated Gastric Cancer RT

- Structures to be included in CTV:
- Tumor bed
- Anastomosis and/or surgical stump
    – Duodenal stump; it is included in cases undergoing partial gastrectomy for distal

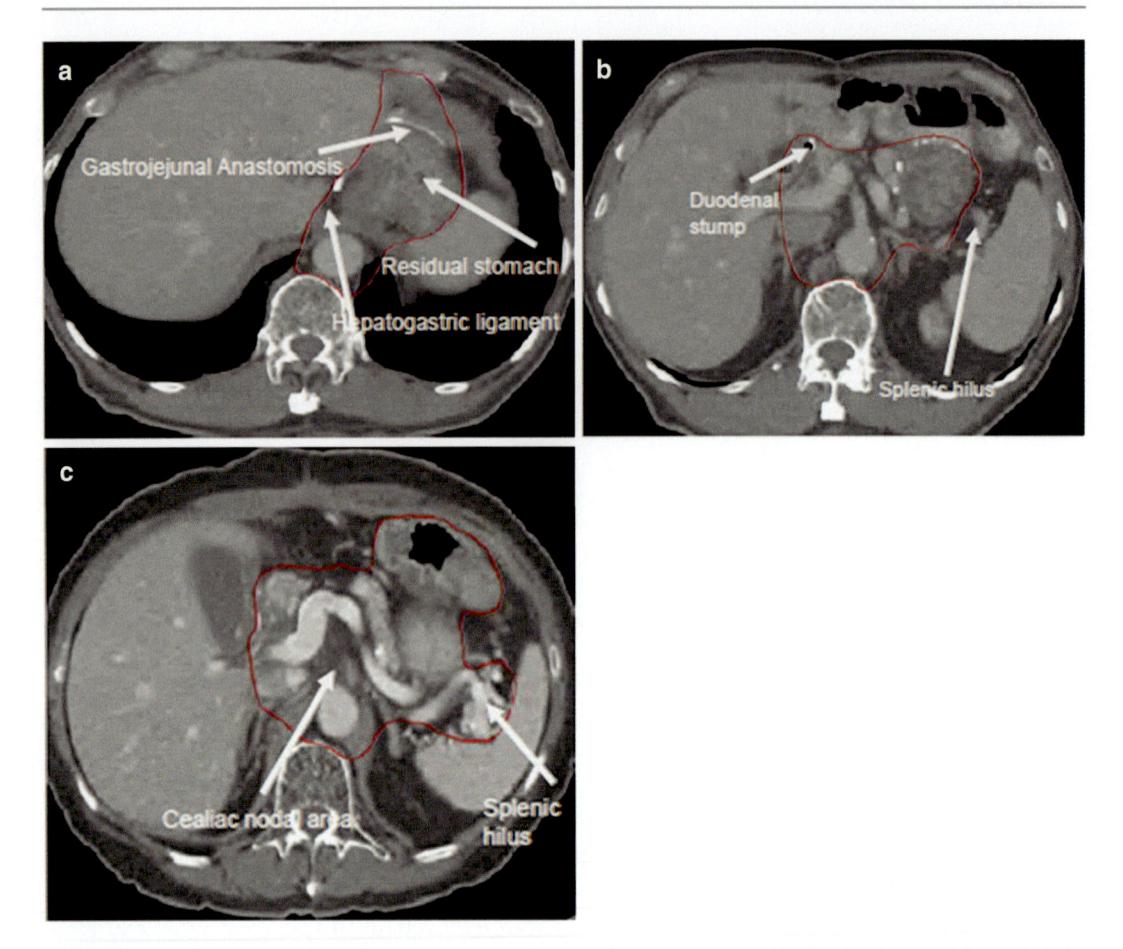

Fig. 10.11 Conformal RT volumes in gastric cancer (**a**) gastrojejunal anastomosis (**b**) duodenal stump (**c**) splenic hilus (from [47])

or antral tumors and not included in proximal/cardia tumors undergoing total gastrectomy.
– Anastomoses
  Gastrojejunal anastomosis is included if partial gastrectomy is performed in distal stomach tumors.
  If partial gastrectomy has been performed in distal stomach tumors, an esophagojejunal anastomosis is included.
– In EGJ tumors, 4–5 cm of the normal esophagus is included.
– Postoperative residual stomach tissue is included in CTV.
– Regional lymphatics; varies according to the location of the primary tumor

In EGJ tumors:
- Perigastric, esophageal, and celiac lymph nodes are included.

In gastric cardia tumors:
- Perigastric, periesophageal, celiac, splenic, pancreaticoduodenal, and porta hepatis lymph nodes are included.

In gastric corpus tumors:
- Perigastric, celiac, splenic, suprapancreatic, pancreaticoduodenal, and porta hepatis lymph nodes are included.

In gastric antrum and pylorus tumors:
- Perigastric, celiac, suprapancreatic, pancreaticoduodenal, and porta hepatis lymph nodes are included.

Splenic lymph nodes are contoured in corpus tumors, but not in antrum/pylorus tumors.
- GTV; it is determined according to CT findings and surgical operation grade.
- CTV; residual stomach + anastomosis/surgical stump + lymph nodes by primary tumor location.
- PTV; CTV + 0.5–1 cm margin is added.
  - PTV boost; GTV ± positive surgical margin zone (areas with surgical margin <5 cm) + 1–1.5 cm margin is added.

Contouring in Neoadjuvant or Definitive Radiotherapy
- Structures to be included in CTV:
  - If the primary tumor is in the fundus:
    Entire stomach (except pylorus and antrum).
    Except for pancreaticoduodenal lymph nodes, other lymphatics are included.
  - If the primary tumor is in the corpus:
    Entire stomach.
    All lymphatics are included.
  - If the primary tumor is in the antrum:
    Entire stomach (except fundus and cardia).
    Except for celiac lymph nodes, other lymphatics are included.
- GTV; it includes primary tumor and involved lymph nodes.
- CTV; GTV + 0.5 cm + lymph nodes are added according to the primary tumor location.
- ITV; created to account for target volume mobility.
  - If 4D simulation was done, it is the sum of the CTVs generated based on the CT images obtained at maximum expiration and inspiration.
  - If a 4D simulation is not done:
  - It is created by giving CTV a margin of 1.5 cm distally, 1 cm radial, and 1 cm proximal.
- PTV; ITV + 0.5 cm margin is added.

Dose/Fractionation:
- 45Gy, 1.8 Gy/fx, 25 fx ± boost (5.4 Gy)

Dose Restrictions (OAR)
- Liver; average dose 30–32 Gy, V30 <33%
- Small intestine; V45 <195 cc
- Heart; mean dose <26 Gy (pericardium), V30 <46% (pericardium), V25 <10% (entire heart)
- Kidney; bilateral kidney mean dose <15–18 Gy
  - Single kidney, V20 <33%
- Lungs; 20% max <20 Gy
- Spinal cord: Dmax is <45 Gy

10.2.6 Selected Publications

Intergroup: INT 0116, 2001 → 556 patients with resected adenocarcinoma of the stomach or gastroesophageal junction were randomly assigned to surgery plus postoperative chemoradiotherapy or surgery alone. The adjuvant treatment was 425 mg/m^2 of fluorouracil plus 20 mg/m^2 of leucovorin (LV) for 5 days, followed by 45 Gy of radiation at 1.8 Gy/day, given 5 days/week for 5 weeks, with modified doses of fluorouracil and LV on the first four and the last three days of radiotherapy. One month after the completion of radiotherapy, two 5-day cycles of fluorouracil (425 mg/m^2) plus LV (20 mg/m^2) were given 1 month apart.

- The median overall survival in the surgery-only group was 27 months compared with 36 months in the chemoradiotherapy group; the hazard ratio for death was 1.35 ($p = 0.005$).
- The hazard ratio for relapse was 1.52 ($p < 0.001$).
- Three patients (1%) died from toxic effects of the chemoradiotherapy; grade 3 toxic effects occurred in 41% of the patients in the chemoradiotherapy group, and grade 4 toxic effects occurred in 32%.
- Postoperative chemoradiotherapy is standard therapy for all patients at high risk for recurrence of adenocarcinoma of the stomach or

gastroesophageal junction who have undergone curative resection.

MacDonald JS et al (2001) Chemoradiotherapy after surgery compared with surgery alone for adenocarcinoma of the stomach or gastroesophageal junction. N Engl J Med 345(10):725–730

Consensus report, 2002 → This consensus report reviewed the following issues:

- Data supporting RT.
- Important clinical and anatomic issues related to RT.
- Details of the practical application of RT to commonly occurring clinical presentations.
- Supportive therapy during and after radiochemotherapy was also discussed in detail.

Smalley SR et al (2002) Gastric surgical adjuvant radiotherapy consensus report: rationale and treatment implementation. Int J Radiat Oncol Biol Phys 52(2):283–293

10.3 Pancreatic Cancer

Pancreatic cancer is the second most common GIS cancer in the USA and has a very poor prognosis [48]. The chances of early diagnosis are very low.

- The pancreas is localized in the posterior abdominal wall, and it is relatively fixed at the second lumbar vertebra.
- It extends transversely from the second part of the duodenal concavity to the splenic hilum.
- It is surrounded by the duodenum, the stomach, the spleen anterosuperiorly, and the duodenum, jejunum, transverse colon, and spleen anteroinferiorly.
- Posterior: Right renal vessels, inferior vena cava, portal vein, crura of diaphragm, aorta, celiac plexus, thoracic ductus, superior mesenteric vessels, splenic vessels, left renal vessels, and left kidney.
- It is 15–20 cm in length, 3.1 cm in width, and 1–1.5 cm thick.
- The pancreas is anatomically divided into five sections: The head, uncinate process, neck, corpus, and tail (Fig. 10.12).

The major pancreatic duct (Wirsung's canal) starts at the tail and passes through the entire gland, producing many branches. It enters the major duodenal papilla from the second part of the duodenum together with the choledoc duct. The pancreatic duct may also separately enter the duodenum (Fig. 10.13).

The accessory pancreatic duct (Santorini's canal), if present, drains the upper part of the

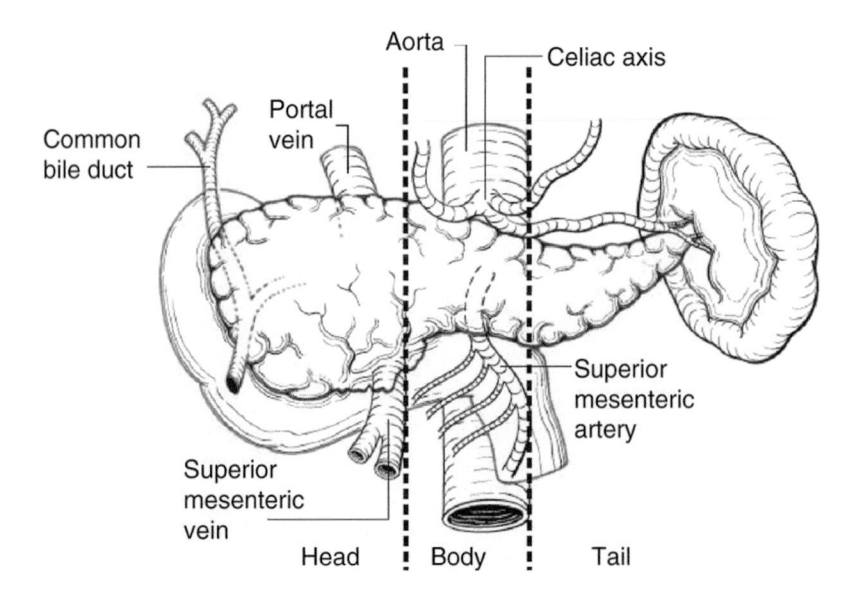

Fig. 10.12 Anatomy of the pancreas (from Compton et al. [49])

10.3 Pancreatic Cancer

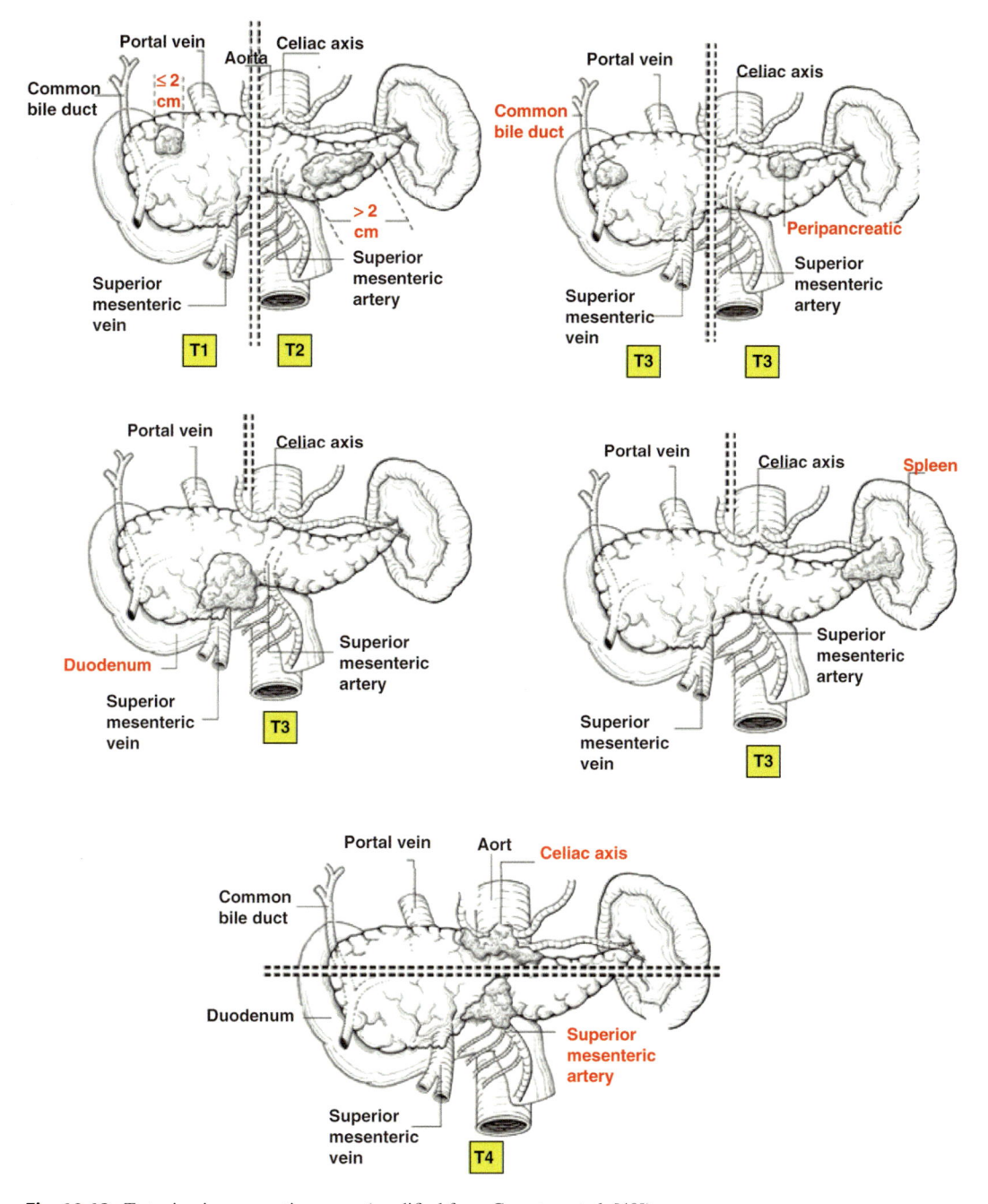

Fig. 10.13 T staging in pancreatic cancer (modified from Compton et al. [49])

pancreas and then enters the duodenum from the minor duodenal papilla (superior to the main pancreatic duct).

Pancreatic Head. This is just on the right of the L2 vertebra. A virtual line passing from the portal vein superiorly and the superior mesenteric vein inferiorly forms the border for the head and neck.

Uncinate Process. This is the part of the head that extends inferiorly. This section is sometimes never present; if it is, it surrounds the superior mesenteric artery and vein completely.

Pancreatic Neck. This covers the portal vein and the superior mesenteric vein; 1.5–2.5 cm in length.

Corpus. This crosses the L1 vertebra.

10.3.1 Pathology

Ductal epithelial cells constitute only 5% of the entire pancreatic tissue. However, more than 90% of all exocrine pancreatic cancers are adenocarcinomas originating from these ductal epithelial cells [50].

Macroscopically, the tumor is an irregularly bordered, firm, pale mass. Bleeding and necrosis are not usually present. Ductal cancers are classified according to glandular shape, mucin production, and epithelial anaplasia into well, moderate, and less differentiated.

- Only 1% of exocrine pancreatic cancers are of the acinic cell type.

Pancreatic cancer is localized in the pancreatic head (60–70%), pancreatic corpus (15–20%), or pancreatic tail (5–10%). Pancreatic cancer is multifocal in 16–30% of cases [50].

10.3.2 General Presentation

The most common symptoms are weight loss, jaundice, pain, anorexia, itching, and occult bleeding.

- Most common triad: Jaundice, pain, weight loss (weight loss and upper abdominal pain, 75% of cases; jaundice depending on the proximity of the lesion to the choledoc canal, 50–90% of cases).

10.3.3 Staging

Pancreas head cancers metastasize frequently to the superior pancreatic and posterior pancreaticoduodenal lymph nodes, and they can also involve inferior pancreatic and anterior pancreaticoduodenal lymph nodes (Table 10.4, Fig. 10.14).

Table 10.4 (A–C) Pancreatic cancer, TNM AJJC 8th edition

(A)

TX	Primary tumor cannot be assessed
T0	No evidence of primary tumor
Tis	Carcinoma in situ this includes high-grade pancreatic intraepithelial neoplasia (PanIn-3), intraductal papillary mucinous neoplasm with high-grade dysplasia, intraductal tubulopapillary neoplasm with high-grade dysplasia, and mucinous cystic neoplasm with high-grade dysplasia
T1	Tumor ≤2 cm in greatest dimension
T1a	Tumor ≤0.5 cm in greatest dimension
T1b	Tumor >0.5 and <1 cm in greatest dimension
T1c	Tumor 1 to 2 cm in greatest dimension
T2	Tumor >2 and ≤4 cm in greatest dimension
T3	Tumor >4 cm in greatest dimension
T4	Tumor involves the celiac axis, superior mesenteric artery, and/or common hepatic artery, regardless of size

The peritoneum and liver are the most frequent metastatic sites. Extra-abdominal metastasis is frequently to lungs.

Primary tumor (T) (Fig. 10.13) [49, 51]
Regional lymph nodes (N)

(B)

NX	Regional lymph nodes cannot be assessed
N0	No regional lymph node metastasis
N1	Metastasis in one to three regional lymph nodes
N2	Metastasis in four or more regional lymph nodes

AJCC stage groups

T	N	M	Stage

(C)

T	N	M	Stage
Tis	N0	M0	0
T1	N0	M0	IA
T1	N1	M0	IIB
T1	N2	M0	III
T2	N0	M0	IB
T2	N1	M0	IIB
T2	N2	M0	III
T3	N0	M0	IIA
T3	N1	M0	IIB
T3	N2	M0	III
T4	Any N	M0	III
Any T	Any N	M1	IV

TNM tumor, node, metastasis, *AJCC* American Joint Committee on Cancer, *UICC* Union for International Cancer Control, *N/A* not applicable. Used with permission of the American College of Surgeons, Chicago, Illinois. The original source for this information is the AJCC Cancer Staging Manual, Eighth Edition (2017) published by Springer International Publishing

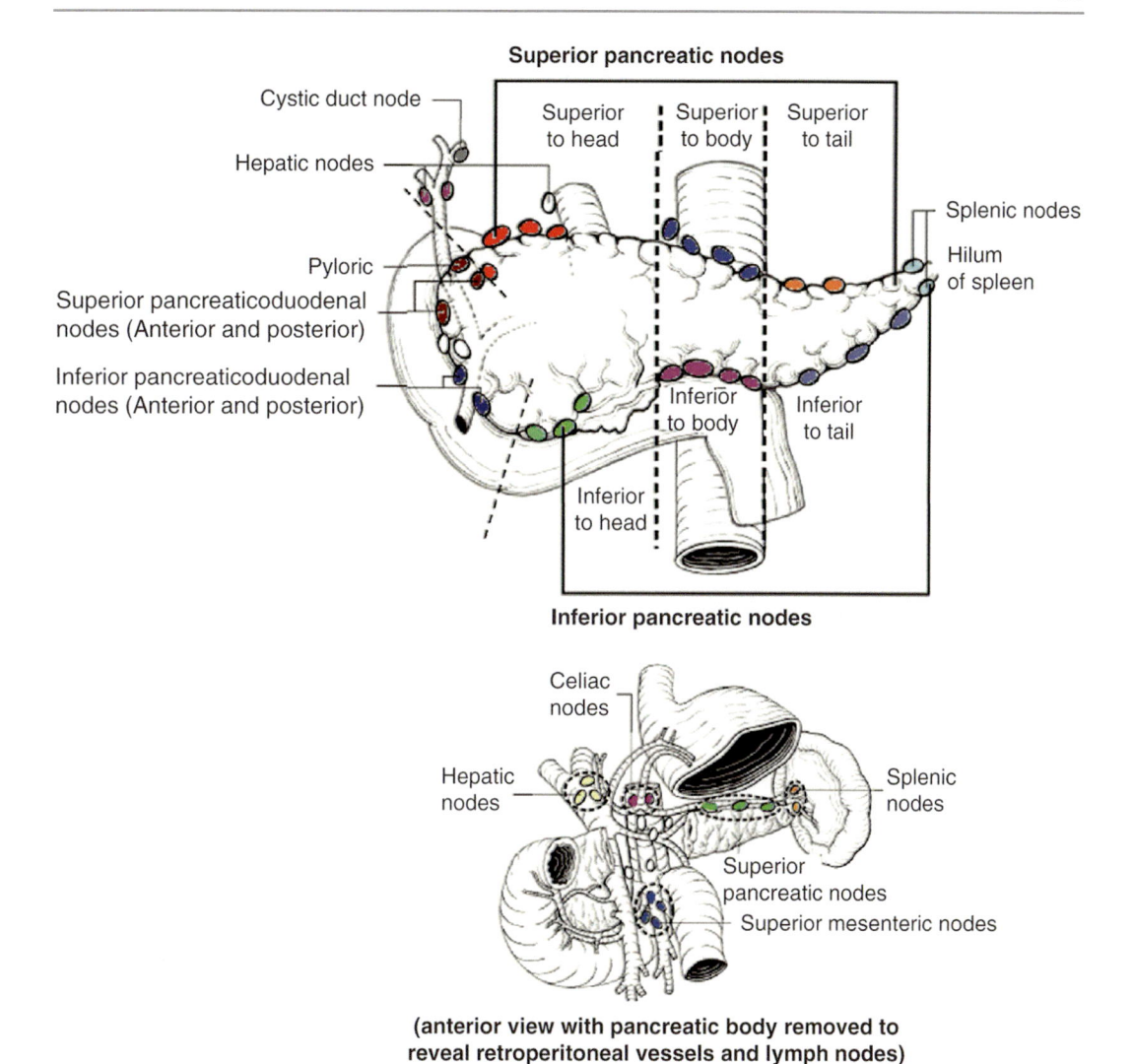

Fig. 10.14 Pancreatic lymphatics (modified from Compton et al. [51])

10.3.4 Treatment Algorithm

Whipple operation [52]: This was defined by Allen Whipple in 1935. This operation is the most common surgical approach for periampullary cancers. Pancreatic head, uncinate process, duodenum, distal stomach, proximal jejunum, and regional lymph nodes are dissected. Truncal vagotomy is generally not performed. Reconstruction is done by pancreatojejunostomy, hepaticojejunostomy, and gastrojejunostomy, respectively. Gastrojejunostomy is done by Roux-en-Y to prevent alkaline reflux.

Stage I–II (resectable) [53]

Pancreaticoduodenectomy (Whipple) + chemotherapy ± chemoradiotherapy or RT alone

- (RT → 45–50.4 Gy)
- (CT → 5-FU/gemcitabine)

Stage III (unresectable) [53]
Chemoradiotherapy or RT alone + CT
Metastatic

Chemotherapy, palliative RT, best supportive care

Resectability Criteria in Pancreatic Cancer:
- Resectable:
  - If there is no distant metastasis.
  - If there is no involvement in the celiac artery, superior mesenteric artery (SMA), a hepatic artery on MRI.
  - If there is <180° involvement in the celiac artery or SMA.
  - If the junction of the superior mesenteric vein (SMV) and portal vein (PV) is open.
  - Borderline resectable:
  - Contact of the solid tumor with the SMA or celiac axis ≤ 180° involvement
  - Common hepatic artery involvement> 180°
  - If the portal vein and SMV junction is occluded
- Anresectable [54]:
  - In pancreatic head/uncinate process lesions:
    If the contact of the solid tumor with the SMA is >180°.
    If the contact of the solid tumor with the celiac axis is >180°.
    If the solid tumor is in contact with the first jejunal branch of the SMA.
    If there is SMV or PV in nonreconstruction condition.
    If there is contact with the jejunal branch closest to the SMV.
  - In pancreatic corpus and tail lesions:
    If the contact of the solid tumor with the SMA or celiac axis is >180°.
    A solid tumor is in contact with SMA and there is aortic involvement.
    If there is SMV or PV in reconstruction unavailable condition.
  - For all regions of the pancreas;
    If there is distant metastasis.
    If there is lymph node metastasis beyond the resection area.

Treatment of Borderline/Locally Advanced Pancreatic Cancer:
- If resection is possible:
  - In the pancreatic head/uncinate process localization, pancreaticoduodenectomy (Whipple, pylor-sparing, subtotal) is applied to appropriate cases [55]. If resection will be delayed, preoperative biliary drainage is performed via an endoscopic stent. Portal/superior mesenteric vein resection and reconstruction may be required.
  - In the corpus/tail localization, distal pancreatectomy is performed [56].
  - Restaging and resection can be performed after neoadjuvant therapy in specialized centers [57].
- If resection is not possible (locally unresectable), the treatment goal is the treatment of occult systemic disease, followed by locoregional tumor and symptom control [58].
  - Endoscopic or surgical palliation of biliary and/or gastric outlet obstruction is performed.
  - Systemic chemotherapy and locoregional chemoradiotherapy are applied.

Radiotherapy in Pancreatic Cancer:
- In early disease:
  - Adjuvant chemoradiotherapy:
    Controversial, but applicable if R1 is resection or N +.
    - Various radiosensitizers can be used. Radiosensitizers such as gemcitabine, 5-FU, capecitabine.
    - CTV; tumor bed, microscopic positive surgical margins, residual disease, and high-risk peripancreatic lymph nodes.
      Celiac, superior mesenteric, porta hepatis, para-aortic lymph nodes.
    - Dose; 45–50.4 Gy is 1.8 Gy/fx.
    - Systemic therapy is added after RT [59, 60].
  - Neoadjuvant chemoradiotherapy:
    CTV; gross tumor and involved lymph nodes + 0.5–1 cm margin is given.
    45–50.4 Gy is 18. Gy / fx.
- Various types of radiosensitizers such as gemcitabine, 5-FU, and capecitabine can be used.
  - In locally advanced disease:
    Chemoradiotherapy is more effective than RT alone, and this efficacy has been demonstrated in the GITSG-9273 and ECOG 4201 studies [61].

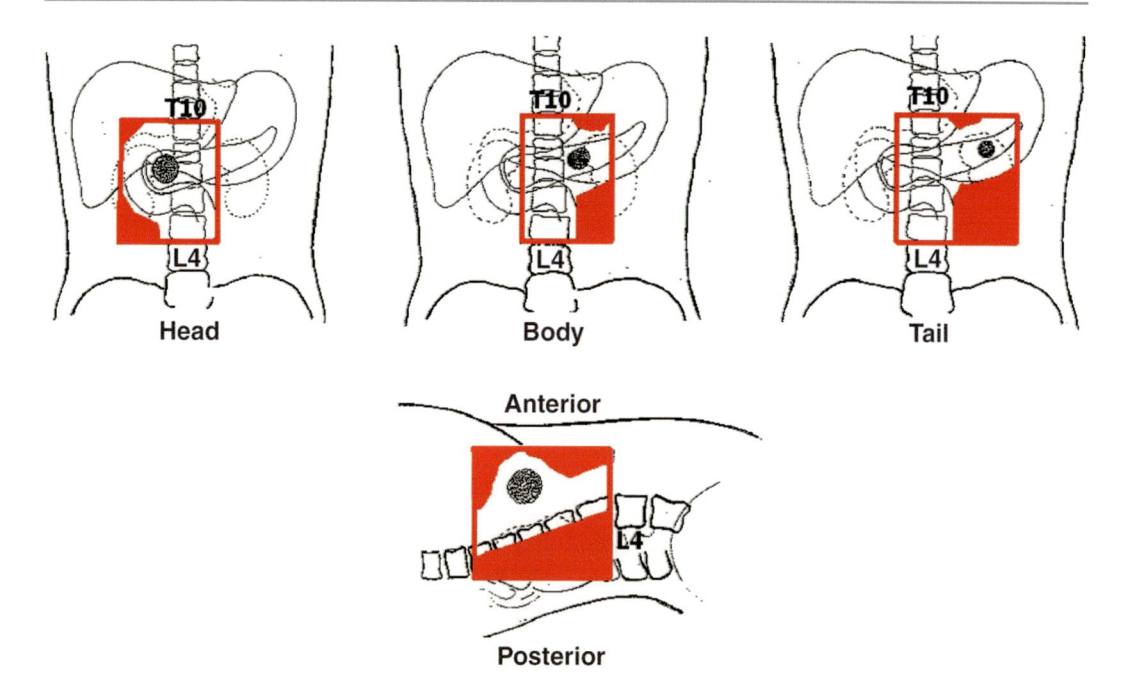

Fig. 10.15 Conventional anterior–posterior fields for different localizations of pancreatic cancer (Courtesy of wikimediacommons.org)

- First induction chemotherapy, then chemoradiotherapy is applied.
- If there is no progression after three or five cycles of gemcitabine induction chemotherapy, concurrent RT with 5-FU is applied.
    Gemcitabine; 1000 mg/m^2 is applied every three weeks.
    5-FU; 250 mg/m^2/day.
- CTV; gross tumor and involved lymph nodes + 0.5–1 cm margin is given.
- 45–50.4 Gy is 1.8 Gy/fx.
    Some centers use hypofractionated regimens.
- Various radiosensitizers can be used.
- SBRT may be an option in selected cases [62].
– Neoadjuvant chemoradiotherapy:
    There are protocols varying from center to center, 30 Gy, 3 Gy/fx; Protocols such as 5-FU, 300 mg/m^2/day can be applied.
– RT alone; can provide locoregional control in selected cases.

10.3.5 Radiotherapy

10.3.5.1 2D Conventional Radiotherapy

Simulation is done in the supine position with arms upon head. Stomach and duodenum is visualized with oral contrast; kidneys with IV contrast. Liver within field should be shielded as much as possible.

The pancreas is between the L1 and L2 vertebrae. Celiac axis is at the level of the T12 vertebra. Superior mesenteric artery is at the level of the L1 vertebra.

Anterior–Posterior Fields (Fig. 10.15):

- Superior: above T11 vertebra
- Inferior: below L3 vertebra
- Left lateral:

Pancreatic head tumor → vertebral corpus + 2 cm

Pancreatic corpus tumor → vertebral corpus + 5 cm

Pancreatic tail tumor → vertebral corpus + 8 cm (left lateral border should be modified according to the tumor/tumor bed + 3 cm if required)

- Right lateral:

Pancreatic head → include duodenum (generally costovertebral joint)

Pancreatic corpus tumor → vertebral corpus + 2 cm

Pancreatic tail tumor → vertebral corpus + 2 cm

Lateral Fields

- Anterior: tumor + 2 cm (includes duodenum)
- Posterior: half of vertebral corpus by sparing the spinal cord
- Superior–inferior: same as anterior–posterior fields

Total dose: 45 Gy/1.8 Gy with high-energy photons. Boost dose can be given by adding lateral or oblique fields (50.4–54 Gy).

10.3.5.2 3D Conformal RT/IMRT
(Fig. 10.16)

CTV: [tumor/tumor bed + 1–2 cm] + regional lymphatics

- Lymphatics:
  - Corpus: upper and lower pancreaticoduodenal, superior and inferior pancreatic, celiac
  - Head: corpus lymphatics + porta hepatis lymphatics
  - Tail: corpus lymphatics (except pancreaticoduodenal LN) + splenic hilum LN

PTV: CTV + 1–1.5 cm

Critical organs: liver, kidneys, small intestines, spinal cord

10.3.6 Selected Publications

ESPAC-1, 2001 → After resection, 541 patients were randomly assigned to adjuvant chemoradiotherapy (20 Gy in ten daily fractions over 2 weeks with 500 mg/m^2 fluorouracil intravenously on days 1–3, repeated after 2 weeks) or chemotherapy (intravenous fluorouracil 425 mg/m^2 and folinic acid 20 mg/m^2 daily for 5 days, monthly for 6 months). Clinicians randomized the patients into a two-by-two factorial design (observation, chemoradiotherapy alone, chemotherapy alone, or both) or into one of the main treatment comparisons (chemoradiotherapy vs. no chemoradiotherapy or chemotherapy vs. no chemotherapy).

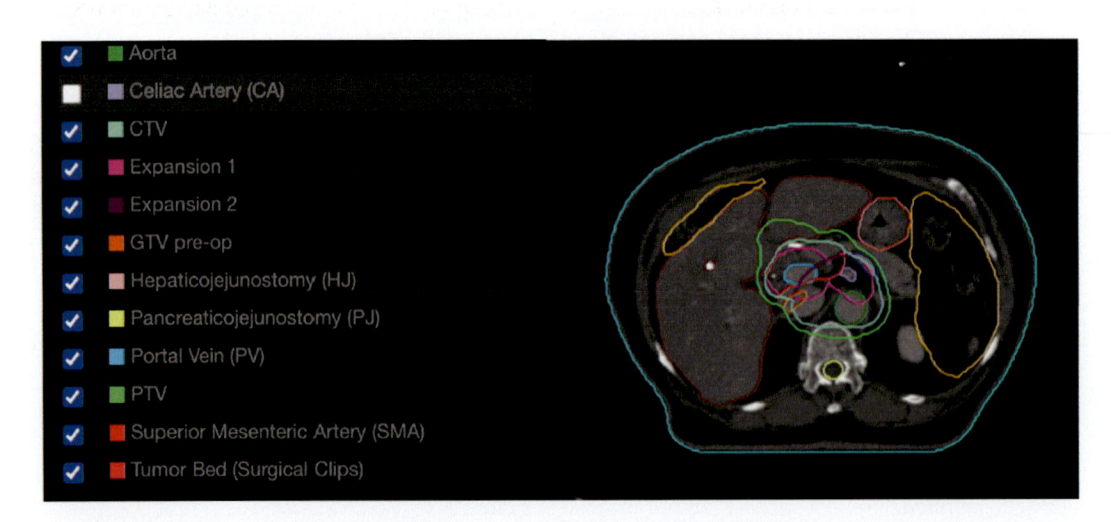

Fig. 10.16 Conformal RT fields in pancreatic cancer (Courtesy of Dr. Cuneyt Ebruli)

- This study showed no survival benefit for adjuvant chemoradiotherapy but revealed a potential benefit for adjuvant chemotherapy.

Neoptolemos JP et al (2001) Adjuvant chemoradiotherapy and chemotherapy in resectable pancreatic cancer: a randomised controlled trial. Lancet 358(9293):1576–1585
→ *2004 Update*: Adjuvant chemotherapy had a significant survival benefit in patients with resected pancreatic cancer, whereas adjuvant chemoradiotherapy had a deleterious effect on survival.
Neoptolemos JP et al (2004) A randomized trial of chemoradiotherapy and chemotherapy after resection of pancreatic cancer. N Engl J Med 350(12):1200–1210
(*Editors' note*. This trial was seriously criticized due its design and several issues related with RT. Therefore, the results of this trial did not affect practice in North American series.)
EORTC, 1999 → 218 patients were randomized to an observation group vs. a treatment group (RT + chemotherapy).

- Adjuvant RT in combination with 5-FU was found to be safe and well tolerated. However, the benefit in this study was small; routine use of adjuvant chemoradiotherapy was not warranted as standard treatment in cancer of the head of the pancreas or periampullary region.

Klinkenbijl JH et al (1999) Adjuvant radiotherapy and 5-fluorouracil after curative resection of cancer of the pancreas and periampullary region: phase III trial of the EORTC Gastrointestinal Tract Cancer Cooperative Group. Ann Surg 230(6):776–782
GITSG, 1985 → 22 patients randomized to no adjuvant treatment and 21 to combined therapy were analyzed. Median survival for the treatment group (20 months) was significantly longer than that observed for the control group (11 months).
Kalser MH et al (1985) Pancreatic cancer. Adjuvant combined radiation and chemotherapy following curative resection. Arch Surg 120(8):899–903

RTOG 97-04/Intergroup, 2006 → 442 patients. Pre- and post-chemoradiotherapy 5-FU vs. pre- and post-chemoradiotherapy gemcitabine. ChemoRT = 50.4 Gy 1.8 Gy/fraction/day with CI 5-FU, 250 mg/m^2/day during RT for all patients.

- The addition of gemcitabine to postoperative adjuvant 5-FU CRT significantly improved survival.

Regine WF et al (2006) RTOG 9704 a phase III study of adjuvant pre and post chemoradiation (CRT) 5-FU vs. gemcitabine (G) for resected pancreatic adenocarcinoma. J Clin Oncol (2006 ASCO Annual Meeting Proceedings Part I) 24(18S (June 20 Suppl)):4007
GITSG 9273, 1981 → 194 patients with locally unresectable adenocarcinoma of the pancreas were randomly assigned to therapy with high-dose (60 Gy) radiation therapy alone, to moderate-dose (40 Gy) radiation + 5-FU, and to high-dose radiation plus 5-FU.

- Median survival with radiation alone was only 5.5 months from the date of diagnosis. Both 5-FU-containing treatment regimens produced a highly significant survival improvement when compared with radiation alone.
- Survival differences between 4,000 rads plus 5-FU and 6,000 rads plus 5-FU were not significant, with an overall median survival of 10 months.

Moertel CG et al (1981) Therapy of locally unresectable pancreatic carcinoma: a randomized comparison of high dose (6000 rads) radiation alone, moderate dose radiation (4000 rads + 5-fluorouracil), and high dose radiation + 5-fluorouracil: The Gastrointestinal Tumor Study Group. Cancer 48(8):1705–1710
GITSG 9283, 1988 → This trial compared multidrug chemotherapy [streptozocin, mitomycin, and 5-FU (SMF)] vs. radiation combined with 5-FU followed by the same three-drug SMF combination.

- An improved median survival for the combined-modality therapy (42 weeks) com-

pared with chemotherapy alone (32 weeks) was demonstrated. Overall survival following this combined-modality treatment program (41% at 1 year) was significantly superior to that following SMF chemotherapy alone (19% at 1 year) ($p < 0.02$).
- Combined-modality therapy was found to be superior to either optimal radiotherapy or chemotherapy alone.

Gastrointestinal Tumor Study Group (1988) Treatment of locally unresectable carcinoma of the pancreas: comparison of combined-modality therapy (chemotherapy plus radiotherapy) to chemotherapy alone. J Natl Cancer Inst 80(10):751–755

FFCD-SFRO, 2006 → Randomization: chemo-RT (60 Gy in 6 weeks, 2 Gy/fraction, concomitant with 5-FU and cisplatin) or gemcitabine (G) (1,000 mg/m² weekly 7q8w) as induction treatment.

- Study was stopped before the planned inclusion due to lower survival with initial chemo-RT when compared to G alone.

Chauffert B (2006) Phase III trial comparing initial chemoradiotherapy (intermittent cisplatin and infusional 5-FU) followed by gemcitabine vs. gemcitabine alone in patients with locally advanced non metastatic pancreatic cancer. J Clin Oncol (2006 ASCO Annual Meeting Proceedings Part I) 24(18S):4008

South Florida, 2008 → Intratumoral (32)P was associated with more serious adverse events and did not improve survival for locally advanced unresectable pancreatic cancer.

Rosemurgy A (2008) (32)P as an adjunct to standard therapy for locally advanced unresectable pancreatic cancer: a randomized trial. J Gastrointest Surg 12(4):682–688

10.4 Rectal Cancer

Colon and rectal cancers are the most common gastrointestinal malignant neoplasms. They are the fourth most common of all cancers, and the second most common cause of cancer-related mortality [63].

The rectum is a part of the large intestine; it starts at the promontorium and exits at the anus inferiorly (Fig. 10.17). The rectum extends anteroinferiorly, following the sacral concavity for 13–15 cm, and reaching 2–3 cm below the coccyx. At this level, it turns back along the canal and passes within the levator muscles before ending at the anus, forming an anal canal that is nearly 3–4 cm in length. The length and diameter

Fig. 10.17 Anatomy of the rectum (from Carmichael and Mills [64], p. 9, Fig. 1.5)

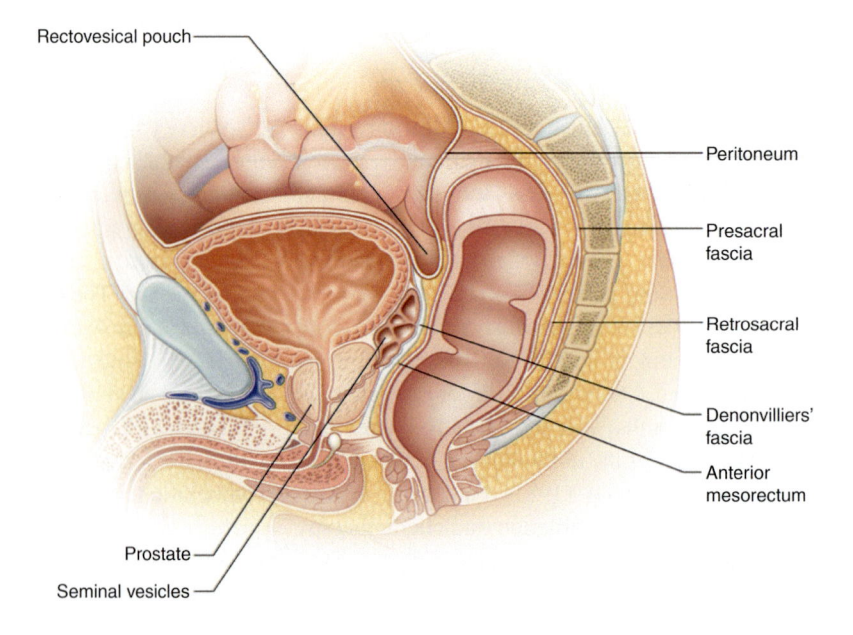

10.4 Rectal Cancer

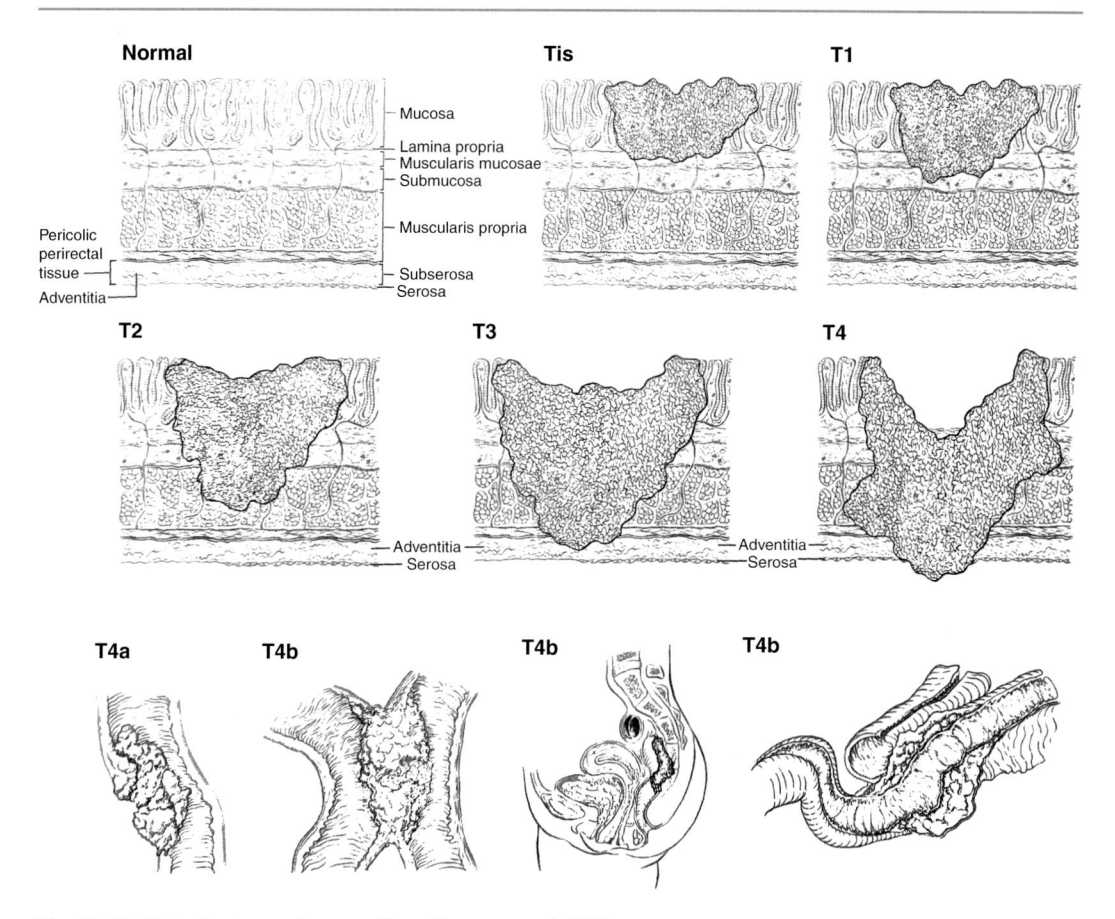

Fig. 10.18 T staging in rectal cancer (from Compton et al. [65])

of the rectum change depending on its fullness (Figs. 10.18 and 10.19).

Rectum–peritoneum relationship:

- The upper third of the rectum is covered with peritoneum on the anterior and lateral surfaces. Only a small part of the thin mesorectum has no peritoneum on the posterior part.
- The middle third of the rectum is covered with peritoneum on its anterior surface; its lateral and posterior surfaces have no peritoneum.
- The peritoneum skips the rectovesical pouch and covers the seminal vesicles and the bladder in males. It forms the rectouterine pouch and passes over the vagina and the uterus in females.
- The lower third of the rectum has no peritoneum.

Colorectal cancer is the most frequent type of cancer after breast and ovary cancer in females, and lung cancer in males.

The rectum is classically divided into three parts: the upper third, the middle third, and the lower third. Each part is practically assumed to be 5 cm in length.

The peritoneal reflection (Douglas' pouch) that turns anteriorly in front of the rectum is 8–9 cm from the anal verge in males and 5–8 cm in females.

10.4.1 Pathology

Most rectal cancers (95%) are adenocarcinomas [21].

Fig. 10.19 T3–4 staging according to peritoneal status in rectal cancer (circumferential margin) (from Compton et al. [65])

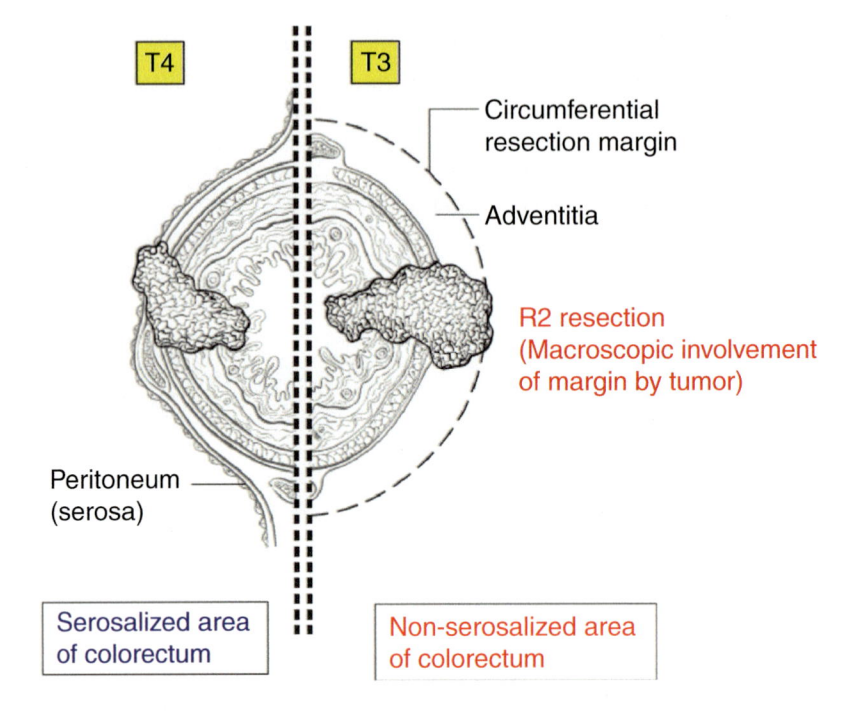

Fig. 10.20 Comparison of the TNM and Astler–Coller rectal cancer staging systems [22–24]

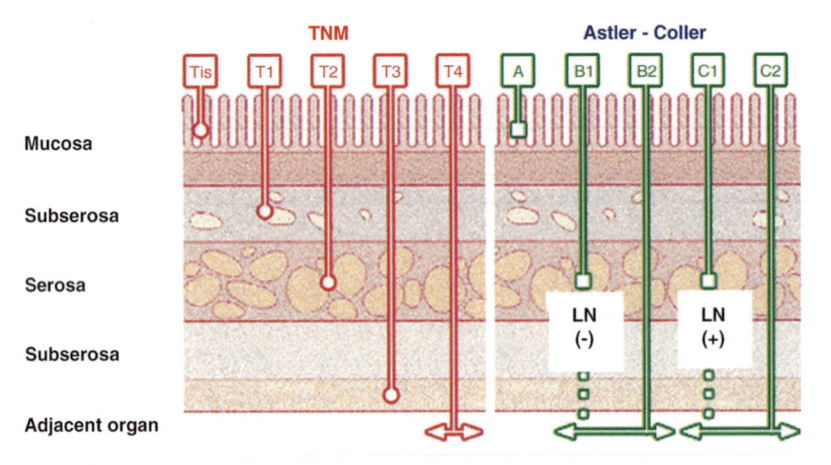

10.4.1.1 Colorectal Cancers According to the WHO Classification

- Adenocarcinoma
- Mucinous adenocarcinoma
- Signet-ring cell carcinoma
- Adenosquamous carcinoma
- Squamous cell carcinoma
- Undifferentiated carcinoma

10.4.2 General Presentation

Frequent defecation, tenesmus, decrease in stool diameter and volume, and fresh blood and mucous discharge in stool may be seen in rectal cancers. Perineal and sacral pain is seen in advanced stages due to surrounding organ invasion.

There are several staging systems in rectal cancer other than the TNM (AJCC) (2002) staging system: Duke's staging (1930) and Astler–Coller staging (modified Duke's) (1954) (Fig. 10.20).

10.4 Rectal Cancer

10.4.2.1 Change in Defecation Habits

This is the most common sign associated with colorectal cancers. Constipation, diarrhea, or constipation–diarrhea may be observed.

- *Bleeding.* The second most common symptom is bleeding, which may be prominent or occult.
- *Rectal mucous discharge.* This may be rectal discharge or it may be mixed with stool or blood. When tumor is close to distal, mucous discharge begins.
- *Morning diarrhea.* Here, mucous and blood collecting in the rectal cancer are discharged as a mixed bloody–mucous material with a foul smell in the morning, termed morning diarrhea. There is no stool in this material.
- *Pain.* Pain is seen in the advanced stages of rectal cancer. The involvement of the sacral plexus causes sacral and sciatic pain.
- A digital rectal exam may reveal a mass in the rectum.
- The carcinoembryonic antigen level may be high, but this should not be used as a diagnostic or screening tool.

10.4.3 Staging (Table 10.5)

10.4.4 Treatment Algorithm

Abdominoperineal resection (APR) (Miles' operation) [66]: Tumoral tissue + entire rectum are dissected and a permanent colostomy is opened.

Lower anterior resection (LAR) [66]: Tumoral tissue + the upper parts of the rectum are dissected and a permanent colostomy is not opened.

Stage I [53]
Lower one-third rectal cancers → APR
Upper/middle one-third rectal cancers → low anterior resection (LAR)
Highly selected T1 tumors with good prognostic factors: local excision alone

- Good prognostic factors: Tumor size <3 cm, circumferential resection line <1/3, maximum distance from anal verge <8 cm, well-differentiated, surgical margin >3 mm, LVSI (−), PNI (−)

- T1 tumors with good prognostic factors (+), surveillance after surgery

Stage II–III (resectable) [53]
Preoperative 5-FU/RT + transabdominal surgery + adjuvant CT
Stage III (unresectable and T4) [53]
5-FU/RT (then surgery if possible) + CT
Stage IV
CT or surgery or RT or combinations of them

Postoperative RT [67]
Advantages:

- There is no overtreatment with radiation due to pathological confirmation of exact stage.
- RT field is accurately designed since tumoral extension is known.

Disadvantages:

- RT efficacy may decrease due to hypoxic medium after surgery.
- Postoperative adhesions may increase intestinal side effects of RT.
- Field size increases after APR due to the inclusion of the perineal scar in the RT portal.

Preoperative RT [68]
Advantages:

- Large and advanced tumors may be downsized during radiation and the chances of resection may also increase.
- Tumor cells that may be implanted into the surgical region are eradicated.
- Tumor cells that may enter circulation are eradicated and distant metastatic risk decreases.
- Tumor cells are more oxic before surgery and sensitive to radiation.

Disadvantages:

- Overtreatment of early or metastatic tumors
- Delayed surgery and risk of stage advancement
- Postoperative perineal complications may increase

Table 10.5 (A–C) Colorectal cancer, TNM AJJC 8th edition

T category	T criteria
A: Primary tumor (T) (Fig. 10.18) [22]	
TX	Primary tumor cannot be assessed
T0	No evidence of primary tumor
Tis	Carcinoma in situ, intramucosal carcinoma (involvement of lamina propria with no extension through muscularis mucosae)
T1	Tumor invades the submucosa (through the muscularis mucosa but not into the muscularis propria)
T2	Tumor invades the muscularis propria
T3	Tumor invades through the muscularis propria into pericolorectal tissues
T4	Tumor invades[a] the visceral peritoneum or invades or adheres[b] to adjacent organ or structure
T4a	Tumor invades[a] through the visceral peritoneum (including gross perforation of the bowel through tumor and continuous invasion of tumor through areas of inflammation to the surface of the visceral peritoneum)
T4b	Tumor directly invades[a] or adheres[b] to adjacent organs or structures

[a]Direct invasion in T4 includes invasion of other organs or other segments of the colorectum as a result of direct extension through the serosa, as confirmed on microscopic examination (e.g., invasion of the sigmoid colon by a carcinoma of the cecum) or, for cancers in a retroperitoneal or subperitoneal location, direct invasion of other organs or structures by virtue of extension beyond the muscularis propria (i.e., respectively, a tumor on the posterior wall of the descending colon invading the left kidney or lateral abdominal wall; or a mid or distal rectal cancer with invasion of prostate, seminal vesicles, cervix, or vagina)

[b]Tumor that is adherent to other organs or structures, grossly, is classified cT4b. However, if no tumor is present in the adhesion, microscopically, the classification should be pT1-4a depending on the anatomical depth of wall invasion. The V and L classification should be used to identify the presence or absence of vascular or lymphatic invasion, whereas the PN prognostic factor should be used for perineural invasion

N category	N criteria
B: Regional lymph nodes (N)	
NX	Regional lymph nodes cannot be assessed
N0	No regional lymph node metastasis
N1	One to three regional lymph nodes are positive (tumor in lymph nodes measuring ≥0.2 mm), or any number of tumor deposits are present and all identifiable lymph nodes are negative
N1a	One regional lymph node is positive
N1b	Two or three regional lymph nodes are positive
N1c	No regional lymph nodes are positive, but there are tumor deposits in the: ■ Subserosa ■ Mesentery ■ Nonperitonealized pericolic or perirectal/mesorectal tissues
N2	Four or more regional nodes are positive
N2a	Four to six regional lymph nodes are positive
N2b	Seven or more regional lymph nodes are positive

Distant metastasis (M)

M category	M criteria
M0	No distant metastasis by imaging, etc.; no evidence of tumor in distant sites or organs (this category is not assigned by pathologists)
M1	Metastasis to one or more distant sites or organs or peritoneal metastasis is identified
M1a	Metastasis to one site or organ is identified without peritoneal metastasis
M1b	Metastasis to two or more sites or organs is identified without peritoneal metastasis
M1c	Metastasis to the peritoneal surface is identified alone or with other site or organ metastases

AJCC stage groups			
T	N	M	Stage
C: Prognostic stage groups			
Tis	N0	M0	0
T1, T2	N0	M0	I
T3	N0	M0	IIA
T4a	N0	M0	IIB
T4b	N0	M0	IIC
T1-T2	N1/N1c	M0	IIIA
T1	N2a	M0	IIIA
T3-T4a	N1/N1c	M0	IIIB
T2-T3	N2a	M0	IIIB
T1-T2	N2b	M0	IIIB
T4a	N2a	M0	IIIC
T3-T4a	N2b	M0	IIIC
T4b	N1-N2	M0	IIIC
Any T	Any N	M1a	IVA
Any T	Any N	M1b	IVB
Any T	Any N	M1c	IVC

TNM tumor, node, metastasis, *AJCC* American Joint Committee on Cancer, *UICC* Union for International Cancer Control, *N/A* not applicable. Used with permission of the American College of Surgeons, Chicago, Illinois. The original source for this information is the AJCC Cancer Staging Manual, Eighth Edition (2017) published by Springer International Publishing

10.4.5 Radiotherapy

Simulation is performed in either the supine or the prone position. The prone position with a full bladder may decrease the small intestinal dose.

- Rectal and vaginal markers are useful during simulation. The perineal scar should be wired after APR.
- Oral contrast may be used to visualize the small intestine, and the bladder is preferably full.

Anterior–Posterior Fields (Fig. 10.21):
- Superior: between levels L5 and S1
- Inferior: preoperative, tumor + 3 cm; LAR, below obturator foramen or tumor + 3 cm; APR, includes perineal scar
- Lateral: bony pelvis +2 cm

Lateral Fields (Fig. 10.22):
- Superior–inferior: same as anterior–posterior fields
- Anterior: ≤T3 tumors, posterior to pubic symphysis; T4 tumors, anterior to pubic symphysis
- Posterior: includes sacrum

Boost field → tumor/tumor bed +2–3 cm. Sacral concavity should be included in lateral fields.

Dose
1.8 Gy/day, total dose: 45 Gy four field-box technique, 5.4 Gy boost may be given.

Conformal RT Volumes [51]
- CTV: tumor/tumor bed + perirectal tissue, presacral region and internal iliac LNs

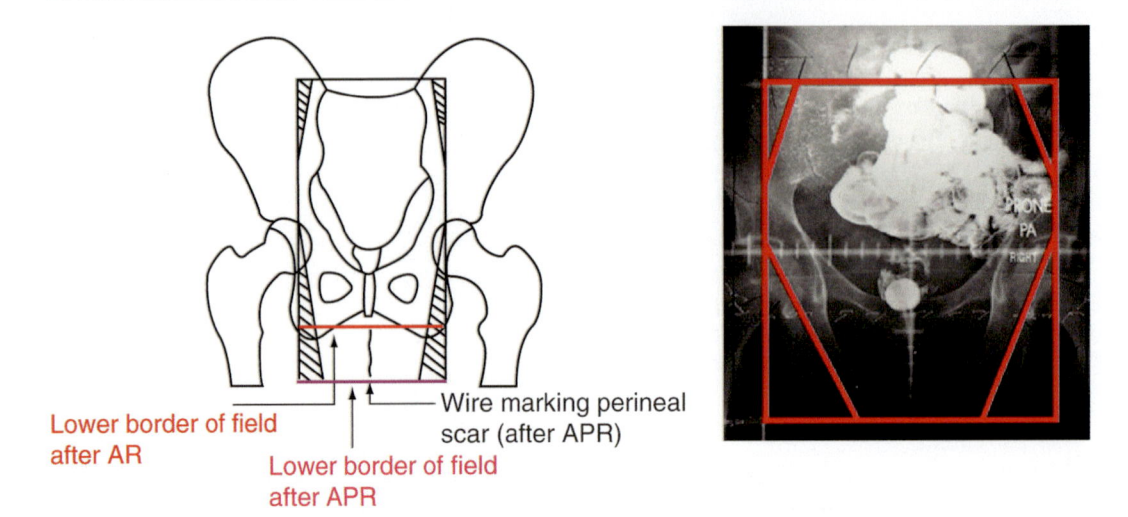

Fig. 10.21 Conventional anterior–posterior RT fields in rectal cancer [57, p 563, Fig. 24.4]

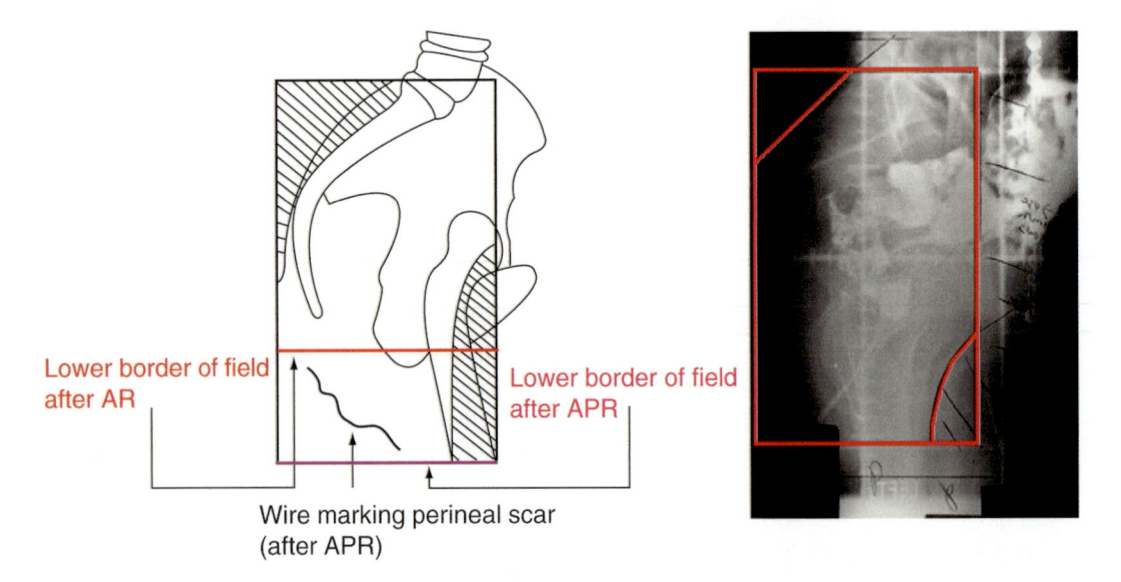

Fig. 10.22 Conventional lateral RT field in rectal cancer [57, p 563, Fig. 24.5]

A craniocaudal 2 cm margin is added to the tumor.

One to two centimeters of the bladder and prostate/vagina are included (particularly in T4 cases).

Anterior wall of the promontorium and sacrum is within the CTV.

- PTV: CTV + 1–1.5 cm

10.4.6 Selected Publications

NSABP R-01, 1988 → The patients were randomized to receive no further treatment (184 patients), postoperative adjuvant chemotherapy with 5-FU, semustine, and vincristine (MOF) (187 patients), or postoperative RT (184 patients).

- Postoperative radiation therapy reduced the incidence of locoregional recurrence, but it failed to affect overall disease-free survival and survival.

Fisher B et al (1988) Postoperative adjuvant chemotherapy or radiation therapy for rectal cancer: results from NSABP protocol R-01. J Natl Cancer Inst 80(1):21–29

NSABP R-02, 2000 → Patients were randomized to postoperative adjuvant chemotherapy alone ($n = 348$) or chemotherapy with postoperative radiotherapy ($n = 346$).

- The addition of postoperative RT to chemotherapy in Dukes' B and C rectal cancer did not alter the subsequent incidence of distant disease, although there was a reduction in locoregional relapse when compared with chemotherapy alone.

Wolmark N et al (2000) Randomized trial of postoperative adjuvant chemotherapy with or without radiotherapy for carcinoma of the rectum: National Surgical Adjuvant Breast and Bowel Project Protocol R-02. J Natl Cancer Inst 92(5):388–396

MRC 3, 1996 → Surgery alone ($n = 235$) vs. surgery followed 4–6 weeks later by radiotherapy ($n = 234$), of 40 Gy in 20 fractions.

- Postoperative radiotherapy delayed and prevented the local recurrence of rectal cancer.

Medical Research Council Rectal Cancer Working Party (1996) Randomised trial of surgery alone versus surgery followed by radiotherapy for mobile cancer of the rectum. Lancet 348(9042):1610–1614

Intergroup INT-0114, 1997 → 1,695 patients were treated with two cycles of chemotherapy followed by chemo-RT and two additional cycles of chemotherapy. Chemotherapy regimens were bolus fluorouracil (5-FU), 5-FU and LV, 5-FU and levamisole, and 5-FU, LV, and levamisole. Pelvic irradiation was given to a dose of 45 Gy to the whole pelvis and a boost to 50.4–54 Gy.

- There was no advantage to LV- or levamisole-containing regimens over bolus 5-FU alone in the adjuvant treatment of rectal cancer when combined with irradiation.

Tepper JE et al (2002) Adjuvant therapy in rectal cancer: analysis of stage, sex, and local control—final report of intergroup 0114. J Clin Oncol 20(7):1744–1750

Intergroup INT-0144, 2006 → After surgery of T3–4, N0, M0 or T1–4, N1, 2M0 rectal adenocarcinoma, 1,917 patients were randomized to bolus FU in two 5-day cycles every 28 days before and after RT plus FU via protracted venous infusion (PVI) during RT; arm 2 (PVI-only arm), PVI 42 days before and 56 days after RT + PVI; or bolus-only arm, with bolus FU + LV in two 5-day cycles before and after XRT, plus bolus FU + LV (levamisole was administered in each cycle before and after RT).

- All arms provided similar relapse-free survivals and OS, along with different toxicity profiles and central catheter requirements.

Smalley SR (2006) Phase III trial of fluorouracil-based chemotherapy regimens plus radiotherapy in postoperative adjuvant rectal cancer: GI INT 0144. J Clin Oncol 24(22):3542–3547

German Rectal Cancer Study, 2004 → Preoperative chemoradiotherapy, as compared with postoperative chemoradiotherapy, improved local control and was associated with reduced toxicity but with no improvement in overall survival.

Sauer R et al (2004) Preoperative versus postoperative chemoradiotherapy for rectal cancer. N Engl J Med 351(17):1731–1740

Netherlands, 2007 → 1861 patients with resectable rectal cancer were randomized between total mesorectal excision (TME) preceded by 5 × 5 Gy or TME alone. No chemotherapy was allowed. Preoperative short-term radiotherapy improved local control in patients with clinically resectable rectal cancer. However, there was no effect on overall survival.

Peeters KC (2007) The TME Trial after a median follow-up of 6 years: increased local control but no survival benefit in irradiated patients with resectable rectal carcinoma. Ann Surg 246(5):693–701

MRC,1996 → Preoperative RT reduced the rate of local recurrence of rectal cancer in patients with locally advanced disease.

Medical Research Council Rectal Cancer Working Party (1996) Randomised trial of surgery alone versus radiotherapy followed by surgery for potentially operable locally advanced rectal cancer. Lancet 348(9042):1605–1610

Swedish Rectal Cancer Trial, 2005 → Preoperative RT with 25 Gy in 1 week before curative surgery for rectal cancer was found to be beneficial to overall and cancer-specific survival and local recurrence rates after long-term follow-up.

Folkesson J et al (2005) Swedish Rectal Cancer Trial: long lasting benefits from radiotherapy on survival and local recurrence rate. J Clin Oncol 23(24):5644–5650

UK, 1994 → 468 patients randomized to RT (3 × 5 Gy over 5 days within 2 days of operation) followed by surgery, or to surgery alone. Long-term survival was unaffected, but long-term local recurrence was reduced by the addition of low-dose RT to surgery. Perioperative mortality was increased.

Goldberg PA (1994) Long-term results of a randomised trial of short-course low-dose adjuvant pre-operative radiotherapy for rectal cancer: reduction in local treatment failure. Eur J Cancer 30A(11):1602–1606

10.5 Anal Cancer

Anal canal cancers constitute 1% of all colorectal cancers and 10% of all rectal cancers. They are usually observed in those aged 60–65 years. Incidence is 1/100,000 in females and 0.5–0.8/100,000 in males [69].

The rectum forms the anal canal, which is 3–4 cm in length and ends at the anus. The anal canal is defined as either the surgical or the anatomical canal (Fig. 10.23).

The epithelium proximal to the dentate line is transitional epithelium, while the epithelium distal to it until the anal verge is a special nonkeratinized squamous epithelium with no hair follicles or sweat glands, also termed anoderm.

Fig. 10.23 Anatomy of the anus (from Compton et al. [70])

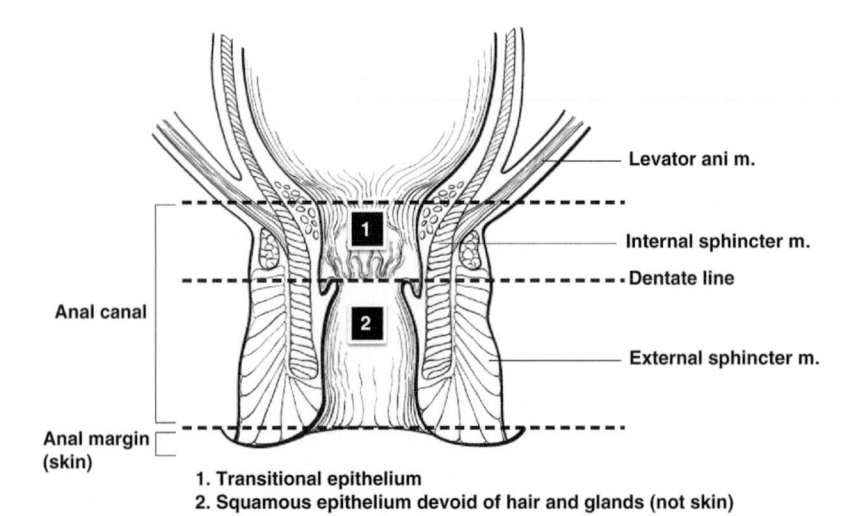

1. Transitional epithelium
2. Squamous epithelium devoid of hair and glands (not skin)

10.5 Anal Cancer

Anatomical Anal Canal: Between the anal verge and the dentate line).

Surgical anal canal:
3–4 cm section localized between the anal verge and the anorectal ring.

Anorectal Ring: This is felt as a muscular ring during a digital exam, and it is the proximal endpoint of the internal sphincter muscle–puborectal muscle complex.

- It forms the border of the rectum–anal canal.

The length of the anal canal shows individual variations (2–8 cm, mean 3.5–4 cm). These variations are important in cancer surgery. For instance, it is theoretically possible to conserve sphincter function when a tumor is palpated at 5 cm in a patient with a 3 cm anal canal. However, the sphincter cannot be conserved in distal rectal tumors that extend into the anal canal and invade the sphincter muscles.

10.5.1 Pathology

Most anal cancers are squamous cell cancers (~60%), followed by transitional cell cancers (~25%) and adeno cancers (~7%) [71]. Less frequent ones are basaloid cell cancers (cloacogenic cancer) and malignant melanomas. Small cell cancers are very rare but have a high risk of distant metastasis.

Anal cancers occur between the anal verge and 2 cm beyond the dentate line; tumors occurring further from the dentate line are called rectal cancers.

10.5.2 General Presentation

Anal cancers usually give symptoms in their early stages. Bleeding, pain around the anus or feelings of compression in the anal canal, anal itching, anal swelling, and changes in bowel habits are the main symptoms.

10.5.3 Staging (Table 10.6)

10.5.4 Treatment Algorithm

10.5.4.1 T1 Tumor (Small and Well Differentiated) [57]

Local excision

- Tm <2 cm, well-differentiated, superficially spreading, and negative surgical margin after excision.
- Sphincter invasion in circumferential excision line >40%, chemoradiotherapy should be used due to risk of incontinence after excision.

Stage I–III [57]
Concurrent chemoradiotherapy (CT → mitomycin + 5-FU)
Stage IV
Palliative therapies and their combinations
Recurrence, salvage therapy
APR, or RT if no previous RT history

10.5.5 Radiotherapy

The anal region may be irradiated with various techniques due to the complex anatomy of this region and the inguinal lymphatics. These techniques include wide anterior field–narrow posterior field, the four-field box technique, 3D conformal radiotherapy, and IMRT.

- The aim when using these techniques is to decrease the doses supplied to the genital organs, bladder, small intestine, and femur heads, and to increase the doses supplied to deep inguinal lymphatics.
- Femur fractures start at 45 Gy and their incidence increases prominently after 50 Gy.
- Deep inguinal lymph nodes are usually located 5–6 cm from the skin.
- Simulation is performed in the prone position with arms on chest.
- Anal marker is used to visualize the anal verge.

Table 10.6 (A–C) Anal cancer, TNM AJJC 8th edition

T	T criteria
A: Primary tumor (T) (Fig. 10.24) [72]	
TX	Primary tumor not assessed
T0	No evidence of primary tumor
Tis	High-grade squamous intraepithelial lesion (previously termed carcinoma in situ, Bowen disease, anal intraepithelial neoplasia II–III, high-grade anal intraepithelial neoplasia)
T1	Tumor ≤2 cm
T2	Tumor >2 cm but ≤5 cm
T3	Tumor >5 cm
T4	Tumor of any size invading adjacent organ(s), such as the vagina, urethra, or bladder

Anal canal lymphatics (Fig. 10.25) From [70]
- Perirectal LN

| Anorectal |
| Perirectal |
| Lateral sacral |

- Internal iliac (hypogastric) LN
- Inguinal LN

| Superficial inguinal |
| Deep femoral |

- Above dentate line: internal iliac LN
- Below dentate line: ingunial LN

N category	N criteria
B: Regional lymph nodes (N) (Fig. 10.26)	
NX	Regional lymph nodes cannot be assessed
N0	No regional lymph node metastasis
N1	Metastasis in inguinal, mesorectal, internal iliac, or external iliac nodes
N1a	Metastasis in inguinal, mesorectal, or internal iliac lymph nodes
N1b	Metastasis in external iliac lymph nodes
N1c	Metastasis in external iliac with any N1a nodes

AJCC stage groupings

T	N	N	Stage
C: Prognostic stage groups			
Tis	N0	M0	0
T1	N0	M0	I
T1	N1	M0	IIIA
T2	N0	M0	IIA
T2	N1	M0	IIIA
T3	N0	M0	IIB
T3	N1	M0	IIIC
T4	N0	M0	IIIB
T4	N1	M0	IIIC
Any T	Any N	M1	IV

TNM: tumor, node, metastasis, *AJCC* American Joint Committee on Cancer, *UICC* Union for International Cancer Control, *N/A* not applicable. Used with permission of the American College of Surgeons, Chicago, Illinois. The original source for this information is the AJCC Cancer Staging Manual, Eighth Edition (2017) published by Springer International Publishing

- Inguinal field may be marked with wires for visualization.
- Palpable inguinal lymph nodes are marked with wires.
- A full bladder is recommended to decrease toxicity to the small intestine.
- The penis is placed cranially to prevent its bolus effect on the scrotum.
- Inguinal electron boost is applied in the supine position.
- High-energy X-rays (6–18 MV) are used.
- Inguinal electron boost is applied with suitable electron energies, and bolus may be used if required.

10.5.5.1 Conventional RT Fields in Anal Cancer (Fig. 10.27)

Anterior field until 36 Gy:

  Superior: above sacroiliac joints
- Inferior: below the anal verge or 3 cm below the tumor
- Lateral: medial side of the trochanter major, including inguinal lymph nodes

Anterior field between 36 and 45 Gy

- Superior: below sacroiliac joints
- Inferior: below the anal verge or 3 cm below the tumor
- Lateral: tips of femur heads

Posterior field until 45 Gy:

- Superior: above sacroiliac joints
- Inferior: below the anal verge or 3 cm below the tumor
- Lateral: tips of femur heads

Anterior–posterior boost field 50.4–54 Gy:

- Tumor/tumor bed + 2.5 cm

Electron boosts are added to inguinal regions (Fig. 10.28).

- LN (−): Total dose is completed to 45 Gy.
- LN (+): Total dose is completed to 50.4–54 Gy.

10.5 Anal Cancer

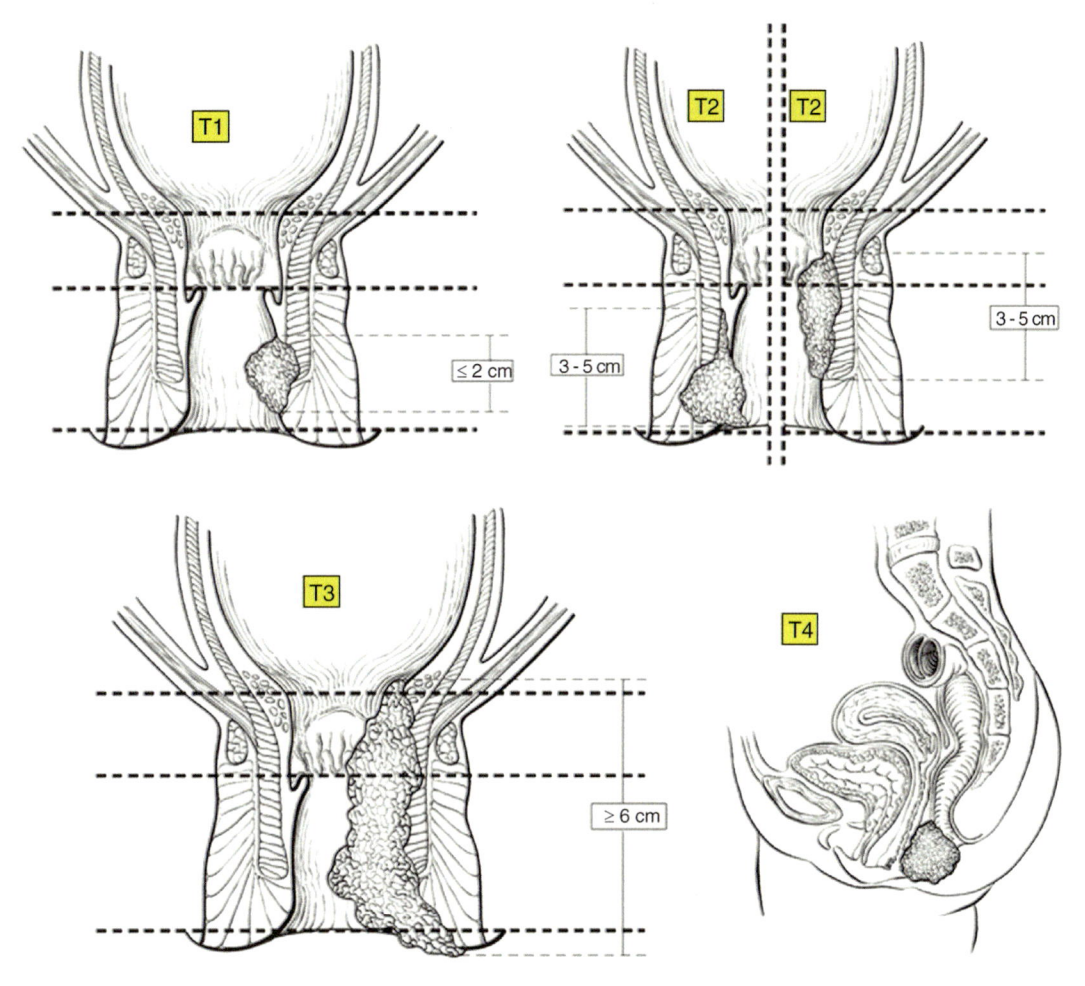

Fig. 10.24 T staging in anal cancer (adapted from [70])

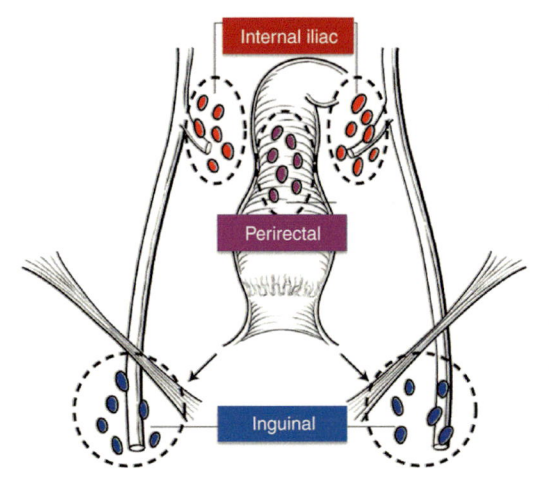

Fig. 10.25 Anal canal lymphatics (from [70], p 121, Fig. 13.2, reproduced with the permission from the American Joint Committee on Cancer)

10.5.5.2 Conformal RT Volumes [51]
CTV: [tumor/tumor bed, pelvic LN, inguinal LN, perianal tissues] + 1–2 cm
  PTV: CTV + 1.5 cm

10.5.6 Selected Publications

EORTC, 1997 → 110 patients were randomized between RT alone (45 Gy given in 5 weeks) and a combination of RT and chemotherapy (750 mg/m² daily fluorouracil as a continuous infusion and a single dose of mitomycin 15 mg/m²). After a rest period of 6 weeks, a boost of 20 or 15 Gy was given in cases of partial or complete response, respectively.

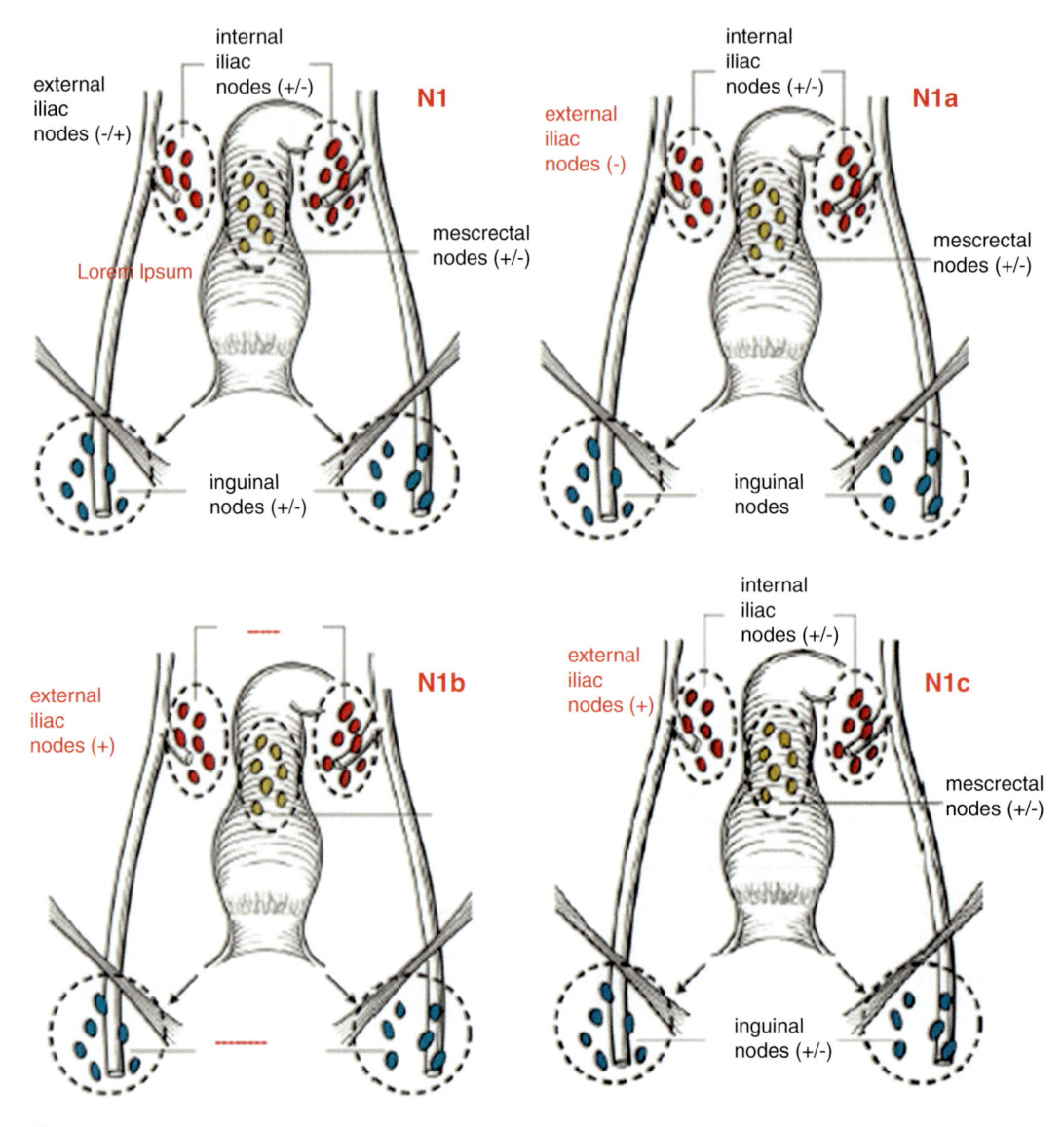

Fig. 10.26 N staging in anal cancer (adapted from [70])

- The concomitant use of RT and chemotherapy resulted in a significantly improved locoregional control rate and a reduction in the need for colostomy in patients with locally advanced anal cancer without a significant increase in late side effects.

Bartelink H (1997) Concomitant radiotherapy and chemotherapy is superior to radiotherapy alone in the treatment of locally advanced anal cancer: results of a phase III randomized trial of the European Organization for Research and Treatment of Cancer Radiotherapy and Gastrointestinal Cooperative Groups. J Clin Oncol 15(5):2040–2049

UKCCCR, 1996 → 585 patients were randomized to receive initially either 45 Gy RT over 4–5 weeks or the same regimen of RT combined with 5-FU by continuous infusion during the first and the final weeks of radiotherapy and mitomycin on day 1 of the first course.

- The standard treatment for most patients with epidermoid anal cancer should be a combina-

10.5 Anal Cancer

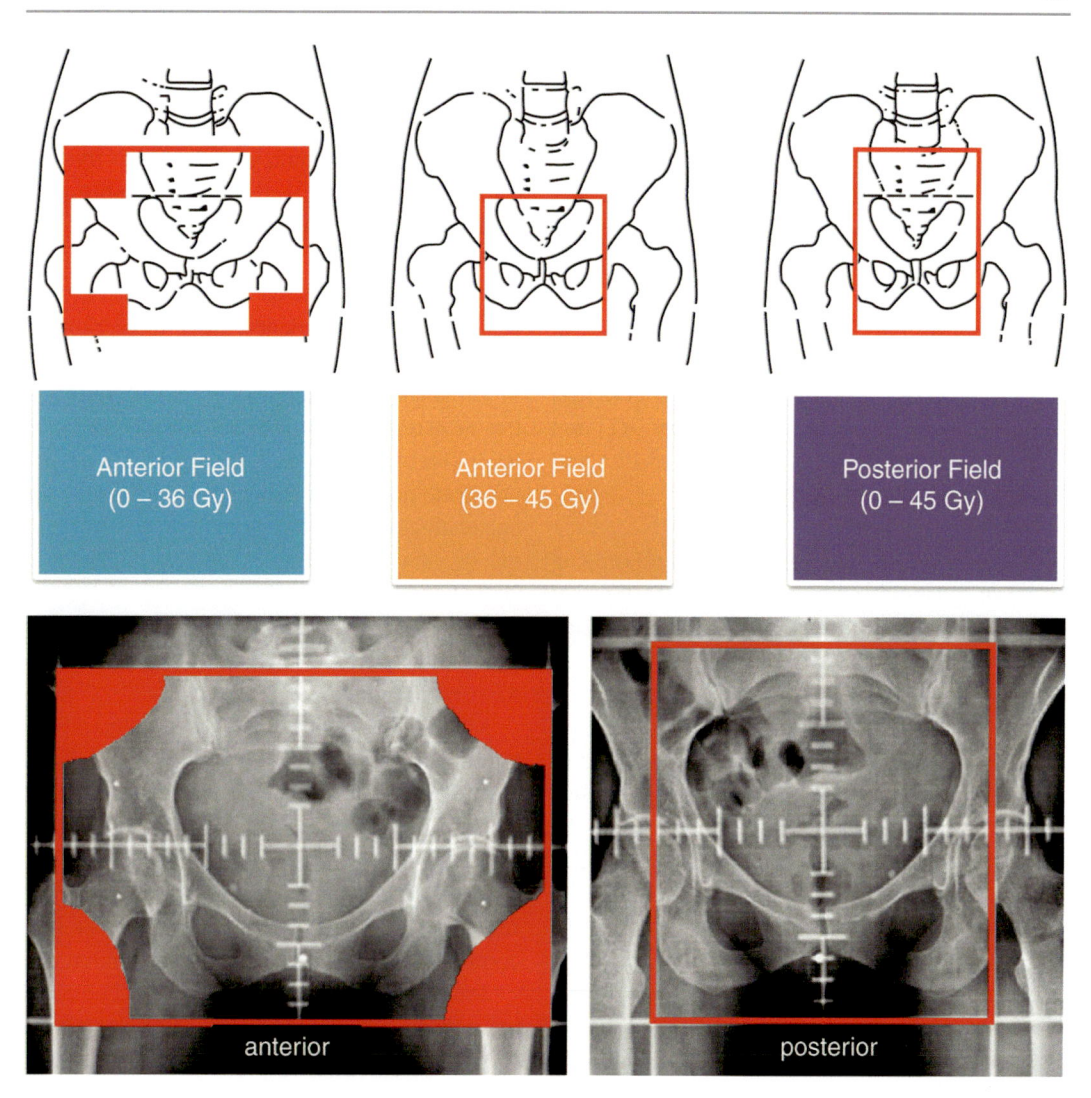

Fig. 10.27 Conventional RT fields in anal cancer (from Fig. 355b–e [73], p 555, Fig. 22.6a–c, reproduced with the permission from the Springer Science and Business Media)

Fig. 10.28 Inguinal boost fields in anal cancer ([74], p 555, Fig. 22.6c])

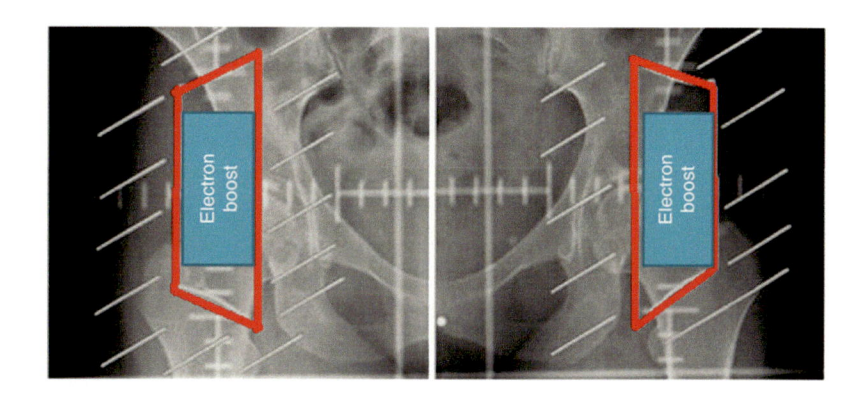

tion of radiotherapy and infused 5-FU and mitomycin, with surgery reserved for those who fail to improve on this regimen.

UK Coordinating Committee on Cancer Research (1996) Epidermoid anal cancer: results from the UKCCCR randomised trial of radiotherapy alone versus radiotherapy, 5-fluorouracil, and mitomycin. UKCCCR Anal Cancer Trial Working Party. Lancet 348(9034):1049–1054

RTOG 87-04, 1996 → 310 patients were randomized to receive either radiotherapy (RT) and fluorouracil (5-FU) or radiotherapy, 5-FU, and mitomycin C.

- Despite greater toxicity, the use of mitomycin C in a definitive chemo-RT regimen for anal cancer was justified, particularly in patients with large primary tumors.

Flam M (1996) Role of mitomycin in combination with fluorouracil and radiotherapy, and of salvage chemoradiation in the definitive nonsurgical treatment of epidermoid carcinoma of the anal canal: results of a phase III randomized intergroup study. J Clin Oncol 14(9):2527–2539

RTOG 83-14, 1989 → 79 patients with any primary tumor stage of anal canal carcinoma were treated by RT combined with mitomycin given by bolus IV injection and 5-FU given by continuous infusion. Radiation was delivered to the perineum and pelvis to a total dose of 40.8 Gy in 4.5–5 weeks. The inguinal nodal areas received 40.8 Gy, calculated at a depth of 3 cm in the center of the nodal area.

- The overall survival rates were 97% at 1 year and 73% at 3 years.
- Combined therapy was effective for patients with anal cancer and allowed preservation of the sphincter and of sexual function.

Sischy B et al (1989) Definitive irradiation and chemotherapy for radiosensitization in management of anal carcinoma: interim report on Radiation Therapy Oncology Group study no. 8314. J Natl Cancer Inst 81(11):850–856

Salama et al., 2007 → 53 patients were treated with concurrent chemotherapy and IMRT for anal cancer. Preliminary outcomes suggested that concurrent chemotherapy and IMRT for anal canal cancers was effective and favorably tolerated compared with historical standards.

Salama JK (2007) Concurrent chemotherapy and intensity-modulated radiation therapy for anal canal cancer patients: a multicenter experience. J Clin Oncol 25(29):4581–4586

University of Chicago, 2005 → IMRT reduced the mean doses to small bowel, bladder, and genitalia. Treatment was well tolerated, with no serious acute nonhematological toxicity. There were no treatment breaks attributable to gastrointestinal or skin toxicity. At a median follow-up of 20.3 months, there were no other local failures. Two-year overall survival, disease-free survival, and colostomy-free survival were 91, 65, and 82%, respectively.

Milano MT et al (2005) Intensity-modulated radiation therapy (IMRT) in the treatment of anal cancer: toxicity and clinical outcome. Int J Radiat Oncol Biol Phys 63(2):354–361

RTOG 92-08, 1996 → Increases in RT dose over those used in conventional chemotherapy regimens for anal cancers did not increase local control when given in a split-course fashion.

John M (1996) Dose escalation in chemoradiation for anal cancer: preliminary results of RTOG 92-08. Cancer J Sci Am 2(4):205–211

References

1. Kubo A, Corley DA (2004) Marked multi-ethnic variation of esophageal and gastric cardia carcinomas within the United States. Am J Gastroenterol 99(4):582–588. https://doi.org/10.1111/j.1572-0241.2004.04131.x
2. Compton CC, Byrd DR, Garcia-Aguilar J, Kurtzman SH, Olawaiye A, Washington MK (2012) Esophagus and esophagogastric junction. In: Compton C, Byrd D, Garcia-Aguilar J, Kurtzman S, Olawaiye A, Washington M (eds) AJCC cancer staging atlas. Springer, New York, NY. https://doi.org/10.1007/978-1-4614-2080-4_10
3. Siegel RL, Miller KD, Jemal A (2017) Cancer statistics, 2017. CA Cancer J Clin 67(1):7–30
4. Rice TW, Kelsen D, Blackstone EH, et al. Esophagus and esophagogastric junction. In: AJCC cancer stag-

ing manual, 8th ed, Amin MB (Ed), AJCC, Chicago 2017. p. 185. Corrected at 4th printing, 2018.
5. Moraca RJ, Low DE (2006) Outcomes and health-related quality of life after esophagectomy for high-grade dysplasia and intramucosal cancer. Arch Surg 141:545
6. Dandara C, Robertson B, Dzobo K et al (2016) Patient and tumour characteristics as prognostic markers for oesophageal cancer: a retrospective analysis of a cohort of patients at Groote Schuur hospital. Eur J Cardiothorac Surg 49:629
7. Puli SR, Reddy JB, Bechtold ML et al (2008) Staging accuracy of esophageal cancer by endoscopic ultrasound: a meta-analysis and systematic review. World J Gastroenterol 14:1479
8. Wani S, Sayana H, Sharma P (2010) Endoscopic eradication of Barrett's esophagus. Gastrointest Endosc 71:147
9. Yau KK, Siu WT, Cheung HY et al (2006) Immediate preoperative laparoscopic staging for squamous cell carcinoma of the esophagus. Surg Endosc 20:307
10. Riedel M, Hauck RW, Stein HJ et al (1998) Preoperative bronchoscopic assessment of airway invasion by esophageal cancer: a prospective study. Chest 113:68
11. van Westreenen HL, Heeren PA, van Dullemen HM et al (2005) Positron emission tomography with F-18-fluorodeoxyglucose in a combined staging strategy of esophageal cancer prevents unnecessary surgical explorations. J Gastrointest Surg 9:54
12. Yendamuri S, Swisher SG, Correa AM et al (2009) Esophageal tumor length is independently associated with long-term survival. Cancer 115:508
13. Gemmill EH, McCulloch P (2007) Systematic review of minimally invasive resection for gastro-oesophageal cancer. Br J Surg 94:1461
14. Dimick JB, Wainess RM, Upchurch GR Jr et al (2005) National trends in outcomes for esophageal resection. Ann Thorac Surg 79:212
15. Liu SL, Xi M, Yang H et al (2016) Is there a correlation between clinical complete response and pathological complete response after neoadjuvant chemoradiotherapy for esophageal squamous cell cancer? Ann Surg Oncol 23:273
16. Zenda S, Kojima T, Kato K et al (2016) Multicenter phase 2 study of cisplatin and 5- fluorouracil with concurrent radiation therapy as an organ preservation approach in patients with squamous cell carcinoma of the cervical esophagus. Int J Radiat Oncol Biol Phys 96:976
17. Stahl M, Stuschke M, Lehmann N et al (2005) Chemoradiation with and without surgery in patients with locally advanced squamous cell carcinoma of the esophagus. J Clin Oncol 23:2310
18. Al-Sarraf M, Martz K, Herskovic A et al (1997) Progress report of combined chemoradiotherapy versus radiotherapy alone in patients with esophageal cancer: an intergroup study. J Clin Oncol 15:277
19. Mariette C, Dahan L, Mornex F et al (2014) Surgery alone versus chemoradiotherapy followed by surgery for stage I and II esophageal cancer: final analysis of randomized controlled phase III trial FFCD 9901. J Clin Oncol 32:2416
20. Adelstein DJ, Rice TW, Rybicki LA et al (2009) Mature results from a phase II trial of postoperative concurrent chemoradiotherapy for poor prognosis cancer of the esophagus and gastroesophageal junction. J Thorac Oncol 4:1264
21. Wong R, Malthaner R (2001) Combined chemotherapy and radiotherapy (without surgery) compared with radiotherapy alone in localized carcinoma of the esophagus. Cochrane Database Syst Rev:CD002092
22. Bédard EL, Inculet RI, Malthaner RA et al (2001) The role of surgery and postoperative chemoradiation therapy in patients with lymph node positive esophageal carcinoma. Cancer 91:2423
23. Adelstein DJ, Rice TW, Rybicki LA et al (2009) Mature results from a phase II trial of postoperative concurrent chemoradiotherapy for poor prognosis cancer of the esophagus and gastroesophageal junction. J Thorac Oncol 4:1264
24. Lou F, Sima CS, Adusumilli PS et al (2013) Esophageal cancer recurrence patterns and implications for surveillance. J Thorac Oncol 8:1558
25. Li C, Wang X, Wang X et al (2019) A multicenter phase III study comparing simultaneous integrated boost (SIB) radiotherapy concurrent and consolidated with S-1 versus SIB alone in elderly patients with esophageal and esophagogastric cancer – the 3JECROG P-01 study protocol. BMC Cancer 19:397. https://doi.org/10.1186/s12885-019-5544-1
26. Blot WJ, Devesa SS, Kneller RW et al (1991) Rising incidence of adenocarcinoma of the esophagus and gastric cardia. JAMA 265(10):1287–1289
27. Compton CC, Byrd DR, Garcia-Aguilar J, Kurtzman SH, Olawaiye A, Washington MK (2012) Stomach. In: Compton C, Byrd D, Garcia-Aguilar J, Kurtzman S, Olawaiye A, Washington M (eds) AJCC cancer staging atlas. Springer, New York, NY. https://doi.org/10.1007/978-1-4614-2080-4_11
28. Werner M, Becker KF, Keller G, Hofler H (2001) Gastric adenocarcinoma: pathomorphology and molecular pathology. J Cancer Res Clin Oncol 127(4):207–216
29. Sayegh ME, Sano T, Dexter S, Katai H, Fukagawa T, Sasako M (2004) TNM and Japanese staging systems for gastric cancer: how do they coexist? Gastric Cancer 7(3):140–148
30. Graham DY, Schwartz JT, Cain GD, Gyorkey F (1982) Prospective evaluation of biopsy number in the diagnosis of esophageal and gastric carcinoma. Gastroenterology 82:228
31. Yoshida S, Tanaka S, Kunihiro K et al (2005) Diagnostic ability of high-frequency ultrasound probe sonography in staging early gastric cancer, especially for submucosal invasion. Abdom Imaging 30:518
32. Power DG, Schattner MA, Gerdes H et al (2009) Endoscopic ultrasound can improve the selection for laparoscopy in patients with localized gastric cancer. J Am Coll Surg 208:173

33. Yoshioka T, Yamaguchi K, Kubota K et al (2003) Evaluation of 18F-FDG PET in patients with advanced, metastatic, or recurrent gastric cancer. J Nucl Med 44:690
34. Lai IR, Lee WJ, Huang MT, Lin HH (2002) Comparison of serum CA72-4, CEA, TPA, CA19-9 and CA125 levels in gastric cancer patients and correlation with recurrence. Hepato-Gastroenterology 49:1157
35. Stoffel EM (2015) Screening in GI cancers: the role of genetics. J Clin Oncol 33:1721–1728
36. Gravalos C, Jimeno A (2008) HER2 in gastric cancer: a new prognostic factor and a novel therapeutic target. Ann Oncol 19(9):1523–1529
37. Bang YJ, Van Cutsem E, Feyereislova A, an ToGA Trial Investigators et al (2010) Trastuzumab in combination with chemotherapy versus chemotherapy alone for treatment of HER2-positive advanced gastric or gastro-oesophageal junction cancer (ToGA): a phase 3, open-label, randomised controlled trial. Lancet 376(9742):687–697
38. Park SH, Sohn TS, Lee J et al (2015) Phase III rial to compare adjuvant chemotherapy with capecitabine and cisplatin versus concurrent chemoradiotherapy in gastric cancer: final report of the adjuvant chemoradiotherapy in stomach tumors trial, including survival and subset analyses. J Clin Oncol 33:3130–3136
39. Roukos DH, Kappas AM (2005) Perspectives in the treatment of gastric cancer. Nat Clin Pract Oncol 2(2):98–107
40. Mocellin S, McCulloch P, Kazi H et al (2015) Extent of lymph node dissection for adenocarcinoma of the stomach. Cochrane Database Syst Rev:CD001964
41. Bozzetti F, Marubini E, Bonfanti G et al (1999) Subtotal versus total gastrectomy for gastric cancer: five-year survival rates in a multicenter randomized Italian trial. Italian Gastrointestinal Tumor Study Group. Ann Surg 230:170
42. Gouzi JL, Huguier M, Fagniez PL et al (1989) Total versus subtotal gastrectomy for adenocarcinoma of the gastric antrum. A French prospective controlled study. Ann Surg 209:162
43. Raju GS, Waxman I (2000) High-frequency US probe sonography-assisted endoscopic mucosal resection. Gastrointest Endosc 52:S39
44. Macdonald JS, Smalley SR, Benedetti J et al (2001) Chemoradiotherapy after surgery compared with surgery alone for adenocarcinoma of the stomach or gastroesophageal junction. N Engl J Med 345:725
45. Stahl M, Walz MK, Stuschke M et al (2009) Phase III comparison of preoperative chemotherapy compared with chemoradiotherapy in patients with locally advanced adenocarcinoma of the esophagogastric junction. J Clin Oncol 27:851
46. Neuhof D, Wenz F (2006) Magenkarzinom. In: Wannenmacher M, Debus J, Wenz F (eds) Strahlentherapie. Springer, Berlin
47. Tey J, Lu JJ (2014) Gastric cancer. In: Lee NY, Riaz N, Lu JJ (eds) Target volume delineation for conformal and intensity-modulated radiation therapy. Medical radiology. Springer, Cham. Fig. 5-6., pp 269–270
48. Greenlee RT, Murray T, Bolden S et al (2000) Cancer statistics, 2000. CA Cancer J Clin 50(1):7–33
49. Compton CC, Byrd DR, Garcia-Aguilar J, Kurtzman SH, Olawaiye A, Washington MK (2012) Exocrine and endocrine pancreas. In: Compton C, Byrd D, Garcia-Aguilar J, Kurtzman S, Olawaiye A, Washington M (eds) AJCC cancer staging atlas. Springer, New York, NY. https://doi.org/10.1007/978-1-4614-2080-4_24
50. Yeo CJ, Yeo TP, Hruban RH et al (2005) Cancer of the pancreas. In: DeVita VT, Hellman S, Rosenberg SA (eds) Cancer: principles & practice of oncology, 7th edn. Lippincott Williams & Wilkins, Philadelphia, pp 947–950
51. Compton CC, Byrd DR, Garcia-Aguilar J, Kurtzman SH, Olawaiye A, Washington MK (2012) Exocrine and endocrine pancreas. In: Compton C, Byrd D, Garcia-Aguilar J, Kurtzman S, Olawaiye A, Washington M (eds) AJCC cancer staging atlas. Springer, New York, p 300
52. Ujiki MB, Talamonti MS (2007) Guidelines for the surgical management of pancreatic adenocarcinoma. Semin Oncol 34(4):311–320
53. Hansen EK, Mack Roach III (2006) Handbook of evidence-based radiation oncology. Springer, US, pp 232–234
54. Al-Hawary MM, Francis IR, Chari ST et al (2014) Pancreatic ductal adenocarcinoma radiology reporting template: consensus statement of the Society of Abdominal Radiology and the American Pancreatic Association. Radiology 270:248
55. Hüttner FJ, Fitzmaurice C, Schwarzer G et al (2016) Pylorus-preserving pancreaticoduodenectomy (pp Whipple) versus pancreaticoduodenectomy (classic Whipple) for surgical treatment of periampullary and pancreatic carcinoma. Cochrane Database Syst Rev 2:CD006053
56. Johnson CD, Schwall G, Flechtenmacher J, Trede M (1993) Resection for adenocarcinoma of the body and tail of the pancreas. Br J Surg 80:1177
57. Suker M, Beumer BR, Sadot E et al (2016) FOLFIRINOX for locally advanced pancreatic cancer: a systematic review and patient-level meta-analysis. Lancet Oncol 17:801
58. Balaban EP, Mangu PB, Khorana AA et al (2016) Locally advanced. unresectable pancreatic cancer: American Society of Clinical Oncology Clinical Practice Guideline. J Clin Oncol 34:2654
59. Neoptolemos JP, Stocken DD, Dunn JA et al (2001) Influence of resection margins on survival for patients with pancreatic cancer treated by adjuvant chemoradiation and/or chemotherapy in the ESPAC-1 randomized controlled trial. Ann Surg 234:758
60. Klinkenbijl JH, Jeekel J, Sahmoud T et al (1999) Adjuvant radiotherapy and 5-fluorouracil after curative resection of cancer of the pancreas and periampullary region: phase III trial of the EORTC gastrointestinal tract cancer cooperative group. Ann Surg 230:776

61. Hammel P, Huguet F, van Laethem JL et al (2016) Effect of Chemoradiotherapy vs chemotherapy on survival in patients with locally advanced pancreatic cancer controlled after 4 months of gemcitabine with or without Erlotinib: the LAP07 randomized clinical trial. JAMA 315:1844
62. Mellon EA, Hoffe SE, Springett GM et al (2015) Long-term outcomes of induction chemotherapy and neoadjuvant stereotactic body radiotherapy for borderline resectable and locally advanced pancreatic adenocarcinoma. Acta Oncol 54:979
63. American Cancer Society (2007) Cancer facts and figures 2007. American Cancer Society, Atlanta, GA
64. Carmichael JC, Mills S (2019) Anatomy and embryology of the colon, rectum, and anus. In: Steele S, Hull T, Hyman N, Maykel J, Read T, Whitlow C (eds) The ASCRS manual of colon and rectal surgery. Springer, Cham, p 9. https://doi.org/10.1007/978-3-030-01165-9_1
65. Compton CC, Byrd DR, Garcia-Aguilar J, Kurtzman SH, Olawaiye A, Washington MK (2012) Colon and Rectum. In: Compton C, Byrd D, Garcia-Aguilar J, Kurtzman S, Olawaiye A, Washington M (eds) AJCC cancer staging atlas. Springer, New York, NY. https://doi.org/10.1007/978-1-4614-2080-4_14
66. Rajput A, Dunn BK (2007) Surgical management of rectal cancer. Semin Oncol 34(3):241–249
67. Tepper JE, O'Connell MJ, Petroni GR, Hollis D, Cooke E, Benson AB III, Cummings B, Gunderson LL, Macdonald JS, Martenson JA (1997) Adjuvant postoperative fluorouracil-modulated chemotherapy combined with pelvic radiation therapy for rectal cancer: initial results of intergroup 0114. J Clin Oncol 15(5):2030–2039
68. Berard P, Papillon J (1992) Role of pre-operative irradiation for anal preservation in cancer of the low rectum. World J Surg 16(3):502–509
69. Johnson LG, Madeleine MM, Newcomer LM et al (2004) Anal cancer incidence and survival: the surveillance, epidemiology, and end results experience, 1973–2000. Cancer 101(2):281–288
70. Compton CC, Byrd DR, Garcia-Aguilar J, Kurtzman SH, Olawaiye A, Washington MK (2012) Anus. In: Compton C., Byrd D., Garcia-Aguilar J., Kurtzman S., Olawaiye A., Washington M. (eds) AJCC cancer staging atlas. Springer, New York, NY. Fig. 15.3-4-5-6, p 207. https://doi.org/10.1007/978-1-4614-2080-4_15
71. Boman BM, Moertel CG, O'Connell MJ et al (1984) Carcinoma of the anal canal. A clinical and pathologic study of 188 cases. Cancer 54:114–125
72. Welton ML, Steele SR, Goodman KA, et al. Anus. In: AJCC cancer staging manual, 8th ed, Amin MB (Ed), AJCC, Chicago 2017. Fig. 21.3, p 279.
73. Martenson JA, Haddock MG, Gunderson LL (2006) Cancers of the colon, rectom and anus. In: Perez C, Vijayakumar S (eds) Technical basis of radiation therapy, 4th, revised edn. Springer, Berlin
74. Levitt SH, Purdy JA, Perez CA, Vijayakumar S (eds) (2006) In: Technical basis of radiation therapy, 4th revised edn. Springer, Berlin

Soft Tissue Sarcoma

11.1 Introduction

Although soft tissue sarcomas (STS) are rarely seen among adult solid tumors, they are considered to be an unusual tumor group due to their different biological behaviors and abundance of histological subtypes. They can present themselves at any site of the body. Nearly 50% of the patients die because of recurrence or distant metastasis within 5 years of diagnosis, even provided with best treatment conditions. Therefore, they compose a group of tumors with challenging treatment and follow-up procedures [1].

Sarcomas are, mesoderm-mesenchyme originated, malignant tumors.

While STSs make up for less than 1% of all adult cancers, they constitute for 12% of pediatric cancers [2].

11.2 Pathology

Even though STSs have more than 100 subtypes according to WHO classification [2], 3/4 of those are undifferentiated pleomorphic sarcomas (previously known as malignant fibrous histiocytoma, MFH), liposarcomas, leiomyosarcomas, synovial sarcomas, and malignant peripheral nerve sheath tumors (MPNST) [3].

STSs are classified according to the soft tissue cell which they originated from. Histological subtypes may be detected by the use of electron microscopic, histochemical, flowcytometric, cytogenetic, and tissue culture analyses. Malignant STSs:

- Angiosarcoma
- Epithelioid sarcoma
- Synovial sarcoma
- Malignant schwannoma
- Clear cell sarcoma
- Malignant fibrous histiocytoma (MFH)
- Neuroectodermal tumor
- Dermatofibrosarcoma
- Liposarcoma
- Leiomyosarcoma (LMS)
- Rhabdomyosarcoma (RMS)
- Hemangiopericytoma
- Myxoid chondrosarcoma
- Synovial chondrosarcoma (Fig. 11.1)

STSs occur 45% in lower extremity, 30% in trunk, 15% in upper extremity, and 9% at head and neck region [4].

Extremities: Liposarcoma, undifferentiated pleomorphic sarcomas (previous MFH), synovial sarcoma, fibrosarcoma, myxoidliposarcoma

Retroperitoneal: Liposarcoma, leiomyosarcoma

Head and neck region: undifferentiated pleomorphic sarcomas (previous MFH)

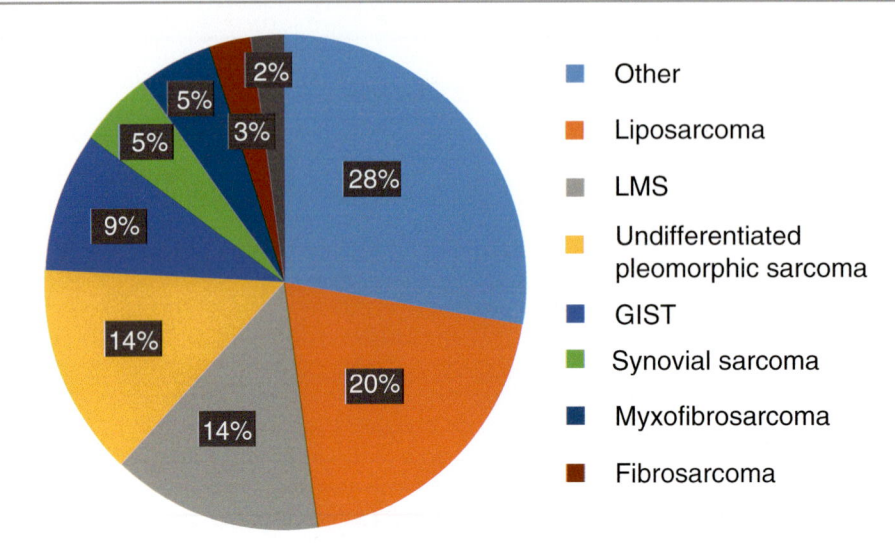

Fig. 11.1 STS subtype distribution. (Management of Soft Tissue Sarcoma, General Description, second ed., 2016, p. 7, Brennan M, Antonescu C, Alektiar K, Maki R, © Springer International Publishing Switzerland 2013, 2016)

- 20–30% of STSs are undifferentiated pleomorphic sarcomas (formerly MFH); 10–20% are liposarcomas.
- Incident rates of leiomyosarcoma, fibrosarcoma, synovial sarcoma, rhabdomyosarcoma, and schwannoma vary between 5 and 10% for each individual subtype [5].

Undifferentiated pleomorphic sarcomas (previous MFH): Most commonly seen at 5th–7th decades, however can be seen in children and young adults as well.

- Most oftenly presents itself with painless mass in the lower extremity, followed by upper extremity and retroperitoneal area. 2/3 of the tumors are intramuscular.
- Can be seen alongside of hematological diseases such as Hodgkin's lymphoma, non-Hodgkin's lymphoma, and multiple myeloma.
- Identified five histological subtypes are (1) storiform/pleomorphic (most common), (2) myxoid, (3) giant cell, (4) inflammatory (generally retroperitoneal), and (5) angiomatoid.
- 10-year survival rate of low-, moderate-, and high-grade patients are 90%, 60%, and 20%, respectively.
- Most common distant metastasis site is lung (90%), bone (8%), and liver (1%) [6].

Liposarcoma : Most commonly seen in adults between the ages of 50 and 65 years, though it can develop in any body part, most oftenly seen in retroperitoneal region and thigh.

- Liposarcoma has three basic forms. Each form has its unique morphology, natural history, and karyotypic and genetic differences.
1. Well-differentiated liposarcomas: Generally, they are deep, painless, and progressively growing masses which can increase to large sizes over the years.
   They are locally agressive tumors without metastasis[7].
2. Myxoid/round cell liposarcomas: they cover 40% of all liposarcomas. They more likely occur at the deep soft tissue in extremities, and 66% of cases appear in the thigh. More than 5% round cell component usually indicates poor prognosis. Pure myxoid ones are considered to be low-grade and have 90% 5-year survival rate, whereas more than 5% round cell component ones have 50%.
   Unlike other liposarcomas, these can cause retroperitoneal and axillary metastasis without pulmonary metastasis.
3. Pleomorphic liposarcomas: these are high-grade tumors. High mitotic activity, bleeding, and necrosis are common. They constitute less

than 5% of all liposarcomas and mostly occur in lower extremity of adults over 50 years of age.

Clinically, 50% of patients experience lung metastasis in early stages [5].

Fibrosarcomas: They have adult and infantile types. Adult type can be seen as painless, deep masses, mostly located in trunk and thigh at the ages between 40 and 60, whereas infantile type occurs generally at first 2 years.

- They are commonly congenital and originates from extremities but rarely metastasizes [6].

Leiomyosarcomas (LMS): They can be seen in all age groups. More than half are located retroperitoneal and intraabdominal regions. Other areas include gastrointestinal system and uterus.

- They are generally resistant to radiotherapy and chemotherapy. 5-year survival rate is around 30%.
- Skin leiomyosarcomas are mostly seen as nodules at extremities.
- Deep extremity leiomyosarcomas may occur mostly at thigh area with middle and large veins.
- The ones localized at retroperitoneal region are usually larger than 10 cm size and tend to cause local recurrence as well as lung and liver metastasis [7].

Synovial sarcomas: They originate from tenosynovial mesothelia and mostly seen in young adults. Typically, these tumors are seen in tendon sheath and paraarticular regions of joints.

- At least 50% of cases are seen at lower extremity (especially knee), the rest are seen at arms.
- They are commonly calcific and have characteristic radiological image.
- Synovial sarcomas can also appear in regions without synovial tissue presence (10% head and neck, less than 10% chest and abdominal wall or intrathoracic region).
- Characteristically synovial sarcomas include chromosomal translocations, t(X;18) (p11.2;q11.2).

Those translocations became gold standard in synovial sarcoma diagnosis. Almost all of them are high grade.

- They have the tendency of high local recurrence rate and lymph node metastasis. 5-year survival rate is nearly 30% [8].

Rhabdomyosarcomas (RMS): These develop from striated muscles and considered to be all high grades in the staging system.

- They are the most common STS of childhood.
- 35% of all cases are localized at head-neck region, other 35% are seen in trunk and extremities, and the final 30% are seen in genitourinary system. .
- These tumors have the tendency for local recurrence.
- They metastasize via venous and lymphatic system.
- They have three subtypes:
1. Embryonal RMS: Affects infants and children.
   Localization rates are 70% head-neck and 15–20% genital region [9].
   Includes botryoid sarcoma.
   5-year survival rate is approximately 70%.
2. Alveolar RMS: They are quite aggressive and caN originate from any region of body. Affected people are typically between 13 and 18 years of age.
   5-year survival rate is approximately 50%.
3. Pleomorphic RMS: These are seen in people over 30 years of age. They are rare and originate from extremities.
   They are commonly anaplastic and may be microscopically mixed with undifferentiated pleomorphic sarcomas. 5-year survival rate is 25%.

Malignant peripheral nerve sheath tumors [MPNST]): They are also referred to as malignant schwannoma, neurofibrosarcoma, and neurogenic sarcoma. They are seen in young and middle-aged adults.

- These ectoderm-based tumors originate from peripheral nerve sheath. Legs and retroperitoneum are the most common localizations.
- Most of MPNSTs are high grade and characteristically stained with S-100 [10].
- 5% of patients with neurofibromatosis (NF-1) may develop MPNST of plexiform neurofibroma. Those patients have the worst prognosis.
- MPNSTs tend to have larger size and higher grade than other types of sarcomas.

Desmoid tumors (aggressive fibromatosis): These locally aggressive and invasive tumors are differentiated from fibrous tissue and show fibroblastic proliferation.

- They have high local recurrence rates after resection.
- Although they are locally invasive, they rarely metastasize [11]. Patients do not typically die of this tumor, but functional morbidity may develop.

Localizations and presentations of desmoid tumors may vary. They may develop in the abdominal walls of pregnant women and cause intraabdominal mesenteric masses.

- In older population, they may form large extremity masses. Abdominal desmoids may be a component of familial adenomatous polyposis syndrome.
- Only 14% of STS are seen in upper extremity,
- Approximately 1/2 of epitheloid sarcomas are observed in front arm and fingers [12].
- Desmoplastic small round cell tumors are seen in adolescent and young adult males and primarily spread to the abdominal cavity and pelvic region [13].

Genetic predispositions that may be seen in STS:
- Li-Fraumeni syndrome
  - Related to TP53 mutation
- Von Recklinghausen disease: neorufibromatosis
- Retinoblastoma
- Gardner syndrome/familial adenomatous polyposis
- Werner syndrome
- Carney diad
  - Gastrointestinal stromal tumors + paraganglioma
- Gorlin syndrome
- Tuberous sclerosis
- Basal nevous syndrome

Histological markers in STSs:
- Desmin is especially important on determining myogenic origin.
  - RMS, leiomyosarcoma
- Presence of S100 and neurofilament indicates tumoral cells originate from neural sheath.
  - But also may be seen in tumors that show melanocytic differentiation.
  - Clear cell sarcomas and perivascular epithelioid cell tumors (PEComa).
- Cytokeratin is rarely expressed in most sarcomas,
  - However can be used to distinguish synovial or epithelioid sarcomas (both express cytokeratin) and fibrosarcoma (does not express cytokeratin).

Factor VIII-related antigen is important for determining whether the tumors are of endothelial originate or not.

11.3 General Presentation

For most STS, the first and only symptom is a painless mass. Besides, some cases may present with compression signs depending on localization. Furthermore, symptoms due to metastasis may occur.

Stewart-Treves syndrome: Chronic upper extremity lymphedema in lymphangiosarcoma [14].

- Typically, sarcomas metastasize by hematogenous route. Lung metastasis is quite common.
  While lung is the primary distant metastatic site for extremity located STS, retroperitoneal and GIS STSs primarily metastasize to the liver.

Retroperitoneal and GIS STSs first metastasis is liver.
- The most frequent lymph node metastases are from clear cell sarcoma, RMS, epithelioid sarcoma, synovial sarcoma, and angiosarcoma [15].
- Liposarcoma (myxoid and well-differentiated type), fibrosarcoma (infantile and well-differentiated type), undifferentiated pleomorphic sarcoma (MFH) (superficial type), and dermatofibrosarcoma protuberans do not generally metastasize.

11.4 Staging

Staged according to AJCC/UICC 8th edition.

In AJCC 8th edition, STSs are staged as two distinct prognostic stage groups as extremity/trunk and retroperitoneal, including similar T, N, and M staging criteria [16, 17].

Furthermore; T, N, and M definitions for thoracic and abdominal visceral STSs have been made while prognostic stage groupings have not [18].

AJCC staging is not commonly used for retroperitoneal STSs because histology and localization which are both major prognostic indicators have not been taken into account.

In addition, AJCC staging of retroperitoneal tumors does not predict survival outcomes; therefore, AJCC recommends the use of nomograms for survival estimation [19, 20] (Tables 11.1, 11.2, 11.3, 11.4, 11.5, and 11.6).

Table 11.1 AJCC 8th edition TNM, grade, differentiation score, and histologic grade definitions

Primary tumor (T)	
T	T criteria
TX	Primary tumor cannot be assessed
T0	No evidence for primary tumor
T1	Tumor's greatest size is ≤5 cm
T2	Tumor's greatest size is >5 cm but ≤10 cm
T3	Tumor's greatest size is >10 cm but ≤15 cm
T4	Tumor's greatest size is >15 cm
N	N criteria
N0	Regional lymph node metastasis is nonexistent or lymph node status unknown
N1	Regional lymph node metastasis exists
M	M criteria
M0	No distant metastasis
M1	Distant metastasis
Grade definitions (G)	
G	G definition
GX	Grade cannot be assessed
G1	Total differentiation, mitotic count, and necrosis score of 2 or 3
G2	Total differentiation, mitotic count, and necrosis score of 4 or 5
G3	Total differentiation, mitotic count, and necrosis score of 6, 7, or 8
Histologic grade	
The FNCLCC (Fédération Nationale des Centres de Lutte Contre Le cancer) grade is determined by three parameters: differentiation, mitotic activity, and extent of necrosis. Each parameter is scored as follows: differentiation (1–3), mitotic activity (1–3), and necrosis (0–2). These scores are added to determine the grade.	
Tumor differentiation	
Differantiation score	Definition
1	Sarcomas closely resembling normal adult mesenchymal tissue (e.g., low-grade leiomyosarcoma)
2	Sarcomas for which histological type is certain (e.g., myxoid/round cell liposarcoma)
3	Embryonal and undifferetiated sarcomas, sarcomas of suspicious type, synovial sarcomas, soft tissue osteosarcoma, Ewing sarcoma/primitive neuroectodermal tumor (PNET) of soft tissue.

Table 11.2 AJCC 8th edition mitotic count score and necrosis score definitions

Mitotic count	
In the most mitotically active area of the sarcoma, 10 successive high-power fields (HPF = one HPF at 400× magnification = 0.1734 mm$^{2)}$) are assessed using a 40× objective	
Mitotic count score	**Definition**
1	0–9 mitoses/10 HPF
2	10–19 mitoses/10 HPF
3	≥20 mitoses/10 HPF

Tumor necrosis	
Evaluated on gross examination and validated with histologic sections	
Necrosis score	Definition
0	No necrosis
1	<%50 tumor necrosis
2	≥%50 tumor necrosis

Table 11.3 AJCC 8th edition histology specific differentiation scores

Histology specific differentiation score	
Histological type	Score
• Atypical lipomatous tumor/well-differantiated liposarcoma	1
• Myxoidliposarcoma	2
• Round cell liposarcoma	3
• Pleomorphicliposarcoma	3
• Dedifferentiated liposarcoma	3
• Fibrosarcoma	2
• Myxofibrosarcoma	2
• Undifferentiated pleomorphic sarcoma (formerly MFH)	3
• Well-differentiated leiomyosarcoma	1
• Classicalleiomyosarcoma	2
• Poorly differentiated/pleomorphic/epitheiloid leiomyosarcoma	3
• Biphasic/ monophasic synovial leiomyosarcoma	3
• Poorly differentiated synovial sarcoma	3
• Pleomorphic rhabdomyosarcoma	3
• Mesenchymal chondrosarcoma	3
• Ekstraskeletal osteosarcoma	3
• Ewing Sarcoma/PNET	3
• Malignanat rhabdoidtumor	3
• Undifferentiated sarcoma (unclassified)	3

Table 11.4 AJCC 8th edition, prognostic stage groups of extremity/trunk STSs

Prognostic stage groups [Extremities /Trunk Soft Tissue Sarcoma]				
T	N	M	Grade	Stage
T1	N0	M0	G1, GX	IA
T2, T3, T4	N0	M0	G1, GX	IB
T1	N0	M0	G2, G3	II
T2	N0	M0	G2, G3	IIIA
T3, T4	N0	M0	G2, G3	IIIB
Any T	N1	M0	Any G	IV
Any T	Any N	M1	Any G	IV

Table 11.5 AJCC 8th edition, prognostic stage groups of retroperitoneal sarcoma

Prognostic stage groups [Retroperitoneal Sarcoma]				
T	N	M	Grade	Stage
T1	N0	M0	G1, GX	IA
T2, T3, T4	N0	M0	G1, GX	IB
T1	N0	M0	G2, G3	II
T2	N0	M0	G2, G3	IIIA
T3, T4	N0	M0	G2, G3	IIIB
Any T	N1	M0	Any G	IIIB
Any T	Any N	M1	Any G	IV

Table 11.6 AJCC 8th edition; TNM classification for STS of abdomen and thoracic visceral organs

Primary tumor (T) (STS of abdomen and thoracic visceral organs)	
T	T criteria
TX	Primary tumor cannot be assessed
T1	Organ confined
T2	Tumor extension into tissue beyond organ
T2a	Tumor invades serosa of visceral peritoneum
T2b	Tumor extends beyond serosa (mesentery)
T3	Tumor invades another organ
T4	Multifocal involvement
T4a	Multifocal, 2 sites
T4b	Multifocal, 3–5 sites
T4c	Multifocal, > 5 sites
N	N criteria
N0	No regional lymph node involvement or unknown lymph node status
N1	Lymph node involvement present
M	M criteria
M0	No metastasis
M1	Metastases present

11.5 Treatment

Diagnostic Procedures Before Treatment
- Imaging before biopsy:
  - CT ± MRI
  - Optional plain graphy
- Biopsy and pathological evaluation:
- Should be performed carefully with probable curative resection in mind
- Incisional biopsy (with minimal disection) or core needle biopsy
- Fine needle aspiration (FNA) biopsy is acceptable if performed in a specialized sarcoma center, otherwise not recommended.
- FNA can be used on suspicion of recurrence.
- Incision must be within probable RT volume or definitive resection edges.
- Grade and histological type must be reported.
- Thorax CT:
- Used for detection of probable lung metastasis.
- PET-CT
- May be beneficial in high-grade STSs but not recommended for primary imaging for benign and low-/medium-grade STSs.
- Exceptionally, can be used for identifying suspicious malignant peripheral nerve sheath tumor (MPNST) in patients with neurofibromatosis [21].
- Myxoid/round cell carcinoma
  - In the presence of suspicious vertebrae metastases, vertebral MRI is required.
  - In the presence of lung metastases, abdominal CT is added for the detection of probable retroperitoneal metastases.
- RMS, MPNST, angiosarcoma;
  - Risk of brain metastasis is greater and CNS imaging is indicated.
- Angiosarcomas;
  - Most reliable method to show multifocal spread is thorough clinical examination.
- Non-pleomorphic RMS;
  - Additional bone narrow investigation and bone scan are required.
  - If located in parameningeal region, lumbar puncture may be required for cerebrospinal fluid examination.

Parameningeal areas are middle ear, nasopharynx, paranasal sinuses, pterygopalatine fossa, and parapharyngeal regions.
- Leptomeningeal metastasis may develop in case of recurrences in these regions.
  - Sentinel lymph node disection may be required if located in extremity (Figs. 11.2 and 11.3).

Diagnostic Markers of Soft Tissue Sarcomas
- PAX3–FOXO1 or PAX7–FOXO1 chromosomal translocations
  - are seen on alveolar subtype of non pleomorphic RMSs [22].
- FUS–DD1T3 (T LS–CHOP) or EWSR1–DD1T3 (E WSR1–CHOP)
  - are seen on myxoid/round cell liposarcomas [23].
- MDM2 amplification [24]
  - is seen on atypical lipomatous tumors and well-differentiated and dedifferentiated liposarcomas.
- ASPL–TFE3
  - is seen on alveolar soft part sarcomas.
- EWSR1–AFT1 or EWSR1–CREB1
  are seen on clear cell sarcomas.
- COLIA1–PDGFB
  - is seen on dermatofibrosarcoma protuberans.
- SS18–SSX1, SSX2, and SSX4 [25]
  - are seen on synovial sarcoma.

Treatment of Soft Tissue Sarcomas
- STSs should be managed in a multidisciplinary clinic which specializes in sarcomas.
- Surgery is the main curative treatment in STSs. However, there are some exceptions.
  - Rhabdomyosarcomas are generally sensitive to both radiotherapy and chemotherapy; nevertheless, surgery is the main treatment modality for selected sites.
  - For advanced stages of face and scalp angiosarcomas, it may be difficult to achieve adequate surgical margins, and RT+surgery is an appropriate option; chemotherapy is also applicable.
  - In surgery, "en bloc resection" should be performed, and surgical margin should be ≥2 cm in every direction. Due to possibil-

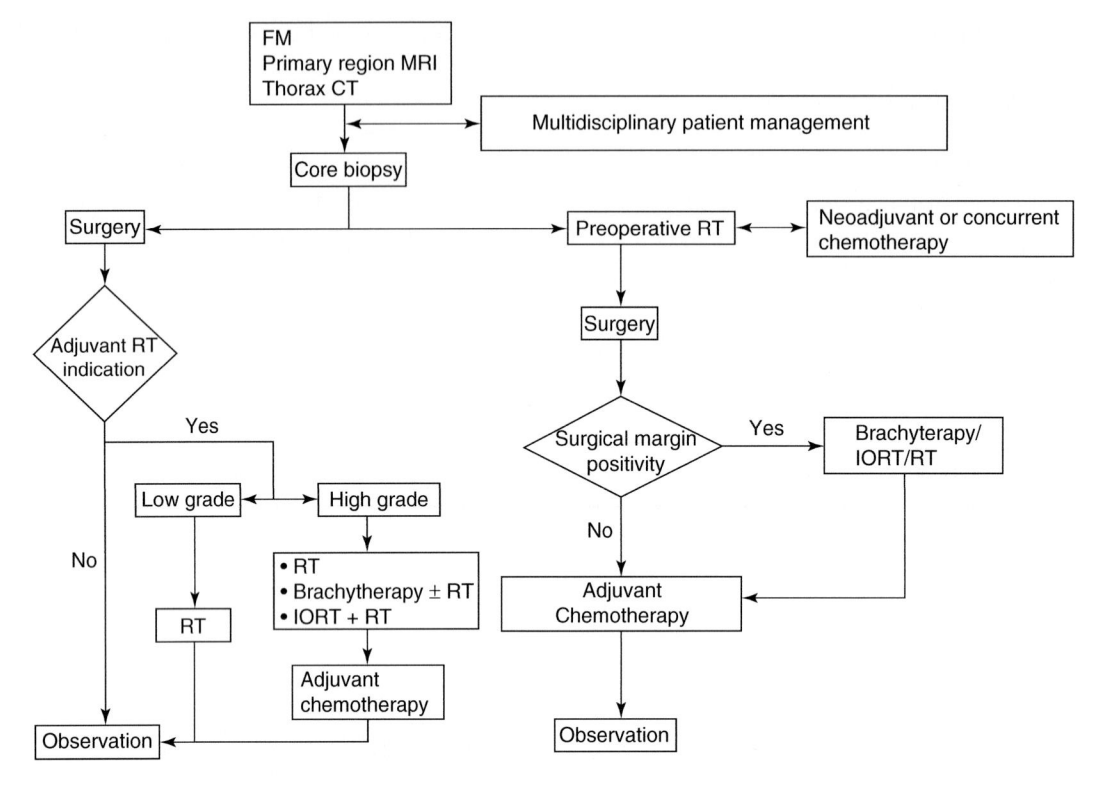

Fig. 11.2 Treatment algorithm for extremity sparing surgery candidates in extremity STSs. (adapted from DiSano J.A., Pameijer C.R. (2019) Management of Soft Tissue Sarcoma. In: Docimo Jr. S., Pauli E. (eds) Clinical Algorithms in General Surgery. Springer, Cham. algorithm 1, p10)

 ity of RT after surgery, clips should be placed.
- Radiotherapy (preoperative or postoperative) reduces the risk of local failure.
- Particularly, deep-seated, high-grade, or larger (>5 cm) tumors.
- In the presence of close or positive surgical margins, re-excision or adjuvant RT should be performed.
- Also for unresectable cases or in the presence of comorbidities, definitive RT is indicated.
- RT combined with surgery increases local control without affecting survival.
- Role of adjuvant chemotherapy has yet to be proven for most subtypes of STSs.
  – Chemotherapy is important on chemosensitive tumors.
  – Extraosseous Ewing sarcoma and rhabdomyosarcoma.

Radiotherapy in Early-Stage STS
- Role:
  – Preoperative or postoperative RT is indicated typically for deep-located and large-sized (>5 cm) sarcomas.
- Indication for adjuvant RT is the presence of close or positive surgical margins.
- Brachytherapy is optional for selected cases.
- Preoperative RT (generally for extremity sarcomas).
  – 50 Gy, 2 Gy/fx.
  – RT volume is usually the volume which includes 3–4 cm. margin longitudinally and 1.5 cm margin radially from the primary tumor.
- Postoperative RT (generally extremity sarcomas).
  – 66 Gy, 2Gy/fx.

11.6 Radiotherapy

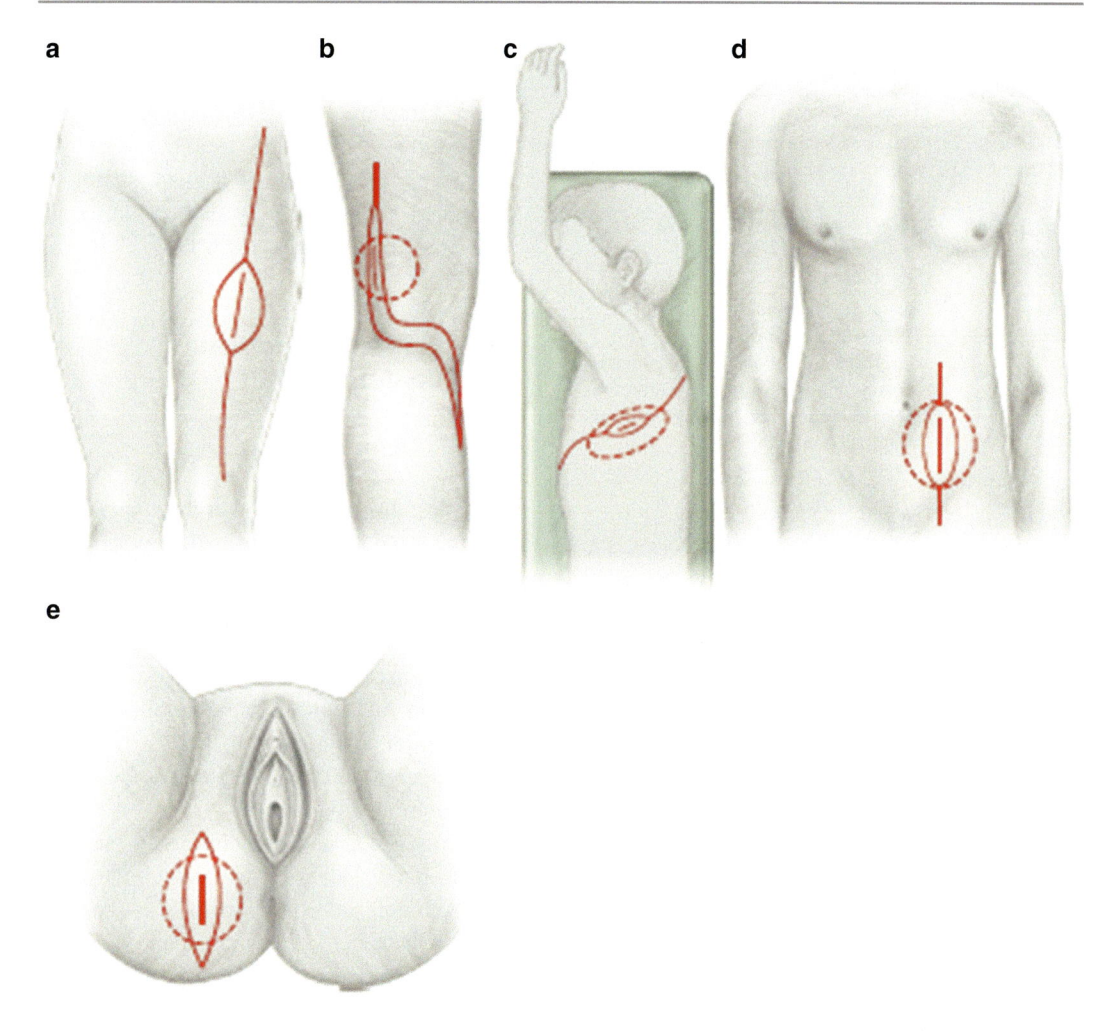

Fig. 11.3 Biopsy for soft tissue sarcomas: principles for planning a biopsy in various anatomical locations. (**a**) Extremity along a limb. (**b**) Extremity over a joint. (**c**) Trunk (chest wall). (**d**) Trunk (abdominal wall). (**e**) Perineum. (from Pheng A.L.H., Rao B. (2019) Soft Tissue Sarcomas. In: Puri P., Höllwarth M. (eds) Pediatric Surgery. Springer Surgery Atlas Series. Springer, Berlin, Heidelberg. Fig. 56.1, p 477)

- 60 Gy is delivered for the cases with low-grade and clear surgical margins.
- Generally, low-risk region is delivered 50 Gy followed by 10–16 Gy boost to primary tumor bed.

Radiotherapy in Advanced Stage Nonmetastatic STSs
- 70 Gy, 2 Gy/fx or equivalent dose RT is to be delivered to unresectable gross residual disease (Tables 11.7 and 11.8).

11.6 Radiotherapy

Two-Dimensional Conventional Radiotherapy
- **RT field:** Tumor/tumor bed, scar, drain zone if exists;
- Until 50 Gy:
- Add 5–7 cm to **craniocaudal**.
- Add 2–3 cm to **mediolateral**.
- After 50 Gy :
- Add 2 cm to **craniocaudal**.
- Add 2 cm to **mediolateral**.

Table 11.7 General treatment approach for extremity-located soft tissue sarcomas

Stage	Treatment
Stage I (extremity located)	Surgery (postoperative RT is added if surgical margin proximity/positivity exists)
Stage II–III (extremity located)	Surgery + postoperative RT or Preoperative RT + surgery
Stage IV	Best care, chemotherapy, palliative RT, ± palliative surgery Surgical resection is to be considered if primary tumor is under control, lung metastases count is ≤4, and /or distant metastasis occured long after diagnosis

Table 11.8 Treatment on retroperitoneum STS and desmoid tumors

Status	Treatment
Retroperitoneal location	Surgery ± IORT (12–15 Gy) + postoperative RT (45–50 Gy)
	Preoperative radiochemotherapy + surgery + IORT boost
Desmoid tumors	Surgery Postoperative RT is added if surgical margin is (+). If inoperable, then RT (56–60Gy)

Postoperative RT:
- If surgical margin is (−) or there is microscopic residue; 60 Gy,
- Surgical margin (+); 66 Gy,
- Gross disease; 70–75 Gy is delivered.
- RT starts 14–20 days after surgery.

Preoperative RT:
- 50 Gy is delivered via 2 Gy/fx.
- Surgery is to be performed 3 weeks after RT.
- For the purpose of preserving lymphatic drainage on skin;
- A band of 1.5 – 2 cm. is purposefully excluded from RT field.
- Epiphyseal plates close prematurely on children at > 20 Gy [26].

- Bone marrow ablation is observed at ≥ 40 Gy [27].
- Bone fracture and delayed healing are observed at ≥ 50 Gy [28].
- Joint intervals within RT field should be excluded after 40–45 Gy if possible, due to fibrotic band formation (Fig. 11.4).

11.6.1 Three-Dimensional Conformal/IMRT

11.6.1.1 Simulation

- Diagnostic imaging workup and pre-simulation data including patient positioning, tumor localization, and compartment involvement should be assessed in the multidisciplinary setting before RT planning.
- Simulation should be performed at the position with the highest reproducibility (generally neutral position).
- Permanent tattoo may be used for facilitation of daily setup control.
- Patient position and immobilization may vary due to tumor localization; however, most suitable immobilization tools are to be used in order to keep the patient in the most stable position.
  Vacuum bed is recommended for cases with pelvic, abdominal, and retroperitoneal sarcoma.
  Head-neck thermoplastic mask is recommended for cases with head-neck sarcoma.

11.6.1.2 Contouring

- **Preoperative cases**
  - Diagnostic MRI images and CT simulation images are registered.
  - GTV and CTV are generated.
- **Postoperative cases**
  - Since a visualized GTV does not exist, GTV is generated by the use of preoperative images.
  CTV

Fig. 11.4 (a–c) RT technique in soft tissue sarcomas. (Courtesy of wikimediacommons.org)

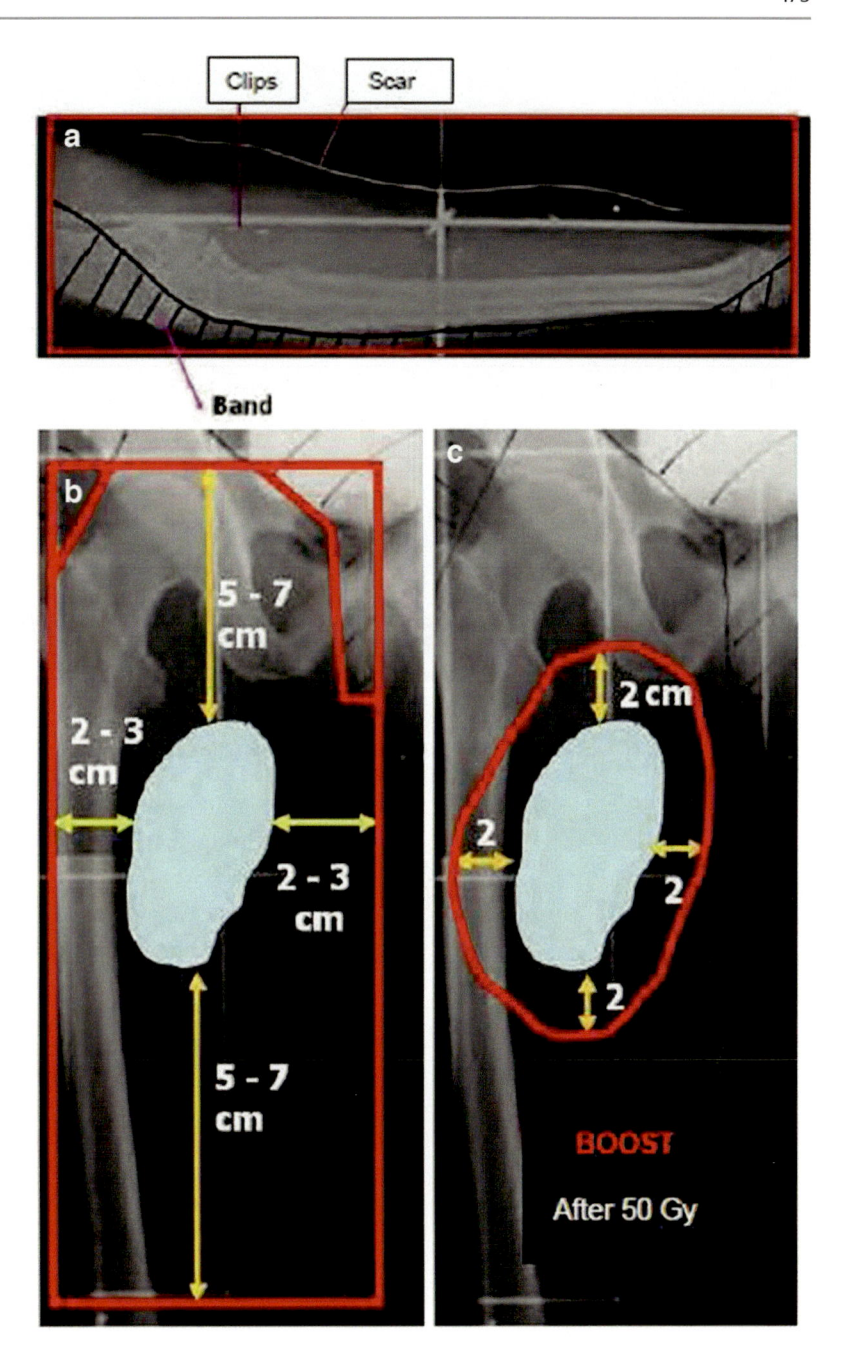

Preoperative cases
- GTV + 3–4 cm longitudinal margin (should include suspicious edema if there is any) is given.
- GTV + 1.5 cm radial margin is given.
- Intact facial barriers, bone, and skin surface are included if they are located within the given margins.
- Suspicious edema can be better visualized on T2 sequences of MRI.

Postoperative cases
- GTV is regenerated by the use of MRI
- CTV1: new GTV + 4 cm longitudinal margin is given.
- New GTV + 1.5 cm radial margin is given.

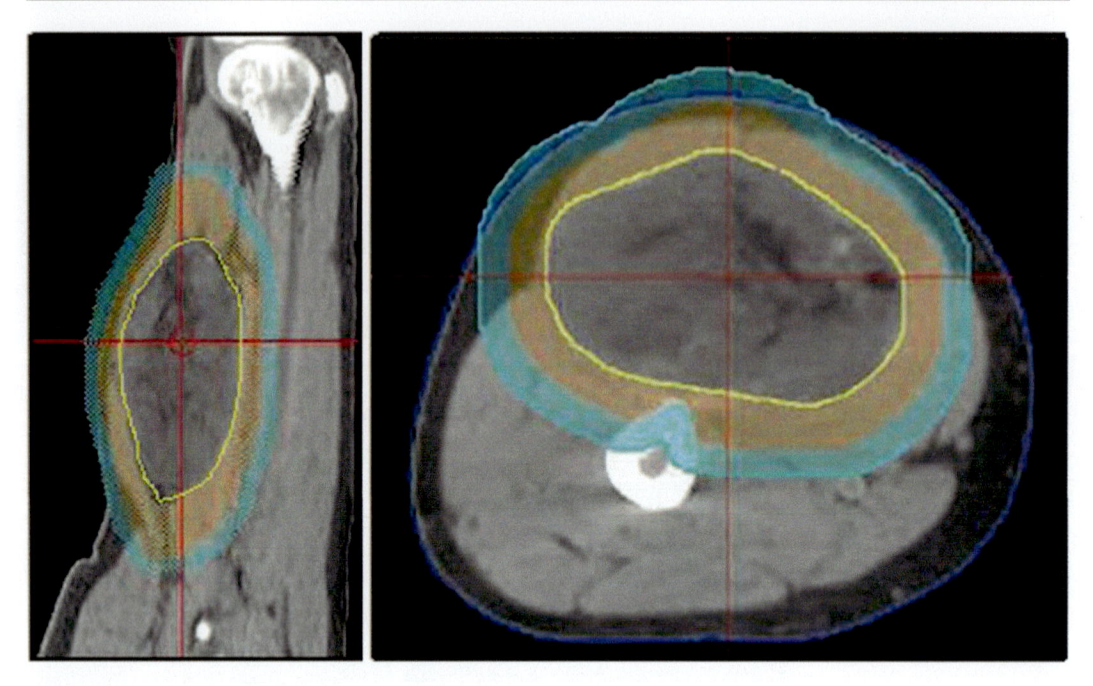

Fig. 11.5 Target volumes on STS RT. (Adapted from Dickie C., O'Sullivan B. (2014) Soft Tissue Sarcoma. In: Lee N.Y., Riaz N., Lu J.J. (eds) Target Volume Delineation for Conformal and Intensity-Modulated Radiation Therapy. Medical Radiology. Springer, Cham. Fig. 2, p 518)

- CTV1, should be within the skin and longer than the scar.
- If there is postoperative residue
- CTV1: $GTV_{residue}$ + 2 cm longitudinal margin is given.
- $GTV_{residue}$ + 1 cm radial margin is given.
  PTV;
- CTV + 5–10 mm margin is given.
- **Dose/Fractionation**
  Preoperative Extremity STS
- PTV50: 50 Gy, 2Gy/fx.
- **Postoperative Extremity STS**
- PTV50: 50 Gy, 2Gy/fx.
- PTV66: 16 Gy, 2 Gy/fx boost.
- If SIB(simultaneous integrated boost) is delivered;
- PTV56; 1.69 Gy/fx and PTV 66; 2 Gy/fx is delivered.
- Surgical margin (-), low-grade tumors planned as PTV54 and PTV 60.
- PTV56;
- [PostopGTV + 4 cm londitudinal / + 1.5 cm radial margin] +[scar/drain region + 1–2 cm margin] + 0.5–1 cm margin is given.

Preoperative Retroperitoneal STS
- **PTV;** 50 Gy, 2 Gy/fx or 50.4 Gy, 1.8Gy/fx.
- PTV;
- [GTV + 2 cm longitudinal margin / + 0.5–2 cm radial margin]±0.5 cm liver tissue depending on proximity to liver] + 0.5 cm margin is given.
- Contralateral kidney should be spared.

Definitive Extremity/Retroperitoneal STS
- Indicated to the cases which are surgically unresectable or inoperable.
- 70 Gy, 2 Gy/fx or equivalent dose is required for local control.

Dose Constraints for OAR
- Bone; mean dose <37 Gy, maximum dose <59 Gy, V40 < 67%
- Spinal cord; maximum dose 45 Gy
- Liver; V30 < 30%
- Small bowel; V45 < 10%
- Lungs; V20 < 20%
- Testis; V3 < 50%
- Kidneys; bilateral kidney V14 < 50%, single kidney V20 < 1/3 (Fig. 11.5).

References

1. Brennan M, Singer S, Maki R et al (2005) Sarcomas of the soft tissues and bone. In: De Vita VT Jr, Hellman S, Rosenberg SA (eds) Cancer: principles and practice of oncology, 7th edn. Lippincott Williams & Wilkins, Philadelphia, pp 1581–1631
2. Siegel RL, Miller KD, Jemal A (2018) Cancer statistics, 2018. CA Cancer J Clin 68:7
3. Fletcher CDM, Bridge JA, Hogendoorn PCW, Mertens F (2013) World Health Organization classification of tumours of soft tissue and bone, 4th edn. IARC Press, Lyon
4. Lawrence W Jr, Donegan WL, Natarajan N et al (1987) Adults of tissue sarcomas. A pattern of care survey of the American College of Surgeons. Ann Surg 205:349
5. Hameed M (2007) Pathology and genetics of adipocytic tumors. Cytogenet Genome Res 118(2–4):138–147. Review
6. Eyden B (2005) The myofibroblast: a study of normal, reactive and neoplastic tissues, with an emphasis on ultrastructure. Part 2—tumours and tumour-like lesions. J Submicrosc Cytol Pathol 37(3–4):231–296. Review
7. Skubitz KM, D'Adamo DR (2007) Sarcoma. Mayo Clin Proc 82(11):1409–1432. Review
8. Siegel HJ, Sessions W, Casillas MA Jr, Said-Al-NaiefN LPH, Lopez-Ben R (2007) Synovial sarcoma: clinicopathologic features, treatment, and prognosis. Orthopedics 30(12):1020–1025; quiz 1026-7. Review
9. Gallego Melcón S, Sánchez de Toledo Codina J (2005) Rhabdomyosarcoma: present and future perspectives in diagnosis and treatment. Clin Transl Oncol 7(1):35–41. Review
10. Ariel IM (1988) Tumors of the peripheral nervous system. Semin Surg Oncol 4(1):7–12. Review
11. Okuno S (2006) The enigma of desmoid tumors. Curr Treat Options in Oncol 7(6):438–443. Review
12. Baratti D, Pennacchioli E, Casali PG et al (2007) Epithelioid sarcoma: prognostic factors and survival in a series of patients treated at a single institution. Ann Surg Oncol 14:3542
13. Levy A, Le Péchoux C, Terrier P et al (2014) Epithelioid sarcoma: need for a multimodal approach to maximize the chances of curative conservative treatment. Ann Surg Oncol 21:269
14. Nakazono T, Kudo S, Matsuo Y, Matsubayashi R, Ehara S, Narisawa H, Yonemitsu N (2000) Angiosarcoma associated with chronic lymphedema (Stewart-Treves syndrome) of the leg: MR imaging. Skeletal Radiol 29(7):413–416. Review
15. Johannesmeyer D, Smith V, Cole DJ et al (2013) The impact of lymph node disease in extremity soft-tissue sarcomas: a population-based analysis. Am J Surg 206:289
16. Yoon SS, Maki RG, Asare EA et al (2017) Soft tissue sarcoma of the trunk and extremities. In: Amin MB (ed) AJCC cancer staging manual, 8th edn. AJCC, Chicago, p 507
17. Pollock RE, Maki RG, Baldini EH et al (2017) Soft tissue sarcoma of the retroperitoneum. In: Amin MB (ed) AJCC cancer staging manual, 8th edn. AJCC, Chicago, p 531
18. Sarcoma of the abdomen and thoracic visceral organs. In: Amin MB (ed) AJCC cancer staging manual, 8th edn. AJCC, Chicago 2017, p 517.
19. Cates JMM (2017) Performance analysis of the American Joint Committee on Cancer 8th edition staging system for retroperitoneal sarcoma and development of a new staging algorithm for sarcoma-specific survival. Ann Surg Oncol 24:3880
20. Gronchi A, Miceli R, Shurell E et al (2013) Outcome prediction in primary resected retroperitoneal soft tissue sarcoma: histology-specific overall survival and disease-free survival nomograms built on major sarcoma center data sets. J Clin Oncol 31:1649
21. Benz MR, Czernin J, Dry SM et al (2010) Quantitative F18-fluorodeoxyglucose positron emission tomography accurately characterizes peripheral nervesheath tumors as malignant or benign. Cancer 116:451
22. Linardic CM, Naini S, Herndon JE 2nd et al (2007) The PAX3-FKHR fusion gene of rhabdomyosarcoma cooperates with loss of p16INK4A to promote bypass of cellular senescence. Cancer Res 67:6691
23. Crozat A, Aman P, Mandahl N, Ron D (1993) Fusion of CHOP to a novel RNA-binding protein in human myxoid liposarcoma. Nature 363:640
24. Oliner JD, Pietenpol JA, Thiagalingam S et al (1993) Oncoprotein MDM2 conceals the activation domain of tumour suppressor p53. Nature 362:857
25. Kadoch C, Crabtree GR (2013) Reversible disruption of mSWI/SNF(BAF) complexes by the SS18-SSX oncogenic fusion in synovial sarcoma. Cell 153:71
26. Margulies BS, Horton JA, Wang Y, Damron TA, Allen MJ (2006) Effects of radiation therapy on chondrocytes in vitro. Calcif Tissue Int 78(5):302–313. Epub 2006 May 9
27. Spadaro JA, Baesl MT, Conta AC, Margulies BM, Damron TA (2003) Effects of irradiation on the appositional and longitudinal growth of the tibia and fibula of the rat with and without radioprotectant. J Pediatr Orthop 23(1):35–40
28. Holt GE, Griffin AM, Pintilie M, Wunder JS, Catton C, O'Sullivan B, Bell RS (2005) Fractures following radiotherapy and limb-salvage surgery for lower extremity soft-tissue sarcomas. A comparison of high-dose and low-dose radiotherapy. J Bone Joint Surg Am 87(2):3159

Non-melanoma Skin Cancers

12.1 Introduction

Non-melanoma skin cancers, affecting approximately 1 million people each year, are the most common cancers, and approximately 2000 people die each year from these cancers [1].

Epidermis: It is a squamous epithelium layer that includes keratinocytes at various stages of differentiation. This layer has no blood vessels. Nutrition and excretion of waste products are achieved by diffusion:

- Epidermis includes three cell types as melanocytes, Langerhans, and Merkel cells.
- Melanin is collected in melanosomes, which are surrounded by keratinocytes.
- Langerhans cells originate from bone marrow and are located in the basal and granular layers of epidermis.
- Merkel cells originated from the mesoderm provide a differentiated function for light touch perception.

Dermis: It consists of two layers, the superficial papillary layer and the deeper reticular dermis layer.

- Papillary dermis which consists of capillaries, elastic fibers, reticular fibers, and collagen feeds the avascular epidermis.
- Reticular layer of the dermis which consists of dense, irregular connective tissue includes fibroblasts, mast, and other connective tissue cells, nerve endings, lymphatics, and epidermal branches.
- The main cell type within the dermis is fibroblasts. Collagen produced by fibroblasts constitutes 70% of the dermis weight.
- There are epidermal appendages containing sebaceous glands, sweat glands, apocrine glands, and hair follicles in the dermis.
- Adipose glands are located on the entire surface of the body except for the sole and dorsum of foot and palmar areas.
- Sweat glands are found extensively in the palmar regions and the dorsum of the foot, in contrast to the adipose glands.

Hypodermis: There is subcutaneous fat and connective tissue which is also called hypodermis under the dermis. These consist of elastic fibers, collagen, and adipose tissue, which can be of different thickness and arrangement according to the regions.

- Connective tissue elements form the superficial and deep fascia systems that generate connections between bone and skin. The spaces between these include adipose tissue, muscle and tendons, nerves, arteries, veins, and lymph vessels.

The skin consists of two different layers as epidermis and dermis (Fig. 12.1).

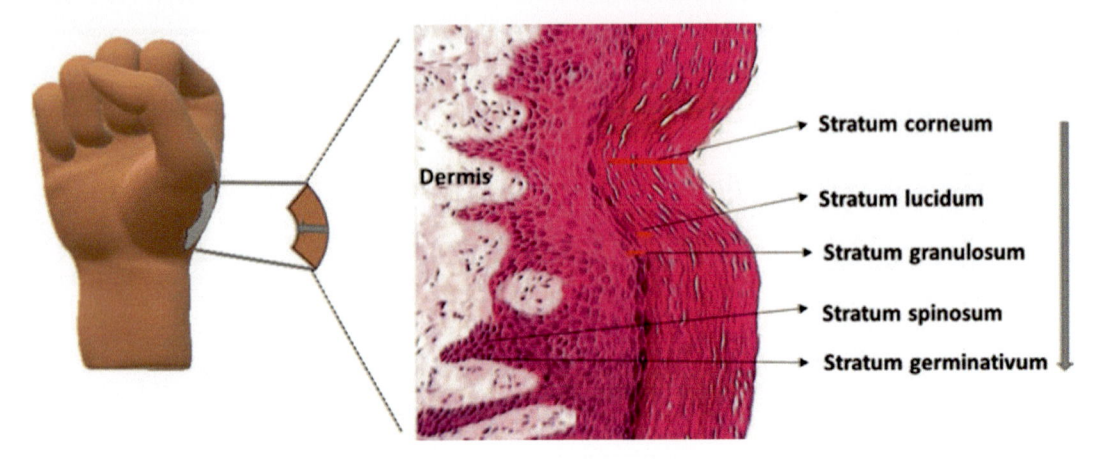

Fig. 12.1 Layers of the skin (from Yadav et al. [2], Fig. 1, p. 3)

The upper layer of the skin is the epidermis with its thickness ranging from 0.4 mm in the eyelid to 1.5 mm in the sole of the foot.

There is a bilayer dermis under the epidermis.

Subcutaneous fat which is also called the hypodermis, and connective tissue is found under the dermis.

12.2 Pathology

Malignant tumors of the skin

- Basal cell carcinoma (BCC) (65–70%)
- Squamous cell carcinoma (SCC) (30–35%)
- Malignant melanoma (1.5%)
- Merkel cell carcinoma
- Malignant sweat gland tumors

Basal Cell Carcinoma (BCC): This is the most common type of cancer according to USA data. BCC originates from pluripotent epithelial cells of the epidermal basal layer or external root sheaths of hair follicles [3].

– BCC rarely metastasizes (<0.1%), but it is a locally aggressive tumor that invades and destroys underlying tissues.
– BCC does not develop from premalignant lesions, unlike SCC.
– The most frequent subtype is noduloulcerative BCC (50%). This ulcerates during growth as a result of tissue destruction. Characteristically, this lesion forms the lesion defined as "rodent ulcer" which is a central ulcer surrounded by a pearl-like periphery appearance.
– The second most common subtype is superficial BCC (33%), which extends into the epidermis without presenting dermal invasion. BCCs are patchy tumors that can be multiple, hypopigmented, and erythematous.
– The most prominent feature of morpheaform (sclerosing) BCC, which is an another of subtype of BCC, is persistent surgical margin positivity after simple surgical excisions.
– Basosquamous BCC, a very rare subtype, usually occurs on the skin of the face, and its metastasis rate is similar to SCC (~5%).

Squamous Cell Cancer (SCC): This is the second most common type of cancer. It originates from the keratinocytes of the epidermis.

- SCC is most commonly seen on the dorsal surfaces of the hands and forearms and the interdigital regions.
- The major etiological factor is solar radiation.
- There are two subtypes of SCC. The slow-growing type is low grade and has a verrucous and exophytic appearance. Verrucous carcinoma often shows a high tendency for metas-

tasis. The second type is nodular and indurated; it grows rapidly and ulcerates early.
- Since SCC has a high potential for metastasis and regional lymph node involvement, a detailed physical exam should be performed.
- While nodal involvement rate is 1% in well-differentiated SCCs, nodal involvement rate increases to 10% in poorly differentiated, recurrent, tumors of size >3 cm and invasion depth >4 mm as well as lip-localized SCC [4].
- SCCs that develop on burn scars and with osteomyelitis show a nodal involvement rate of 10–30%.
- The rate of distant metastasis is 2%, and the most frequent metastatic regions are the lungs, liver, and bones.

Perineural Invasion
- Approximately 1% in BCC; it is generally seen in recurrent or locally advanced cases.
- Two to fifteen percent in SCC; frequently related with nodal involvement and the cranial base.
  - Recurrent SCC involves perineural invasion unless proven by pathological evaluation.
  - Perineural invasion is often asymptomatic; however, patients may occasionally have paresthesia, pain, dysesthesia, and paralysis.

Bowen's Disease
- This is a preinvasive form of SCC and is also known as skin carcinoma in situ [5].
- Epidermal full thickness dysplasia is seen without the presence of invasion histologically. It is a clinically well-circumscribed, erythematous, scaly, or irregularly circumscribed plaque.
- Treatment: surgery, cryotherapy, topical 5-FU or 40 Gy RT at 2 Gy/day.

Queyrat Disease/Erythroplasia
- This is Bowen's disease of the penis and is characterized by erythematous, shiny plaques caused by human papillomavirus (HPV) [6].

Marjolin's Ulcer
- This is an SCC that develops at the base of a burn scar [7].

Merkel Cell Carcinoma: Merkel cell carcinoma is a rare tumor that originates in the basal layer of the skin and is related to the terminal axons.

- Merkel cell carcinoma is a neuroendocrine carcinoma of the skin.
- Merkel cell carcinoma is an aggressive tumor and has ~75% local recurrence and mortality rate of ~35%. This mortality rate is higher than that of malignant melanoma.
- Lymphatic involvement is 25%, and 50–60% of patients develop distant metastases within 10 months of diagnosis [8].
- Treatment includes sentinel lymph node dissection followed by regional lymph node dissection and extensive surgical excision. Adjuvant RT (60–66 Gy) ± KT is added to the treatment.

Malignant Sweat Gland Tumor: These are aggressive tumors that usually metastasize to regional lymph nodes and hematogenously metastasize to distant organs [9].

- Malignant eccrine sweat gland tumors are locally aggressive and destructive tumors. They are often seen as slow-growing, painless nodular lesions on the palmar surface of the hand.
- Treatment includes the evaluation of metastatic disease, aggressive resection, and therapeutic lymphadenectomy for clinically positive lymph nodes. The roles of radiotherapy and chemotherapy are not clear due to the rarity of these tumors.

12.3 General Presentation

Basal Cell Carcinoma
- BCC is associated with UV light exposure.
- 70% of BCCs are seen on the skin of the head and face. The most common forms are nodular and superficial forms.
- **Basal cell nevus syndrome:** Also known as Gorlin syndrome and it is a rare autosomal dominant syndrome [10].

- It is observed as a result of germline mutation in the PTCH gene.
- Both developmental anomalies and postnatal tumors (especially multiple BCCs) are seen in patients.

Squamous Cell Cancer
- Often occurs on skin exposed to sun exposure (in people with white skin).
- Location:
  - Head and neck (55%)
    In AJCC 8th edition, only head-neck SCC staging is available [11].
  - Dorsum of hand and forearm (18%)
  - Leg (13%)
  - Arm (3%)
  - Shoulder/back (4%)
  - Thorax/abdomen (4%)
- SCC development is generally less common in those with skin that is not exposed to the sun and those with dark skin.
- In people with dark skin, possibly, epidermal melanin has protection against the carcinogenic effects of UV rays.
  In Black persons, SCC is usually seen in the legs, anus, and areas of scar/inflammation [12].
  SCC associated with chronic scars constitutes 20–40% of SCCs in Black patients (Figs. 12.2, 12.3, and 12.4).

12.4 Staging

In AJCC/UICC 8th edition (Tables 12.1 and 12.2),

- Only head-neck SCC staging is available [11].
- Merkel cell carcinoma has also been staged separately [17].

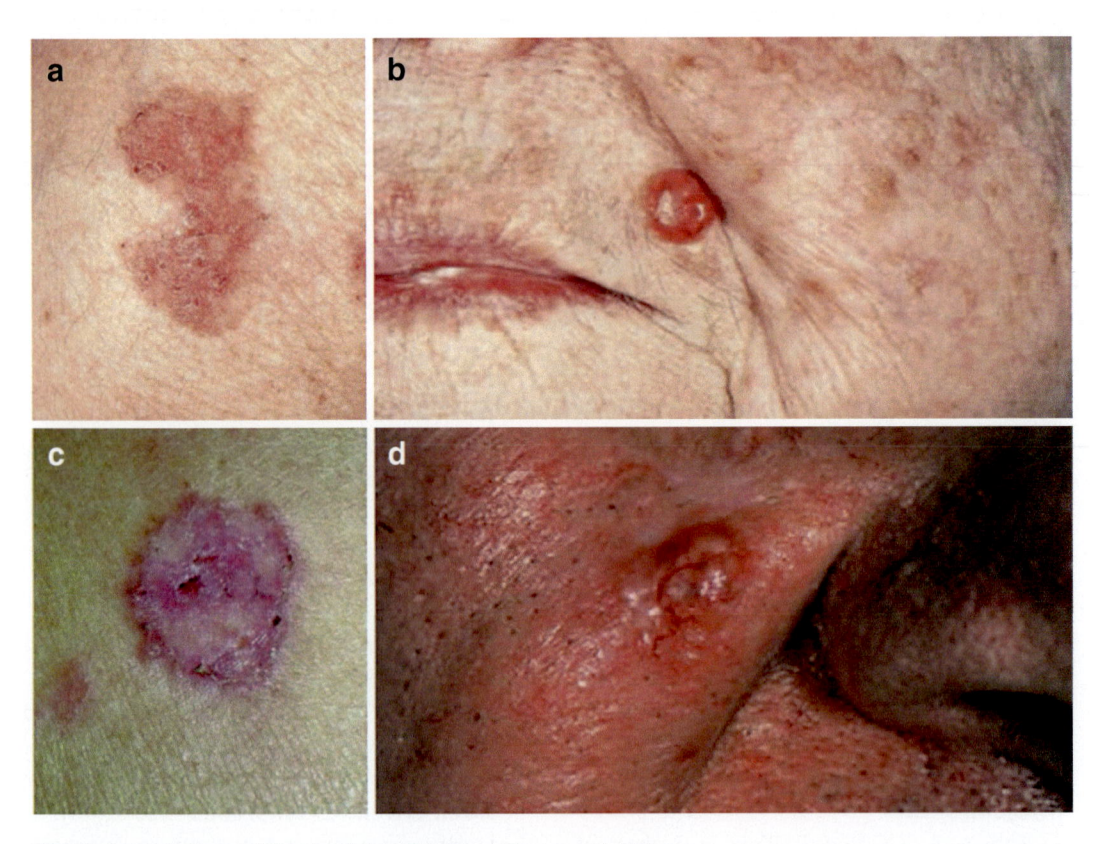

Fig. 12.2 (a–d) Superficial and nodular BCC (from Gloster et al. [13])

12.5 Treatment

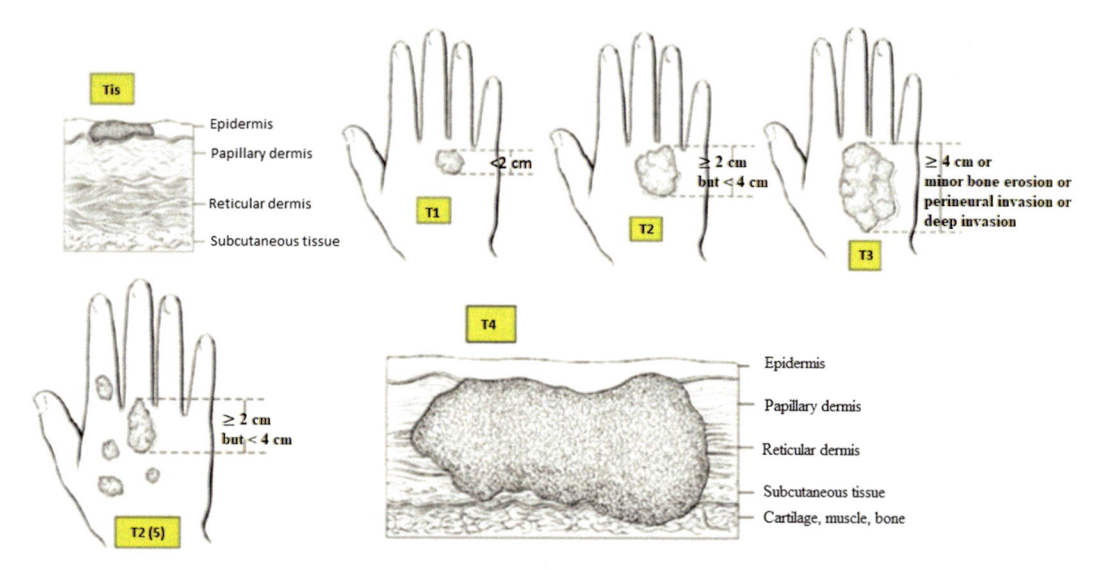

Fig.12.3 T staging in nonmelanoma skin cancers (modified from [14, 15])

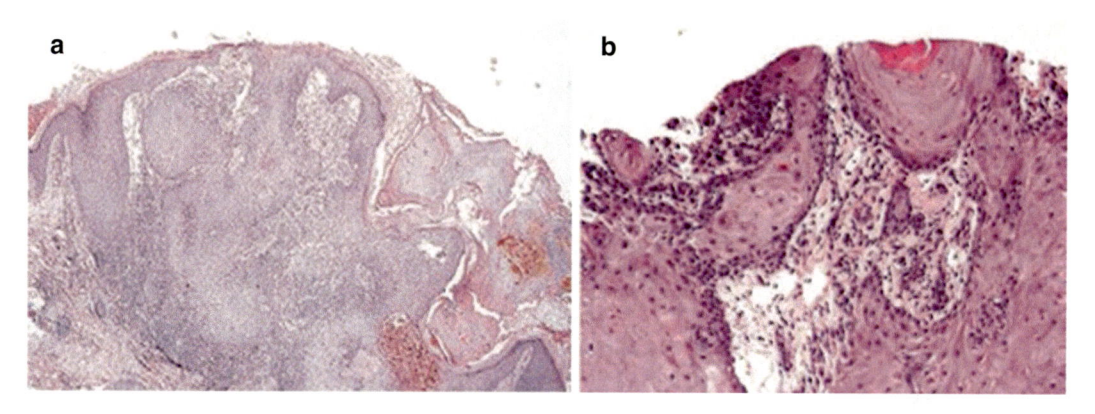

Fig. 12.4 (a, b) SCC histology (from Morgan [16], Figs. 6.15 and 6.16, p. 67)

12.5 Treatment

The major therapeutic approaches utilized in the management of nonmelanoma skin cancers are cryotherapy, curettage + electrodesiccation, chemotherapy, surgical excision, Mohs micrographic surgery, and radiotherapy.

Cryotherapy [11]
- ≤2 cm, superficial BCC therapy.
- Well-differentiated SCC therapy (margin: <3 cm tumors, 0.5–1 cm; ≥3 cm tumors, 1 cm margin).
- Cryogenic liquid nitrogen is most commonly used. The temperature should be at least −30°C to kill malignant cutaneous tissues.
- Advantages: high cure rate, ability to protect tissue, ability to treat multiple lesions.
- Disadvantages: wound care, long period of scar healing, hypopigmentation, poor scar tissue [18].

Curettage and Electrodesiccation
- If the location of skin cancers which have cryotherapy indication is over scar, cartilage and bone, and recurrent tumors.

Table 12.1 AJCC 8th edition TNM classification

Primary tumor (T)	
T	T criteria
TX	No evidence of primary tumor
Tis	Carcinoma in situ
T1	Tumor; <2 cm
T2	Tumor; ≥2 cm but <4 cm
T3	Tumor; ≥4 cm or minor bone erosion or perineural invasion or deep invasion[a]
T4	Tumor; gross cortical bone/bone marrow, skull base invasion, and/or skull base foramen invasion
T4a	Tumor; gross cortical bone/bone marrow invasion
T4b	Tumor; skull base invasion and/or skull base foramen invasion

[a]Deep invasion: invasion beyond subcutaneous fat tissue or >6 mm invasion (measured from the granular layer of the adjacent normal epidermis to the tumor base)
Perineural invasion in the T3 classification, the presence of tumor cells in the nerve sheath deeper in the dermis or clinical/radiological presence of nerve involvement with 0.1 mm or greater in caliber or without skull base invasion or without border violation (transgression)

Regional lymph nodes (N)	
N	N criteria
Clinical N (cN)	
NX	Regional lymph nodes cannot be assessed
N0	No regional lymph node metastasis
N1	Metastasis in a single ipsilateral lymph node, 3 cm or less in greatest dimension and ENE (−)
N2	Metastasis in a single ipsilateral lymph node, more than 3 cm but not more than 6 cm in greatest dimension and ENE (−), or metastasis in multiple ipsilateral lymph nodes, none more than 6 cm in greatest dimension or metastasis in bilateral or contralateral lymph nodes, none more than 6 cm in greatest dimension
N2a	Metastasis in a single ipsilateral lymph node, more than 3 cm but not more than 6 cm in greatest dimension and ENE (−)
N2b	Metastasis in multiple ipsilateral lymph nodes, none more than 6 cm in greatest dimension and ENE (−)
N2c	Metastasis in bilateral or contralateral lymph nodes, none more than 6 cm in greatest dimension and ENE (−)
N3	Metastasis in a lymph node, more than 6 cm in greatest dimension and ENE (−) or any metastatic lymph node clinically ENE (+)
N3a	Metastasis in a lymph node, more than 6 cm in greatest dimension and ENE (−)
N3b	Any metastatic lymph node clinically ENE (+)

Regional lymph nodes (N)	
N	N criteria
Pathological N (pN)	
NX	Regional lymph nodes cannot be evaluated
N0	No regional lymph node metastasis
N1	Metastasis in a single ipsilateral lymph node, 3 cm or less in greatest dimension and ENE (−)
N2	Metastasis in a single ipsilateral lymph node, 3 cm or less in greatest dimension and ENE (+) or Metastasis in a single ipsilateral lymph node, more than 3 cm but not more than 6 cm in greatest dimension and ENE (−), or metastasis in multiple ipsilateral lymph nodes, none more than 6 cm in greatest dimension and ENE (−), or metastasis in bilateral or contralateral lymph nodes, none more than 6 cm in greatest dimension and ENE (−)
N2a	Metastasis in a single ipsilateral lymph node, 3 cm or less in greatest dimension and ENE (+) or Metastasis in a single ipsilateral lymph node, more than 3 cm but not more than 6 cm in greatest dimension and ENE (−)
N2b	Metastasis in multiple ipsilateral lymph nodes, none more than 6 cm in greatest dimension and ENE (-)
N2c	Metastasis in bilateral or contralateral lymph nodes, none more than 6 cm in greatest dimension and ENE (−)
N3	Lymph node metastasis more than 6 cm and ENE (−), or a single ipsilateral lymph node metastasis larger than 3 cm and ENE (+), or multiple ipsilateral/contralateral/bilateral lymph node metastases and ENE (+) in either, or a single contralateral lymph node metastasis smaller than 3 cm and ENE (+)
N3a	Lymph node metastasis more than 6 cm and ENE (−) or a single ipsilateral lymph node metastasis larger than 3 cm and ENE (+)
N3b	Multiple ipsilateral/contralateral/bilateral lymph node metastases and ENE (+) in either or a single contralateral lymph node metastasis smaller than 3 cm and ENE (+)

Note: If the lymph node metastasis is above the lower limit of the cricoid cartilage (U) and below the lower limit of the cricoid (L) indicator is used

- This type of therapy is associated with a high incidence of recurrence, and no samples are taken for pathological evaluation.

Topical 5-FU
- This is only used for tumors that are limited to the epidermis (superficial BCC and Bowen's disease).

Table 12.2 AJCC 8th edition stage groupings

Distant metastasis (M)	
M	M criteria
M0	No distant metastasis
M1	Distant metastasis

Stage groupings

T	N	M	Stage
Tis	N0	M0	0
T1	N0	M0	I
T2	N0	M0	II
T3	N0	M0	III
T1	N1	M0	III
T2	N1	M0	III
T3	N1	M0	III
T1	N2	M0	IV
T2	N2	M0	IV
T3	N2	M0	IV
Any T	N3	M0	IV
T4	Any N	M0	IV
Any T	Any N	M1	IV

- It can be used in the management of premalignant lesions associated with SCC, but it is contraindicated in invasive SCC with nodular or infiltrative BCC [19].

Surgical Excision
- This provides total excision of the lesion and histopathological evaluation.
- It is usually preferred for tumors that are >5 mm in size.
- Surgical margin: 0.5–1 cm for tumors <3 cm; 1 cm for tumors ≥3 cm.
- Reconstructive surgery may also be performed according to the localization and size of the tumor.
- Cure rate ~90%.

Mohs' Micrographic Surgery
- In this technique, a fixative (zinc chloride paste) is applied to the tumor, and serial excisions are performed after fixation. The tumor map is generated by histological examination of the removed tumor, and re-excision is made in areas of residual tumor. Thus, while the tumor is removed in a controlled manner, protection of normal tissues is maximally provided. The cure rate is between 96% and 99% [20].

- Indications; high-risk localized BCC (postauricular, midface and eyelids, etc.), morpheaform or infiltrative type BCC, recurrent BCC, verrucous carcinoma, malignant sweat gland tumors.

High Risk Factors in Cutaneous SCC (Conditions with High Risk of Recurrence)
- By location/size:
  - Tumors on the extremity and body with size of ≥20 mm (excluding pretibia, hands, fingers, nail and appendages, ankles)
  - Cheeks, forehead, scalp, neck, or pretibial tumors with size of ≥10 mm
  - Mask areas with size ≥6 mm
     Central face, eyelids, eyebrows, periorbital nose, lips (cutaneous and Vermilion), chin, mandible, preauricular and postauricular skin, ear, genital area, hands, and fingers

It is performed:

- In lesions with irregular borders
- In recurrent tumors
- In immunosuppressed patients
- In areas with previous RT or chronic inflammation history
- Fast-growing tumors
- In the presence of neurological symptoms
- Moderately or poorly differentiated tumors
- Adenoid (acantholytic), adenosquamous, or desmoplastic histologies
- Depth of invasion for tumors ≥2 mm
- In the presence of perineural or vascular involvement [21]

High Risk Factors for Merkel Cell Carcinoma
- If the size is >10 mm
- If the tumor thickness is >10 mm
- If the excision margin is <5 mm
- Presence of perineural invasion
- Immunosuppressed conditions

Surgical treatment is the major treatment for high-risk cutaneous SCC patients.

– In surgically resectable tumors, Mohs micrographic surgery or surgical excision in which

surgical margins (radial and deep) are fully evaluated is recommended [22].
- Postsurgical RT indications
  - If the surgical margin is negative, adjuvant RT indications are controversial.

    There is an indication for severe perineural invasion and/or obvious nerve involvement.

    Postoperative adjuvant RT is recommended for patients with multiple risk factors for recurrence (especially if the surgical margin is positive or suspicious, rapid infiltrative growth pattern, and/or poorly differentiated histology) [23].

- **Treatment in N (+) cases**
  - Purpose is curative treatment.
  - Surgery + adjuvant RT is usually indicated.
  - Postoperative RT indications:

    If there is multiple N + and/or the involved lymph node is ≥3 cm

    If there is extranodal extension (ENE)

    If there is residual disease

    If there is involvement in the surrounding tissue (bone, nerve, orbita, etc.)

    If the patient is immunosuppressed

12.6 Radiotherapy

12.6.1 2D Conventional Radiotherapy

Radiotherapy can be used in the treatment of primary or recurrent non-melanoma skin cancers.

Primary radiotherapy in skin cancers
- Face skin: lesions >5 mm (particularly eyelids, tip of nose, nose wings, lips) [24]
- Ear and forehead: lesions >2 cm
- Hairy scalp

RT is planned with 0.5–2 cm safety margin according to tumor size and histology

- Recurrent or morpheaform BCC: 0.5–1 cm margin
- High-risk SCC (>3 cm, poorly differentiated or infiltrative–ulcerative SCC): 2 cm margin
- Regional lymph nodes may be included in the RT portal for high-risk SCCs

Photons or electrons with suitable energy are used according to tumor depth. Single RT field is used in the management of skin cancers.

- Dose prescription: Dmax dose for photons, 95% isodose for electrons
- Total dose: 50 Gy for BCC, 60–66 Gy for SCC

Bolus may be used to increase the surface dose (Fig. 12.5).

- Minimum bolus thickness for 6 and 9 MeV: 1 cm
- Minimum bolus thickness for 12 MeV: 0.5 cm

3D Conformal Radiotherapy/IMRT
- According to the location and size of the tumor and the nodal involvement
- Superficial therapy, orthovoltage therapy, electron therapy, 3D conformal RT, and IMRT can be used.

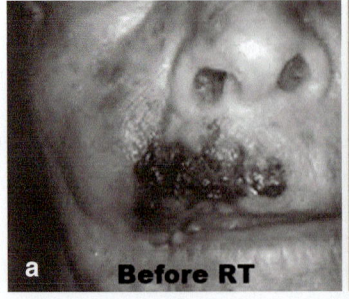

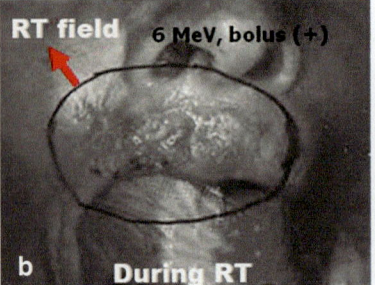

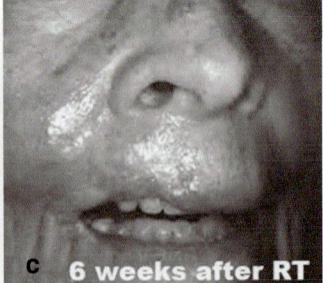

Fig. 12.5 (a–c) Appearance before, during, and after RT in non-melanoma skin cancers

Simulation
- It varies according to the location of the tumor, size, nodal involvement, and the selected treatment modality.

Contouring
- It varies according to the location of the tumor, size, nodal involvement, and the selected treatment modality.
- CTV is generated by adding a minimum 0.5 cm margin to the GTV, but it can be reduced according to the condition of the adjacent organs.
- 1 cm margin is added in high-risk cases and for all SCCs.
- For PTV, usually CTV + 0.5 cm margin is used.

Dose/Fractionation
- Definitive RT
- 64–66 Gy, 32–33 fx or 55 Gy, 20 fx or 45–50 Gy, 15 fx or 35 Gy, 5 fx
- Postoperative
  - 60 Gy, 30 fx or 50 Gy, 20 fx

As the fraction dose decreases, cosmetic result improves, and risk of late side effects decreases (Table 12.3).

- Nodal RT
  - If dissection is not performed; 66–70 Gy, 33–35 fx
  - If dissection has been performed
  - If head and neck located and ENE (+): 60–66 Gy in 30–33 fx

Table 12.3 Dose/fractionation recommendations in skin cancers

Primary tumor		
Tumor diameter	Margin	Dose/fractionation
<2 cm	1–1.5 cm	64 Gy, 32 fx
		55 Gy, 20–30 fx
		50 Gy, 15–20 fx
		35 Gy, 5 fx
≥2 cm	1.5–2 cm	64-66 Gy, 32–33 fx
		60 Gy, 30 fx
		55 Gy, 20 fx
Postoperative		60-66 Gy, 30–33 fx
		50 Gy, 20 fx
		40 Gy, 15 fx
		36 Gy, 12 fx
If cosmesis will be ignored		40 Gy, 10 fx
		30 Gy, 6 fx
		18–20 Gy, 1 fx
After lymph node dissection pN+		
Head-neck, ENE (+)		60–66 Gy, 30–33 fx
Head-neck, ENE (−)		56 Gy ; 28 fx
Axilla or inguinal, ENE(+)		60 Gy, 30 fx
Axilla or inguinal, ENE(−)		54 Gy, 27 fx
Lymph node dissection not performed, cN + and subclinical cN0		
cN0 but subclinical disease at-risk		50 Gy, 25 fx
Head-neck, cN+		66–70 Gy, 33–35 fx
Axilla or inguinal cN+		66 Gy, 33 fx

References

1. Wagner RF, Casciato DA (2000) Skin cancers. In: Casciato DA, Lowitz BB (eds) Manual of clinical oncology, 4th edn. Lippincott, Williams, and Wilkins, Philadelphia, PA, pp 336–373
2. Yadav N, Parveen S, Chakravarty S, Banerjee M (2019) Skin anatomy and morphology. In: Dwivedi A, Agarwal N, Ray L, Tripathi A (eds) Skin aging & cancer. Springer, Singapore
3. Rowe DE, Carroll RJ, Day CL Jr (1989) Long-term recurrence rates in previously untreated (primary) basal cell carcinoma: implications for patient follow-up. J Dermatol Surg Oncol 15(3):315–328
4. Committee on Guidelines of Care (1993) Guidelines of care for cutaneous squamous cell carcinoma. Task force on cutaneous squamous cell carcinoma. J Am Acad Dermatol 28(4):628–631
5. Cox NH, Eedy DJ, Morton CA (2007) Therapy guidelines and audit subcommittee, British Association of Dermatologists. Guidelines for management of Bowen's disease: 2006 update. Br J Dermatol 156(1):11–21
6. von Krogh G, Horenblas S (2000) Diagnosis and clinical presentation of premalignant lesions of the penis. Scand J Urol Nephrol Suppl 205:201–214
7. Phillips TJ, Salman SM, Bhawan J, Rogers GS (1998) Burn scar carcinoma. Diagnosis and management. Dermatol Surg 24(5):561–565
8. Eng TY, Boersma MG, Fuller CD, Goytia V, Jones WE 3rd, Joyner M, Nguyen DD (2007) A comprehensive review of the treatment of Merkel cell carcinoma. Am J Clin Oncol 30(6):624–636
9. Crowson AN, Magro CM, Mihm MC (2006) Malignant adnexal neoplasms. Mod Pathol 19(Suppl 2):S93–S126
10. Cohen MM Jr (1999) Nevoid basal cell carcinoma syndrome: molecular biology and new hypotheses. Int J Oral Maxillofac Surg 28:216
11. Amid MB (2017) Part II head and neck. In: AJCC cancer staging manual, 8th edn. Springer, New York, p 53
12. Gloster HM Jr, Neal K (2006) Skin cancer in skin of color. J Am Acad Dermatol 55:741
13. Gloster HM, Gebauer LE, Mistur RL (2016) Basal cell carcinoma. In: Absolute dermatology review. Springer, Cham, pp 391–392
14. Greene FL, Page DL, Fleming ID et al (2002) American joint committee on cancer. AJCC cancer staging manual, 6th edn. Springer, New York
15. Compton CC, Byrd DR, Garcia-Aguilar J, Kurtzman SH, Olawaiye A, Washington MK (2012) Cutaneous squamous cell carcinoma and other cutaneous carcinomas. In: Compton C, Byrd D, Garcia-Aguilar J, Kurtzman S, Olawaiye A, Washington M (eds) AJCC cancer staging atlas. Springer, New York, NY. https://doi.org/10.1007/978-1-4614-2080-4_29
16. Morgan MB (2018) Squamous cell carcinoma: variants and challenges. In: Morgan M, Spencer J, Hamill J Jr, Thornhill R (eds) Atlas of mohs and frozen section cutaneous pathology. Springer, Cham
17. Califano JA, Lydiatt WM, Nehal KS et al (2017) Cutaneous squamous cell carcinoma of the head and neck. In: Amin MB (ed) AJCC cancer staging manual, 8th edn. Springer, New York, p 171
18. Morton C, Horn M, Leman J, Tack B, Bedane C, Tjioe M, Ibbotson S, Khemis A, Wolf P (2006) Comparison of topical methyl aminolevulinate photodynamic therapy with cryothera py or fluorouracil for treatment of squamous cell carcinoma in situ: results of a multicenter randomized trial. Arch Dermatol 142(6):729–735
19. Szeimies RM, Karrer S (2006) Towards a more specific therapy: targeting nonmelanoma skin cancer cells. Br J Dermatol 154(Suppl 1):16–21
20. Lang PG Jr (2004 Jan) The role of Mohs' micrographic surgery in the management of skin cancer and a perspective on the management of the surgical defect. Clin Plast Surg 31(1):5–31
21. Haisma MS, Plaat BE, Bijl HP et al (2016) Multivariate analysis of potential risk factors for lymph node metastasis in patients with cutaneous squamous cell carcinoma of the head and neck. J Am Acad Dermatol 75:722
22. Howle JR, Hughes TM, Gebski V, Veness MJ (2012) Merkel cell carcinoma: an Australian perspective and the importance of addressing the regional lymph nodes in clinically node-negative patients. J Am Acad Dermatol 67:33
23. Harrington C, Kwan W (2014) Outcomes of Merkel cell carcinoma treated with radiotherapy without radical surgical excision. Ann Surg Oncol 21:3401
24. Lovett RD, Perez CA, Shapiro SJ et al (1990) External irradiation of epithelial skin cancer. Int J Radiat Oncol Biol Phys 19(2):235–242

Lymphomas and Total Body Irradiation

13

Lymphomas originate from immune system cells in various stages of differentiation. They cause several morphological, immunological, and clinical situations according to their origins.

> All lymphoid cell's originate from hematopoietic progenitor cells. These progenitor cells are divided into two subgroups: lymphoid and myeloid precursor cells. Lymphoid stem cells differentiate into B and T lymphocytes, which are the final products.
>
> - B cell origin: 75% of lymphoid leukemias and 90% of all lymphomas [1, 2]

Lymphoma classification changes frequently (see, e.g., Rapaport in 1966, Lukes/Collins in 1974, Working Formulation in 1982, REAL in 1994, WHO in 2001, 2008, and revised in 2016). The WHO classification is the most recent valid system for lymphomas (Table 13.1) [3].

13.1 Hodgkin's Lymphoma

Hodgkin's lymphoma (HL) is histologically characterized by the presence of various numbers of diagnostic multinucleated giant cells (Reed–Sternberg cells) within a mixed inflammatory infiltrate and originates from the lymphoid system. This disease was first defined by Thomas Hodgkin in 1832. Old terms for this disease are lymphogranulomatous lymphadenoma and malignant granuloma [4].

- HL constitutes 10% of all lymphomas and 1% of all malignancies.

13.1.1 Pathology/General Presentation

It is pathognomonic to find multinucleated or multilobulated giant cells called Reed–Stenberg (CD15+, CD30+) cells in biopsy material. Similar cells may be seen in infectious mononucleosis and non-Hodgkin's lymphoma (NHL). These cells are surrounded by normal lymphocytes, plasma cells, and eosinophils.

HL is classified into two groups according to the presence of fibrosis, collagen bands, necrosis, and malignant reticular cells (WHO classification) [3].

1. Nodular lymphocyte-predominant HL (2–5%)
2. Classical HL (~95%):
    - Nodular sclerosing HL (~65%)
    - Lymphocyte-rich HL (~10%)
    - Mixed cellular HL (~20%)
    - Lymphocyte-depleted HL (<5%)

Table 13.1 WHO lymphoma classification [3]

B cell NHL	T and NK cell NHL	HL
Chronic lymphocytic leukemia/small lymphocytic lymphoma	T cell prolymphocytic leukemia	Nodular lymphocyte-predominant Hodgkin's lymphoma
Monoclonal B cell lymphocytosis	T cell large granular lymphocytic leukemia	Classical Hodgkin's lymphoma
B cell prolymphocytic leukemia	Aggressive NK cell leukemia	▪ Nodular sclerosis classical Hodgkin's lymphoma
Splenic marginal zone lymphoma	Systemic EBV+ T cell lymphoma of childhood	▪ Lymphocyte-rich classical Hodgkin's lymphoma
Hairy cell leukemia	Hydroa vacciniforme-like lymphoproliferative disorder	▪ Mixed cellularity classical Hodgkin's lymphoma
Lymphoplasmacytic lymphoma	Adult T cell leukemia/lymphoma	▪ Lymphocyte-depleted classical Hodgkin's lymphoma
▪ Waldenström macroglobulinemia	Extranodal NK/T cell lymphoma, nasal type	
Monoclonal gammopathy of undetermined significance (MGUS), IgM	Enteropathy-associated T cell lymphoma	
μ heavy chain disease	Monomorphic epitheliotropic intestinal T cell lymphoma	
γ heavy chain disease	Hepatosplenic T cell lymphoma	
α heavy chain disease	Subcutaneous panniculitis-like T cell lymphoma	
Monoclonal gammopathy of undetermined significance (MGUS), IgG/A	Mycosis fungoides	
Plasma cell myeloma	Sézary syndrome	
Solitary plasmacytoma of bone	Primary cutaneous CD30+ T cell lymphoproliferative disorders	
Extraosseous plasmacytoma	▪ Lymphomatoid papulosis	
Monoclonal immunoglobulin deposition diseases	▪ Primary cutaneous anaplastic large cell lymphoma	
Extranodal marginal zone lymphoma of mucosa-associated lymphoid tissue (MALT lymphoma)	Primary cutaneous γδ T cell lymphoma	
Nodal marginal zone lymphoma	Peripheral T cell lymphoma, NOS	
Follicular lymphoma	Angioimmunoblastic T cell lymphoma	
▪ In situ follicular neoplasia	Anaplastic large cell lymphoma, ALK+	
▪ Duodenal-type follicular lymphoma	Anaplastic large cell lymphoma, ALK−	
Pediatric-type follicular lymphoma		
Primary cutaneous follicle center lymphoma		
Mantle cell lymphoma		
▪ In situ mantle cell neoplasia		
Diffuse large B cell lymphoma (DLBCL), NOS		
▪ Germinal center B cell type		
▪ Activated B cell type		
T cell/histiocyte-rich large B cell lymphoma		
Primary DLBCL of the central nervous system (CNS)		
Primary cutaneous DLBCL, leg type		
EBV+ DLBCL, NOS		
DLBCL associated with chronic inflammation		
Lymphomatoid granulomatosis		
Primary mediastinal (thymic) large B cell lymphoma		
Intravascular large B cell lymphoma		
ALK+ large B cell lymphoma		
Plasmablastic lymphoma		
Primary effusion lymphoma		
Burkitt lymphoma		
High-grade B cell lymphoma, with *MYC* and *BCL2* and/or *BCL6* rearrangements		
High-grade B cell lymphoma, NOS		
B cell lymphoma, unclassifiable, with features intermediate between DLBCL and classical Hodgkin's lymphoma		

NHL non-Hodgkin's lymphoma, *HL* Hodgkin's lymphoma

13.1.2 Clinical Signs

Lymphadenopathy: lymph nodes are rubber-like and painless. They grow slowly. The most common site for LAP is the cervical region (80%). Axilla and inguinal region involvement is rare. An anterior mediastinal mass is frequently seen (50%). Hepatosplenomegaly is rare.

B symptoms:

- Fever (due to increases in interleukins 1 and 2; should be of unknown origin and above 38°C)
- Drenching night sweats (due to increases in interleukins 1 and 2)
- Weight loss (>10% in the last 6 months; due to TNF increase)

In addition, itching and pain after alcohol intake may be observed. The most common symptom presented is a cervical mass (80%) [5]. Axillary or inguinal adenopathy are less frequently seen, found in approximately 15% and 10% of patients, respectively. B symptoms are present in one-third of all cases. Splenic involvement is seen in nearly all cases with bone marrow and hepatic involvement [5].

- *Nodular lymphocyte predominant HL*: CD15 (−), CD30 (−), CD45 (+), CD20 (+). Characterized with "popcorn cells" [1, 5].
- This type is seen after 40 years of age. Mediastinal involvement is usually not observed. The most common presentation is lymphadenopathies in peripheral lymph node regions. Generally, late relapses are observed. Survival is excellent.
- *Nodular sclerosing HL*: Mediastinal involvement is common.
- *Mixed cellular HL*: Such cases are usually at advanced stages, and even early-stage supradiaphragmatic cases commonly have microscopic abdominal disease.
- *Lymphocyte-rich HL*: Male predominance and older age. It may resemble nodular lymphocyte predominant HL morphologically. Prognosis is slightly better than other subtypes of classical HL.
- *Lymphocyte-depleted HL*: These cases are rarely seen and usually occur in older patients. Most of them have B symptoms. This type has poor prognosis and is associated with HIV.

13.1.3 Staging

Lugano Classification for Lymphomas (Derived from Ann Arbor Staging with Cotswolds Modifications [6])

Limited
  Stage I: One node or a group of adjacent nodes; IE is defined as single extranodal lesions without nodal involvement.
  Stage II: Two or more nodal groups on the same side of the diaghragm; IIE is defined as stage I or II by nodal extent with limited contiguous extranodal involvement.
  Stage II bulky: Stage II as above, with "bulky disease."
Advanced
  Stage III: Nodes on both sides of the diaphragm; nodes above the diaphragm with spleen involvement.
  Stage IV: Additional noncontiguous extranodal involvement.
  Note
  A: No symptoms.
  B: Fever (temperature > 38°C), drenching night sweats, unexplained weight loss >10% of body weight within the prior 6 months.
  Bulk: Defined as a single nodal mass of 10 cm or greater than one-third of the transthoracic diameter at any level of thoracic vertebrae as determined by computed tomography; record the longest measurement by computed tomography scan, with the term "X" no longer necessary (Tables 13.2 and 13.3).

Table 13.2 Risk classification for clinical stages I–II Hodgkin's lymphoma [5]

Risk factor	EORTC	GHSG	NCIC
Age	>50	–	≥40
ESR or B symptoms	No B symptoms with ESR≥50 or B symptoms with ESR ≥30	No B symptoms with ESR≥ 50 or B symptoms with ESR ≥ 30	ESR ≥50
Histology	–	–	Other than LP or NS
Large mediastinal adenopathy	Bulky mediastinal involvement	Large mediastinal mass	–
Number of nodal sites	≥4 sites of involvement	≥3 sites of involvement	≥3 sites of involvement
Extranodal lesions	–	Any	–

EORTC European Organization for Research and Treatment of Cancer, *GHSG* German Hodgkin Study Group, *NCIC* National Institute of Canada, *ESR* erythrocyte sedimentation rate, *LP* lymphocyte predominance, *NS* nodular sclerosis

Table 13.3 International Prognostic Score (IPS) for advanced stage Hodgkin's lymphoma [7]

Prognostic score[a]	Five-year freedom from progression (%)	Five-year overall survival (%)
0	84	89
1	77	90
2	67	81
3	60	78
4	51	61
5	42	56

[a]Each of the following factors carries a score of 1: hypoalbuminemia (albumin <4 g/dL), anemia (Hgb <10.5 g/dL), male sex, age ≥45 years, stage IV disease, lymphocytopenia (lymphocyte <8% or <600 μL

13.1.4 Treatment Algorithm

Treatment Algorithm for HL [8]
Favorable stage IA–IIA (bulky (−), <10 cm adenopathy, involvement of ≤3 LN regions, ESR <50, no B symptoms):

- Combined chemoradiotherapy: ABVD × 2–4c and involved-site radiation therapy (ISRT) 20–30 Gy
- Chemotherapy alone: ABVD × 3–4c (if PET-negative after 2–3 cycles, i.e., Deuville 1–2)
- Stanford V × 8 weeks and ISRT 30 Gy

Unfavorable stage IA–IIA:
- Combined chemoradiotherapy: ABVD × 4c and ISRT 30 Gy
- ABVD × 6c
- Stanford V × 12 weeks and ISRT 30–36 Gy

Stage III–IV:
- ABVD × 6c and consider ISRT to initially bulky or select PET+ sites
- Escalated BEACOPP × 6c and ISRT to residual >2.5 cm PET +

Treatment Algorithm for Nodular Lymphocyte-Predominant HL
Favorable stage IA–IIA:
- ISRT 30 Gy alone (consider +6 Gy boost for bulky disease)

Unfavorable stage IA–IIA:
- Chemotherapy and ISRT ± rituximab

Stage III–IV:
- Chemotherapy + rituximab ± ISRT
- Local radiotherapy for palliation only

Treatment Algorithm for Refractory/Relapsed HL
- High-dose chemotherapy and autologous stem cell rescue (for primary refractory disease)
- Chemotherapy or IFRT or extended-field RT alone as salvage only in highly selected cases otherwise same algorithm as primary refractory (for relapsed disease)

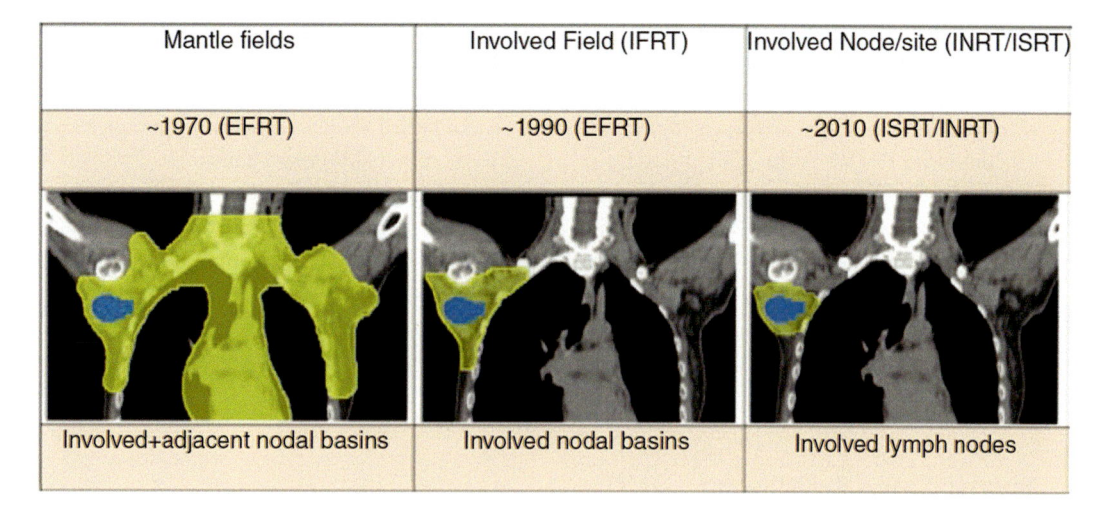

Mantle fields	Involved Field (IFRT)	Involved Node/site (INRT/ISRT)
~1970 (EFRT)	~1990 (EFRT)	~2010 (ISRT/INRT)
Involved+adjacent nodal basins	Involved nodal basins	Involved lymph nodes

Fig. 13.1 Evolution of radiotherapy fields (from [1], p. 740)

ABVD → Adriamycin (doxorubicin), bleomycin, vinblastine, dacarbazine.

Stanford V → Nitrogen mustard, doxorubicin, vinblastine, vincristine, bleomycin, etoposide, prednisone. Quicker treatment (8–12 weeks vs. 16–24 weeks for 4–6 cycles of ABVD).

Risks of sterility and secondary malignancy are lower than with the MOPP regimen (MOPP indicates mechlorethamine, vincristine, procarbazine, and prednisolone).

Ninety percent of children and 80% of adults at all anatomical stages and with all histological subtypes are cured with modern therapeutic approaches.

13.1.5 Radiotherapy

Treatment for Hodgkin's lymphoma includes small field radiotherapy with chemotherapy. Involved-node radiotherapy (INRT)/ISRT has replaced involved-field radiotherapy (IFRT) and extended-field radiotherapy (EFRT) (Fig. 13.1).

Involved-Node Radiotherapy (INRT) [9, 10]
- All patients must be examined by radiation oncologist before chemotherapy.
- All patients must have pre- and post-chemotherapy CT scans that are performed in the treatment position, as well as the prechemotherapy PET-CT.
- Fusion possibilities are also strongly recommended.
- CT simulation, modern radiation techniques such as IMRT or espiratory-gated radiotherapy, and immobilization devices are recommended for proper implementation of INRT.
- Imaging abnormalities before any intervention that might affect lymphoma volume should be outlined on the simulation study, and these volumes should be included n the clinical target volume (CTV) (CTV = initial volume of the lymph nodes before chemotherapy).
- Normal structures that were clearly uninvolved should be excluded from the CTV.
- The internal target volume (ITV) is defined as the CTV plus a margin taking into account uncertainties in size, shape, and position of the CTV.
- For patients with partial response, the gross tumor volume (GTV) is the lymph node remnant alone and should be con-

toured first. The CTV is the initial volume of the lymph nodes before chemotherapy Two PTVs should be contoured [PTV1 is the CTV including the GTV (initial tumor and remnant lymph node) with a margin to take into account organ movement and set-up variations. The boost PTV (PTV2) is the GTV alone with a margin for organ movement and set-up variations].

Involved-Field Radiotherapy (IFRT) [9, 10]
- IFRT is no longer routinely used.
- It is now limited to salvage treatment in patients in whom chemotherapy is unsuccessful and who are unable to embark on more intensive salvage treatment schedules.
- IFRT encompasses a region, not an individual lymph node.
- Involve-field regions are detailed in Table 13.4.

Involved-Site Radiotherapy (INRT) [9, 10] [Fig. 13.2].
- In both INRT and ISRT, the prechemotherapy GTV determines the CTV.
- When optimal pre-chemotherapy PET-CT imaging is not available, ISRT is used with clinical judgment to contour a larger CTV in order to eliminate uncertainties in defining the prechemotherapy GTV.

Neck should be at maximum extension (Table 13.5).
- Extension should be in a position where the chin is in the same plane as the mastoid process and external occipital protuberance (Fig. 13.3).
- This ensures the exclusion of the oral cavity and teeth from the RT fields, and decreases the dose to the mandible.

Table 13.4 Lymphatic regions for involved field radiotherapy

Lymph node	Radiotherapy fields (IFRT)
Preauricular and submandibular	Preauricular, cervical, supraclavicular LN, Waldeyer's field LN
Upper cervical	Bilateral cervical/preauricular, supraclavicular LN
Lower cervical, supraclavicular, infraclavicular	Bilateral cervical/supraclavicular/infraclavicular
Axilla	Ipsilateral axilla/supraclavicular/infraclavicular
Cubital	Cubital and ipsilateral axilla LN
Epitrochlear	Epitrochlear and ipsilateral axilla LN
Mediastinal	Hilar, mediastinal
Hilar	Hilar, mediastinal
Para-aortic	Para-aortic, splenic, portal LN
Iliac	Ipsilateral iliac/inguinal
Inguinal/femoral	Ipsilateral inguinal/femoral
Popliteal	Ipsilateral popliteal/inguinal/femoral LN

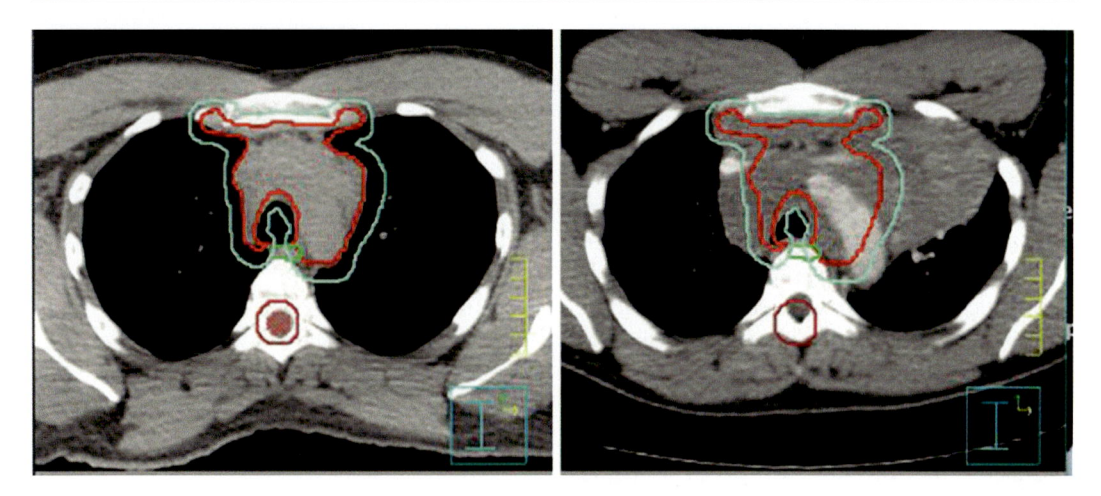

Fig. 13.2 Images depicting the CTV (red) and the PTV (green). The lift side is the CT scan, the right side is the fused pre-chemotherapy image (from [11], pp. 1–28)

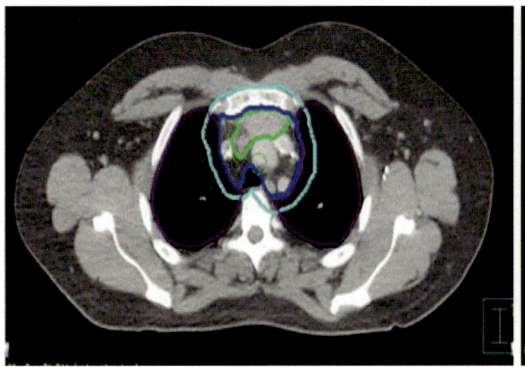

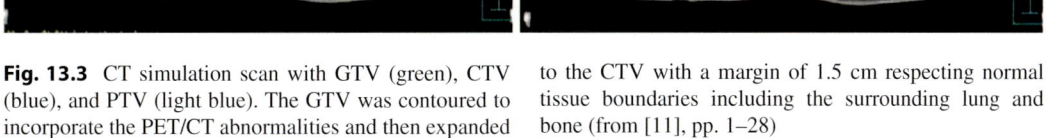

Fig. 13.3 CT simulation scan with GTV (green), CTV (blue), and PTV (light blue). The GTV was contoured to incorporate the PET/CT abnormalities and then expanded to the CTV with a margin of 1.5 cm respecting normal tissue boundaries including the surrounding lung and bone (from [11], pp. 1–28)

Table 13.5 Deauville criteria [12]

Score	Definition
Deauville 1	No uptake
Deauville 2	Uptake ≤ mediastinum
Deauville 3	Uptake > mediastinum but ≤ liver
Deauville 2	Uptake moderately higher than liver
Deauville 2	Uptake markedly higher than liver and/or new lesions
X	New areas of uptake unlikely to be related to lymphoma

13.2 Selected Publications

13.2.1 Early-Stage HL

H7 **(Noordijk EM et al (2006) Combined-modality therapy for clinical stage I or II Hodgkin's lymphoma: long-term results of the European Organisation for Research and Treatment of Cancer H7 randomized controlled trials. J Clin Oncol 24(19):3128–**

3135) → Patients with stage I or II HL were stratified into two groups, favorable and unfavorable, based on the following four prognostic factors: age, symptoms, number of involved areas, and mediastinal–thoracic ratio. The experimental therapy consisted of six cycles of epirubicin, bleomycin, vinblastine, and prednisone (EBVP) followed by IFRT. It was randomly compared, in favorable patients, to subtotal nodal irradiation (STNI) and, in unfavorable patients, to six cycles of mechlorethamine, vincristine, procarbazine, prednisone, doxorubicin, bleomycin, and vinblastine (MOPP/ABV hybrid) and IFRT. The median follow-up time of the 722 patients included was 9 years.

- In 333 favorable patients, the 10-year event-free survival rates (EFS) were 88% in the EBVP arm and 78% in the STNI arm ($p = 0.0113$), with similar 10-year overall survival (OS) rates (92 vs. 92%, respectively; $p = 0.79$).
- In 389 unfavorable patients, the 10-year EFS rate was 88% in the MOPP/ABV arm compared with 68% in the EBVP arm ($p < 0.001$), leading to 10-year OS rates of 87 and 79%, respectively ($p = 0.0175$).
- A treatment strategy for early-stage HL based on prognostic factors leads to high OS rates in both favorable and unfavorable patients. In favorable patients, the combination of EBVP and IFRT can replace STNI as standard treatment. In unfavorable patients, EBVP is significantly less efficient than MOPP/ABV.

H8, 2007 (Ferme C (2007) Chemotherapy plus involved-field radiation in early-stage Hodgkin's disease. N Engl J Med 357(19):1916–1927) → 1538 patients (age, 15–70 years) who had untreated stage I or II supradiaphragmatic Hodgkin's disease with favorable prognostic features (the H8-F trial) or unfavorable features (the H8-U trial). In the H8-F trial, three cycles of mechlorethamine, vincristine, procarbazine, and prednisone (MOPP) combined with doxorubicin, bleomycin, and vinblastine (ABV) plus involved-field radiotherapy were compared with subtotal nodal radiotherapy (STNI) alone (reference group). In the H8-U trial, three regimens were compared: six cycles of MOPP-ABV plus involved-field radiotherapy (reference group), four cycles of MOPP-ABV plus involved-field radiotherapy, and four cycles of MOPP-ABV plus subtotal nodal radiotherapy. The median follow-up was 92 months.

- In the H8-F trial, the estimated 5-year EFS rate was significantly higher after three cycles of MOPP-ABV plus involved-field radiotherapy than after subtotal nodal radio therapy alone (98 vs. 74%, $p < 0.001$). The 10-year OS estimates were 97 and 92%, respectively ($p = 0.001$).
- In the H8-U trial, the estimated 5-year EFS rates were similar in the three treatment groups: 84% after six cycles of MOPP-ABV plus involved-field radiotherapy, 88% after four cycles of MOPP-ABV plus involved-field radiotherapy, and 87% after four cycles of MOPP-ABV plus STNI. The 10-year OS estimates were 88, 85, and 84%, respectively.
- Chemotherapy plus involved-field radiotherapy should be the standard treatment for Hodgkin's disease with favorable prognostic features.
- In patients with unfavorable features, four courses of chemotherapy plus involved-field radiotherapy should be the standard treatment.

German HD8, 2003 (Engert A et al (2003) Involved-field radiotherapy is equally effective and less toxic com- pared with extended-field radiotherapy after four cycles of chemotherapy in patients with early-stage unfavorable Hodgkin's lymphoma: results of the HD8 trial of the German Hodgkin's Lymphoma Study Group. J Clin Oncol 21(19):3601–3608) → 1204 patients with newly diagnosed early-stage unfavorable HD were randomly assigned to receive cyclophosphamide, vincristine, procarbazine, and prednisone (COPP) + doxorubicin, bleomycin, vinblastine, and dacarbazine (ABVD) for two cycles followed by radiotherapy of 30 Gy EF + 10 Gy to bulky disease (arm A) or 30 Gy IF + 10 Gy to bulky disease (arm B). The median follow-up time was 54 months.

- Survival rates at 5 years after start of radiotherapy revealed no differences between arms A and B, respectively, in terms of FFTF (85.8 and 84.2%) and OS at 5 years (90.8 and 92.4%).
- There were no differences between arms A and B, respectively, in terms of complete remission (98.5 and 97.2%), progressive disease (0.8 and 1.9%), relapse (6.4 and 7.7%), death (8.1 and 6.4%), and secondary neoplasia (4.5 and 2.8%).
- Radiotherapy volume size reduction from EF to IF after COPP + ABVD chemotherapy for two cycles produces similar results and less toxicity in patients ith early-stage unfavorable HD.

Italy, 2007 (Picardi M (2007) Randomized comparison of consolidation radiation versus observation in bulky Hodgkin's lymphoma with post-chemotherapy negative positron emission tomography scans. Leuk Lymphoma 48(9):1721–1727) → Among 260 patients treated with induction chemotherapy for bulky HL, 160 patients achieved negative residual masses at 2-[18F]fluoro-2-deoxy-D-glucose positron emission tomography (FDG-PET) scans. They were randomly divided into two well-matched groups to receive either 32 Gy radiotherapy to bulky area or no further therapy. The median follow-up was 40 months.

- Histology showed a malignancy in 14% of patients in the chemotherapy-only group and in 4% of patients in the chemotherapy + radiotherapy group ($p = 0.03$). All the relapses in the chemotherapy-only group involved the bulky site and the contiguous nodal regions. Thus, the overall diagnostic accuracy of FDG-PET to exclude future relapses in the patients not protected by radiotherapy was 86%, with a false-negative rate of 14%.
- The addition of irradiation helps improve EFS in HL patients with postchemotherapy FDG-PET-negative residual masses.

NCI Canada/ECOG, 2005 (Meyer RM et al (2005) Randomized comparison of ABVD chemotherapy with a strategy that includes radiation therapy in patients with limited-stage Hodgkin's lymphoma: National Cancer Institute of Canada Clinical Trials Group and the Eastern Cooperative Oncology Group. J Clin Oncol 23(21):4634–4642) → 399 patients with nonbulky clinical stage I–IIA Hodgkin's lymphoma were stratified into favorable and unfavorable risk cohorts. Patients allocated to radiation-containing therapy received subtotal nodal radiation if favorable risk or combined-modality therapy if unfavorable risk. Patients allocated to ABVD received 4–6 treatment cycles. The median follow-up was 4.2 years.

- Five-year freedom from disease progression was superior in patients allocated to radiation therapy ($p = 0.006$; 93 vs. 87%); no differences in EFS ($p = 0.06$; 88 vs. 86%) or OS ($p = 0.4$; 94 vs. 96%) were detected.
- In a subset analysis comparing patients stratified into the unfavorable cohort, freedom from disease progression was superior in patients allocated to combined-modality treatment ($p = 0.004$; 95 vs. 88%); no difference in OS was detected ($p = 0.3$; 92 vs. 95%).
- In patients with limited-stage Hodgkin's lymphoma, no difference in OS was detected between patients randomly assigned to receive treatment that includes radiation therapy or ABVD alone. Although 5-year freedom from disease progression was superior in patients receiving radiation therapy, this advantage is offset by deaths due to causes other than progressive Hodgkin's lymphoma or acute treatment-related toxicity.

NCIC/ECOG HD6 (Meyer et al (2012) ABVD alone versus radiation-based therapy in limited-stage Hodgkin's lymphoma. NEJM 366(5):399–408) → 405 patients, stage IA–IIA nonbulky randomized to receive ABVD chemotherapy alone or STNI with or without ABVD chemotherapy. In ABVD-only group, patients with both favorable and unfavorable risk factors received 4–6 cycles of ABVD. In STNI group, patients with favorable risk factors received STNI alone and those with unfavorable risk factors received two cycles of ABVD plus STNI. The median follow-up was 11.3 years.

- In favorable group, no difference in OS or EFS at 12 years.
- In unfavorable group, ABVD + STNI improved 12-year freedom from disease progression (94% vs. 86%) but decreased 12-year OS (81% vs. 92%) due to more on-cancer deaths.
- More secondary cancers (23 vs. 10) and cardiac events (26 vs. 16) with STNI.

GHSG HD11 (Eich et al (2011) Intensified chemotherapy and dose-reduced involved-field radiotherapy in patients with early unfavorable Hodgkin's lymphoma: final analysis of the German Hodgkin Study Group HD11 trial. J Clin Oncol 28(27):4199–4206) → 1395 patients with early-stage unfavorable HL randomized to ABVD × 4c +30 Gy IFRT vs. ABVD × 4c +20 Gy IFRT vs. BEACOPP × 4c + 30 Gy IFRT vs. BEACOPP × 4c + 20 Gy IFRT. The median follow-up was 82 months.

- No difference in overall freedom from treatment failure or OS between ABVD vs. BEACOPP or 20 Gy vs. 30 Gy
- There was more toxicity with BEACOPP
- More relapses in the 20 Gy arm requiring salvage

GHSG HD10 (Engert et al (2010) **Reduced treatment intensity in patients with early-stage Hodgkin's lymphoma. NEJM 363(7):640–652)** → 1370 patients with favorable stage I–II randomized to ABVD × 2c vs. × 4c followed by IFRT 20 vs. 30 Gy. The median follow-up was 7.5 years.

- No statistically significant difference in treatment failure or OS between any of the arms.
- Two cycles of ABVD followed by 20 Gy of IFRT is as effective as, and less toxic than, four cycles of ABVD followed by 30 Gy of involved-field radiation therapy.

Updated in 2017 (Sasse et al (2017) Long-Term Follow-Up of Contemporary Treatment in Early-Stage Hodgkin Lymphoma: Updated Analyses of the German Hodgkin Study Group HD7, HD8, HD10, and HD11 Trials. J Clin Oncol 35(18):1999–2007)
The median follow-up was 98 months. PET-CT was not used to assess response.

- No significant difference between either randomization.
- Noninferiority was confirmed for both (10-year PFS of ABVD × 4c + 30 Gy vs. ABVD × 2c + 20 Gy was 87.4% vs. 87.2%).

EORTC H10 (Raemaekers et al (2014) Omitting radiotherapy in early positron emission tomography-negative stage I/II Hodgkin lymphoma is associated with an increased risk of early relapse: Clinical results of the preplanned interim analysis of the randomized EORTC/LYSA/FIL H10 trial. J Clin Oncol 32(12):1188–1194) → 1137 patients. This study evaluated whether INRT could be omitted without compromising progression-free survival in patients attaining a negative early PET scan after two cycles of ABVD as compared with standard combined-modality treatment. In H10F (favorable group), patients were randomized to PET-adapted therapy versus standard treatment, then received ABVD × 2c followed by PET. In standard arm, patients received one additional cycle of ABVD with INRT to 30 Gy (6 Gy boost allowed for residual disease). In experimental PET-adapted arm, patients received two additional cycles of ABVD (total four) if PET was negative (Deauville 1–2). If positive, patients received escalated BEACOPP × 2 cycles and INRT to 30 Gy (6 Gy boost allowed for residual). Randomization to PET-adapted therapy was stopped early as noninferiority was unlikely.

Updated in 2017 (Andre et al (2017) Early Positron Emission Tomography Response-Adapted Treatment in Stage I and II Hodgkin Lymphoma: Final Results of the Randomized EORTC/LYSA/FIL H10 Trial. J Clin Oncol 35(16):1786–1794)

- 1950 patients were recruited.
- Noninferiority of ABVD alone could not be established (H10F 5-year PFS 99% vs. 87.1%, HR 15.8, 95% CI: 3.8–66.1, noninferiority margin was 3.2).
- Escalation to BEACOPP improved 5-year PFS from 77.4% (ABVD + INRT) to 90.6% (BEACOPP + INRT, p = .002).
- Even in patients with excellent PET response, omission of INRT is associated with increased risk of progression (but no difference in OS).

UK RAPID (Radford et al (2015) Results of a trial of PET-directed therapy for early-stage Hodgkin's lymphoma. NEJM 372(17):1598–1607) → 602 patients, stage I–IIA treated with ABVD × 3c followed by PET. 426 PET-negative (Deauville 1–2) patients randomized to IFRT 30 Gy or no further treatment. If PET-positive, patients received ABVD × 4c and 30 Gy IFRT. The median follow-up was 60 months.

- Three-year PFS was 94.6% in RT group vs. 90.8% in ABVD alone (difference –3.8% (95% CI: –8.8% to 1.3%).
- The results did not show the noninferiority of the strategy of no further treatment after chemotherapy.
- Nevertheless, patients with early-stage Hodgkin's lymphoma and negative PET findings after three cycles of ABVD had a very good prognosis either with or without consolidation radiotherapy.

EORTC H9F (Thomas et al (2018) Comparison of 36 Gy, 20 Gy, or No Radiation Therapy After 6 Cycles of EBVP Chemotherapy and Complete Remission in Early-Stage Hodgkin Lymphoma Without Risk Factors: Results of the EORT-GELA H9-F Intergroup Randomized Trial. Int J Rad Biol Phys 100(5):1133–1145) → 783 patients with early-stage HL received 6c × EBVP and randomized to no IFRT, 20 Gy IFRT, and 36 Gy IFRT. No-IFRT arm was prematurely stopped due to toxicity, treatment modification, early relapse or death.

- 20-Gy arm (5-year RFS, 84.2%) was not inferior to those in the 36-Gy arm (5-year RFS, 88.6%) (difference, 4.4%; 90% confidence interval [CI] –1.2% to 9.9%).
- A difference of 16.5% (90% CI 8.0–25.0%) in 5-year RFS estimates was observed between the no-RT arm (69.8%) and the 36-Gy arm (86.3%); the hazard ratio was 2.55 (95% CI 1.44–4.53; P<.001).
- The 5-year overall survival estimates ranged from 97% to 99%.
- In patients with early-stage HL without risk factors in complete remission after EBVP chemotherapy, the RT dose may be limited to 20 Gy without compromising disease control.
- Omitting RT in these patients may jeopardize the treatment outcome

13.2.2 Advanced-Stage HL

GHSG HD3 (German Hodgkins Study Group), 1995 **(Diehl V et al (1995) Further chemotherapy versus low-dose involved-field radiotherapy as consolidation of complete remission after six cycles of alternating chemotherapy in patients with advanced Hodgkin's disease. German Hodgkins' Study Group (GHSG). Ann Oncol 6(9):901–910)** → 288 patients with stage IIIB or IV HD received induction chemotherapy with 3× (COPP + ABVD). Patients achieving CR were eligible for randomization to either 20 Gy radiotherapy to initially involved fields (RT-arm) or to an additional 1× (COPP + ABVD) (CT arm). Patients with nodal PR were allocated to more intense radiotherapy (IRT arm: 20 Gy IF, 40 Gy to persisting tumor).

- 171 (59%) achieved CR after induction chemotherapy. Of these, 100 patients were successfully randomized to RT or CT. In the CT arm, relapses were observed in 10 of 49 patients compared with 13 of 51 patients in the RT arm (p = n.s.).
- No differences in treatment efficacy were detected between 20 Gy IF radiotherapy and 1× (COPP + ABVD) chemotherapy following

CR after six cycles of alternating chemotherapy in patients with advanced-stage HD.

***SWOG 7808, 1994* (Fabian CJ et al (1994) Low-dose involved field radiation after chemotherapy in advanced Hodgkin disease. A Southwest Oncology Group randomized study. Ann Intern Med 120(11):903–912)** → 278 adults with clinical or pathological stage III or IV Hodgkin's disease, who achieved complete responses after six cycles of MOP-BAP (nitrogen mustard, vincristine, prednisone, bleomycin, doxorubicin, and procarbazine).

- Remission duration, relapse-free survival, and OS were similar for the two groups ($p = 0.09$, $p > 0.2$, and $p = 0.14$, respectively).
- Low-dose radiation improved remission duration in the subgroups of patients with nodular sclerosis and bulky disease. For all patients with bulky disease, the 5-year remission duration estimate was 75% for the low-dose radiation group and 57% for the no further treatment group ($p = 0.05$).
- No difference in OS was noted between low-dose radiation and no further treatment in all patients or major subgroups.
- Low dose involved field radiation after MOP-BAP chemotherapy in patients with stage III or IV Hodgkin's disease did not prolong remission duration or OS in randomized patients.

***EORTC 20884, 2007* (Aleman BM et al (2007) Involved-field radiotherapy for patients in partial remission after chemotherapy for advanced Hodgkin's lymphoma. Int J Radiat Oncol Biol Phys 67(1):19–30)** → 739 stage III–IV HL cases were treated with 6–8 cycles of mechlorethamine, vincristine, procarbazine, prednisone/doxorubicin, bleomycin, and vinblastine hybrid chemotherapy. Patients in complete remission (CR) after chemotherapy were randomized to no further treatment and 24 Gy IFRT. Those in PR after six cycles received IFRT (30 Gy to originally involved nodal areas and 18–24 Gy to extranodal sites with or without a boost). The median follow-up was 7.8 years.

- The 8-year EFS and OS rates for the 227 patients in PR who received IFRT were 76% and 84%, respectively. These rates were not significantly different from those for CR patients who received IFRT (73 and 78%) or from those in CR who did not receive IFRT (77 and 85%).
- Patients in PR after six cycles of mechlorethamine, vincristine, procarbazine, prednisone/doxorubicin, bleomycin, and vinblastine treated with IFRT had 9-year EFS and OS rates similar to those of patients in CR, suggesting a definite role for RT in these patients.

***Germany, 1998* (Loeffler M (1998) Meta-analysis of chemotherapy versus combined modality treatment trials in Hodgkin's disease. International Database on Hodgkin's Disease Overview Study Group. J Clin Oncol 16(3):818–829)** → Data on 1,740 patients treated in 14 different trials that included 16 relevant comparisons were analyzed.

- Additional RT showed an 11% overall improvement in tumor control rate after 10 years ($p = 0.0001$). No difference could be detected with respect to OS ($p = 0.57$).
- When combined-modality treatment was compared with CT alone in the parallel-design trials, no difference could be detected in tumor control rates ($p = 0.43$), but OS was significantly better after 10 years in the group that did not receive RT ($p = 0.045$).
- There were significantly fewer fatal events among patients in continuous complete remission (relative risk, 1.73; $p = 0.005$) if no RT was given.
- Combined-modality treatment in patients with advanced-stage Hodgkin's disease overall has a significantly inferior long-term survival outcome than CT alone if CT is given over an appropriate number of cycles. The role of RT in this setting is limited to specific indications.

***GHSG HD12, 2011* (Borchmann et al (2011) Eight cycles of escalated-dose BEACOPP compared with four cycles of escalated-dose**

BEACOPP followed by four cycles of baseline-dose BEACOPP with or without radiotherapy in patients with advanced-stage Hodgkin's lymphoma: final analysis of the HD12 trial of the German Hodgkin Study Group. J Clin Oncol 29(32):4234–4242) → 1670 patients with stage IIB/IIIA and risk factors or stage IIIB/IV randomized to escalated BEACOPP × 8c with 30 Gy IFRT to initial bulky or residual vs. escalated BEACOPP × 4c and standard BEACOPP × 4c with 30 Gy IFRT to initial bulky or residual vs. no RT to residual for both arms.

- At 5 years, no statistical difference between any of the four arms.
- RT improved freedom from treatment failure (90.4% vs. 87%; difference −3.4%; 95% CI − 6.6% to −0.1%), particularly in patients who had residual disease after chemotherapy (difference, −5.8%; 95% CI, −10.7% to −1.0%), but not in patients with bulk in complete response after chemotherapy (difference, −1.1%; 95% CI, −6.2% to 4%).
- RT did not improve OS. Because RT was given to 11% of patients in "non-RT" arms, equivalency of a non-RT strategy cannot be proved.

GHSG HD15, 2017 **(Engert et al (2017) Reduced-Intensity Chemotherapy in Patients With Advanced-Stage Hodgkin Lymphoma. Updated Results of the Open-Label, International, Randomised Phase 3 HD15 Trial by the German Hodgkin Study Group. Hemasphere 1(1):e5)** → 2182 patients with advanced HL. Patients with partial response with PET+ residual disease >2.5 cm were treated with 6–8 c BEACOPP and PET-guided 30 Gy RT. PET-negative patients received no additional RT. Negative predictive value of PET was 94%. The median follow-up was 102 months.

- Four-year PFS was 92% for PET-negative CT-persistent residual disease not irradiated, suggesting consolidative RT may be omitted.
- Four-year PFS for PET+ partial response patients who were irradiated was 86%, suggesting consolidative RT effective.

- Treatment with six cycles of BEACOPP$_{escalated}$ followed by PET-guided radiotherapy was more effective in terms of freedom from treatment failure and less toxic than eight cycles of the same chemotherapy regimen.

UK RATHL, 2016 **(Johnson et al (2016) Adapted Treatment Guided by Interim PET-CT Scan in Advanced Hodgkin's Lymphoma. NEJM 374(25):2419–2429)** → 1214 advanced-stage HL patients underwent a baseline PET-CT scan, received two cycles of ABVD, and underwent an interim PET-CT scan. Patients with negative PET findings after two cycles were randomly assigned to continue ABVD (ABVD group) or omit bleomycin (AVD group) in cycles 3–6. Those with positive PET findings after two cycles received BEACOPP. Radiotherapy was not recommended for patients with negative findings on interim scans. The median follow-up was 41 months.

- The omission of bleomycin from the ABVD regimen after negative findings on interim PET resulted in a lower incidence of pulmonary toxic effects than with continued ABVD but not significantly lower efficacy.

ECOG E2496, 2013 **(Gordon et al (2013) Randomized phase III trial of ABVD versus Stanford V with or without radiation therapy in locally extensive and advanced-stage Hodgkin lymphoma: an intergroup study coordinated by the Eastern Cooperative Oncology Group (E2496). J Clin Oncol 31(6):684–691)** → 794 advanced-stage HL patients randomized to 6–8c × ABVD followed by 36 Gy IFRT to massive mediastinal disease vs. 12 weeks of Stanford V regimen followed by 36 Gy IFRT to sites > 5 cm in maximum transverse dimension plus spleen if involved on CT. The median follow-up was 6.4 years.

- There was no significant difference in the overall response rate between the two arms, with complete remission and clinical com-

plete remission rates of 73% for ABVD and 69% for Stanford V.
- There was no difference in FFS: 74% for ABVD and 71% for Stanford V at 5 years ($P = 0.32$).
- ABVD with consolidation RT sites of pretreatment bulky disease remains the standard of care for patients with advanced Hodgkin's lymphoma.

13.2.3 Ongoing Trials

GHSG HD16 trial randomizes low-risk early-stage patients to two cycles of ABVD followed by 30-Gy IFRT irrespective of PET results after chemotherapy vs. two cycles of ABVD followed by 30 Gy IFRT for only patients with positive PET after chemotherapy and no further therapy if PET scan is negative after chemotherapy.

13.2.4 Nodular Lymphocyte-Predominant HL

GHSG (*Eichenauer et al (2015)* **Long-Term Course of Patients With Stage IA Nodular Lymphocyte-Predominant Hodgkin Lymphoma: A Report From the German Hodgkin Study Group. J Clin Oncol 33:2857–2862**) → 256 patients with stage IA NLPHL received combined-modality treatment (CMT), extended-field radiotherapy (EFRT), IFRT, or four weekly standard doses of rituximab. The median follow-up was 91 months.

- Eight-year PFS and OS were 88.5% and 98.6% for CMT 84.3% and 95.7% for EFRT, and 91.9% and 99% for IFRT, respectively.
- Patients treated with rituximab had 4-year PFS and OS rates of 81% and 100%, respectively.
- Tumor control rates were equivalent with CMT, EFRT, and IFRT.
- IFRT is considered the standard of care due to the lowest risk of toxic effects.

GHSG (*Eichenauer et al (2020)* **Long-Term Follow-Up of Patients With Nodular Lymphocyte-Predominant Hodgkin Lymphoma Treated in the HD7 to HD15 Trials: A Report From the German Hodgkin Study Group. J Clin Oncol 38(7):698–705**) → 471 patients with NLPHL (early stages, $n = 251$; intermediate stages, $n = 76$; advanced stages, $n = 144$) had received stage-adapted first-line treatment in the randomized GHSG HD7 to HD15 studies were investigated. Treatment consisted of radiotherapy alone, chemotherapy alone, or combined-modality approaches. The median follow-up was 9.2 years.

- At 10 years, progression-free survival and overall survival estimates were 75.5% and 92.1% (early stages, 79.7% and 93.3%; intermediate stages, 72.1% and 96.2%; advanced stages, 69.8% and 87.4%), respectively.

13.3 Non-Hodgkin's Lymphoma

NHLs are malignancies of the lymphoid–reticular system, like HL. They are usually observed in older patients and show more extranodal involvement than HLs. They also do not show contagious LN extension patterns, unlike HL. NHLs exhibit different responses to treatment modalities [5].

> Extranodal involvement is commonly seen in the head–neck region, particularly in the Waldeyer's ring, nasal cavity, paranasal sinus, oral cavity, larynx, and orbita. The most commonly involved extralymphatic localizations are the liver, lungs, and bone marrow.

13.3.1 Pathology/General Presentation

The Rappaport, Lukes/Collins, International Formulation, and Kiel classifications are not used anymore. The current NHL classification system is the WHO system, which was revised (4th edition) in 2017 [3].

> - *B cell NHLs constitute 85% of all NHLs* [1] Diffuse B cell lymphoma (DLBCL), 33% Follicular lymphoma, 20%
> - MALT-type extranodal marginal zone B cell lymphoma, 5–10%
> - B-CLL/small lymphocytic lymphoma, 5–10%
> - Mantle cell lymphoma, 5%
> - *T cell NHLs constitute 15% of all NHLs* [1]
> - T/NK cell lymphoma, peripheral T cell lymphoma, 6% Mycosis fungoides/Sézary syndrome, <1% Anaplastic large cell lymphoma, 2%
> - *Low-grade NHL* [1].
> - Follicular lymphoma (grade I–II)
> - B-CLL/small lymphocytic lymphoma
> - MALT-type extranodal marginal zone B cell lymphoma Mycosis fungoides/Sézary syndrome
> - *Moderate-grade NHL* [1]
> - Follicular lymphoma (grade III)
> - Mantle cell lymphoma
> - Diffuse large B cell lymphoma (DLBCL)
> - T/NK cell lymphoma, peripheral T cell lymphoma Anaplastic large cell lymphoma
> - High-grade NHL [1]
> - Lymphoblastic lymphoma Burkitt's lymphoma

- DLBCL is presented at stage I–II in 30–40% of cases and is commonly associated with extranodal disease.
- Follicular lymphoma is presented at stage IV in 60% of cases (21% in stage I–II, 19% in stage III).
- MALT-type extranodal marginal zone B cell lymphoma is generally seen in the stomach, ocular adnexae, skin, thyroid, parotids, lungs, and breasts. Most cases are at stage I–II (65–70%).
- Mantle cell lymphoma is generally presented as disseminated disease, and together with spleen, bone marrow, and GIS involvement.

B symptoms are more common in HL.

13.3.2 Staging

Staging is the same as HL [6].

Note: Sites that are extranodal, but not extralymphatic (therefore, not classified as E): Waldeyer's ring, thymus, and spleen.

13.3.2.1 Prognostic Factors in Non-Hodgkin's Lymphoma [13]

Adverse features include diffuse large cell and mantle cell lymphoma, T/NK cell rather than B cell phenotype, tumor bulk >10 cm, Ann Arbor Stage III–IV, B symptoms, high % S-phase Ki-67 > 80%, age of ≥60 years, abnormal level of LDH, high level of β-2 microglobulin, absent HLA-DR and BCL-6 rearrangement, high CD-44 expression, presence of c-MYC translocation, low level of BCL-2 protein, involvement of brain and testis.

> **International Prognostic Index (IPI) for NHL** [13]
> - IPI estimates the prognosis in intermediate- to high-grade NHL.
> - It consists of five elements:
>   - Age: ≥60
>   - Serum LDH level above normal
>   - ECOG performance status ≥2
>   - Stage III–IV
>   - Number of extranodal disease sites >1
> - One point is given for each of the above characteristics present in the patient.
> - Five-year OS: IPI 0–1, 73%; IPI 2, 51%; IPI 3, 43%; IPI, 4–5 26%

Follicular Lymphoma International Prognostic Index-2 [14]
- It consists of 10 elements:
  - Beta-2 microglobulin >upper limit of normal
  - Bone marrow involvement
  - Nodes >6 cm in greatest diameter
  - Number of involved nodal and extra nodal sites
  - B-symptoms
  - Age >60 years
  - Stage III/IV
  - Hemoglobin level <12 g/dL
  - Number of nodal areas (>4)
  - Elevated LDH
- Five-year OS for low-, intermediate-, and high-risk patients was 98%, 88%, and 77%, respectively.

Mantle Cell Lymphoma International Prognostic Index [15]
- For advanced-stage mantle cell lymphoma
- It consists of four elements:
  - Age (<50 = 0, 50–59 = 1, 60–69 = 2, ≥70 = 3)
  - Preformance status (ECOG ≥2 = 2)
  - LDH (<0.67 = 0, 0.67–0.99 = 1, 1–1.49 = 2, ≥1.5 = 3)
  - Leukocyte count (<6.7 = 0, 6.7–9.9 = 1, 10–14.9 = 2, ≥15 = 3)
- Five-year OS for low- (0–3), intermediate- (4–5), and high- (6–11) risk patients was 70%, 45%, and 10%, respectively.

13.3.3 Treatment Algorithm

Treatment Algorithm for NHL [1, 8]
Low-Grade B Cell Lymphoma
Limited stage
ISRT (24–30 Gy at 1.8–2 Gy/fx)
Gastric MALT: 30 Gy/15 fx

Median OS → 10–15 years Ten-year DFS → 40–50%
Transformation rate to diffuse B cell lymphoma → 10–15%
Advanced stage
Asymptomatic → surveillance
Symptomatic → rituximab ± chemotherapy (CHOP, CVP, or bendamustine), radioimmunotherapy, or palliative local RT (24–30 Gy/12–15 fx or 1 × 4 Gy or 2 × 2 Gy)
Median survival 8–9 years
Recurrence
High-dose CT + stem cell transplantation, or radioimmunotherapy, or rituximab ± chemotherapy
Transformed disease
Treated as intermediate-grade NHL
Radioimmunotherapy

Radioimmunotherapy → The application of monoclonal antibodies (MAb) conjugated with radioisotopes.

- *Drugs used:* I-131-anti CD 20 MAb (tositumomab; Bexxar®) and yttrium-90 anti-CD 20 MAb (ibritumomab; Zevalin®).

Moderate-Grade B Cell Lymphoma [2, 15]
Limited stage (30%)
IPI 0–1 → three cycles of R-CHOP + ISRT (30–36 Gy), or
six cycles of R-CHOP
IPI ≥2 → six cycles of R-CHOP ± ISRT (30–36 Gy), or
three cycles of R-CHOP + ISRT (30–36 Gy)
Advanced stage (70%)
6–8 cycles of R-CHOP
Consider ISRT to initially bulky disease
Recurrence
Second-line chemotherapy ± high-dose CT + stem cell transplantation

If not a candidate for further chemotherapy, consider RT alone (40–55 Gy)
Palliation
Solitary recurrence → RT (20 Gy in 10 fractions or 4 Gy in 2 fractions) Diffuse recurrence → CT (rituximab, etoposide, etc.)
Nasal NK/T cell lymphoma → RT is recommended as primary therapy (RT dose >50 Gy); consolidation CT is used for stage IIE and IPI >2
High-Grade B Cell Lymphoma [2, 15]
Stage I–II
Chemotherapy ± RT (bulky disease, residual disease)
Stage III–IV
Chemotherapy
Gastric MALT-Type Extranodal Marginal Zone B Cell Lymphoma (MALTOMA)
Stage I–II
For *H. pylori*-positive patients, 2 weeks of proton pump inhibitor + bismuth salicylate + tetracycline + metronidazole
Complete response → 97–99%
Stage III–IV
Induction chemoimmunotherapy or ISRT for symptoms
Recurrence or resistance to antibiotics
t(11:18) is a predictor for lack of response to antibiotic therapy, and these patients should be considered for RT. If disease persists despite antibiotic therapy or *H. pylori* negative, RT (30 Gy in 20 fx) to entire stomach and perigastric nodes should be considered. If RT is contraindicated, rituximab may be considered.

13.3.4 Radiotherapy

RT planning in NHL is similar to that in HL.

13.3.5 Selected Publications

UK multicenter trial (Lowry et al (2011) **Reduced dose radiotherapy for local control in non-Hodgkin lymphoma: a randomised phase III trial. Radiother Oncol 100(1):86–92)** → 361 sites of indolent lymphoma and 640 sites of aggressive lymphoma randomized to different RT doses [40–45 Gy in 20–23 fx. Vs. 24 Gy in 12 fx. (indolent) vs. 30 Gy in 15 fx. (aggressive)].

- For indolent group, no difference in local control at 5 years (79% high dose vs. 76% low dose).
- For aggressive group, no difference in local control at 5 years (84% high dose vs. 82% low dose).
- No significant difference in PFS and OS at 5 years for both indolent and aggressive lymphoma.

MD Anderson, 2003 → Low-dose IFRT (4 Gy) was found to be a valuable option in the management of relapsed disease in both indolent and aggressive lymphoma, and should be considered to palliate symptoms in patients with recurrent and/or chemotherapy refractory disease.

13.3.5.1 Low-Grade Lymphoma

JAROG; 2007 (Isobe K (2007) **A multicenter phase II study of local radiation therapy for stage IEA mucosa-associated lymphoid tissue lymphomas: a preliminary report from the Japan Radiation Oncology Group (JAROG). Int J Radiat Oncol Biol Phys 69(4):1181–1186)** → 37 patients with MALT lymphoma. The median tumor dose was 30.6 Gy (range, 30.6–39.6 Gy). The median follow-up was 37.3 months.

- Moderate-dose RT was highly effective in achieving local control with acceptable morbidity in 37 patients with MALT lymphoma.

British Columbia (Campbell et al (2010) **Long-term outcomes for patients with limited stage follicular lymphoma: involved regional radiotherapy versus involved node radiotherapy. Cancer 116(16):3797–3806)** → 237 patients with stage I–II follicular lymphoma treated with RT alone. INRT vs. involved regional radiotherapy (obtained using INRT with up to 5 cm margins)

- Ten-year OS and PFS was 66% and 49%, respectively.
- No difference in OS and PFS between INRT and involved regional RT groups.

***UK FORT* (Hoskin et al (2014) 4 Gy versus 24 Gy radiotherapy for patients with indolent lymphoma (FORT): a randomised phase 3 non-inferiority trial. Lancet Oncol 15(4):457–463)** → Patients with follicular or marginal zone lymphoma. 614 sites randomized to receive 4 Gy in 2 fx. vs. 24 Gy in 12 fx.

- Higher response rate with 24 Gy (overall response 91% vs. 81%; CR 68% vs. 49%)
- Shorter time to progression with 4 Gy
- No difference in survival

***NCDB, 2015* (Vargo et al (2015) What is the optimal management of early-stage low-grade follicular lymphoma in the modern era? Cancer 121(18):3325–3334)** → 35,961 patients with follicular lymphoma were retrospectively evaluated.

- Patients who received RT had improved 5- and 10-year OS vs. those who did not (86%/68% vs. 74%/54%).

13.3.5.2 Limited Stage Intermediate-Grade Lymphoma

***British National Lymphoma Investigation, 2004* (Spicer J (2004) Long-term follow-up of patients treated with radiotherapy alone for early-stage histologically aggressive non-Hodgkin's lymphoma. Br J Cancer 90(6): 1151–1115)** → 377 adults treated with RT alone for early-stage diffuse large cell lymphoma.

- Ten-year cause-specific survival in patients older than 60 years was poor and significantly inferior to that in younger patients (47 and 75%, respectively; $p < 0.001$).
- Short-course chemotherapy, with or without RT, is superior to RT alone in early-stage aggressive NHL in elderly as well as in younger patients.
- Increased age alone should not exclude patients from systemic treatment for early-stage aggressive NHL.

***SEER Data, 2008* (Ballonoff A et al (2008) Outcomes and effect of radiotherapy in patients with stage I or II diffuse large B-cell lymphoma: a surveillance, epidemiology, and end results analysis. Int J Radiat Oncol Biol Phys 72(5):1465–1471)** → 13,420 patients with localized diffuse large B cell lymphoma (DLBCL). The Surveillance, Epidemiology, and End Results database was queried for all patients diagnosed with stage I, IE, II, or IIE DLBCL between 1988 and 2004. 5,547 (41%) had received RT and 7,873 (59%) had not.

- RT was associated with a significant DSS (hazard ratio, 0.82, $p < 0.0001$) and OS benefit that persisted during the 15 years of follow-up.
- Elderly patients, defined as either >60 or >70 years old, had significantly improved DSS and OS associated with RT.
- On multivariate analysis, RT was significantly associated with increased DSS and OS.
- This analysis presents the largest detailed dataset of stage I–II DLBCL patients. RT is associated with a survival advantage in patients with localized DLBCL, a benefit that extends to elderly patients.

***Italy, 2007* (Mazzarotto R (2007) Primary mediastinal large B-cell lymphoma: results of intensive chemotherapy regimens (MACOP-B/VACOP-B) plus involved field radiotherapy on 53 patients. A single institution experience. Int J Radiat Oncol Biol Phys 68(3): 823–829)** → 53 patients with primary mediastinal large B cell lymphoma. Planned treatment consisted of induction chemotherapy [I-CT; prednisone, methotrexate, doxorubicin, cyclophosphamide, etoposide–mechlorethamine, vincristine, procarbazine, prednisone (ProMACE-MOPP) in the first two patients, MACOP-B in the next 11, and VACOP-B in the last 40] followed by IFRT. The median follow-up was 93.9 months.

- The response rates after I-CT were complete response (CR) in 20 (37.73%) and partial response (PR) in 30 (56.60%); three patients (5.66%) were considered nonresponders.

  Among patients in PR after chemotherapy, 92% showed a CR after IFRT.
- IFRT had a pivotal role in inducing CR in patients in PR after chemotherapy.

GELA LNH93-1, 2005 (Reyes F (2005) ACVBP versus CHOP plus radiotherapy for localized aggressive lymphoma. N Engl J Med 352(12):1197–1205) → Patients less than 61 years old with localized stage I or II aggressive lymphoma and no adverse prognostic factors according to the International Prognostic Index were randomly assigned to three cycles of cyclophosphamide, doxorubicin, vincristine, and prednisone (CHOP) plus involved-field radiotherapy or chemotherapy alone with dose-intensified doxorubicin, cyclophosphamide, vindesine, bleomycin, and prednisone (ACVBP) plus sequential consolidation.

- Event-free and overall survival rates were significantly higher in the group given chemotherapy alone than in the group given CHOP plus radiotherapy ($p < 0.001$ and $p = 0.001$, respectively).
- In patients under 61 years of age, chemotherapy with three cycles of ACVBP followed by sequential consolidation is superior to three cycles of CHOP plus radiotherapy for the treatment of low-risk localized lymphoma.

GELA LNH93-4, 2007 (Bonnet C (2007) CHOP alone compared with CHOP plus radiotherapy for localized aggressive lymphoma in elderly patients: a study by the Groupe d'Etude des Lymphomes de l'Adulte. J Clin Oncol 25(7):787–792) → Patients older than 60 years with localized stage I or II histologically aggressive lymphoma and no adverse prognostic factors of the International Prognostic Index were randomly assigned to receive either four cycles of cyclophosphamide, doxorubicin, vincristine, and prednisone (CHOP) plus involved-field radiotherapy or chemotherapy alone with four cycles of CHOP.

- Event-free and overall survival did not differ between the two treatment groups ($p = 0.6$).
- CHOP plus radiotherapy did not provide any advantage over CHOP alone for the treatment of low-risk localized aggressive lymphoma in elderly patients.

SWOG 8736, 1998 (Miller TP (1998) Chemotherapy alone compared with chemotherapy plus radiotherapy for localized intermediate- and high-grade non-Hodgkin's lymphoma. N Engl J Med 339(1):21–26) → 200 patients were randomly assigned to receive 3c × CHOP plus IFRT (50–50 Gy), and 201 received 8c × CHOP alone.

- Patients treated with three cycles of CHOP plus radiotherapy had significantly better progression-free survival ($p = 0.03$) and OS ($p = 0.02$) than patients treated with CHOP alone.
- Three cycles of CHOP followed by involved-field radiotherapy are superior to eight cycles of CHOP alone for the treatment of localized intermediate- and high-grade non-Hodgkin's lymphoma.

Updated in 2016 (Stephens et al (2016) Continued Risk of Relapse Independent of Treatment Modality in Limited-Stage Diffuse Large B-Cell Lymphoma: Final and Long-Term Analysis of Southwest Oncology Group Study S8736. J Clin Oncol 34(25):2997–3004) → 401 patients were included. The median follow-up was 17.7 years.

- 7-, 10- and 12-year OS and PFS results no longer show difference between groups.

ECOG 1484, 2004 (1984–1992) (Horning SJ (2004) Chemotherapy with or without radiotherapy in limited-stage diffuse aggressive non-Hodgkin's lymphoma: Eastern Cooperative Oncology Group Study 1484. J Clin Oncol 22(15):3032–303) → Stage I (with risk factors) and II adults with diffuse aggressive lymphoma in CR after eight cycles of cyclophosphamide, doxorubicin, vincristine, and prednisone (CHOP) were randomly assigned to 30 Gy IFRT or OBS. PR patients received 40 Gy RT.

- For patients in CR after CHOP, low-dose RT prolonged DFS and provided local control, but no survival benefit was observed.

Role of RT in Limited-Stage Intermediate-Grade Lymphoma in Rituximab Era

Lyse/Goelams Group 02-03 Trial (Lamy et al (2017) R-Chop +/-Radiotherapy In Non-Bulky Limited-Stage Diffuse Large B-Cell Lymphoma (Dlbcl): Final Results Of The Prospective Randomized Phase II 02-03 Trial From The Lysa/Goelams. Hematological Oncology. (Oral Presentations)) → 334 patients with nonbulky, limited-stage DLBCL randomized to 4–6c × R-CHOP followed by or not 40 Gy RT.

- No difference in OS and EFS between the groups.
- RT was recommended for all patients with residual PET-avid disease after four cycles of R-CHOP.

NCDB, 2015 (Vargo et al (2015) Treatment Selection and Survival Outcomes in Early-Stage Diffuse Large B-Cell Lymphoma: Do We Still Need Consolidative Radiotherapy? J Clin Oncol 33(32):3710–3717) → 59,255 patients with nonbulky, limited-stage DLBCL were evaluated retrospectively. Median follow-up 60 months.

- Addition of RT improved 5- and 10-year OS (82%/64% vs. 75%/55%).

SEER-Medicare database, 2015 (Odejide (2015) Limited stage diffuse large B-cell lymphoma: comparative effectiveness of treatment strategies in a large cohort of elderly patients. Leuk Lymphoma 56(3):716–724) → 874 patients with limited-stage DLBCL were evaluated retrospectively. 359 received abbreviated R-CHOP with radiation, and 515 received a full course of R-CHOP.

- Similar OS results in both groups
- Lower risk of second-line therapy and febrile neutropenia with abbreviated R-CHOP and RT than 6–8 cycles of R-CHOP

13.3.5.3 Advanced-Stage Intermediate-Grade Lymphoma

MiNT (Pfreundschuh et al (2006) CHOP-like chemotherapy plus rituximab versus CHOP-like chemotherapy alone in young patients with good-prognosis diffuse large-B-cell lymphoma: a randomised controlled trial by the MabThera International Trial (MInT) Group. Lancet Oncol 7(5):379–391) → 824 patients <60 years with stage II–IV, good prognosis DLBCL randomized to 6c × CHOP-like or 6c × CHOP-like + rituximab.

- Addition of rituximab improved 3-year EFS (79% vs. 59%) and 3-year OS (93% vs. 84%)

Updated in 2011 (Pfreundschuh et al (2011) CHOP-like chemotherapy with or without rituximab in young patients with good-prognosis diffuse large-B-cell lymphoma: 6-year results of an open-label randomised study of the MabThera International Trial (MInT) Group. Lancet Oncol 12(11):1013–1022.

- Addition of rituximab improved 6-year EFS (74.3% vs. 55.8%) and 6-year OS (90.1% vs. 80%)

RICOVER-60 **(Pfreundschuh et al (2008) Six versus eight cycles of bi-weekly CHOP-14 with or without rituximab in elderly patients with aggressive CD20+ B-cell lymphomas: a randomised controlled trial (RICOVER-60). Lancet Oncol 9(2):105–116)** → 1222 patients 61–80 years with stage I–IV DLBCL randomized to 6 vs. 8 cycles of CHOP-14 ± rituximab. Patients with bulky disease or extranodal involvement received 36 Gy RT.

- Addition of rituximab to 6c × CHOP improved 3-year EFS (66% vs. 47%) and OS (78% vs. 68%).
- No benefit of two more cycles to six cycles of R-CHOP was observed.

Updated in 2014 (Held et al (2014) Role of radiotherapy to bulky disease in elderly patients with aggressive B-cell lymphoma. J Clin Oncol 32:1112–1118)

- Addition of RT for bulky disease or extranodal involvement improved 3-year EFS (80% vs. 54%), PFS (88% vs. 62%), and OS (90% vs. 65%).

UNFOLDER (Pfreundschuh et al (2018) Radiotherapy (RT) to bulky (B) and extra-lymphatic (E) disease in combination with 6xR-CHOP-14 or R-CHOP-21 in young good-prognosis DLBCL patients: Results of the 2x2 randomized UNFOLDER trial of the DSHNHL/GLA. J Clin Oncol (Abstract) → 285 patients with good prognosis DLBCL randomized to R-CHOP-21 or R-CHOP-14 with or without RT.

- RT improved 3-year EFS (84% vs. 68%)
- No difference in 3-year OS (93% vs. 93%) and 3-year PFS (89% vs. 81%) between groups with and without RT
- No difference in outcome between R-CHOP-14 and R-CHOP-21

13.4 Cutaneous Lymphoma

Cutaneous lymphoma is a heterogeneous group of NHLs: 30% B cell, 70% T cell. They are rarely seen, with an incidence of 1–1.5/100,000. Cutaneous lymphomas are usually observed in older patients and are twice as common in males. They are usually presented with dermal lesions, and the period between symptoms and diagnosis is usually 5 years [16]. The finding of atypical T lymphocytes, termed Sézary cells, in peripheral blood is pathognomonic. Cutaneous lymphomas constitute 2% of all NHLs.

- Sézary syndrome = erythroderma + Sézary cells in peripheral blood
- Primary cutaneous lymphomas were classified by WHO-EORTC in 2018

Cutaneous T Cell Lymphoma (70%) [17]
Mycosis fungoides (MF) (39%) MF variants

- Folliculotropic MF (5%)
- Pagetoid reticulosis (<1%)
- Granulomatous slack skin (<1%)
- Sézary syndrome (2%)
- Adult T-cell leukemia/lymphoma (<1%)

Primary cutaneous CD30 (+) LPDs

- C-ALCL (8%)
- LyP (12%)
- Subcateous panniculitis-like T cell lymphoma (1%)
- Extranodal NK/T cel lymphoma, nasal type (<1%)
- Chronic active EBV infection (<1%)

Primary cutaneous peripheral T-cell lymphoma, rare subtypes

- Primary cutaneous γ/δ T-cell lymphoma (<1%)
- CD8+AECTCL (<1%)
- Primary cutenous CD4+ small/medium T-cell lymphoproliferative disorder (6%)
- Primary cutenous acral CD8+ T-cell lymphoma
- Primary cutaneous peripheral T-cell lymphoma, NOS

Cutaneous B Cell Lymphoma (30%) [17]
- Primary cutaneous marginal zone lymphoma (9%)
- Primary cutaneous follicular center lymphoma (12%)
- Primary cutaneous diffuse large B-cell, leg type (4%)
- EBV+ mucocutaneous ulcer (<1%)
- Intravascular large B-cell lymphoma (<1%)

AJCC Staging System 8th Edition, 2017 [18]
Primary Cutaneous B-Cell/T-Cell Lymphoma (Non-MF/SS)
T stage
T1: Solitary skin involvement
T1a: Solitary lesion <5 cm
T1b: Solitary lesion ≥5 cm
T2: Regional skin involvement: multiple lesions limited to one body region or two contigious body regions
T2a: All diseases encompassing in a <15 cm circular area
T2b: All diseases encompassing in a ≥15 cm and <30 cm circular area
T2c: All diseases encompassing in a >30 cm circular area
T3: Generalized skin involvement
N Stage
NX: Regional lymph nodes cannot be assessed
N0: No clinical or pathological lymph node involvement
N1: Involvement of one peripheral node region that drains an area of current or prior skin involvement
N2: Involvement of two or more peripheral node regions or involvement of any lymph node region that does not drain an area of current or prior skin involvement
N3: Involvement of central nodes
M Stage
M0: No evidence of extracutaneous non-lymph node disease
M1: Extracutaneous non-lymph node disease present

Mycosis Fungoides and Sézary Syndrome
T Stage
T1: Limited patches, papules, and/or plaques covering <10% of the skin surface
T1a: Patch only
T1b: Plaque ± patch
T2: Patches, papules, or plaques covering ≥10% of the skin surface
T2a: Patch only
T2b: Plaque ± patch
T3: One or more tumors
T4: Confluence of erythema covering ≥80% of body surface area
N Stage
NX: Clinically abnormal peripheral lymph nodes; no histological confirmation
N0: No clinically abnormal peripheral lymph nodes; biopsy not required
N1: Clinically abnormal peripheral lymph nodes; histopathologically Dutch grade 1 or National Cancer Institute (NCI) LN0-2
N1a: Clone negative
N1b: Clone positive
N2: Clinically abnormal peripheral lymph nodes; histopathologically Dutch grade 2 or National Cancer Institute (NCI) L3
N2a: Clone negative
N2b: Clone positive
N3: Clinically abnormal peripheral lymph nodes; histopathologically Dutch grade 3–4 or National Cancer Institute (NCI)
LN4; clone positive or negative
Continued...
M Stage
M0: No visceral organ involvement
M1: Visceral involvement (must have pathology confirmation, and organ involved should be specified)
Peripheral Blood Involvement (B)
B0: Absence of significant blood involvement: ≥5% of peripheral blood lymphocytes are atypical (Sézary) cells
B0a: Clone negative

B0b: Clone positive

B1: Low blood tumor burden: >5% of peripheral blood lymphocytes are atypical (Sézary) cells, but do not meet the criteriaof B2

B1a: Clone negative

B1b: Clone positive

B2: High blood tumor burden: ≥ 1,000/ μL Sézary cells with positive clone

13.4.1 Treatment Algorithm [1]

Cutaneous Marginal Zone and Follicular Center Lymphomas

T1–2 disease: Local RT (24–30 Gy with electron) and/or excision; in selected cases, observation or topical agents

T3 generalized disease: Observation or rituximab or RT for palliation (4 Gy in 2 fx or 30 Gy in 10–15 fx) or systemic therapy

Cutaneous Diffuse Large B-Cell, Leg Type

Solitary disease: R-CHOP followed by local RT (36–40 Gy with electron) or RT ± rituximab

Cutaneous Anaplastic Large-Cell Lymphoma

Localized disease: Similar with cutaneous marginal zone lymphoma

Multifocal disease: Methotrexate, systemic retinoids, pralatrexate, brentuximab, or observation

Subcutaneous Panniculitis-Like T-Cell Lymphoma

Solitary lesions: >40 Gy RT with electron

Cutaneous NK/T-Cell Lymphoma, Nasal Type

Localized disease: 50 Gy with 5–10 Gy boost for residual disease

Lymphomatoid Papulosis: T Cell

Observation or PUVA, methotrexate, interferon, topical steroids, and topical bexarotene for palliation

Mycosis Fungoides

T1 disease: Topical steroids, imiquimod, retinoids, chemotherapy, PUVA, or local RT (20–30 Gy with >2 cm margin)

T2 disease: Same local treatment options as above, or total skin irradiation (TSI) (10–36 Gy with electron)

T3 disease:

If limited; Local RT (8–12 Gy for small lesions, 20–30 Gy for larger or thicker lesions) or systemic treatment ± local therapies or TSI if refractory

If generalized: TSI or systemic treatment ± local therapies

T4 disease:

Skin-directed therapy if no blood involvement

Systemic therapy if blood involvement

Sézary syndrome: Systemic therapise and RT for local control

RT in local therapies, 24–36 Gy with 6–9 MeV electron energies (with 0.5 cm bolus). Palliative RT: 4 Gy in 2 fx or 8 Gy in 2 fx or 7–8 Gy in 1 fx or 12 Gy in 3–4 fx or 20–30 Gy in 10 fx.

TSI: 12–36 Gy with 1.5–2 Gy delivered per 2-day cycle, 4 days per week with 2–7 MeV electron energies (usually 6 MeV). In cases of prominent skin edema, TSI is stopped for 1 week.

13.4.2 Total Skin Irradiation (TSI)

The aim of TSI is the homogeneous irradiation of all of the skin on the body. Plaques of cutaneous lymphomas that can be found in skin locally or extensively are localized in the first 5 mm of skin depth. Therefore, electron energies should be used. In addition, palpable or visible tumors (nodules) are irradiated with a boost dose.

TSI Is Applied Using Two Main Techniques
- The treatment couch is moved manually or automatically while the patient is in the supine position.
- Three to six meters are left between the therapy machine and the patient (the Stanford 6 technique).

Stanford 6 Technique (Fig. 13.4) [20]
The body is divided into six treatment fields:
  Anterior, posterior, two posterior oblique, and two anterior oblique fields.
  On odd days (first day, third day, ...) during the TSI course, the anterior, left posterior oblique, and right posterior oblique fields are irradiated.
  On even days (second day, fourth day, ...) during the TSI course, the posterior, left anterior oblique, and right anterior oblique fields are irradiated.

- A dose of 2 Gy is provided to the entirety of skin during one cycle.
- The entire body is therefore covered every 2 days (one cycle = 2 days of TSI).
- TSI is applied 4 days each week, with a duration of 30 min for each fraction.
- SSD = 320 cm, gantry angle: 15–18°.
- The patient is treated on a special formica frame.
- Skin doses are measured with TLDs in all therapy fields or in the phantom prior to radiotherapy.
- A transparent plaque is placed approximately 30 cm away from the patient skin to block scattered electrons.

13.4.3 Selected Publications

13.4.3.1 Multifocal Cutaneous Lymphoma

Netherland, 1999 (1985–1997) (Bekkenk MW et al (1999) **Treatment of multifocal primary cutaneous B-cell lymphoma: a clinical follow-up study of 29 patients. J Clin Oncol 17(8):2471–2478**) → They evaluated the clinical behavior of and results of treatment for multifocal CBCL in 29 patients and formulated therapeutic guidelines. The study group included 16 patients with primary cutaneous follicular center cell lymphoma (PCFCCL), 8 with primary cutaneous immunocytoma (PCI), and 5 with primary cutaneous large B cell lymphoma presenting on the legs (PCLBCL of the leg). Radiotherapy directed toward all skin lesions was as effective as multiagent chemotherapy. Patients with PCLBCL of the leg had a more unfavorable prognosis, particularly patients presenting with multifocal skin lesions. This last group should always be treated with multiagent chemotherapy.

13.4.3.2 Primary Cutaneous Diffuse B Cell Lymphoma

Europe, Multicentric, 2001 (1979–1998) (Grange F et al (2001) **Prognostic factors in primary cutaneous large B-cell lymphomas: a European multicenter study. J Clin Oncol 19(16):3602–3610**) → 48 patients had a PCLBCL of the leg and 97 had a PCFCCL. Data from both groups were compared. Compared with PCFCCLs, PCLBCLs of the leg were characterized by an older age of onset, a more recent history of skin lesions, a more frequent predominance of tumor cells with round nuclei and positive bcl-2 staining, and a poorer 5-year disease-specific survival rate (52 vs. 94%; $p < 0.0001$). Round-cell morphology was an adverse prognostic factor in both PCLBCLs of the leg and in PCFCCLs, whereas multiple skin lesions were

13.4 Cutaneous Lymphoma

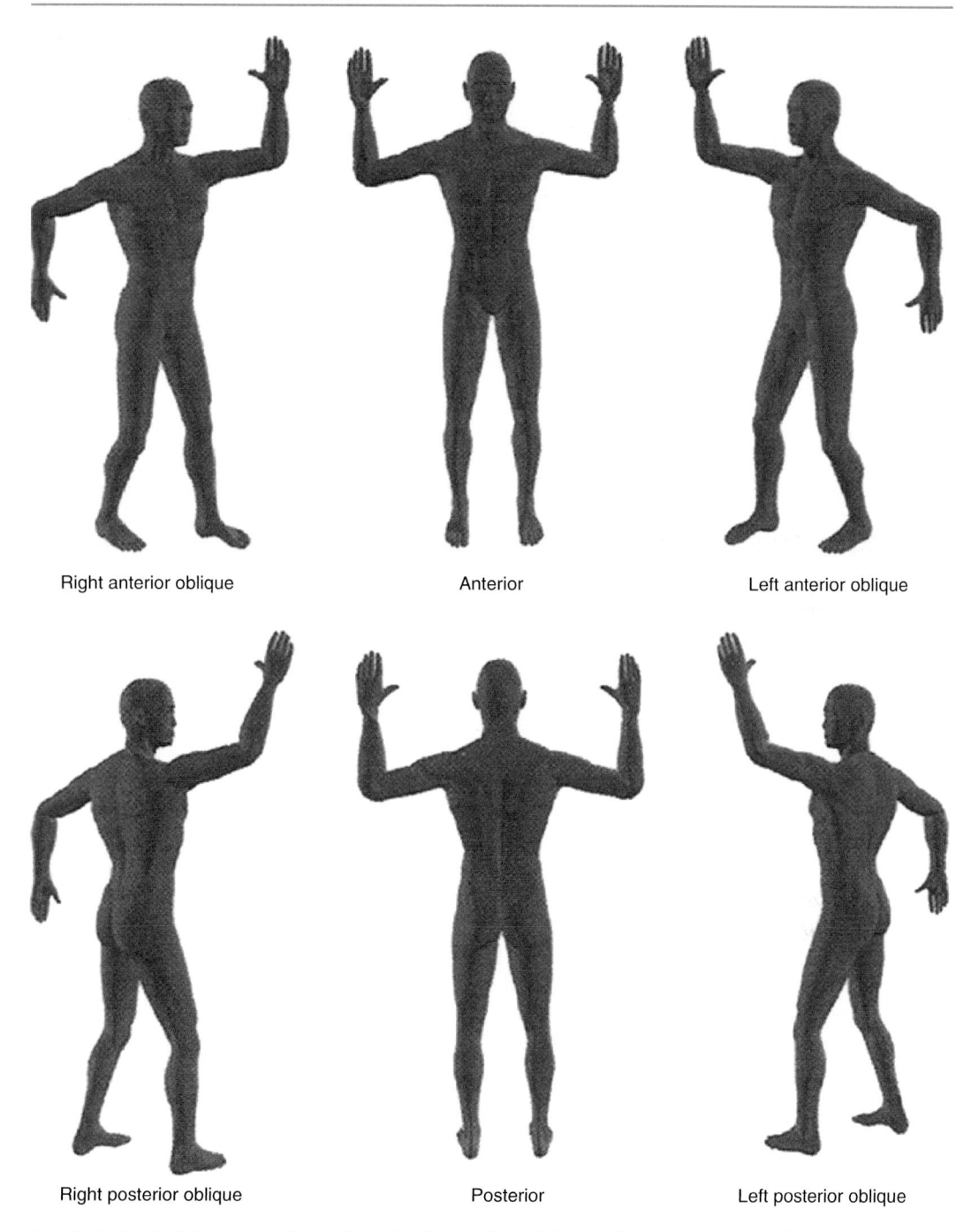

Right anterior oblique   Anterior   Left anterior oblique

Right posterior oblique   Posterior   Left posterior oblique

Fig. 13.4 Stanford 6 technique for total skin irradiation (from [19], p. 818, Fig. 33, reproduced with the permission from Springer Science and Business Media)

associated with a poor prognosis only in patients with PCLBCLs of the leg.

13.4.3.3 Primary Cutaneous T-Cell Lymphoma

Thomas et al (2013) Outcome of patients treated with a single-fraction dose of palliative radiation for cutaneous T-cell lymphoma. Int J Radiat Oncol Biol Phys 85(3):747–753. → Data of 58 patients with mycosis fungoides and treated with a single fraction of palliative radiotherapy were collected. The majority of patients (97%) were treated with ≥7 Gy.

- Complete response and partial response was observed in 94% and 4% of patients, respectively.
- A single fraction of 700–800 cGy provides excellent palliation for CTCL lesions and is cost effective and convenient for the patient.

RT + CT in Cutaneous Lymphoma
MD Anderson, 2001 **(Sarris AH et al (2001) Primary cutaneous non-Hodgkin's lymphoma of Ann Arbor stage I: preferential cutaneous relapses but high cure rate with doxorubicin-based therapy. Clin Oncol 19(2):398–405)** → 46 patients, 27 males, with median age of 57 years. Treatment was radiotherapy in 10 patients, doxorubicin-based therapy in 33 patients that was followed by radiotherapy in 25 patients, and another combination with radiotherapy in 1 patient. The complete response rate was 95%. PCNHL is rare, and its first relapse is exclusively cutaneous in 50% of patients. Patients with diffuse large B cell lymphomas are curable with doxorubicin-based regimens but not with radiotherapy.

13.4.3.4 Cutaneous Lymphoma Staging (Review)

EORTC/ISCL, 2007 **(Olsen E et al (2007) Revisions to the staging and classification of mycosis fungoides and Sézary syndrome: a proposal of the International Society for Cutaneous Lymphomas (ISCL) and the cutaneous lymphoma task force of the European Organization of Research and Treatment of Cancer (EORTC). Blood 110(6):1713–1722)** → These revisions were made to incorporate advances related to tumor cell biology and diagnostic techniques as pertains to mycosis fungoides (MF) and Sézary syndrome (SS) since the 1979 publication of the original guidelines, to clarify certain variables that currently impede effective interinstitution and interinvestigator communication and/or the development of standardized clinical trials in MF and SS, and to provide a platform for tracking other variables of potential prognostic significance.

13.4.3.5 Total Skin Irradiation

EORTC, 2002 **(Jones GW (2002) Total skin electron radiation in the management of mycosis fungoides: consensus of the European Organization for Research and Treatment of Cancer (EORTC) Cutaneous Lymphoma Project Group. Am Acad Dermatol 47(3):364–370)** → The European Organization for Research and Treatment of Cancer Cutaneous Lymphoma Project Group, in association with experts from radiotherapy centers in North America, reached a consensus on acceptable methods and clinical indications for TSEB in the treatment of MF.

Stanford, 2015 **(Hoppe et al (2015) Low-dose total skin electron beam therapy as an effective modality to reduce disease burden in patients with mycosis fungoides: results of a pooled analysis from 3 phase-II clinical trials. J Am Acad Dermatol 72(2):286–292.)** → Data of 33 patients with mycosis fungoides from three clinical trial using low dose (12 Gy) were pooled.

- Overall response rate was 88%.
- Median time to response was 7.6 weeks.
- Toxicities were mild.
- Low-dose TSEBT provides reliable and rapid reduction of disease burden in patients with mycosis fungoides.

13.5 Total Body Irradiation (TBI)

TBI is used for a conditioning regimen prior to stem cell transplantation [21].

13.5 Total Body Irradiation (TBI)

Diseases Requiring Stem Cell Transplantation
- Acute myeloid leukemia (AML) Acute lymphoblastic leukemia Chronic myeloid leukemia (CML)
- Chronic lymphocytic leukemia
- Myelodysplasia
- Lymphoma
- Multiple myeloma
- Aplastic anemia
- Idiopathic
- Fanconi
- Paroxysmal nocturnal hemoglobinuria
- Congenital immune deficiencies
- Autoimmune diseases
- Rheuomatoidarthiritis
- SLE
- Osteopetrosis
- Leukoencephalopathy
- Hurler syndrome
- Sickle cell anemia
- Thalassemia

Advantages of TBI Compared to Other Conditioning Regimens [22]
- TBI affects sanctuary organs (testis, brain).
- TBI does not cause a requirement for blood transfusion. TBI is independent of hepatic and renal functions.
- TBI does not cause cross-resistance with other agents.
- TBI does not require metabolization, unlike chemotherapeutic agents.
- Tumor cells are more sensitive to radiation since transplantation cases usually do not have a previous history of radiotherapy.
- The cyclophosphamide + TBI conditioning regimen has a reduced toxicity profile compared to the busulfan + cyclophosphamide regimen.
- **Disadvantages of Stem Cell Transplantation:**
- Potential late side effects (sterility, cataract, growth retardation, neurological toxicity).

Conditioning Regimens for Stem Cell Transplantation [22]

- Standard intensity regimens for leukemia

  Cy/TBI (cyclophosphamide + TBI) Bu/Cy (busulfan + cyclophosphamide)

- Standard intensity regimens for lymphoma

  BEAM (BCNU/etoposide/ARA-C/melphalan) CBV (BCNU/etoposide/cyclophosphamide)

- Standard intensity regimens for multiple myeloma

  High-dose melphalan

TBI is used for bone marrow ablation, immune system suppression, and eradication of genetically damaged cells prior to stem cell transplantation.

- TBI is usually applied after immunosuppressive chemotherapy. Cyclophosphamide is the most frequently used chemotherapeutic agent for a fractionated TBI regimen. Etoposide use is rapidly increasing in leukemia.
- TBI is now commonly applied with linear accelerators, 1–6 MV photon energies with a single fraction (8 Gy) or in fractionated doses. Immunosuppression and tumor cell eradication requires 8–12 Gy radiation doses.
- The thickness of the patient along the beam axis is the dominant factor dictating the choice of beam energy; typically, lateral field arrangements require higher beam energies than AP/PA field arrangements; however, this ultimately depends on patient size, beam spoilers, and source–axis distance.

Hydration and sedation should be done 2 h before fractionated TBI and 12 h before single-dose TBI (steroid and phenobarbital).

TBI Technique (Fig. 13.5) [22]
- Maximum field size for a SSD of 100 cm in most therapy machines is 40 × 40 cm. However, a field size of 120 × 120 can be achieved with an SSD of 300 cm due to divergence.
- The AP/PA TBI technique is now employed most frequently.
- Patients can be irradiated in the prone (pediatric), supine, or leaning positions.

A compensator is required for the lungs, head–neck, and the lower extremities in the leaning position (Fig. 13.6).
TBI in the standing position: lung blocks are required (Fig. 13.6).

- Lung is the dose-limiting organ (maximum 10 Gy).
- Newer techniques such as IMRT with TomoTherapy or VMAT appear dosimetrically advantageous because they allow targeting and radiation dose escalation in the bone marrow. IMRT can be delivered as beams spiral down the axis of a patient supine on the treatment couch over a maximum couch travel length of 172.5 cm. The beams can be planned to deliver dose to the bones and bone marrow, liver, and spleen as well as major nodal groups and to relatively spare the lungs and kidneys. However, sparing the circulating hematopoietic cells may defeat the purpose of delivering radiation to the entire body (Fig. 13.7).

- In anterior–posterior TBI, bone marrow within costae under lung blocks is irradiated with the electron boost (6 Gy at D_{max}) [23].
- An electron boost is also given to the testicles in all males (4 Gy at D_{max}, 90% isodose covers the posterior surface of the scrotum) [23].
- TLD measurements should be done in the phantom and in the patient prior to the first fraction.
- After TBI, transplanted stem cells start proliferating at 2–3 weeks.

TBI Doses in the Standing Position
First day (first, second, and third fractions): 1.65 Gy/fraction with 6 MV, and 3 Gy electron boost to the testicles at the first fraction.
Second day (fourth, fifth, and sixth fractions): 1.65 Gy/fraction with 6 MV.
Third day (seventh, eighth, and ninth fractions): 1.65 Gy/fraction with 6 MV, and 3 Gy electron boost to the chest wall at the seventh fraction.
Fourth day (tenth and eleventh fractions): 1.65 Gy/fraction with 6 MV, and 3 Gy electron boost to chest wall at the tenth fraction.

TBI Doses in the Leaning Position
First day (first and second fractions): 1.65 Gy/fraction with 18–25 MV. Second day (third and fourth fractions): 1.65 Gy/fraction with 18–25 MV. Third day (fifth and sixth fractions): 1.65 Gy/fraction with 18–25 MV. Fourth day (seventh and eighth fractions): 1.65 Gy/fraction with 18–25 MV.

13.5.1 Selected Publications

13.5.1.1 TBI (Randomized Trials)
Fred Hutchinson, 1994 (1988–1992) **(Clift RA (1994) Marrow transplantation for chronic myeloid leukemia: a randomized study comparing cyclophosphamide and total body irradiation with busulfan and cyclophosphamide. Blood 84(6):2036–2043)** → 69 patients received 60 mg/kg of cyclophosphamide on each of two successive days, followed by six fractions of TBI each of 2.0 Gy (CY-TBI), and 73 patients received 16 mg/kg of busulfan delivered over 4 days followed by 60 mg/kg CY on each of two successive days (BU-CY). The BU-CY regimen was better tolerated than, and associated with, survival and relapse probabilities that compare favorably with the CY-TBI regimen.

1999 Update: Clift RA (1999). Long-term follow-up of a randomized study comparing cyclophosphamide and total body irradiation with busulfan and cyclophosphamide for patients

13.5 Total Body Irradiation (TBI)

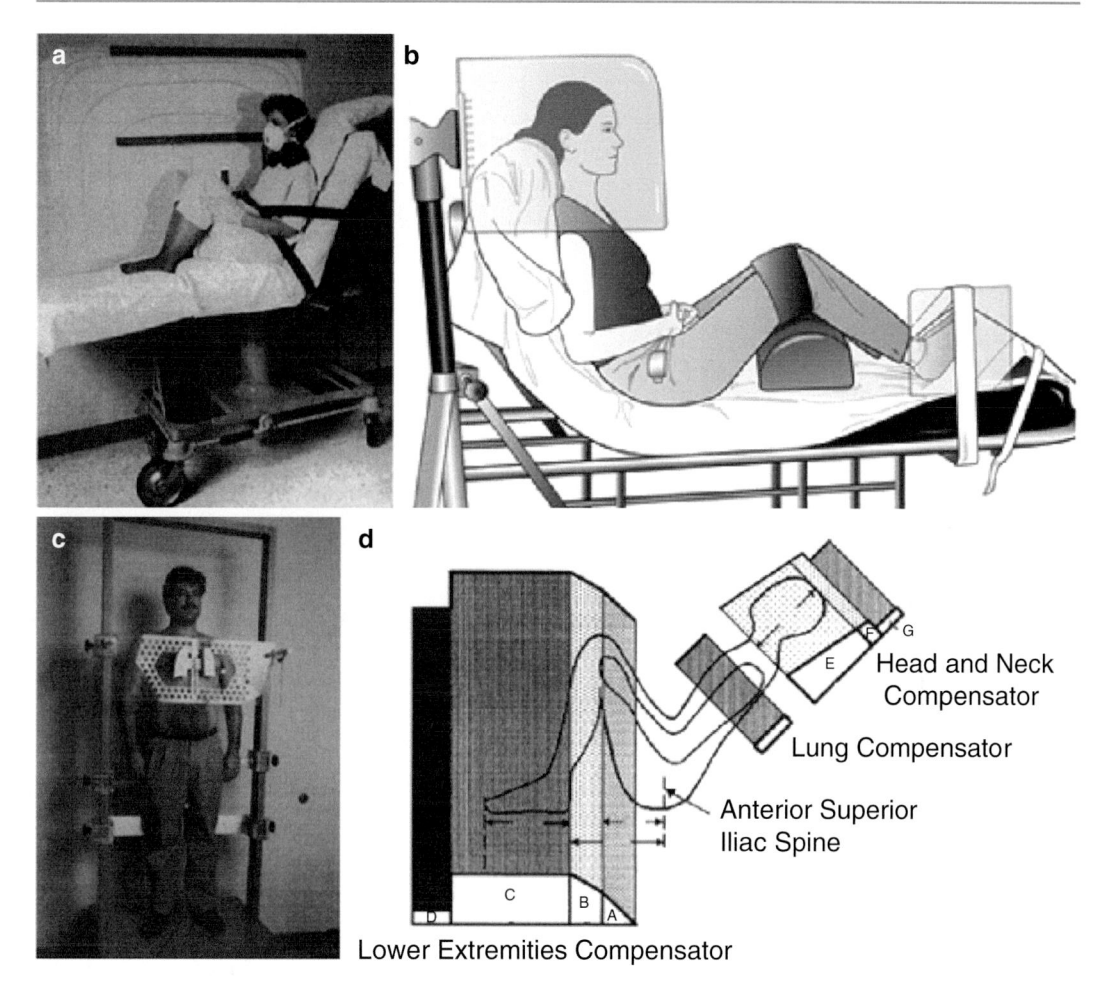

Fig. 13.5 (a–d) Patient positions in total body irradiation (from [23], pp. 1077–1098, Figs. 1, 5, and 12 reproduced with the permission from Springer Science and Business Media, with permission)

receiving allogenic marrow transplants during chronic phase of chronic myeloid leukemia. Blood 94(11):3960–3962. → Final results support original findings.

Nordic BMTG, 1994 (1988–1992) **(Ringden O (1994) A randomized trial comparing busulfan with total body irradiation as conditioning in allogeneic marrow transplant recipients with leukemia: a report from the Nordic Bone Marrow Transplantation Group. Blood 83(9):2723–2730)** → 167 patients with leukemia receiving marrow transplants from HLA-identical donors and conditioned with cyclophosphamide (120 mg/kg) were randomized to additional treatment with either busulfan (16 mg/kg, $n = 88$) or TBI (TBI; $n = 79$). Patients treated with busulfan had more early toxicity and increased transplant-related mortality in patients with advanced disease. TBI was therefore the treatment of choice, especially in adults and patients with advanced disease. However, busulfan was an acceptable alternative for patients with early disease and for those in whom TBI is not feasible.

1999 Update: Ringden O (1999). Increased risk of chronic graft-versus-host disease, obstructive bronchiolitis, and alopecia with busulfan versus total body irradiation: long-term results of a randomized trial in allogeneic marrow recipients with leukemia. Nordic Bone Marrow Transplantation Group. Blood 93(7):2196–2201

Seven-year leukemia-free survival (LFS) in patients with more advanced disease was 17% in

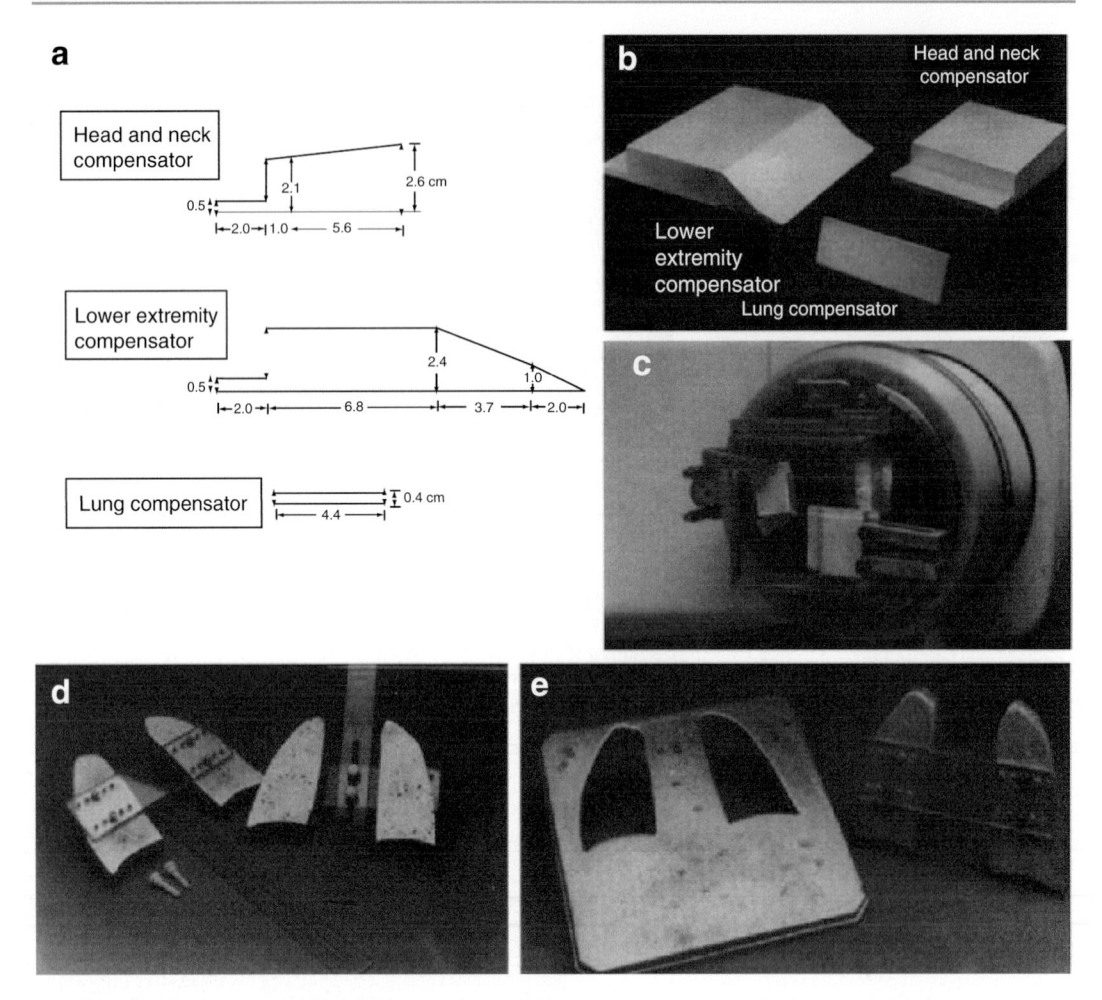

Fig. 13.6 Compensator (**a–c**) and lung blocks (**d, e**) used in TBI (from [23], pp. 1077–1098, Figs. 6, 7, and 11 reproduced with the permission from Springer Science and Business Media)

the busulfan group vs. 49% in the TBI group ($p < 0.01$). In patients with CML in the first chronic phase, 7-year LFS was 72% vs. 83% in the two groups, respectively.

GEGMO, 1991 (1987–1990) (Blaise D (1992) Allogeneic bone marrow transplantation for acute myeloid leukemia in first remission: a randomized trial of a busulfan–Cytoxan versus Cytoxan–TBI as preparative regimen: a report from the Group d'Etudes de la Greffe de Moelle Osseuse. Blood 79(10):2578–2582) → 101 patients with AML were randomized to be transplanted in first complete remission (CR1). A preparative regimen including cytoxan (120 mg/kg) with TBI (CYTBI) ($n = 50$) or busulfan (16 mg/kg) (BUSCY) ($n = 51$) was followed by allogeneic BMT from an HLA-identical sibling. Besides the antileukemic effect of preparative regimens, this trial pointed out the progress accomplished in BMT management (transplant mortality = 8% in CYTBI) over the last 20 years, as well as the effectiveness of transplant in early first CR after CYTBI (DFS = 72% at 2 years).

2001 Update: **Blaise D (2001) Long-term follow-up of a randomized trial comparing the combination of cyclophosphamide with total body irradiation or busulfan as conditioning regimen for patients receiving HLA-identical marrow grafts for acute myeloblastic leukemia in first complete remission. Blood 97(11):3669–3671**

Fig. 13.7 The isodose distribution achieved with TomoTherapy. Since this is delivered locally, an extremely conformal dose distribution can be achieved, thus delivering the dose only to the narrow spaces, while avoiding irradiation of sensitive structures (lung, kidney, liver) (from [23], pp. 1077–1089, Fig. 16)

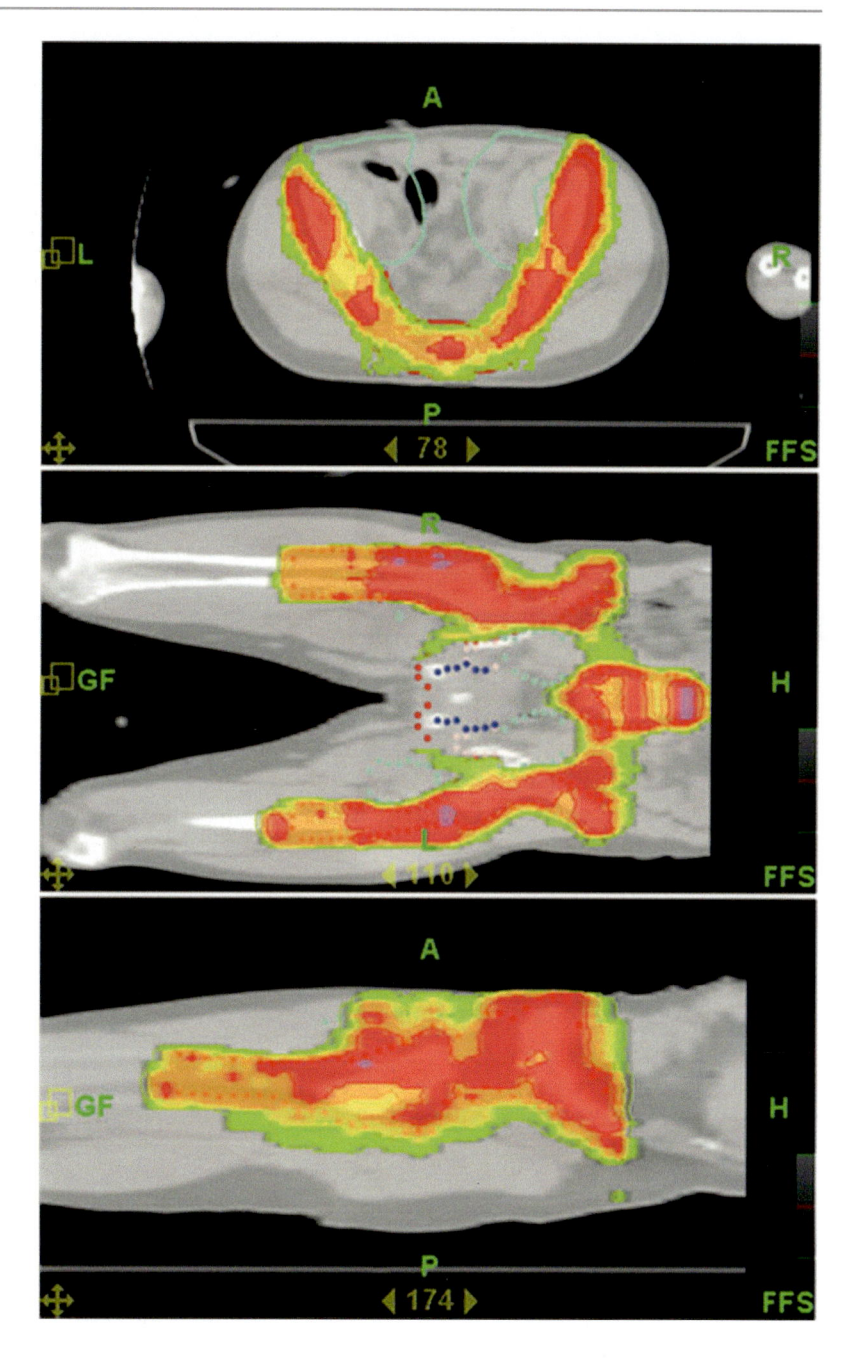

- The use of BUCY (120 mg/kg of CY) is associated with a poorer outcome than a standard CYTBI regimen.

SWOG 8612, 1993 (1987–1991) **(Blume KG (1993) A prospective randomized comparison of total body irradiation– etoposide versus busulfan–cyclophosphamide as preparatory regimens for bone marrow transplantation in patients with leukemia who were not in first remission: a Southwest Oncology Group study. Blood 81(8):2187–2193)** → Two regimens consisting either of fractionated TBI and etoposide (FTBI/VP-16) or high-dose busulfan with cyclophosphamide (BU/CY). Both regimens were well tolerated with no regimen-related deaths encountered during the 6-week period after BMT. The leading cause of treatment failure was leukemic relapse (45 of the 114 BMT recipients suffered a recurrence of their leukemia), whereas 38 patients died without evidence of relapse.

Girinsky et al (2000) **Prospective randomized comparison of single-dose versus hyperfractionated total-body irradiation in patients with hematologic malignancies. J Clin Oncol 18(5):981–986** → 160 patients with various hematological malignancies randomized to 10 Gy single-dose TBI vs. 14.85 Gy hyperfractionated TBI.

- Hyperfractionated TBI improved 8-year OS and cancer-specific survival (CSS) (45% vs. 38% and 77% vs. 63.5%, respectively).
- Similar toxicity results between two groups.

Bunin et al (2003) **Randomized trial of busulfan vs total body irradiation containing conditioning regimens for children with acute lymphoblastic leukemia: a Pediatric Blood and Marrow Transplant Consortium study. Bone Marrow Transplant 32(6):543–548.** → 43 pediatric patients with ALL undergoing allogeneic stem cell transplant randomized to busulfan (Bu) versus TBI. Conditioning included either Bu or TBI, with etoposide 40 mg/kg and cyclophosphamide 120 mg/kg.

- Three-year EFS was higher in TBI arm than Bu arm (58% vs. 29%).
- Relapses were similar in both arms.
- Bu is inferior to TBI for pediatric patients with ALL undergoing allogeneic SCT.

13.5.1.2 TBI (meta-analyses)

Socie G **(Socie G (2001) Busulfan plus cyclophosphamide compared with total-body irradiation plus cyclophosphamide before marrow transplantation for myeloid leukemia: long-term follow-up of 4 randomized studies. Blood 98(13):3569–3574** → This study analyzed the long-term outcomes of 316 patients with CML and 172 patients with AML who participated in these four trials, with a mean follow-up of more than 7 years.

Bu-CY and CY-TBI provided similar probabilities of cure for patients with CML. In patients with AML, a nonsignificant 10% lower survival rate was observed after Bu-CY. Late complications occurred equally after both conditioning regimens (except for increased risk of cataract after CY-TBI and of alopecia with Bu-CY).

Hartman AR **(Hartman AR (1998) Survival, disease-free survival and adverse effects of conditioning for allogeneic bone marrow transplantation with busulfan/cyclophosphamide vs. total body irradiation: a meta-analysis. Bone Marrow Transplant 22(5):439–443)** → The OSs, the disease-free survivals, and the toxicities of BUCY vs. TBI-based regimens were compared by conducting a meta-analysis of five published, randomized, prospective trials comparing these regimens. Survival and disease-free survival were better with TBI-based regimens than with BUCY, but these differences were not statistically significant. TBI-based regimens cause less VOD than BUCY and were at least as good for survival and disease-free survival.

Shi-Xia et al (2010) **Total body irradiation plus cyclophosphamide versus busulphan with cyclophosphamide as conditioning regimen for patients with leukemia undergoing allogeneic stem cell transplantation: a meta-analysis. Leuk Lymphoma 51(1):50–60.** → 3172 patients from 18 trials were evaluated. TBI + cyclophosphamide (CY) versus busulfan + CY.

- For patients with acute leukemia (ALL and AML), the TBI/CY regimen leads to lower rates of leukemia relapse, lower transplant-related mortality (TRM), and higher DFS, while for chronic myeloid leukemia (CML), the TBI/CY regimen had a higher rate of leukemia relapse, lower TRM, and similar DFS.

The TBI/CY regimen was associated with similar occurrence of engraftment, acute and chronic graft-versus-host disease (GVHD), but with higher rates of interstitial pneumonitis, later growth, or development problems. BU/CY regimen was associated with higher rates of complications like liver veno-occlusive, hemorrhagic cystitis, and TRM.

References

1. Jason Chan SEB (2018) Lymphomas and myeloma. In: Hansen EK (ed) Handbook of evidence-based radiation oncology, 3rd edn. Springer, Cham, pp 725–749
2. Mann RB, Jaffe ES, Berard CW (1979) Malignant lymphomas--a conceptual understanding of morphologic diversity. A review. Am J Pathol. 94:105–192
3. Swerdlow SH, Campo E, Pileri SA, Harris NL, Stein H, Siebert R et al (2016) The 2016 revision of the World Health Organization classification of lymphoid neoplasms. Blood. 127:2375–2390
4. Diehl VHN, Mauch PM (2005) Hodgkin's lymphoma. In: VT DV Jr, Rosenberg SA (eds) Cancer: principles and practice of oncology, 7th edn. Lippincott Williams & Wilkins, Philadelphia, pp 2021–2026
5. Andrea K, Ng ASL (2021) Hodgkin Lymphoma. In: Tepper JE, Michalski JM (eds) Gunderson and Tepper's clinical radiation oncology, 5th edn. Elsevier, Philadelphia, pp 1518–1521
6. Cheson BD, Fisher RI, Barrington SF, Cavalli F, Schwartz LH, Zucca E et al (2014) Recommendations for initial evaluation, staging, and response assessment of Hodgkin and non-Hodgkin lymphoma: the Lugano classification. J Clin Oncol. 32:3059–3068
7. Moccia AA, Donaldson J, Chhanabhai M, Hoskins PJ, Klasa RJ, Savage KJ et al (2012) International Prognostic Score in advanced-stage Hodgkin's lymphoma: altered utility in the modern era. J Clin Oncol. 30:3383–3388
8. Senthilkumar Gandhidasan MCW, Shah C (2018) Hematologic. In: Ward MC, Videtic GMM (eds) Essentials of clinical radiation oncology. Springer, Cham, pp 450–479
9. Girinsky T, van der Maazen R, Specht L, Aleman B, Poortmans P, Lievens Y et al (2006) Involved-node radiotherapy (INRT) in patients with early Hodgkin lymphoma: concepts and guidelines. Radiother Oncol. 79:270–277
10. Specht L, Yahalom J, Illidge T, Berthelsen AK, Constine LS, Eich HT et al (2014) Modern radiation therapy for Hodgkin lymphoma: field and dose guidelines from the international lymphoma radiation oncology group (ILROG). Int J Radiat Oncol Biol Phys. 89:854–862
11. Stephanie Terezakis JPP (2017) Hodgkin Lymphoma. In: Bouthaina Shbib Dabaja AKN (ed) Radiation therapy for hematologic malignancies. Springer, Switzerland, pp 1–28
12. Barrington SF, Mikhaeel NG, Kostakoglu L, Meignan M, Hutchings M, Mueller SP et al (2014) Role of imaging in the staging and response assessment of lymphoma: consensus of the International Conference on Malignant Lymphomas Imaging Working Group. J Clin Oncol. 32:3048–3058
13. International Non-Hodgkin's Lymphoma Prognostic Factors P (1993) A predictive model for aggressive non-Hodgkin's lymphoma. N Engl J Med. 329:987–994
14. Federico M, Bellei M, Marcheselli L, Luminari S, Lopez-Guillermo A, Vitolo U et al (2009) Follicular lymphoma international prognostic index 2: a new prognostic index for follicular lymphoma developed by the international follicular lymphoma prognostic factor project. J Clin Oncol. 27:4555–4562
15. Hoster E, Dreyling M, Klapper W, Gisselbrecht C, van Hoof A, Kluin-Nelemans HC et al (2008) A new prognostic index (MIPI) for patients with advanced-stage mantle cell lymphoma. Blood. 111:558–565
16. Fungoides M (2006) In: Michaelis S, Kempf W (eds) Cutaneous lymphomas unusual cases 2, 1st edn. Steinkopff, pp 2–12
17. Willemze R, Cerroni L, Kempf W, Berti E, Facchetti F, Swerdlow SH et al (2019) The 2018 update of the WHO-EORTC classification for primary cutaneous lymphomas. Blood. 133:1703–1714
18. Steven T, Rosen ADZ, Jaffe ES, Leonard AJP (2017) Primary cutaneous lymphomas. In: Meyer LR (ed) AJCC cáncer staging manual. Springer, Chicago, pp 967–972
19. Gerbi BJ (2006) Clinical applications of high-energy electrons. In: Levitt SH, Purdy JA, Perez CA, Vijayakumar S (eds) Technical basis of radiation therapy, 4th edn. Springer, Berlin
20. James E, Hansen YHK, Hoppe RT, Lynn D, Wilson CB (2019) Primary cutaneous lymphomas. In: Halperin EC, Perez CA, Brady LW (eds) Principles and practice of radiation oncology, 7th edn. Wolters Kluwer, China
21. Christopher Andrew Barker JYCW, Yahalom J (2021) Total body irradiation. In: Tepper JE, Michalski JM (eds) Gunderson and Tepper's clinical radiation oncology, 5th edn. Elsevier, Philadelphia, pp 388–407.e10
22. Sarah Jo Stephens KBR, Chen ZJ, Seropian SE, Kelsey CR (2019) Stem cell transplantation and total-body irradiation. In: Halperin EC, Brady LW (eds) Principles and practice of radiation oncology, 7th edn. Wolters Kluwer, China
23. Dusenbery KEGB (2012) Total body irradiation conditioning regimens in stem cell transplantation. In: Levitt SH, Perez CA, Poortmans P (eds) Technical basis of radiation therapy, 5th edn. Springer, Berlin

Index

A

Abdominoperineal resection (APR) (Miles' operation), 447
Abscopal effect, 93
Absorbed dose, 14–16, 25, 32, 41, 42, 59, 92, 366
Accelerated fractionation, 89, 90, 172, 241–243
Acinic cell tumors, 223, 229
Adenoid cystic carcinoma, 192, 204, 211, 215, 222, 228, 229, 280, 402
α/β ratio, 68, 69, 73
α-fetoprotein (AFP), 337
ALARA, 91
Alpha decay, 5, 6
Amifostine, 73
Anal cancers, 452, 453, 456–458
Anaplastic astrocytoma (AA), 127, 134–136, 145
Anaplastic oligodendroglioma (AO), 127, 134, 136, 147
Androgen ablation, 318, 324
A point, 369
Atom, 1, 6, 7, 9, 11
Avalanche phenomenon, 93

B

Back scatter factor (BSF), 33–34
Basal cell carcinoma (BCC), 402, 478, 479
Beam modifiers, 37
BED, 68–70, 90, 94
Bergonie & Tribondeau law, 75, 88
Beta decays, 5
Beta human chorionic gonadotropin (β-hCG), 337
Bladder cancer, 345, 348–354
Bladder conserving surgery, 349
Bolus, 37, 38, 109, 143, 404, 451, 454, 458, 484, 509
Boron neutron capture therapy, 101
Bowen disease, 454
B point, 369
Brachytherapy, 33, 71, 79, 81, 101, 308, 318, 323, 324, 361, 365–372, 375, 377, 378, 387–388, 390–393, 398, 400–402, 470
Bragg peak, 44, 115
Breast conserving surgery, 302, 305
Bremsstrahlung X-rays, 4, 11, 12
B symptoms, 489, 490, 501, 502

Build-up region, 26, 28
Bulky disease, 170, 490, 494, 498, 500, 502, 503, 506, 507
Bystander effect, 93

C

Carcinogenesis, 47, 52, 53, 55, 77, 78, 94
Cell cycle, 49–53, 57, 70, 76, 84–87, 92, 94
Cell survival curve, 62, 63, 70, 71, 73, 86, 87, 94
Central lung distance
Cervical intraepithelial neoplasias (CIN), 357
Cervical cancer, 357, 359–365, 369, 370, 372–379, 385
Characteristic X-rays, 6, 11, 12
Chassagne point, 368
Chromosomes, 47–51, 54, 55, 71, 72, 78, 94
Cobalt–Gray equivalent dose (CGE), 114, 221
Cobalt-60 teletherapy unit (Co-60), 17–19, 37
Coherent effect, 11, 41
Collimator angle, 104, 106, 144
Collimator scattering factor, 35–37
Compensating filters, 37
Compton effect, 9–12, 26, 41
Concomitant boost, 90, 184, 221, 241, 243, 307, 353, 376, 377
Conformal planning, 109, 142, 143
Contouring procedure, 485
Conventional fractionation, 89, 90, 172, 180, 200, 308, 326, 328, 329, 332, 335
Conventional planning, 108, 109, 143
Conventional simulation, 103, 141
Cotswolds lymphoma staging, 489
Craniospinal RT (CSRT), 140, 143, 149, 154
Critical volume model, 81, 82
CT simulation, 106, 107, 120, 147, 258, 265, 267, 350, 362, 472, 493
Cutaneous lymphoma, 507, 509, 510, 512
Cyclins, 53

D

Deauville criteria, 493
Delta electrons, 42
Denonvillier's fascia, 313, 322

© The Editor(s) (if applicable) and The Author(s), under exclusive license to Springer Nature Switzerland AG 2022
M. Beyzadeoglu et al., *Basic Radiation Oncology*, https://doi.org/10.1007/978-3-030-87308-0

Desmoid tumors (aggressive fibromatosis), 466
Detector types, 23
Deterministic effects, 76, 77, 94
Direct effect, 55–57, 66, 70, 71, 90
Dose equivalent, 59–60
Dose rate, 17, 23, 32, 37, 43, 55, 59, 65, 71, 72, 79, 81, 92, 94, 119, 366, 368, 375
Ductal carcinoma in situ (DCIS), 280, 281, 301, 302

E

Effective dose, 44, 60, 68, 91
Electrodesiccation, 481
Electromagnetic radiation, 1–3, 9, 11, 16, 58
Electromagnetic spectrum, 1, 3
Electron, 1–12, 14–18, 20–22, 24, 25, 28–32, 34, 37, 41–43, 56–58, 73, 101, 108–110, 113, 298, 299, 308, 311, 454, 484, 509, 512, 514
Electron capture phenomenon, 6, 7
Ellis model, 69
Endometrial cancer, 300, 378–385, 387–394
Ependymoma, 125, 130, 139, 154
Equivalent dose, 15, 41, 59, 60, 62, 81, 91, 94, 371, 471, 474
Equivalent uniform dose (EUD), 81
Erythroplakia, 203, 204
Esophageal cancer, 411, 413–416, 422–425
Esthesioneuroblastoma, 214
Eukaryotic cells, 47–51, 53
Exponential survival curves, 65
Exposure unit, 41
Extensive stage SCLC, 271–274
External radiotherapy, 33, 42, 93, 101, 141–143, 235, 237, 308, 318, 362–365, 398, 404

F

Fibrosarcoma, 129, 463–467
Film dosimeters, 24, 42
5R of radiotherapy, 90
Flexible tissues, 75, 93
Flexure dose, 73
Follicular thyroid cancer, 231
Forgue sign, 283
Fractionation, 68, 79, 81, 82, 86, 88–90, 92, 94, 153, 171, 185, 200, 221, 238, 241, 242, 274, 307, 308, 328–329, 390, 422, 435, 474, 485

G

Gamma emission, 6, 7, 18
Gamma (γ) rays, 4
Gap calculation, 144
Gastric cancer, 425–433, 435
Geiger–Muller counter, 23, 24
Gleason score, 316, 318, 321–323, 331–333
Glioblastoma (GBM), 127, 131, 134, 135, 145, 148, 153, 154

H

Half value layer (HVL), 13, 26, 27, 29, 39, 42
HDR brachytherapy, 369
Head and neck lymphatics, 159
Helmet field, 142
Hierarchical tissues, 75
High-grade B cell lymphoma, 488, 503
Hodgkin lymphoma, 488, 490, 491, 496, 497, 499, 500, 505
Horner syndrome, 252, 253, 256
H points, 369
Hurthle cell thyroid cancer, 238
Hyperfractionation (HF), 73, 89, 90, 153, 180, 200, 242, 272, 274
Hyperthermia, 101
Hypofractionation, 89, 90, 94, 299, 307, 308, 326, 328, 329, 332
Hypopharyngeal cancer, 183, 186–192, 202

I

Image guided radiotherapy (IGRT), 79, 81, 101, 119–121, 147, 325, 326, 422
Immobilization, 102–104, 179, 258, 398, 420, 433, 472
Indirect effect, 55–58, 66, 70, 71, 87, 93
Inflammatory carcinoma, 281, 283, 290, 291
Intensity modulated radiotherapy (IMRT), 79, 81, 82, 101, 115, 117, 133, 135, 138, 145–149, 161, 163, 168, 171, 174, 180, 182, 183, 186, 208, 209, 217, 221, 226, 242, 258, 264, 308, 311, 325, 332, 335, 336, 362, 376–378, 385, 386, 393, 398, 401, 404, 420, 433, 453, 458, 472–475, 484, 514
Internal margin (IM), 101, 111, 112, 324
Internal target volume (ITV), 111, 112, 264, 365, 404, 422, 435
International Commission on Radiation Units and Measurements (ICRU), 110–113, 115, 116, 365, 370
International prognostic index (IPI), 502, 503, 505
Interphase, 49, 51, 52
Intracavitary brachytherapy application, 371–372
Intraoperative radiotherapy (IORT), 101, 308, 472
Invasive ductal carcinoma
Invasive lobular carcinoma, 280
Inverse square law, 33, 35, 92
Ionization chamber, 23, 31
Ionizing electromagnetic radiation, 3
Ionizing radiation units, 14, 16
Irish nodule, 426
Isolated tumor cells (ITC), 290

J

Japanese staging system for gastric cancer, 426

K

Kernohan grading system, 126

Index

Kinetic energy released in the medium (KERMA), 14, 16, 42, 366
Klystron, 20, 21, 42
Krukenberg tumor, 426

L

Laryngeal cancer, 84, 99, 192–194, 197, 198, 200–202
Leiomyosarcoma (LMS), 129, 334, 395, 463–467
Lethal damage, 66, 70, 85
Leukoplakia, 203, 204
Lichen planus, 204
Limited stage SCLC, 272, 273
Linear accelerator (LINAC), 4, 11, 14, 17, 18, 20–22, 32, 35, 39, 42, 101, 117, 120
Linear energy transfer (LET), 55, 58, 59, 61, 70, 71, 79, 81, 94
Linear quadratic model (LQ model), 67–70, 73, 86, 90, 371
Liposarcomas, 129, 463–465, 467, 469
Lobular carcinoma in situ (LCIS), 280, 282, 294, 298, 300
Low anterior resection (LAR), 447, 449
Low grade astrocytoma, 151
Low-grade B cell lymphoma, 130, 502
Lung lymphatics, 251, 258, 263, 477
Lyman model, 81, 82
Lymphatic trapezoid, 368
Lymphoma, 130, 136, 174, 204, 277, 357, 413, 464, 487, 488, 493, 495, 501–505, 507, 513

M

Magnetron, 20, 21, 42
Major salivary gland tumors, 221–223, 228–230
Malignant fibrous histiocytoma (MFH), 129, 395, 463, 464, 467
Malignant schwannoma, 463, 465
Malignant sweat gland tumors, 478, 479, 483
MALT-type extranodal marginal zone B cell lymphoma (MALTOMA), 503
Mantle radiotherapy
Marjolin's ulcer, 479
Measurement of ionizing radiation, 22–25
Medullary carcinoma, 231, 232, 280
Merkel cell carcinoma, 478–480, 483
Metastasis, 54, 134, 149, 159, 166, 168, 175, 193, 205, 214, 219, 222, 231, 233, 235, 237, 251–254, 257, 262, 263, 269–274, 281, 316, 319, 322, 338, 359, 379, 383, 403, 408, 414, 416, 438, 448, 454, 463–469, 472, 478–479, 482, 483
Michalowski tissue sensitivity classification, 75
Mini mantle
Mitosis, 47, 49–53, 55, 70, 85, 88, 93
Modified radical mastectomy (MRM), 294–297, 304, 305
Moh's micrographic surgery, 481, 483
Monitor unit, 118, 119
Multileaf collimator (MLC), 21, 22, 32, 39, 41, 101, 107, 109, 118, 119, 258

Mycosis fungoides (MF), 488, 501, 507–509, 512

N

Nasopharyngeal cancer, 99, 159, 163–174, 214
Neurovascular bundle, 267, 314, 315
Neutron, 1, 2, 5–8, 14, 18, 20, 23–25, 30, 58, 101, 135, 230
Neutron dosimeter, 25
Non-Hodgkin lymphoma (NHL), 136, 487, 488, 500–504, 507
Non-small cell lung cancer (NSCLC), 251–255, 258, 260–263, 267–272
Normal tissue complication probability (NTCP), 78–82, 311
Normalized total dose, 94, 95

O

Oligodendroglioma, 127, 133, 139, 146, 149, 151, 154
Oncogenes, 52, 53
Oral cavity cancers, 159, 202–206, 210, 243
Oropharyngeal cancer, 174–180, 182–186, 244
Orton–Ellis model, 69
Overkill effect, 61, 63
Oxygen enhancement ratio (OER), 71

P

Pair production, 9, 10, 12, 26, 41
Pancoast syndrome, 256, 257
Pancreatic cancer, 436
Papillary thyroid cancers, 231
Papilloma, 127, 204, 280, 348
Parallel organs, 77, 95
Particulate radiation, 1, 2, 7, 16, 42, 58
Pelvic plexus (PP), 313
Penumbra
    geometrical penumbra, 33
    physical penumbra, 33
Percentage depth dose (PDD), 18, 27, 36, 37, 43
Phantom, 25, 29, 31–35, 43, 514
Phantom scattering factor, 35–37
Pharyngeal cancers, 163, 165–168, 170–180, 182, 184–186, 188, 190–192
Photoelectric effect, 9, 12, 26, 41, 56
Photon, 2, 3, 9–15, 18, 26, 28, 30, 32, 34, 37, 41–43, 58, 110, 114–116, 143–145, 221, 230, 263, 299, 307, 311, 336, 398, 442, 484, 513
Pleomorphic adenoma, 222
Posterior fossa syndrome, 140, 154
Potentially lethal damage, 70, 72
P point, 369
Prokaryotic cells, 47
Prophylactic cranial RT, 269, 272
Prostate cancer, 94, 95, 110, 313
Prostate-specific antigen (PSA), 313, 317, 318, 321–323, 327, 328, 330, 331, 333–336
Prostatic intraepithelial neoplasia (PIN), 315

Proton, 1, 2, 5–7, 14, 18, 20, 42, 44, 58, 59, 93, 110, 113–115, 137, 149, 221, 336, 503
PSA bounce phenomenon, 318
PSA density (PSAD), 317, 322
PSA relapse definition, 327–328
PSA velocity (PSAV), 317

Q
Quality assurance (QA), 115, 118
Queyrat disease/erythroplasia, 479

R
Radiation dosimetry, 25–27, 31, 33, 34
Radiation hormesis, 92, 93
Radiation protection, 15, 59, 91, 92
Radiation recall phenomenon, 93, 94
Radiation weighting factor, 15, 58, 59
Radical hysterectomy, 360, 361, 373, 374, 376, 379, 397, 398
Radical prostatectomy, 323, 333, 334, 336
Radioactivity, 16
Radioimmunotherapy, 502
Radionuclides, 1, 5, 6, 366
Radioprotective agents, 72, 73
Radiosensitivity, 49, 59, 64, 67, 70, 71, 73, 75, 76, 78, 79, 81, 84, 86–88, 92–94, 139
Radiosensitizers, 73, 79, 87, 88, 440, 441
Radiotherapy generators, 16
RAI indications, 235
Rectal cancers, 444–453
Redistribution, 84, 86, 87, 89, 92, 94
Reed–Stenberg cells, 487
Reference air kerma, 16
Relative biological effect (RBE), 61, 79, 81, 114
Remote afterloading, 365
Reoxygenation, 84, 86, 87, 89, 92
Repair, 49, 52–54, 57, 62, 67, 70–72, 84–88, 92, 94, 101, 381
Repopulation, 68, 71, 75, 84–90, 92, 94, 241
Retromolar trigone, 161, 179, 202, 203, 205, 206, 210
Retroperitoneal sarcoma, 314, 316, 337, 420, 428, 448, 463–469, 472, 474
Rhabdomyosarcoma (RMS), 129, 395, 463–467, 469, 470
Roach formulas, 318
Robotic radiosurgery, 220, 261, 327
Roentgen, 3, 14, 16, 19
Rogers, Will phenomenon, 428
Rubin and Casarett tissue sensitivity classification, 76

S
SAD techniques, 26, 27
Santorini canal, 436
Santorini's plexus, 313
Scatter air ratio (SAR), 35
Seminoma, 337, 340, 341, 343, 344
Serial organs, 76, 81, 95
Setup, 38, 103, 111, 115, 116, 298, 299, 472
Setup margin (SM), 101, 111
Sézary syndrome, 488, 501, 507–509, 512
Simulation, 38, 102, 108, 115, 116, 144, 260, 264, 267, 325–327, 398, 404, 420, 433, 441, 449, 453, 472, 485
Sinonasal cancers, 210, 212–216, 219, 221
Sister Mary Joseph nodule, 426
Skin cancers, 14, 16, 18, 69, 99, 477–485
Small cell lung cancer (SCLC), 100, 251–256, 271–274
Soft tissue sarcoma (STS), 463–467, 469–474
Spinal tumors, 126, 130, 132, 139, 141, 144, 145
Split-course, 90, 171, 180, 192, 243, 244, 406, 424, 458
Squamous cell cancer (SCC), 163, 164, 174, 180, 186, 192, 204, 211, 240, 245, 345, 357, 374, 395, 402, 406, 412, 413, 424, 453, 478–481, 483, 484
SSD technique, 33, 36
Stanford 6 technique, 511
Stereotactic radiotherapy, 90, 94, 101, 138, 260
Stewart-Treves syndrome, 466
Stochastic effects, 61, 77, 78
Strandqvist model, 69
Sublethal damage (SLD), 62, 71
Subtotal nodal irradiation (STNI), 494, 496
Surgical anal canal, 453
Surviving fraction (SF), 62, 63, 65, 68, 69, 80–82
Synovial sarcoma, 463–465, 467, 469

T
Target theory, 61, 65
Target volume definitions, 109, 190
Temozolamide (TMZ), 133, 134, 148
Testicular cancers, 336–345
Therapeutic index, 78, 79
Therapeutic ratio, 82
Thermoluminescence dosimeters (TLD), 23–25, 42
Three-dimensional conformal RT (3D-CRT), 101, 152, 325, 362
Thyroid cancer, 230–238, 246
Tissue air ratio (TAR), 34, 35, 42, 43
Tissue and organ response to radiation, 73, 75, 76
Tissue maximum ratio (TMR), 31, 34–36
Tomotherapy, 101, 121, 258, 517
Total body irradiation (TBI), 77, 100
Total nodal irradiation, 494
Total skin irradiation (TSI), 113, 509–512
Treatment planning, 31, 82, 106–108, 116, 118, 134, 371, 398, 404
Trimmer bars, 42
Tumor control probability (TCP), 78–82
Tyratron, 42

U
Uncinate process, 436, 437, 439, 440
Unknown primary head-neck cancers, 238–241

Index

V
Vaginal cancer, 394
Van Nuys prognostic index (VPNI), 281
Virchow node, 426
Vulvar carcinoma, 406

W
Waldeyer field RT, 492
Wang staging for nasal vestibule
Warthin tumor, 222
Wedge filter, 21, 37
Whipple operation, 439

Whole brain RT, 99, 136, 141, 142, 147, 152, 153
Wirsung canal, 436

X
X-rays, 1, 3, 4, 6, 8, 9, 11–14, 16, 18, 20, 23–25, 28–30, 41–44, 56, 58, 61, 83, 92, 101, 104, 109, 253, 358, 454

Y
Y applicator, 388

Printed by Printforce, United Kingdom